Springer Complexity

Springer Complexity is an interdisciplinary program publishing the best research and academic-level teaching on both fundamental and applied aspects of complex systems - cutting across all traditional disciplines of the natural and life sciences, engineering, economics, medicine, neuroscience, social and computer science.

Complex Systems are systems that comprise many interacting parts with the ability to generate a new quality of macroscopic collective behavior the manifestations of which are the spontaneous formation of distinctive temporal, spatial or functional structures. Models of such systems can be successfully mapped onto quite diverse "real-life" situations like the climate, the coherent emission of light from lasers, chemical reaction-diffusion systems, biological cellular networks, the dynamics of stock markets and of the internet, earthquake statistics and prediction, freeway traffic, the human brain, or the formation of opinions in social systems, to name just some of the popular applications.

Although their scope and methodologies overlap somewhat, one can distinguish the following main concepts and tools: self-organization, nonlinear dynamics, synergetics, turbulence, dynamical systems, catastrophes, instabilities, stochastic processes, chaos, graphs and networks, cellular automata, adaptive systems, genetic algorithms and computational intelligence.

The two major book publication platforms of the Springer Complexity program are the monograph series "Understanding Complex Systems" focusing on the various applications of complexity, and the "Springer Series in Synergetics", which is devoted to the quantitative theoretical and methodological foundations. In addition to the books in these two core series, the program also incorporates individual titles ranging from textbooks to major reference works.

Understanding Complex Systems

Founding Editor: J.A. Scott Kelso

Future scientific and technological developments in many fields will necessarily depend upon coming to grips with complex systems. Such systems are complex in both their composition - typically many different kinds of components interacting simultaneously and nonlinearly with each other and their environments on multiple levels - and in the rich diversity of behavior of which they are capable.

The Springer Series in Understanding Complex Systems series (UCS) promotes new strategies and paradigms for understanding and realizing applications of complex systems research in a wide variety of fields and endeavors. UCS is explicitly transdisciplinary. It has three main goals: First, to elaborate the concepts, methods and tools of complex systems at all levels of description and in all scientific fields, especially newly emerging areas within the life, social, behavioral, economic, neuroand cognitive sciences (and derivatives thereof); second, to encourage novel applications of these ideas in various fields of engineering and computation such as robotics, nano-technology and informatics; third, to provide a single forum within which commonalities and differences in the workings of complex systems may be discerned, hence leading to deeper insight and understanding.

UCS will publish monographs, lecture notes and selected edited contributions aimed at communicating new findings to a large multidisciplinary audience.

Leandro Pardo,
Narayanaswamy Balakrishnan,
and María Ángeles Gil (Eds.)

Modern Mathematical Tools and Techniques in Capturing Complexity

Springer

Editors

Leandro Pardo
Departamento de Estadística e I. O.
Facultad de Matemáticas
Universidad Complutense de Madrid
28040-Madrid Spain
E-mail: lpardo@mat.ucm.es
http://www.ucm.es/dir/gi001.htm

Narayanaswamy Balakrishnan
Department of
Mathematics and Statistics
McMaster University Hamilton,
Ontario Canada L8S 4K1
E-mail: bala@univmail.cis.mcmaster.ca
http://www.math.mcmaster.ca/bala/

María Ángeles Gil
Departamento de Estadística e I. O. y D.M.
Universidad de Oviedo
C/ Calvo Sotelo, s/n 33071 Oviedo
Spain
E-mail: magil@uniovi.es
http://bellman.ciencias.uniovi.es/SMIRE

ISBN 978-3-642-20852-2 e-ISBN 978-3-642-20853-9

DOI 10.1007/978-3-642-20853-9

Understanding Complex Systems ISSN 1860-0832

Library of Congress Control Number: 2011928064

Typeset & Cover Design: Scientific Publishing Services Pvt. Ltd., Chennai, India.

Printed on acid-free paper

9 8 7 6 5 4 3 2 1

springer.com

This book is prepared as a tribute to
Professor María Luisa Menéndez
who was not only an exceptional researcher and teacher,
but was also a person with many notable skills and
and outstanding human qualities

«Teaching is the highest form of understanding » (Aristotle)

Preface

> "The essence of mathematics is not to make simple things complicated, but to make complicated things simple."
>
> Stanley Gudder

Real-life problems are often quite complicated and, for centuries, Mathematics has provided the necessary tools and ideas to formulate these problems mathematically and then help in solving them either exactly or approximately.

This book aims to gather a collection of papers dealing with several different problems arising from many disciplines and some modern mathematical approaches to handle them. In this respect, the book offers a wide overview on many of the current trends in Mathematics as valuable formal techniques in capturing and exploiting the complexity involved in real-world situations.

The need for this volume arose in a natural way as many mathematicians wanted to pay tribute and recognition to their beloved and admired friend and colleague María Luisa Menéndez, who passed away in late 2009 after fighting cancer courageously for two years. Marisa has been, and will continue to be, a constant source of inspiration for all of us because of her sweet and strong personality and her high scientific reputation.

Professor Maria Luisa Menéndez was born on March 15, 1956, in Cuenca, Spain. Her parents, Mr. Manuel Menéndez and Mrs. Vicenta Calleja, had two daughters, María Dolores and María Luisa. Marisa lived in Cuenca and studied at the public high school "Lorenzo Hervás y Panduro". After getting the access to the University, she came to the Complutense University of Madrid where she studied Mathematics with the specialization in Statistics. Once she got her Bachelor in Sciences (Mathematics) in 1978, she joined the Master in Statistics program also at the Complutense University of Madrid.

In October 1978, she joined the Technical University of Madrid in the Higher Technical School of Architecture, and started to teach also at the high school where she got a permanent position in 1979. In 1986 she submitted her PhD Dissertation in the Computer Science School at the Technical University of Madrid. In 1988, she got a permanent position as Associate Professor in the Department of Applied Mathematics in the Higher Technical School of Architecture and subsequently in 1997 she was promoted to the rank of Full Professor. In 1990, the General Foundation of the Technical University of Madrid awarded Professor Menéndez the Prize for the best research developed by young professors (under 35 years).

In July 1981, Professor Menéndez started her own family when she married the mathematician Leandro Pardo. They have had two children: María Luisa, who was born in 1983 and graduated in Business Administration some years ago, and Leandro, who was born in 1989 and studying currently in the fourth year of the 6 years high degree in Civil Engineering.

Professor Menéndez was an excellent teacher and an exemplary mentor, and was genuinely interested in the students and their achievements. Her office door was always open to students. Some of her last students paid her a warm tribute on 29 October 2010 during the homage that was organized by the Higher Technical School of Architecture. Details of this homage can be seen in the web site at `http://dma.aq.upm.es/actividades/2010_hmm`.

Professor Menéndez was an excellent researcher as is clearly evident from the international databases. She published nearly 100 papers in different scientifically recognized journals; she presented many papers in international conferences, and participated in numerous national and international research projects. Her primary research interest was on Statistical Information Theory, a topic in which she created jointly with Professor Leandro Pardo an internationally renowned and active research group.

María Luisa Menéndez, in addition to being an exceptional researcher and teacher, was one with with many notable skills and outstanding human qualities. She was humble and modest, and always made all the students, colleagues and friends at ease and felt welcome. Marisa was a great woman, a foremost researcher, and a great inspiration to those around and those who came into contact with her. With a deep humanistic view and attitude, she faced other people as friends and relatives, and carefully cultivated friendship with many. Gentle and polite with a distinctive and delicate humor, she was a source of great joy to anyone who interacted with her. This volume is a small but sincere gesture from all of us to show how much we appreciated her and how sorely we miss her!

Madrid-Ontario-Oviedo, February 2011

Leandro Pardo
Narayanaswamy Balakrishnan
María Ángeles Gil

Contents

Part III: Distribution Theory and Applications

Part IV: Divergence Measures and Statistical Applications

Part V: Modelling in Engineering Problems

Part I

Advanced Methods in Statistics

A General View of the Goodness-of-Fit Tests for Statistical Models*

Wenceslao González-Manteiga and Rosa M. Crujeiras

Department of Statistics and Operations Research, Faculty of Mathematics, University of Santiago de Compostela, Spain
wenceslao.gonzalez@usc.es, rosa.crujeiras@usc.es

Summary. Goodness-of-Fit tests have been a major topic of research in Statistics since their introduction by Pearson at the beginning of the last century. This type of tests has been applied to remarkable curves such as the distribution, the density or the regression function. In this paper, we provide a unified approach of the Goodness-of-Fit testing theory based on the formulation of the test statistic as a functional of a certain empirical process. We will focus our attention in the regression function, revising also some new testing approaches based on the likelihood ratio test and the empirical distribution of the residuals. Finally, we also collect some ideas for the extension of Goodness-of-Fit tests to other settings, such as incomplete information or dependent data.

1 Introduction: From Distribution to Regression

The term *Goodness-of-Fit* (GoF) was introduced by Pearson at the beginning of the 20th century and it refers to tests that check how a distribution fits to a data set in an omnibus way. Since then, many papers have been devoted to the χ^2 test, the Kolmogorov-Smirnov test, the Cramer-von-Mises test and other related methods.

The basic idea consists in comparing a nonparametric pilot estimator for the unknown distribution F or the density f, with a consistent parametric estimator under the null hypothesis. In these cases, the pilot function used for testing is the empirical distribution or, a density estimator, since the seminal paper by [4].

It should be also mentioned that [11] and [4] settled the beginnings of the mathematical developments for GoF tests, based on the estimation of the distribution and the density function, respectively.

The general statement of the problem is as follows: given a random sample $\{X_1, \ldots, X_n\}$ of X (i.i.d., independent and identically distributed) with distribution function F, the goal is to test if

* We would like to thank the Editors of this book for giving us the opportunity to honor Professor Marisa Menéndez.

$$H_0: F \in \mathcal{F} = \{F_\theta\}_{\theta \in \Theta \subset \mathbb{R}^q}, \quad \text{vs.} \quad H_a: F \notin \mathcal{F}.$$

The test statistic is based on the discrepancy between a pilot estimator of the distribution or the density function and the corresponding estimator under H_0. Hence, for a distribution test, we may consider:

$$T_n = T(F_n, F_{\hat{\theta}}) = T(\alpha_n) \tag{1}$$

where F_n is the empirical distribution given by $F_n(x) = \dfrac{\#\{j|X_j \leq x\}}{n}$ and $F_{\hat{\theta}}$ is a parametric estimator under H_0, where $\hat{\theta}$ is usually a $\sqrt{n}$-consistent estimator.

In expression (1), α_n denotes the empirical process with estimated parameter $\hat{\theta}$, specifically:

$$\alpha_n(x) = \sqrt{n}(F_n(x) - F_{\hat{\theta}}(x)).$$

This general statement encloses the well-known Kolmogorov-Smirnov test:

$$T(F_n, F_{\hat{\theta}}) = \sup_x \sqrt{n}|F_n(x) - F_{\hat{\theta}}(x)| = \sup_x |\alpha_n(x)|$$

and the Cramer-von-Mises test:

$$T(F_n, F_{\hat{\theta}}) = n \int (F_n(x) - F_{\hat{\theta}}(x))^2 dF_n(x) = \int \alpha_n^2(x) dF_n(x).$$

In fact, a more general approach is obtained taking T_n as any continuous functional of α_n. In this setting, for deriving the asymptotic distribution of the test statistic, once should note that the limit convergence of α_n is given by a Gaussian process where the covariance structure of $\hat{\theta}$ is directly involved and tabulation is needed. For a detailed analysis of the empirical process α_n, see [11].

Following the ideas for the GoF tests for the distribution function, in order to design a testing procedure for the density function, test statistics usually follow this structure:

$$T_n = T(\widetilde{\alpha}_n) = T\left(\sqrt{nh}(f_{nh}(\cdot) - \mathbb{E}_{\hat{\theta}} f_{nh}(\cdot))\right),$$

where $f_{nh}(x) = (nh)^{-1} \sum_{i=1}^{n} K\left(\dfrac{x - X_i}{h}\right)$ is the kernel density estimator, with K the kernel function and h smoothing parameter (c.f. [32] and [33]). Here, $\mathbb{E}_{\hat{\theta}} f_{nh}(x)$ represents the expected value

$$\mathbb{E}_\theta f_{nh}(x) = \int h^{-1} K(h^{-1}(x - u)) dF_\theta(u),$$

with θ replaced by a $\sqrt{n}$-consistent estimator.

Besides, $\widetilde{\alpha}_n$ is the empirical process associated to the density function and its limit convergence is given by a Gaussian process where the covariance structure of $\hat{\theta}$ is not involved (c.f. [34]).

The ideas of Bickel and Rosenblatt where extended to the p-dimensional case in the nineties (see [13] and [15]). For instance, consider the test statistic:

$$T_n = \int \widetilde{\alpha}_n^2(x)\omega(x)dx = \int \left[\sqrt{nh^p}\left(f_{nh}(x) - \mathbb{E}_{\hat{\theta}}(f_{nh}(x))\right)\right]^2 \omega(x)dx,$$

where ω is a weight function included for mitigating the edge-effect in curve estimation. It holds that:

$$h^{-p/2}\left(T_n - \int K^2(x)dx \cdot \int f(x)\omega(x)dx\right) \xrightarrow{d} N\left(0, 2\int (K * K)^2(x)dx \int f^2(x)\omega^2(x)dx\right)$$

where $K * K$ denotes the convolution, and the limit is obtained as $h \equiv h_n \to 0$ and $nh_n^p \to \infty$. This convergence can be derived with arguments involving Gaussian processes (see [34]) or using the Central Limit Theorem for U-statistics with kernels varying with n(see, for instance, [23]).

Also in the nineties, these ideas for GoF testing were extended to the general case of a regression model, such as in [21] and [19]. Consider, for example, the regression model with random design:

$$Y = m(X) + \varepsilon$$

with $\{(X_i, Y_i)\}_{i=1}^n$ an i.i.d. sample of $(X, Y) \in \mathbb{R}^{p+1}$. The goal is to obtain a specification test for:

$$H_0 : m \in \mathcal{M} = \{m_\theta\}_{\theta \in \Theta \subset \mathbb{R}^q}, \quad \text{vs.} \quad H_a : m \notin \mathcal{M}$$

where $m(x) = \mathbb{E}(Y|X = x)$ is the regression function of Y over X. In this context, the nuisance functions $\sigma^2(x) = \text{Var}(Y|X = x)$ and f appear, where f denotes the density of the explanatory variable (if exists).

From the initial works of [21] and [19], the statistical literature on this topic is explosive. The different alternatives are based on the nonparametric estimator proposed for the regression function. For instance, a Nadaraya-Watson estimator ([28], [42]):

$$m_{nh}(x) = \sum_{i=1}^n W_{ni}(x)Y_i, \quad W_{ni}(x) = \frac{K\left(\frac{x - X_i}{h}\right)Y_i}{\sum_{j=1}^n K\left(\frac{x - X_j}{h}\right)}$$

being W_{ni} the Nadaraya-Watson weights, or more generally, a local polynomial estimator (c.f. [14]):

$$m_{nh}(x) = \hat{\beta}_0(x) = \sum_{j=1}^n W_{n,\bar{q}}\left(\frac{x - X_i}{h}\right)Y_i,$$

where $\hat{\beta}(x) = (\hat{\beta}_0(x), \ldots, \hat{\beta}_{\bar{q}}(x))^t$, is the minimizer of:

$$\sum_{i=1}^{n} \left(Y_i - \sum_{r=0}^{\bar{q}} \beta_r (x - X_i)^r \right)^2 K\left(\frac{x - X_i}{h} \right),$$

and $W_{n,\bar{q}} = u^t (X^T W X)^{-1} (1, ht, \ldots, h^{\bar{q}} t^{\bar{q}}) \frac{K(t)}{h}$, with $u^t = (1, 0, \ldots, 0) \in \mathbb{R}^{\bar{q}+1}$, $X = ((x - X_i)^j)_{1 \le i \le n, 1 \le j \le \bar{q}}$, $W = \operatorname{diag}\left(K\left(\frac{x - X_i}{h}\right)\right)$.

Following the spirit of the previous tests for the distribution and the density function, the initial empirical process for this regression problem, with a p-dimensional explanatory variable, is given by:

$$\begin{aligned} \overline{\alpha_n}(x) &= \sqrt{nh^p} \left(m_{nh}(x) - \mathbb{E}_{\hat{\theta}}(m_{nh}(x)) \right) \\ &= \sqrt{nh^p} \sum_{i=1}^{n} W_{ni}(x) \left(Y_i - m_{\hat{\theta}}(X_i) \right) \\ &= \sqrt{nh^p} \sum_{i=1}^{n} W_{ni}(x) \hat{\varepsilon}_i \end{aligned}$$

where $\mathbb{E}_{\hat{\theta}}$ is the estimation of $\mathbb{E}_{\theta_0}$ (with θ_0 theoretical parameter under H_0) and $\hat{\theta}$ is a $\sqrt{n}$-consistent estimator of θ_0 (for instance, least squares or maximum likelihood).

We may consider several possible tests based on $\overline{\alpha_n}$, such as:

$$T_n = \int \overline{\alpha_n}^2(x) \omega(x) dx \tag{2}$$

as a test statistic, which in the case of testing a polynomial regression model:

$$H_0 : m(x) = \sum_{j=1}^{q} \theta_j x^{j-1}, \quad \text{vs.} \quad H_a : m(x) \neq \sum_{j=1}^{q} \theta_j x^{j-1}$$

verifies that (see [1]):

$$h^{-1/2}(T_n - c_1) \xrightarrow{d} N(0, c_2)$$

where

$$c_1 = \int \widetilde{K}^2(x) dx \int \frac{\sigma^2(x)\omega(x)}{f(x)} dx, \quad c_2 = 2 \int (\widetilde{K} * \widetilde{K})^2(x) dx \int \frac{\sigma^4(x)\omega^2(x)}{f^2(x)} dx,$$

being $\widetilde{K}$ the equivalent kernel corresponding to a q-th order local polynomial estimate of m.

The test statistic in (2) can be seen as a consistent estimator, under H_0, of the expected value

$$\mathbb{E}(C_1) = \mathbb{E}[\mathbb{E}^2(\varepsilon_0|X)\omega(X)] = \mathbb{E}[(m(X) - m_{\theta_0}(X))^2\omega(x)],$$

where $\varepsilon_0 = Y - m_{\theta_0}(X)$. This expected value is equal to zero if and only if the null hypothesis holds (see [21]). On the other hand, H_0 is true if and only if $\mathbb{E}(C_2) = \mathbb{E}[\varepsilon_0\mathbb{E}(\varepsilon_0|X)f(X)\omega(X)] = 0$, and this quantity can be estimated by:

$$\frac{1}{n}\sum_{i=1}^{n}(Y_i - m_{\hat{\theta}}(X_i))(m_{nh}(X_i) - \mathbb{E}_{\hat{\theta}}(m_{nh}(X_i)))f_{nh}(X_i)\omega(X_i)$$

which gives rise to the Zheng's test (see [44]). Following similar arguments, H_0 is true if $\mathbb{E}(C_3) = \mathbb{E}[(\varepsilon_0 - (\varepsilon_0 - \mathbb{E}(\varepsilon_0|X))^2)\omega(X)] = 0$, and this can be estimated by:

$$\frac{1}{n}\sum_{i=1}^{n}(Y_i - m_{\hat{\theta}}(X_i))^2\omega(X_i) - \frac{1}{n}\sum_{i=1}^{n}(Y_i - m_{nh}(X_i))^2\omega(X_i),$$

which produces the variance difference test (see [9], [3]).

All these approaches share some common drawbacks. More specifically, the bandwidth choice, the slow rate of convergence of T_n to its Gaussian limit and the estimation of unknown curves involved in the test statistic. This idea for testing regression models takes as a methodological reference the previous work for density testing.

Similar to the developments for the GoF distribution tests, we may also consider the integrated regression function:

$$I(x) = \int_{-\infty}^{x} m(t)dF(t) = \mathbb{E}(Y \cdot \mathbb{I}(X \leq x)),$$

where $\mathbb{I}$ is the indicator. $I(x)$ can be empirically estimated by:

$$I_n(x) = \frac{1}{n}\sum_{i=1}^{n} Y_i \cdot \mathbb{I}(X_i \leq x)$$

with associated empirical process:

$$\overline{\overline{\alpha_n}}(x) = \sqrt{n}(I_n(x) - \mathbb{E}_{\hat{\theta}}(I_n(x))) = \frac{1}{\sqrt{n}}\sum_{i=1}^{n}\mathbb{I}(X_i \leq x)\hat{\varepsilon}_i.$$

This empirical process is the basis for a broad class of test statistics. For instance, the Cramer-von-Mises test $T_n = \int \overline{\overline{\alpha_n}}^2(x)dF_n(x)$ can be analyzed taking into account the convergence properties of the empirical process $\overline{\overline{\alpha_n}}$ (see [36] for details).

2 Calibration, Size and Power

In the general testing problem:

$$H_0 : g \in \mathcal{G} = \{g_\theta\}_{\theta\in\Theta}, \quad \text{vs.} \quad H_a : g \notin \mathcal{G} = \{g_\theta\}_{\theta\in\Theta}$$

with test statistic $T_n = T(g_n, g_{\hat{\theta}})$ where g may be F_θ, f_θ, m_θ or I_θ, calibration of critical points is crucial. The estimation of c_α such that $\mathbb{P}_{H_0}(T_n \geq c_\alpha) = \alpha$ can be done using the asymptotic normality of the test statistic with known g_n (for nonparametric density or regression estimators) or approximating the distribution of the corresponding empirical process, typically used in the calibration of tests for the distribution or the integrated regression function. In this last situation, the tabulation of the Gaussian limit process (or a continuous functional) is required.

In [35], an approximation of the critical point c_α by resampling methods is proposed. Specifically, if $T_n^* = T(F_n^*, F_{\hat{\theta}^*})$, c_α is estimated by $\hat{c}_\alpha$ such that:

$$\mathbb{P}^*_{H_0}(T_n^* \geq \hat{c}_\alpha) = \alpha$$

where $\mathbb{P}^*$ is the probability under resampling and F_n^* and $F_{\hat{\theta}^*}$ are obtained with the Bootstrap samples $\{X_1^*, \ldots, X_n^*\}$ (iid from $X^* \sim F_{\hat{\theta}}$). This idea has been applied in other tests, such as in [21] for $T_n = \displaystyle\int \overline{\overline{\alpha_n}}^2(x)\omega(x)dx$, in regression models with kernel estimator; in [1] for the local linear estimator or in [37], for empirical regression processes.

The key idea in all these approaches lies in the resample $\{(X_i^*, Y_i^*)\}_{i=1}^n$, which comes from the regression model $Y_i^* = m_{\hat{\theta}}(X_i^*) + \varepsilon_i^*$. The error term ε_i^* can be obtained from the empirical distribution of the residuals $\{\hat{\varepsilon}_i = Y_i - m_{\hat{\theta}}(X_i)\}$ as follows:

1. Construct the parametric residuals:

$$\hat{\varepsilon}_i = Y_i - m_{\hat{\theta}}(X_i), \, i = 1, 2, \ldots, n.$$

2. Recenter the previous residuals:

$$\overline{\hat{\varepsilon}}_i = \hat{\varepsilon}_i - \overline{\hat{\varepsilon}}, \, i = 1, 2, \ldots, n, \text{ where } \overline{\hat{\varepsilon}} = \frac{\sum_{i=1}^n \hat{\varepsilon}_i}{n}.$$

3. Draw bootstrap versions of the residuals, ε_i^*, from the empirical cumulative distribution function of the $\left\{\overline{\hat{\varepsilon}}_i\right\}_{i=1}^n$.
4. Compute $Y_i^* = m_{\hat{\theta}}(X_i) + \varepsilon_i^*$, $i = 1, 2, \ldots, n$ (no resampling of the X's).

Another alternative is to use Wild Bootstrap ([43], [26], [21]):

1. Construct the parametric residuals:

$$\hat{\varepsilon}_i = Y_i - m_{\hat{\theta}}(X_i), \, i = 1, 2, \ldots, n.$$

2. Draw independent random variables $V_1^*, V_2^*, \ldots, V_n^*$ (also independent of the observed sample) satisfying

$$E^*(V_i^*) = 0, \; E^*\left(V_i^{*2}\right) = 1, \; E^*\left(V_i^{*3}\right) = 1$$

and construct the $\varepsilon_i^* = \hat{\varepsilon}_i V_i^*$.
3. Compute $Y_i^* = m_{\hat{\theta}}(X_i) + \varepsilon_i^*$, $i = 1, 2, \ldots, n$ (no resampling of the X's).

The resampling approximation of the test statistic distribution is consistent, so the problems of the asymptotic approach (for instance, the slow rate of convergence) are overcome. Related to the calibration problem, other alternatives to Bootstrap have been also proposed, such as the martingale transform (see [37], [24]) or Monte Carlo methods (see [45]).

Once the test has been calibrated for a certain level α, another problem arises when choosing the test that maximizes the power. In the regression context, the comparison between tests can be done with Pitman alternatives:

$$H_0 : \, m \in \mathcal{M} = \{m_\theta\}_{\theta \in \Theta \subset \mathbb{R}^1}, \quad \text{vs.} \quad H_{an} : \, m_n(\cdot) = m_\theta(\cdot) + c_n d(\cdot)$$

where $c_n \to 0$ and d denotes the direction of the alternative. A test with level α and with power higher than α, and with the fastest c_n tending to zero, will be the most powerful.

As an example, the F-test in multiple linear regression models with parametric alternatives of higher dimension shows $c_n \sim n^{-1/2}$. On the other hand, a test for linearity $H_0 : \; m_\theta(x) = \theta^t x, \, \theta \in \mathbb{R}^q$ based on a smooth estimation of the residuals via $T_n = \int \overline{\alpha_n}^2(x)\omega(x)dx$ has $c_n \sim n^{-1/2}h^{-p/4}$, which comprises the price to pay for a nonparametric alternative. Finally, a test based on the empirical regression process with $T_n = \int \overline{\overline{\alpha_n}}^2(x)dF_n(x)$ has $c_n \sim n^{-1/2}$. From these results, we may conclude that the tests based on the empirical regression process are the best ones in practice. However, this is not true, as it is shown is simulation studies with finite samples (as in [27]). Even though the tests based on the estimators exhibit the curse of dimensionality, this is not clear in practice, and some modifications of the test statistic allow to get $c_n \sim n^{-1/2}(\log\log n)^{1/2}$, as in [22], where the following transformation is proposed:

$$T_n = \max_{h \in H_n} \frac{\int \overline{\alpha_{nh}}^2(x)\omega(x)dx - \widehat{\mathbb{E}_{H_0}}(\int \overline{\alpha_{nh}}^2(x)\omega(x)dx)}{\widehat{\mathrm{Var}_{H_0}}(\int \overline{\alpha_{nh}}^2(x)\omega(x)dx)}$$

with H_n being a family of smoothing parameters.

Nevertheless, there are other issues that make the comparison even more difficult. The non-existance of a dominant class of tests justifies the variety of alternatives, finding in the literature tests based on smoothing methods, with different nonparametric pilot estimators, such as kernels, splines, orthogonal expansions, etc.

3 Recent GoF Tests for Regression Models

In the last decade, some alternatives procedures for testing a regression model have been proposed. In this section, we will give some ideas of the extensions of the likelihood ratio tests, based on the empirical likelihood, as well as other new tests based on the empirical distribution of the residuals.

3.1 The Generalized Likelihood Ratio Test

In a regression model, assume that $\varepsilon \sim N(0, \sigma^2)$. The Generalized Likelihood Ratio Test statistic is given by:

$$T_n = l(m_{nh}, \hat{\sigma}) - l(m_{\hat{\theta}}, \hat{\sigma}_0)$$

where $l(m, \sigma)$ is the Gaussian log-likelihood:

$$l(m, \sigma) = -n \log(\sqrt{2\pi\sigma^2}) - \frac{1}{2\sigma^2} \sum_{i=1}^{n} (Y_i - m(X_i))^2$$

and

$$\hat{\sigma}_0^2 = \frac{1}{n} \sum_{i=1}^{n} (Y_i - m_{\hat{\theta}}(X_i))^2, \quad \hat{\sigma}^2 = \frac{1}{n} \sum_{i=1}^{n} (Y_i - m_{nh}(X_i))^2$$

giving the loglikelihood under H_0 and the estimated loglikelihood with nonparametric estimator m_{nh} of the regression function.

This procedure is studied in detail in [16] and more recently, in [17], and it represents and extension of the natural idea of the likelihood ratio, where under the alternative, the likelihood is evaluated on a nonparametric estimator of the regression function, which is not the maximum likelihood estimator, in a general setting.

It should be also noticed that:

$$T_n = \frac{n}{2} \log \frac{RSS_0}{RSS_1} \approx \frac{n}{2} \frac{RSS_0 - RSS_1}{RSS_0},$$

being RRS_0 the residual sum of squares under H_0 and RSS_1 the residual sum of squares under the alternative, which resembles the F-test expression. In the numerator, we have the variance difference test, mentioned in Section 1 and this general view allows to establish connection with other tests.

Once again, the calibration can be done by an asymptotic Gaussian approximation, by resampling or with an intermediate approach using:

$$r_k T_n \sim \chi^2_{\sqrt{n}}, \quad \text{with } r_k = \frac{1}{h} \frac{(K(0) - 1/2\|K\|^2)^2}{\|K - 1/(2K * K)\|^2} \mu(Supp(X))$$

where μ denotes the Lebesgue measure and $Supp(X)$ is the support of the explanatory variable. This is an extension of the Wilks theorem for parametric alternative hypothesis (see [16] for details).

3.2 The Local Empirical Likelihood Ratio Test

With the view described in this section, the test statistics are now based on the local version of the empirical likelihood (see [30]). A possible test statistic is the integrated empirical likelihood ratio test:

$$T_n = \int \left(-2 \log(L_n(\tilde{m}(x, \hat{\theta}))n^n)\right) \omega(x)dx,$$

where

$$L_n(\tilde{m}(x, \hat{\theta})) = \max \prod_{i=1}^{n} p_i(x),$$

subject to $\displaystyle\sum_{i=1}^{n} p_i(x) = 1$ and

$$\sum_{i=1}^{n} p_i(x) K\left(\frac{x - X_i}{n}\right)(Y_i - \tilde{m}(x, \hat{\theta})) = 0, \quad \text{with } \tilde{m}(x, \hat{\theta}) = \mathbb{E}_{\hat{\theta}}(m_{nh}(x)).$$

See [6] and [7] for more details.

For $H_0 : m \in \mathcal{M} = \{m_\theta\}_{\theta \in \Theta \subset \mathbb{R}^p}$, this test verifies that:

$$h^{-p/2}(T_n - c_1) \xrightarrow{d} N(0, c_2),$$

where $c_1 = 1$ and

$$c_2 = 2 \frac{\int (K * K)^2(x)dx}{\int K^2(x)dx} \int \omega(x)dx.$$

An interesting property is that the limit is distribution free.

3.3 The Empirical Process Based on the LRT

A recent study in [41] combines the empirical likelihood ideas with the empirical regression process view. Taking into account that:

$$H_0 : m \in \{m_\theta\}_{\theta \in \Theta} \Leftrightarrow \mathbb{E}\left(\mathbb{I}(X \leq x)(Y - m_{\theta_0}(X))\right) = 0,$$

for $\theta_0 \in \theta$ and $x \in Supp(X)$, and using the empirical likelihood for $p = 1$:

$$L(\mathbb{F}) = \prod_{i=1}^{n} \left(\mathbb{F}(X_i, Y_i) - \mathbb{F}(X_i^-, Y_i) - \mathbb{F}(X_i, Y_i^-) + \mathbb{F}(X_i^-, Y_i^-)\right)$$

with $\{(X_i, Y_i)\}_{i=1}^n$ iid from (X, Y) with joint distribution $\mathbb{F}$, the test statistic can be built based on the empirical likelihood ratio:

$$\begin{aligned}\Lambda_n(x) &= \frac{\sup\{L(\mathbb{F});\ \mathbb{E}_{\mathbb{F}}\left(\mathbb{I}(X \leq x)(Y - m_{\hat{\theta}}(X))\right) = 0\}}{\sup L(\mathbb{F})} \\ &= \sup\left\{n^n \prod_{i=1}^{n} p_i;\ p_i \geq 0,\ i = 1, \ldots, n,\ \sum_{i=1}^{n} p_i = 1, \right. \\ &\qquad \left. \sum_{i=1}^{n} p_i \mathbb{I}(X_i \leq x)(Y_i - m_{\hat{\theta}}(X_i)) = 0\right\}.\end{aligned}$$

The test statistic T_n may be any continuous functional of $\Lambda_n(\cdot)$. The test statistic is associated with the process $-2\log\Lambda_n(x)$, and its calibration can be done by Bootstrap, Monte Carlo methods or using the asymptotic distribution.

3.4 Test Based on the Empirical Distribution of the Residuals

Consider the ideas of the location-scale regression model, in a nonparametric framework:

$$Y = m(X) + \sigma(X)\varepsilon,$$

where $m(x) = \mathbb{E}(Y|X = x)$ is a smooth regression function, $\sigma^2(x) = \mathrm{Var}(Y|X = x)$ is the variance function and ε denotes the error term (independent of the covariate) with distribution function:

$$F_\varepsilon(y) = \mathbb{P}(\varepsilon \leq y) = \mathbb{P}\left(\frac{Y - m(X)}{\sigma(X)} \leq y\right).$$

Assume that $\{(X_i, Y_i)\}_{i=1}^n$ are i.i.d. observations from (X, Y). In this context, [2] proposed estimating the error distribution by the empirical distribution of estimated residuals:

$$\hat{F}_\varepsilon(y) = \frac{1}{n}\sum_{i=1}^{n} \mathbb{I}\left(\frac{Y_i - m_{nh}(X_i)}{\hat{\sigma}(X_i)} \leq y\right),$$

where $m_{nh}(x)$ and $\hat{\sigma}^2(x)$ are Nadaraya-Watson estimators of m and σ^2:

$$\hat{\sigma}^2(x) = \sum_{i=1}^{n} W_{ni}(x) Y_i^2 - \hat{m}^2(x),$$

where W_{ni} denote the Nadaraya-Watson weights. During the last years, the estimation of the error distribution has been used to test several hypothesis about the regression model. The basic idea of the tests consists of getting a nonparametric estimator of the error distribution, namely $\hat{F}_{\varepsilon}$ and a parametric estimator under the null hypothesis, $\hat{F}_{\varepsilon 0}$. These two estimators can be compared through a certain criterion and Bootstrap can be used to approximate the critical values of the test.

An important example is the GoF test for parametric regression models, which is studied in [40] for $p = 1$ and [29], for $p \geq 1$.

4 GoF in Other Contexts

All the previous GoF tests can be extended in several ways, for instance, extending the null hypothesis (testing partially linear models, significance tests or testing additivity) as well as considering incomplete and dependent data. In this final section, we will give some references on the extensions of the studied GoF tests to the incomplete and dependent data settings.

4.1 Incomplete Data

A particular case of incomplete information arises when data are censored and/or truncated. The GoF testing problem for conditional models have been studied in [5].

In [20], a GoF test adapted to the situation where the Y variable may be missing is proposed. In the missing data case, we may not observe Y_i for some index i, which implies that we have to deal with: (X_i, Y_i) if Y_i is observed and $(X_i, \cdot)$, otherwise. To control whether an observation is complete or not, a new variable δ is introduced, as an indicator of missing observations.

Other situations that can be treated in the incomplete data context are, for example, length biased data and double censored or truncated data.

4.2 Dependent Data

Dependent data usually appear when samples are collected along time and/or over space. For the time dependent setting, consider $\{(X_t, Y_t\}_{1 \leq t \leq T}$ a sequence of observations from a joint stationary density function, $f(x, y)$ corresponding to (X, Y), a $(d+1)$-dimensional random vector.

In order to describe the behaviour of (X, Y), we may consider the conditional density of Y given $X = x$, denoted by $f(y|x)$ and the conditional moments $m_j(x) = \mathbb{E}(Y^j | X = x)$, for $j = 1, 2, \ldots$. For $j = 1$, we have the conditional expectation and m_1 is simply denoted by m. Consider the model:

$$Y_t = m(X_t) + \varepsilon_t, \quad t = 1, \ldots, T$$

where $\{\varepsilon_t\}$ is an iid sequence with $\mathbb{E}(\varepsilon_t) = 0$ and $\mathbb{E}(\varepsilon_t^2) = \sigma^2 < \infty$ and $\{X_t\}$ is strictly stationary. If we are interested in testing the null hypothesis of m

belonging the a certain parametric family, it is possible to generalize to this context the test introduced in [21], among others, which have already been described for the iid case (see [18]).

In addition, the empirical regression process introduced in Section 1 can be also generalized to a more general framework. For instance, if $X_t = Y_{t-1}$, [25] studied the problem of testing linearity (AR(1) assumption). The same problem is also tackled in [10] taking $X_t = (Y_{t-1}, \ldots, Y_{t-s})$, and in this same case, [39] introduced a test for the link of a linear model.

In [12], a nonlinear model for $m(x) = \mathbb{E}(Y|X = x)$ is tested. The author introduces a general empirical regression process given by:

$$R_{n,\omega}(x, \hat{\theta}) = \frac{1}{\sqrt{n}} \sum_{t=1}^{n} (Y_t - m_{\hat{\theta}}(X_t))\omega(X_t, x).$$

Some particular cases taking $\omega(X_t, x) = \mathbb{I}(X_t \le x)$, where $\mathbb{I}$ denotes the indicator function, or $\omega(X_t, x) = \mathbb{I}(\hat{\beta}^t X_t \le x)$ are also studied.

In the spatial context, the interest may be focused on the trend function collecting the large scale variation, or in the dependence structure, being this last one crucial for prediction. This small scale variability structure is capture by the covariogram C. For second order stationary processes, the covariogram C admits a Fourier transform given by the so-called spatial spectral density f. Hence, the problem of testing a parametric family for the covariogram can be written in terms of the spectral density. Besides, the spectral density can be nonparametrically estimated by the spatial periodogram, providing an alternative statement of the testing problem in terms of f or its logarithm $\log f$. Actually, the log-spectral density can be seen as a regression function of the log-periodogram values over the different frequencies.

This alternative statement of the testing problem allows for the application of the GoF testing methods in regression models introduced in the previous section, with suitable modifications. In [8], the authors extend the tests of [16] and [31] to the spatial setting, providing also calibration techniques.

Acknowledgement. The authors would like to acknowledge the financial support of Project MTM2008-03010, from the Spanish Ministry of Science and Innovation.

References

1. Alcalá, T., Cristóbal, J.A., González-Manteiga, W.: Goodness of fit test for linear models based on local polynomials. Statist. Prob. Lett. 42, 39–46 (1999)
2. Akritas, M.G., Van Keilegom, I.: Non-parametric estimation of the residual distribution. Scand. J. Statist. 28, 549–567 (2001)
3. Azzalini, A., Bowman, A.W.: On the use of nonparametric regression for checking linear relationships. J. Royal Statist. Soc. Ser. B. Method 55, 549–557 (1993)
4. Bickel, P.J., Rosenblatt, M.: On some global measures of the deviations of density function estimates. Ann. Statist. 1, 1071–1095 (1973)

5. Cao, R., González-Manteiga, W.: Goodness-of-fit tests for conditional models under censoring and truncation. J. Econometrics 143, 166–190 (2008)
6. Chen, S.X., Cui, H.: An extended empirical likelihood for generalized linear models. Statistica Sinica 13, 69–81 (2003)
7. Chen, S.X., Van Keilegom, I.: A review on empirical likelihood methods for regression. TEST 18, 415–447 (2009)
8. Crujeiras, R., Fernández-Casal, R., González-Manteiga, W.: Goodness-of-fit tests for the spatial spectral density. Stoch Environ. Res. Risk Asses. 24, 67–79 (2010)
9. Dette, H.: A consistent test for the functional form of a regression based on a difference of variance estimators. Ann. Statist. 27, 1012–1040 (1999)
10. Domínguez, M.A., Lobato, I.N.: Testing the martingale difference hypothesis. Economet. Rev. 22, 351–377 (2003)
11. Durbin, J.: Weak convergence of the sample distribution function when parameters are estimated. Ann. Statist. 1, 279–290 (1973)
12. Escanciano, J.C.: Model checks using residual marked empirical processes. Statistica Sinica 17, 115–138 (2007)
13. Fan, Y.: Testing the goodness of fit of a parametric density function by the kernel method. Economet. Theory 10, 316–356 (1994)
14. Fan, J., Gijbels, I.: Local polynomial modelling and its applications. Monographs on Statistics and Applied Probability, vol. 66. Chapman & Hall, London (1996)
15. Fan, Y.: Goodness-of-fit tests based on kernel density estimators with fixed smoothing parameters. Economet. Theory 14, 604–621 (1998)
16. Fan, J., Zhang, C., Zhang, J.: Generalised likelihood ratio statistics and Wilks phenomenon. Ann. Statist. 29, 153–193 (2001)
17. Fan, J., Jiang, J.: Nonparametric inference with generalized likelihood ratio tests. TEST 16, 409–444 (2007)
18. Gao, J.: Nonlinear Time Series. Semiparametric and Nonparametric Methods. Chapman & Hall, London (2007)
19. González-Manteiga, W., Cao, R.: Testing the hypothesis of a general linear model using nonparametric regression estimation. TEST 2, 161–188 (1993)
20. González-Manteiga, W., Pérez-González, A.: Goodness-of-fit tests for linear regression models with missing response data. Canad. J. Statist. 34, 1–22 (2006)
21. Härdle, W., Mammen, E.: Comparing nonparametric versus parametric regression fits. Ann. Statist. 21, 1926–1947 (1993)
22. Horowitz, J., Spokoiny, V.: An adaptive, rate-optimal test of a parametric mean-regression model against a nonparametric alternative. Econometrica 69, 599–631 (2001)
23. de Jong, P.: A central limit theorem for generalized quadratic forms. Prob. Theor. Rel. Fields 75, 261–277 (1987)
24. Khmadladze, E.V., Koul, H.L.: Martingale transforms goodness-of-fit tests in regression models. Ann. Statist. 37, 995–1034 (2004)
25. Koul, H.L., Stute, W.: Nonparametric model checks for time series. Ann. Statist. 27, 204–236 (1999)
26. Liu, R.: Bootstrap procedures under some non-i.i.d. models. Ann. Statist. 16, 1696–1708 (1988)
27. Miles, D., Mora, J.: On the performance of nonparametric specification test in regression models. Comp. Statist. Data Anal. 42, 477–490 (2002)

28. Nadaraya, E.A.: On estimating regression. Theor. Prob. and its Appl. 10, 186–196 (1964)
29. Neumeyer, N., Van Keilegom, I.: Estimating the error distribution in nonparametric multiple regression with applications to model testing. J. Multiv. Anal. 101, 1067–1078 (2010)
30. Owen, A.B.: Empirical Likelihood. Chapman & Hall, London (2001)
31. Paparoditis, E.: Spectral density based goodness-of-fit tests for time series models. Scand. J. Statist. 27, 143–176 (2000)
32. Parzen, E.: On estimation of a probability density function and mode. Ann. Math. Statist. 33, 1065–1076 (1962)
33. Rosenblatt, M.: Remarks on some nonparametric estimates of a density function. Ann. Math. Statist. 27, 832–837 (1956)
34. Rosenblatt, M.: Stochastic Curve Estimation. Institute of Mathematical Statistics, Hayward (1991)
35. Stute, W., González-Manteiga, W., Presedo-Quindimil, M.: Bootstrap based goodness-of-fit tests. Metrika 40, 243–256 (1993)
36. Stute, W.: Nonparametric model checks for regression. Ann. Statist. 25, 613–641 (1997)
37. Stute, W., González-Manteiga, W., Presedo-Quindimil, M.: Bootstrap approximations in model checks for regression. J. Amer. Statist. Assoc. 93, 141–149 (1998)
38. Stute, W., Thies, S., Zhu, L.X.: Model checks for regression: an innovation process approach. Ann. Statist. 26, 1916–1934 (1998)
39. Stute, W., Presedo-Quindimil, M., González-Manteiga, W., Koul, H.L.: Model checks of higher order time series. Statist. Prob. Lett. 76, 1385–1396 (2006)
40. Van Keilegom, I., González-Manteiga, W., Sánchez-Sellero, C.A.: Goodness-of-fit tests in parametric regression based on the estimation of the error distribution. TEST 17, 401–415 (2008)
41. Van Keilegom, I., Sánchez-Sellero, C.A., González-Manteiga, W.: Empirical likelihood based testing for regression. Elect. J. Statist. 2, 581–604 (2008)
42. Watson, G.S.: Smooth regression analysis. Sankhyā Ser. A 26, 359–372 (1964)
43. Wu, C.F.J.: Jackknife, bootstrap and other resampling methods in regression analysis. Ann. Statist. 14, 1261–1350 (1986)
44. Zheng, J.X.: A consistent test of functional form via nonparametric estimation techniques. J. Economet. 75, 263–289 (1996)
45. Zhu, L.: Nonparametric Monte Carlo Tests and their Applications. Lect. Notes Statist. 182 (2005)

Linear Properties in a Continuous Time Linear Model*

Pilar Ibarrola[1] and Ana Pérez-Palomares[2]

[1] Departamento de Estadística e Investigación Operativa,
Facultad de Matemáticas, Universidad Complutense, 28040, Madrid, Spain
ibarrola@mat.ucm.es

[2] Departamento de Métodos Estadísticos, Universidad de Zaragoza,
Facultad de Ciencias, C/ Pedro Cerbuna 12, 50009 Zaragoza, Spain
anapp@unizar.es

Summary. This work provides further contribution to the linear properties in a continuous time linear model. We deal with linear sufficiency and linear completeness properties, together with the linear admissibility property. These concepts were originally introduced and characterized in a discrete time context and subsequently were extended by the authors of the present paper to a continuous time linear model. Our objective is to study in depth these properties showing a general unified context where the classical linear model appears as a particular case.

1 Introduction

The problem of linear estimation in a linear model has been extensively studied in the literature. In particular, linear properties have been characterized by different authors in the last years. The concepts of linear sufficiency and linear completeness were introduced by Baksalary and Kala [1] and by Drygas [6], respectively, in the following way: in a linear model $(Y, X\beta, V), (E[Y] = X\beta, Var(Y) = V)$, a linear combination AY is a linear sufficient estimator if each BLUE can be written as BAY, for a matrix B. A linear combination AY is a linear complete estimator if the unique estimator BAY unbiased for 0 is the 0 estimator. These properties have been studied in [2], [6] and [10], among others.

The problem of finding and characterizing admissible estimators is an important topic in this context. In particular, many works have been published characterizing the linear admissibility property, see [23], [2] and [3], for instance.

Recently, we have defined and characterized these properties in a continuous time linear model. Our former works considered integral type estimators and the results were obtained for this class of estimators. The objective of

* In memory of Marisa.

the present paper is to show in a more general context the linear properties above mentioned extending some results given in those papers.

The paper is organized as follows: next section establishes the context in which we develop this work. In addition, we present the construction of BLUE estimators together with a representation of the deterministic part of the process. Section 3 deals with the linear sufficiency and linear completeness properties. The main result of this section is Theorem 3. Finally, in Section 4 we study the linear admissibility property.

2 Framework

From now on, let $(Z_t, t \in T)$, with T a compact set of $\mathbb{R}$, be a stochastic process with distribution P_0 in $(\mathbb{R}^T, \mathcal{F}_T)$ where $\mathcal{F}_T$ is the $\sigma-$algebra generated by $Z_t, t \in T$. Let E_0 be the mathematical expectation with respect to P_0. Suppose that $E_0[Z_t] = 0$, $t \in T$ and $E_0[Z_s Z_t] = B(s,t)$, $s, t \in T$ is a known continuous function in $T \times T$. For each $\theta \in \mathbb{R}^p$, we denote by P_θ the distribution of the process $(X_t, t \in T)$ which is defined as $X_t = A(t)\theta + Z_t$, $t \in T$, where $A(t)'$ is a vector in $\mathbb{R}^p$, with known continuous components in T. Let μ be the normal distribution on $(\mathbb{R}^p, \mathcal{B}(\mathbb{R}^p))$ with zero mean and covariance matrix I. $\bar{P}$ denotes the measure defined on $(\mathbb{R}^T, \mathcal{F}_T)$ as

$$\bar{P}(A) = \int_{\mathbb{R}^p} P_\theta(A) d\mu(\theta), \quad A \in \mathcal{F}_T.$$

The mathematical expectation with respect to $\bar{P}$ will be denoted by $\bar{E}$ and with respect to P_θ by E_θ. Thus, the process $(X_t, t \in T)$ is an element of $L^2(\mathbb{R}^T, \mathcal{F}_T, \bar{P})$ with

$$\bar{E}[X_t] = 0 \text{ and } \bar{E}[X_s X_t] = B(s,t) + A(s)A(t)', \quad s, t \in T. \tag{1}$$

We denote by $\bar{\mathcal{L}}(X_t, t \in T)$ the closure on $L^2(\mathbb{R}^T, \mathcal{F}_T, \bar{P})$ of the set of finite linear combinations of type $\sum_{i=1}^n c_i X_{t_i}$, $c_i \in \mathbb{R}$, $t_i \in T$. $\bar{\mathcal{L}}(X_t, t \in T)$ is a Hilbert space with the inner product $< Y, Z >= \bar{E}[YZ]$. We are interested in estimators constructed from the observed paths of the process $(X_t, t \in T)$ in a linear way, that is, we are interested in estimators belonging to the class $\bar{\mathcal{L}}(X_t, t \in T)$. We refer to this class of estimators as linear estimators. With this framework, we can essentially use the same concepts that in a discrete time context with some technical differences. From now on, the terms minimum variance, unbiased and uncorrelated estimators are all referred to the measure P_θ, $\theta \in \mathbb{R}^p$. When the measure $\bar{P}$ is involved, it will be mentioned explicitly.

Now, we define, for each $j = 1, \ldots, p$, the operator L_j as $L_j(\sum_{i=1}^n c_i X_{t_i}) = \sum_{i=1}^n c_i A^j(t_i)$, where A^j is the j-th component of A. These operators can be extended to $\bar{\mathcal{L}}(X_t, t \in T)$ in the following way. For each $Y \in \bar{\mathcal{L}}(X_t, t \in T)$, there exists a sequence $Y_n = \sum_{i=1}^n c_i X_{t_i}$ which converges to Y in

$L^2(\mathbb{R}^T, \mathcal{F}_T, \bar{P})$. We can write $Y_n = \int_T V_n(dt)X_t$. From (1) it is easy to see that

$$\bar{E}[(Y_n - Y_m)^2] = E_0[(Y_n - Y_m)^2] + \sum_{j=1}^{p} \left(\int_T (V_n - V_m)(dt)A^j(t) \right)^2 .$$

Since $(Y_n, n \geq 1)$ is a Cauchy sequence in $L^2(\mathbb{R}^T, \mathcal{F}_T, \bar{P})$, then $(Y_n, n \geq 1)$ and $\int_T V_n(dt)A(t)$ are Cauchy sequences, the first as an element in $L^2(\mathbb{R}^T, \mathcal{F}_T, P_0)$ and the second as a vector of real numbers. This implies that if $(Y_n, n \geq 1)$ converges to Y in $L^2(\mathbb{R}^T, \mathcal{F}_T, \bar{P})$ then it converges to Y in $L^2(\mathbb{R}^T, \mathcal{F}_T, P_0)$ and $L_j(Y_n) = \int_T V_n(dt)A^j(t)$ is convergent. Thus, we can define $L_j(Y) = \lim_{n\to\infty} L_j(Y_n)$. Therefore, for each j, L_j is a continuous linear operator and, applying the Riesz representation theorem, we can assure the existence of an element $\hat{\theta}_j \in \bar{\mathcal{L}}(X_t, t \in T)$ such that $L_j(Y) =< Y, \hat{\theta}_j >$, $Y \in \bar{\mathcal{L}}(X_t, t \in T)$. Defining $\hat{\theta}_B = (\hat{\theta}_1, \cdots, \hat{\theta}_p)'$ and taking $Y = X_t$, we have

$$A(t) =< X_t, \hat{\theta}'_B >= \bar{E}[X_t \hat{\theta}'_B], \quad t \in T. \tag{2}$$

It is immediate that

$$E_\theta[Y] = \bar{E}[Y\hat{\theta}'_B]\theta, \quad Y \in \bar{\mathcal{L}}(X_t, t \in T), \tag{3}$$

and, from (1),

$$\bar{E}[YZ] = E_0[YZ] + \bar{E}[Y\hat{\theta}'_B]\bar{E}[\hat{\theta}_B Z], \quad Y, Z \in \bar{\mathcal{L}}(X_t, t \in T). \tag{4}$$

As we shall see throughout the paper, equality (1) and therefore (10) and, (11), will play a relevant role in this work. In a discrete time linear model $(Y, X\beta, V), (E[Y] = X\beta, Var(Y) = V)$, the characterizations of the linear properties have been given in terms of the matrix $W = V + XX'$ whose analogous element in continuous time is $\bar{E}[X_s X_t]$. It is easy to verify that $X = WC$ for some matrix C but, as we have checked, it is not so immediate in this approach. In fact, in the problem of estimating the expectation of a process in a continuous time context, most of papers ([5] and [4], see among others) impose equality (1) with E_0 instead of $\bar{E}$, in order to construct optimum estimators. In this paper the unique assumption on the expectation and on the covariance of the process is their continuity, for that reason we have defined the measure $\bar{P}$ and we have proved equality (1).

Finally, we present the following notation to be used throughout the paper: $\Sigma = \bar{E}[\hat{\theta}_B \hat{\theta}'_B]$ and $C = E_0[\hat{\theta}_B \hat{\theta}'_B] = (I - \Sigma)\Sigma$. M^- denotes a g-inverse of a matrix M, that is, $MM^-M = M$.

2.1 BLUE Estimators

In this section we show a characterization of the BLUE estimators. Moreover, we give a representation of the deterministic part of each linear estimator.

From now on, an estimable linear combination is a linear combination of θ which can be unbiasedly estimated by elements of $\bar{\mathcal{L}}(X_t, t \in T)$. We say that a linear estimator is the BLUE for an estimable linear combination of θ if it is of minimum variance among all linear estimators unbiased for the linear combination.

First, note that (10) implies $\bar{E}[Y\hat{\theta}'_B] = 0$ for each linear estimator Y unbiased for 0 and using (11) we have $E_0[Y\hat{\theta}'_B] = 0$. This means that $\hat{\theta}_B$ is uncorrelated with all linear estimators unbiased for 0, so $\hat{\theta}_B$ is the BLUE for its expectation. Thus, $A(t)\Sigma^-\hat{\theta}_B$ is the BLUE for $A(t)\theta$, $t \in T$, and the estimator $\hat{\theta}_B$ generates all BLUE estimators.

As we have commented, it is possible to find deterministic elements in $\bar{\mathcal{L}}(X_t, t \in T)$, since we have not imposed restrictions on the covariance function B. In fact, an estimable linear combination $M\theta$, which is completely determined by observing $(X_t, t \in T)$, can be characterized as follows: let Y be a linear estimator with $M = \bar{E}[Y\hat{\theta}'_B]$ and null variance. We have that $Y = M\Sigma^-\hat{\theta}_B$ a.s. and $0 = E_0[Y\hat{\theta}_B] = M\Sigma^-C$. Moreover $M\Sigma^-C = 0 \Leftrightarrow M\Sigma^-C\Sigma^- = MP^- = 0$, where $P = \Sigma C^-\Sigma$ and $P^- = \Sigma^-C\Sigma^-$. Thus, each estimable linear combination $M\theta$ which generates a deterministic BLUE verifies $MP^- = 0$, that is, $M' = (I - PP^-)R$ for some R matrix. Thereby, the main deterministic component of the process can be represented as:

$$C_t = A(t)(I - P^-P)\Sigma^-\hat{\theta}_B, t \in T.$$

This component coincides with that given by the authors in [9, Lemma 1]. Obviously, we can write

$$X_t = C_t + X_t^*, \quad t \in T,$$

with $X_t^* = X_t - C_t$. Each linear estimator $\hat{\theta}$ can be factorized as

$$\hat{\theta} = C_{\hat{\theta}} + \hat{\theta}^*,$$

where $C_{\hat{\theta}} = \bar{E}[Y\hat{\theta}'_B](I - P^-P)\Sigma^-\hat{\theta}_B$ is a deterministic estimator and $\hat{\theta}^* = \hat{\theta} - C_{\hat{\theta}} \in \bar{\mathcal{L}}(X_t^*, t \in T)$.

The following lemma gives some straightforward properties of the process $(X_t^*, t \in T)$.

Lemma 1. *The process $(X_t^*, t \in T)$ satisfies*

(i) $E_\theta[X_t^*] = A(t)P^-P\theta$, $t \in T$, *and*

(ii) There exists an estimator $\hat{\theta}_B^* \in \bar{\mathcal{L}}(X_t^*, t \in T)$ *such that* $E_0[X_t^*\hat{\theta}_B^{*'}] = A(t)P^-P$, $t \in T$. *Thus,* $\bar{\mathcal{L}}(X_t^*, t \in T)$ *does not contain deterministic linear estimators.*

Proof. (i) It is immediate by the definition of X_t^*. (ii) Consider $\hat{\theta}_B^* = P\Sigma^-\hat{\theta}_B$. On the one hand $C_{\hat{\theta}_B^*} = P(I - P^-P)\Sigma^-\hat{\theta}_B = 0$ so $\hat{\theta}_B^* \in \bar{\mathcal{L}}(X_t^*, t \in T)$.

On the other hand, $E_0[X_t^* \hat{\theta}_B^{*\prime}] = E_0[X_t \hat{\theta}_B'] \Sigma^- P = A(t)(I - \Sigma)\Sigma^- P = A(t)\Sigma^- C \Sigma^- P = A(t)P^- P$, $t \in T$. □

Lemma 1 proves that $\hat{\theta}_B^*$ verifies condition (1) with E_0 instead of $\bar{E}$ in the process $(X_t^*, t \in T)$, so $\hat{\theta}_B^*$ generates the BLUE estimators for this model. Some results of the present paper are given in terms of $\bar{E}$ and in its equivalent form in terms of E_0. In most cases, we shall omit the proof of this last part.

We conclude this section giving a characterization of BLUE estimators.

Lemma 2. *Let $Y \in \bar{\mathcal{L}}(X_t, t \in T)$ and $c\theta$ an estimable linear combination. Then, Y is the BLUE for $c\theta$ if and only if*

$$\bar{E}[Y X_t] = c\Sigma^- A(t)', \quad t \in T, \tag{5}$$

equivalently

$$E_0[Y X_t] = cP^- A(t)', \quad t \in T.$$

Proof. Suppose that Y is the BLUE for $c\theta$. Then, $Y = c\Sigma^- \hat{\theta}_B$ a.s. from the uniqueness of a BLUE estimator. Thus, using (11) we obtain the only if part. Conversely, (5) and (11) imply

$$\bar{E}[(Y - c\Sigma^- \hat{\theta}_B) X_t] = 0, \quad t \in T.$$

Then, $Y = c\Sigma^- \hat{\theta}_B$, a.s. and the conclusion follows. The lemma is proved. □

3 Linear Sufficiency and Linear Completeness

The concept of linear sufficiency in this context is straightforward. Let K be a compact subset of $\mathbb{R}$. We consider a family $(\hat{\theta}_r, r \in K)$ of elements in $\bar{\mathcal{L}}(X_t, t \in T)$. If K is not a finite set then we shall suppose that $(\hat{\theta}_r, r \in K)$ is continuous in square mean sense. We denote by $\bar{\mathcal{L}}(\hat{\theta}_r, r \in K)$ the closure in $L^2(\mathbb{R}^T, \mathcal{F}_T, \bar{P})$ of the linear combinations of $(\hat{\theta}_r, r \in K)$. Then $(\hat{\theta}_r, r \in K)$ is linearly sufficient if the BLUE of each estimable linear combination belongs to $\bar{\mathcal{L}}(\hat{\theta}_r, r \in K)$. The following characterization is an extension of that given in [1].

Theorem 1. *$(\hat{\theta}_r, r \in K)$ is linearly sufficient if and only if there exists $Y \in \bar{\mathcal{L}}(\hat{\theta}_r, r \in K)$ such that*

$$\bar{E}[Y X_t] = A(t)', \quad t \in T, \tag{6}$$

equivalently, there exists $Y^* \in \bar{\mathcal{L}}(\hat{\theta}_r^*, r \in K)$ such that

$$E_0[Y^* X_t^*] = PP^- A(t)', \quad t \in T.$$

Proof. First we prove the right implication. Let Y be the BLUE for $\Sigma\theta$. By hypothesis, $Y \in \bar{\mathcal{L}}(\hat{\theta}_r, r \in K)$ and applying Lemma 2, we obtain $\bar{E}[YX_t] = \Sigma\Sigma^- A(t)' = A(t)'$, which is (6).

Conversely, if $Y \in \bar{\mathcal{L}}(\hat{\theta}_r, r \in K)$ satisfies (6), Lemma 2 assures that Y is the BLUE for $\Sigma\theta$ and therefore $c\Sigma^- Y$ is the BLUE for each estimable linear combination $c\theta$. The theorem is proved. □

Now, we consider the concept of linear completeness whose definition is the following: $(\hat{\theta}_r, r \in K)$ is linearly complete if for each $Y \in \bar{\mathcal{L}}(\hat{\theta}_r, r \in K)$ such that $E_\theta[Y] = 0$, $\theta \in \mathbb{R}^p$, we have that $Y = 0$, a.s.

Theorem 2. $(\hat{\theta}_r, r \in K)$ *is linearly complete if and only if*

$$\bar{E}[\hat{\theta}_r X_t] = \bar{E}[\hat{\theta}_r \hat{\theta}'_B] h(t), \quad t \in T, \quad r \in K, \tag{7}$$

for a vector h of continuous functions on T, equivalently

$$E_0[\hat{\theta}_r X_t] = E_0[\hat{\theta}_r \hat{\theta}^{*\prime}_B] g(t), \quad t \in T, \quad r \in K,$$

for a vector g of continuous functions on T.

Proof. First, we prove the left implication. Suppose that $Y \in \bar{\mathcal{L}}(\hat{\theta}_r, r \in K)$ verifies $E_\theta[Y] = 0$, for all θ, equivalently $\bar{E}[Y\hat{\theta}'_B] = 0$. The construction of $\bar{\mathcal{L}}(\hat{\theta}_r, r \in K)$ and (7) give $\bar{E}[YX_t] = \bar{E}[Y\hat{\theta}'_B]h(t) = 0$, $t \in T$. Since $Y \in \bar{\mathcal{L}}(X_t, t \in T)$, we obtain $Y = 0$, a.s. proving the left implication.

For the right implication, first we consider that K is a finite set $K = \{r_1, \ldots, r_n\}$. Define the row vector $g(r_i) = \bar{E}[\theta_{r_i} \hat{\theta}'_B]$, $i = 1, \ldots, n$ and the $n \times n$ matrix $G = (g(r_1)', \ldots, g(r_n)')$. Consider the estimator

$$\bar{\theta}_{r_i} = g(r_i) H^- G \begin{pmatrix} \theta_{r_1} \\ \vdots \\ \theta_{r_n} \end{pmatrix},$$

where $H = GG'$. Then,

$$\bar{E}[\bar{\theta}_{r_i} \hat{\theta}'_B] = g(r_i)(GG')^- GG' = g(r_i) H^- H = g(r_i), \quad i = 1, \ldots, n.$$

On the one hand, the estimator $\hat{\theta}_{r_i} - \bar{\theta}_{r_i}$ has 0 expectation and, on the other hand, $\hat{\theta}_{r_i} - \bar{\theta}_{r_i} \in \bar{\mathcal{L}}(\hat{\theta}_r, r \in K)$. Due to the linear completeness of $(\hat{\theta}_r, r \in K)$, we conclude $\hat{\theta}_{r_i} = \bar{\theta}_{r_i}$, $i = 1, \ldots, n$, and then

$$\bar{E}[\hat{\theta}_{r_i} X_t] = \bar{E}[\hat{\theta}_{r_i} \hat{\theta}_B] h(t),$$

with $h(t) = H^- G(\bar{E}[\hat{\theta}_{r_1} X_t], \ldots, E[\hat{\theta}_{r_n} X_t])'$, $t \in T$. If K is not a finite set, the proof is the same with $H = \int_K g(s)' g(s) ds$ and $\bar{\theta}_r = g(r) H^- \int_K g(s)' \theta_s ds$, $r \in K$. □

Note that in the proof of the previous theorem we have obtained the following results.

Corollary 1. *$(\hat{\theta}_r, r \in K)$ is linearly complete if and only if $\hat{\theta}_r = g(r)\bar{\theta}$, a.s. $r \in K$, where $\bar{\theta} \in \bar{\mathcal{L}}(\hat{\theta}_r, r \in K)$ and $g(r) = \bar{E}[\hat{\theta}_r \hat{\theta}'_B]$, $r \in K$.*

Corollary 2. *$(\hat{\theta}_r, r \in K)$ is linearly complete if and only if*

$$\hat{\theta}_r = C_{\hat{\theta}_r} + g^*(r)\bar{\theta}^*, a.s. \quad r \in K,$$

where $C_{\hat{\theta}_r}$ is its deterministic component, $\bar{\theta}^* \in \bar{\mathcal{L}}(\hat{\theta}^*_r, r \in K)$ and $g^*(r) = E_0[\hat{\theta}_r \hat{\theta}'_B]$, $r \in K$.

To conclude this section, we introduce the concept of linear minimal sufficiency. A linear estimator $(\hat{\theta}_r, r \in K)$ is linear minimal sufficient if it is linear sufficient and for each linear sufficient estimator $(\bar{\theta}_s, s \in S)$ we have $\hat{\theta}_r \in \bar{\mathcal{L}}(\bar{\theta}_s, s \in S)$, $r \in K$.

Let $(\hat{\theta}^{(1)}_r, r \in K)$ and $(\hat{\theta}^{(2)}_s, s \in S)$ be two linear estimators. We consider the functions

$$g_Y(s) = \bar{E}[Y\hat{\theta}^{(2)}_s], \quad s \in S, \quad Y \in \bar{\mathcal{L}}(\hat{\theta}^{(1)}_r, r \in K). \tag{8}$$

Thus, we define

$$\mathcal{H}_{\hat{\theta}^{(1)}}(\hat{\theta}^{(2)}) = \{g_Y, Y \in \bar{\mathcal{L}}(\hat{\theta}^{(1)}_r, r \in K)\},$$

with g_Y defined in (8). In order to abridge notation we write $\mathcal{H}(\hat{\theta}^{(2)})$ when $(\hat{\theta}^{(1)}_r, r \in K) = (X_t, t \in T)$ and $\mathcal{H}_{\hat{\theta}^{(1)}}$ when $(\hat{\theta}^{(2)}_s, s \in S) = (X_t, t \in T)$. These elements allow us to characterize the linear properties in a unified context in the following way:

Theorem 3. *Let $(\hat{\theta}_r, r \in K)$ be a linear estimator. Then,*

(i) $(\hat{\theta}_r, r \in K)$ is linearly sufficient $\Leftrightarrow \mathcal{H}_{\hat{\theta}_B} \subseteq \mathcal{H}_{\hat{\theta}}$.
(ii) $(\hat{\theta}_r, r \in K)$ is linearly complete $\Leftrightarrow \mathcal{H}(\hat{\theta}) \subseteq \mathcal{H}_{\hat{\theta}_B}(\hat{\theta})$.
(iii) $(\hat{\theta}_r, r \in K)$ is linearly sufficient and linearly complete $\Leftrightarrow \mathcal{H}_{\hat{\theta}_B} = \mathcal{H}_{\hat{\theta}}$.
(iv) $(\hat{\theta}_r, r \in K)$ is linearly sufficient and linearly complete $\Leftrightarrow (\hat{\theta}_r, r \in K)$ is linearly minimal sufficient.

Proof. (i) First, we observe that $\mathcal{H}_{\hat{\theta}_B} = \{A(t)c, t \in T, c \in \mathbb{R}^p\}$. On the other hand, $(A(t), t \in T) \in \mathcal{H}_{\hat{\theta}}$ is equivalent to condition (6). Then, applying Theorem 1 we obtain the claim.

(ii)$\mathcal{H}(\hat{\theta}) \subseteq \mathcal{H}_{\hat{\theta}_B}(\hat{\theta})$ means that for $t \in T$, $(\bar{E}[X_t\hat{\theta}_r], r \in K) \subseteq \mathcal{H}_{\hat{\theta}_B}(\hat{\theta})$, that is, there exists a vector $h(t) \in \mathbb{R}^p$ such that $\bar{E}[X_t\hat{\theta}_r] = h(t)'\bar{E}[\hat{\theta}_B\hat{\theta}_r]$, $r \in K$, which is, from Theorem 2, the property of linear completeness.

(iii) First, we prove the left implication. From part (i) of the present theorem, the inclusion $\mathcal{H}_{\hat{\theta}_B} \subseteq \mathcal{H}_{\hat{\theta}}$ is equivalent to the linear sufficiency of

$(\hat{\theta}_r, r \in K)$. On the other hand, $\mathcal{H}_{\hat{\theta}} \subseteq \mathcal{H}_{\hat{\theta}_B}$ is equivalent to the following: for each $r \in K$, there exists a vector c_r such that

$$\bar{E}[\hat{\theta}_r X_t] = c_r' \bar{E}[\hat{\theta}_B X_t], \quad t \in T.$$

Then, $\bar{E}[\hat{\theta}_r \hat{\theta}_B'] = c_r' \bar{E}[\hat{\theta}_B \hat{\theta}_B'] = c_r' \Sigma$ and, therefore,

$$\bar{E}[\hat{\theta}_r X_t] = c_r' \bar{E}[\hat{\theta}_B X_t] = c_r' \Sigma^- \Sigma \bar{E}[\hat{\theta}_B X_t] = \bar{E}[\hat{\theta}_r \hat{\theta}_B'] h(t), \tag{9}$$

with $h(t) = \Sigma^- \bar{E}[\hat{\theta}_B X_t]$, $t \in T$. Equality (9) is equivalent to the linear completeness of the estimator and it concludes the proof of the only if part.

Conversely, if $\hat{\theta}_r$ is linearly sufficient and linear complete, it is the BLUE for its expectation. By applying Lemma 2, we obtain

$$\bar{E}[\hat{\theta}_r X_t] = \bar{E}[\hat{\theta}_r \hat{\theta}_B'] \Sigma^- \bar{E}[\hat{\theta}_B X_t],$$

which yields $\mathcal{H}_{\hat{\theta}} \subseteq \mathcal{H}_{\hat{\theta}_B}$. The other inclusion follows from part (i) of the present theorem. The proof is complete.

(iv) Suppose that $(\hat{\theta}_r, r \in K)$ is linearly sufficient and linearly complete. Let $(\bar{\theta}_s, s \in S)$ be a linear sufficient estimator. By definition of this property, the BLUE for $(E_\theta[\hat{\theta}_r], r \in K)$ belongs to the class $\bar{\mathcal{L}}(\bar{\theta}_s, s \in S)$. On the other hand, $(\hat{\theta}_r, r \in K)$ is the BLUE for its expectation because it is linearly sufficient and complete, therefore $\hat{\theta}_r \in \bar{\mathcal{L}}(\bar{\theta}_s, s \in S)$, proving the linear minimal sufficiency of $(\hat{\theta}_r, r \in K)$.

For the converse implication, we have only to prove the linear completeness of $(\hat{\theta}_r, r \in K)$. Taking into account that $\hat{\theta}_B$ is a linear sufficient estimator, we can write $\hat{\theta}_r = c_r \hat{\theta}_B'$, for a row vector c_r. Since $\hat{\theta}_B$ is a linear complete estimator we conclude the proof of the theorem. □

Note 1. In a discrete linear model, Drygas [4] obtained the following characterization:

(i) AY is linearly sufficient $\Leftrightarrow Im(X) \subseteq Im(WA')$.
(ii) AY is linearly complete $\Leftrightarrow Im(AW) \subseteq Im(AX)$.
(iii) AY is linearly sufficient and linearly complete $\Leftrightarrow Im(X) = Im(WA')$.
(iv) AY is linearly sufficient and linearly complete $\Leftrightarrow$ AY is linearly minimal sufficient,

where $Im(A)$ denotes the space generated by the columns of the matrix A.

If we substitute the estimator $(\hat{\theta}_r, r \in K)$ by AY, it is easy to check that $\mathcal{H}_{\hat{\theta}_B} = Im(X)$, $\mathcal{H}_{\hat{\theta}} = Im(WA')$, $\mathcal{H}(\hat{\theta}) = Im(AW)$ and $\mathcal{H}_{\hat{\theta}_B}(\hat{\theta}) = Im(AX)$. Thus, Theorem 3 is a generalization of the characterizations in discrete time.

4 Linear Admissibility

In this section we deal with the problem of admissibility with respect to the square mean error for parametric functions $K\theta$, where K is a

$k \times p$-matrix. From now on, we shall suppose that $K\theta$ is an estimable function. More precisely, let $\hat{\theta} \in \bar{\mathcal{L}}(X_t, t \in T)$ be an estimator (a vector of k components), it is said that $\hat{\theta}$ is linearly admissible for $K\theta$ if there does not exist an estimator $\bar{\theta} \in \bar{\mathcal{L}}(X_t, t \in T)$ such that

$$E_\theta[(\bar{\theta} - K\theta)'(\bar{\theta} - K\theta)] \leq E_\theta[(\hat{\theta} - K\theta)'(\hat{\theta} - K\theta)], \quad \theta \in \mathbb{R}^p,$$

with strict inequality for at least one value of θ.

The objective of this section is to establish necessary and sufficient conditions for linear admissibility by means of analogous criteria to those given in [23] and in [2, 3].

In order to establish the main result, first we give the following lemma.

Lemma 3. *Let $\hat{\theta} \in \mathcal{L}(X_t, t \in T)$ be an estimator with $E_\theta[\hat{\theta}] = M\theta$. Then,*

$$E_\theta[(\bar{\theta} - K\theta)'(\bar{\theta} - K\theta)] = E_\theta[(\bar{\theta} - M\Sigma^-\hat{\theta})'(\bar{\theta} - M\Sigma^-\hat{\theta})] + E_\theta[(M\Sigma^-\hat{\theta} - K\theta)'(M\Sigma^-\hat{\theta} - K\theta)].$$

Theorem 4. *Let $\hat{\theta}$ be a linear estimator with $M = \bar{E}[\hat{\theta}\hat{\theta}']$. $\hat{\theta}$ is linearly admissible for $K\theta$ if and only if*

(i) $\bar{E}[\hat{\theta}X_t] = RA(t)'$, $t \in T$, for a matrix R.
(ii) $M(I - \Sigma)\Sigma^- K'$ is a symmetric matrix.
(iii) $M(I - \Sigma)\Sigma^-(M - K)'$ is a non negative definite matrix.
(iv) There exists a matrix R such that $(M - K) = (M - K)(I - \Sigma)R$.

Proof. Suppose that $\hat{\theta}$ is a linear admissible estimator for $K\theta$. By Lemma 3, $\hat{\theta} = M\Sigma^-\hat{\theta}_B$, a.s. and $\bar{E}[\hat{\theta}X_t] = \bar{E}[M\Sigma^-\hat{\theta}_B X_t] = M\Sigma^- A(t)'$, $t \in T$, which is condition (i). On the other hand, since $\hat{\theta}$ is linearly admissible for $K\theta$ in the model $(\hat{\theta}_B, \Sigma\theta, \Sigma(I - \Sigma))$, we can apply [1, Theorem in p.352] with $A = M\Sigma^-$, $V = W = \Sigma(I - \Sigma)$ and $C = K\Sigma^-$ obtaining the following conditions:

(ii) $AVC' = M\Sigma^-\Sigma(I-\Sigma)\Sigma^-K' = M(I-\Sigma)\Sigma^-K'$ is a symmetric matrix,
(iii) $AV(C - A) = M\Sigma^-\Sigma(I - \Sigma)(K\Sigma^- - M\Sigma^-)' = M(I - \Sigma)\Sigma^-(M - K)'$ is a non negative definite matrix, and
(iv) $(M - K) = (M - K)(I - \Sigma)R$, for a matrix R.

It concludes the only if part of this theorem.

Conversely, condition (i) is equivalent, from Lemma 2, to that $\hat{\theta}$ is the BLUE for $M\theta$, therefore $\hat{\theta} = M\Sigma^-\hat{\theta}_B$, a.s. Lemma 3 shows that in order to find linearly admissible estimators of $K\theta$, we need to consider only linear functions of $\hat{\theta}_B$. So, if $M\Sigma^-\hat{\theta}_B$ is admissible for $K\theta$ among linear combinations of $\hat{\theta}_B$, $\hat{\theta}$ will be linearly admissible. Applying again [1, Theorem] it is enough to check that conditions (ii) to (iv) imply the linear admissibility of $M\Sigma^-\hat{\theta}_B$ in the model $(\hat{\theta}_B, \Sigma\theta, \Sigma(I - \Sigma))$. The proof is complete. □

Note 2. It is immediate that conditions (ii) to (iv) can be rewritten in terms of P in the following way:

(ii) $MP^{-}K'$ is a symmetric matrix.
(iii) $MP^{-}(M-K)'$ is a non negative definite matrix.
(iv) There exists a matrix R such that $(M-K)=(M-K)P^{-}PR$.

By considering the deterministic part of the process, we can establish the following result:

Theorem 5. *Let $\hat{\theta} \in \bar{\mathcal{L}}(X_t, t \in T)$ with $M = \bar{E}[\hat{\theta}\hat{\theta}'_B]$. Then, $\hat{\theta}$ is a linearly admissible estimator for $K\theta \Leftrightarrow \hat{\theta}^*$ is linearly admissible for $KP^{-}P\theta$ in the model $\bar{\mathcal{L}}(X_t^*, t \in T)$ and $(M-K)=(M-K)P^{-}PR$.*

Proof. First, we establish the equivalent conditions to the linear admissibility of $\hat{\theta}^*$ for $KP^{-}P\theta$ in $\bar{\mathcal{L}}(X_t^*, t \in T)$. For that, we use [7, Theorem 5] with the matrix P instead of Σ. Thus, $\hat{\theta}^*$ is a linear admissible estimator for $KP^{-}P\theta$ in $\bar{\mathcal{L}}(X_t^*, t \in T)$ if and only if:

(i) $\bar{E}_0[X_t\hat{\theta}] = E_0[X_t^*\hat{\theta}^*] = A(t)P^{-}PR = A(t)S$, $t \in T$, for a matrix S.
(ii) $MP^{-}PP^{-}PP^{-}K' = MP^{-}K'$ is a symmetric matrix.
(iii) $MP^{-}(M-K)'$ is a no negative definite matrix.

By applying Theorem 4 of the present paper we obtain the equivalence of this theorem. □

Note 3. Since $\hat{\theta} - K\theta = (M-K)(I-P^{-}P)\Sigma^{-}\hat{\theta}_B + (\hat{\theta}^* - KP^{-}P\theta) = a + (\hat{\theta}^* - KP^{-}P\theta)$, condition (iv) is equivalent to a belongs to the space generated by the columns of $(M-K)P^{-}P$.

To conclude this paper we show two characterizations of the linear admissibility which appear in [7]. These results assume that condition (1) is verified with E_0 and that θ is linearly estimable.

Theorem 6. *$\hat{\theta}$ is linearly admissible for θ if and only if*

(i) $E_0[\hat{\theta}'X_t] = A(t)R$, $t \in T$, *with R a symmetric $p \times p$-matrix, and*
(ii) $M\Sigma^{-1}M' \leq M\Sigma^{-1}$.

Theorem 7. *$\hat{\theta}$ is linearly admissible for θ if and only if*

(iii) $A(s)E_0[\hat{\theta}'X_t] := H(s,t)$, $s,t \in T$ *is a symmetric function, that is,* $H(s,t) = H(t,s)$.
(iv) The spectrum of M belongs to $[0,1]$.

Acknowledgement. Research supported by research projects MTM2010-15972 and DGA E22.

References

1. Baksalary, J.K., Kala, R.: Linear transformations preserving best linear unbiased estimators in a general Gauss-Markoff model. Ann. Statist. 9, 913–916 (1981)
2. Baksalary, J.K., Markiewicz, A.: Admissible linear estimators in restricted linear models. Linear Alg. Appl. 70, 9–19 (1985)
3. Baksalary, J.K., Markiewicz, A.: Admissible linear estimators in the general Gauss-Markov model. J. Statist. Plann. Infer. 19, 349–359 (1988)
4. Berlinet, A., Thomas-Agnan, C.: RKHS in Probability and Statistics. Kluwer Acad. Pub., Boston (2004)
5. Del Pino, G.E.: On restricted linear estimation for regression in stochastic processes. Statist. Prob. Lett. 3, 9–13 (1985)
6. Drygas, H.: Sufficiency and completeness in the general. Gauss-Markov model Sankhya Ser. A 45, 88–98 (1983)
7. Ibarrola, P., Pérez-Palomares, A.: Linear sufficiency and linear admissibility in a continuous time. Gauss-Markov model J. Multiv. Anal. 87, 315–327 (2003)
8. Ibarrola, P., Pérez-Palomares, A.: Linear completeness in a continuous time. Gauss-Markov model Statist. Prob. Lett. 69, 143–149 (2004)
9. Ibarrola, P., Pérez-Palomares, A.: A decomposition of a linear model. Statist. Prob. Lett. 4, 101–1024 (2009)
10. Müller, J.: Sufficiency and completeness in the linear model. J. Multivariate Anal. 21, 312–323 (1987)
11. Rao, C.R.: Estimation of parameters in a linear model. Ann. Statist. 4, 1023–1037 (1976)

δ-Outliers and Their Identification in Real-Valued Continuous Random Fields*

María Macarena Muñoz-Conde[1], Joaquín Muñoz-García[2], and María Dolores Jiménez-Gamero[2]

[1] Instituto de Estadística de Andalucía, Pabellón de Nueva Zelanda, C/ Leonardo da Vinci, 21, 41071, Sevilla, Spain
mmacarena.munoz@juntadeandalucia.es

[2] Dpto. de Estadística e Investigación Operativa, Universidad de Sevilla, 41012 Sevilla, Spain
joaquinm@us.es, dolores@us.es

Summary. Due to the fact that the spatial outlier observations depend on the neighborhood where they are located, a definition of δ-outlier is given, δ being the diameter of the neighborhood. Two methods to identify δ-outliers are proposed. One of them for continuous random fields and the other for Gaussian continuous random fields.

1 Introduction

The treatment of outlier observations in any statistical data analysis is a crucial step if one want the conclusions drawn from the analysis not to be affected by such observations. In this article we deal with the case of spatial outlier observations. These observations are characterized by their local nature, that is to say, they are classified as outliers just in a neighborhood in which they are located. In this line, we propose a definition of spatial outlier observations, which takes into account the diameter of the neighborhood in which they are located. So, an outlier observation, in the sense of our definition, may not be an outlier observation for the whole sample.

Once the concept of δ-outlier is introduced, we propose two statistical techniques to identify such observations. One for the general case of having a real-valued continuous random field and the other for Gaussian fields. In both cases, with the aim of developing formally the techniques, we assume that the random field has locally homogeneous p-expected differences, for some $p > 0$. This condition is less restrictive than those usually assumed in other techniques to identify spatial outlier observations. For example, isotropy of the field is supposed in many graphical techniques of identification of spatial outliers (see Wackernagel [11]). Another example is the method proposed by

---

* This paper is dedicated to Marisa, who will certainly be missed by her many students, colleagues and friends, and will always be affectionately remembered, not only for her research career, but also for her sterling human qualities.

Cerioli and Riani [4], which requires practically a intrinsically stationary field throughout the parameter space.

2 Real Continuous Random Fields

Let $\mathbb{T} \subset \mathbb{R}^k$ and let $(\mathbb{T}, d_{\mathbb{T}})$ be a compact metric space, with $d_{\mathbb{T}}$ the euclidean metric. Let $(\mathbb{T}, \mathcal{T})$ and $(\mathbb{R}, \mathcal{B}(\mathbb{R}))$ be two measurable spaces, where $\mathcal{T}$ and $\mathcal{B}(\mathbb{R})$ are the σ-algebra of Borel sets in $\mathbb{T}$ and $\mathbb{R}$, respectively. Let $B(\mathbb{T})$ be the space of bounded real functions defined on $\mathbb{T}$, and let $C(\mathbb{T})$ be the space of functions in $B(\mathbb{T})$ which are continuous on $\mathbb{T}$, endowed with the supremum norm.

The functions of the space $C(\mathbb{T})$ are characterized by its modulus of continuity, Billingsley [3, p. 80]

$$w_X^*(\delta) = \sup_{\{t,s \in \mathbb{T}/d_{\mathbb{T}}(t,s) \leq \delta\}} |X(t) - X(s)|.$$

These functions satisfy the following.

Proposition 1. *Let $X \in B(\mathbb{T})$. Then, $X \in C(\mathbb{T})$ if and only if*

$$\lim_{\delta \to 0} w_X^*(\delta) = \lim_{\delta \to 0} \sup_{\{t,s \in \mathbb{T}/d_{\mathbb{T}}(t,s) \leq \delta\}} |X(t) - X(s)| = 0.$$

We will assume that $(\Omega, \mathcal{A}, P)$ is a complete probability space and defined on it we consider the functions $X \in C(\mathbb{T})$, defined on Ω, with the adequate σ-algebra (see Billingsley [3, p. 84]). The functions X generate a real-valued random field which will be assumed separable and almost surely continuous, i.e.,

$$P\left[\omega \in \Omega \Big/ \lim_{s \to t} |X(\omega, s) - X(\omega, s)| \to 0 \quad \forall t \in \mathbb{T}\right] = 1.$$

Next, we define a random field with locally homogeneous p-expected differences. This concept is crucial for our subsequent developments.

Definition 1. *Let $p > 0$. Let $(\mathbb{T}, d_{\mathbb{T}})$ be a compact metric space, with $\mathbb{T} \subset \mathbb{R}^k$, and let $X \in C(\mathbb{T})$ be a real-valued random field. It is said that X is a random field with locally homogeneous p-expected differences if $\exists r \in \mathbb{R}$, $r > 0$, such that $\forall t, s \in \mathbb{T}$ with $d_{\mathbb{T}}(t,s) \leq r$, it is satisfied that*

$$E\left[|X(t) - X(s)|^p\right] = V_p\{d_{\mathbb{T}}(t,s)\}$$

is a function of the distance between the points t and s.

3 Outlier Observations

The data depuration is one of the most important parts in any statistical data analysis. In this step it is essential the treatment of outlier observations. The early treatments of such observations are placed by various authors between the mid-eighteenth century and early nineteenth centuries, as referenced in Barnett and Lewis [2].

Many definitions of outlier observations have been provided in the statistical literature. The paper by Muñoz-García, Moreno-Rebollo and Pascual-Acosta [7] collects a large number of them. All definitions share that outlier observations deviate markedly from the general behaviour of the available observations. Some definitions specify that such deviations are with respect to the statistical analysis to be applied to the data or with respect to some characteristics of the data.

When the data consist of the realization of a stochastic process, the existing definitions of outlier observations still share the peculiarity of "deviate markedly" but, due to the dependence structure, new considerations must be added. For example, if outlier observations are studied in the context of time series, terms such as "additive outliers" and "innovation outliers" appear in the book by Barnett and Lewis [2, p. 396]. The paper by Tsay [9] characterizes in the time series setting the "level change", the "transient change" and the "variance change". Also in this setting, Wu, Hosking and Ravishankara [12] propose the "reallocation outliers".

In the context of spatial data, many definitions of outlier observations have been also proposed. Next we present the definition given by Wartenberg [10] because it has some features which are shared with many other definitions given by other authors. Wartenberg [10] defines the following three types of outlier observations:

Outlying locations: in some data sets, one (or a few) observations are situated far away from the others in the geographic space. This positional anomaly will affect their influence on spatially weighted statistics. In short, these are isolated observations.

Aspatial outliers: these are observations that are different from all the others in a data set. This type of observations can be seen as general outliers, since they do not depend on the parameter space. In this sense, they may also be called global outliers.

Spatial outliers: these are observations that are different from all the others in a neighborhood of the parametric space $\mathbb{T}$. They influence statistics that asses spatial pattern of variate values because, in comparison with their neighbors, they show large differences.

In this last definition, we must highlight the local nature that characterizes the spatial outlier observations. Wartenberg emphasizes that these spatial outlier observations can be not classified as global outliers, but their behaviour is quite different from those in a neighborhood.

A more recent definition is given by Shekkar, Lu and Zhang [8]. They gave a definition of spatial outlier observation or S-outlier, as they name it, where in addition to the reference to the neighborhood of the observation, they include also a reference to the criterion used to identify this type of observations.

In these definitions of spatial outliers, it can be seen that they depend on the neighborhood where they are situated. This forces us to label this type of observations according to such neighborhood. Observe that the diameter of the neighborhood where the observation is located can be crucial to classify it as an outlier. Therefore we propose the definition of δ-outlier, δ being the diameter of the considered neighborhood.

Therefore the definition of δ-outlier that is proposed for real-valued continuous random field, is given by

Definition 2. *Let* $\{t_1, t_2, ..., t_n\} \in \mathbb{T}$ *be the locations of the observed data of the real-valued continuous random field* $X \in C(\mathbb{T})$, $\{X(t_1), X(t_2), ..., X(t_n)\}$. *Let* $\alpha \in (0,1)$ *be fixed. It is said that* $X(t_i)$ *is a* δ*-outlier, for some* $i \in \{1, 2, ..., n\}$, *if*

$$\max_{\substack{j \in \{1,2,\ldots,n\} \setminus \{i\} \\ d_{\mathbb{T}}(t_i, t_j) \le \delta}} |X(t_i) - X(t_j)| \ge x_\alpha,$$

where x_α *satisfies*

$$P\left[\sup_{\substack{s,t \in \mathbb{T} \\ d_{\mathbb{T}}(s,t) \le \delta}} |X(s) - X(t)| \ge x_\alpha\right] \le \alpha.$$

Remark 1. Strictly speaking, it should be named an (α, δ)-outlier, because it also depends on α. Because in practice α is taken to be small (0.01-0.05), that is, it always ranges in a narrow interval, he have suppressed the dependence of α in the name.

4 The α-Percentile, x_α

Since $(\mathbb{T}, d_{\mathbb{T}})$ is a compact metric space, for any $\delta_m > 0$, $m = 1, 2, \ldots$, with $\delta_m \to 0$ when $m \to \infty$, it is possible to have a finite δ_m-net, N_{δ_m}, and the δ_m-net family $\{N_{\delta_m}\}$ satisfies that $\mathbb{T}_N = \cup_m N_{\delta_m}$ is a countable dense set in $\mathbb{T}$ (see for example Kolmogorov and Fomin [6, p. 113]). This fact, together with the condition of separability of the continuous random field, is important because the continuity of the field X is reduced to equivalent problems on dense and countable sets.

Let $\boldsymbol{B}_m$ be a finite collection of balls covering $\mathbb{T}$ according to the δ_m-net N_{δ_m}. For each $t \in \mathbb{T}$ there is an $B_m \in \boldsymbol{B}_m$ such that $t \in B_m$. This ball will be represented by $B_m(t)$. Following to Khoshnevisan [5, pp. 160–161], $B_m(t) = B_m(s)$ if $s \in B_m(t)$. It can be always selected $r_m(t) \in B_m(t) \cap B_{m+1}(t)$ satisfying $d_{\mathbb{T}}(r_m(t), t) \le \delta_m$, $\forall t \in \mathbb{T}$. These points generate the finite sets

$$R_m = \{r_m(t), t \in \mathbb{T}\}.$$

Thus, given two points $t, s \in \mathbb{T}$ with $d_{\mathbb{T}}(t,s) \leq \delta_m$, there exists two sequences $\{r_p(t)\}_{p\geq m}$ and $\{r_p(s)\}_{p\geq m}$, such that $r_m(t) = r_m(s)$ and

$$d_{\mathbb{T}}(r_p(t), t) \leq \delta_p, \quad d_{\mathbb{T}}(r_p(s), s) \leq \delta_p,$$

with $\delta_p \leq \delta_m$. Thus, we can always construct two sequences of balls $\{B(r_p(t))\}_{p\geq m}$ and $\{B(r_p(s))\}_{p\geq m}$, which allow us the use a sort of chaining reasoning in the inequalities that are obtained below.

Proposition 2. *Let $(\mathbb{T}, d_{\mathbb{T}})$ be a compact metric space and let X be a real-valued continuous random field. Then, $\forall t, s \in \mathbb{T}$ with $d_{\mathbb{T}}(t,s) \leq \delta_m$ and $k > m$, it is satisfied*

$$P[|X(t) - X(s)| \geq x'] \leq$$
$$P\Bigg[|X(t) - X(r_{m+k}(t))| + \sum_{j=1}^{k} |X(r_{m+j}(t)) - X(r_{m+j-1}(t))|$$
$$+ \sum_{j=1}^{k} |X(r_{m+j}(s)) - X(r_{m+j-1}(s))| + |X(s) - X(r_{m+k}(s))| \geq x'\Bigg].$$

Proof. Since $X(r_m(t)) = X(r_m(s))$

$$\begin{aligned} |X(t) - X(s)| &= |X(t) - X(r_m(t)) + X(r_m(s)) - X(s)| \\ &\leq |X(t) - X(r_m(t))| + |X(r_m(s)) - X(s)| \end{aligned}$$

Now, by applying the triangular inequality we obtain the desired result. □

From this proposition it is immediately obtained the following corollary, by using the finiteness of the sets R_m.

Corollary 1. *Let $(\mathbb{T}, d_{\mathbb{T}})$ be a compact metric space and let X be a real-valued continuous random field. Then, $\forall t, s \in \mathbb{T}$ with $d_{\mathbb{T}}(t,s) \leq \delta_m$ and $k > m$, it is satisfied*

$$P\left[\sup_{\substack{t,s\in\mathbb{T} \\ d_{\mathbb{T}}(t,s)\leq\delta_m}} |X(t) - X(s)| \geq x\right]$$
$$\leq P\left[2\sum_{j=1}^{k} \max_{\tau\in R_{m+j}} |X(r_{m+j}(\tau)) - X(\tau)| \geq x\right]$$

The next proposition is the basis for the method that will be proposed in order to identify δ-outliers.

Proposition 3. *Let $(\mathbb{T}, d_{\mathbb{T}})$ be a compact metric space and let X be a real-valued continuous random field. Then, $\forall t, s \in \mathbb{T}$ with $d_{\mathbb{T}}(t,s) \leq \delta_m$ and $k > m$, it is satisfied*

$$P\left[\sup_{\substack{t,s\in\mathbb{T}\\ d_{\mathbb{T}}(t,s)\leq\delta_m}} |X(t)-X(s)|\geq x\right]$$
$$\leq \sum_{j=1}^{k} \#R_{m+j} \frac{\max_{\tau\in R_{m+j}} E\left[|X(r_{m+j}(\tau))-X(\tau)|^p\right]}{\left(\beta_j \frac{x}{2}\right)^p},$$

for $p>0$ and $0\leq\beta_j\leq 1$ satisfying $\sum_{j=1}^{k}\beta_j=1$, where $\#A$ is the number of elements in A.

Proof.

$$P\left[\sup_{\substack{t,s\in\mathbb{T}\\ d_{\mathbb{T}}(t,s)\leq\delta_m}} |X(t)-X(s)|\geq x\right]$$
$$\leq P\left[2\sum_{j=1}^{k} \max_{\tau\in R_{m+j}} |X(r_{m+j}(\tau))-X(\tau)|\geq x\right]$$
$$\leq \sum_{j=1}^{k} \sum_{\tau\in R_{m+j}} P\left[|X(r_{m+j}(\tau))-X(\tau)|\geq \beta_j\frac{x}{2}\right]$$

with $\{\beta_j\}\in\mathbb{R}$, $0\leq\beta_j\leq 1$ satisfying $\sum_{j=n+1}^{k}\beta_j=1$. Now, by applying the Markov inequality we get

$$P\left[\sup_{\substack{t,s\in\mathbb{T}\\ d_{\mathbb{T}}(t,s)\leq\delta_m}} |X(t)-X(s)|\geq x\right]$$
$$\leq \sum_{j=1}^{k} \#R_{m+j} \frac{\max_{\tau\in R_{m+j}} E\left[|X(r_{m+j}(\tau))-X(\tau)|^p\right]}{\left(\beta_j \frac{x}{2}\right)^p}. \qquad \square$$

For the calculation of x_α it can be assumed in the previous inequality the condition of stationarity or homogeneity of the real-valued continuous random field. This condition is widely used both in theory and in practice, but this condition is rarely verified in many applications. However, it is also known that the replication of the field is required in order to calculate averages. To circumvent this problem, we assume that the field has locally homogeneous p-expected differences, for some $p>0$, which is not a strong restriction. This allows us to formulate the following corollaries, which are immediate consequences of the previous proposition.

Corollary 2. *Let $\mathbb{T}\subset\mathbb{R}^k$, let $(\mathbb{T},d_{\mathbb{T}})$ be a compact metric space and let $X\in C(\mathbb{T})$ be a real-valued random field with locally homogeneous p-expected differences, for some $p>0$. Then, $\forall t,s\in\mathbb{T}$, with $d_{\mathbb{T}}(t,s)\leq\delta_m$, it is satisfied*

$$P\left[\sup_{\substack{t,s\in\mathbb{T}\\ d_{\mathbb{T}}(t,s)\leq\delta_m}} |X(t)-X(s)|\geq x\right] \leq \sum_{j=1}^{k} \#R_{m+j}\frac{E\left[|X(r_{m+j}(\tau))-X(\tau)|^p\right]}{\left(\beta_j\frac{x}{2}\right)^p},$$

for $\{\beta_j\}\in\mathbb{R}$, $0\leq\beta_j\leq 1$, *satisfying* $\sum_{j=1}^{k}\beta_j=1$.

Corollary 3. *Let* $\mathbb{T}\subset\mathbb{R}^k$, *let* $(\mathbb{T},d_{\mathbb{T}})$ *be a compact metric space and let* $X\in C(\mathbb{T})$ *be a real-valued random field with locally homogeneous p-expected differences, for some* $p>0$. *Let* $\alpha\in(0,1)$ *be fixed. Then,* $\forall t,s\in\mathbb{T}$, *with* $d_{\mathbb{T}}(t,s)\leq\delta_m$, *it is satisfied*

$$x_\alpha=\sqrt[p]{\frac{1}{\alpha}\sum_{j=1}^{k}\#R_{m+j}\frac{E\left[|X(r_{m+j}(\tau))-X(\tau)|^p\right]}{\left(\frac{\beta_j}{2}\right)^p}}$$

5 The Gaussian Case

Now the condition of Gaussian is imposed on the real-valued continuous random field. This condition will allow us to get a more accurate result than that obtained previously for real continuous random fields, since in this case, we will use a suitable inequality for Gaussian fields such as the inequality of Borel-Tsirelson, Ibragimov and Sudakov, which is shown in Adler and Taylor [1, p. 50]. Next we give this inequality.

Theorem 1. *Let* $Z(y)$ *be a centered Gaussian field, almost surely bounded on* $\mathbb{Y}$. *Then*

$$E\left[\sup_{y\in\mathbb{Y}}Z(y)\right]<\infty,$$

and for all $u>0$

$$P\left[\sup_{y\in\mathbb{Y}}Z(y)-E\left[\sup_{y\in\mathbb{Y}}Z(y)\right]>u\right]\leq\exp\left(-\frac{u^2}{2\sigma_{\mathbb{Y}}^2}\right)$$

where $\sigma_{\mathbb{Y}}^2=\sup_{y\in\mathbb{Y}}E\left[(Z(y))^2\right]$.

From the above theorem the following corollary is immediately obtained.

Corollary 4. *Let* $Z(y)$ *be a centered Gaussian field, almost surely bounded on* $\mathbb{Y}$. *Then,* $\forall u>E\left[\sup_{y\in\mathbb{Y}}Z(y)\right]$ *it is satisfied*

$$P\left[\sup_{y\in\mathbb{Y}} Z(y) > u\right] \leq \exp\left(-\frac{\left(u - E\left[\sup_{y\in\mathbb{Y}} Z(y)\right]\right)^2}{2\sigma_{\mathbb{Y}}^2}\right).$$

These results are now applied to the module of continuity for the random field $X \in C(\mathbb{T})$.

Proposition 4. *Let $\mathbb{T} \subset \mathbb{R}^k$, let $(\mathbb{T}, d_{\mathbb{T}})$ be a compact metric space and let $X \in C(\mathbb{T})$ be a centered Gaussian real-valued random field with locally homogeneous p-expected differences, for $p = 2$. Then, $\forall t, s \in \mathbb{T}$, with $d_{\mathbb{T}}(t,s) \leq \delta_m$ and*

$$\forall x > E\left[\sup_{\substack{t,s\in\mathbb{T} \\ d_{\mathbb{T}}(t,s)\leq\delta_m}} (X(t) - X(s))\right],$$

it is satisfied

$$P\left[\sup_{\substack{t,s\in\mathbb{T} \\ d_{\mathbb{T}}(t,s)\leq\delta_m}} |X(t) - X(s)| > x\right]$$
$$\leq 2\exp\left(-\frac{\left(x - E\left[\sup_{\substack{t,s\in\mathbb{T} \\ d_{\mathbb{T}}(t,s)\leq\delta_m}} (X(t) - X(s))\right]\right)^2}{2\sigma^2_{\{t,s\in\mathbb{T},d_{\mathbb{T}}(t,s)\leq\delta_m\}}}\right),$$

with

$$\sigma^2_{\{t,s\in\mathbb{T},d_{\mathbb{T}}(t,s)\leq\delta_m\}} = \sup_{\substack{t,s\in\mathbb{T} \\ d_{\mathbb{T}}(t,s)\leq\delta_m}} E\left[(X(t) - X(s))^2\right]$$
$$= \sup_{\substack{t,s\in\mathbb{T} \\ d_{\mathbb{T}}(t,s)\leq\delta_m}} V_2(d_{\mathbb{T}}(t,s)).$$

Proof. The result follows from Corollary 4 and the following inequality (see Adler and Taylor [1, p. 51]),

$$P\left[\sup_{\substack{t,s\in\mathbb{T} \\ d_{\mathbb{T}}(t,s)\leq\delta_m}} |X(t) - X(s)| > x\right]$$
$$\leq 2P\left[\sup_{\substack{t,s\in\mathbb{T} \\ d_{\mathbb{T}}(t,s)\leq\delta_m}} (X(t) - X(s)) > x\right].$$

□

Finally, for fixed α we have the following.

Corollary 5. *Let $\mathbb{T} \subset \mathbb{R}^k$, let $(\mathbb{T}, d_{\mathbb{T}})$ be a compact metric space and let $X \in C(\mathbb{T})$ be a centered Gaussian real-valued random field with locally homogeneous p-expected differences, for $p = 2$. Let $\alpha \in (0,1)$ be fixed. Then, $\forall t, s \in \mathbb{T}$, with $d_{\mathbb{T}}(t,s) \leq \delta_m$, it is satisfied*

$$x_\alpha = E\left[\sup_{\substack{t,s\in\mathbb{T}\\ d_{\mathbb{T}}(t,s)\leq\delta_m}} (X(t) - X(s))\right] + \sqrt{2\sigma^2_{\{t,s\in\mathbb{T}, d_{\mathbb{T}}(t,s)\leq\delta_m\}}\left(-\ln\frac{\alpha}{2}\right)}.$$

Acknowledgement. The research in this paper has been partially supported by grant MTM2008-00018 (Spain).

References

1. Adler, R.J., Taylor, J.E.: Random Fields and Geometry. Springer, New York (2007)
2. Barnett, V., Lewis, T.: Outliers in Statistical Data, 3rd edn. Wiley, New York (1994)
3. Billingsley, P.: Convergence of Probability Measures, 2nd edn. Wiley, New York (1999)
4. Cerioli, A., Riani, M.: The ordering of spatial data and the detection of multiple outliers. J. Comp. Graph Statist. 8, 239–258 (1999)
5. Khoshnevisan, D.: Multiparameter Processes. An Introduction to Random Fields. Springer, New York (2002)
6. Kolmogorov, A.N., Fomin, S.V.: Elementos de la Teoría de Funciones y del Análisis Funcional. In: MIR, Moscow (1975)
7. Muñoz-García, J., Moreno-Rebollo, J.L., Pascual-Acosta, A.: Outliers: A formal approach. Int. Sta. Rev. 58, 215–226 (1990)
8. Shekhar, S., Lu, C.-T., Zhang, P.: A unified approach to detecting spatial outliers. Geoinformatica 7, 139–166 (2003)
9. Tsay, R.S.: Outliers, level shifts, and variance changes in time series. J. Forecasting 7, 1–20 (1988)
10. Wartenberg, D.: Exploratory Spatial Analyses: Outliers, Leverage Points, and Influence Functions. In: Griffith, D.A. (ed.) Spatial Statistics. Past, Present and Future, pp. 131–156. Institute of Mathematical Geography, University of Michigan (1990)
11. Wackernagel, H.: Multivariate Geostatistics. An Introduction with Applications, 3rd edn. Springer, New York (2003)
12. Wu, L.S.-Y., Hosking, J.R.M., Ravishanker, N.: Reallocation outliers in time series. App. Stat.-J. Roy St. C 42, 301–313 (1993)

Capability Index for Multivariate Processes That Are Non-stable over Time*

Miquel Salicrú[1], Juan J. Barreiro[2], and Sergi Civit-Vives[1]

[1] Department of Statistics. University of Barcelona, Spain
msalicru@ub.edu, svives@ub.edu
[2] Department of Statistics and Operational Research. Vigo University, Spain
jjbi@uvigo.es

Summary. A key element of the management process is to pay close attention to the strategies that minimize total production cost. One of the ways of doing this is to minimize stocks. However, planning smaller lot sizes can lead to a shorter time for process set-up and pre-adjustment, which often gives rise to processes that are not stable over time. These processes are characterized by their dual variability: variability between batches (largely attributable to differences in tuning) and variability within the batch (attributable to uncontrolled process factors). In order to build the understanding to this "management process" reality, a new process capability index $C_{(u,v)}(p_o, \sigma_o)$ based on the proportion of conformance of the process and applications at an inferential level are developed. For illustrative purposes, a case study of the manufacture of car hoods and the results of the new methodology are introduced.

Keywords: capability index, process non stable over time.

1 Introduction

The growing interest of companies in meeting customers' expectations has entailed, among other things, the need to reduce the costs associated with the production process. In this respect, the operative actions have been aimed at the reduction of stocks, in order to minimize financing costs and to guarantee and minimize production costs, on the basis of reducing the cost of raw material, reducing non-conformities, reducing energy costs, reducing movements or increasing the productivity of workers.

Planning and producing smaller sized batches requires reducing preset time (tuning of equipment) and this often gives rise to manufacturing with

* Since that autumn sun's morning, we remember Marisa for all the shared experiences and the values she conveyed. Her courage and drive to improve in all respects have inspired and guided our work towards the continuous improvement and optimization.

processes which are unstable over time. These processes are characterized by their dual variability: variability between batches (largely attributable to differences in tuning) and variability within the batch (attributable to uncontrolled process factors).

The improvement of processes to reduce and guarantee operative costs can be approached from different perspectives. One of these is focused on minimizing variabilities (between and within batches) and on focuses the process on the objective value. In this context, it is important to ensure the correct characterization of the capacity index, the analysis of the process to identify the key elements and the root causes of the non-conformity and the contrast of alternatives to identify the best production practices.

On the other hand, processes which need to guarantee multiple variables are increasingly common. In this field, various approaches are possible: the multi-variant approach (powerful but very difficult to characterize with mixed variables or with non-normal background distribution), variable to variable control (costly and with many false alarms when there are many variables), or characterization based on the percentage of non-conformities (less informative but often useful).

1.1 Aims and Scope

Optimization of real-world industrial processes require, the guarantee of multiple variables, with the manufacture of smaller sized batches and control of the percentage of non-conformities. In this scenario, we have proposed

- Introducing a new capacity index that suitably characterizes this experimental situation.
- Providing tools aimed at identifying the alternatives that optimize capacity. In other words, we introduce statistics and decision criteria. Specifically a hypothesis test and confidence intervals.
- Applying the methodology developed to a case study. We shows the performance of the methodology for statistical process control of a car hood manufacturing application. We identify the best option combining technology and materials.

1.2 Notations and Preliminary Results

The observation of n batches with independent samples of sizes $m_1, m_2, \ldots, m_n$, provides an information table with x_{ij} values (see Table 1). These x_{ij} values are characterized by p_i parameter Bernoulli distributions, being p_i the mean percentage of non-conformities corresponding to batch i.

For processes which are stable over time, the p_i $(i = 1, \ldots, n)$ parameters are all the same, whereas for processes that are non stable over time, the p_i parameters are associated with a normal distribution with mean p_o (the mean number of non-conformities produced) and variance σ_o^2 (the variance between batches).

Table 1. Data structure

Batch	1	2	...	n
Data	x_{11}	x_{21}	...	x_{n1}
	x_{12}	x_{22}	...	x_{n2}
	$\vdots$	$\vdots$	$\vdots$	$\vdots$
	x_{1m_1}	x_{2m_2}	...	x_{nm_n}
Mean	$\hat{p}_1$	$\hat{p}_2$	...	$\hat{p}_n$

Regardless of the background distribution, the algebraic decomposition of the total variability (between batches and within batches):

$$m_+ \cdot \hat{p}_o(1-\hat{p}_o) = \sum\nolimits_{i=1}^{n} m_i \cdot (\hat{p}_i - \hat{p}_o)^2 + \sum\nolimits_{i=1}^{n} m_i \cdot \hat{p}_i(1-\hat{p}_i)$$

where $m_+ = m_1 + m_2 + \ldots + m_n$ and

$$\hat{p}_o = \frac{m_1\hat{p}_1 + \ldots + m_n\hat{p}_n}{m_1 + m_2 + \ldots + m_n} \tag{1}$$

the estimation of variance between batches can be obtained from

$$\hat{\sigma}_o^2 = \max\left\{0, \frac{1}{n-1}\sum\nolimits_{i=1}^{n} \frac{m_i}{\bar{m}} \cdot (\hat{p}_i - \hat{p}_o)^2 - \frac{1}{n \cdot (\bar{m}-1)}\sum\nolimits_{i=1}^{n} \frac{m_i}{\bar{m}} \cdot \hat{p}_i(1-\hat{p}_i)\right\} \tag{2}$$

where $\bar{m} = m_+/n$ (see details Sect. 4).

However, processes which are non stable over time with control of the percentage of non-conformities (see Figure 1) requires the consideration of compound distributions and could be done according the following approach:

$$\hat{p}_i \approx \frac{1}{m_i} B(m_i, \mathrm{p}) \underset{p}{\wedge} N(p_o, \sigma_o) \approx N(p, \sigma_1) \underset{p}{\wedge} N(p_o, \sigma_o) \approx N(p_o, \sqrt{\sigma_1^2 + \sigma_o^2}) \tag{3}$$

where $\hat{p}_i$ is the $i-th$ trial as a success obtained with a sample of sizes m_i (which is characterized by the binomial law $B(m_i, p_i)$).

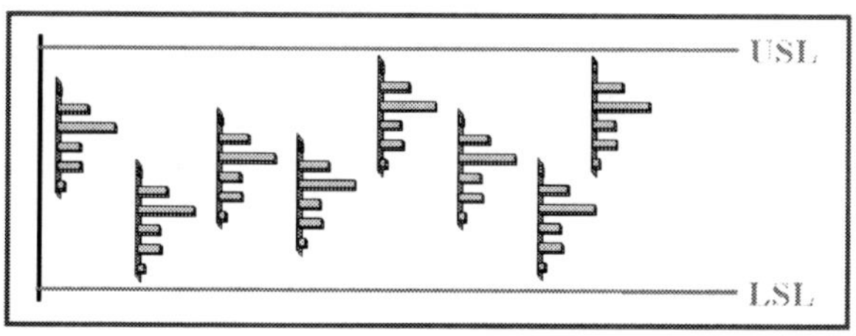

Fig. 1. Process behaviour

1.3 Revisiting the Literature

In process management, the capability of a manufacturing process is measured using process capability index ratios. Such capability indices are widely used to evaluate the ability of process to manufacture product that meets specifications. The analytical formulation of these indices is generally easy to understand and straightforward to apply. A large number of indices they have been described in the literature since [8] define the simplest and most common measure used to describe the performance of a process relative to the specification limits as:

$$C_p = \frac{USL - LSL}{6\sigma} \tag{4}$$

where USL and LSL denote the upper and lower specification limits for the characteristic X to be measured, and σ denote the process mean and standard deviation values of the process. The C_p index does not require knowledge of the process location μ and for this reason it can be viewed as a measure of the capability of an optimally centered process. It basically reflects changes in the amount of product that exceeds the specification limits and does not consider the target value of the process, which is a critical quantity when assessing process performance. The adaptation of (4) to different experimental situations has given rise to new indices (C_{pk}, C_{pm}, C_{pmk} among others) and investigations into the properties of existing indices ([6], [9], [3], [13], [1], [19], [20], [4], [21], [23], [16], [11], [12]).

To facilitate global studies, [20] introduced a general index for the most widely used indices or their generalizations described in the literature:

$$C_{pa}(u,v) = \frac{d - |\mu - M| - u|\mu - T|}{n_{\alpha/2}\sqrt{\sigma^2 + v(\mu - T)^2}}, \tag{5}$$

where $d = (USL - -LSL)/2$, $M = (USL + LSL)/2$, T is the target value, μ is the production mean, $n_{\alpha/2}$ is quantile $\alpha/2$ corresponding to a standard normal distribution and u, v are non-negative parameters. Revisions of capability indices have been reported by [10], [17], [18] and [14]. However and taking into account another point of view, [22] introducing a process capability index based on the proportion of conformance of the process. The performance of this capability index is based on the relationship between the non-conformity ratio accepted by the customer, p_c (p_c^L or p_c^U correspond to the lower and upper control limits accepted by the customer), and the non-conformity ratio of the process, p_o (p_o^L or p_o^U correspond to the process lower and upper control limits):

$$C_{YB} = \frac{p_c}{p_o} \qquad\qquad C_f = \min\{\frac{p_c^L}{p_o^L}, \frac{p_c^U}{p_o^U}\}$$

According to this new point of view and under distributional assumptions, point estimators for the process capability index based on the proportion of conformance were suggested by [2] and [15]. Also, in this environment, [7] proposed a process capability index based on Taguchi quadratic quality loss function. This approach is very attractive when wants to be guaranteed the quality in stable processes in the time. However, this assumption is not appropriate for most industrial processes where the strategies are to minimize stocks in order to make manufacturing processes more flexible.

2 Capability Index: Statistical Behavior

Taking into account the improvement/optimization process, we characterize the capacity of the non-stable processes over time (in the same way as in continuous model) and we determine its statistical behavior (according to the Taylor approximation results).

2.1 Capacity Process: Characterization

For a guarantee level of $1-\alpha$ for the production process, adaptation of (5) to variables that describe the percentage of non-conformities requires: the mean and variance for the percentage non-conformity of the production process, the percentage non-conformity accepted by the customer, and the target. In this framework, when only non-conformity percentages higher than the target are penalized, the index is defined as

$$C_{(u,v)}(p_o, \sigma_o) = \frac{p_c - u \cdot \max\{0, p_o - T\}}{p_o + n_{\alpha/2}\sqrt{\sigma_o^2 + v \cdot [\max\{0, \mathrm{p}_o - T\}]^2}} \tag{6}$$

where p_o and $\sigma_o = \sigma(p_1, p_2, \ldots, p_n, \ldots)$ are the mean and standard deviation of the non-conformity percentages over time.

Under this general expression, many other indices, including that of [22] for processes stable over time, can be considered a particular case:

$$C_{(0,0)}(p_o, \sigma_o = 0) = \frac{p_c}{p_o} = C_{YB}$$

and could also be considered a reasonable approximation to a process that is not stable over time, $\mathrm{C}_{(0,0)}\ (p_o, \sigma_o)$.

Given a random sample from the process, the capability index can be obtained by replacing p_o and σ_o with their estimates (1) and (2):

$$\hat{C}_{(u,v)}(\hat{p}_o, \hat{\sigma}_o) = \frac{p_c - u \cdot \max\{0, \hat{p}_o - T\}}{\hat{p}_o + n_{\alpha/2}\sqrt{\hat{\sigma}_o^2 + v \cdot [\max\{0, \hat{p}_o - T\}]^2}} \tag{7}$$

2.2 Capacity Process: Asymptotic Distribution

Based on the first-order Taylor expansion of $\hat{C}_{(u,v)}(\hat{p}_o, \hat{\sigma}_o)$ around the true value of the parameter (p_o, σ_o),

$$\hat{C}_{(u,v)}(\hat{p}_o, \hat{\sigma}_o) = C_{(u,v)}(p_o, \sigma_o) + (t_1, t_2)(\hat{p}_o - p_o, \hat{\sigma}_o - \sigma_o)^t + R_2$$

being

$$t_1 = \frac{\partial\, C_{(u,v)}(p_o,\, \sigma_o)}{\partial\, p_o}, \quad t_2 = \frac{\partial\, C_{(u,v)}(p_o,\, \sigma_o)}{\partial\, \sigma_o}$$

and the asymptotic distribution:

$$n^{1/2}(\hat{p}_o - p_o,\, \hat{\sigma}_o - \sigma_o)^t \approx N(\mu_1,\, \Sigma_1)$$

where $(\mu_1,\, \Sigma_1)$ are the values defined in (15) and (16) (see Sect. 4), the following result can be obtained.

Theorem 1. *For sample size n, if n is sufficiently large, the asymptotic distribution of the capability index $\hat{C}_{(u,v)}(\hat{p}_o,\, \hat{\sigma}_o)$ defined in (7) satisfies*

$$n^{1/2}[\hat{C}_{(u,v)}(\hat{p}_o,\, \hat{\sigma}_0) - C_{(u,v)}(p_o,\, \sigma_0)] \approx N(\mu,\, \sigma) \approx N(\hat{\mu},\, \hat{\sigma})$$

where (μ, σ) are the values defined in (21)and (22)(see Sect. 4), and $(\hat{\mu},\, \hat{\sigma})$ the respectively estimates according to (1) and (2).

3 Inferential Applications

In the analysis given so far, we have dealt only with the definition of some point estimators of the true value of the capability index. However, knowing these values is insufficient to measure the capability of a process. It would therefore be useful to construct confidence intervals for the true values as well. In fact, given the asymptotic distribution obtained in Theorem (see Sect. 2), it is possible to build general tests and confidence intervals for the index and for the difference of indices. In this section we demonstrate how to build confidence intervals and a hypothesis test.

3.1 One-Sample Hypothesis Test

$$H_o :\ \mathrm{C}_{(u,v)}(p_o, \sigma_o) = C^o_{(u,v)}$$

In this experimental situation, the statistic to be used is

$$Z_{\exp} = \frac{n^{1/2}[\hat{C}_{(u,v)}(\hat{p}_o, \hat{\sigma}_o) - C^o_{(u,v)} + \hat{B}]}{\hat{\sigma}} \approx N(0,\, 1),$$

where $\hat{\sigma}$ is the estimator defined in (22), $B = -n^{1/2}\mu$ the bias estimation of $C_{(u,v)}(p_o, \sigma_o)$ and μ the value defined in (21).

The decision rule consists of rejection of the null hypothesis if

$$|Z_{\exp}| > z_{\varepsilon/2} \qquad H_1 :\ \mathrm{C}_{(u,v)}(p_o, \sigma_o) \neq C^o_{(u,v)}$$

$$Z_{\exp} > z_{\varepsilon} \qquad H_1 :\ \mathrm{C}_{(u,v)}(p_o, \sigma_o) > C^o_{(u,v)}$$

$$Z_{\exp} < -z_{\varepsilon} \qquad H_1 :\ \mathrm{C}_{(u,v)}(p_o, \sigma_o) < C^o_{(u,v)}$$

The asymptotic power of the test is obtained by considering the distribution of the statistic described in Theorem (Sect. 2)

3.2 Multiple Testing

$$H_o: \mathrm{C}^1_{(u,v)}(p_{o(1)}, \sigma_{o(1)}) = \mathrm{C}^2_{(u,v)}(p_{o(2)}, \sigma_{o(2)}) = \ldots = \mathrm{C}^s_{(u,v)}(p_{o(s)}, \sigma_{o(s)})$$

For the experimental situation with **s** independent populations, the statistic to be used is

$$\chi^2_{\text{exp}} = \sum_{j=1}^{s} \frac{n_j[\hat{C}^j_{(u,v)}(\hat{p}_{o(j)}, \hat{\sigma}_{o(j)}) - \bar{C}^*_{(u,v)} + \hat{B}_j]^2}{\hat{\sigma}^2_j} \approx \chi^2_{s-1} \qquad (8)$$

where n_j, $\hat{\sigma}^2_j$ are the sample size and variance estimator associated with $C^j_{(u,v)}(p_{o(j)}, \sigma_{o(j)})$, respectively, and $\bar{C}^*_{(u,v)}$ is the statistic defined by

$$\bar{C}^*_{(u,v)} = [\sum_{j=1}^{s} \frac{n_j[\hat{C}^j_{(u,v)}(\hat{p}_{o(j)}, \hat{\sigma}_{o(j)}) + \hat{B}_j]}{\hat{\sigma}^2_j}] / [\sum_{j=1}^{s} \frac{n_j}{\hat{\sigma}^2_j}]$$

where B_j is the bias estimation of $C^j_{(u,v)}(p_{o(j)}, \sigma_{o(j)})$.

Note that the χ^2 distribution of the statistic defined in (8) is determined as follows.

Denoting $C^o_{(u,v)}$ as the common value of the capability indices, we obtain

$$\sum_{j=1}^{s} \frac{n_j[\hat{C}^j_{(u,v)}(\hat{p}_{o(j)}, \hat{\sigma}_{o(j)}) + \hat{B}_j - \mathrm{C}^o_{(u,v)}]^2}{\hat{\sigma}^2_j}$$

$$= \sum_{j=1}^{s} \frac{n_j[(\hat{C}^j_{(u,v)}(\hat{p}_{o(j)}, \hat{\sigma}_{o(j)}) + \hat{B}_j - \bar{C}^*_{(u,v)}) + (\bar{C}^*_{(u,v)} - C^o_{(u,v)})]^2}{\hat{\sigma}^2_j}$$

$$= \sum_{j=1}^{s} \frac{n_j[\hat{C}^j_{(u,v)}(\hat{p}_{o(j)}, \hat{\sigma}_{o(j)}) + \hat{B}_j - \bar{C}^*_{(u,v)}]^2}{\hat{\sigma}^2_j} + \sum_{j=1}^{s} \frac{n_j[\bar{C}^*_{(u,v)} - C^o_{(u,v)}]^2}{\hat{\sigma}^2_j}$$

$$+ 2\sum_{j=1}^{s} \frac{n_j[\hat{C}^j_{(u,v)}(\hat{p}_{o(j)}, \hat{\sigma}_{o(j)}) + \hat{B}_j - \bar{C}^*_{(u,v)}][\bar{C}^*_{(u,v)} - C^o_{(u,v)}]}{\hat{\sigma}^2_j}$$

Taking into account the definition of $\bar{C}^*_{(u,v)}$, the following result is obtained:

$$\sum_{j=1}^{s} \frac{n_j[\hat{C}^j_{(u,v)}(\hat{p}_{o(j)}, \hat{\sigma}_{o(j)}) + \hat{B}_j - C^o_{(u,v)}]^2}{\hat{\sigma}^2_j} =$$

$$= \sum_{j=1}^{s} \frac{n_j[\hat{C}^j_{(u,v)}(\hat{p}_{o(j)}, \hat{\sigma}_{o(j)}) + \hat{B}_j - \bar{C}^*_{(u,v)}]^2}{\hat{\sigma}^2_j}$$

$$+ \sum_{j=1}^{s} \frac{n_j[\bar{C}^*_{(u,v)} - C^o_{(u,v)}]^2}{\hat{\sigma}^2_j}$$

where

$$\sum_{j=1}^{s} \frac{n_j[\hat{C}^j_{(u,v)}(\hat{p}_{o(j)}, \hat{\sigma}_{o(j)}) + \hat{B}_j - C^o_{(u,v)}]^2}{\hat{\sigma}^2_j} \approx \chi^2_s$$

and

$$\sum_{j=1}^{s} \frac{n_j[\bar{C}^*_{(u,v)} - C^o_{(u,v)}]^2}{\hat{\sigma}^2_j} = \sum_{j=1}^{s} \left[\frac{\sum_{j=1}^{s} \frac{n_j[\hat{C}^j_{(u,v)}(\hat{p}_{o(j)}, \hat{\sigma}_{o(j)}) + \hat{B}_j - C^o_{(u,v)}]}{\hat{\sigma}^2_j}}{(\sum_{j=1}^{s} \frac{n_j}{\hat{\sigma}^2_j})^{1/2}} \right]^2 \approx \chi^2_1$$

In particular, we obtain the statistical $\chi^2_{\text{exp}}$by application of Cochran's theorem:

$$\chi^2_{\text{exp}} = \sum_{j=1}^{s} \frac{n_j[\hat{C}^j_{(u,v)}(\hat{p}_{o(j)}, \hat{\sigma}_{o(j)}) - \bar{C}^*_{(u,v)} + \hat{B}_j]^2}{\hat{\sigma}^2_j} \approx \chi^2_{s-1}$$

3.3 Confidence Intervals

According to the results obtained for Theorem (Sect. 2) confidence intervals for the capability index and for the differences of capability indices can be obtained. Thus, the confidence interval associated with $C_{(u,v)}(p_o, \sigma_o)$ is

$$(\hat{C}_{(u,v)}(\hat{p}_o, \hat{\sigma}_o) + \hat{B} - n_{\varepsilon/2}\frac{\hat{\sigma}}{n^{1/2}}, \hat{C}_{(u,v)}(\hat{p}_o, \hat{\sigma}_o) + \hat{B} + n_{\varepsilon/2}\frac{\hat{\sigma}}{n^{1/2}}) \quad (9)$$

and for the difference $C^i_{(u,v)}(p_{o(i)}, \sigma_{o(i)}) - C^j_{(u,v)}(p_{o(j)}, \sigma_{o(j)})$, the confidence interval is expressed as

$$[\hat{C}^i_{(u,v)}(\hat{p}_{o(i)}, \hat{\sigma}_{o(i)}) - \hat{C}^j_{(u,v)}(\hat{p}_{o(j)}, \hat{\sigma}_{o(j)}) + (\hat{B}_i - \hat{B}_j)] \pm z_{\varepsilon/2}\sqrt{\frac{\hat{\sigma}^2_i}{n_i} + \frac{\hat{\sigma}^2_j}{n_j}}$$

In order to improve the accuracy in estimating capability indices, it is interesting to explore confidence intervals that combine analytical results with computational techniques. A good approach is bootstrap-t confidence intervals with a pivot:

$$\hat{P}_i = \frac{[\hat{C}^*_{(u,v)}(\hat{p}_o, \hat{\sigma}_o)_i - \hat{C}_{(u,v)}(\hat{p}_o, \hat{\sigma}_o)]}{\hat{\sigma}^*_i / n^{1/2}},$$

where $\hat{C}_{(u,v)}(\hat{p}_o, \hat{\sigma}_o)$ is the estimated value of the capability index, $\hat{C}^*_{(u,v)}(\hat{p}_o, \hat{\sigma}_o)_i$ are the values obtained from resampling and σ^*_i is the standard deviation described in Theorem (see Sect. 2).

4 A Case Study

The automotive industry is involved in the design, development, manufacture, marketing and sale of motor vehicles. As the whole product life cycle should be regarded in an integrated perspective, representatives from advance development, design, production, marketing, purchasing and project management should work together on the *ecodesign* of further developments or new products to predict the overall effects of changes in the product and their environmental impact.

Environmental aspects that should be analyzed for every stage of the life cycle include consumption of resources (energy, materials, water or land area), emissions to air or water and human health impacts such as noise and vibration, among others.

In this section, for illustrative purposes we demonstrate the performance of the methodology for statistical process control of a car hood manufacturing application, for which a variety of thermoplastics have been used to obtain weight reduction, design flexibility and reduced noise transmission.

To meet passenger comfort requirements and to comply with legislation, polymer-based acoustic hood coverings are widely accepted in the automotive market. The most environmentally friendly option is molding of badges obtained from prebaked recycled fibers and resins. Standard processes are usually designed for a specific product and may require cost-intensive equipment that is profitable only for high-volume production. In the automotive field, where the scale of manufacture means that high production rates are very difficult, establishing satisfactory performance at the time of commercialization is critical. Thus, the aim is to obtain the highest flexibility and to fulfill customer needs in a fast and reliable process which minimize stocks and consequently minimize cost. A flexible method for producing a small series of high-quality components can be achieved by using flexibility for different molds, fast adaptation to layout changes and shorter periods for adapting equipment to the normal process flow. For these reasons and because of the intrinsic changeability of recycled fibers, the production process is not stable over time.

The intrinsic characteristics of the product (fireproof function and absorption of acoustics and vibrations) are guaranteed by the components, the thickness and the weight/m^2, whereas the esthetic characteristics (no resin spots >2 mm in diameter in the weave, no wrinkles or loose threads, inflexibility, etc.) are subject to self-control and/or final control because they affect the non-conformity accepted by the customer.

Process capability was determined without withdrawing pieces at any time, under the following conditions:

Condition 1. Short fibers+30% resin;
Condition 2. Intermediate fibers+28% resin; and
Condition 3. Short fibers+28% resin+2% glass wool.

For these experimental conditions, the non-conformity percentages observed for sample sizes ($n = 60, 100$) are collected in Table 1. We can consider $p_c = 0.02$ (a common assumption in most "real situations") as the minimum acceptable proportion of non-conformity. According to Table 1, the $\mathrm{C}_{(0,0)}(p_o, \sigma_o)$ values with $n_{\alpha/2} = 3$ are:

$$\mathrm{C}^1_{(0,0)}(p_{o(1)}, \sigma_{o(1)}) = 0.85 \quad \mathrm{C}^2_{(0,0)}(p_{o(2)}, \sigma_{o(2)}) = 1.18 \quad \mathrm{C}^3_{(0,0)}(p_{o(3)}, \sigma_{o(3)}) = 1.25$$

However, knowledge of these values does not suffice for measuring the capability of a process. Keeping in mind that multiple testing theory provides a framework for defining and controlling appropriate error rates in order to protect against wrong conclusions the hypothesis to be tested is

$$\begin{array}{ll} H_o: & \mathrm{C}^1_{(0,0)}(p_{o(1)}, \sigma_{o(1)}) = \mathrm{C}^2_{(0,0)}(p_{o(2)}, \sigma_{o(2)}) = C^3_{(0,0)}(p_{o(3)}, \sigma_{o(3)}) \\ H_1: & \text{There are differences} \end{array} \tag{10}$$

where the experimental statistic (8) had a value

$$\chi^2_{\mathrm{exp}} = \sum_{j=1}^{s} \frac{n_j [\hat{C}^j_{(u,v)}(\hat{p}_{o(j)}, \hat{\sigma}_{o(j)}) - \bar{C}^*_{(u,v)} + \hat{B}_j]^2}{\hat{\sigma}^2_j} = 6.79$$

greater than the significance bound, thus leading to the rejection of the null hypothesis of homogeneity (p-value $=0.034$).

In addition, a selection of results for the capability process is given below. Figure 2 shows the confidence intervals (confidence level 95% according to (9) for the experimental conditions), while Table 3 summarizes some descriptive results of the process, obtained from Table 2.

Thus, as a final practical conclusion, we summarized the results:

First of all, talking in plain language, the experimental conditions suggested do not improve the process (capacity process >1.33). Some more work is necessary to set more clearly the proper experimental conditions and more flexible methodology for producing high-quality components. However,

1. In terms of multiple testing and confidence intervals, the results show differences between the experimental conditions examined. In fact, experimental conditions 2 and 3 show similar $\mathrm{C}_{(u,v)}(p, \sigma)$ values, in both cases greater than experimental condition 1.
2. On non-stability over time, characterized by σ_o (see Table 3), the results show that short fibers (associated with experimental conditions 1, 3) perform worse than intermediate fibers (experimental condition 2). However, the non-conformity mean p_o (see Table 3), associated with the glass wool effect, suggested better results among the short fibers experimental conditions.

In addition, a pilot study applying intermediate fibers+resin+glass wool was considered and the total cost was evaluated. Unfortunately, the results indicated that this new experimental condition was less cost-efficient than the original one (intermediate fibers+resin).

Table 2. Non conformity values obtained for every experimental condition

Condition 1		Condition 2		Condition 3	
NC*	Sample size	NC	Sample size	NC	Sample size
1	100	1	60	1	60
0	100	1	60	0	60
2	100	2	60	2	60
1	100	0	60	1	60
1	100	1	60	2	60
1	100	1	60	1	60
0	100	2	10	1	60
1	100	1	100	1	60
2	100	2	100	2	100
2	100	2	100	1	100
0	100	3	100	1	100
1	100	1	100	1	100
3	100	1	100	2	100
3	100	2	100	1	100
3	100	2	100	1	100
3	100	1	100	3	100
4	100	2	100	1	100
5	100	2	100	2	100
3	100	1	100	1	100
3	100			1	100
				3	100

* Non Conformity values

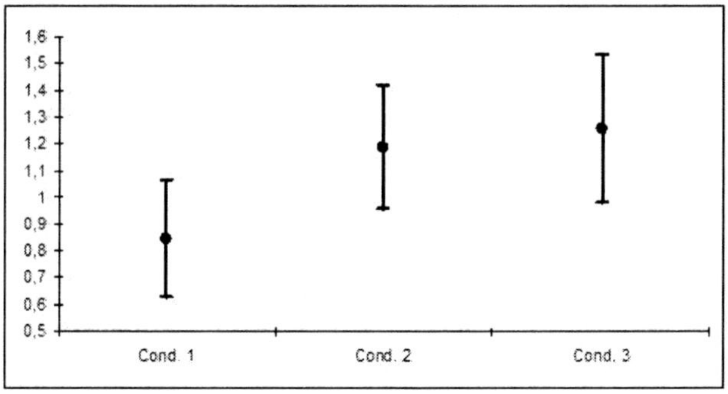

Fig. 2. Experimental conditions. Confidence intervals for a $100(1-\alpha)\% = 95\%$ confidence coefficient.

Table 3. Non conformity proportion: Descriptive results

	Condition 1	Condition 2	Condition 3
Non-conformity mean p_o	0.0190	0.0169	0.0160
Non-conformity standard deviation σ_o	0.0014	0.0000	0.0000

References

1. Boyles, R.A.: Process capability with asymmetric tolerances. Comm. Statist. -Simul. Comp. 23, 615–643 (1994)
2. Borges, W., Ho, L.L.: A fraction defective capability index. Qual. Reliab. Engin. Int. 17, 447–458 (2001)
3. Chan, L.K., Cheng, S.W., Spiring, F.A.: A new measure of process capability: Cpm. J. Qual. Tech. 20, 162–175 (1998)
4. Chen, S.M., Pearn, W.L.: The asymptotic distribution of the estimated process capability index Cpk. Comm. Statist.-Theor. Meth. 26, 2489–2497 (1997)
5. Dik, J.J., Gunst, M.C.M.: The distribution of general. quadratic forms in normal variables. Statistica Neerlandica 39, 14–26 (1985)
6. Hsiang, T.C., Taguchi, G.A.: Handbook of ASA Annual Meeting. Las Vegas, Nevada (1985)
7. Hsieh, K.L., Tong, L.I.: Manufacturing performance evaluation for IC products. The Int. J. Adv. Manuf. Tech. 27, 1217–1222 (2006)
8. Juran, J.M.: Juran's quality control handbook, 3rd edn. McGraw-Hill, New York (1974)
9. Kane, V.E.: Process capability indices. J. Qual. Tech. 18, 41–52 (1986)
10. Kotz, S., Johnson, N.L.: Process capability indices – A review, 1992 –2000. J. Qual. Tech. 34, 2–53 (2002)
11. Mathew, T., Sebastian, G., Kurian, K.M.: Generalized confidence intervals for process capability indices. Qual. Reliab. Engin. Int. 23, 471–481 (2007)
12. Parchami, A., Mashinchi, M.: Fuzzy estimation for process capability indices. Inform. Sci. 177, 1452–1462 (2007)
13. Pearn, W.L., Kotz, S., Johnson, N.L.: Distribution and inferential properties of capability indices. J. Qual. Tech. 24, 216–231 (1992)
14. Pearn, W.L., Kotz, S.: Encyclopedia and Handbook of Process Capability Indices. World Scientific, Singapore (2006)
15. Perakis, M., Xekalaki, E.: A process capability index that is based on the proportion of conformance. J. Statist. Comp. Simul. 72, 707–718 (2002)
16. Salicrú, M., Vaamonde, A.: Confidence intervals for capability indices. Adv. Appl. in Statist. 4, 129–137 (2004)
17. Singpurwalla, N.D.: The stochastic control of process capability indices. TEST 7, 1–74 (1988)
18. Spiring, F., Leung, B., Cheng, S., Yeung, A.: A bibliography of process capability papers. Qual. Reliab. Engin. Int. 19, 445–460 (2003)
19. Vannman, K.: Distribution and moments in simplified form for a general class of capability indices. Comm. Statist.-Theor. Meth. 26, 159–179 (1997)

20. Vannman, K.: A general class of capability indices in the case of tolerances. Comm. Statist.-Theor. Meth. 26, 2049–2072 (1997)
21. Wright, P.A.: The probability density function of process capability index Cpmk. Comm. Statist.-Theor. Meth. 27, 1781–1789 (1998)
22. Yeh, A.B., Battacharya, S.: A robust process capability index. Comm. Statist.-Simul. Comp. 27, 565–589 (1998)
23. Zimmer, L.S., Hubele, N.F., Zimmer, W.J.: Confidence intervals and sample size determination for Cpm. Qual. Reliab. Engin. Int. 17, 51–68 (2001)

Appendix 1

Exploiting algebraic structure of sum of squares:

$$\sum_{i=1}^{n}\sum_{j=1}^{m_i}(x_{ij}-\hat{p}_o)^2$$

$$\begin{aligned}
&=\sum_{i=1}^{n}\sum_{j=1}^{m_i}((x_{ij}-\hat{p}_i)+(\hat{p}_i-\hat{p}_o))^2\\
&=\sum_{i=1}^{n}\sum_{j=1}^{m_i}(x_{ij}-\hat{p}_i)^2+2\cdot\sum_{i=1}^{n}(\hat{p}_i-\hat{p}_o)\sum_{j=1}^{m_i}(x_{ij}-\hat{p}_i)\\
&+\sum_{i=1}^{n}\sum_{j=1}^{m_i}(\hat{p}_i-\hat{p}_o)^2\\
&=\sum_{i=1}^{n}\sum_{j=1}^{m_i}(x_{ij}-\hat{p}_i)^2+\sum_{i=1}^{n}m_i\cdot(\hat{p}_i-\hat{p}_o)^2
\end{aligned}$$

we can obtain

$$m_+\cdot\hat{p}_o(1-\hat{p}_o)=\sum_{i=1}^{n}m_i\cdot(\hat{p}_i-\hat{p}_o)^2+\sum_{i=1}^{n}m_i\cdot\hat{p}_i(1-\hat{p}_i)$$

being $m_+ = m_1 + m_2 + \ldots + m_n$.

On the other hand and according to the degrees of freedom for each component, we can obtain the intra-batch and inter-batch variance estimations respectively.

$$\hat{\sigma}_1^2=\frac{1}{n\cdot(\bar{m}-1)}\sum_{i=1}^{n}m_i\cdot\hat{p}_i(1-\hat{p}_i)$$

$$\hat{\sigma}_o^2=\max\left\{0,\ \frac{1}{n-1}\sum_{i=1}^{n}\frac{m_i}{\bar{m}}\cdot(\hat{p}_i-\hat{p}_o)^2-\frac{1}{n\cdot(\bar{m}-1)}\sum_{i=1}^{n}\frac{m_i}{\bar{m}}\cdot\hat{p}_i(1-\hat{p}_i)\right\}$$

Appendix 2

To determine the statistical distribution of the capacity, we obtain the first-order Taylor expansion of $\hat{C}_{(u,v)}(\hat{p}_o,\hat{\sigma}_o)$ around the true value of the parameter (p_o,σ_o).

According to the Taylor expansion of

$$\hat{C}_{(u,v)}(\hat{p}_o,\hat{\sigma}_o)=C_{(u,v)}(p_o,\sigma_o)+(t_1,t_2)(\hat{p}_o-p_o,\hat{\sigma}_o-\sigma_o)^t+R_2,$$

the following approximation can be obtained:

$$n^{1/2}[\hat{C}_{(u,v)}(\hat{p}_o,\hat{\sigma}_o)-C_{(u,v)}(p_o,\sigma_o)]\approx(t_1,t_2)\cdot n^{1/2}(\hat{p}_o-p_o,\hat{\sigma}_o-\sigma_o)^t$$

where

$$t_1=\frac{\delta_1\,(p_o+n_{\alpha/2}\delta_2)-[1+n_{\alpha/2}\cdot V(\frac{\delta_4}{\delta_2})]\delta_3}{[p_o+n_{\alpha/2}\delta_2]^2}$$

$$\begin{aligned}
t_2 &= \frac{-n_{\alpha/2} \cdot V(\frac{\delta_3 \cdot \sigma_o}{\delta_2})}{[p_o + n_{\alpha/2}\delta_2]^2} \\
\delta_1 &= -u \cdot \max\{0,\, \mathrm{sgn}(p_o - T)\} && (11) \\
\delta_2 &= [\sigma_o^2 + v \cdot (\max\{0,\, \mathrm{p}_o - T\})^2]^{1/2} && (12) \\
\delta_3 &= p_c - u \cdot \max\{0,\, \mathrm{p}_o - T\} && (13) \\
\delta_4 &= v \cdot \max\{0,\, p_o - T\,\mathrm{sgn}(p_o - T) && (14) \\
V(\tfrac{X}{Y}) &= \begin{cases} 0 & \text{if}\, Y = 0 \\ \frac{X}{Y} & \text{if}\, Y \neq 0 \end{cases} \\
\mathrm{sgn}(x) &= x/|x|,
\end{aligned}$$

and R_2 is the second-order remainder of the Taylor expansion.

On the other hand, by taking into account the asymptotic distribution

$$n^{1/2}(\hat{p}_o - p_o,\, \hat{\sigma}_o - \sigma_o)^t \approx N(\mu_1,\, \Sigma_1),$$

where

$$\begin{aligned}
\mu_1 &= (0,\, \lambda_1 \sigma_o), && (15) \\
\Sigma_1 &= \begin{pmatrix} \sigma_{ob}^2 & 0 \\ 0 & \lambda_2 \cdot \sigma_o^2 \end{pmatrix}, && (16) \\
\lambda_1 &= n^{1/2} \cdot (c_4 - 1) && (17) \\
\sigma_{ob}^2 &= \frac{1}{(n-1)\bar{m}} \sum_{i=1}^{n} m_i \, (\hat{p}_i - \hat{p}_o)^2 && (18) \\
\lambda_2 &= n - n \cdot c_4^2 && (19) \\
c_4 &= \frac{E(\hat{\sigma}_o)}{\sigma_o} = \sqrt{\frac{2}{n-1}} \frac{\Gamma(n/2)}{\Gamma((n-1)/2)} && (20)
\end{aligned}$$

the following result can be obtained.

For sample size n, if n is sufficiently large, the asymptotic distribution of the capability index $\hat{C}_{(u,v)}(\hat{p}_o,\, \hat{\sigma}_o)$ defined in (7) satisfies:

$$n^{1/2}[\hat{C}_{(u,v)}(\hat{p}_o,\, \hat{\sigma}_0) - C_{(u,v)}(p_o,\, \sigma_0)] \approx N(\mu,\, \sigma) \approx N(\hat{\mu},\, \hat{\sigma}),$$

where

$$\mu = \frac{-n_{\alpha/2}\, \lambda_1\, \sigma_o}{[p_o + n_{\alpha/2}\delta_2]^2} V(\tfrac{\delta_3 \cdot \sigma_o}{\delta_2}) \qquad (21)$$

$$\sigma = \frac{1}{[p_o + n_{\alpha/2}\delta_2]^2} \Big[[\delta_1 \, (p_o + n_{\alpha/2}\delta_2) - [1 + n_{\alpha/2} \cdot V(\tfrac{\delta_4}{\delta_2})]\delta_3]^2 \sigma_{ob}^2$$

$$+ [n_{\alpha/2} \cdot V(\tfrac{\delta_3 \cdot \sigma_o}{\delta_2})]^2 \lambda_2 \sigma_o^2 \Big]^{1/2} \qquad (22)$$

where δ_1, δ_2, δ_3, δ_4, λ_1, and λ_2 are the values defined in (11), (12), (13), (14), (17), (19), respectively, and $(\hat{\mu},\, \hat{\sigma})$ the respective estimates according to (1, 2).

Remark 1. The remainder R_2 of the Taylor expansion converges in probability to zero and is given by

$$R_2 = [(\hat{p}_o - p_o,\, \hat{\sigma}_o - \sigma_o) \cdot C \cdot (\hat{p}_o - p_o,\, \hat{\sigma}_o - \sigma_o)^t],$$

where C is the matrix of the second partial derivatives of C_p with respect to (p_o, σ_o), and

$$\begin{aligned} n^{1/2}\, R_2 &= n^{-1/2}[n^{1/2}(\hat{p}_o - p_o,\, \hat{\sigma}_o - \sigma_o) \cdot C \cdot n^{1/2}(\hat{p}_o - p_o,\, \hat{\sigma}_o - \sigma_o)^t] \\ &\approx n^{-1/2} \sum \beta_i \chi^2_{1,i} \overset{n\to\infty}{\longrightarrow} 0, \end{aligned}$$

where β_i are the eigenvalues of the matrix $C\Sigma_1$ and $\chi^2_{1,i}$ are the independent χ^2 distributions with one degree of freedom.

The results that allow us to obtain the distribution of the quadratic forms resulting from the second-order term of the Taylor expansion can be found in [5].

Remark 2. The asymptotic distribution

$$n^{1/2}(\hat{p}_o - p_o,\, \hat{\sigma}_o - \sigma_o) \approx N(\mu_1,\, \Sigma_1)$$

results from taking into account the independence between the mean and the standard deviation in a simple random sample, the approximation based on Taylor expansion of the function $f(x) = x^{1/2}$

$$\hat{\sigma}_o - \sigma_o = \sqrt{\hat{\sigma}_o^2} - \sqrt{\sigma_o^2} \approx \frac{1}{2 \cdot \sigma_o}(\hat{\sigma}_o^2 - \sigma_o^2),$$

the convergence of the distribution χ^2 to the normal distribution, and the relation $E(\hat{\sigma}_o) = c_4 \sigma_o$ to adjust the expected value and the variance.

Part II

Applied Mathematics

Hopf Bifurcation and Bifurcation from Constant Oscillations to a Torus Path for Delayed Complex Ginzburg-Landau Equations*

Alfonso Casal[1], Jesús Ildefonso Díaz[2], Michael Stich[1,3], and José Manuel Vegas[2]

[1] Dpto. de Matemática Aplicada, E.T.S. Arquitectura, Universidad Politécnica de Madrid, 28040 Madrid, Spain
alfonso.casal@upm.es

[2] Dpto de Matemática Aplicada, Facultad de Matemáticas, Universidad Complutense de Madrid, 28040 Madrid, Spain
ji_diaz@mat.ucm.es, JM_Vegas@mat.ucm.es

[3] Centro de Astrobiología (CSIC-INTA), Ctra de Ajalvir, km. 4, 28850 Torrejón de Ardoz (Madrid), Spain
stichm@inta.es

Summary. We consider the complex Ginzburg-Landau equation with feedback control given by some delayed linear terms (possibly dependent on the past spatial average of the solution). We prove several bifurcation results by using the delay as parameter. We start proving a Hopf bifurcation result for the equation without diffusion (the so-called Stuart-Landau equation) when the amplitude of the delayed term is suitably chosen. The diffusion case is considered first in the case of the whole space and later on a bounded domain with periodicity conditions. In the first case a linear stability analysis is made with the help of computational arguments (showing evidence of the fulfillment of the delicate transversality condition). In the last section the bifurcation takes place starting from an uniform oscillation and originates a path over a torus. This is obtained by the application of an abstract result over suitable functional spaces.

Keywords: delayed complex Ginzburg-Landau equations, Hopf bifurcation, torus bifurcation, linearization, uniform oscillations.

* To the memory of Maria Luisa Menéndez: excellent mathematician, admirable colleague and great person.

1 Introduction

1.1 Reaction-Diffusion Equations and the Complex Ginzburg-Landau Equation

The evolution of a chemical system consisting of n species which are reacting with each other and allowed to diffuse in a spatially extended medium, is generally described by a n-component reaction-diffusion equation for the $n-$concentrations $\mathbf{c}(x,t)$

$$\partial_t \mathbf{c} = \mathbf{F}(\mathbf{c};p) + \mathbf{D}\Delta\mathbf{c}, \tag{1}$$

where $\mathbf{F}$ denotes the typically nonlinear reaction term representing chemical kinetics, $\mathbf{D}\Delta\mathbf{c}$ the diffusion term (being $\mathbf{D}$ the diffusion matrix) and p a scalar control parameter. We assume that this system has a homogeneous, stationary solution $\mathbf{c_s}$ which undergoes a *Hopf bifurcation* at $p = p_0$: i.e., for $p \in (p_0, p_0 + \varepsilon)$ the stationary solution $\mathbf{c_s}$ becomes a time-periodic solution, at least for $\varepsilon > 0$ small enough.

It has been shown by Kuramoto and others that the dynamics of any reaction-diffusion system (1) in the vicinity of a Hopf bifurcation is described, by means of suitable parametrizations, by a nonlinear parabolic equation with complex coefficients, the so-called *complex Ginzburg-Landau equation* (CGLE), see, e.g., [12, 8]. The relation between reaction-diffusion systems and the CGLE has been treated in many texts, here we will follow the presentation of [10].

After a convenient choice of variables $\mathbf{X} = \mathbf{c} - \mathbf{c_s}$ (the *concentration deviations*) and $\epsilon = p - p_0$, the system can be reformulated as

$$\partial_t \mathbf{X} = \mathbf{J}\mathbf{X} + \mathbf{f}(\mathbf{x}, \epsilon) + \mathbf{D}\Delta\mathbf{X},$$

where $\mathbf{J}$ is the Jacobian matrix for the homogeneous system evaluated at $\mathbf{X_s} = \mathbf{0}$, i.e. $\mathbf{F}(\mathbf{c};p) - \mathbf{F}(\mathbf{c}_s;p_0) = \mathbf{J}\mathbf{X} + \mathbf{f}(\mathbf{x},\epsilon)$. At the *bifurcation point*, $\mathbf{J}$ has two imaginary eigenvalues $\pm i\omega_0$, being ω_0 the so-called *Hopf frequency*. The corresponding *right eigenvectors* $\mathbf{e}_1$ and $\mathbf{e}_2 = \bar{\mathbf{e}}_1$ (normalized with left eigenvectors $\mathbf{e}_i^+$ according to $\mathbf{e}_i^+\mathbf{e}_j = \delta_{ij}$) span the *center subspace* E^c of the homogeneous solution. The *center manifold* W^c is tangent to E^c at $\mathbf{X} = \mathbf{0}$, $\epsilon = 0$. The other $n-2$ eigenvalues are all assumed to be large and negative. This assures that a homogeneous solution converges fast toward W^c provided that $\mathbf{X}$ and ϵ are sufficiently small (for details and further references see [10]).

This allows us to express the concentration deviations $\mathbf{X}$ in terms of *amplitude coordinates* $\mathbf{Y} \in E^c$ by

$$\mathbf{X} = \mathbf{Y} + \mathbf{h}(\mathbf{Y}, \epsilon).$$

This equation describes a mapping from coordinates in the center subspace E^c onto the center manifold W^c. The function $\mathbf{h}(\mathbf{Y},\epsilon)$ is selected in such a way to successively eliminate as many nonlinear terms as possible from the kinetic equations starting from the lowest order [10]. Each kind of bifurcation is characterized by the specific terms which cannot be eliminated (the so-called *resonant terms*). In this way we obtain a general equation valid for all reaction-diffusion equations undergoing a given bifurcation. In the case of the Hopf bifurcation, neglecting the diffusion term, to third order we obtain the so-called *Stuart-Landau equation*

$$\frac{\mathrm{d}Y}{\mathrm{d}t} = (\mathrm{i}\omega_0 + \sigma_1\epsilon)Y - g|Y|^2 Y,$$

where Y is a complex amplitude given by $\mathbf{Y} = Y\mathbf{e}_1 + \overline{Y}\mathbf{e}_2$. The parameters σ_1 and g are complex and given by solutions of lengthy equations given in [10]. The Stuart-Landau equation represents the *normal form* of a homogeneous system close to a Hopf bifurcation. Performing a similar derivation, but including diffusion, we arrive at

$$\partial_t Y = (\mathrm{i}\omega_0 + \sigma_1\epsilon)Y - g|Y|^2 Y + d\Delta Y,$$

with $d = \mathbf{e}_1^+ \cdot \mathbf{D}\mathbf{e}_1$. After rescaling of space, time, and introducing A for Y, we finally arrive at the rescaled complex Ginzburg-Landau equation

$$\partial_t A = (1 - \mathrm{i}\omega)A - (1 + \mathrm{i}\alpha)|A|^2 A + (1 + \mathrm{i}\beta)\Delta A, \tag{2}$$

where A is the *complex oscillation amplitude*, ω the *linear frequency parameter*, α the *nonlinear frequency parameter*, and β the *linear dispersion coefficient*. All reaction-diffusion systems sufficiently close to a Hopf bifurcation are described by the complex Ginzburg-Landau equation. The specific details of the original system are incorporated in the parameter values. If one wishes to express the solution of the CGLE in the original variables, to first order the concentrations of the chemical species are expressed by

$$\mathbf{c} = \mathbf{c_s} + \sqrt{\epsilon}(Y(x,t)\mathbf{e}_1 + \overline{Y}(x,t)\mathbf{e}_2).$$

Different scalings of the CGLE are considered in the literature [3]. Here, we assume that the Hopf frequency is not scaled out, and hence contributes to ω in Eq. (2). We also send the reader to Appendix B of [12] for the detailed derivation of the CGLE associated to the Brusselator model.

1.2 On Feedback Control Using Delayed Terms

Over the decades, the complex Ginzburg-Landau equation has been studied intensively because of its frequent appearance in different contexts of science, and its rich repertoire of different spatio-temporal wave patterns like plane waves, spiral waves, or localized hole solutions [3]. Remarkable, even if the Hopf bifurcation is supercritical, and hence the limit cycle a stable solution of the Stuart-Landau equation, the oscillations in the spatially-extended system may be unstable. The resulting states of spatiotemporal chaos appear if the

Benjamin-Feir-Newell criterion $1 + \alpha\beta < 0$ is fulfilled, a phenomenon that is induced by the diffusive coupling and that is therefore genuine to a system with spatial degrees of freedom.

Considerable efforts have been made to understand this type of chaotic behavior and to apply methods to suppress this kind of turbulence and replace it by regular dynamics. In the context of the reaction-diffusion systems, the introduction of forcing terms or global feedback terms have been shown to be efficient ways to control turbulence [13, 11]. Still, control of chaotic states in nonlinear systems is a wide field of research that we cannot review here [15].

Global feedback methods, where a spatially independent quantity (or, e.g., a spatial average of a space-dependent quantity) is coupled back to the system dynamics, have attracted much attention since in many cases the models are simpler and easier to be carried out experimentally. Nevertheless, local methods have gained interest in recent years since they allow to access other solutions of the systems and may also be implemented, such as in the light-sensitive BZ reaction or in neurophysiological experiments [13].

Feedback methods with an explicit time delay amplify the range of possibilities of control that can be applied to the system and provide the researcher with an additional adjustable parameter. On the level of the mathematical description, the model equations become delay differential equations [9, 4]. Obviously, time delay feedback can be applied to any solution of the dynamics, not necessarily to a chaotic one.

1.3 Main Results

In this paper we analyze several bifurcation effects produced by the delay time in the behavior of solutions of the complex Ginzburg-Landau equation with this type of feedback.

In Section 2 we prove a Hopf bifurcation result for the equation without diffusion (the Stuart-Landau equation) when the amplitude of the delayed term is suitably chosen. This simplified formulation has the advantage that closed analytical solutions are possible and the necessary eigenvalue computations can be carried out in full. The diffusion case is considered firstly in the case of the whole space (Section 3) and later on a bounded domain with periodicity conditions (Section 4).

In the case in which the space is the whole $\mathbb{R}$ (we consider here the one-dimensional case) we performed a linear stability analysis of uniform oscillations with respect to spatiotemporal perturbations following the treatment made in [16]: we express the complex oscillation amplitude A as the superposition of a homogeneous mode H (corresponding to uniform oscillations) with spatially inhomogeneous perturbations,

$$A(x,t) = H(t) + A_+(t)\mathrm{e}^{\mathrm{i}\kappa x} + A_-(t)\mathrm{e}^{-\mathrm{i}\kappa x}\,.$$

With the help of computational arguments we get several bifurcation diagrams where, besides the delay time it is possible to use the feedback magnitude term. Among many other detailed informations, we obtain numerical evidence of the fulfillment of the delicate transversality condition.

The paper ends by analyzing the case in which the bifurcation takes place starting from an uniform oscillation and originating a path over a torus. This time the study is carried out in two spatial dimensions over a rectangle in which we impose periodic boundary conditions. We show the applicability of an abstract result ([22]) to our formulation thanks to a suitable choice of the involved functional spaces. In this way, the spatial perturbations can be considered in their greatest generality.

The presentation of this chapter is very condensed due to limited space. A more detailed study will be published elsewhere.

2 Hopf Bifurcation for the Stuart-Landau Equation with a Time Delay Feedback

For the purposes of clarity and ease of understanding, we start by considering in this section a very simplified version of the general model to be given later which has the advantage that closed analytical solutions are possible and the necessary eigenvalue computations can be carried out in full. Unfortunately, such precise calculations are not available for the general model and a fairly complete graphical-numerical study will be given in exchange.

Equation (2) reads

$$\partial_t A = (1 - \mathrm{i}\omega)A - (1 + \mathrm{i}\alpha)\,|A|^2\,A + (1 + \mathrm{i}\beta)\Delta A.$$

In the Stuart-Landau equation, the diffusion term is absent, which amounts to restricting our study to the *spatially homogeneous* solutions (which always satisfy periodic boundary conditions as it will be formulated in Section 4). On the other hand, we assume that a *delayed linear feedback term* is added, so the equation under study in this section will be

$$\partial_t A = (1 - \mathrm{i}\omega)A - (1 + \mathrm{i}\alpha)\,|A|^2\,A + m_1 A + m_3 A(t - \tau). \tag{3}$$

More general control terms will be considered in the remaining sections of the paper. The change of variables $\mathbf{w}(t) = \mathrm{e}^{-\mathrm{i}\phi t} A(t)$ gives

$$\partial_t \mathbf{w} = (1 - \mathrm{i}\omega - \mathrm{i}\phi)\mathbf{w} - (1 + \mathrm{i}\alpha)\,|\mathbf{w}|^2\,\mathbf{w} + m_1 \mathbf{w} + m_3 \mathrm{e}^{-\mathrm{i}\phi\tau}\mathbf{w}(t - \tau). \tag{4}$$

We now choose $\phi = -\alpha - \omega$ and $m_3 = -\mathrm{e}^{\mathrm{i}\phi\tau} m_1$ and denote the stationary solution of

$$\partial_t \mathbf{w} = (1 + \mathrm{i}\alpha)(\mathbf{w} - |\mathbf{w}|^2\,\mathbf{w}) + m_1\left[\mathbf{w} - \mathbf{w}(t - \tau)\right]. \tag{5}$$

by $\mathbf{w}_0$.

In order to check if at some critical value of the delay $\tau = \tau^*$ a Hopf bifurcation takes place, we linearize the equation around $\mathbf{w}_0 = 1$ and check whether a pair of complex eigenvalues $\lambda(\tau) = a(\tau) \pm ib(\tau)$ of the linearization cross transversally the imaginary axis away from the origin, i.e., they satisfy $a(\tau^*) = 0,\ b(\tau^*) \neq 0$ and $a'(\tau^*) \neq 0$ (see, e.g., [22]).

Observe now that the complex term $|v|^2 v$, although perfectly differentiable from the real point of view (in fact, the complex map $z \longmapsto |z|^2 z = z^2\bar{z}$ is *real-analytic*), is *not an analytic* (or holomorphic) function from the complex viewpoint. Therefore it becomes convenient at this point to abandon the complex notation and write the system in real form ($\mathbf{w} = u + iv$) as follows

$$\partial_t \begin{pmatrix} u \\ v \end{pmatrix} = \begin{pmatrix} 1 & -\alpha \\ \alpha & 1 \end{pmatrix} (1 - (u^2 + v^2)) \begin{pmatrix} u \\ v \end{pmatrix} + m_1 \begin{pmatrix} u - u(t-\tau) \\ v - v(t-\tau) \end{pmatrix}.$$

Let us fix our attention to the stationary solution $\mathbf{w}_0 = (u_0, v_0) = (1, 0)$. The linearization around $\mathbf{w}_0$ is given by

$$\partial_t \begin{pmatrix} U \\ V \end{pmatrix} = \begin{pmatrix} 1 & -\alpha \\ \alpha & 1 \end{pmatrix} \begin{pmatrix} -2 & 0 \\ 0 & 0 \end{pmatrix} \begin{pmatrix} U \\ V \end{pmatrix} + m_1 \begin{pmatrix} U - U(t-\tau) \\ V - V(t-\tau) \end{pmatrix} \tag{6}$$

and the eigenvalue-eigenvector pairs associated to this vector equation are the solutions of (6) of the special form $U(t) = \mathrm{e}^{\lambda t} U_0,\ V(t) = \mathrm{e}^{\lambda t} V_0$ where $\lambda \in \mathbb{C}$ and U_0, V_0 are (possibly complex) constant (nonzero) 2-vectors. One thus easily finds

$$\lambda \begin{pmatrix} U_0 \\ V_0 \end{pmatrix} = \begin{pmatrix} -2 + m_1 & 0 \\ -2\alpha & m_1 \end{pmatrix} \begin{pmatrix} U_0 \\ V_0 \end{pmatrix} - m_1 \mathrm{e}^{-\lambda\tau} \begin{pmatrix} U_0 \\ V_0 \end{pmatrix},$$

thus arriving to the *characteristic equation*

$$\begin{vmatrix} \lambda + 2 - m_1 + m_1 \mathrm{e}^{-\lambda\tau} & 0 \\ 2\alpha & \lambda - m_1 + m_1 \mathrm{e}^{-\lambda\tau} \end{vmatrix} = 0.$$

This means that we have a double collection of eigenvalues: those satisfying $\lambda - m_1 + m_1 \mathrm{e}^{-\lambda\tau} = 0$ and those satisfying $\lambda + 2 - m_1 + m_1 \mathrm{e}^{-\lambda\tau}$. Denoting $\lambda = a + ib$, we identify two classes of eigenvalues:

$$\lambda - m_1 + m_1 \mathrm{e}^{-\lambda\tau} = 0 \Longleftrightarrow \begin{cases} a - m_1 + m_1 \mathrm{e}^{-a\tau} \cos b\tau = 0 \\ b - m_1 \mathrm{e}^{-a\tau} \sin b\tau \end{cases} \quad \text{(Class 1)}$$

$$\lambda + 2 - m_1 + m_1 \mathrm{e}^{-\lambda\tau} = 0 \Longleftrightarrow \begin{cases} a + 2 - m_1 + m_1 \mathrm{e}^{-a\tau} \cos b\tau = 0 \\ b - m_1 \mathrm{e}^{-a\tau} \sin b\tau \end{cases} \quad \text{(Class 2)}$$

We now look for values $\tau = \tau^*$ for which $a = 0$ and $b \neq 0$. We find no eigenvalues of this kind for Class 1, since $-1 + \cos b\tau = 0$ implies $\sin b\tau = 0$, and hence $b = 0$ from the second equation.

However, Class 2 does give us some useful values:

$$2 - m_1 + m_1 \cos b\tau = 0 \Longrightarrow \cos b\tau = \frac{m_1 - 2}{m_1},$$

$$b - m_1 \sin b\tau = 0 \Longrightarrow \sin b\tau = \frac{b}{m_1}.$$

Thus,

$$1 = \cos^2 b\tau + \sin^2 b\tau = \left(\frac{m_1 - 2}{m_1}\right)^2 + \frac{b^2}{m_1^2} \Longrightarrow b^2 = m_1^2 - (m_1 - 2)^2 = 4(m_1 - 1).$$

Hence, if $m_1 > 1$, we have

$$\cos b\tau = \frac{m_1 - 2}{m_1} \Longrightarrow b\tau = \arccos\left(\frac{m_1 - 2}{m_1}\right)$$

which is well defined for every $m_1 > 1$.

Summarizing, the set of values

$$b^* = 2\sqrt{m_1 - 1},\ \tau^* = \frac{1}{b^*}\left[\arccos\left(\frac{m_1 - 2}{m_1}\right) + 2k\pi\right]$$

corresponds to a (possible) bifurcation point of Hopf type. For instance, for $m_1 = 2$ we have $b^* = 2$ and $\tau^* = k\pi + \pi/4$.

We now need to compute the derivative $a'(\tau^*)$. It is easier now to go back to the complex formulation of Class 2 eigenvalues

$$\lambda + 2 - m_1 + m_1 \mathrm{e}^{-\lambda\tau} = 0,$$

and find $\mathrm{d}\lambda/\mathrm{d}\tau$ by implicit differentiation:

$$\frac{\mathrm{d}\lambda}{\mathrm{d}\tau} + m_1 \mathrm{e}^{-\lambda\tau}\left(-\frac{\mathrm{d}\lambda}{\mathrm{d}\tau}\tau - \lambda\right) = 0 \Longrightarrow \frac{\mathrm{d}\lambda}{\mathrm{d}\tau} = \frac{\lambda \mathrm{e}^{-\lambda\tau}}{1 - m_1 \mathrm{e}^{-\lambda\tau}\tau} = \frac{\lambda}{1 - m_1 \mathrm{e}^{\lambda\tau}\tau}.$$

Concentrating on the specific values $b^* = 2$ and $\tau^* = \pi/4$ we find, at the bifurcation values τ^*, $\lambda^* = \mathrm{i}b^*$, that

$$\left.\frac{\mathrm{d}\lambda}{\mathrm{d}\tau}\right|_{(\tau^*,\lambda^*)} = \frac{\mathrm{i}b^*}{1 - m_1 \mathrm{e}^{\mathrm{i}b^*\tau^*}\tau^*} = -\frac{4\pi}{\pi^2 + 4} + \frac{8}{\pi^2 + 4}\mathrm{i}.$$

Hence

$$\frac{\mathrm{d}a}{\mathrm{d}\tau}(\tau^*) = -\frac{4\pi}{\pi^2 + 4} \neq 0$$

and the transversality condition is satisfied. Therefore, a Hopf bifurcation occurs, and a periodic orbit of approximate period

$$T \simeq \frac{2\pi}{b(\tau^*)} = \pi$$

exists for delay values τ near τ^*.

Remark 1. To decide the sub- or supercritical character of the bifurcation a much longer analysis is necessary. On the other hand, for $\tau > 1/2$ there are always positive *real* eigenvalues coming from the first class, which means that the stationary point has become already unstable *before* the delay reaches $\tau^* = \pi/4$ value. Hence the periodic orbit cannot capture the stability lost by the stationary point, since that stability was already lost.

3 Hopf Bifurcation for the Complex Ginzburg-Landau Equation on the Whole Space and with Delayed Time Feedback

We come back to the consideration of the complex Ginzburg-Landau equation subjected to a time-delay feedback with local and global terms but now for the case of a spatial domain given by the whole space:

$$\begin{aligned} \partial_t A &= (1 - \mathrm{i}\omega)A - (1 + \mathrm{i}\alpha)|A|^2 A + (1 + \mathrm{i}\beta)\partial_{xx} A + F, \\ F &= \mu \mathrm{e}^{\mathrm{i}\xi} \left[m_1 A + m_2 \langle A \rangle + m_3 A(t - \tau) + m_4 \langle A(t - \tau) \rangle \right], \end{aligned} \tag{7}$$

where

$$\langle A \rangle = \frac{1}{L} \int_0^L A(x, t)\, \mathrm{d}x$$

denotes the spatial average of A over a one-dimensional medium of length L. There are many previous works in the literature dealing with such type of formulations: [6, 7, 17, 16].

Extensive simulations [17] and an analytical stability analysis [16] for a special case representing a Pyragas-type feedback [14] ($m_3 = -m_1 = m_l$, $m_4 = -m_2 = m_g$) showed the range of patterns that can be stabilized as function of the local and global feedback terms. If the feedback is global, uniform oscillations can be stabilized for a large range of feedback parameters, while as the contribution of the local feedback term becomes larger, the parameter regions increase where the homogeneous fixed point solution, standing waves and traveling waves are found.

Uniform oscillations $A(t) = \rho_0 \exp(-\mathrm{i}\theta t)$ are a solution of Eqs. (7) with amplitude and frequency given by

$$\begin{aligned} \rho_0 &= \sqrt{1 + \mu(m_g + m_l)(\cos(\xi + \theta\tau) - \cos\xi)}, \\ \theta &= \omega + \alpha + \mu(m_g + m_l)\left[\alpha(\cos(\xi + \theta\tau) - \cos\xi) - (\sin(\xi + \theta\tau) - \sin\xi)\right]. \end{aligned}$$

In [16], we performed a linear stability analysis of uniform oscillations with respect to spatiotemporal perturbations. There, we expressed the complex oscillation amplitude A as the superposition of a homogeneous mode H (corresponding to uniform oscillations) with spatially inhomogeneous perturbations,

$$A(x, t) = H(t) + A_+(t)e^{\mathrm{i}\kappa x} + A_-(t)e^{-\mathrm{i}\kappa x}. \tag{8}$$

Notice that here we are using the fact that the equation takes place on the whole space, which allows the justification of the spatially inhomogeneous perturbations of the form $A_+(t)e^{\mathrm{i}\kappa x} + A_-(t)e^{-\mathrm{i}\kappa x}$. Inserting Eq. (8) into Eq. (7), and assuming that the amplitudes $A_\pm$ are small, we obtain a set of equations for H, A_+, and A_-^* (see [16] for details of this derivation). To investigate linear stability of uniform oscillations with respect to spatiotemporal perturbations, we make the ansatz

$$
\begin{aligned}
A_+ &= A_+^0 \exp(-\mathrm{i}\theta t)\exp(\lambda t),\\
A_-^* &= A_-^{*0} \exp(\mathrm{i}\theta t)\exp(\lambda t),
\end{aligned}
\tag{9}
$$

where $\lambda = \lambda_1 + \mathrm{i}\lambda_2$ is a complex eigenvalue. Using ansatz (9), we arrive at the following eigenvalue equation:

$$
F = (A + \mathrm{i}B - \mathrm{i}\lambda_2 + D_1 + \mathrm{i}D_2)(A - \mathrm{i}B - \mathrm{i}\lambda_2 + C_1 + \mathrm{i}C_2), \tag{10}
$$

where we have defined

$$
\begin{aligned}
F &= (1+\alpha^2)\rho_0^4,\\
A &= 1 - \lambda_1 - 2\rho_0^2 - \kappa^2,\\
B &= \theta - \omega - 2\alpha\rho_0^2 - \beta\kappa^2,\\
C_1 &= \mu m_l e^{-\lambda_1\tau}\cos(\xi + \theta\tau + \lambda_2\tau) - \mu m_l \cos\xi,\\
C_2 &= -\mu m_l e^{-\lambda_1\tau}\sin(\xi + \theta\tau + \lambda_2\tau) + \mu m_l \sin\xi,\\
D_1 &= \mu m_l e^{-\lambda_1\tau}\cos(\xi + \theta\tau - \lambda_2\tau) - \mu m_l \cos\xi,\\
D_2 &= \mu m_l e^{-\lambda_1\tau}\sin(\xi + \theta\tau - \lambda_2\tau) - \mu m_l \sin\xi.
\end{aligned}
$$

We point out that the above eigenvalue equation can be obtained also by a formal linearization argument involving the Fréchet derivatives as in the next section. There is no general analytic solution to Eq. (10) for $\lambda_{1,2}$. Thus, Eq. (10) must be solved numerically for a given set of parameters. We keep the CGLE parameters α, β, ω and the feedback parameters m_l, m_g, and ξ constant and solve Eq. (10) with the FindRoot routine of the Mathematica package [21]. We then find, for each point in the (τ, μ)-space, the functional dependence of λ_1 and λ_2 on κ. Notice that if we assume $\kappa = 0$ the study can be applied to the case of the Stuart-Landau equation, as in Section 2.

In general, Eq. (10) has multiple solutions, reflected by multiple branches in the dispersion relation. Stability is determined by the sign of λ_1. The curves $\lambda_1(\kappa)$ either lie below $\lambda_1 = 0$, so that uniform oscillations are stable, or they display an interval of κ-values, where $\lambda_1 > 0$, so that uniform oscillations are unstable. At criticality, we have $\lambda_1 = 0$, $\partial_\epsilon\lambda_1 \neq 0$, where ϵ stands for either μ or τ. For the critical wavenumber κ_c, there are two possibilities: $\kappa_c = 0$ or $\kappa_c \neq 0$ ($\pm\kappa_c$ are solutions, although below, we consider only $\kappa_c > 0$ without loss of generality).

Two instabilities are particularly important in our system: the first one is associated with $\kappa_c > 0$ and $\lambda_2(\kappa_c) = 0$, and the second one with $\kappa_c = 0$ and $\lambda_2(\kappa_c) \neq 0$. In Figure 1, we show as an example the control diagram in (μ, τ)-space for $m_l = 0.4$, $m_g = 0.6$. Stable uniform oscillations are observed above the solid curve and to the right of the dotted curve. At the solid curve, uniform oscillations become unstable with respect to perturbations with $\kappa_c > 0$ and $\lambda_2(\kappa_c) = 0$, at the dotted curve, with $\kappa_c = 0$ and $\lambda_2(\kappa_c) \neq 0$. In Figure 2(a,b), the dispersion relations $\lambda_{1,2} = \lambda_{1,2}(\kappa)$ are shown for three τ values close to criticality, demonstrating clearly the nature of the underlying

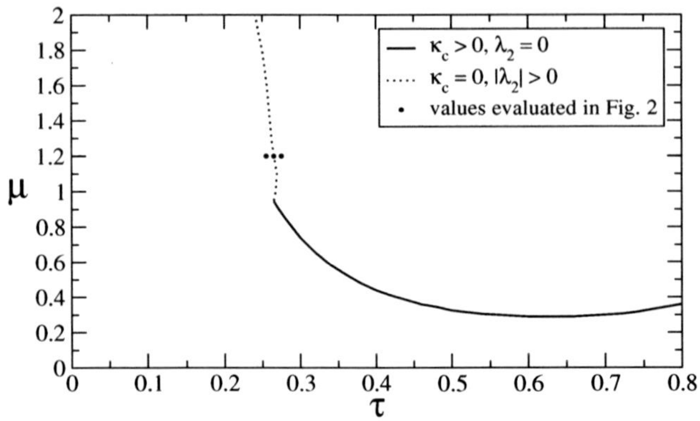

Fig. 1. Control diagram in (μ, τ)-space for $m_l = 0.4$, $m_g = 0.6$. The other parameters are $\alpha = -1.4$, $\beta = 2$, $\omega = 2\pi - \alpha$, $\xi = \pi/2$. At the solid curve, uniform oscillations become unstable with respect to perturbations with $\kappa_c > 0$ and $\lambda_2(\kappa_c) = 0$, at the dotted curve, with $\kappa_c = 0$ and $\lambda_2(\kappa_c) \neq 0$. The dots indicate parameter values further studied in Figure 2.

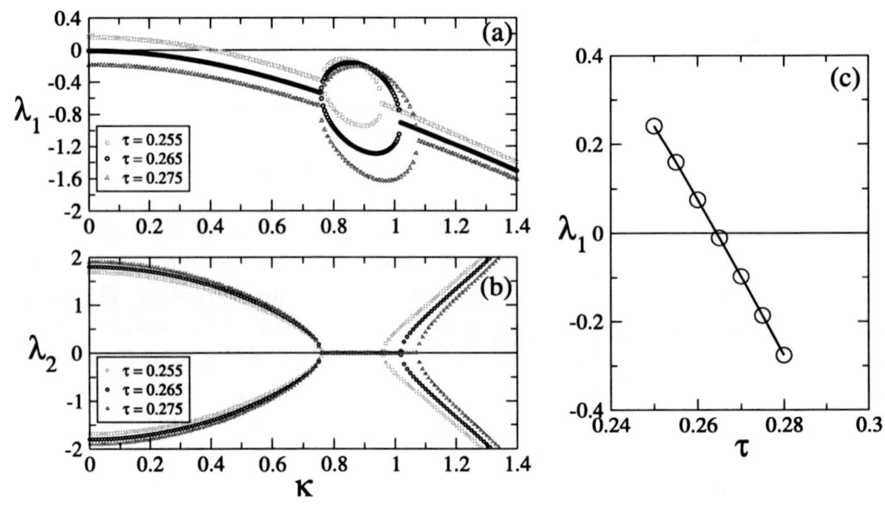

Fig. 2. Dispersion relations for three parameter sets close to criticality: $\tau = 0.255$ (light grey squares), $\tau = 0.265$ (black circles), $\tau = 0.275$ (dark grey triangles). (a) Real part of the eigenvalue as function of the wavenumber κ. (b) Imaginary part of the eigenvalue. The instability is characterized by $\kappa_c = 0$ and $\lambda_2(\kappa_c) \neq 0$ and occurs for $\mu = 1.2$ at $\tau = 0.264399$. (c) Real part of the eigenvalue as function of τ, demonstrating transversality.

instability. In Figure 2(c), we show that λ_1 crosses $\lambda_1 = 0$ as τ is varied, hence demonstrating transversality. As the uniform oscillations become unstable with respect to a mode with complex conjugated eigenvalues and since ρ_0 remains finite, we infer the presence of a secondary Hopf bifurcation.

4 Hopf Bifurcation for the Delayed CGLE in a Bounded Domain

In this section we consider the case of two spatial dimensions varying on the domain $\Omega = (0, L_1) \times (0, L_2)$ (note a slight change of notation with respect to Sect. 3). Our goal is to show a bifurcation phenomenon near uniform oscillations for the CGLE in terms of the delay term as parameter. We define the faces of the boundary

$$\Gamma_j = \partial\Omega \cap \{x_j = 0\}\,, \Gamma_{j+2} = \partial\Omega \cap \{x_j = L_j\}\,,\ j = 1, 2,$$

on which we assume periodic boundary conditions and, hence, the problem under study can be formulated as

$$(P_1)\begin{cases} \partial_t \mathbf{u} - (1 + \mathrm{i}\beta)\Delta\mathbf{u} = (1 - \mathrm{i}\omega)\mathbf{u} - (1 + \mathrm{i}\alpha)|\mathbf{u}|^2\mathbf{u} \\ \qquad\qquad\qquad\qquad +\mu \mathrm{e}^{\mathrm{i}\xi}\mathbf{F}(\mathbf{u}, t, \tau) & \Omega \times (0, \infty), \\ \left.\begin{array}{l} \mathbf{u}|_{\Gamma_j} = \mathbf{u}|_{\Gamma_{j+2}}, \\ \left(-\frac{\partial\mathbf{u}}{\partial\mathbf{n}}\Big|_{\Gamma_j} =\right) \frac{\partial\mathbf{u}}{\partial x_j}\Big|_{\Gamma_j} = \frac{\partial\mathbf{u}}{\partial x_j}\Big|_{\Gamma_{j+2}} \left(= \frac{\partial\mathbf{u}}{\partial\mathbf{n}}\Big|_{\Gamma_{j+2}}\right) \end{array}\right\} & \partial\Omega \times (0, \infty), \\ \mathbf{u}(x, s) = \mathbf{u}_0(x, s) & \Omega \times [-\tau, 0], \end{cases}$$

where $\mathbf{n}$ is the outpointing normal unit vector, and

$$\mathbf{F}(\mathbf{u}, t, \tau) = [m_1\mathbf{u}(x, t) + m_2\langle\mathbf{u}(t)\rangle + m_3\mathbf{u}(x, t - \tau) + m_4\langle\mathbf{u}(t - \tau)\rangle]$$

with

$$\langle\mathbf{u}(s)\rangle = \frac{1}{|\Omega|}\int_\Omega \mathbf{u}(x, s)\mathrm{d}x.$$

Again, the parameters $\alpha, \beta, \omega, \mu, \xi, m_i$ and τ are real, while $\mathbf{u}(x, t) = \mathbf{u}_1(x, t) + \mathrm{i}\mathbf{u}_2(x, t)$ is complex.

We study the stability of *uniform oscillations*, i.e., solutions of (P_1) of the form $\mathbf{v}_{\mathrm{uo}}(t) = \rho_0\mathrm{e}^{-\mathrm{i}\theta t}$ which determines completely ρ_0 and θ. We are interested in the Hopf bifurcation close to $\mathbf{v}_{\mathrm{uo}}(t)$ which gives rise to some paths on a suitable torus (for a different study dealing with invariant tori see [18]).

In order to avoid the application of very sophisticated techniques (dealing with periodic solutions), we can reduce the study to the Hopf bifurcation near a stationary solution of some auxiliary problem by introducing the change of unknown $\mathbf{z}(x, t) = \mathbf{v}(x, t)\mathrm{e}^{\mathrm{i}\theta t}$ where $\mathbf{v}(x, t)$ is a solution of (P_1). Thus, $\mathbf{z}(x, t)$ satisfies

$$(P_2)\begin{cases} \partial_t \mathbf{z} - (1+\mathrm{i}\beta)\Delta\mathbf{z} = (1+\mathrm{i}\theta)\mathbf{z} - (1+\mathrm{i}\alpha)|\mathbf{z}|^2\mathbf{z} + \mu\mathrm{e}^{\mathrm{i}\xi}\times \\ \times[m_1\mathbf{z} + m_2\langle\mathbf{z}\rangle + \mathrm{e}^{\mathrm{i}(\omega+\theta)\tau}(m_3\mathbf{z}(t-\tau) + m_4\langle\mathbf{z}(t-\tau)\rangle)] & \Omega\times(0,\infty), \\ \left.\begin{array}{l} \mathbf{z}|_{\Gamma_j} = \mathbf{z}|_{\Gamma_{j+2}}, \\ \left(-\frac{\partial\mathbf{z}}{\partial\mathbf{n}}\Big|_{\Gamma_j} =\right) \frac{\partial\mathbf{z}}{\partial x_j}\Big|_{\Gamma_j} = \frac{\partial\mathbf{z}}{\partial x_j}\Big|_{\Gamma_{j+2}} \left(= \frac{\partial\mathbf{z}}{\partial\mathbf{n}}\Big|_{\Gamma_{j+2}}\right) \end{array}\right\} & \partial\Omega\times(0,\infty), \\ \mathbf{z}(x,s) = \mathbf{u}_0(x,s)\mathrm{e}^{\mathrm{i}(\omega-\theta)s} & \Omega\times[-\tau,0]. \end{cases}$$

Now, $\mathbf{v}_{\mathrm{uo}}(t) = \rho_0\mathrm{e}^{-\mathrm{i}\theta t}$ is an uniform oscillation if and only if $\mathbf{z}(x,t) = \mathbf{v}_{\mathrm{uo}}(t)\mathrm{e}^{\mathrm{i}\theta t} = \mathbf{z}_\infty = \rho_0$ is an stationary solution of (P_2), i.e.,

$$\mathbf{0} = (1+\mathrm{i}\theta)\mathbf{z}_\infty - (1+\mathrm{i}\alpha)\left|\mathbf{z}_\infty\right|^2\mathbf{z}_\infty + \mu\mathrm{e}^{\mathrm{i}\xi}\left[m_1 + m_2 + \mathrm{e}^{\mathrm{i}(\omega+\theta)\tau}(m_3+m_4)\right]\mathbf{z}_\infty.$$

4.1 The Abstract Hopf Bifurcation Theorem for Semilinear Functional Equations

We shall apply to our setting an abstract result due to J. Wu (see [22], Theorem 2.1) stated for problems of the type

$$\begin{cases} \frac{du}{dt}(t) + Au(t) = L(\mu, u_t(.)) + g(u_t(.)) & in\ X, \\ u(s) = u_0(s) & s\in[-\tau,0]. \end{cases}$$

on a Banach space X, where $u_t : [-\tau,0] \to X$, under the following list of conditions:

(H_1) A generates an analytic compact semigroup $\{T(t)\}_{t\geq 0}$;

(H_2) The point spectrum of A consists of a sequence of real number $\{\mu_k\}_{k\geq 1}$ with the corresponding eigenspace M_k and the projection $P_k : X \to M_k$. Moreover, if $\sum_{k=1}^{\infty} x_k = 0$ for $x_k \in M_k$ then each x_k must be zero;

(H_3) Every $x \in D(A)$ has a unique expression $x = \sum_{k=1}^{\infty} P_k x$ and $Ax = \sum_{k=1}^{\infty} \mu_k P_k x$;

(H_4) The mapping $L : \mathbb{R}\times C \to X$ (with $C := C([-\tau,0] : X)$) is C^k-smooth ($k\geq 4$) and is given by

$$L(\mu,\phi) = \int_{-\tau}^{0} \phi(\theta)\mathrm{d}\eta(\mu,\theta)$$

for any $(\mu,\phi)\in\mathbb{R}\times C$, for a function $\eta(\mu,.)$: $[-\tau,0]\to B(X,X)$ of bounded variation. Moreover, $L(\mu,P_k\phi)\in M_k$, $k\geq 1$, $\phi\in C$ and $L(\mu,\sum_{k=1}^{\infty} P_k\phi) = \sum_{k=1}^{\infty} L(\mu,P_k\phi)$ for any $\phi\in C$ such that $\sum_{k=1}^{\infty} P_k\phi\in C$, where $P_k\phi$ is defined by $(P_k\phi)(\theta) = P_k\phi(\theta)$ for $\theta\in[-\tau,0]$;

(H_5) $g : \mathbb{R}\times C \to X$ has k-th-continuous Fréchet derivatives with $g(\mu,0) = 0$ and $Dg(\mu,0) = 0$ for $\mu\in\mathbb{R}$;

(H_6) There exists $\mu_0\in\mathbb{R}$ and $\omega_0 > 0$ such that $\pm\mathrm{i}\omega_0$ are simple characteristic values of the linear equation

$$\dot{u}(t) + Au(t) = L(\mu_0, u_t(.)) \tag{12}$$

and all other characteristic values have negative real parts;

(H_7) *Transversality condition.* If μ is near μ_0 the eigenvalues of the corresponding problem (12) are given by $\lambda(\mu) = \alpha(\mu) + \mathrm{i}\omega(\mu)$, $\lambda(\mu_0) = \mathrm{i}\omega_0$, $\lambda(\mu)$ is C^k-smooth in μ and

$$\alpha'(\mu_0) \neq 0.$$

Remark 2. A careful reading of the proof of Theorem 2.1 of [22] allows to see that the use of the same notation u_t in the terms $L(\mu, u_t(.))$ and $g(u_t(.))$ does not needs that the kernels envolved in each of the possible nonlocal terms be exactly the same. So, in particular, the conclusion remains valid in the special case in which $g(u_t(.)) = g(u(.))$, i.e., without delay or neutral term.

4.2 Applications of the Abstract Result to the Delayed CGLE on a Bounded Domain

Motivated by the special form of the nonlinear term of the equation in (P_2) we shall take $X = \mathbf{L}^4(\Omega)$ and $Y = \mathbf{L}^{4/3}(\Omega)$. A detailed analysis of the associated diffusion operator is consequence of some previous results in the literature: see, e.g., Amann [1]. Notice that the operator $A\mathbf{u}$ can be formulated matricially as

$$\begin{pmatrix} u_1 \\ u_2 \end{pmatrix} \to \begin{pmatrix} \Delta & -\beta\Delta \\ \beta\Delta & \Delta \end{pmatrix} \begin{pmatrix} u_1 \\ u_2 \end{pmatrix}.$$

So, if $\beta \neq 0$ the diffusion matrix has a nonzero antisymmetric part. In particular, A is the generator of a semigroup of contractions $\{T(t)\}_{t \geq 0}$ on X and the compactness of the semigroup is consequence of the compactness of the inclusion $D(A) \subset X$ (notice that, since $N = 2$, $\mathbf{W}^{1,4}(\Omega) \subset \mathbf{W}^{1,4/3}(\Omega) \subset \mathbf{C}(\overline{\Omega})$ with compact imbedding) and some regularity results for nonsymmetric systems. A study of the eigenvalues of A can be found, e.g., in Temam [19].

Concerning the rest of the terms of the equation in (P_2), we define $g(\mathbf{u}) = -(1 + \mathrm{i}\alpha) |\mathbf{u}|^2 \mathbf{u}$ with $D(g) = \mathbf{L}^{12}(\Omega)$. By using the characterization of the semi inner-braket $[,]$ for the spaces $L^p(\Omega)$ (see, e.g., Benilan, Crandall and Pazy [5]) it is easy to see that $\mathbf{B} = -\mathbf{g}$ is an accretive operator on X, which is dominated by A; i.e.,

$$D_X(A) \subset D_X(B) \text{ and } |Bu| \leq k \left|A^0 u\right| + \sigma(|u|)$$

for any $u \in D_X(A)$, some $k < 1$ and some continuous function $\sigma : \mathbb{R} \to \mathbb{R}$.

Here and in what follows, $|.|$ denotes the norm in the space X (in contrast to the norm in space C which will be denoted by $\|.\|$ if there is no ambiguity, when handling two spaces X and Y the corresponding norms will be indicated), $\left|A^0 u\right| := \inf\{|\xi| : \xi \in Au\}$ for $u \in D_X(A)$. In particular, the operator $A + B$ is also an accretive operator on X.

In order to calculate the Fréchet differential of Nemitsky operator $g(\mathbf{u})$, it is useful to start analyzing the Gateaux derivative of the complex function $\mathbf{h}(\mathbf{z}) := \|\mathbf{z}\|^2 \mathbf{z}$ in the direction of an arbitrary vector $\mathbf{v}$ of $\mathbb{C}$

$$\lim_{\substack{\beta\in\mathbb{R}\\|\beta|\to 0}} \frac{\mathbf{h}(\mathbf{z}_0+\beta\mathbf{v})-\mathbf{h}(\mathbf{z}_0)}{|\beta|} = \mathbf{z}_0^2\overline{\mathbf{v}} + 2\left\|\mathbf{z}_0\right\|^2 \mathbf{v}.$$

Then, we identify the Fréchet differential of operator $g(\mathbf{u})$ as

$$D\mathbf{B}(\mathbf{y})\mathbf{v} = (1+\mathrm{i}\alpha)[\mathbf{y}^2\overline{\mathbf{v}} + 2\left\|\mathbf{y}\right\|^2 \mathbf{v}]. \tag{13}$$

Since we have $\|D\mathbf{B}(\mathbf{y})\| \leq c\left\|\mathbf{y}\right\|^2$, by the results on the Fréchet differentiability of Nemitsky operators (see Theorem 2.6 (with $p = 4$) of Ambrosetti and Prodi [2]) we get that, if we take $Y = \mathbf{L}^{4/3}(\Omega)$, then exists $\delta^B > 0$ such that $\mathbf{B}$ is Fréchet differentiable as function from $B_{\delta^B}(w) = \left\{z \in D(\mathbf{B});\ |w-z| < \delta^B\right\}$ into Y, and that the Fréchet derivative is locally Lipschitz continuous.

The nonlocal term is defined by

$$\begin{aligned} F(\mathbf{u}_t) &= (1+\mathrm{i}\theta)\mathbf{u}(t) \\ &\quad + \mu\mathrm{e}^{\mathrm{i}\xi}\left[m_1\mathbf{u}(t) + m_2\langle\mathbf{u}(t)\rangle + \mathrm{e}^{\mathrm{i}(\omega+\theta)\tau}(m_3\mathbf{u}(t-\tau) + m_4\langle\mathbf{u}(t-\tau)\rangle)\right], \end{aligned}$$

is locally Lipschitz continuous and its Fréchet derivative is given by

$$\begin{aligned} \mathrm{D}F(\widehat{\mathbf{y}})\ \mathbf{v}(t) &= -(1+\mathrm{i}\theta)\mathbf{v}(t) \\ &\quad - \mu\mathrm{e}^{\mathrm{i}\xi}\left[m_1\mathbf{v}(t) + m_2\langle\mathbf{v}(t)\rangle - \mathrm{e}^{\mathrm{i}(\omega+\theta)\tau}(m_3\mathbf{v}(t-\tau) - m_4\langle\mathbf{v}(t-\tau)\rangle)\right]. \end{aligned}$$

In consequence, the operator $y \to Ay + \mathrm{D}B(w)y - \mathrm{D}F(\widehat{w})\left(e^{\omega^*\cdot}y\right)$ belongs to $\mathcal{A}(\omega^* : Y)$, for some $\omega^* \in \mathbb{C}$ with $\mathrm{Re}\omega^* = \gamma^* < 0$. This means that the operator $y \to Ay + \mathrm{D}B(w)y - \mathrm{D}F(\widehat{w})\left(e^{\omega^*\cdot}y\right) + \omega^* y$ is accretive in $Y = \mathbf{L}^{4/3}(\Omega)$. We recall (see Ambrosetti and Prodi [2]) that this differentiability of B does not hold if we take $X = Y = \mathbf{L}^2(\Omega)$.

We also recall that in [6] the existence (and uniqueness) of a mild solution of problem (P_2) was obtained through a pseudolinearization argument near a stationary solution $\widehat{w}$:

Theorem 1 ([6]). *Assume $(H_1)-(H_7)$. Then there exists $\alpha > 0$, $\beta > 0$ and $M \geq 1$ such that if $u_0 \in B_\beta^X(\widehat{w})$, $u_0(s) \in D_X(B)$ for any $s \in [-\tau, 0]$ then the solution $u(\cdot : u_0)$ of (12) exists on $[-\tau, +\infty)$ and*

$$|u(t:u_0) - w| \leq M\mathrm{e}^{-\alpha t}\left\|u_0 - \widehat{w}\right\|,\ \textit{for any } t > 0.$$

Moreover, there exists $\alpha^* > 0$, $\beta^* \in (0,\beta]$ and $M^* \geq 1$ such that if $u_0 \in B_{\beta^*}^{X\cap Y}(\widehat{w})$, $u_0(s) \in D_X(B) \cap D_Y(B)$ for any $s \in [-\tau, 0]$ then, for any $t > 0$,

$$|u(t:u_0) - w|_X + |u(t:u_0) - w|_Y \leq M^*\mathrm{e}^{-\alpha^* t}(\|u_0 - \widehat{w}\|_X + \|u_0 - \widehat{w}\|_Y).$$

We can get better a priori estimates on the sup norm of the solution $\mathbf{u}$ if we assume more regular initial data in such a way that $u_0 \in B_{\beta^*}^{X\cap Y}(\widehat{w})$,

$u_0(s) \in D(A) \cap D_X(B) \cap D_Y(B)$ for any $s \in [-\tau, 0]$. Indeed, the solution can be found (after technical arguments) as a fixed point for the application $f \to Q_1(Q_2(f))$, with $w = Q_2 f$ (for $f \in W^{1,1}(0, T : X)$, for any arbitrary $T > 0$) being the solution of the problem

$$\begin{cases} \frac{dw}{dt}(t) + Aw(t) + B(w(t)) = f(t) \text{ in } X, \\ w(0) = w_0, \end{cases}$$

and Q_1 a suitable operator (see [20], Theorem 5.3.1). Since X is a reflexive Banach space, we know (see, e.g., [5], Lemma 7.8) that $w_0 \in D(A) \cap D_X(B)$ implies that $w(t) \in D(A) \cap D_X(B)$ for a.e. $t \in (0, T)$ and that

$$\|Aw(t)\|_X \leq C(\|Aw_0\|_X + \|B(w_0)\|_X, \|f\|_{W^{1,1}(0,T:X)}).$$

Thus, by the Sobolev imbedding theorems we know that

$$\|w(t)\|_{\mathbf{C}(\overline{\Omega})} \leq M$$

for a.e. $t \in (0, T)$ with $M = M(\|Aw_0\|_X + \|B(w_0)\|_X, \|f\|_{W^{1,1}(0,T:X)})$. In particular, this property remains true for the fixed point of $Q_1(Q_2(f))$ (see [20], Theorem 5.3.1) and thus

$$\|\mathbf{u}(t)\|_{\mathbf{C}(\overline{\Omega})} \leq M^*$$

for a suitable $M^* = M * (\|Au_0\|_{C([-\tau,0];X)} + \|B(w_0)\|_{C([-\tau,0];X)}, F)$. In consequence, without any loss of generality we can replace function $\mathbf{g}$ by the truncated one $\mathbf{g}_{M^*}(\mathbf{u})$:

$$\mathbf{g}_{M^*}(\mathbf{u}) = \begin{cases} -(1 + \mathrm{i}\alpha) |\mathbf{u}|^2 \mathbf{u} \quad \text{if } |\mathbf{u}| \leq M^*, \\ \\ -2(1 + \mathrm{i}\alpha) (2M^*)^2 \mathbf{u} \text{ if } |\mathbf{u}| \geq M^*, \end{cases}$$

and with $\mathbf{g}_{M^*}(\mathbf{u})$ a C^k-smooth function generating an accretive operator $\mathbf{B}_{M^*} = -\mathbf{g}_{M^*}$ on X dominated by A as before. This proves that, at least for regular initial data, $\mathbf{u}$ coincides with the solution of

$$\begin{cases} \frac{du}{dt}(t) + Au(t) = L(\mu, u_t(.)) + g_{M^*}(u_t(.)) & \text{in } X, \\ u(s) = u_0(s) & s \in [-\tau, 0]. \end{cases}$$

Thanks to this argument we can verify now the assumption (H_5) since by the results of Ambrosetti and Prodi (see [2], Sect. 3, Chap. 1) we know that the Nemitsky operator associated to g_{M^*} has k-th-continuous Fréchet derivatives on any $\mathbf{L}^p(\Omega)$, $p > 1$.

Remark 3. By introducing the representation operator $\mathbf{P} : \mathbb{R}^2 \to \mathbb{C}$, $\mathbf{P}(\rho,\phi) = \rho e^{\mathrm{i}\phi}$ it is clear that the quasilinear operator $A\mathbf{P}(\mathbf{q})$ obtained from the operator $A\mathbf{u} = -(1 + \mathrm{i}\beta)\Delta\mathbf{u}$ satisfies also condition $A \in \mathcal{A}(\omega)$ (since $\mathbf{P}$ is merely a change of variables). We point out that

$$AP(\mathbf{q}) = -(1+\mathrm{i}\beta)[\Delta\rho - \rho\,|\nabla\phi|^2 + \mathrm{i}(2\nabla\rho\cdot\nabla\phi + \rho\Delta\phi)]\mathrm{e}^{\mathrm{i}\phi}.$$

Then, the formal linearization of the operator $\mathbf{E}(\mathbf{q}) := A\mathbf{P}(\mathbf{q})$ at $\mathbf{q}^*(x,y) := \mathbf{y} \equiv \rho_0$ becomes

$$D\mathbf{E}(\mathbf{q}^*)(\rho\mathrm{e}^{\mathrm{i}\phi}) = -(1+\mathrm{i}\beta)[\Delta\rho + \mathrm{i}\rho_0\Delta\phi]\mathrm{e}^{\mathrm{i}\phi}.$$

Notice that the linearization of $\mathbf{C}(\mathbf{q})^{-1}A\mathbf{P}(\mathbf{q})$ needs a slight modification of the above linear expression. Nevertheless by applying the representation operator $\mathbf{P}$, after the linearization used in the abstract theorem, we get a curious result relating two nonlinear problems which are closed (in some sense) in the same spirit as the *pseudo-linearization principle* obtained in [6].

4.3 Some Comments on the Associated Transversality Assumption

Concerning problem (P_2), we give an outline of the study of eigenvalues and its implications on the associated transversality condition. The eigenvalue equation can be obtained by a linearization argument involving the Fréchet derivative of the nonlinear part, as in the preceding section.

As usual, the linear structure of the equation leads to the search of non-trivial solutions $\mathbf{z}(x)$ of the form $\mathbf{A}_{\mathbf{k}}w^j_{\mathbf{k}}(x)$, with $j=1,2$, where $w^j_{\mathbf{k}}(x)$ are the eigenfunctions for the usual Laplacian operator Δ with periodic boundary conditions on $\Omega = (0,L_1)\times(0,L_2)$. The eigenvalues of this problem are given by

$$\lambda^0_0 = 0,\ \ \lambda^0_{\mathbf{k}} = 4\pi\left(\frac{k_1^2}{L_1^2} + \frac{k_2^2}{L_2^2}\right);\quad k_1,k_2\in\mathbb{N}$$

with the associate eigenfunctions

$$w_0 = \frac{1}{\sqrt{|\Omega|}},\, w^1_{\mathbf{k}} = \sqrt{\frac{2}{|\Omega|}}\cos 2\pi\mathbf{k}\mathbf{x},\, w^2_{\mathbf{k}} = \sqrt{\frac{2}{|\Omega|}}\sin 2\pi\mathbf{k}\mathbf{x},\ \text{ with } |\Omega| = L_1L_2,$$

where we have written $\mathbf{k}\mathbf{x} := \left(\frac{k_1}{L_1}x_1 + \frac{k_2}{L_2}x_2\right)$. This study can be found in Temam [19]. We introduce the notation $\lambda_{\mathbf{k}} = a_{\mathbf{k}} + ib_{\mathbf{k}}$ for the real and imaginary parts of the eigenvalues of the problem, and taking into account Fréchet derivative of the nonlinear part (13), the eigenvalue equations for the problem (P_2) are

$$\begin{cases} \qquad (a_{\mathbf{k}} + ib_{\mathbf{k}})[v_r + \mathrm{i}v_i] - (1+\mathrm{i}\beta)(-\lambda_{\mathbf{k}})[v_r + \mathrm{i}v_i] \\ \quad = (1+\mathrm{i}\theta)[v_r + \mathrm{i}v_i] - (1+\mathrm{i}\alpha)[3\rho_0^2 v_r + \mathrm{i}\rho_0^2 v_i] \\ +\mu\mathrm{e}^{\mathrm{i}\xi}\left[m_1 + m_2\delta_{0\mathbf{k}} + \mathrm{e}^{-a\tau + \mathrm{i}(\omega+\theta-b)\tau}(m_3 + m_4\delta_{0\mathbf{k}})\right][v_r + \mathrm{i}v_i], \end{cases}$$

where v_r and v_i are the real and imaginary parts of the linearization $\mathbf{v}$, and $\delta_{0\mathbf{k}}$ denotes the Kronecker delta function. We arrive at

$$
\begin{cases}
a_{\mathbf{k}} v_r - b_{\mathbf{k}} v_i = -\lambda^0_{\mathbf{k}} v_r + \beta \lambda^0_{\mathbf{k}} v_i + \left(\left[1 - 3\rho_0^2\right] v_r + \left[\alpha \rho_0^2 - \theta\right] v_i\right) \\
\qquad + \mu(m_1 + m_2 \delta_{0k}) \left[v_r \cos\xi - v_i \sin\xi\right] + \{\mu \mathrm{e}^{-a_{\mathbf{k}}\tau}(m_3 + m_4 \delta_{0k}) \\
\qquad \left[\cos(\xi + (\omega + \theta - b_{\mathbf{k}})\tau) v_r - \sin(\xi + (\omega + \theta - b_{\mathbf{k}})\tau) v_i\right]\}, \\
b_{\mathbf{k}} v_r + a_{\mathbf{k}} v_i = -\beta \lambda^0_{\mathbf{k}} v_r + \lambda^0_{\mathbf{k}} v_i + (v_i + \theta v_r) - \left[\rho_0^2 v_i - 3\alpha \rho_0^2 v_r\right] \\
\qquad + \mu(m_1 + m_2 \delta_{0k}) \left[v_r \sin\xi + v_i \cos\xi\right] + \{\mu \mathrm{e}^{-a_{\mathbf{k}}\tau}(m_3 + m_4 \delta_{0k}) \\
\qquad \left[\sin(\xi + (\omega + \theta - b_{\mathbf{k}})\tau) v_r + \cos(\xi + (\omega + \theta - b_{\mathbf{k}})\tau) v_i\right]\}
\end{cases}
$$

To show the procedure, without loss of generality, we consider the case

$$
m_3 + m_4 \delta_{0\mathbf{k}} = 0. \tag{14}
$$

This represents a special, and important, choice of the combination of instantaneous and delayed terms in the global feedback, none of them necessarily zero. The equations for the eigenvalues become

$$
\begin{cases}
a_{\mathbf{k}} v_r - b_{\mathbf{k}} v_i = -\lambda^0_{\mathbf{k}} v_r + \beta \lambda^0_{\mathbf{k}} v_i + \left(\left[1 - 3\rho_0^2\right] v_r + \left[\alpha \rho_0^2 - \theta\right] v_i\right) \\
\qquad + \mu(m_1 + m_2 \delta_{0k}) \cos\xi v_r - \mu(m_1 + m_2 \delta_{0k}) \sin\xi v_i \\
b_{\mathbf{k}} v_r + a_{\mathbf{k}} v_i = -\beta \lambda^0_{\mathbf{k}} v_r + \lambda^0_{\mathbf{k}} v_i + (v_i + \theta v_r) - \left[\rho_0^2 v_i - 3\alpha \rho_0^2 v_r\right] \\
\qquad + \mu(m_1 + m_2 \delta_{0k}) \sin\xi v_r + \mu(m_1 + m_2 \delta_{0k}) \cos\xi v_i
\end{cases}
$$

If we call

$$
C_1\left(\mu, m_1, m_2, \xi, \lambda^0_{\mathbf{k}}\right) = 1 - \lambda^0_{\mathbf{k}} - \mu(m_1 + m_2 \delta_{0k}) \cos\xi,
$$

$$
C_2\left(\mu, m_1, m_2, \xi, \lambda^0_{\mathbf{k}}\right) = 1 + \lambda^0_{\mathbf{k}} + \mu(m_1 + m_2 \delta_{0k}) \cos\xi,
$$

$$
D\left(\beta, \mu, m_1, m_2, \xi, \lambda^0_{\mathbf{k}}\right) = -\beta \lambda^0_{\mathbf{k}} + \mu(m_1 + m_2 \delta_{0k}) \sin\xi,
$$

we obtain

$$
\begin{cases}
\left(a_{\mathbf{k}} - \left[C_1 - 3\rho_0^2\right]\right) v_r - \left(b_{\mathbf{k}} + \left[\alpha \rho_0^2 - \theta - D\right]\right) v_i = 0 \\
\left(b_{\mathbf{k}} - \left[-3\alpha \rho_0^2 + \theta + D\right]\right) v_r + \left(a_{\mathbf{k}} - \left[C_2 - \rho_0^2\right]\right) v_i = 0
\end{cases}
$$

The compatibility of this system implies

$$
\det \begin{pmatrix} a_{\mathbf{k}} - \left[C_1 - 3\rho_0^2\right] & -b_{\mathbf{k}} - \left[\alpha \rho_0^2 - \theta - D\right] \\ b_{\mathbf{k}} - \left[-3\alpha \rho_0^2 + \theta + D\right] & a_{\mathbf{k}} - \left[C_2 - \rho_0^2\right] \end{pmatrix} = 0,
$$

that is

$$
\begin{cases}
\left(a_{\mathbf{k}} - \left[C_1 - 3\rho_0^2\right]\right) \left(a_{\mathbf{k}} - \left[C_2 - \rho_0^2\right]\right) = \\
\left(b_{\mathbf{k}} - \left[-3\alpha \rho_0^2 + \theta + D\right]\right) \left(b_{\mathbf{k}} + \left[\alpha \rho_0^2 - \theta - D\right]\right).
\end{cases} \tag{15}
$$

This expression is of the same type as (10) and, similarly, there is no general analytic solution for $a_{\mathbf{k}}$ and $b_{\mathbf{k}}$. Thus, Eq. (15) must also be solved numerically for a given set of parameters, to find the numerical values of the eigenvalues as in the equation (10). One of the relevant parameter spaces of the representation is the one of (τ, μ) because they are the parameters of the perturbation.

Although the explicit analytical representation of the functions $a_{\mathbf{k}}$ and $b_{\mathbf{k}}$ is not possible, we can still say something analytic in the study of the transversality, already proved by the numerical computation of Sect. 3. From the equation (15), it is possible to find the implicit derivative

$$\left[\frac{\mathrm{d}}{\mathrm{d}\tau} a_{\mathbf{k}}\right]_{a_{\mathbf{k}}=0}.$$

The analytic computation is rather involved. We show how to proceed in a simpler, and still very important example

$$m_1 + m_2 \delta_{0\mathbf{k}} = 0, \tag{16}$$

where a remark similar as the one made for the expression (14) remains valid, in this case for the local part of the perturbation. For the case (16), we have

$$C_1\left(\mu, m_1, m_2, \xi, \lambda_{\mathbf{k}}^0\right) = 1 - \lambda_{\mathbf{k}}^0,$$
$$C_2\left(\mu, m_1, m_2, \xi, \lambda_{\mathbf{k}}^0\right) = 1 + \lambda_{\mathbf{k}}^0,$$
$$D\left(\beta, \mu, m_1, m_2, \xi, \lambda_{\mathbf{k}}^0\right) = -\beta\lambda_{\mathbf{k}}^0.$$

If we expand Eq. (15) for this case,

$$\begin{cases} a_{\mathbf{k}}^2 - 2\left[1 - 2\rho_0^2\right] a_{\mathbf{k}} + \left(\left[1 - \lambda_{\mathbf{k}}^0 - 3\rho_0^2\right]\left[1 + \lambda_{\mathbf{k}}^0 - \rho_0^2\right]\right) = \\ -b_{\mathbf{k}}^2 + 2\left[-\beta\lambda_{\mathbf{k}}^0 + \alpha\rho_0^2 + \theta\right] b_{\mathbf{k}} + \left(\left[-\beta\lambda_{\mathbf{k}}^0 + 3\alpha\rho_0^2 + \theta\right]\left[+\beta\lambda_{\mathbf{k}}^0 + \alpha\rho_0^2 - \theta\right]\right), \end{cases}$$

and differentiate implicitly

$$\begin{cases} 2a_{\mathbf{k}}\frac{\mathrm{d}}{\mathrm{d}\tau}a_{\mathbf{k}} - 2\left[1 - 2\rho_0^2\right]\frac{\mathrm{d}}{\mathrm{d}\tau}a_{\mathbf{k}} - a_{\mathbf{k}}\frac{\mathrm{d}}{\mathrm{d}\tau}\left(2\left[1 - 2\rho_0^2\right]\right) \\ +\frac{\mathrm{d}}{\mathrm{d}\tau}\left(1 - \left(\lambda_{\mathbf{k}}^0\right)^2 - 2\left[2 + \lambda_{\mathbf{k}}^0\right]\rho_0^2 + 3\rho_0^4\right) \\ = -2b_{\mathbf{k}}\frac{\mathrm{d}}{\mathrm{d}\tau}b_{\mathbf{k}} + 2\left[-\beta\lambda_{\mathbf{k}}^0 + \alpha\rho_0^2 + \theta\right]\frac{\mathrm{d}}{\mathrm{d}\tau}b_{\mathbf{k}} - b_{\mathbf{k}}\frac{\mathrm{d}}{\mathrm{d}\tau}\left(2\left[-\beta\lambda_{\mathbf{k}}^0 + \alpha\rho_0^2 + \theta\right]\right) \\ +\frac{\mathrm{d}}{\mathrm{d}\tau}\left(\left[-\beta\lambda_{\mathbf{k}}^0 + 3\alpha\rho_0^2 + \theta\right]\left[+\beta\lambda_{\mathbf{k}}^0 + \alpha\rho_0^2 - \theta\right]\right). \end{cases}$$

The derivative of the real part $a_{\mathbf{k}}$ in the value $a_{\mathbf{k}} = 0$ can be written as

$$\begin{cases} \left[-2(1 - 2\rho_0^2)\frac{\mathrm{d}}{\mathrm{d}\tau}a_{\mathbf{k}}\right]_{a_{\mathbf{k}}=0} \\ = \left[-\frac{\mathrm{d}}{\mathrm{d}\tau}\left(1 - \left(\lambda_{\mathbf{k}}^0\right)^2 - 2\left[2 + \lambda_{\mathbf{k}}^0\right]\rho_0^2 + 3\rho_0^4\right)\right]_{a_{\mathbf{k}}=0} \\ +2\left[-b_{\mathbf{k}}\frac{\mathrm{d}}{\mathrm{d}\tau}b_{\mathbf{k}} + \left[-\beta\lambda_{\mathbf{k}}^0 + \alpha\rho_0^2 + \theta\right]\frac{\mathrm{d}}{\mathrm{d}\tau}b_{\mathbf{k}} - b_{\mathbf{k}}\frac{\mathrm{d}}{\mathrm{d}\tau}\left(\left[-\beta\lambda_{\mathbf{k}}^0 + \alpha\rho_0^2 + \theta\right]\right)\right]_{a_{\mathbf{k}}=0} \\ + \left[\frac{\mathrm{d}}{\mathrm{d}\tau}\left(\left[-\beta\lambda_{\mathbf{k}}^0 + 3\alpha\rho_0^2 + \theta\right]\left[+\beta\lambda_{\mathbf{k}}^0 + \alpha\rho_0^2 - \theta\right]\right)\right]_{a_{\mathbf{k}}=0}. \end{cases}$$

The coefficient of the derivative of $a_{\mathbf{k}}$,

$$-2(1-2\rho_0^2) = -2\left[1-2(1+\mu\cos\xi)\right] = 2(1+2\mu\cos\xi)$$

does not vanish either for stability reasons as can be seen, e.g., in [6] and references therein.

Acknowledgement. The research of AC, JID and JMV was partially supported by the project ref. MTM200806208 of the DGISPI (Spain) and the Research Group MOMAT (Ref. 910480) supported by UCM. The research of JID has received funding from the ITN *FIRST* of the Seventh Framework Programme of the EU (grant agreement number 238702). MS acknowledges support from Spanish MICIIN through project FIS2008-05273 and from Comunidad Autónoma de Madrid, project MODELICO (S2009/ESP-1691). AC acknowledges support also from the research group MMNL from UPM.

References

1. Amann, H.: Dynamic theory of quasilinear parabolic equations: II. reaction-diffusion systems. Diff. Int. Equ. 3, 13–75 (1990)
2. Ambrosetti, A., Prodi, G.: A Primer of Nonlinear Analysis. Cambridge University Press, Cambridge (1993)
3. Aranson, I.S., Kramer, L.: The world of the complex Ginzburg-Landau equation. Rev. Mod. Phys. 74, 99–143 (2002)
4. Atay, F.M. (ed.): Complex Time-Delay Systems. Springer, Berlin (2010)
5. Benilan, P., Crandall, M.G., Pazy, A.: Nonlinear Evolution Equations in Banach Spaces. Book in preparation
6. Casal, A.C., Díaz, J.I.: On the complex Ginzburg-Landau equation with a delayed feedback. Math. Mod. Meth. App. Sci. 16, 1–17 (2006)
7. Casal, A.C., Díaz, J.I., Stich, M.: On some delayed nonlinear parabolic equations modeling CO oxidation. Dyn. Contin. Discret. Impuls Syst. A 13(supp. S), 413–426 (2006)
8. Cross, M.C., Hohenberg, P.C.: Pattern formation outside of equilibrium. Rev. Mod. Phys. 65, 851–1112 (1993)
9. Erneux, T.: Applied Delay Differential Equations. Springer, New York (2009)
10. Ipsen, M., Kramer, L., Sørensen, P.G.: Amplitude equations for description of chemical reaction-diffusion systems. Phys. Rep. 337, 193–235 (2000)
11. Kim, M., Bertram, M., Pollmann, M., von Oertzen, A., Mikhailov, A.S., Rotermund, H.H., Ertl, G.: Controlling chemical turbulence by global delayed feedback: pattern formation in catalytic CO oxidation reaction on Pt(110). Science 292, 1357–1360 (2001)
12. Kuramoto, Y.: Chemical Oscillations, Waves, and Turbulence. Springer, Berlin (1984)
13. Mikhailov, A.S., Showalter, K.: Control of waves, patterns and turbulence in chemical systems. Phys. Rep. 425, 79–194 (2006)
14. Pyragas, K.: Continuous control of chaos by self–controlling feedback. Phys. Lett. A 170, 421–428 (1992)

15. Schöll, E., Schuster, H.G. (eds.): Handbook of Chaos Control. Wiley-VCH, Weinheim (2007)
16. Stich, M., Beta, C.: Control of pattern formation by time-delay feedback with global and local contributions. Physica D 239, 1681–1691 (2010)
17. Stich, M., Casal, A.C., Díaz, J.I.: Control of turbulence in oscillatory reaction-diffusion systems through a combination of global and local feedback. Phys. Rev. E 76, 036209, 1–9 (2007)
18. Takáč, P.: Invariant 2-tori in the time-dependent Ginzburg-Landau equation. Nonlinearity 5, 289–321 (1992)
19. Temam, R.: Infinite-Dimensional Dynamical Systems in Mechanics and Physics. Springer, New York (1988)
20. Vrabie, II: Compactness Methods for Nonlinear Evolutions. Pitman Monographs and Surveys in Pure and Applied Mathematics, vol. 75, 2nd edn. Longman, Harlow (1995)
21. Wolfram Research, Inc.: Mathematica edition 5.2, Wolfram Research, Inc. Champaign (2005)
22. Wu, J.: Theory and Applications of Partial Differential Equations. Springer, New York (1996)

Invariant Lagrangians on the Vertically Adapted Linear Frame Bundle*

Jeffrey K. Lawson[1] and M. Eugenia Rosado-María[2]

[1] Department of Mathematics & Computer Science,
Western Carolina University,
Cullowhee, NC 28723, USA
jlawson@wcu.edu

[2] Departamento de Matemática Aplicada,
Escuela Técnica Superior de Arquitectura, U.P.M.,
Avda. Juan de Herrera 4,
28040-Madrid, Spain
eugenia.rosado@upm.es

Summary. For the bundle of linear frames LM of a manifold M, diffeomorphism invariance on the vertically adapted linear frame bundle $L_\pi(LM)$ and its infinitesimal counterpart, invariance under the natural representation of vector fields of M, are analyzed. Furthermore, the structure of the vector-field-invariant Lagrangians on $L_\pi(LM)$ is determined.

1 Introduction

The principal fiber bundle approach is ideally suited for covariant field theories, such as $U(1)$ electromagnetism, Yang-Mills theory, Weinberg-Salam electroweak theory, and even the Higgs mechanism for mass acquisition via spontaneous symmetry breaking. This technique is "forced upon [physicists] by their own perception of nature." [2, p. xiv] Indeed, it lays out the geometry and topology of the spacetime M in a way that, once a reference frame for spacetime is chosen (choice of gauge), the transformation to another valid reference frame (change of gauge) is effected by a unique translation in the fiber that is equivariant with respect to the group symmetry. This provides a scenario in which the global structure for the physical situation is readily manipulated, and there is a straightforward way to view phenomenology in spacetime from any one of a collection of valid reference frames. The variational principle for the field theory now can be expressed in terms of varying a gauge-invariant action defined on the first jet bundle (the bundle of field configurations and "velocities") of the principal fiber bundle. Dynamics are then computed by a choice of gauge with a declared time parameter. Furthermore, physical observables such as field strength are readily computable on the principal fiber bundle.

* Dedicated in honour of Professor Marisa Menéndez.

If we view general relativity with an eye toward covariance, the gauge group is the infinite-dimensional group of diffeomorphisms of M, but the induced smooth changes of coordinates are effected by the action of the finite-dimensional general linear group $GL(n, \mathbb{R})$ on the local coordinate frames. Hence, the linear frame bundle, the principal fiber bundle over M with fiber $GL(n, \mathbb{R})$, seems to be the natural framework viewing gravity as an affine gauge theory (see [2]) and for formulating diffeomorphism-invariant problems in gauge-theoretical gravity (see [6, 19, 20]). Diffeomorphism invariance also plays an important role in supersymmetry and gauge theories (see, for example, [1, 12, 17]), as it allows the formulation of the principle of general covariance in each of these contexts.

Let M be a smooth connected m-dimensional manifold, and let Diff$M$ denote the group of diffeomorphisms of M. Let $\pi\colon LM \to M$ be the bundle of linear frames of M. The variational problems defined by Lagrangian densities on the bundle of first-order jets $J^1(LM)$ are the simplest of the diffeomorphism-invariant variational problems that appear in G-structure theory. The more important G-structure is obtained when a closed subgroup G of the general linear group $GL(m, \mathbb{R})$ is considered. In this case, the G-structures are classified by sections of the quotient fiber bundle $LM/G \to M$. Such an approach has the advantage of separating diffeomorphism invariance from G-invariance.

The Hamiltonian structure of variational problems defined by the natural basis $\mathcal{L}^i_{jk}$ of diffeomorphism-invariant Lagrangians on the first jet bundle of $\pi\colon LM \to M$ has been studied in [14]. From the structural point of view, the invariant Lagrangians $\mathcal{L}^i_{jk}$ represent the most elementary diffeomorphism-invariant variational problems. Hence, although they are too simple to be of immediate application to field theory, they provide interesting geometric models, precisely due to their simple properties. In fact, each of the Lagrangians proposed as a relativistic model on LM can be written as a function of the basic Lagrangians.

Variational problems for a covariant classical field theory may be expressed using the affine construct of ***multisymplectic geometry***, as developed in [4] and [5]. The field configurations are represented by a fiber bundle $\pi\colon E \to M$, where the fibers of E represent the field values. The variational problems are defined geometrically on the first jet bundle J^1E by pulling back the *canonical* covariant one-form, that is, the tautological one-form on the affine dual to J^1E, via a Legendre transformation induced by a Lagrangian $L : J^1E \to \mathbb{R}$.

Subsequently, it has been shown in [9, 10, 11, 13, 16] that the multisymplectic geometry for a covariant field theory may be generalized using a geometry possessed by the vertically adapted linear frame bundle $L_\pi E$. The bundle $L_\pi E$ is a symmetry-broken subbundle of the bundle of linear frames LE in which only the frames adapted to the fibration of E are allowed. If we apply Norris' ***n-symplectic geometry*** [15], a vector-valued symplectic geometry on a linear frame bundle in which the soldering (tautological) one-form is treated as a vector-valued canonical one-form, the subbundle $L_\pi E$

naturally inherits from LE a generalized symplectic geometry adapted to the particular field theory.

The vertically adapted frame bundle $L_{\pi}E$ is a principal bundle over J^1E (see [13]), and the generalized symplectic geometry of $L_{\pi}E$ forms a covering theory (in the sense of [3]) for the multisymplectic geometry of J^1E. Furthermore, a Hamiltonian theory on $L_{\pi}E$ reproduces the Poisson brackets of momentum observables for a covariant classical field, but in a manner that removes algebraic obstructions to closure, allowing us to define a full Poisson algebra of momentum observables for covariant fields (see [3, 9]). For a covariant field theory with symmetry, the geometry of the vertically adapted frame bundle produces vector-valued momentum mappings and, thusly, field conservation laws. This sets the stage for a finite-dimensional approach to reduction by symmetry for field theories (see [10]).

The purpose of the present paper is to describe the structure of the diffeomorphism-invariant Lagrangians $\mathcal{L}$ defined on the vertically adapted linear frame bundle $L_{\pi}(LM)$ of $\pi\colon LM \to M$. This would allow the study of the generalized geometric structure of a Lagrangian gauge field theory, including the Poisson algebra of field momentum observables. The G-structure then could be identified readily in this paradigm by an appropriate symmetry breaking, and additional symmetries could be used to determine conserved field quantities.

The outline of the paper is as follows. In §2, we review the construction of the $\operatorname{Diff} M$-invariant Lagrangians on the first jet bundle in terms of the finite basis of natural Lagrangians. In §3 we review the geometric structure of the vertically adapted linear frame bundle of a fiber bundle $\pi : E \to M$. In §4, we introduce the coordinates on $L_{\pi}(LM)$ and the soldering form on $L_{\pi}(LM)$. We define the $\operatorname{Diff} M$-invariance on $L_{\pi}(LM)$ and its infinitesimal counterpart; *i.e.*, invariance under the natural representation of vector fields of M into $L_{\pi}(LM)$. Both definitions are not exactly equivalent due to some global topological obstructions on M, although they are essentially equivalent. We use the infinitesimal definition of invariance as it allows us to employ tools such as vector distributions, involutiveness, Frobenius theorem, etc. In this section we also present the basis of $\operatorname{Diff} M$-invariant Lagrangians in $L_{\pi}(LM)$. In §5 we state and prove the main result.

2 Diffeomorphism-Invariant Lagrangians on $J^1(LM)$

Let $\pi\colon LM \to M$ be the bundle of linear frames of M. Let $\widetilde{\varphi}\colon LM \to LM$ be the natural automorphism induced from a diffeomorphism $\varphi\colon M \to M$ (see, for example, [8]) and let $\widetilde{\varphi}^{(1)}\colon J^1(LM) \to J^1(LM)$ be its natural jet prolongation (see, for example, [18]). A Lagrangian function $\mathcal{L}\colon J^1(LM) \to \mathbb{R}$ is said to be invariant under $\operatorname{Diff} M$ if $\mathcal{L} \circ \widetilde{\varphi}^{(1)} = \mathcal{L}$ for every $\varphi \in \operatorname{Diff} M$.

Let $\mathfrak{X}(M)$ be the set of all differentiable vector fields of M. If $X \in \mathfrak{X}(M)$ is the infinitesimal generator of a one-parameter group φ_t of diffeomorphisms, then $\widetilde{\varphi}_t$ induces a vector field $\widetilde{X}$ on LM (see [8]). Let $\widetilde{X}^{(1)}$ denote the first jet prolongation of $\widetilde{X}$. From the invariance condition for $\mathcal{L}$ we conclude that $\widetilde{X}^{(1)}\mathcal{L} = 0$. The Lagrangians satisfying the latter condition are said to be $\mathfrak{X}(M)$-invariant. Hence invariance under diffeomorphisms implies $\mathfrak{X}(M)$-invariance. The converse is also true (see [14]), except when M is orientable and admits an orientation reversing diffeomorphism onto itself. As a consequence, throughout this paper we prefer to use the notion of $\mathfrak{X}(M)$-invariance in order to formulate invariance under diffeomorphisms.

Let $\mathcal{L}^i_{jk} \colon J^1(LM) \to \mathbb{R}$, $j < k$, be the functions defined by

$$\mathcal{L}^i_{jk}(j^1_x s) = \omega^i([\bar{X}_j, \bar{X}_k](x)) ,$$

where $(\bar{X}_1, \ldots, \bar{X}_m)$ is the linear frame attached to the local section s of π, with dual coframe $(\omega^1, \ldots, \omega^m)$. Such functions are invariant under $\operatorname{Diff} M$, and, in addition, every $\mathfrak{X}(M)$-invariant function on $J^1(LM)$ can be written locally as a differentiable function of $\{\mathcal{L}^i_{jk}\}^i_{j<k}$. Since $\{\mathcal{L}^i_{jk}\}^i_{j<k}$ is functionally independent and has a geometric meaning, we can say that these Lagrangians are a natural basis of $\mathfrak{X}(M)$-invariant functions; we refer the reader to [14] for the details.

3 The Vertically Adapted Linear Frame Bundle

Let $\pi \colon E \to M$ be an arbitrary fiber bundle over M with fiber dimension k. The vertically adapted linear frame bundle $L_\pi E$ is a subbundle of LE consisting of the linear frames (X_i, Y_A), in which $i = 1, 2, \ldots, m$ and $A = m+1, m+2, \ldots, m+k$, and the last k vectors $\{Y_A\}$ are vertical with respect to π. Note that throughout the paper the convention is that lower case indices run from 1 to m, and the upper case indices run from $m+1$ to $m+k$. The structure group $G_v \subset GL(m+k, \mathbb{R})$ of $\lambda \colon L_\pi E \to E$ is the group of nonsingular block lower triangular matrices (see [7, 13]),

$$G_v = \left\{ \begin{pmatrix} A & 0 \\ C & B \end{pmatrix} , \; A \in GL(m, \mathbb{R}), B \in GL(k, \mathbb{R}), C \in \hom(\mathbb{R}^m, \mathbb{R}^k) \right\} .$$

The vertically adapted linear frame bundle $L_\pi E$ is also a principal bundle over $J^1 E$ with structure group $GL(m, \mathbb{R}) \times GL(k, \mathbb{R})$; its projection $\rho \colon L_\pi E \to J^1 E$, is given by $\rho(e, (X_i, Y_A)) = \tau_e$, where $\tau_e \in \hom(T_{\pi(e)} M, T_e E)$ is defined by $\tau_e(\pi_*(X_i)) = X_i$ (see [13]).

4 Lifting Field Theories to $L_\pi(LM)$

4.1 Lifting to $L_\pi(LM)$

If the fiber bundle E is the linear frame bundle LM, the vertically adapted linear frame bundle $\lambda \colon L_\pi(LM) \to M$ is a principal bundle with structure group $G_v \subset GL(m+m^2, \mathbb{R})$. The bundle $L_\pi(LM)$ is also a principal bundle over $J^1(LM)$ with structure group $G = GL(m, \mathbb{R}) \times GL(m^2, \mathbb{R})$. Using the principal bundle $\rho\colon L_\pi(LM) \to J^1(LM)$, every Lagrangian $\mathcal{L}\colon J^1(LM) \to \mathbb{R}$ can be lifted to a Lagrangian $L_\pi(LM) \to \mathbb{R}$ defined by $L := \rho^*(\mathcal{L})$.

Proposition 1. *For every diffeomorphism φ of M the induced automorphism $\widetilde{\widetilde{\varphi}}\colon\ L(LM) \to L(LM)$ maps an adapted frame (X_i, X^i_j) at $e \in LM$ onto the adapted frame $\widetilde{\widetilde{\varphi}}(X_i, X^i_j) = (\widetilde{\varphi}_* X_i, \widetilde{\varphi}_* X^i_j)$ at $\widetilde{\varphi}(e) \in LM$, making the following diagram commutative:*

$$\begin{array}{ccc} L_\pi(LM) & \xrightarrow{\widetilde{\widetilde{\varphi}}} & L_\pi(LM) \\ \downarrow \rho & & \downarrow \rho \\ J^1(LM) & \xrightarrow{\widetilde{\varphi}^{(1)}} & J^1(LM) \end{array}$$

where $\widetilde{\varphi}^{(1)}$ is the first jet prolongation of $\widetilde{\varphi}$.

Proof. Let us denote by $\tau'_{\widetilde{\varphi}(e)}$ the point in $J^1(LM)$ defined as

$$\tau'_{\widetilde{\varphi}(e)}(\pi_*(\widetilde{\varphi}_* X_i)) = \widetilde{\varphi}_* X_i$$

(see [13]), we have

$$\begin{aligned} \widetilde{\varphi}_* \tau_e(\pi_* X_i) &= \widetilde{\varphi}_* X_i \\ &= \tau'_{\widetilde{\varphi}(e)}(\pi_*(\widetilde{\varphi}_* X_i)) \\ &= \tau'_{\widetilde{\varphi}(e)}((\pi \circ \widetilde{\varphi})_* X_i) \\ &= \tau'_{\widetilde{\varphi}(e)}((\varphi \circ \pi)_* X_i) = \tau'_{\widetilde{\varphi}(e)}\big(\varphi_*(\pi_* X_i)\big). \end{aligned}$$

Hence, $\tau'_{\widetilde{\varphi}(e)} = \widetilde{\varphi}_* \circ \tau_e \circ (\varphi_*)^{-1}$, and therefore

$$\widetilde{\varphi}^{(1)}(\rho(X_i, X^i_j)) = \widetilde{\varphi}^{(1)}(\tau_e) = \widetilde{\varphi}_* \circ \tau_e \circ (\varphi_*)^{-1} = \tau'_{\widetilde{\varphi}(e)}\,. \qquad \square$$

Definition 1. *A function $f\colon L_\pi(LM) \to \mathbb{R}$ is said to be* Diff$M$***-invariant*** *if $f \circ \widetilde{\widetilde{\varphi}} = f$ for every $\varphi \in$ Diff$M$. If $X \in \mathfrak{X}(M)$ is the infinitesimal generator of a one-parameter group φ_t of diffeomorphisms, according to Proposition 1, then $\widetilde{\widetilde{\varphi}}_t\colon L_\pi(LM) \to L_\pi(LM)$ induces a vector field $\widetilde{\widetilde{X}}$ on $L_\pi(LM)$. A function $f\colon L_\pi(LM) \to \mathbb{R}$ is said to be* ***$\mathfrak{X}$(M)-invariant*** *if $\widetilde{\widetilde{X}}(f) = 0$ for every $X \in \mathfrak{X}(M)$.*

We may now begin the process to determine the structure of the $\mathfrak{X}(M)$-invariant functions on $L_\pi(LM)$. Let $u = (X_i, X^i_j)$ be a linear frame in $L_\pi(LM)$ at a point $e = (\bar{X}_i) \in LM$ over $x \in M$. Let $\mathcal{L}^h_i\colon L_\pi(LM) \to \mathbb{R}$ be the functions defined by

$$(\bar{X}_1, \ldots, \bar{X}_m) = (\pi_* X_1, \ldots, \pi_* X_m) \begin{pmatrix} \mathcal{L}_1^1(u) & \cdots & \mathcal{L}_m^1(u) \\ \vdots & \ddots & \vdots \\ \mathcal{L}_1^m(u) & \cdots & \mathcal{L}_m^m(u) \end{pmatrix}. \tag{1}$$

Similarly, let $\mathcal{L}_{ir}^{sj} \colon L_\pi(LM) \to \mathbb{R}$ be the functions defined by

$$(E_j^{i*})_e = \mathcal{L}_{ir}^{sj}(u) X_s^r \tag{2}$$

(using index notation here and throughout the paper), or in matrix form

$$(E_j^{i*})_e = (X_s^r)(\mathcal{L}_{ir}^{sj}(u)),$$

where (E_j^{i*}) is the global basis of $V(LM)$ associated with the standard basis (E_j^i) of $\mathfrak{gl}(m, \mathbb{R})$; *i.e.*, $(E_j^i)_k^h = \delta_i^h \delta_k^j$ (see, for example, [8]), where $V(LM)$ denotes the vertical bundle of LM.

4.2 Coordinates on $L_\pi(LM)$

Each coordinate system (x^i) on an open domain $U \subseteq M$ induces a local coordinate system, $(x^i, \bar{x}_j^i)$ on $V = \pi^{-1}(U) \subset LM$, by setting

$$e = ((\partial/\partial x^1)_x, \ldots, (\partial/\partial x^m)_x) \cdot (\bar{x}_j^i(e)), \tag{3}$$

where $x = \pi(e)$, and $(x^i, \bar{x}_j^i)$ induces a coordinate system $(x^i, \bar{x}_j^i, x_{j,k}^i)$ on $J^1(V)$ by setting

$$x^i(j_x^1 s) = x^i(x), \quad \bar{x}_j^i(j_x^1 s) = \bar{x}_j^i(s(x)), \text{ and } \quad x_{j,k}^i(j_x^1 s) = \frac{\partial(x_j^i \circ s)}{\partial x^k}(x).$$

The soldering form $\bar{\theta}$ on LM is the $\mathbb{R}^m$-valued one-form defined on LM by

$$\bar{\theta}(\bar{X}) = e^{-1}(\pi_*(\bar{X})), \quad \forall \bar{X} \in T_e(LM).$$

With respect to the standard basis $(\bar{r}_i) = (0, \ldots, \overset{(i)}{1}, \ldots, 0)$ for $\mathbb{R}^m$, we write: $\bar{\theta} = \bar{\theta}^i \otimes \bar{r}_i$, where $\bar{\theta}^i$ is a differential one-form on LM. In local coordinates, $\bar{\theta}^i = \bar{x}_j{}^i dx^j$.

Each adapted coordinate system $(x^i, \bar{x}_j^i)$ on $V = \pi^{-1}(U) \subset LM$ induces a coordinate system $(x^i, \bar{x}_j^i, v_j^i, v_{jk}^i, v_{jk}^{ih})$ on $\lambda^{-1}(V) \subset L_\pi(LM)$, setting

$$u = ((X_j)_e, (X_k^h)_e) = \left(v_j^i(u) \left.\frac{\partial}{\partial x^i}\right|_e + v_{kj}^i(u) \left.\frac{\partial}{\partial \bar{x}_k^i}\right|_e, \; v_{jk}^{ih}(u) \left.\frac{\partial}{\partial \bar{x}_j^i}\right|_e \right), \tag{4}$$

$e = \lambda(u)$. We will consider the Lagrangian coordinate system $(x^i, \bar{x}^i_j, x^i_j, x^i_{jk}, x^{ih}_{jk})$ defined on $\lambda^{-1}(V)$ from the above coordinates by

$$x^i_j = v_j{}^i, \; x^i_{jk} = v_k{}^s v^i_{js}, \text{ and } v^{ih}_{jk} x^{bs}_{ar} = \delta^h_a \delta^b_k \delta^i_r \delta^s_j \tag{5}$$

where $(v_j{}^i) = (v^i_j)^{-1}$ (see [13]).

The soldering form θ on $L_\pi(LM)$ is the $\mathbb{R}^{m+m^2}$-valued one-form defined on $L_\pi(LM)$ by

$$\theta(X) = u^{-1}(\lambda_*(X)), \quad \forall X \in T_u(L_\pi(LM)).$$

We write

$$\theta = \theta^i \otimes r_i + \theta^i_j \otimes r^j_i,$$

where θ^i, θ^i_j are differential one-forms on $L_\pi(LM)$ and (r_i, r^j_i) is the standard basis for $\mathbb{R}^{m+m^2}$. In Lagrangian coordinates (5) we have

$$\theta^i = x^i_j dx^j \text{ and } \theta^i_j = -x^{il}_{jk} x^k_{lh} dx^h + x^{il}_{jk} d\bar{x}^k_l. \tag{6}$$

4.3 Lifting Vector Fields to $L_\pi(LM)$

Let us first compute the local expression of the vector field $\tilde{\tilde{X}} \in \mathfrak{X}(L_\pi(LM))$ for every $X \in \mathfrak{X}(M)$. We begin by computing the local expression of the natural lift to $L_\pi(LM)$ of a vector field $X \in \mathfrak{X}(LM)$.

Locally a vector field on LM is written as

$$X = f^i \frac{\partial}{\partial x^i} + \bar{f}^i_j \frac{\partial}{\partial \bar{x}^i_j}, \quad f^i, \bar{f}^i_j \in C^\infty(V).$$

Hence, the local expression of a vector field $\widetilde{X}$ on $L_\pi(LM)$ that projects onto X is given by

$$\widetilde{X} = f^i \frac{\partial}{\partial x^i} + \bar{f}^i_j \frac{\partial}{\partial \bar{x}^i_j} + f^i_j \frac{\partial}{\partial x^i_j} + f^i_{jk} \frac{\partial}{\partial x^i_{jk}} + f^{il}_{jk} \frac{\partial}{\partial x^{il}_{jk}}.$$

Let us see that the functions f^i_j, f^i_{jk}, f^{il}_{jk} can be uniquely determined by using the assumption that $L_{\widetilde{X}}\theta = 0$. Applying the local expressions of (θ^i, θ^i_j) in (6), it follows that

$$0 = L_{\widetilde{X}}\theta^i = L_{\widetilde{X}}(x^i_j dx^j) = \left(\widetilde{X}(x^i_k) + x^i_j \frac{\partial f^j}{\partial x^k}\right) dx^k.$$

Hence,

$$\widetilde{X}(x^i_k) = -x^i_j \frac{\partial f^j}{\partial x^k}. \tag{7}$$

Also,

$$\begin{aligned}
0 &= L_{\widetilde{X}}\theta^i_j \\
&= L_{\widetilde{X}}(-x^{il}_{jk}x^k_{lh}dx^h + x^{il}_{jk}d\bar{x}^k_l) \\
&= \widetilde{X}(-x^{il}_{jk}x^k_{lh})dx^h - x^{il}_{jk}x^k_{lh}L_{\widetilde{X}}(dx^h) + \widetilde{X}(x^{il}_{jk})d\bar{x}^k_l + x^{il}_{jk}L_{\widetilde{X}}(d\bar{x}^k_l) \\
&= -\widetilde{X}(x^{il}_{jk})x^k_{lh}dx^h - x^{il}_{jk}\widetilde{X}(x^k_{lh})dx^h - x^{il}_{jk}x^k_{lh}\frac{\partial f^h}{\partial x^r}dx^r \\
&\quad + \widetilde{X}(x^{il}_{jk})d\bar{x}^k_l + x^{il}_{jk}\left(\frac{\partial f^k}{\partial x^h}d\bar{x}^h_l + \bar{x}^h_l\frac{\partial^2 f^k}{\partial x^h\partial x^r}dx^r\right) \\
&= \left(-\widetilde{X}(x^{il}_{jk})x^k_{lr} - x^{il}_{jk}\widetilde{X}(x^k_{lr}) - x^{il}_{jk}x^k_{lh}\frac{\partial f^h}{\partial x^r} + x^{il}_{jk}\bar{x}^h_l\frac{\partial^2 f^k}{\partial x^h\partial x^r}\right)dx^r \\
&\quad + \left(\widetilde{X}(x^{il}_{jh}) + x^{il}_{jk}\frac{\partial f^k}{\partial x^h}\right)d\bar{x}^h_l.
\end{aligned}$$

Hence,

$$0 = -\widetilde{X}(x^{il}_{jk})x^k_{lr} - x^{il}_{jk}\widetilde{X}(x^k_{lr}) - x^{il}_{jk}x^k_{lh}\frac{\partial f^h}{\partial x^r} + x^{il}_{jk}\bar{x}^h_l\frac{\partial^2 f^k}{\partial x^h\partial x^r}, \tag{8}$$

$$0 = \widetilde{X}(x^{il}_{jh}) + x^{il}_{jk}\frac{\partial f^k}{\partial x^h}. \tag{9}$$

Thus, substituting the expression of $\widetilde{X}(x^{il}_{jh})$ in (9) into (8) we obtain

$$x^{il}_{jh}\widetilde{X}(x^h_{lr}) = x^{il}_{jk}\frac{\partial f^k}{\partial x^h}x^h_{lr} - x^{il}_{jk}x^k_{lh}\frac{\partial f^h}{\partial x^r} + x^{il}_{jk}\bar{x}^h_l\frac{\partial^2 f^k}{\partial x^h\partial x^r},$$

and multiplying by v^{cj}_{di}, we have

$$\delta^c_h\delta^l_d\widetilde{X}(x^h_{lr}) = \delta^c_k\delta^l_d\frac{\partial f^k}{\partial x^h}x^h_{lr} - \delta^c_k\delta^l_d x^k_{lh}\frac{\partial f^h}{\partial x^r} + \delta^c_k\delta^l_d\bar{x}^h_l\frac{\partial^2 f^k}{\partial x^h\partial x^r}.$$

Therefore,

$$\widetilde{X}(x^c_{dr}) = \frac{\partial f^c}{\partial x^h}x^h_{dr} - x^c_{dh}\frac{\partial f^h}{\partial x^r} + \bar{x}^h_d\frac{\partial^2 f^c}{\partial x^h\partial x^r}. \tag{10}$$

Hence, from (9) and (10) we have

$$\begin{aligned}
\widetilde{X} &= f^i\frac{\partial}{\partial x^i} + \bar{f}^i_j\frac{\partial}{\partial\bar{x}^i_j} - x^i_k\frac{\partial f^k}{\partial x^j}\frac{\partial}{\partial x^i_j} \\
&\quad + \left(\bar{x}^h_j\frac{\partial^2 f^i}{\partial x^h\partial x^k} + x^h_{jk}\frac{\partial f^i}{\partial x^h} - x^i_{jh}\frac{\partial f^h}{\partial x^k}\right)\frac{\partial}{\partial x^i_{jk}} - x^{il}_{jr}\frac{\partial f^r}{\partial x^k}\frac{\partial}{\partial x^{il}_{jk}}.
\end{aligned} \tag{11}$$

Proposition 2. *Let* $\widetilde{X}^{(1)}$ *be the first jet prolongation of a vector field* X *on* M (see [14]). *There exists a unique vector field* $\widetilde{\widetilde{X}}$ *on* $L_\pi(LM)$ *that satisfies the following two conditions:*

(i) $L_{\widetilde{\widetilde{X}}}\theta = 0$, *where* θ *is the soldering form on* $L_\pi(LM)$;

(ii) $\widetilde{\widetilde{X}}$ *is* ρ*-projectable onto* $\widetilde{X}^{(1)}$; *that is,* $\rho_*(\widetilde{\widetilde{X}}) = \widetilde{X}^{(1)}$.

Proof. The local expression of the natural lift $\widetilde{X}$ of a vector field

$$X = f^i \frac{\partial}{\partial x^i}, \quad f^i \in C^\infty(U),$$

on M is given by

$$\widetilde{X} = f^i \frac{\partial}{\partial x^i} + \bar{x}^h_j \frac{\partial f^i}{\partial x^h} \frac{\partial}{\partial \bar{x}^i_j}.$$

Hence, from (11) we have

$$\begin{aligned} \widetilde{\widetilde{X}} &= f^i \frac{\partial}{\partial x^i} + \bar{x}^h_j \frac{\partial f^i}{\partial x^h} \frac{\partial}{\partial \bar{x}^i_j} - x^i_k \frac{\partial f^k}{\partial x^j} \frac{\partial}{\partial x^i_j} \\ &+ \left(\bar{x}^h_j \frac{\partial^2 f^i}{\partial x^h \partial x^k} + x^h_{jk} \frac{\partial f^i}{\partial x^h} - x^i_{jh} \frac{\partial f^h}{\partial x^k} \right) \frac{\partial}{\partial x^i_{jk}} - x^{il}_{jr} \frac{\partial f^r}{\partial x^k} \frac{\partial}{\partial x^{il}_{jk}}. \end{aligned} \tag{12}$$

We conclude taking into account the local expression of $\widetilde{X}^{(1)}$ (see [14]) and

$$\rho_* \left(\frac{\partial}{\partial x^i} \right) = \frac{\partial}{\partial x^i}, \quad \rho_* \left(\frac{\partial}{\partial \bar{x}^i_j} \right) = \frac{\partial}{\partial \bar{x}^i_j},$$

$$\rho_* \left(\frac{\partial}{\partial x^i_j} \right) = 0, \quad \rho_* \left(\frac{\partial}{\partial x^i_{jk}} \right) = \frac{\partial}{\partial x^i_{j,k}}, \quad \rho_* \left(\frac{\partial}{\partial x^{il}_{jk}} \right) = 0,$$

where $(x^i, \bar{x}^i_j, x^i_{j,k})$ are the local coordinate system induced on $J^1(V)$ introduced in §4.2. □

4.4 Local Expression for $\mathcal{L}^r_h$ and $\mathcal{L}^{kl}_{hr}$

Let us compute now the local expression for the functions $\mathcal{L}^h_i, \mathcal{L}^{sj}_{ir}$. Let $u = ((X_i)_e, (X^i_j)_e) \in L_\pi(LM)$, where $e = (\bar{X}_i) \in LM$ over $x \in M$. Using the definition of the functions $\mathcal{L}^j_i$ (see formula (1)) and the local expressions of the vector fields $\bar{X}_i$ and X_j (see formulas (3) and (4)), we obtain

$$\begin{aligned} \bar{x}^k_i(e) \left. \frac{\partial}{\partial x^k} \right|_x &= \mathcal{L}^j_i(u) \pi_*(X_j) = \mathcal{L}^j_i(u) \pi_* \left(v^k_j(u) \left. \frac{\partial}{\partial x^k} \right|_e + v^h_{kj}(u) \left. \frac{\partial}{\partial \bar{x}^h_k} \right|_e \right) \\ &= \mathcal{L}^j_i(u) v^k_j(u) \left. \frac{\partial}{\partial x^k} \right|_x. \end{aligned}$$

Therefore

$$\mathcal{L}^h_i = \bar{x}^k_i v_k{}^h.$$

In Lagrangian coordinates (see (5))

$$\mathcal{L}_i^h = \bar{x}_i^k x_k^h. \tag{13}$$

Similarly, from the definition of the functions $\mathcal{L}_{ir}^{sj}$ (see formula (2)) and taking into account the local expression of E_j^{i*}, that is, $E_j^{i*} = \bar{x}_i^h \dfrac{\partial}{\partial \bar{x}_j^h}$ (see [14]), and of X_s^r (see formula (4)), we have

$$\bar{x}_i^h(e)\delta_b^j \left.\frac{\partial}{\partial \bar{x}_b^h}\right|_e = \mathcal{L}_{ir}^{sj}(u)(X_s^r)_e = \mathcal{L}_{ir}^{sj}(u) v_{bs}^{hr}(u) \left.\frac{\partial}{\partial \bar{x}_b^h}\right|_e .$$

Therefore, multiplying by x_{dh}^{cb} and taking into account the definition of the Lagrangian coordinates we obtain

$$\bar{x}_i^h \delta_b^j x_{dh}^{cb} = \mathcal{L}_{ir}^{sj} v_{bs}^{hr} x_{dh}^{cb} = \mathcal{L}_{ir}^{sj} \delta_s^c \delta_d^r ;$$

that is,

$$\mathcal{L}_{ir}^{sj} = \bar{x}_i^h x_{rh}^{sj} . \tag{14}$$

5 Invariant Lagrangians on $L_\pi(LM)$

5.1 Decomposition into Basic Lagrangians

Once the basic Lagrangians have been introduced, we may state the main result as follows:

Theorem 1. *Every smooth $\mathfrak{X}(M)$-invariant function on $L_\pi(LM)$ can be written as a differentiable function of the Lagrangians $\mathcal{L}_{jk}^i \circ \rho$, $\mathcal{L}_i^h$ and $\mathcal{L}_{ir}^{sj}$.*

Proof. From the local expression of $\tilde{\tilde{X}}$ (see equation (12)) we conclude that a function $\mathcal{L} \in C^\infty(L_\pi(LM))$ is $\mathfrak{X}(M)$-invariant if and only if it satisfies the following system of $m + m^2 + \frac{1}{2}m^2(m+1)$ PDEs:

$$0 = \frac{\partial \mathcal{L}}{\partial x^i}, \tag{15}$$

$$0 = \bar{x}_h^j \frac{\partial \mathcal{L}}{\partial \bar{x}_h^i} - x_i^h \frac{\partial \mathcal{L}}{\partial x_j^h} + x_{hr}^j \frac{\partial \mathcal{L}}{\partial x_{hr}^i} - x_{hi}^r \frac{\partial \mathcal{L}}{\partial x_{hj}^r} - x_{hi}^{rl} \frac{\partial \mathcal{L}}{\partial x_{hj}^{rl}}, \tag{16}$$

$$0 = \bar{x}_h^j \frac{\partial \mathcal{L}}{\partial x_{hk}^i} + \bar{x}_h^k \frac{\partial \mathcal{L}}{\partial x_{hj}^i}, \tag{17}$$

where $j \le k$. Let us consider the distribution $\mathcal{D}$ generated by the vector fields:

$$X_i = \frac{\partial}{\partial x^i},$$
$$X_i^j = \bar{x}_h^j \frac{\partial}{\partial \bar{x}_h^i} - x_i^h \frac{\partial}{\partial x_j^h} + x_{hr}^j \frac{\partial}{\partial x_{hr}^i} - x_{hi}^r \frac{\partial}{\partial x_{hj}^r} - x_{hi}^{rl} \frac{\partial}{\partial x_{hj}^{rl}},$$
$$X_i^{jk} = \bar{x}_h^j \frac{\partial}{\partial x_{hk}^i} + \bar{x}_h^k \frac{\partial}{\partial x_{hj}^i}, \qquad j \leq k.$$

It is readily checked that: $[X_i, X_a] = 0$, $[X_i, X_a^b] = 0$, $[X_i, X_a^{bc}] = 0$, $[X_i^{jk}, X_a^{bc}] = 0$, $[X_a^b, X_i^{jk}] = \delta_a^j X_i^{bk} + \delta_a^k X_i^{bj} - \delta_i^b X_a^{jk}$ and $[X_a^b, X_i^j] = \delta_i^b X_a^j - \delta_a^j X_i^b$, and therefore the distribution $\mathcal{D}$ is involutive. Hence, the number of invariant functions is equal to

$$\dim L_\pi(LM) - \mathrm{rk}(\mathcal{D}) = m^2 + \frac{m^2(m-1)}{2} + m^4.$$

Taking into account the local expressions of the functions $\mathcal{L}_h^r$, $\mathcal{L}_{hr}^{kl}$, $\mathcal{L}_{jk}^i \circ \rho$, $j < k$ (see formulas (13), (14) and ([14, formula 6])), it is easy to prove that these functions are functionally independent and satisfy the system (15)-(17). Therefore, every function $\mathcal{L} \in C^\infty(L_\pi(LM))$ satisfying $\widetilde{\widetilde{X}}(\mathcal{L}) = 0$ can be written locally as

$$\mathcal{L} = \Phi(\mathcal{L}_h^r, \mathcal{L}_{jk}^i \circ \rho, \mathcal{L}_{hr}^{kl}),$$

where $\Phi \in C^\infty(\mathbb{R}^N)$, $N = m^2 + \frac{1}{2}m^2(m-1) + m^4$. □

5.2 Invariant Functions with Structure Group Symmetry

The smooth $\mathfrak{X}(M)$-invariant functions that admit structure group symmetry may be further classified.

Corollary 1. *The smooth $\mathfrak{X}(M)$-invariant functions on $L_\pi(LM)$ which are also invariant under the structure group of the projection $\rho\colon L_\pi(LM) \to J^1(LM)$ are the differentiable functions of $\mathcal{L}_{jk}^i \circ \rho$.*

Proof. The $\mathfrak{X}(M)$-invariant function $\mathcal{L} = \Phi(\mathcal{L}_h^r, \mathcal{L}_{jk}^i \circ \rho, \mathcal{L}_{hr}^{kl})$ is invariant under the action of the structure group of the principal bundle $\rho\colon L_\pi(LM) \to J^1(LM)$ if

$$A^*\mathcal{L} = 0 \ \forall\, A = \begin{pmatrix} a_j^i & 0 \\ 0 & a_{jl}^{ik} \end{pmatrix} \in \mathfrak{gl}(m, \mathbb{R}) \times \mathfrak{gl}(m^2, \mathbb{R}),$$

where A^* is the infinitesimal generator of $R_{\exp(tA)}$. Let u_h^r, u_{jk}^i, u_{hr}^{kl}, $j < k$, be the coordinates in $\mathbb{R}^N$. For every $u \in L_\pi(LM)$,

$$\left.\frac{d}{dt}\right|_{t=0} x_j^i(u \cdot \exp(tA)) = x_h^i(u) a_j^h \text{ and } \left.\frac{d}{dt}\right|_{t=0} x_{jl}^{ik}(u \cdot \exp(tA)) = x_{rs}^{ik}(u) a_{jl}^{rs},$$

and therefore

$$A^*\mathcal{L} = \left(x_h^i a_j^h \frac{\partial}{\partial x_j^i} + x_{rs}^{ik} a_{jl}^{rs} \frac{\partial}{\partial x_{jl}^{ik}} \right)(\mathcal{L}) = x_h^i a_j^h \frac{\partial \Phi}{\partial u_b^a} \frac{\partial \mathcal{L}_b^a}{\partial x_j^i} + x_{rs}^{ik} a_{jl}^{rs} \frac{\partial \Phi}{\partial u_{cd}^{ab}} \frac{\partial \mathcal{L}_{cd}^{ab}}{\partial x_{jl}^{ik}}$$

$$= x_h^i a_j^h \frac{\partial \Phi}{\partial u_b^a} \bar{x}_b^s \delta_i^a \delta_s^j + x_{rs}^{ik} a_{jl}^{rs} \frac{\partial \Phi}{\partial u_{cd}^{ab}} \bar{x}_c^m \delta_i^a \delta_k^b \delta_d^j \delta_m^l$$

$$= x_h^a a_s^h \bar{x}_b^s \frac{\partial \Phi}{\partial u_b^a} + x_{rt}^{ab} a_{ds}^{rt} \bar{x}_c^s \frac{\partial \Phi}{\partial u_{cd}^{ab}} .$$

Hence, $\mathcal{L}$ is $(GL(m, \mathbb{R}) \times GL(m^2, \mathbb{R}))$-invariant if and only if

$$x_h^a \bar{x}_b^s \frac{\partial \Phi}{\partial u_b^a} = 0 \quad \text{and} \tag{18}$$

$$x_{rt}^{ab} \bar{x}_c^s \frac{\partial \Phi}{\partial u_{cd}^{ab}} = 0 . \tag{19}$$

Multiplying (18) by ${x_i}^h$ and by ${\bar{x}_s}^j$ and adding in h and s, we have

$$0 = x_h^a {x_i}^h \bar{x}_b^s {\bar{x}_s}^j \frac{\partial \Phi}{\partial u_b^a} = \frac{\partial \Phi}{\partial u_j^i} ,$$

and multiplying (19) by v_{ji}^{tr} and by ${\bar{x}_s}^k$ and adding in rt and s, it follows that

$$0 = x_{rt}^{ab} v_{ji}^{tr} \bar{x}_c^s {\bar{x}_s}^k \frac{\partial \Phi}{\partial u_{cd}^{ab}} = \frac{\partial \Phi}{\partial u_{kd}^{ij}} .$$

Therefore, if $\mathcal{L}$ is $(GL(m, \mathbb{R}) \times GL(m^2, \mathbb{R}))$-invariant, then $\mathcal{L} = \Psi(\mathcal{L}_{jk}^i \circ \rho)$, where $\Psi \in C^\infty(\mathbb{R}^{m^2(m-1)/2})$. □

Acknowledgement. Supported by Ministerio de Ciencia y Tecnología of Spain, under grant BFM2002(00141). Partial funding from the Lecturers and Visiting Scholars Fund at Trinity University, San Antonio, Texas, USA.

References

1. Baez, J.C.: Diffeomorphism-invariant generalized measures on the space of connections modulo gauge transformations. In: Proc. of the Conference on Quantum Topology, Manhattan, KS 1993, pp. 21–43. World Sci. Pub., River Edge (1994)
2. Bleecker, D.: Gauge Theory and Variational Principles. Addison-Wesley, Reading (1981)
3. Fulp, R.O., Lawson, J.K., Norris, L.K.: Generalized symplectic geometry as a covering theory for the Hamiltonian theories of classical particles and fields. J.Geom. Phys. 20, 195–206 (1996)

4. Gotay, M.: A multisymplectic framework for classical field theory and the calculus of variations. II. Space + time decomposition. In: Francaviglia, M. (ed.) Mechanics, Analysis and Geometry: 200 Years after Lagrange, pp. 203–235. North Holland, Amsterdam (1991)
5. Gotay, M.J., Isenberg, J.A., Marsden, J.E.: Momentum maps and classical fields, I: Covariant field theory. `arXiv.org` preprint Physics/9801019 (1997)
6. Gronwald, F.: Metric-affine gauge theory of gravity. I. Fundamental structure and field equations. Int. J. Mod. Phys. D 6, 263–303 (1997)
7. Kobayashi, S.: Transformation Groups in Differential Geometry. Springer, Heidelberg (1995)
8. Kobayashi, S., Nomizu, K.: Foundations of Differential Geometry, vol. 1. Wiley, New York (1963)
9. Lawson, J.K.: A frame bundle generalization of multisymplectic geometries. Rep. Math. Phys. 45, 183–205 (2000)
10. Lawson, J.K.: A frame bundle generalization of multisymplectic momentum mappings. Rep. Math. Phys. 53, 19–37 (2004)
11. de León, M., McLean, M., Norris, L.K., Rey-Roca, A.C., Salgado, M.: Geometric structures in field theory. `arXiv.org` preprint math-ph/0208036 (2002)
12. Luo, Y., Shao, M.X., Zhu, Z.Y.: Diffeomorphism invariance of geometric descriptions of Palatini and Ashtekar gravity. Phys. Lett. B 419, 37–39 (1998)
13. McLean, M., Norris, L.K.: Covariant field theory on frame bundles of fibered manifolds. J. Math. Phys. 41, 6808–6823 (2000)
14. Muñoz Masqué, J., Rosado, M.E.: Invariant variational problems on linear frame bundles. J. Phys. A 35, 2013–2036 (2002)
15. Norris, L.K.: Generalized symplectic geometry on the frame bundle of a manifold. In: Proc. Symp. Pure Math., vol. 54(2), pp. 435–465 (1993)
16. Norris, L.K.: n-symplectic algebra of observables in covariant Lagrangian field theory. J. Math. Phys. 42, 4827–4845 (2001)
17. Piguet, O.: Ghost equations and diffeomorphism-invariant theories. Class Quantum Grav. 17, 3799–3806 (2000)
18. Saunders, D.J.: The Geometry of Jet Bundles. Cambridge University Press, Cambridge (1989)
19. Sławianowski, J.J.: Field of linear frames as a fundamental self-interacting system. Rep. Math. Phys. 22, 323–371 (1985)
20. Sławianowski, J.J.: $GL(n, R)$ as a candidate for fundamental fymmetry in Field Theory. Nuovo. Cimento. B 11(106), 645–668 (1991)

On the Computation of Differential Resultants*

Sonia L. Rueda

Dpto de Matemática Aplicada, E.T.S. Arquitectura,
Universidad Politécnica de Madrid, Avda. Juan de Herrera 4,
28040-Madrid, Spain
soniall.uisa.rueda@upm.es

Summary. The definition of the differential resultant of a set of ordinary differential polynomials is reviewed and its computation via determinants is revisited, using a modern language. This computation is also extended to differential homogeneous resultants of homogeneous ordinary differential polynomials. A numeric example is included and an example of the application of elimination theory to biological modelling is revisited, in terms of differential resultants.

1 Introduction

The resultant of two algebraic polynomials is a well known tool in elimination theory (dating back to Euler and Bezout). Given two polynomials $f_1, f_2 \in \mathbb{C}[x, y]$ of nonzero degree in y the resultant of f_1 and f_2 with respect to y is a polynomial $\mathrm{Res}_y(f_1, f_2)$ in $\mathbb{C}[x]$. It is also well known that $\mathrm{Res}_y(f_1, f_2)$ can be computed as the determinant of the Sylvester matrix (see [31], §5.8 or [11], Chapter 3), making it very useful in computer algebra and allowing its implementation in numerous computer algebra systems.

The resultant of a set of n homogeneous algebraic polynomials in n variables was defined by Macaulay in [23], where its computation by means of determinants was also explained (see also [32]). It is called multipolynomial resultant in [11]. In fact, the multipolynomial resultant of a set of generic polynomials can be computed as a quotient of two determinants. Unfortunately, multipolynomial resultants often vanish after specialization and a relevant line of research has been conducted on the study of other resultants, for instance the so called sparse resultants or toric resultants (for an overview on the subject see [11] and [17]). All of these resultants have proved to be powerful tools in computer algebra and in particular in elimination theory, leading the way to numerous applications of algebraic geometry.

The differential resultant problem was first studied for differential operators by Ore [26], Berkovich and Tsirulik [2], Carra'Ferro [7], Chardin [10]

* To our beloved Marisa with great admiration.

and Li [10]. The differential resultant of two differential polynomials in one variable was studied by Ritt in [27], p.47, under some hypothesis on the differential polynomials. It was G. Carra'Ferro, who gave the definition of a differential resultant for a set of n ordinary differential polynomials in $n-1$ differential variables in [9], and for $n = 2$ in [8]. The differential resultant defined by Carra'Ferro is based on the algebraic resultant of Macaulay, and it was explained in [9] how, for generic differential polynomials, it can be computed as a quotient of two determinants. Apparently forgotten for some years, Carra'Ferro's definition has been used recently by Rueda and Sendra to approach the linear ordinary differential implicitization problem in [30]. The differential resultant of a set of partial differential operators was used also by Kasman and Previato in [19], suggesting a revival of the subject (of differential resultants in general). We may say that the theory of differential resultants is rather incomplete and offers a wide field of research, in many different directions. Very recently, Gao *et al.* in [16] gave a more complete definition of the differential resultant of n differential polynomials in $n-1$ variables, in terms of the generalized differential Chow form.

Differential resultants are over all differential elimination tools. Differential elimination methods are commonly applied in control theory and, more precisely, to the identifiability study of differential systems. There is a broad literature available, of which some examples are [1], [13], [14],[15], [22], [25], [33]. In particular, there are some interesting applications of differential elimination techniques in cellular biology developed by Boulier *et al.* in [6] and [5].

In this paper, the definition of the differential resultant of a set of ordinary differential polynomials is revisited, together with its computation via determinants in a modern language. This computation is then extended to the differential homogeneous resultant, which we defined in [30]. Two examples of this computation are included: a numeric example and the computation via differential resultants of an example of application (of differential elimination) in biological modelling, provided by Boulier in [4].

1.1 The Sylvester Matrix

Let us begin with the construction of the Sylvester matrix $\mathrm{Syl}(f_1, f_2)$, of the polynomials f_1 and f_2, to motivate the exposition of this paper. Let $\mathbb{D} = \mathbb{C}[x]$, then $f_1, f_2 \in \mathbb{D}[y]$ and they have degree in y greater than zero, $d_i = \deg(f_i) > 0$, $i = 1, 2$. Then $\mathrm{Syl}(f_1, f_2)$ is the coefficient matrix of the polynomials in the set (as polynomials in y)

$$\{y^{d_2-1}f, \dots, yf, f, y^{d_1-1}g, \dots, y\,g, g\}.$$

If $f_1(y) = a_{d_1}y^{d_1} + \cdots + a_1y + a_0$ and $f_2(y) = b_{d_2}y^{d_2} + \cdots + b_1y + b_0$ with coefficients in $\mathbb{D}$ then

$$
\mathrm{Syl}(f_1,f_2) = \begin{bmatrix}
a_{d_1} & \cdots & a_1 & a_0 & 0 & \cdots & 0 \\
0 & a_{d_1} & \cdots & a_1 & a_0 & \cdots & 0 \\
\vdots & & \ddots & & \ddots & \ddots & \vdots \\
0 & \cdots & 0 & a_{d_1} & \cdots & a_1 & a_0 \\
b_{d_2} & \cdots & b_1 & b_0 & 0 & \cdots & 0 \\
0 & b_{d_2} & \cdots & b_1 & b_0 & \cdots & \vdots \\
\vdots & & \ddots & & \ddots & \ddots & 0 \\
0 & \cdots & 0 & b_{d_2} & \cdots & b_1 & b_0
\end{bmatrix}
\begin{matrix} d_2 \text{ rows} \\ \\ d_1 \text{ rows} \end{matrix} .
$$

Observe that the columns of the previous $(d_1 + d_2) \times (d_1 + d_2)$ Sylvester matrix are indexed by the components of the vector $(y^{d_1+d_2-1}, \ldots, y, 1)$. Therefore, if (f_1, f_2) is the ideal generated by f_1 and f_2 in $\mathbb{C}[x, y]$ then $\mathrm{Res}_y(f_1, f_2) = \det(\mathrm{Syl}(f_1, f_2))$ belongs to the elimination ideal $(f_1, f_2) \cap \mathbb{C}[x]$ (see [11], Chapter 3 and references therein). It is also well known that in some extension field $\mathbb{E}$ of $\mathbb{D}$ it holds:

$$\{f_1 = 0, f_2 = 0\} \text{ has a solution in } \mathbb{E} \Leftrightarrow \partial\mathrm{Res}(f_1, f_2) = 0.$$

1.2 Notation

Let $\mathbb{D}$ be an ordinary differential domain with derivation ∂. Let $U = \{u_1, \ldots, u_{n-1}\}$ be a set of differential indeterminates over $\mathbb{D}$. For $k \in \mathbb{N}$ we denote by u_{jk} the k-th derivative of u_j, $j = 1, \ldots, n-1$. We denote by $\{U\}$ the set of derivatives of the elements of U, and by $\mathbb{D}\{U\}$ the ring of differential polynomials in the differential indeterminates $u_1, \ldots, u_{n-1}$, that is

$$\mathbb{D}\{U\} = \mathbb{K}[u_j, u_{jk} \mid j = 1, \ldots, n-1,\ k \in \mathbb{N}].$$

For further concepts and results on differential algebra we refer to [20] and [28].

2 The Differential Resultant of Carra'Ferro

Let us consider the ordinary differential polynomial f_i in $\mathbb{D}\{U\}$ of order o_i, $i = 1, \ldots, n$.

Definition 1. *The* ***differential resultant*** *(of Carra'Ferro)* $\partial\mathrm{Res}(f_1, \ldots, f_n)$*, of* $f_1, \ldots, f_n$*, is the Macaulay's algebraic resultant of the differential polynomial set*

$$\mathcal{P}(f_1, \ldots, f_n) = \{f_i, \partial f_i, \ldots, \partial^{N-o_i} f_i \mid i = 1, \ldots, n, N = \sum_{i=1}^{n} o_i\}.$$

The Macaulay's algebraic resultant of a set of homogeneous algebraic polynomials is defined by Macaulay in [23] as the greatest common divisor of a set of determinants. For nonhomogeneous algebraic polynomials, the same construction can be used introducing a homogenizing variable. This was the approach used by Carra'Ferro in [9], to compute $\partial\text{Res}(f_1,\ldots,f_n)$ via determinants.

2.1 Computation via Determinants

In this section, the construction of the matrices used to compute the differential resultant in [9] is revisited, but using the notation introduced in [19] (to define differential resultants for partial differential operators) and the construction given in [11], Chapter 3, §4 for multipolynomial resultants. The definition of multipolynomial resultant of a set of algebraic homogeneous polynomials given in [11], Chapter 3 is intrinsic to its properties, and it is also proved afterwards that is can be computed as the greatest common divisor of a set of determinants.

Observe that $\mathcal{P}(f_1,\ldots,f_n)$ is a set with $L=\sum_{i=1}^{n}(N-o_i+1)$ polynomials in the set of $L-1$ variables

$$\mathcal{V}=\{u_j,u_{j1},\ldots,u_{jN}\mid j=1,\ldots,n-1\},$$

that is, $\mathcal{P}$ is included in the polynomial ring $\mathbb{D}[\mathcal{V}]$. We define the L component vector

$$Y=(y_1,\ldots,y_L)$$

where y_l belongs to $\mathcal{V}\cup\{1\}$, $l=1,\ldots,L$. By writing it as a vector, we are supposing that the variables $y_1,\ldots,y_L$ have an ordering, although the particular ordering chosen is not important at this point of the discussion. We call the variable $y_{l_0}=1$, $l_0\in\{1,\ldots,L\}$ the ***homogenizing variable***, and it will allow us to use the construction given in [11], Chapter 3, §4 to compute $\partial\text{Res}(f_1,\ldots,f_n)$ via determinants.

We also impose an ordering on the polynomials in $\mathcal{P}(f_1,\ldots,f_n)$ (and again the particular ordering chosen is not important so far). We denote by $\text{PS}(f_1,\ldots,f_n)$ the L component vector

$$\text{PS}(f_1,\ldots,f_n)=(P_1,\ldots,P_L)$$

where P_l, $l=1,\ldots,L$ belongs to $\mathcal{P}(f_1,\ldots,f_n)$.

Let d_i be the degree of f_i, we assume $d_i>0$. We denote by D the positive integer

$$D=\sum_{l=1}^{L}(\deg(P_l)-1)+1=\sum_{i=1}^{n}(N-o_i+1)(d_i-1)+1.$$

We define the positive integers

$$\mathbb{L} = \binom{L-1+D}{L-1}, \quad \mathbb{L}_{d_i} = \binom{L-1+D-d_i}{L-1}, \; i = 1,\ldots,n.$$

Given $\alpha = (\alpha_1,\ldots,\alpha_L) \in \mathbb{N}^L$, we denote by y^α the monomial $y_1^{\alpha_1} \cdot \ldots \cdot y_L^{\alpha_L}$ and by $|\alpha| = \sum_{l=1}^{L} \alpha_l$. The set

$$\mathcal{M}^D = \{y^\alpha \mid |\alpha| = D\}$$

has cardinality $\mathbb{L}$. Observe that since $y_{l_0} = 1$ then the monomials in $\mathcal{M}^D$ have degree less than or equal to D as monomials in $\mathbb{D}[\mathcal{V}]$. We establish an ordering of the elements of $\mathcal{M}^D$ induced by the ordering of the components of Y. We denote by $\mathbb{Y}_D$ the $\mathbb{L}$-component vector

$$\mathbb{Y}_D = (w_1^D,\ldots,w_{\mathbb{L}}^D),$$

whose components run over the elements of $\mathcal{M}^D$ in the order chosen. For all $i = 1,\ldots,n$ we denote by $\mathcal{M}^{D-d_i}$ the set

$$\mathcal{M}^{D-d_i} = \{y^\alpha \mid |\alpha| = D - d_i\},$$

which has cardinality $\mathbb{L}_{d_i}$. Observe that since $y_{l_0} = 1$ then the monomials in $\mathcal{M}^{D-d_i}$ have degree less than or equal to $D - d_i$ as monomials in $\mathbb{D}[\mathcal{V}]$, $i = 1,\ldots,n$. We denote by $\mathbb{Y}_{D-d_i}$ the $\mathbb{L}_{d_i}$-component subvector of $\mathbb{Y}_D$

$$\mathbb{Y}_{D-d_i} = (w_1^{D-d_i},\ldots,w_{\mathbb{L}_{d_i}}^{D-d_i}),$$

whose components run over the elements of $\mathcal{M}^{D-d_i}$.

Given a differential polynomial $P \in \mathbb{D}[\mathcal{V}]$ of degree less than or equal to D, we denote by $\overline{v}(P)$ the $\mathbb{L}$ component vector whose l-th entry is the coefficient of w_l^D in P, $l = 1,\ldots,\mathbb{L}$. Thus $P = \overline{v}(P)\mathbb{Y}_D^t$ where $\mathbb{Y}_D^t$ is the transpose of $\mathbb{Y}_D$. Observe that $\deg(P_l) \in \{d_1,\ldots,d_n\}$ for all $l = 1,\ldots,L$.

Definition 2. *1. We denote by $M(f_1,\ldots,f_n)$ the $(\sum_{i=1}^{n}(N-o_i+1)\mathbb{L}_{d_i})\times\mathbb{L}$ matrix whose rows are*

$$\overline{v}(w_k^{D-\deg(P_l)} \cdot P_l), \; l = 1,\ldots,L, \; k = 1,\ldots,\mathbb{L}_{\deg(P_l)}.$$

2. By [23], page 4 then $\partial\mathrm{Res}(f_1,\ldots,f_n)$ is the greatest common divisor of a set of polynomials in $\mathbb{D}$, namely

$$\gcd\{\det(M) \mid M \text{ is an } \mathbb{L}\times\mathbb{L} \text{ submatrix of } M(f_1,\ldots,f_n)\}.$$

We define next an $\mathbb{L}\times\mathbb{L}$ submatrix M of $M(f_1,\ldots,f_n)$ such that, if f_i is a generic polynomials of order o_i and degree d_i for $i = 1,\ldots,n$ then

$$\det(M) = \partial\mathrm{Res}(f_1,\ldots,f_n) \cdot \text{extraneous factor}.$$

Furthermore, the extraneous factor can be computed by means of determinants. In fact, if the differential resultant $\partial\mathrm{Res}(f_1,\ldots,f_n)$ is not zero after specialization, it can me computed as the quotient of two determinants (see [23] and [11], Chapter 3, §4).

Given $l\in\{1,\ldots,L\}$, a monomial $y^\alpha\in\mathcal{M}^D$ is said to be ***reduced in*** y_l if $y_l^{\deg(P_l)}$ does not divide y^α. Observe that, if $y_{l_0}^{\deg(P_{l_0})}$ divides y^α then, as a monomial in $\mathbb{D}[\mathcal{V}]$, the degree of y^α is less than $D-\deg(P_{l_0})$. We say that a monomial $y^\alpha\in\mathcal{M}^D$ is ***reduced*** if $y_{l_\alpha}^{\deg(P_{l_\alpha})}$ divides y^α, for exactly one $l_\alpha\in\{1,\ldots,L\}$. Let us call Ω the set of all reduced monomials in $\mathcal{M}^D$.

We define the sets M_l for $l=1,\ldots,L$ as follows:

$$\begin{aligned}M_1&=\{y^\alpha/y_1^{\deg(P_1)}\mid y^\alpha\in\mathcal{M}^D \text{ is not reduced in } y_1\},\\ M_l&=\{y^\alpha/y_l^{\deg(P_l)}\mid y^\alpha\in\mathcal{M}^D \text{ is not reduced in } y_l,\\ &\qquad\qquad \text{and it is reduced in } y_k, k=1,\ldots,l-1\}.\end{aligned}$$

for $l=2,\ldots,L$. The sum of the cardinalities c_l of the sets M_l, $l=1,\ldots,L$ equals $\mathbb{L}$. We denote by $\mathbb{Y}_l$ the subvector of $\mathbb{Y}_D$

$$\mathbb{Y}_l=(w_1^l,\ldots,w_{c_l}^l),$$

whose components run over the elements of $\mathcal{M}_l$, $l=1,\ldots,L$, thus in the order established in $\mathcal{M}^D$.

Definition 3. *Let* $\mathrm{PS}=\mathrm{PS}(f_1,\ldots,f_n)$.

1. *We define the* $\mathbb{L}\times\mathbb{L}$ *matrix* $M(Y,\mathrm{PS})$ *whose rows are*
$$\overline{v}(w_k^l\cdot P_l),\ l=1,\ldots,L,\ k=1,\ldots,c_l.$$
2. *We define the submatrix* $A(Y,\mathrm{PS})$ *of* $M(Y,\mathrm{PS})$ *obtained by removing:*
   a) The columns corresponding to the monomials in Ω.
   b) The rows $\overline{v}(w_k^l\cdot P_l)$ *such that* $w_k^l\cdot y_l^{\deg(P_l)}\in\Omega$, $l=1,\ldots,L$,$k=1,\ldots,c_l$.

It was proved by Macaulay [23] that, if $f_1,\ldots,f_n$ are generic polynomials then

$$\det(M(Y,\mathrm{PS}))=\partial\mathrm{Res}(f_1,\ldots,f_n)\det(A(Y,\mathrm{PS})).$$

Thus, if $\det(A(Y,\mathrm{PS}))$ is non zero after specialization of $f_1,\ldots,f_n$ then

$$\partial\mathrm{Res}(f_1,\ldots,f_n)=\frac{\det(M(Y,\mathrm{PS}))}{\det(A(Y,\mathrm{PS}))}.$$

There are $L!$ ways of ordering the elements in $\mathcal{V}\cup\{1\}$ to write Y. For some specialization of $f_1,\ldots,f_n$, let us suppose that $\det(A(Y,\mathrm{PS}))=0$ for a choice of Y, then we can permute the components of Y to get Y' such that $\det(A(Y',\mathrm{PS}))\neq 0$.

The next properties of $\det(M(Y,\mathrm{PS}))$ make of $\partial\mathrm{Res}(f_1,\ldots,f_n)$ an elimination tool. Let $[f_1,\ldots,f_n]$ be the differential ideal in $\mathbb{D}\{U\}$ generated by the differential polynomials $f_1,\ldots,f_n$.

Proposition 1. *1.* $\det(M(Y, \mathrm{PS}))$ *belongs to* $[f_1, \ldots, f_n] \cap \mathbb{D}$.

2. If $\{f_1 = 0, \ldots, f_n = 0\}$ *has a solution in a differential extension field* $\mathbb{E}$ *of* $\mathbb{D}$ *then* $\det(M(Y, \mathrm{PS})) = 0$.

Proof. The first claim follows from [9], Theorem 12. The second claim was proved in [9], Proposition 11(i).

If f_i are polynomials of order $o_i = 0$, $i = 1, \ldots, n$ then $\partial\mathrm{Res}(f_1, \ldots, f_n)$ is the Macaulay's algebraic resultant of a set of algebraic polynomials. Only for $n = 2$, the second statement of Proposition 2 is an equivalence.

As mentioned by Gao *et al.* in [16], the definition of the differential resultant given by Carra'Ferro seems to be incomplete. They gave recently a new definition of the differential resultant, of a set of n differential polynomials in $n-1$ variables, as a generalized differential Chow form in [16].

A Maple a package of functions that allows the computation of the differential resultant defined by Carra'Ferro in [9] is available at [29]. The implementation of the Macaulay's algebraic resultant, available at [24], could be also used to compute differential resultants.

2.2 Example

Let $\mathbb{D} = \mathbb{C}(t)$ and $\partial = \frac{\partial}{\partial t}$. The differential resultant $\partial\mathrm{Res}(f_1, f_2)$, of the differential polynomials

$$f_1(u_1) = t - 4{u_{11}}^2 - 4u_1^2 - tu_{11}u_1 - 5u_{11} - 4u_1,$$
$$f_2(u_1) = t - u_1 - 3u_{11}$$

in $\mathbb{D}\{u_1\}$ ($u_{1i} = \partial^i u_1/\partial t^i$), is the Macaulay's algebraic resultant of the set $\mathrm{PS}(f_1, f_2) = \{\partial f_1, f_1, \partial f_2, f_2\}$. We can compute $\partial\mathrm{Res}(f_1, f_2)$ as the quotient of two determinants. The numerator is the determinant of a matrix of order 20. The rows of this matrix are the coefficients of the polynomials in $\mathrm{PS}(f_1, f_2)$, written in decreasing order using first the degree and then the lexicographic order with $u_{12} < u_1 < u_{11}$, that is, $Y = (u_{11}, u_1, u_{12}, 1)$. The columns of the matrix $M(Y, \mathrm{PS})$ are indexed by the components of the vector

$$\mathbb{Y}_D = ({u_{11}}^3, {u_{11}}^2 u_1, {u_{11}}^2 u_{12}, u_{11}u_1^2, u_{11}u_1u_{12}, u_{11}{u_{12}}^2, u_1^3, u_1^2 u_{12}, u_1{u_{12}}^2, {u_{12}}^3,$$
$${u_{11}}^2, u_{11}u_1, u_{11}u_{12}, u_1^2, u_1u_{12}, {u_{12}}^2, u_{11}, u_1, u_{12}, 1).$$

The rows of the matrix $M(Y, \mathrm{PS})$ are the coefficients of the following differential polynomials:

rows $1\ldots4$ $\rightarrow \{u_{11}\partial f_1, u_1\partial f_1, u_{12}\partial f_1, \partial f_1\}$
rows $5\ldots8$ $\rightarrow \{u_{11}f_1, u_1 f_1, u_{12}f_1, f_1\}$
rows $9\ldots16$ $\rightarrow \{u_{11}u_1\partial f_2, u_{11}u_{12}\partial f_2, u_1u_{12}\partial f_2, {u_{12}}^2\partial f_2,$
$u_{11}\partial f_2, u_1\partial f_2, u_{12}\partial f_2, \partial f_2\}$
rows $17\ldots20$ $\rightarrow \{u_{11}u_1 f_2, u_{11}f_2, u_1 f_2, f_2\}$.

The matrix $A(Y, \text{PS})$ is the submatrix of $M(Y, \text{PS})$ obtained by removing rows 1, 2, 5, 6, 10, 11, 12, 17, 18, 19, 20 and columns 1, 2, 4, 5, 6, 7, 9, 10, 12, 17, 18, 20 of $M(Y, \text{PS})$.

Finally, the differential resultant verifies

$$\det(M(Y, \text{PS})) = \partial\text{Res}(f_1, f_2)\det(A(Y, \text{PS}))$$

where

$$\begin{aligned}\det(M(Y, \text{PS})) =& -3888t(3t-8)(15t^7 - 227t^6 - 771t^5 + 7260t^4 + 39993t^3 \\ &+ 17497t^2 + 23115t + 6078),\\ \det(A(Y, \text{PS})) =&\ 432t(3t-8),\end{aligned}$$

that is

$$\begin{aligned}\partial\text{Res}(f_1, f_2) =& \frac{-3888}{432}(15t^7 - 227t^6 - 771t^5 + 7260t^4 + 39993t^3 + 17497t^2 \\ &+ 23115t + 6078).\end{aligned}$$

The matrix $M(Y, \text{PS})$, obtained using [29], is equal to

$$\begin{bmatrix}
-t & -9 & -8 & 0 & -t & 0 & 0 & 0 & 0 & 0 & -4 & 0 & -5 & 0 & 0 & 0 & 1 & 0 & 0 & 0\\
0 & -t & 0 & -9 & -8 & 0 & 0 & -t & 0 & 0 & 0 & -4 & 0 & 0 & -5 & 0 & 0 & 1 & 0 & 0\\
0 & 0 & -t & 0 & -9 & -8 & 0 & 0 & -t & 0 & 0 & 0 & -4 & 0 & 0 & -5 & 0 & 0 & 1 & 0\\
0 & 0 & 0 & 0 & 0 & 0 & 0 & 0 & 0 & 0 & -t & -9 & -8 & 0 & -t & 0 & -4 & 0 & -5 & 1\\
-4 & -t & 0 & -4 & 0 & 0 & 0 & 0 & 0 & 0 & -5 & -4 & 0 & 0 & 0 & 0 & t & 0 & 0 & 0\\
0 & -4 & 0 & -t & 0 & 0 & -4 & 0 & 0 & 0 & 0 & -5 & 0 & -4 & 0 & 0 & 0 & t & 0 & 0\\
0 & 0 & -4 & 0 & -t & 0 & 0 & -4 & 0 & 0 & 0 & 0 & -5 & 0 & -4 & 0 & 0 & 0 & t & 0\\
0 & 0 & 0 & 0 & 0 & 0 & 0 & 0 & 0 & 0 & -4 & -t & 0 & -4 & 0 & 0 & -5 & -4 & 0 & t\\
0 & -1 & 0 & 0 & -3 & 0 & 0 & 0 & 0 & 0 & 0 & 1 & 0 & 0 & 0 & 0 & 0 & 0 & 0 & 0\\
0 & 0 & -1 & 0 & 0 & -3 & 0 & 0 & 0 & 0 & 0 & 0 & 1 & 0 & 0 & 0 & 0 & 0 & 0 & 0\\
0 & 0 & 0 & 0 & -1 & 0 & 0 & 0 & -3 & 0 & 0 & 0 & 0 & 0 & 1 & 0 & 0 & 0 & 0 & 0\\
0 & 0 & 0 & 0 & 0 & -1 & 0 & 0 & 0 & -3 & 0 & 0 & 0 & 0 & 0 & 1 & 0 & 0 & 0 & 0\\
0 & 0 & 0 & 0 & 0 & 0 & 0 & 0 & 0 & 0 & -1 & 0 & -3 & 0 & 0 & 0 & 1 & 0 & 0 & 0\\
0 & 0 & 0 & 0 & 0 & 0 & 0 & 0 & 0 & 0 & 0 & -1 & 0 & 0 & -3 & 0 & 0 & 1 & 0 & 0\\
0 & 0 & 0 & 0 & 0 & 0 & 0 & 0 & 0 & 0 & 0 & 0 & -1 & 0 & 0 & -3 & 0 & 0 & 1 & 0\\
0 & 0 & 0 & 0 & 0 & 0 & 0 & 0 & 0 & 0 & 0 & 0 & 0 & 0 & 0 & 0 & -1 & 0 & -3 & 1\\
0 & -3 & 0 & -1 & 0 & 0 & 0 & 0 & 0 & 0 & 0 & t & 0 & 0 & 0 & 0 & 0 & 0 & 0 & 0\\
0 & 0 & 0 & 0 & 0 & 0 & 0 & 0 & 0 & 0 & -3 & -1 & 0 & 0 & 0 & 0 & t & 0 & 0 & 0\\
0 & 0 & 0 & 0 & 0 & 0 & 0 & 0 & 0 & 0 & 0 & -3 & 0 & -1 & 0 & 0 & 0 & t & 0 & 0\\
0 & 0 & 0 & 0 & 0 & 0 & 0 & 0 & 0 & 0 & 0 & 0 & 0 & 0 & 0 & 0 & -3 & -1 & 0 & t
\end{bmatrix}.$$

3 Differential Resultants of Ordinary Homogeneous Differential Polynomials

If the polynomials $f_1, \ldots, f_n \in \mathbb{D}\{U\}$ are homogeneous then $\partial\text{Res}(f_1, \ldots, f_n)$ is zero, because the column indexed by $y_{l_0} = 1$ in the matrix $M(f_1, \ldots, f_n)$

is a column of zeros. Therefore, the previous definition is not appropriate for homogeneous ordinary differential polynomial. We gave a definition of the homogeneous differential resultant in [30] that is included next.

Let $h_i \in \mathbb{D}\{U\}$ be an ordinary differential homogeneous polynomial of order o_i, $i = 1, \ldots, n$, with $N = \sum_{i=1}^{n} o_i \geq 1$.

Definition 4. *The* ***differential homogenous resultant*** $\partial\text{Res}^h(h_1, \ldots, h_n)$*, of the homogeneous differential polynomials* $h_1, \ldots, h_n$*, is the Macaulay's algebraic resultant of the differential polynomial set*

$$\text{PS}^h(h_1, \ldots, h_n) = \{\partial^{N-o_i-1}h_i, \ldots, \partial h_i, h_i \mid i = 1, \ldots, n, \ \ N - o_i - 1 \geq 0\}.$$

A differential homogeneous resultant was defined also by Carra'Ferro in [8] for $n = 2$. In addition, when the homogeneous polynomials have degree one and $n = 2$ the differential homogeneous resultant coincides with the differential resultant of two differential operators, studied by Berkovitch-Tsirulik in [2] and by Chardin in [10].

Observe that $\mathcal{P}^h(h_1, \ldots, h_n)$ is a set with $L^h = \sum_{i=1}^{n}(N - o_i)$ polynomials in the set of L^h variables

$$\mathcal{V}^h = \{u_j, u_{j1}, \ldots, u_{j\,N-1} \mid j = 1, \ldots, n-1\},$$

that is $\mathcal{P}^h \subset \mathbb{D}[\mathcal{V}^h]$. We define the L^h component vector

$$Y^h = (y_1^h, \ldots, y_{L^h}^h),$$

where y_l^h belongs to $\mathcal{V}^h$, $l = 1, \ldots, L^h$. By writing it as a vector, we are supposing that the variables $y_1^h, \ldots, y_{L^h}^h$ have an ordering, although the particular ordering chosen is not important at this point of the explanation. Let us also impose an ordering on the polynomials in $\mathcal{P}^h(h_1, \ldots, h_n)$. We denote by $\text{PS}^h(h_1, \ldots, h_n)$ the L^h component vector

$$\text{PS}^h(h_1, \ldots, h_n) = (P_1^h, \ldots, P_{L^h}^h),$$

where P_l^h, $l = 1, \ldots, L^h$ belongs to $\mathcal{P}^h$.

Let $\text{PS}^h = \text{PS}^h(h_1, \ldots, h_n)$, now definitions 2 and 3 apply to $M(Y^h, \text{PS}^h)$ and $A(Y^h, \text{PS}^h)$. Consequently, for generic polynomials, the differential homogeneous resultant verifies

$$\det(M(Y^h, \text{PS}^h)) = \partial\text{Res}^h(h_1, \ldots, h_n) \cdot \det(A(Y^h, \text{PS}^h)).$$

The next properties of $\det(M(Y^h, \text{PS}^h))$ make of $\partial\text{Res}^h(h_1, \ldots, h_n)$ an elimination tool.

Proposition 2. *1.* $\det(M(Y^h, \text{PS}^h))$ *belongs to* $[h_1, \ldots, h_n] \cap \mathbb{D}$.

2. If $\{h_1 = 0, \ldots, h_n = 0\}$ *has a solution in a differential extension field* $\mathbb{E}$ *of* $\mathbb{D}$ *then* $\det(M(Y^h, \text{PS}^h)) = 0$.

Proof. 1. The proof is analogous to the one of [9], Theorem 12.
2. It follows by [18], Proposition 5.

Unfortunately, the condition $\det(M(Y^h, \mathrm{PS}^h)) = 0$ is only sufficient for the existence of nonzero solutions of the system $\{h_1 = 0, \ldots, h_n = 0\}$ in the cases: $n = 2$ and $o_i = 0$, $i = 1, 2$ ($M(Y^h, \mathrm{PS}^h)$ is the Sylvester matrix); $n = 2$ and $d_i = 1$, $i = 1, 2$ (by [2], Theorem 3.1). Counterexamples can be found in: [8], Example 4 for $n = 2$, $o_i \neq 0$, $d_i > 1$, $i = 1, 2$; and [30], Remark 6 for $n = 3$, $d_i = 1$ and $o_i > 0$, $i = 1, 3$.

4 An Example of Application to Biological Modelling

We revisit an example of application of differential elimination to biological modelling. This example was presented by F. Boulier in [4] to illustrate how differential elimination can contribute to the following problem designed in [12]: Estimate parameter values of parametric ordinary differential systems whose dependent variables are not all observed. The method developed in [12] relies on differential elimination combined with a final numerical treatment, thus it is a hybrid symbolic-numeric method. Some references to the work on the application of this method to cellular biology developed by Boulier *et al.* are [6] and [5].

The differential elimination part of the example was carried out using the Rosenfeld-Gröebner algorithm developed by Boulier in [3]. We explain how the differential elimination part in the example may be computed using differential resultants.

We include next the description of the compartmental model (for two compartments, the blood and the organ) given by Boulier in [4], §3.1.

> A medical product is injected in the blood at $t = 0$. It can go from the blood to the organ and conversely. It may also get degraded and exit from the system. In order to write the corresponding differential system, some hypotheses must be made on the nature of the exchanges: exchanges between the two compartments are assumed to be linear i.e. that, over every small enough interval of time, the amount of product going from compartment i to compartment j is proportional to the concentration of product in compartment i. The proportionality constant is denoted k_{ij} . The degradation is assumed to follow a Michaelis-Menten law. This law is a bit more difficult to explain. It can be derived from the modelling of an enzyme-catalyzed reaction by means of some model reduction. Two parameters are associated to this degradation: a maximal speed V_e and another constant k_e.

Dependent variables x and u_1 are associated to compartments 1 (the blood) and 2 (the organ), which represent the concentrations of product present in these compartments. Let $x_i = \frac{\partial^i x}{\partial t^i}$ and $u_{1i} = \frac{\partial^i u_1}{\partial t^i}$, $i \in \mathbb{N}$. A system of parametric ordinary differential equations is obtained where parameters and dependent variables are positive real numbers.

$$x_1 = -k_{12}x + k_{21}u_1 - \frac{V_e x}{k_e + x},$$

$$u_{11} = k_{12}x - k_{21}u_1.$$

It is assumed that some extra information is available: parameters k_{12} and k_{21} are completely unknown, an interval of possible values $70 \le V_e \le 110$ is known for V_e and $k_e = 7$. Compartment 1 is assumed to be observed (a file of measures is assumed to be available for x) and compartment 2 is assumed to be non observed. In this situation, the goal is to estimate the values of the three unknown parameters: V_e, k_{12} and k_{21}. A classical numerical solution relies on the use of a numerical nonlinear least squares solver i.e. a Newton method. Differential elimination gets involved in the process to help solving the most difficult part of the Newton method: guessing the starting point. The idea is to eliminate the non observed variables of the model.

The denominator of the first equation cannot vanish so we can view the system as a system of differential polynomial equations:

$$\begin{cases} f_1(u_1) = xx_1 + k_{12}x^2 + k_e x_1 + (k_{12}k_e + V_e)x - k_{21}(k_e + x)u_1 = 0, \\ f_2(u_1) = -k_{12}x + k_{21}u_1 + u_{11} = 0, \end{cases}$$

where f_1 and f_2 are differential polynomials in $\mathbb{D}\{u_1\}$, with differential domain $\mathbb{D} = \mathbb{C}(t)\{x\}$ and derivation $\frac{\partial}{\partial t}$. Observe that x and u are differential variables over $\mathbb{C}(t)$. Let us call $a_0(x) = xx_1 + k_{12}x^2 + k_e x_1 + (k_{12}k_e + V_e)x$ the independent term in $f_1(u_1)$.

Observe that f_1 and f_2, both have degree one and order one so $D = 1$ and $\mathbb{L} = L = 4$. Let $Y = (u_{12}, u_{11}, u_1, 1)$ and $\mathrm{PS} = (\partial f_1, f_1, \partial f_2, f_2)$, then $M(Y, \mathrm{PS})$ is a 4×4 matrix whose columns are indexed by the monomials in $Y = \mathbb{Y}_D$. In this situation

$$M(Y, \mathrm{PS}) = \begin{bmatrix} 0 & -k_{21}(k_e + x) & -k_{21}x_1 & \frac{\partial a_0(x)}{\partial t} \\ 0 & 0 & -k_{21}(k_e + k_{21})x & a_0(x) \\ 1 & k_{21} & 0 & -x_1 k_{12} \\ 0 & 1 & k_{21} & -xk_{12} \end{bmatrix}.$$

The differential resultant $\partial\mathrm{Res}(f_1, f_2) = \det(M(Y, \mathrm{PS}))$ in this case and equals

$$\begin{aligned} p(x) = & k_{21}(k_e^2 k_{21} x_1 + k_{21} x_1 x^2 + k_{21} k_e V_e x + 2k_{21} k_e x x_1 + k_{21} V_e x^2 + 2k_e x_2 x \\ & + 2k_e k_{12} x x_1 + k_{12} x^2 x_1 + k_e^2 x_2 + k_e^2 k_1 2 x_1 + x^2 x_2 + V_e k_e x_1), \end{aligned}$$

which is $k_{12}b(x)$ with $b(x)$ the differential polynomial provided in [4], §3.3, where the non observed variable u_1 has been eliminated. The polynomial $b(x)$ is used in [4], §3.3 to estimate the values of V_e, k_{12} and k_{21}.

Acknowledgement. Supported by the Spanish "Ministerio de Ciencia e Innovación" under the Project MTM2008-04699-C03-01.

References

1. Audoly, S., Bellu, G., D'Angio, L., Saccomani, M.P., Cobelli, C.: Global Identifiability of Nonlinear Models of Biological Systems. IEEE Trans. Biomed. Engin. 48, 55–65 (2001)
2. Berkovich, L.M., Tsirulik, V.G.: Differential resultants and some of their applications. Differ. Equat. 22, 750–757 (1986)
3. Boulier, F.: Étude et implantation de quelques algorithmes en algèbre différentielle. PhD Thesis, Université Lille I, 59655, Villeneuve d'Ascq, France (1994)
4. Boulier, F.: Differential elimination and biological modelling. Radon Series Comp. Appl. Math. 2, 111–139 (2007), `http://hal.archives-ouvertes.fr/hal-00139364`
5. Boulier, F., Lemaire, F., Sedoglavic, A., Ürgüplü, A.: Towards an automated reduction method for polynomial ODE models in cellular biology. Math. Comp. Sci. Special Issue on Symbolic Computation in Biology 2(3), 443–464 (2009)
6. Boulier, F., Lefranc, M., Lemaire, F., Morant, P.E.: Model Reduction of Chemical Reaction Systems using Elimination. Presented at the International Conference MACIS (2007), `http://hal.archives-ouvertes.fr/hal-00184558`
7. Carra'Ferro, G.: A resultant theory of systems of linear partial differential equations. Proc. Modern Group Anal. 1, 47–55 (1994)
8. Carra'Ferro, G.: A resultant theory for systems of two ordinary algebraic differential equations. Appl. Alg. Engin. Commun. Comp. 8, 539–560 (1997)
9. Carra'Ferro, G.: A resultant theory for ordinary algebraic differential equations. In: Mattson, H.F., Mora, T. (eds.) AAECC 1997. LNCS, vol. 1255, pp. 55–65. Springer, Heidelberg (1997)
10. Chardin, M.: Differential resultants and subresultants. In: Budach, L. (ed.) FCT 1991. LNCS, vol. 529, pp. 180–189. Springer, Heidelberg (1991)
11. Cox, D., Little, J., O'Shea, D.: Using Algebraic Geometry. Springer, New York (1998)
12. Denis-Vidal, L., Joly-Blanchard, G., Noiret, C.: System identifiability (symbolic computation) and parameter estimation (numerical computation). Numer. Algor. 34, 282–292 (2003)
13. Diop, S.: Elimination in Control Theory. Math. Contr. Sign. Syst. 4, 17–42 (1991)
14. Diop, S.: Differential algebraic decision methods and some application to system theory. Theor. Comp. Sci. 98, 137–161 (1992)
15. Diop, S., Michel, F.: Nonlinear observability, identifiability, and persistent trajectories. In: Proc. 30th CDC, Brighton, pp. 714–719 (1991)
16. Gao X S, Li W, Yuan C M (2010) Intersection theory for generic differential polynomials and differential Chow form. `arXiv:1009.0148v1`
17. Gelfand, I., Kapranov, M., Zelevinsky, A.: Discriminants, Resultants and Multidimensional Determinants. Birkhäuser, Basel (1994)
18. Gonzalez-Vega, L.: Une théorie des sous-résultants pour les polynômes en plusieurs variables. CR Acad. Sci. Paris Ser. I 313, 905–908 (1991)
19. Kasman, A., Previato, E.: Commutative partial differential operators. Physica D 152-153, 66–77 (2001)
20. Kolchin, E.R.: Differential Algebra and Algebraic Groups. Academic Press, London (1973)

21. Li, Z.: A subresultant theory for linear ordinary differential polynomials. RISC-Linz Techn. Report Series No. 95–35 (1995)
22. Ljung, L., Glad, S.T.: On global identifiability for arbitrary model parametrisations. Automatica 30, 265–276 (1994)
23. Macaulay, F.S.: The Algebraic Theory of Modular Systems. Proc. Camb. Univ. Press, Cambridge (1916)
24. Minimair, M.: MR: Macaulay Resultant Package for Maple (2005), http://www.minimair.org/mr
25. Ollivier, F.: Le problème de l'identifiabilité structurelle globale: approche théorique, méthodes effectives et bornes de complexité. PhD Thesis, École Polytechnique, Palaiseau, France (1990)
26. Ore, O.: Formale theorie der linearen differentialgleichungen. J. Reine. Angew. Math. 167, 221–234 (1932)
27. Ritt, J.F.: Differential Equations from the Algebraic Standpoint. Amer. Math. Soc., New York (1932)
28. Ritt, J.F.: Differential Algebra. Amer. Math. Soc. Colloquium (1950)
29. Rueda, S.L.: DiffRes: Differential Resultant Package for Maple (2008), http://www.aq.upm.es/Departamentos/Matematicas/srueda/srueda_archi-vos/DiffRes/DiffRes.htm
30. Rueda, S.L., Sendra, J.R.: Linear complete differential resultants and the implicitization of linear DPPEs. J. Symbol Comput. 45, 324–341 (2010)
31. van der Waerden, B.L.: Algebra, vol. 1. Springer, Heidelberg (2003)
32. van der Waerden, B.L.: Algebra, vol. II. Springer, Heidelberg (2003)
33. Walter, E.: Identifiability of State Space Models. Lect. Notes Biomath., vol. 46. Springer, Heidelberg (1982)

Part III

Distribution Theory and Applications

Convolution of Heterogeneous Bernoulli Random Variables and Some Applications*

Narayanaswamy Balakrishnan[1] and Maochao Xu[2]

[1] Department of Mathematics and Statistics, McMaster University, Hamilton, Ontario, Canada
bala@mcmaster.ca
[2] Department of Mathematics, Illinois State University, Normal, IL, USA
mxu2@ilstu.edu

Summary. In this paper, we review some recent results on stochastic comparisons of convolutions of heterogeneous Bernoulli random variables. We then highlight some new applications of these results in the context of statistical inference, reliability theory and software testing.

Keywords: Bernoulli variables, heterogeneous variables, likelihood ratio order, majorization, order statistics, reversed hazard rate order, inference, reliability theory, software testing

1 Introduction

Bernoulli distribution is one of the most fundamental distributions in statistics, and has found key applications in statistics, actuarial science, operation research and reliability theory. Models with independent heterogeneous Bernoulli variables come up in a wide array of applied problems. Specifically, let $X_{p_1}, \ldots, X_{p_n}$ be a sequence of independent Bernoulli random variables with parameters $p_1, \ldots, p_n$, respectively. It is well known that if $p_1 = \ldots = p_n = p$, then $\sum_{i=1}^{n} X_i$ is a Binomial(n, p) random variable. However, when p_i's are not all equal, there is no simple distributional form for the convolution of heterogeneous Bernoulli random variables.

In this paper, we will review some recent results on stochastic comparisons of convolutions of heterogeneous Bernoulli random variables. We will then highlight some new applications of these results in statistical inference, reliability theory and software testing. The rest of this paper is organized as follows. In Section 2, we describe all the preliminary notions and concepts that are pertinent to subsequent developments. In Section 3, we present the main results concerning stochastic comparisons of convolutions of heterogeneous Bernoulli variables. Finally, we detail in Section 4 some interesting

* It is our distinct honor to present this work to the memory of our beloved lost friend María Luisa Menéndez whose ever smiling face, kindness and generosity are deeply missed.

applications of these results in statistical inference, reliability theory and software testing.

2 Preliminaries

Let us first recall some notions of stochastic orders and majorization orders which are most pertinent to the main results detailed in the following section.

Definition 1. *For two discrete random variables X and Y with common support on integers $\mathbb{N}_0 = \{0, 1, 2, \ldots\}$, let $f(k)$ and $g(k)$ denote their respective probability mass functions, $F(k) = P(X \leq k)$ and $G(k) = P(Y \leq k)$ their respective distribution functions, and $\bar{F}(k) = P(X \geq k)$ and $\bar{G}(k) = P(Y \geq k)$ the corresponding survival functions. Then, X is said to be smaller than Y in the*

1. *likelihood ratio order, denoted by $X \leq_{lr} Y$, if $g(k)/f(k)$ is increasing in $k \in \mathbb{N}_0$;*
2. *hazard rate order, denoted by $X \leq_{hr} Y$, if $\bar{G}(k)/\bar{F}(k)$ is increasing in $k \in \mathbb{N}_0$;*
3. *reversed hazard rate order, denoted by $X \leq_{rh} Y$, if $G(k)/F(k)$ is increasing in $k \in \mathbb{N}_0$;*
4. *usual stochastic order, denoted by $X \leq_{st} Y$, if $\bar{F}(k) \leq \bar{G}(k)$ for all $k \in \mathbb{N}_0$.*

The following implications are well known in the literature:

$$X \leq_{lr} Y \Rightarrow X \leq_{hr(rh)} Y \Rightarrow X \leq_{st} Y.$$

For a comprehensive discussion on various stochastic orders and their applications, one may refer to the book by Shaked and Shanthikumar [16].

We shall also be using the concept of majorization in our discussion. Let $x_{(1)} \leq x_{(2)} \leq \cdots \leq x_{(n)}$ be the increasing arrangement of components of the vector $\mathbf{x} = (x_1, x_2, \ldots, x_n)$.

Definition 2. *For vectors $\mathbf{x}, \mathbf{y} \in \mathbb{R}^n$, $\mathbf{x}$ is said to be*

- *majorized by $\mathbf{y}$ (denoted by $\mathbf{x} \preceq_{\mathrm{m}} \mathbf{y}$) if*

$$\sum_{i=1}^{j} x_{(i)} \geq \sum_{i=1}^{j} y_{(i)}$$

for $j = 1, \ldots, n-1$, and $\sum_{i=1}^{n} x_{(i)} = \sum_{i=1}^{n} y_{(i)}$;

- *weakly supmajorized by $\mathbf{y}$ (denoted by $\mathbf{x} \overset{\mathrm{w}}{\preceq} \mathbf{y}$) if*

$$\sum_{i=1}^{j} x_{(i)} \geq \sum_{i=1}^{j} y_{(i)}$$

for $j = 1, \ldots, n$.

For extensive and comprehensive details on the theory of majorization orders and their applications, one may refer to the book by Marshall and Olkin [13].

3 Main Results

Convolutions of heterogeneous Bernoulli random variables have a long history, and was first considered by Hoeffding [7] who established the following result.

Theorem 1. *Let $X_{p_1}, \ldots, X_{p_n}$ be independent Bernoulli random variables with parameters $p_1, \ldots, p_n$, respectively. Then,*

$$P\left(\sum_{i=1}^{n} X_{p_i} \le k\right) \le \sum_{j=0}^{k} \binom{n}{j} \bar{p}^j (1-\bar{p})^{n-j} \quad \textit{for} \quad 0 \le k \le n\bar{p}-1,$$

and

$$P\left(\sum_{i=1}^{n} X_{p_i} \le k\right) \ge \sum_{j=0}^{k} \binom{n}{j} \bar{p}^j (1-\bar{p})^{n-j} \quad \textit{for} \quad n\bar{p} \le k \le n,$$

where $\bar{p} = \frac{1}{n} \sum_{i=1}^{n} p_i$.

As a consequence, it follows that, for any two integers b and c such that $0 \le b \le n\bar{p} \le c \le n$,

$$P\left(b \le \sum_{i=1}^{n} X_{p_i} \le c\right) \ge \sum_{j=b}^{c} \binom{n}{j} \bar{p}^j (1-\bar{p})^{n-j}.$$

Using the concept of majorization, Gleser [6] refined the result of by Hoeffding [7] as follows:

Theorem 2. *Let $X_{p_1}, \ldots, X_{p_n}$ be independent Bernoulli random variables with parameters $p_1, \ldots, p_n$, respectively. If*

$$(p_1, \ldots, p_n) \succeq_{\mathrm{m}} (p_1^*, \ldots, p_n^*), \tag{1}$$

then

$$P\left(\sum_{i=1}^{n} X_{p_i} \le k\right) \le P\left(\sum_{i=1}^{n} X_{p_i^*} \le k\right) \quad \textit{for} \quad 0 \le k \le n\bar{p}-2,$$

and

$$P\left(\sum_{i=1}^{n} X_{p_i} \le k\right) \ge P\left(\sum_{i=1}^{n} X_{p_i^*} \le k\right) \quad \textit{for} \quad n\bar{p}+2 \le k \le n.$$

This result was subsequently partly extended by Boland and Proschan [2] as follows.

Theorem 3. *Let $X_{p_1},\dots,X_{p_n}$ be independent Bernoulli random variables with parameters $p_1,\dots,p_n$, respectively. If condition (1) holds, then*

$$P\left(\sum_{i=1}^{n} X_{p_i} \le k\right) \le P\left(\sum_{i=1}^{n} X_{p_i^*} \le k\right) \quad \textit{for} \;\; p_i \ge \frac{k}{n-1},$$

and

$$P\left(\sum_{i=1}^{n} X_{p_i} \le k\right) \ge P\left(\sum_{i=1}^{n} X_{p_i^*} \le k\right) \quad \textit{for} \;\; p_i \le \frac{k}{n-1}.$$

In the context of reliability theory, Proschan and Sethuraman [15] proved the following interesting result:

Theorem 4. *Let $X_{p_1},\dots,X_{p_n}$ be independent Bernoulli random variables with parameters $p_1,\dots,p_n$, respectively. Then,*

$$(-\log(p_1),\dots,-\log(p_n)) \succeq_{\mathrm{m}} (-\log(p_1^*),\dots,-\log(p_n^*)) \Rightarrow \sum_{i=1}^{n} X_{p_i} \ge_{st} \sum_{i=1}^{n} X_{p_i^*} \tag{2}$$

and

$$\left(\frac{1-p_1}{p_1},\dots,\frac{1-p_n}{p_n}\right) \succeq_{\mathrm{m}} \left(\frac{1-p_1^*}{p_1^*},\dots,\frac{1-p_n^*}{p_n^*}\right) \Rightarrow \sum_{i=1}^{n} X_{p_i} \ge_{st} \sum_{i=1}^{n} X_{p_i^*}. \tag{3}$$

This result has been recently strengthened by Xu and Balakrishnan [21], who established the following result:

Theorem 5. *Let $X_{p_1},\dots,X_{p_n}$ be independent Bernoulli random variables with parameters $p_1,\dots,p_n$, respectively. Then,*

$$(-\log(p_1),\dots,-\log(p_n)) \overset{\mathrm{w}}{\succeq} (-\log(p_1^*),\dots,-\log(p_n^*)) \Rightarrow \sum_{i=1}^{n} X_{p_i} \ge_{rh} \sum_{i=1}^{n} X_{p_i^*} \tag{4}$$

and

$$\left(\frac{1-p_1}{p_1},\dots,\frac{1-p_n}{p_n}\right) \overset{\mathrm{w}}{\succeq} \left(\frac{1-p_1^*}{p_1^*},\dots,\frac{1-p_n^*}{p_n^*}\right) \Rightarrow \sum_{i=1}^{n} X_{p_i} \ge_{lr} \sum_{i=1}^{n} X_{p_i^*}. \tag{5}$$

Remark 1. Boland *et al.* [3], [4] studied the comparison of convolutions of heterogeneous and homogeneous Bernoulli random variables. Suppose $Y_1,\dots,Y_n$ are i.i.d. Bernoulli random variables with parameter p. They then showed that

$$\sum_{i=1}^{n} X_{p_i} \ge_{st} \sum_{i=1}^{n} Y_i \Longleftrightarrow p \le p_g \tag{6}$$

and

$$\sum_{i=1}^{n} X_{p_i} \le_{st} \sum_{i=1}^{n} Y_i \Longleftrightarrow p \ge p_{cg},$$

where $p_g = \sqrt[n]{\prod_{i=1}^{n} p_i}$ is the geometric mean of p_i's, and $p_{cg} = 1 - \sqrt[n]{\prod_{i=1}^{n}(1-p_i)}$ is the complement of the geometric mean of $(1-p_i)$'s. Furthermore, they proved that

$$\sum_{i=1}^{n} X_{p_i} \ge_{hr} \sum_{i=1}^{n} Y_i \Longleftrightarrow \sum_{i=1}^{n} X_{p_i} \ge_{lr} \sum_{i=1}^{n} Y_i \Longleftrightarrow p \le p_h \tag{7}$$

and

$$\sum_{i=1}^{n} X_{p_i} \le_{hr} \sum_{i=1}^{n} Y_i \Longleftrightarrow \sum_{i=1}^{n} X_{p_i} \le_{lr} \sum_{i=1}^{n} Y_i \Longleftrightarrow p \ge p_{ch},$$

where $p_h = \dfrac{n}{\sum_{i=1}^{n} \frac{1}{p_i}}$ is the harmonic mean of p_i's, and $p_{ch} = 1 - \dfrac{n}{\sum_{i=1}^{n} \frac{1}{1-p_i}}$ is the complement of the harmonic mean of $(1-p_i)$'s. It is easy to see that the results in (4) and (5) generalize the corresponding ones in (6) and (7). Interested readers may also refer to Boland [5] for an overview of various developments on this topic.

One of the interesting properties of a convolution of heterogeneous Bernoulli random variables is its variability. Wang [17] considered the variance of the convolution of independent Bernoulli random variables and proved the following result.

Theorem 6. *Let $X_{p_1}, \ldots, X_{p_n}$ be independent Bernoulli random variables with parameters $p_1, \ldots, p_n$, respectively. Then,*

$$(p_1, \ldots, p_n) \succeq_{\mathrm{m}} (p_1^*, \ldots, p_n^*) \Rightarrow \mathrm{Var}\left(\sum_{i=1}^{n} X_{p_i}\right) \le \mathrm{Var}\left(\sum_{i=1}^{n} X_{p_i^*}\right),$$

which means that the variance of the convolution increases as the components of $(p_1, \ldots, p_n)$ become *more* homogeneous. Actually, a stronger version of this result has been established by Karlin and Novikoff [10] by using convex transforms; see also Ma [12] for a general result.

Theorem 7. *Let $X_{p_1}, \ldots, X_{p_n}$ be independent Bernoulli random variables with parameters $p_1, \ldots, p_n$, respectively. Then,*

$$(p_1, \ldots, p_n) \succeq_{\mathrm{m}} (p_1^*, \ldots, p_n^*) \Rightarrow \mathrm{E}\phi\left(\sum_{i=1}^{n} X_{p_i}\right) \le \mathrm{E}\phi\left(\sum_{i=1}^{n} X_{p_i^*}\right),$$

where ϕ is any convex function.

One may also refer to Hu and Ruan [9] for a multivariate extension of this result.

4 Applications

In this section, we will highlight some interesting applications of Theorem 5 in statistical inference, reliability theory and software testing.

4.1 Statistical Inference

Suppose a person bought n stocks, and for each stock the probability of a loss is p_i, $i = 1, \ldots, n$. In practice, p_i's will not all be equal, for $i = 1, \ldots, n$. The person would then wish to know whether the average probability of a loss $\bar{p} = \frac{1}{n}\sum_{i=1}^n p_i$ is less than or equal to some fixed personal risk level p_0 (a natural choice is, of course, $p_0 < 0.5$). Equivalently, we may express it as a hypothesis testing problem

$$H_0 : \bar{p} = p_0 \quad \text{vs.} \quad H_a : \bar{p} < p_0.$$

Then, as described in Hoeffding [7], Theorem 3 is directly applicable in this case. However, if the individual is confident that the average probability of a loss is at most p^*, then we may use Theorem 5 to arrive at the inequality

$$P\left(\sum_{i=1}^n X_i \geq t \,\middle|\, \sum_{i=1}^n X_i \leq np^*\right) \geq P\left(\sum_{i=1}^n Y_i \geq t \,\middle|\, \sum_{i=1}^n Y_i \leq np^*\right),$$

where Y_i is a Binomial(n, p) random variable, for $p \leq p_g$, the geometric mean of p_i's; equivalently, we have

$$P\left(\sum_{i=1}^n X_i \geq t \,\middle|\, \sum_{i=1}^n X_i \leq np^*\right) \geq \frac{\sum_{x=t^*}^{n^*} \binom{n}{x} p^x (1-p)^{n-x}}{\sum_{x=0}^{n^*} \binom{n}{x} p^x (1-p)^{n-x}},$$

where $t^* = \lceil t \rceil$ and $n^* = \lfloor np^* \rfloor$.

4.2 Reliability Theory

In reliability theory, a system of n components is said to be a k-out-of-n system if it functions as long as at least k components work. Let $X_1, \ldots, X_n$ be non-negative random variables representing the lifetimes of the components, and let $X_{1:n} \leq X_{2:n} \leq \ldots \leq X_{n:n}$ denote ordered lifetimes of these components. Then, evidently $X_{n-k+1:n}$ corresponds to the lifetime of a k-out-of-n system. Now, for $i = 1, \ldots, n$, let

$$Y_{p_i} = \begin{cases} 1, & X_i > t, \\ 0, & \text{otherwise}, \end{cases}$$

where

$$P(Y_{p_i} = 1) = P(X_i > t) = p_i.$$

The lifetime of the k-out-of-n system could then be represented as

$$P(X_{n-k+1:n} > t) = P\left(\sum_{i=1}^{n} Y_{p_i} \geq k\right).$$

By using Theorem 5 and the fact that the likelihood ratio order implies the hazard rate order, the following result then holds.

Theorem 8. *Let $X_1, \ldots, X_n$ be independent non-negative random variables with survival functions $\bar{F}_1(t), \ldots, \bar{F}_n(t)$, and $X_1^*, \ldots, X_n^*$ be another set of independent non-negative random variables with survival functions $\bar{F}_1^*(t), \ldots, \bar{F}_n^*(t)$. If*

$$\left(\frac{1}{\bar{F}_1(t)}, \ldots, \frac{1}{\bar{F}_n(t)}\right) \overset{\mathrm{w}}{\succeq} \left(\frac{1}{\bar{F}_1^*(t)}, \ldots, \frac{1}{\bar{F}_n^*(t)}\right),$$

then, for any $1 \leq m \leq k$,

$$P\left(X_{n-k+1:n} > t \,|X_{n-m+1:n} > t\right) \geq P\left(X_{n-k+1:n}^* > t \,\middle|X_{n-m+1:n}^* > t\right).$$

Remark 2. Properties of conditional order statistics have been studied extensively in the literature; one may refer to Kochar and Xu [11] and Balakrishnan *et al.* [1] and the references therein for further details. The result in Theorem 8 gives a new insight into the conditions for stochastically comparing the lifetimes of two k-out-of-n systems with heterogeneous components.

Remark 3. Theorem 8 also has an interesting application in estimating the probability associated with number of surviving components. For example, in the case of a parallel system with n components, if the system is still working at the end of the experiment, then we may be interested in the number of surviving components. Theorem 8 would provide lower bounds for associated probabilities in this case.

Since

$$R_i(t) = -\log p_i,$$

where $R_i(t)$ is the cumulative hazard function, by Theorem 5, we readily obtain the following result.

Theorem 9. *Let $X_1, \ldots, X_n$ be independent non-negative random variables with survival functions $\bar{F}_1(t), \ldots, \bar{F}_n(t)$, and $X_1^*, \ldots, X_n^*$ be another set of independent non-negative random variables with survival functions $\bar{F}_1^*(t), \ldots, \bar{F}_n^*(t)$. If*

$$(R_1(t), \ldots, R_n(t)) \overset{\mathrm{w}}{\succeq} (R_1^*(t), \ldots, R_n^*(t)),$$

where $R_i(t) = -\log \bar{F}_i(t)$, then, for any $1 \leq k < l \leq n$,

$$P\left(X_{n-k+1:n} > t \,|X_{n-l+1:n} \leq t\right) \geq P\left(X_{n-k+1:n}^* > t \,\middle|X_{n-l+1:n}^* \leq t\right).$$

Remark 4. The condition in Theorem 9 has also been used by Pledger and Proschan [14] and Hu [8] for the stochastic comparison of two k-out-of-n systems, but the result obtained here is a stronger one.

Let us now consider the special case of proportional hazard rates model, which is commonly used in survival and reliability analyses. Under this model, the cumulative hazard rates may be expressed as

$$R_i(t) = \lambda_i R(t), \quad R_i(t) = \lambda_i^* R(t),$$

where $R(t)$ is some baseline cumulative hazard function. Then, the following result holds.

Corollary 1. *Let X_i $(i = 1, \ldots, n)$ be independent non-negative random variables with survival functions $\bar{F}_i(t) = e^{-\lambda_i R(t)}$, and X_i^* $(i = 1, \ldots, n)$ be another set of independent non-negative random variables with survival functions $\bar{F}_i^*(t) = e^{-\lambda_i^* R(t)}$. Then, if*

$$(\lambda_1, \ldots, \lambda_n) \overset{\text{w}}{\succeq} (\lambda_1^*, \ldots, \lambda_n^*),$$

we have, for any $1 \le k < l \le n$,

$$P\left(X_{n-k+1:n} > t \,|X_{n-l+1:n} \le t\right) \ge P\left(X^*_{n-k+1:n} > t \,\middle|X^*_{n-l+1:n} \le t\right).$$

4.3 Software Testing

While testing softwares for faults, it would be impractical to test all possible inputs, which necessitates the selection of a sample for testing purposes. Two common sampling schemes are stratified and simple random sampling. It has been shown by Boland *et al.* [3], [4], that, under certain conditions, the number of faults discovered under stratified sampling is greater than that under random sampling in terms of various stochastic orders. In this section, we shall use our results from the preceding section to provide some further discussion in this direction.

Assume that the testing domain of the software is divided into k non-overlapping subdomains. Let d_i denote the size of the ith subdomain and F_i denote the number of failure causing inputs within the ith subdomain. Then,

$$\theta_i = \frac{F_i}{d_i}$$

corresponds to the failure rate in the ith subdomain $(i = 1, \ldots, k)$. Let X_i be the number of faults found in a random sample of size n_i taken with replacement from the ith subdomain, for $i = 1, \ldots, k$. Evidently, X_i is a Binomial(n_i, θ_i) random variable, for $i = 1, \ldots, k$, and then $X = \sum_{i=1}^k X_i$ is the total number of faults found under the stratified sampling. Next, let Y denote the number of faults found in a random sample of size $n = \sum_{i=1}^k n_i$

taken with replacement from the entire input domain. In this situation, the distribution of Y is clearly Binomial(n,θ).

Then, Boland *et al.* [3] showed that

$$X \geq_{lr} Y \Longleftrightarrow X \geq_{hr} Y \Longleftrightarrow \theta \leq \bar{\theta}_h, \tag{8}$$

$$X \leq_{lr} Y \Longleftrightarrow X \leq_{hr} Y \Longleftrightarrow \theta \geq \bar{\theta}_{ch},$$

where $\bar{\theta}_h = \dfrac{n}{\sum_{i=1}^k \frac{n_i}{\theta_i}}$ and $\bar{\theta}_{ch} = 1 - \dfrac{n}{\sum_{i=1}^k \frac{n_i}{1-\theta_i}}$. They also established that

$$X \geq_{st} Y \Longleftrightarrow \theta \leq \bar{\theta}_g, \tag{9}$$

$$X \leq_{st} Y \Longleftrightarrow \theta \geq \bar{\theta}_{cg},$$

where $\bar{\theta}_g = \left(\prod_{i=1}^k \theta_i^{n_i}\right)^{1/n}$ and $\bar{\theta}_{cg} = 1 - \left(\prod_{i=1}^k (1-\theta_i)^{n_i}\right)^{1/n}$.

Using Theorem 5, we can establish the following two theorems, which generalize and strengthen the results in (8) and (9) due to Boland *et al.* [3].

Theorem 10. *Let X_i be Binomial(n_i,θ_i) random variables and X_i^* be another set of independent Binomial(n_i^*,θ_i^*) random variables, for $i=1,\dots,k$. If there exists some permutation π to make the vectors $\mathbf{n} = (n_1,\dots,n_k)$, $\mathbf{n}^* = (n_1^*,\dots,n_k^*)$, $\boldsymbol{\theta} = (\theta_1,\dots,\theta_k)$ and $\boldsymbol{\theta}^* = (\theta_1^*,\dots,\theta_k^*)$ all to be in the descending order, then*

$$\left(\frac{1}{\theta_1},\dots,\frac{1}{\theta_k}\right) \stackrel{\text{w}}{\succeq} \left(\frac{1}{\theta_1^*},\dots,\frac{1}{\theta_k^*}\right)$$

and

$$(n_1,\dots,n_k) \succeq_{\text{m}} (n_1^*,\dots,n_k^*)$$

imply

$$X \geq_{lr} X^*,$$

where $X = \sum_{i=1}^k X_i$ as before and $X^* = \sum_{i=1}^k X_i^*$.

Proof. Assume that $n_1 \geq n_2 \geq \ldots \geq n_k$ ($n_1^* \geq n_2^* \geq \ldots \geq n_k^*$). Since

$$\left(\frac{1}{\theta_1},\dots,\frac{1}{\theta_k}\right) \stackrel{\text{w}}{\succeq} \left(\frac{1}{\theta_1^*},\dots,\frac{1}{\theta_k^*}\right)$$

implies

$$\Bigg(\underbrace{\frac{1}{\theta_1},\dots,\frac{1}{\theta_1}}_{n_1},\dots,\underbrace{\frac{1}{\theta_k},\dots,\frac{1}{\theta_k}}_{n_k}\Bigg) \stackrel{\text{w}}{\succeq} \Bigg(\underbrace{\frac{1}{\theta_1^*},\dots,\frac{1}{\theta_1^*}}_{n_1},\dots,\underbrace{\frac{1}{\theta_k^*},\dots,\frac{1}{\theta_k^*}}_{n_k}\Bigg),$$

for $\theta_1 \geq \theta_2 \geq \ldots \geq \theta_n$ ($\theta_1^* \geq \theta_2^* \geq \ldots \geq \theta_n^*$) (see Lemma 2.2 of Zhao and Balakrishnan [19]), and

$$(n_1, \ldots, n_k) \succeq_{\mathrm{m}} (n_1^*, \ldots, n_k^*)$$

implies

$$\left(\underbrace{\frac{1}{\theta_1^*}, \ldots, \frac{1}{\theta_1^*}}_{n_1}, \ldots, \underbrace{\frac{1}{\theta_k^*}, \ldots, \frac{1}{\theta_k^*}}_{n_k}\right) \overset{\mathrm{w}}{\succeq} \left(\underbrace{\frac{1}{\theta_1^*}, \ldots, \frac{1}{\theta_1^*}}_{n_1^*}, \ldots, \underbrace{\frac{1}{\theta_k^*}, \ldots, \frac{1}{\theta_k^*}}_{n_k^*}\right),$$

by Lemma 2.5 of Zhao and Balakrishnan [19], the required result follows from Theorem 5. □

Using similar arguments, the following result can also be established.

Theorem 11. *Let X_i be a set of independent Binomial(n_i, θ_i) random variables and X_i^* be another set of independent Binomial(n_i^*, θ_i^*) random variables, for $i = 1, \ldots, k$. If there exists some permutation π to make the vectors $\mathbf{n} = (n_1, \ldots, n_k)$, $\mathbf{n}^* = (n_1^*, \ldots, n_k^*)$, $\boldsymbol{\theta} = (\theta_1, \ldots, \theta_k)$ and $\boldsymbol{\theta}^* = (\theta_1^*, \ldots, \theta_k^*)$ all to be in the descending order, then*

$$(-\log(\theta_1), \ldots, -\log(\theta_k)) \overset{\mathrm{w}}{\succeq} (-\log(\theta_1^*), \ldots, -\log(\theta_k^*))$$

and

$$(n_1, \ldots, n_k) \succeq_{\mathrm{m}} (n_1^*, \ldots, n_k^*)$$

imply

$$X \geq_{rh} X^*.$$

We shall now present some examples to illustrate these results.

Example 1. Suppose there exist different strata with parameters as listed in Table 1.

Table 1. Parameters of different strata for Example 1

	(n_1, n_2, n_3, n_4)	$(\theta_1, \theta_2, \theta_3, \theta_4)$
X	(20, 30, 40, 50)	(0.05, 0.10, 0.20, 0.25)
Y	(25, 25, 40, 50)	(0.05, 0.10, 0.10, 0.20)
Z	(35, 35, 35, 35)	(0.09, 0.09, 0.09, 0.09)

Observe that

$$\left(\frac{1}{0.05}, \frac{1}{0.10}, \frac{1}{0.20}, \frac{1}{0.25}\right) \overset{\mathrm{w}}{\succeq} \left(\frac{1}{0.05}, \frac{1}{0.10}, \frac{1}{0.10}, \frac{1}{0.20}\right)$$

and

$$(20, 30, 40, 50) \succeq_{\mathrm{m}} (25, 50, 25, 40).$$

Then, from Theorem 1, it follows that

$$X \geq_{lr} Y.$$

Similarly, since

$$\left(\frac{1}{0.05}, \frac{1}{0.10}, \frac{1}{0.10}, \frac{1}{0.20}\right) \overset{\mathrm{w}}{\succeq} \left(\frac{1}{0.09}, \frac{1}{0.09}, \frac{1}{0.09}, \frac{1}{0.09}\right)$$

and

$$(25, 25, 40, 50) \succeq_{\mathrm{m}} (35, 35, 35, 35),$$

we also have from Theorem 1 that

$$Y \geq_{lr} Z.$$

Example 2. Suppose there exist different strata with parameters as listed in Table 2. For convenience, we shall use the same values of n's as given in Table 1, but with difference θ's.

Table 2. Parameters of different strata for Example 2

	(n_1, n_2, n_3, n_4)	$(\theta_1, \theta_2, \theta_3, \theta_4)$
X	(20, 30, 40, 50)	$(0.05, 0.10, 0.20, 0.25)$
Y	(25, 25, 40, 50)	$(0.05, 0.125, 0.20, 0.20)$
Z	(35, 35, 35, 35)	$(0.125, 0.125, 0.125, 0.125)$

Since

$$\begin{aligned}&(-\log(0.05), -\log(0.10), -\log(0.20), -\log(0.25))\\ &\quad \overset{\mathrm{w}}{\succeq} (-\log(0.05), -\log(0.125), -\log(0.20), -\log(0.20)),\end{aligned}$$

by Theorem 11, it follows that

$$X \geq_{rh} Y.$$

Similarly, since

$$\begin{aligned}&(-\log(0.05), -\log(0.125), -\log(0.20), -\log(0.20))\\ &\quad \overset{\mathrm{w}}{\succeq} (-\log(0.125), -\log(0.125), -\log(0.125), -\log(0.125)),\end{aligned}$$

Theorem 11 readily implies that

$$Y \geq_{rh} Z.$$

References

1. Balakrishnan, N., Belzunce, F., Hami, N., Khaledi, B.-H.: Univariate and multivariate likelihood ratio ordering of generalized order statistics and associated conditional variables. Probab. Engin. Inform. Sci. 24, 441–455 (2010)
2. Boland, P.J., Proschan, F.: The reliability of k-out-of-n systems. Ann. Probab. 11, 760–764 (1983)
3. Boland, P.J., Singh, H., Cukic, B.: Stochastic orders in partition and random testing of software. J. Appl. Probab. 39, 555–565 (2002)
4. Boland, P.J., Singh, H., Cukic, B.: The stochastic precedence ordering with applications in sampling and testing. J. Appl. Probab. 41, 73–82 (2004)
5. Boland, P.J.: The probability distribution for the number of successes in independent trials. Comm. Statist.-Theor. Meth. 36, 1327–1331 (2007)
6. Gleser, L.: On the distribution of the number of successes in independent trials. Ann. Probab. 3, 182–188 (1975)
7. Hoeffding, W.: On the distribution of the number of successes in independent trials. Ann. Math. Statistics 27, 713–721 (1956)
8. Hu, T.: Monotone coupling and stochastic ordering of order statistics. Syst. Sci. & Math. Sci (English Series) 8, 209–214 (1995)
9. Hu, T., Ruan, L.: A note on multivariate stochastic comparisons of Bernoulli random variables. J. Statist. Plann. Infer. 126, 281–288 (2004)
10. Karlin, S., Novikoff, A.: Generalized convex inequalities. Pacif. J. Math. 13, 1251–1279 (1963)
11. Kochar, S., Xu, M.: On residual lifetimes of k-out-of-n systems with nonidentical components. Probab. Engin. Inform. Sci. 24, 109–127 (2010)
12. Ma, C.: Convex orders for linear combinations of random variables. J. Statist. Plann. Infer. 84, 11–25 (2000)
13. Marshall, A., Olkin, I.: Inequalities: Theory of Majorization and Its Applications. Academic Press, New York (1979)
14. Pledger, G., Proschan, F.: Comparisons of order statistics and of spacings from heterogeneous distributions. In: Rustagi, J.S. (ed.) Optimizing Methods in Statistics, pp. 89–113. Academic Press, New York (1971)
15. Proschan, F., Sethuraman, J.: Stochastic comparisons of order statistics from heterogeneous populations, with applications in reliability. J. Multivar. Anal. 6(4), 608–616 (1976)
16. Shaked, M., Shanthikumar, J.G.: Stochastic Orders and Their Applications. Springer, New York (2007)
17. Wang, Y.H.: On the number of successes in independent trials. Statist. Sinica 3, 295–312 (1993)
18. Xu, M., Balakrishnan, N.: On the convolution of heterogeneous Bernoulli random variables. Tech. Rep. Illinois State University, Normal, Illinois (2011)
19. Zhao, P., Balakrishnan, N.: Ordering properties of convolutions of heterogeneous Erlang and Pascal random variables. Statist. Probab. Lett. 80, 969–974 (2010)

The Effect of Non-normality in the Power Exponential Distributions

Miguel A. Gómez-Villegas[1], Eusebio Gómez-Sánchez-Manzano[1], Paloma Maín[1], and Hilario Navarro[2]

[1] Fac. de C.C. Matemáticas, Dpto. de Estadística e I.O., Universidad Complutense de Madrid, Pza. de Ciencias 3, 28040-Madrid, Spain
ma_gv@mat.ucm.es, eusebio_gomez@mat.ucm.es, pmain@mat.ucm.es
[2] Fac. de Ciencias, Dpto. de Estadística, I.O. y Calc. Num., UNED, Paseo Senda del Rey 9, 28040-Madrid, Spain
hnavarro@ccia.uned.es

Summary. As an alternative to the multivariate normal distribution we have dealt with a wider class of distributions, including the normal, that considers slightly different tail behavior than the normal tail. This is the multivariate exponential power family of distributions with a kurtosis parameter to give the possible forms of the distributions. To measure distribution deviations the Kullback-Leibler divergence will be used as an asymmetric dissimilarity measure from an information-theoretic basis. Thus, a local quantitative description of the non-normality could be established for joint distributions in this family as well as the impact this perturbation causes in the marginal and conditional distributions.

1 Introduction

The multivariate normal distribution is traditionally used as a model for multivariate data in applications. However, this assumption may be doubtful in many real data analysis and it demands a wider class of distributions than the normal to be handled. Our choice is the multivariate exponential power family of distributions presented in [5] as a generalization of the multivariate normal family in that a new parameter, β, is introduced, as an exponent (see (1) below), which governs the kurtosis, and so the sharpness, of the distribution; for $\beta = 1$ we have the normal distribution, thus this parameter represents the disparity of an exponential power distribution from the normal distribution.

The multivariate exponential power family is also a generalization of the univariate one (see [15] and [1, p. 157]) and can be included in the class of Kotz type distributions (see [4, p. 69] and [13]), which, in its turn, is a subset of the more general class of elliptical distributions (see a survey on these in [7]) Also, a matrix generalization of the exponential power distribution can be found in [6].

This distribution can be used to modelize multidimensional random phenomena with distributions having higher or lower tails than those of the normal distribution. Besides, the use of this distribution can robustify many multivariate statistical procedures. The multivariate exponential power distribution has been used to obtain robust models for nonlinear repeated measurements [10], to modeling dependencies among responses, as an alternative to models based upon the multivariate t distribution, and also to obtain robust models for the physiology of breathing. [2] use the multivariate exponential power distribution, as a heavy tailed distribution, in the field of speech recognition.

In this paper we evaluate the effect of this source of non-normality on the joint distributions and the corresponding marginal and conditional distributions for a specific partition. To measure distribution deviations, the Kullback-Leibler (KL) divergence will be used as an asymmetric dissimilarity measure from an information-theoretic basis. Thus, a local quantitative description of the non-normality can be established for joint distributions in this family as well as the impact this perturbation causes in the marginal and conditional distributions. This approach could be useful in problems where, given a model for the joint distribution, the interest is focussed in the distribution of a subset of variables given some values of the remaining ones. Such situations occur, among others, when we deal with Gaussian Bayesian networks for which the output is the conditional distribution of the variables of interest given fixed values of the evidential variables and a sensitivity analysis to non-normality is performed to prove the robustness and accuracy of the inferences.

The paper is organized as follows. In Section 2 the multivariate exponential power family is presented, highlighting some probabilistic characteristics to be handled in later sections. Section 3 is devoted to describe the impact of non-normality on the probabilistic structures of a random vector. The paper ends with conclusions in Section 4.

2 On the Multivariate Exponential Power Distributions

Next, we summarize the most important features of this family of distributions. An absolutely continuous random vector $\mathbf{X} = (X_1, \ldots, X_n)'$ is said to have a power exponential distribution if its density has the form

$$f(\mathbf{x}; \boldsymbol{\mu}, \boldsymbol{\Sigma}, \beta) = k \left|\boldsymbol{\Sigma}\right|^{-\frac{1}{2}} \exp\left\{-\frac{1}{2}\left((\mathbf{x}-\boldsymbol{\mu})' \boldsymbol{\Sigma}^{-1} (\mathbf{x}-\boldsymbol{\mu})\right)^{\beta}\right\}, \qquad (1)$$

with $k = \frac{n\Gamma\left(\frac{n}{2}\right)}{\pi^{\frac{n}{2}}\Gamma\left(1+\frac{n}{2\beta}\right)2^{1+\frac{n}{2\beta}}}$, where $(\boldsymbol{\mu}, \boldsymbol{\Sigma}, \beta) \in (\mathbb{R}^n, S, (0,\infty))$, S being the set of $(n \times n)$ positive definite symmetric matrices, then, we write $\mathbf{X} \sim EP_n(\boldsymbol{\mu}, \boldsymbol{\Sigma}, \beta)$. The parameters $\boldsymbol{\mu}$ and $\boldsymbol{\Sigma}$ are location and scale parameters. The parameter β is a shape parameter, as the kurtosis depends only on it. Figures 1-3 show the graphs of the density $EP_2(\mathbf{0}, \mathbf{I}_2, \beta)$ for the values 6, 1, 1/2 of β.

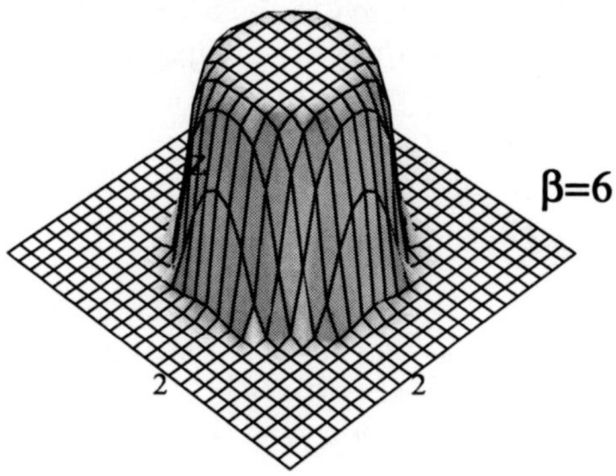

Fig. 1. EP_2 density for $\beta = 6$

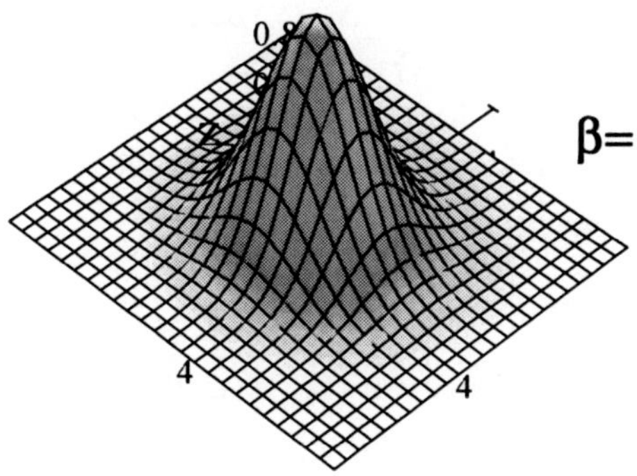

Fig. 2. Multivariate Normal density function, $\beta = 1$

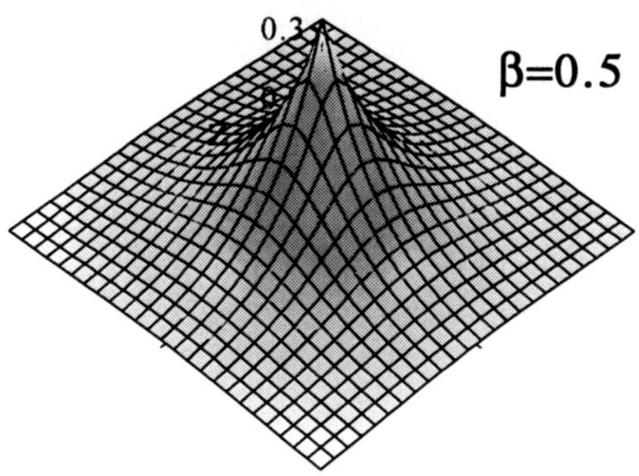

Fig. 3. Multivariate Double Exponential density, $\beta = \frac{1}{2}$

It can be pointed that as β increases the sharpness diminishes; for β going to infinity, (1) tends to be uniform in the ellipsoid $(\mathbf{x}-\boldsymbol{\mu})'\,\boldsymbol{\Sigma}^{-1}\,(\mathbf{x}-\boldsymbol{\mu})$ and also when β goes to 0 the pick narrows infinitely and (1) tends to the improper density constant in $\mathbb{R}^n$.

2.1 Some Related Distributions

Let $\mathbf{X}\sim EP_n\left(\boldsymbol{\mu},\boldsymbol{\Sigma},\beta\right)$. If $\beta=1$, then $\mathbf{X}$ has a normal distribution: $\mathbf{X}\sim N_n\left(\boldsymbol{\mu},\boldsymbol{\Sigma}\right)$. In any case, $\mathbf{X}$ has an elliptical distribution: $\mathbf{X}\sim E_n(\boldsymbol{\mu},\boldsymbol{\Sigma},g)$ (in the sense given in [7]) with $g\left(t\right)=\exp\left\{-\frac{1}{2}t^{\beta}\right\}$.

An exponential power distribution $EP_n\left(\boldsymbol{\mu},\boldsymbol{\Sigma},\beta\right)$ is a scale mixture of normal distributions (see [8]) in the strict sense (namely, with respect to a probability distribution function) if $\beta\in(0,1]$. If we exclude the normal case, that is, if $\beta\in(0,1)$, then

$$f(\mathbf{x};\boldsymbol{\mu},\boldsymbol{\Sigma},\beta)=\int_0^{\infty}N_n\left(\mathbf{x};\boldsymbol{\mu},v^2\boldsymbol{\Sigma}\right)\,dH_\beta(v), \tag{2}$$

where $N_n\left(\mathbf{x};\boldsymbol{\mu},v^2\boldsymbol{\Sigma}\right)$ is the normal density with mean $\boldsymbol{\mu}$ and covariance matrix $v^2\boldsymbol{\Sigma}$, and H_β is the distribution function having density

$$h_\beta(v)=\frac{2^{1+\frac{n}{2}-\frac{n}{2\beta}}\Gamma\left(1+\frac{n}{2}\right)}{\Gamma\left(1+\frac{n}{2\beta}\right)}v^{n-3}S_\beta\left(v^{-2};2^{1-\frac{1}{\beta}}\right),$$

where $S_\beta\left(\,\cdot\,;\sigma\right)$ means the density of the (positive) stable distribution having characteristic function (see [14, p. 8])

$$\varphi(t)=\exp\left\{-\sigma^{\beta}\left|t\right|^{\beta}e^{-i\frac{\pi}{2}\beta sign(t)}\right\}.$$

For $\beta=1$ (the normal case) (2) holds, of course, H_β being the distribution function degenerate in 1. For $\beta\in(1,\infty)$, the exponential power distribution $EP_n\left(\boldsymbol{\mu},\boldsymbol{\Sigma},\beta\right)$ is a scale mixture of normal distributions too, as all the elliptical distributions are (see [3]), but only in a wider sense, since in this case function H_β in (2) is like a distribution function in $(0,\infty)$, but it is not a nondecreasing function.

2.2 Probabilistic Characteristics

If $\mathbf{X}\sim EP_n\left(\boldsymbol{\mu},\boldsymbol{\Sigma},\beta\right)$, its characteristic function is

$$\varphi_{\mathbf{X}}(\mathbf{t})=\frac{n}{\Gamma\left(1+\frac{n}{2\beta}\right)2^{\frac{n}{2\beta}}}\exp\left(i\mathbf{t}'\boldsymbol{\mu}\right)\int_0^{\infty}\Psi_n\left(r\sqrt{\mathbf{t}'\boldsymbol{\Sigma}\mathbf{t}}\right)r^{n-1}\exp\left\{-\frac{1}{2}r^{2\beta}\right\}dr,$$

where $\Psi_1(x)=\cos x$ and $\Psi_n(x)=\frac{\Gamma(\frac{n}{2})}{\pi^{\frac{1}{2}}\Gamma(\frac{n-1}{2})}\int_0^{\pi}\exp\left\{ix\cos\theta\right\}\sin^{n-2}\theta d\theta$, for $n>1$. Besides,

$$E[\mathbf{X}] = \boldsymbol{\mu},$$

$$Var[\mathbf{X}] = \frac{2^{\frac{1}{\beta}}\Gamma\left(\frac{n+2}{2\beta}\right)}{n\Gamma\left(\frac{n}{2\beta}\right)}\boldsymbol{\Sigma},$$

$$\gamma_1[\mathbf{X}] = 0,$$

$$\gamma_2[\mathbf{X}] = n^2\frac{\Gamma\left(\frac{n+4}{2\beta}\right)\Gamma\left(\frac{n}{2\beta}\right)}{\left(\Gamma\left(\frac{n+2}{2\beta}\right)\right)^2} - n(n-2),$$

where γ_1 and γ_2 are the asymmetry and kurtosis coefficients as shown in [12, p. 31].

Figure 4 shows the kurtosis coefficient as a function of β for $n = 1$ (dotted line), 2, 3, 5 and 7, supporting the previous comments about the monotony relation between kurtosis and the non-normality coefficient in this family.

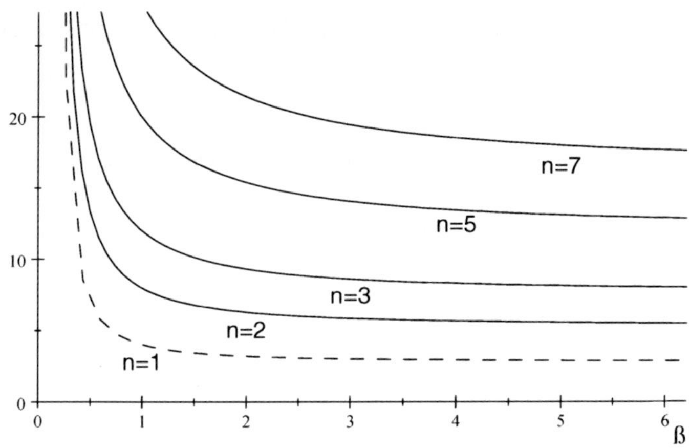

Fig. 4. Kurtosis coefficient as a function of β

2.3 Marginal and Conditional Distributions and Regression

The marginal and conditional distributions are elliptical. But the regression function is linear, as in the normal case. Specifically, let $\mathbf{X} \sim EP_n(\boldsymbol{\mu}, \boldsymbol{\Sigma}, \beta)$ and make $\mathbf{X} = (\mathbf{X}'_{(1)}, \mathbf{X}'_{(2)})'$, with $\mathbf{X}_{(1)} = (X_1, \ldots, X_p)'$ and $\mathbf{X}_{(2)} = (X_{p+1}, \ldots, X_n)'$, with $p < n$; analogously make $\boldsymbol{\mu} = (\boldsymbol{\mu}'_{(1)}, \boldsymbol{\mu}'_{(2)})'$ and $\boldsymbol{\Sigma} = \begin{pmatrix} \boldsymbol{\Sigma}_{11} & \boldsymbol{\Sigma}_{12} \\ \boldsymbol{\Sigma}_{21} & \boldsymbol{\Sigma}_{22} \end{pmatrix}$, where $\boldsymbol{\Sigma}_{11}$ is a $(p \times p)$ matrix. Then $\mathbf{X}_{(1)}$ has an elliptical distribution: $\mathbf{X}_{(1)} \sim E_p(\boldsymbol{\mu}_{(1)}, \boldsymbol{\Sigma}_{11}, g_{(1)})$, where

$$g_{(1)}(t) = \int_0^\infty w^{\frac{n-p}{2}-1} \exp\left\{-\frac{1}{2}(t+w)^\beta\right\} dw.$$

The distribution of $\mathbf{X}_{(2)}$ conditional to $\mathbf{X}_{(1)} = \mathbf{x}_{(1)}$ is elliptical too. $\left(\mathbf{X}_{(2)} \middle| \mathbf{X}_{(1)} = x_{(1)}\right) \sim E_{n-p}\left(\boldsymbol{\mu}_{(2.1)}, \boldsymbol{\Sigma}_{22.1}, g_{(2.1)}\right)$, with

$$\begin{aligned}
\boldsymbol{\mu}_{(2.1)} &= \boldsymbol{\mu}_{(2)} + \boldsymbol{\Sigma}_{21}\boldsymbol{\Sigma}_{11}^{-1}\left(\mathbf{x}_{(1)} - \boldsymbol{\mu}_{(1)}\right), \\
\boldsymbol{\Sigma}_{22.1} &= \boldsymbol{\Sigma}_{22} - \boldsymbol{\Sigma}_{21}\boldsymbol{\Sigma}_{11}^{-1}\boldsymbol{\Sigma}_{12}, \\
g_{(2.1)}(t) &= \exp\left\{-\frac{1}{2}\left(t + \mathbf{q}_{(1)}\right)^\beta\right\},
\end{aligned}$$

where $\mathbf{q}_{(1)} = \left(\mathbf{x}_{(1)} - \boldsymbol{\mu}_{(1)}\right)' \boldsymbol{\Sigma}_{11}^{-1} \left(\mathbf{x}_{(1)} - \boldsymbol{\mu}_{(1)}\right)$.

3 The Effects of Deviations from Normality

Now, we are interested in the effects of small changes in the parameter β of the $EP_n(\boldsymbol{\mu}, \boldsymbol{\Sigma}, \beta)$ distribution taking as a reference the one with $\beta_0 = 1$, that, as pointed above, corresponds to a normal distribution with parameters $\boldsymbol{\mu}$ and $\boldsymbol{\Sigma}$. When β is close to $\beta_0 = 1$, that is, $\beta = \beta_0 + \delta$ with δ representing a small deviation from normality, the Taylor expansion leads to the approximation

$$D_{KL}(f, f^{(\delta)}) \approx \frac{1}{2} F_\beta(1)\delta^2 , \tag{3}$$

being f the normal density, $f^{(\delta)}$ the perturbed density and $F_\beta(1)$ the Fisher information with respect to β in $\beta_0 = 1$. The same problem can be formulated in terms of the marginal and conditional distributions for a fixed partition of the random vector $\mathbf{X}$. From now on, our goal is both analytical and graphical description of the function (3).

3.1 Joint Distributions

Let $f(\mathbf{x})$ be a density function of the family $EP_n(\boldsymbol{\mu}, \boldsymbol{\Sigma}, \beta_0 = 1)$, that is a normal density $N_n(\boldsymbol{\mu}, \boldsymbol{\Sigma})$ and $f^{(\delta)}(\mathbf{x})$ be the perturbed density $EP_n(\boldsymbol{\mu}, \boldsymbol{\Sigma}, \beta = 1 + \delta)$, then the KL divergence between these densities can be calculated using that, if $\mathbf{X} \sim N_n(\boldsymbol{\mu}, \boldsymbol{\Sigma})$, the quadratic form $(\mathbf{X} - \boldsymbol{\mu})' \boldsymbol{\Sigma}^{-1}(\mathbf{X} - \boldsymbol{\mu})$ is distributed as a chi-square distribution with n degrees of freedom. Specifically, since

$$D_{KL}(f, f^{(\delta)}) = E_f\left[\log \frac{f(\mathbf{X})}{f^{(\delta)}(\mathbf{X})}\right]$$

it follows

$$D_{KL}(f, f^{(\delta)}) = \log \frac{2^{\frac{n}{2(1+\delta)}} \Gamma\left(\frac{n}{2(1+\delta)}\right)}{2^{\frac{n}{2}} \Gamma\left(\frac{n}{2}\right)(1+\delta)} - \frac{1}{2}\{E_f\left[(\mathbf{X}-\boldsymbol{\mu})' \boldsymbol{\Sigma}^{-1} (\mathbf{X}-\boldsymbol{\mu})\right] - E_f\left[\left((\mathbf{X}-\boldsymbol{\mu})' \boldsymbol{\Sigma}^{-1} (\mathbf{X}-\boldsymbol{\mu})\right)^{1+\delta}\right]\}$$

that is

$$D_{KL}(f, f^{(\delta)}) = \log \frac{2^{\frac{n}{2(1+\delta)}} \Gamma\left(\frac{n}{2(1+\delta)}\right)}{2^{\frac{n}{2}} \Gamma\left(\frac{n}{2}\right)(1+\delta)} - \frac{1}{2}\left(n - \frac{2^{(1+\delta)} \Gamma\left(\frac{n}{2} + (1+\delta)\right)}{\Gamma\left(\frac{n}{2}\right)}\right) \quad (4)$$

According to this result, the divergence between joint densities depends on the dimension of the random vector n and the perturbation δ applied to the reference normal distribution. Figure 5 illustrates the relation (4) when δ is small and, consequently, the approximation (3) holds. Observe that there is a monotone behavior with respect to the dimension n with a faster growth for high dimensions. From a local point of view, Figure 5 confirms the approximation (3).

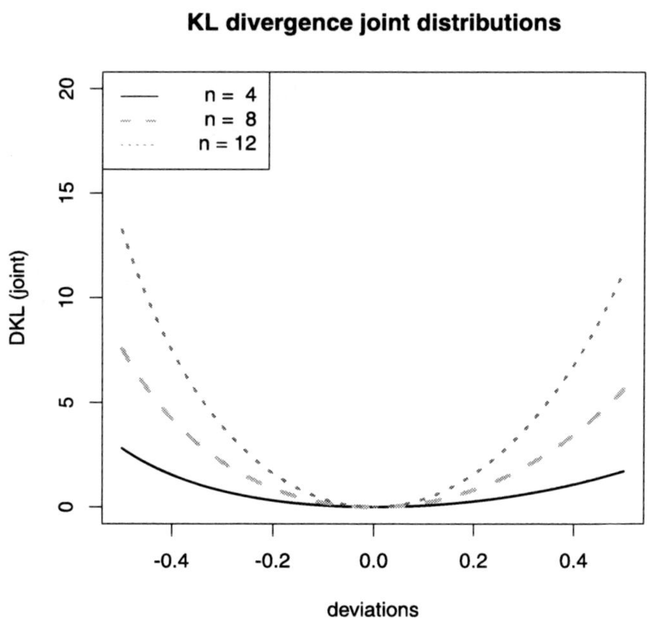

Fig. 5. KL divergence of the joint distributions for $n = 4, 8, 12$

3.2 Conditional Distributions

Now we focus on the analysis of conditional distributions sensitivity to small perturbations of the parameter β. Using previous notation it follows

$$f_{2.1}^{(\delta)}\left(\mathbf{x}_{(2)} \mid \mathbf{x}_{(1)}\right) =$$

$$= k_1 \left|\boldsymbol{\Sigma}_{22.1}\right|^{-\frac{1}{2}} \exp -\frac{1}{2}\left\{\left[\left(\mathbf{x}_{(2)} - \boldsymbol{\mu}_{(2.1)}\right)' \boldsymbol{\Sigma}_{22.1}^{-1}\left(\mathbf{x}_{(2)} - \boldsymbol{\mu}_{(2.1)}\right) + \mathbf{q}_1\right]^{(1+\delta)}\right\},$$

being

$$k_1 = \frac{\Gamma\left(\frac{n-p}{2}\right)}{\pi^{\frac{n-p}{2}} \int_0^\infty t^{\frac{n-p}{2}-1} \exp\left\{-\frac{1}{2}\left(t + \mathbf{q}_1\right)^{(1+\delta)}\right\} dt}$$

and consequently the KL divergence is [11]

$$D_{KL}\left(f_{2.1}, f_{2.1}^{(\delta)}\right) = \log \frac{\int_0^\infty t^{\frac{n-p}{2}-1} \exp\left\{-\frac{1}{2}\left(t + \mathbf{q}_1\right)^{(1+\delta)}\right\} dt}{2^{\frac{n-p}{2}} \Gamma\left(\frac{n-p}{2}\right)} - \frac{1}{2}\left[n - p - \frac{\mathbf{q}_1^{(1+\delta)+\frac{n-p}{2}}}{2^{\frac{n-p}{2}}} U(a, b, x)\right],$$

where $U(a, b, x)$ is the *Confluent Hypergeometric Function* calculated in

$$a = \frac{n-p}{2}, \; b = 2 + \delta + \frac{n-p}{2}, \; x = \frac{\mathbf{q}_1}{2}.$$

Figure 6 shows the KL divergence, as a function of δ, for the conditional distributions corresponding to selected values of the $\mathbf{q}_1$ distribution: the mean and the $10th$ and $90th$ quantiles. In this setting the divergence is affected by the dimension of $\mathbf{X}$, the dimension of the conditioning random vector $\mathbf{X}_{(1)}$ and the particular value of the conditioning variables through the Mahalanobis distance to its mean. As it was expected, the KL divergence has a quadratic appearance compatible with Equation (3) for δ close to zero. From a statistical point of view, a larger variability is found for distributions with lighter tails than the normal. Also, for the chosen values (mean and quantiles) of the $\mathbf{q}_1$ distribution, the KL divergence functions are monotone and their relative position are directly related to the ratio p/n.

3.3 Marginal Distributions

A similar approach holds for the case of marginal distributions. However, deriving an exact expression for the KL divergence using analytical methods

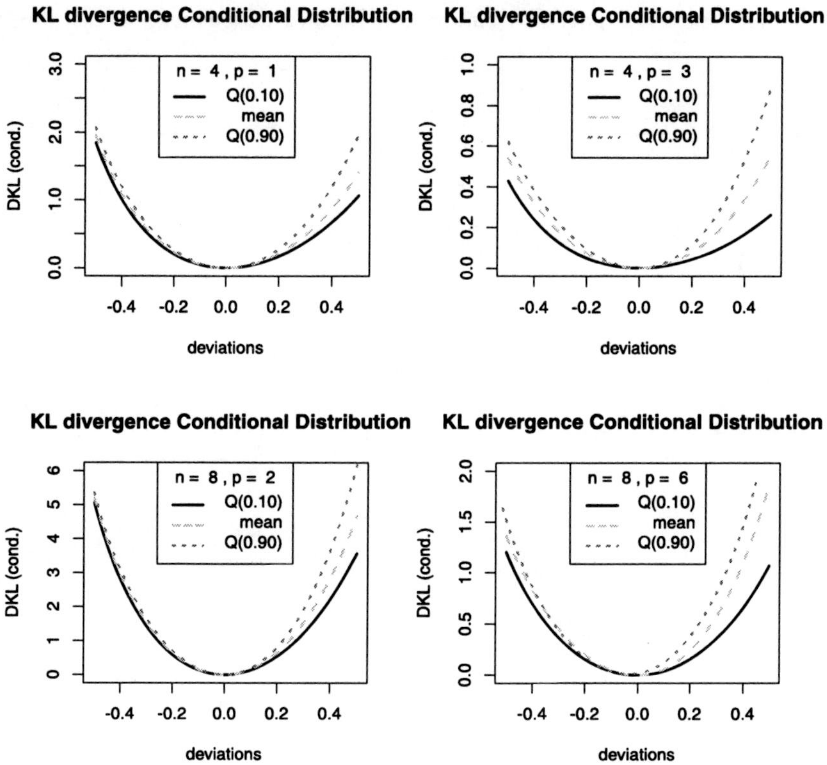

Fig. 6. KL divergence of the conditional distributions: $n = 4, 8$; $p/n = 0.25, 0.75$

appears to be a complicated task. Here, Monte Carlo simulation data were used to approximate the value of this measure, under a variety of conditions.

$$D_{KL}\left(f_1, f_1^{(\delta)}\right) = \log \frac{2^{\frac{n}{2(1+\delta)} - \frac{p}{2}} \Gamma\left(\frac{n}{2(1+\delta)}\right) \Gamma\left(\frac{n-p}{2}\right)}{\Gamma\left(\frac{n}{2}\right)(1+\delta)\exp\left(\frac{p}{2}\right)} - \log E_{\chi_p^2}\left[\int_0^\infty w^{\frac{n-p}{2}-1} e^{-\frac{w}{2}} e^{-\frac{1}{2}\left(\chi_p^2 + w\right)^\delta} dw\right]$$

Figure 7 shows the simulation results obtained for different values of n, p and δ, using 50,000 replications of the random variable χ_p^2 for each case. In general the behavior observed is similar to that of the previous sections.

On the other hand, it is well known that the divergences between the different distributions we have considered are related as follows

$$D_{KL}(f, f^{(\delta)}) = E_{f_1}\left[D_{KL}\left(f_{2.1}, f_{2.1}^{(\delta)}\right)\right] + D_{KL}\left(f_1, f_1^{(\delta)}\right) \tag{5}$$

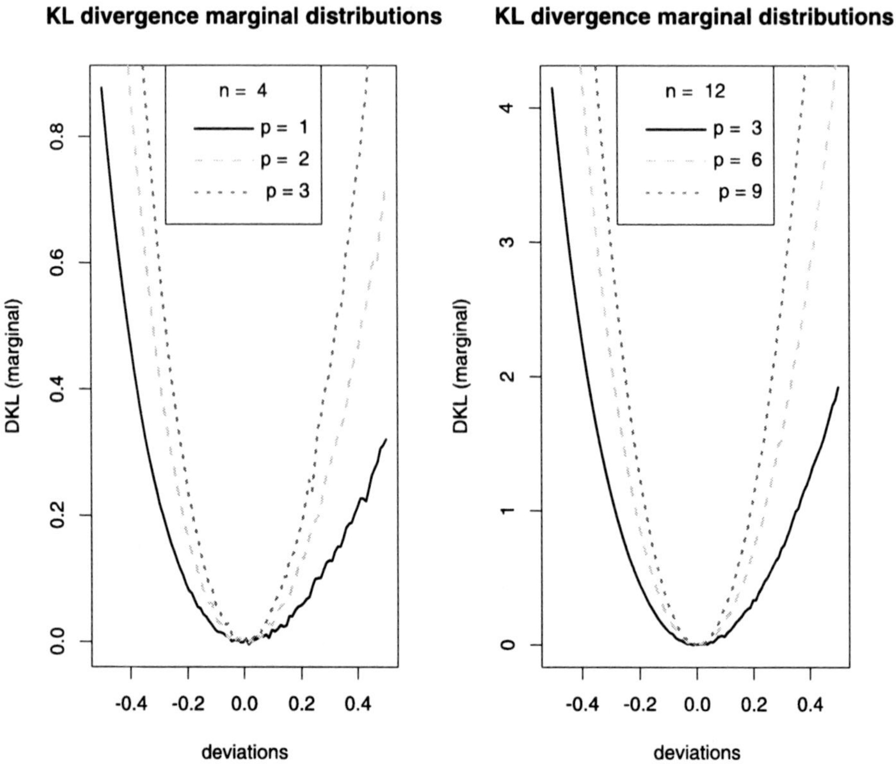

Fig. 7. KL divergence of the marginal distributions: $n = 4, 8$; $p/n = 0.25, 0.50, 0.75$

and therefore the divergence between marginal densities would also be approximated from the previous identity with Monte Carlo simulations to estimate the conditional KL divergence mean.

Finally, Equation (5) suggests the definition of a relative divergence measure for the conditional and marginal distributions in terms of the ratios

$$\frac{E_{f_1}\left[D_{KL}\left(f_{2.1}, f_{2.1}^{(\delta)}\right)\right]}{D_{KL}(f, f^{(\delta)})}, \frac{D_{KL}\left(f_1, f_1^{(\delta)}\right)}{D_{KL}(f, f^{(\delta)})}.$$

A recent approach to this problem is presented in [9].

4 Conclusions

In this paper we considered the multivariate exponential power family of distributions as an alternative model when normality assumption was doubtful. The Kullback-Leibler divergence measure is used as a tool for exploring the influence of deviations from multivariate normal in joint, conditional and marginal distributions. The obtained expressions for divergence measures provide quadratic sensitivity functions both globally and locally. Moreover, it results that this effect depends on the dimension of the vectors involved as well as the values of the conditioning variables through the Mahalanobis distance to its mean, for the case of conditionals.

References

1. Box, G., Tiao, G.: Bayesian Inference in Statistical Analysis. Addison-Wesley, Reading (1973)
2. Basu, S., Miccheli, C.A., Olsen, P.: Power exponential densities for the training and classification of acoustic feature vectors in speech recognition. J. Comput. Graph Statist. 10(1), 158–184 (2001)
3. Chu, K.C.: Estimation and decision for linear systems with elliptical random processes. IEEE Trans. Automat. Control 18, 499–505 (1973)
4. Fang, K.T., Kotz, S., Ng, K.W.: Symmetric Multivariate and Related Distributions. Chapman and Hall, London (1990)
5. Gómez, E., Gómez-Villegas, M.A., Marín, J.M.: A multivariate generalization of the power exponential family of distributions. Comm. Statist.-Theor. Meth. 27, 589–600 (1998)
6. Gómez, E., Gómez-Villegas, M.A., Marín, J.M.: A matrix variate generalization of the power exponential family of distributions. Comm. Statist.-Theor. Meth. 31(12), 2167–2182 (2002)
7. Gómez, E., Gómez-Villegas, M.A., Marín, J.M.: A survey on continuous elliptical vector distributions. Rev. Mat. Complut. 16, 345–361 (2003)
8. Gómez, E., Gómez-Villegas, M.A., Marín, J.M.: Multivariate exponential power distributions as mixtures of normal distributions with Bayesian applications. Comm. Statist.-Theor. Meth. 37(6), 972–985 (2008)
9. Gómez-Villegas, M.A., Main, P., Navarro, H., Susi, R.: Relative sensitivity of conditional distributions to kurtosis deviations in the joint model. Procedia.-Soc. Behav. Sci. 2, 7664–7665 (2010)
10. Lindsey, J.K.: Multivariate elliptically contoured distributions for repeated measurements. Biometrics 56, 1277–1280 (1999)
11. Main, P., Navarro, H.: Analyzing the effect of introducing a kurtosis parameter in Gaussian Bayesian networks. Reliab. Eng. Syst. Safety 94, 922–926 (2009)
12. Mardia, K.V., Kent, J.T., Bibby, J.M.: Multivariate Analysis. Academic Press, London (1979)
13. Nadarajah, S.: The Kotz-type distribution with applications. Statistics 37(4), 341–358 (2003)
14. Samorodnitsky, G., Taqqu, M.S.: Stable Non-Gaussian Random Processes. Chapman and Hall, Boca Raton (2000)
15. Subbotin, M.: On the law of frequency of error. Mathematicheskii Sbornik 31, 296–301 (1923)

Characterization Results for the Skewed Double Exponential Distributions

Keshav Jagannathan[1], Arjun K. Gupta[2], and Truc T. Nguyen[2]

[1] Coastal Carolina University, Conway, SC, USA
kjaganna@coastal.edu
[2] Bowling Green State University, Bowling Green, OH, USA
gupta@bgsu.edu, tnguyen@bgsu.edu

Summary. In defining the skew-normal distribution, [1] introduced a method of modifying symmetric distributions to obtain their skewed counterparts. In this paper, the authors present moment properties of the distribution obtained by adding skewness to the double exponential distribution, i.e. the Skewed Double Exponential(SDE) distribution ([6]). The authors also provide characterization results of distributions in the SDE family of distributions and present several interesting corollaries of the characterization results.

1 Introduction

There are many methods of introducing skewness in statistical models. [2] introduced the skew-normal distribution, i.e $f_Y(y) = 2\phi_X(y)\Phi_X(\lambda y)$ where ϕ and Φ represent the standard normal probability density function (p.d.f.) and cumulative distribution function (c.d.f.) respectively. Although credit is given to Azzalini, [8] had already used this distribution in studying twin data. [5] have given a characterization of this distribution. Azzalini however, introduced a way to "skew" symmetric distributions by considering similar products of p.d.f.'s and c.d.f.'s of symmetric distributions. [4] and [3] use this approach to skew a host of symmetric distributions. [6] introduced the Skewed Double Exponential(SDE) family of distributions and provided a stochastic representation for the SDE family of distributions. In this paper, we present moment properties and provide characterizations for the SDE family of distributions.

2 Definition and Moment Results

The definition of a SDE distribution is given based on the following result.

Lemma 1. *Let f be a density function symmetric about 0, and G an absolutely continuous distribution function such that G' is symmetric about 0. Then,*

$$2f(y)G(\lambda y) \quad (-\infty < y < \infty) \tag{1}$$

is a density function for any real λ.

Definition 1. *A random variable Y is said to have a skewed double exponential distribution with parameter* λ, *denoted* $SDE(\lambda)$, *if for* $-\infty < y < \infty$, *its probability density function is given by*

$$g(y,\lambda) = 2f(y)F(\lambda y) \tag{2}$$

where f and F are respectively the density and the c.d.f of a $DE(0,1)$ *distribution.*

Using Definition 1, we have that, for $\lambda > 0$, the density of a $SDE(\lambda)$ distribution is

$$g(y,\lambda) = \begin{cases} \frac{1}{2}e^{(1+\lambda)y} & \text{when } y < 0 \\ e^{-y} - \frac{1}{2}e^{-(1+\lambda)y} & \text{when } y \geq 0 \end{cases}$$

and, for $\lambda < 0$, is

$$g(y,\lambda) = \begin{cases} e^{y} - \frac{1}{2}e^{(1-\lambda)y} & \text{when } y < 0 \\ \frac{1}{2}e^{-(1-\lambda)y} & \text{when } y \geq 0 \end{cases}$$

[4] gave a simple formula for this density. It is given by

$$g(y,\lambda) = \frac{1}{2}e^{-|y|}\left(1 + sign(\lambda y)(1 - e^{-|\lambda y|}\right), y \in \mathbb{R}.$$

Note 1. Since the distribution function F is symmetric, we notice that $F(-\lambda x) = 1 - F(\lambda x)$ and hence, we can restrict our attention to the case when $\lambda > 0$.

Graphs of the $SDE(\lambda)$ density function for different values of λ are given in Figure 1 and Figure 2.

2.1 Moment Generating Function

In this section, we give the moment generating function (m.g.f) of the $SDE(\lambda)$ model defined above and also an expression for the kth moment.

If the random variable Y is distributed as $Y \sim SDE(\lambda)$, with $\lambda > 0$ and $-(1+\lambda) < t < \min(1, (1+\lambda))$, then it has m.g.f given by

$$M_Y(t) = \frac{1}{2(1+\lambda+t)} + \frac{1}{1-t} - \frac{1}{2(1+\lambda-t)}, \tag{3}$$

and it's kth moment is given by

$$\mu_k = \frac{(-1)^k k!}{2(1+\lambda)^{(k+1)}} + k! - \frac{k!}{2(1+\lambda)^{(k+1)}}, \; k = 1, 2, \ldots \tag{4}$$

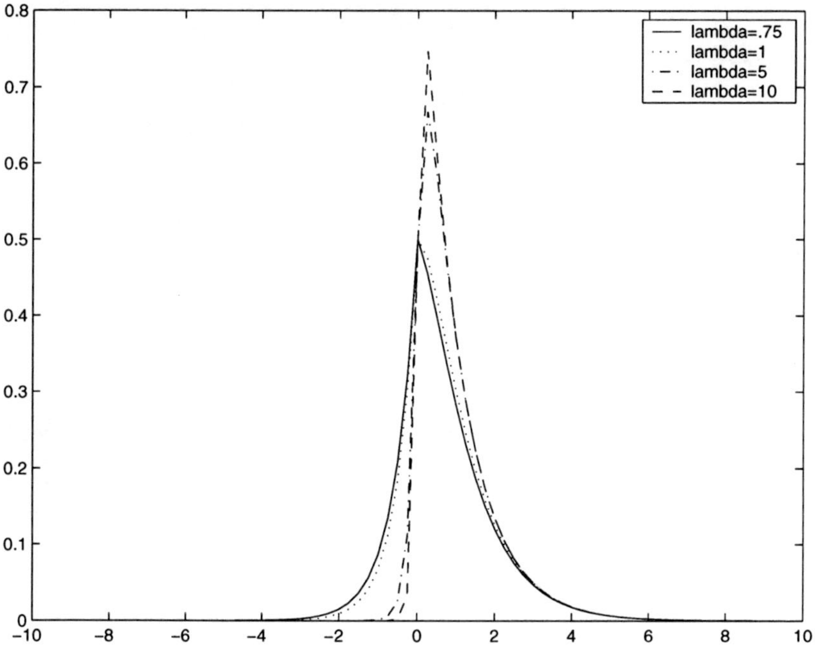

Fig. 1. Graphs of the $SDE(\lambda)$ density function for $\lambda = 0, 0.01, 0.1, 0.5$

Table 1. Expected and simulated values of skewness of $SDE(\lambda)$ distribution in 10,000 repetitions

	Sample size n						
λ	10	50	100	500	1000	5000	True Value
0.01	0.0232	0.0303	0.0315	0.0437	0.0363	0.0413	0.0410
0.1	0.1356	0.2573	0.2717	0.3031	0.3129	0.3132	0.3151
0.5	0.3807	0.6703	0.7429	0.8116	0.8203	0.8287	0.8294
1	0.5058	0.9184	1.0151	1.1131	1.1270	1.1411	1.1423
5	0.8641	1.4963	1.6378	1.8036	1.8242	1.8399	1.8462
10	0.9311	1.5956	1.7389	1.9039	1.9247	1.9479	1.9516

2.2 Skewness and Kurtosis

We calculate the skewness and kurtosis of a SDE random variable in this section using the moment generating function and moments of all orders obtained in the previous section.

If a random variable Y has a $SDE(\lambda)$ distribution with $\lambda > 0$, then from the expression for the kth moment, the skewness(ν_1) and kurtosis(ν_2) are given by

$$\nu_1 = \frac{2((\lambda+1)^6 - 1)}{((1+\lambda)^4 + 2(1+\lambda)^2 - 1)^{3/2}}, \tag{5}$$

and

$$\nu_2 = \frac{9(1+\lambda)^8 + 4(1+\lambda)^6 + 6(1+\lambda)^4 - 4(1+\lambda)^2 - 1}{((1+\lambda)^4 + 2(1+\lambda)^2 - 1)^2}. \tag{6}$$

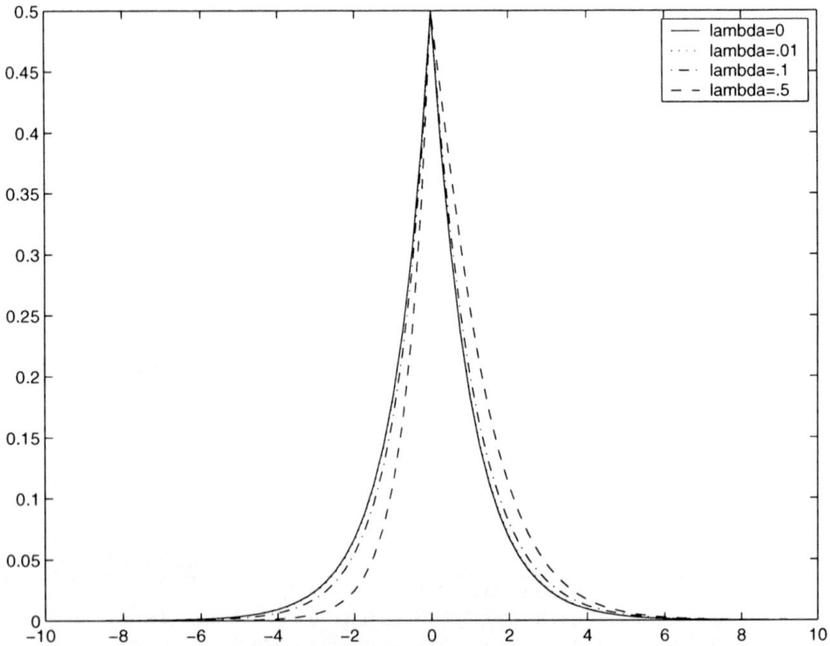

Fig. 2. Graphs of the $SDE(\lambda)$ density function for $\lambda = 0.75, 1, 5, 10$

Table 2. Expected and simulated values of kurtosis of $SDE(\lambda)$ distribution in 10,000 repetitions

	Sample size n						
λ	10	50	100	500	1000	5000	True Value
0.01	3.0378	4.8104	5.2656	5.8865	5.9307	5.9965	5.9989
0.1	3.0784	4.8276	5.1501	5.7175	5.8584	5.9082	5.9371
0.5	2.9908	4.7905	5.2820	5.6482	5.7464	5.8242	5.8442
1	2.9341	4.8349	5.5630	6.0072	6.1129	6.2686	6.2552
5	3.0932	5.8098	6.6780	7.9184	8.1481	8.3000	8.4012
10	3.1244	6.0746	7.0944	8.3197	8.5705	8.8287	8.8080

We make a few observations about the skewness and kurtosis based on their derivatives.

1. The skewness is an increasing function in λ.
2. The kurtosis is also an increasing function in λ irrespective of the sign of λ as long as $|\lambda| > \sqrt{5}/2 - 1$.

We conduct a simulation study to verify the rate of convergence of the sample skewness and kurtosis to the respective population parameters. The results of that study are summarized in Table 1 and Table 2.

The above tables show that the sample skewness and kurtosis converge to their population parameters at a rate of $\sqrt{n}$.

3 Characterization Results for the *SDE*(λ) Family of Distributions

From the definition of a $SDE(\lambda)$ distribution given by (2) in definition 1, the density of $|Y|$ is

$$\begin{aligned} g_{|Y|}(y) &= 2f(y)[F(-\lambda y) + F(\lambda y)] \\ &= 2f(y) \\ &= e^{-|y|}, \end{aligned} \tag{7}$$

since the $DE(0,1)$ distribution is symmetric about the origin. Hence, $|Y|$ has an $Exp(1)$ distribution.

Now consider a large family $\mathcal{P}$ of distributions that contain the family of $SDE(\lambda)$ distributions for all $\lambda \in \mathbb{R}$. A random variable Y is said to have a distribution belonging to $\mathcal{P}$ if $|Y|$ has an $Exp(1)$ distribution. It is trivial to see that Y has a distribution in $\mathcal{P}$ if and only if Y has a density given by

$$g_Y(y) = e^{|y|}\, h(y)\,,\ -\infty < y < \infty,\ h(y) \geq 0, \tag{8}$$

such that $h(y) + h(-y) = 1$. ([9])

As shown above, the family $SDE(\lambda)$, $\lambda \in \mathbb{R}$ is a subfamily of $\mathcal{P}$. If Y has a $SDE(\lambda)$ distribution, then $h(y) = F(\lambda y)$.

If X and Y are two independent random variables with densities $f_X(x) = e^{-|x|}\, h_1(x)$, $h_1(x) \geq 0$ and $f_Y(y) = e^{-|y|}\, h_2(y)$, $h_2(y) \geq 0$ of $\mathcal{P}$, then $|X|$ and $|Y|$ are i.i.d. $Exp(1)$ and $|X| + |Y|$ has a $Gam(2,1)$ distribution.

The joint distribution of X and Y is

$$f_{X,Y}(x,y) = e^{-(|x|+|y|)}\, h_1(x)\, h_2(y) \quad -\infty < x, y < \infty \tag{9}$$

and the joint distribution of X and $T = |X| + |Y|$ is

$$\begin{aligned} f_{X,T}(x,t) &= e^{-t}\, h_1(x)\, h_2(t - |x|) + e^{-t}\, h_1(x)\, h_2(-t + |x|) \\ &= e^{-t}\, h_1(x) \quad -t < x < t\,, t > 0. \end{aligned}$$

Therefore, the conditional density of X given $T = t$ is

$$f_{X|T}(x|t) = \frac{h_1(x)}{t}, \quad -t < x < t, \tag{10}$$

with $h_1(x) + h_1(-x) = 1$ and $h_1(x) \geq 0$. The following result is a characterization of a distribution in $\mathcal{P}$.

Theorem 1. *Let X and Y be two independent random variables with finite second moment with $\mathbb{E}[|X|] = 1$. Then the distributions of X and Y are in $\mathcal{P}$ if and only if the conditional density of X given $T = |X| + |Y| = t$ is*

$$f_{X\,||X|+|Y|=t}(x|t) = \frac{h_1(x)}{t}, \text{ for } -t < x < t, h_1(x) + h_1(-x) = 1,\ h_1(x) \geq 0. \tag{11}$$

Proof. We need only to prove the reverse implication. Assume that the conditional density of $X|T = t$ is given by (11). Then,

$$\begin{aligned} \mathbb{E}[X|T = t] &= \int_{-t}^{t} |x| \, \frac{h_1(x)}{t} \, dx \\ &= \int_0^t x \, \frac{h_1(x) + h_1(-x)}{t} \, dx \\ &= \frac{t}{2} \end{aligned} \tag{12}$$

and

$$\begin{aligned} \mathbb{E}[X^2|T = t] &= \mathbb{E}[|X|^2|T = t] \\ &= \int_{-t}^{t} x^2 \, \frac{h_1(x)}{t} \, dx \\ &= \frac{t^2}{3}. \end{aligned} \tag{13}$$

From (12) and (13), multiply both sides by e^{isT} and taking expectations, the following system of differential equations in characteristic functions of $\varphi_{|X|}(s)$ and $\varphi_{|Y|}(s)$ is obtained.

$$\mathbb{E}[|X|e^{isT}] = \mathbb{E}[\frac{t}{2}e^{isT}] \tag{14}$$

$$\mathbb{E}[|X|^2e^{isT}] = \mathbb{E}[\frac{t^2}{3}e^{isT}] \tag{15}$$

(14) gives

$$\varphi'_{|X|}(s)\, \varphi_{|Y|}(s) = \varphi_{|X|}(s)\, \varphi'_{|Y|}(s) \tag{16}$$

(15) gives

$$2\varphi''_{|X|}(s)\,\varphi_{|Y|}(s) = 2\varphi'_{|X|}(s)\,\varphi'_{|Y|}(s) + \varphi_{|X|}(s)\,\varphi'_{|Y|}(s) \tag{17}$$

From (16), we get that $\varphi_{|X|}(s) = \varphi_{|Y|}(s) = \varphi(s)$. Substituting in (17), we obtain

$$\varphi''(s)\,\varphi(s) = 2\varphi'^2(s). \tag{18}$$

The general solution to (18) is

$$\varphi(s) = \frac{1}{D - Cs}$$

From $\varphi(0) = 1$ and $\varphi'(0) = \mathbb{E}[|X|] = 1$, we get that $D = 1, C = i$. Hence,

$$\varphi(s) = \frac{1}{1 - is}, \quad -\infty < s < \infty,$$

and $|X|$ and $|Y|$ are i.i.d. according to an $Exp(1)$ distribution. Therefore X has the density $e^{-|x|}\,h_1(x)$ and Y has the density $e^{-|y|}\,h_2(y)$ where $h_1(x)$ is given by the hypothesis of the theorem and $h_2(y)$ is an arbitrary function with $h_2(y) \geq 0$, $h_2(y) + h_2(-y) = 1$.

Theorem 2. *Let $Y_1, \ldots, Y_n$ be i.i.d. according to a distribution F with a finite second moment and $\mathbb{E}[|Y_1|] = 1$. Then F is a distribution of $\mathcal{P}$ with density $e^{-|y|}\,h(y)$, $-\infty < y < \infty$, $h(y) + h(-y) = 1$ if and only if*

$$f_{Y_1|T}(y_1|t) = \frac{(t - |y_1|)^{n-2}}{t^{n-1}}\,h(y_1), \quad -t < y_1 < t$$

where $T = \sum_{i=1}^{n} |Y_i|$.

Proof. It is trivial that $\mathbb{E}[|Y_1||T = t] = \dfrac{t}{n}$ and

$$\begin{aligned}
\mathbb{E}|Y_1^2|T = t] &= \mathbb{E}[|Y|^2|T = t] \\
&= \int_{-t}^{t} (n-1)|y_1|^2\,\frac{(t - |y_1|)^{n-2}}{t^{n-1}}\,h(y_1)\,dy_1 \\
&= \int_{0}^{t} (n-1)y^2\,\frac{(t - y)^{n-2}}{t^{n-1}}\,(h(y_1) + h(-y_1))\,dy \\
&= \int_{0}^{t} (n-1)y^2\,\frac{(t - y)^{n-2}}{t^{n-1}}\,dy \\
&= \frac{2t^2}{n(n+1)}
\end{aligned}$$

Using the same technique as in the proof of Theorem 1, the same differential equation in the characteristic function φ of $|Y|$ is obtained,

$$\varphi''\varphi = 2\varphi'^2.$$

Hence, $|Y_1|, \ldots, |Y_n|$ are i.i.d. according to an $Exp(1)$ distribution and therefore F has the density $e^{-|y|}\, h(y)$, $-\infty < y < \infty$, $h(y) + h(-y) = 1$ with $h(y) \geq 0$.

The following results are obtained directly from Theorem 2.

Corollary 1. *Let $Y_1, \ldots, Y_n$ be i.i.d. according to a distribution F with a finite second moment and $\mathbb{E}[|Y_j|] = 1$. Then F has the density $e^{-|y|}\, h(y)$, $-\infty < y < \infty$, $h(y) + h(-y) = 1$, $h(y) \geq 0$ if and only if*

$$f_{Y_1,\ldots,Y_{n-1}|T}(y_1, \ldots, y_{n-1}|t) = \frac{(n-1)! \prod_{i=1}^{n} h(y_i)}{t!}, \; : \; 0 < \sum_{i=1}^{n} |Y_i| < t,$$

where $T = \sum_{j=1}^{n} |Y_j|$.

Corollary 2. *Let $Y_1, \ldots, Y_n$ be i.i.d. according to a distribution F with a finite second moment and $\mathbb{E}[|Y_j|] = 1$. Then F is a $SDE(\lambda)$ distribution if and only if*

$$f_{Y_1|T}(y_1|t) = (n-1)\frac{(t - |y_1|)^{n-2}}{t^{n-1}}\Big[\frac{1 + sign(\lambda y_1)(1 - e^{-|\lambda y_1|}}{2}\Big],$$

where $T = \sum_{j=1}^{n} |Y_j|$, $-t < y_1 < t$.

Corollary 3. *Let $Y_1, \ldots, Y_n$ be i.i.d. according to a distribution F with a finite second moment and $\mathbb{E}[|Y_j|] = 1$. Then F is a $SDE(\lambda)$ distribution if and only if*

$$f_{Y_1,\ldots,Y_{n-1}|T}(y_1, \ldots, y_{n-1}|t) = \frac{(n-1)!}{t!} \prod_{j=1}^{n-1} \Big[\frac{1 + sign(\lambda y_j)(1 - e^{-|\lambda y_j|}}{2}\Big],$$

where $T = \sum_{j=1}^{n} |Y_j|$ *and* $0 < \sum_{j=1}^{n} |Y_j| < t$.

4 Conclusion

[6] introduced the SDE family of distributions and provided certain basic properties as well as stochastic representations of the SDE family of distributions. [7] provides estimation results for the SDE family of distributions.

This paper presents moment properties of the SDE family of distributions along with some interesting results based on the characterization of the SDE family of distributions. The starting point for the SDE family of distributions is the $DE(0,1)$ distribution. We can extend the SDE family of distributions to a location-scale family of distributions by starting with the $DE(\eta, \theta)$ distribution.

References

1. Azzalini, A.: A class of distributions that includes the normal ones. Scand. J. Statist. 12, 171–178 (1985)
2. Azzalini, A.: Further results on a class of distributions which includes the normal ones. Statistica 56, 199–208 (1986)
3. Gupta, A.K., Chang, F.C.: Multivariate skew symmetric distributions. Appl. Math. Lett. 16, 643–646 (2003)
4. Gupta, A.K., Chang, F.C., Huang, W.J.: Some skew symmetric models. Random Oper. Stoc. Equ. 20, 89–103 (2002)
5. Gupta, A.K., Nguyen, T.T., Sanqui, J.A.T.: Characterization of the skew-normal distribution. Ann. Inst. Statist. Math. 56(2), 351–360 (2004)
6. Jagannathan, K., Gupta, A.K., Nguyen, T.T.: Skewed double exponential distribution and its stochastic representation. Eur. J. Pure Appl. Math. 2(1), 1–20 (2009)
7. Jagannathan, K., Gupta, A.K., Nguyen, T.T.: Parameter Estimation of a Skewed Double Exponential Distribution. J. Probab. Stat. Sci. (2010) (to appear)
8. Roberts, C.: A correlation model useful in the study of twins. J. Amer. Statist. Assoc. 61, 1184–1190 (1966)
9. Roberts, C.: On a distribution of random variables whose m-th power is gamma. Sankhyã Ser. A 33, 229–232 (1971)

Generalized Beta Generated-II Distributions*

Kostas Zografos

University of Ioaninna, Department of Mathematics, 451 10 Ioaninna, Greece
kzograf@uoi.gr

Summary. A family of univariate distributions, generated by beta random variables, has been proposed by Jones [9]. This broad family of univariate distributions has received considerable attention in the recent literature since it possesses great flexibility while fitting symmetric as well as skewed models with varying tail weights. This paper introduces and studies a new broad class of univariate distributions which is defined by means of a generalized beta distribution and includes Jones family as a particular case. Some properties of the proposed class of distributions are discussed. These properties include its moments, generalized moments, representation and relationship with other distributions, expressions for Shannon entropy. Two examples are given and the paper is completed with some conclusions.

Keywords: beta generated distributions, income distributions, Shannon entropy, maximum entropy principle.

1 Introduction

The last decade is characterized by an increasing effort of the researchers to introduce and study broad classes of univariate distributions which compose useful characteristics of two or more univariate distributions. These models extend well known univariate distributions and provide great flexibility in modelling data in practice. The beta-normal distribution, introduced by Eugene *et al.* [6], is the initial effort in this direction, to the best of our knowledge. Two years later, Jones proposed, in a discussion paper in TEST [9], the class of "beta-generated distributions", which is defined as follows. For a continuous distribution function F with density f, the family of univariate distributions generated by F, and the parameters $\alpha, \beta > 0$, has its pdf as [9]

$$g_F^{(B)}(x;\alpha,\beta) = \frac{1}{B(\alpha,\beta)} f(x) \{F(x)\}^{\alpha-1} \{1 - F(x)\}^{\beta-1}, \tag{1}$$

* This paper is devoted to the memory of Maria Luisa Menéndez, an exceptional scientist and an outstanding human personality. The long standing friendship and research collaboration with Marisa it was an inexhaustible source of knowledge and of humaneness to me.

for $\alpha > 0$ and $\beta > 0$, where $B(\alpha, \beta) = \int_0^1 t^{\alpha-1}(1-t)^{\beta-1}dt$ is the complete beta function. If the parameters α and β are positive integers, the beta-generated model in (1) is the distribution of the i-th order statistic in a random sample of size n from distribution F, where $i = \alpha$ and $n = \alpha+\beta-1$. The above family of distributions can be, moreover, obtained by means of a transformation of an initial random variable Y with beta distribution $Beta(\alpha, \beta)$, $\alpha > 0$ and $\beta > 0$. In particular, if $Y \sim Beta(\alpha, \beta)$, then the density of the random variable $X = F^{-1}(Y)$ is given by (1). This representation of X helps to generate random numbers from (1) while the case $\alpha = \beta = 1$ corresponds to the well-known quantile function representation $X = F^{-1}(U)$, where $U \sim U(0, 1)$, which is used in order to generate data from a distribution F. The family (1) has been recently studied by Zografos and Balakrishnan [24]. Following the terminology of Arnold in the discussion of Jones' [9] paper, the distribution F will be referred to as the "parent distribution", in the sequel.

It is clear that special choices of the parent model F, lead to specific models generated by the classic beta distribution. Hence, if F is the c.d.f. of the normal distribution, then (1) leads to the beta normal distribution of Eugene *et al.* [6], if F is the c.d.f. of the exponential distribution, then (1) is the beta exponential distribution of Nadarajah and Kotz [19] and so on. The next table summarizes the models obtained from (1) for special choices of the parent c.d.f. F, where the second ingredient in the name of the distribution refers to the parent model F.

Table 1. Specific beta generated distributions

Name of the distribution	**Authors / Year**
Beta-Normal	Eugene *et al.* [6]
Beta-Logistic	Brown *et al.* [5] and Olapade [20]
Beta-Frechet	Nadarajah and Gupta [17]
Beta-Gumbel	Nadarajah and Kotz [18]
Beta-Exponential	Nadarajah and Kotz [19]
Beta-Gamma	Kong *et al.* [12]
Beta-Weibull	Lee *et al.* [13] and Zografos [23]
Beta-Pareto	Akinsete *et al.* [3]
Beta-Power	Zografos and Balakrishnan [24]
Beta-Generalized Half Normal	Pescim *et al.* [22]
Beta-Generalized Exponential	Barreto-Souza *et al.* [4]
Beta-Burr XII	Paranaiba *et al.* [21]

Several models can be obtained as particular cases of the beta generated distributions of the above table. We mention, as an example, the Kumaraswamy distribution which is introduced recently by Jones [10]. This is obtained as a particular case of the beta-power function distribution with density $\frac{1}{B(\alpha,\beta)} k\theta^{k\alpha} x^{k\alpha-1} \left\{1 - (\theta x)^k\right\}^{\beta-1}$, $0 < x < \frac{1}{\theta}$, for $\theta = 1$ and $\alpha = 1$

(cf. Zografos and Balakrishnan [24]). Another example is the exponentiated exponential distribution of Gupta and Kundu [7], which is a particular case of the beta-exponential distribution of Nadarajah and Kotz [19].

This paper concentrates on generalized beta generated distributions which include as special or limiting cases all of the above models. The generalized beta generated distributions are defined in Section 2 in a manner quite similar to that of the definition of the beta generated distributions. To be precise, the classic beta distribution which is the kernel of the beta generated distribution is replaced by generalized beta models, considered in the work of McDonald and his colleagues (cf. McDonald [14], McDonald and Xu [15], McDonald and Ransom [16]). This work is motivated by the fact that generalized beta distributions include, as particular cases, models that are suitable to formulate econometric data with possibly skewed and leptokurtic error distributions. Hence, it is expected that the introduced here generalized beta distributions, generated by a parent distribution F, will provide great flexibility in modelling data in practice. Section 3 studies some properties of the introduced model while some specific examples are presented in the Section 4. The final section provides with some conclusions and problems for a future work.

2 The Model: Distribution and Density Functions

Generalized beta distributions have been discussed in the papers by McDonald and his colleagues, mentioned above, as income distributions and they include as particular cases several univariate distributions. A nice presentation of the generalized beta distributions is provided in the book by Kleiber and Kotz [11]. Following McDonald and Xu [15], a random variable Y is described by a generalized beta distribution (hereafter referred to as GB) if its density is given by

$$g_{GB}(y) = \frac{\alpha y^{\alpha p-1}[1-(1-c)(y/b)^{\alpha}]^{q-1}}{b^{\alpha p}B(p,q)[1+c(y/b)^{\alpha}]^{p+q}}, \; 0 < y^{\alpha} < \frac{b^{\alpha}}{1-c}, \tag{2}$$

and zero otherwise, with $0 \leq c \leq 1$ and α, b, p, q positive.

Following terminology in Kleiber and Kotz [11], for $c = 0$, we obtain the generalized beta distribution of the first kind (hereafter referred to as GB1) with density

$$g_{GB1}(y) = \frac{\alpha y^{\alpha p-1}[1-(y/b)^{\alpha}]^{q-1}}{b^{\alpha p}B(p,q)}, \; 0 < y < b. \tag{3}$$

For $c = 1$, the GB distribution is reduced to the generalized beta of the second kind (GB2) with density

$$g_{GB2}(y) = \frac{\alpha y^{\alpha p-1}}{b^{\alpha p}B(p,q)[1+(y/b)^{\alpha}]^{p+q}}, 0 < y < \infty. \tag{4}$$

If $\alpha = b = 1$ and $c = 0$, then the GB distribution is reduced to the classic beta distribution $Beta(p,q)$.

In order to define the generalized beta generated distributions by parent F, consider a parent distribution function F with respective density f and let Y denotes a random variable with generalized beta distribution and density, given by (2). Then it can be easily seen that the density function of the random variable $X = F^{-1}(Y)$, is the following:

$$g_{GBG}^{II}(x) = \frac{\alpha}{b^{\alpha p} B(p,q)} f(x) \frac{[F(x)]^{\alpha p-1}[1-(1-c)(F(x)/b)^{\alpha}]^{q-1}}{[1+c(F(x)/b)^{\alpha}]^{p+q}}, \quad (5)$$

for $F^{-1}(0) < x < F^{-1}\left(\frac{b}{(1-c)^{1/\alpha}}\right)$ and zero otherwise, with $0 \leq c \leq 1$ and α, b, p, q positive. The above distribution will be named Generalized Beta Generated distribution by parent F and it will be hereafter referred to as GBG-II. It can be seen, by using successive suitable transformation on the resulting integrals, that $g_{GBG}^{II}(x)$ is indeed a density function, that is, it integrates to one.

Remark 1. Two particular Generalized Beta Generated distributions by parent F can be obtained from g_{GBG}^{II}, given in (5), depending on whether $c = 0$ or $c = 1$. For $c = 0$, the density (5) leads to the Generalized Beta Generated distributions of the first kind (GBG-1) by parent F, with density,

$$g_{GBG}^{(1)}(x) = \frac{\alpha}{b^{\alpha p} B(p,q)} f(x)[F(x)]^{\alpha p-1}[1-(F(x)/b)^{\alpha}]^{q-1}, \quad (6)$$

for $F^{-1}(0) < x < F^{-1}(b)$. If $b = 1$, then $g_{GBG}^{(1)}$ is the distribution introduced recently by Alexander and Sarabia [2]. For $c = 1$ we obtain the Generalized Beta Generated distributions of the second kind (hereafter referred to as GBG-2) by parent F, with density,

$$g_{GBG}^{(2)}(x) = \frac{\alpha}{b^{\alpha p} B(p,q)} f(x) \frac{[F(x)]^{\alpha p-1}}{[1+(F(x)/b)^{\alpha}]^{p+q}}, \quad (7)$$

for $F^{-1}(0) < x < F^{-1}(\infty)$.

The cumulative distribution function (c.d.f.), denoted by G_{GBG}^{II}, of the GBG-II distribution with density (5) is defined by the integral

$$G_{GBG}^{II}(x) = \int_{F^{-1}(0)}^{x} g_{GBG}^{II}(t)dt.$$

Using the transformation $y = F(t)$ in the above integral and then the transformation $z = (y/b)^{\alpha}$, in the resulting integral, we get

$$G_{GBG}^{II}(x) = \frac{1}{B(p,q)} \int_0^{(F(x)/b)^{\alpha}} z^{p-1}[1-(1-c)z]^{q-1}(1+cz)^{-(p+q)}dz. \quad (8)$$

Following exactly the same procedure, it can be shown that the c.d.f. G_{GB} of the generalized beta distribution of McDonald and Xu [15], with density (2), is given by

$$G_{GB}(x) = \frac{1}{B(p,q)} \int_0^{(x/b)^\alpha} z^{p-1}[1-(1-c)z]^{q-1}(1+cz)^{-(p+q)}dz.$$

Observe that

$$G_{GBG}^{II}(x) = G_{GB}(F(x)). \tag{9}$$

Remark 2.

(i) It has been just proved that the c.d.f. of the GBG-II distribution coincides with the respective distribution function of the GB distribution, at the point $F(x)$. Exactly the same behavior it is true between Jones' [9] distribution and the classic beta distribution.

(ii)The c.d.f. of the GB distribution with density (2) is not available in a closed form or, at least, in the form of a series representation, to the best of our knowledge. The same is also true for the GBG-II distribution, taking into account equation (9).

(iii) Although it is not possible to obtain the c.d.f. of the GBG-II distribution in a closed form or in the form of a series representation, it can be easily obtained, by using (8), the c.d.f. of the GBG-1 and GBG-2 distributions with densities given by (6) and (7), respectively. Indeed, for $c = 0$, equation (8) becomes,

$$G_{GBG}^{II}(x) = G_{GBG}^{(1)}(x) = \frac{1}{B(p,q)} \int_0^{(F(x)/b)^\alpha} z^{p-1}(1-z)^{q-1}dz.$$

Taking into account that the incomplete beta function $B_r(p,q)$ is defined by the integral $B_r(p,q) = \int_0^r z^{p-1}(1-z)^{q-1}dz$, it is immediate to see that

$$\begin{aligned} G_{GBG}^{(1)}(x) &= \tfrac{1}{B(p,q)} B_{(F(x)/b)^\alpha}(p,q) \\ &= \tfrac{(F(x)/b)^{\alpha p}}{pB(p,q)}\ {}_2F_1(p, 1-q; p+1; (F(x)/b)^\alpha), \end{aligned}$$

where ${}_2F_1$ is used to denote the hypergeometric function (cf. Abramowitz and Stegun ([1], pp. 263, 556)). It can be also written as an incomplete beta function ratio

$$G_{GBG}^{(1)}(x) = I_z(p,q),\ z = (F(x)/b)^\alpha,$$

and $I_z(p,q) = B_z(p,q)/B(p,q)$, the incomplete beta function ratio. It is interesting to note that for $b = 1$, the c.d.f. $G_{GBG}^{(1)}$ coincides with the similar one of Alexander and Sarabia [2].

In a similar manner, for $c = 1$, equation (8) leads to the c.d.f. of the GBG-2 distribution, with density given by (7). This is given by

$$G_{GBG}^{(2)}(x) = \frac{1}{B(p,q)} \int_0^{(F(x)/b)^\alpha} z^{p-1}(1+z)^{-(p+q)}dz.$$

Based on equation (6.3) of Kleiber and Kotz ([11], p. 184),

$$I_t(p,q) = \frac{1}{B(p,q)} \int_0^t z^{p-1}(1+z)^{-(p+q)} dz, t > 0$$

and by analogy with the GBG-1 case, discussed above, it can be written as an incomplete beta function ratio

$$G_{GBG}^{(2)}(x) = I_z(p,q), \ z = (F(x)/b)^{\alpha}.$$

3 Properties of the GBG-II Distribution

This section is devoted to the study of some properties of the GBG-II distribution, namely, its moments, generalized moments, representation and relationship with other distributions, the Shannon entropy.

3.1 Moments

If the random variable X follows the GBG-II distribution, then for k a positive integer, the moments are defined by

$$E\left(X^k\right) = \int_{F^{-1}(0)}^{F^{-1}\left(b/(1-c)^{1/\alpha}\right)} x^k g_{GBG}^{II}(x) dx$$

and using successive transformations: $y = F(x)$, $z = (y/b)^{\alpha}$ and $\omega = (1-c)z$, $c \neq 1$, to the respective integrals, we obtain that

$$\begin{aligned} E\left(X^k\right) &= \tfrac{1}{B(p,q)(1-c)^{p-1}} \int_0^1 \left(F^{-1}\left(\tfrac{b\omega^{1/\alpha}}{(1-c)^{1/\alpha}}\right)\right)^k \omega^{p-1}(1-\omega)^{q-1} \\ &\times \left(1 + \tfrac{c}{1-c}\omega\right)^{-(p+q)} d\omega, \end{aligned}$$

for $c \neq 1$ and k a positive integer. This integral is not simplified further. However, it is possible to obtain the moments of the GBG-II distribution in a closed form, in some specific cases. For instance, if $c = 0$ and $b = 1$ the above formula is reduced to the moments of GBG-1 distribution and they are derived in formula (10) of Alexander and Sarabia [2].

3.2 Generalized Moments

For positive integers k and ℓ we are interested in generalized moments of a random variable X with GBG-II distribution which are defined by the expected value

$$M_{k,\ell} = E_{g_{GBG}^{II}} \left\{ (F(X))^k [1 - (1-c)(F(X)/b)^{\alpha}]^{\ell} \right\}. \tag{10}$$

Using the successive transformations: $y = F(x)$, $z = (y/b)^\alpha$ and $\omega = (1-c)z$, $c \neq 1$, to the respective integrals, we obtain that

$$M_{k,\ell} = \frac{b^k}{(1-c)^{p+(k/\alpha)}B(p,q)} \times \int_0^1 \omega^{p+(k/\alpha)-1}(1-\omega)^{q+\ell-1}\left(1+\frac{c}{1-c}\omega\right)^{-(p+q)} d\omega, \tag{11}$$

for $0 \leq c < 1$. Using the definition of the hypergeometric function (cf. Abramowitz and Stegun ([1], p. 558)), we obtain

$$M_{k,\ell} = \frac{b^k B(p+(k/\alpha), q+\ell)}{(1-c)^{p+(k/\alpha)}B(p,q)}\, {}_2F_1(p+q, p+(k/\alpha); p+(k/\alpha)+q+\ell; c/(1-c)).$$

It can be easily seen, from (11), that for $c = 0$ we can have in a closed form the generalized moments. In particular,

$$M_{k,\ell} = \frac{b^k}{B(p,q)} B(p+(k/\alpha), q+\ell).$$

The above equation, for $b = 1$, leads to the equation (13) of Alexander and Sarabia [2].

Remark 3.

(*i*) Although the classic moments are not available in an explicit form, we can have in an analytic form, at least for $c = 0$, the generalized moments defined by the formula (10). This permits the development of an alternative to the method of moments estimation procedure, as it is described in Zografos and Balakrishnan [24].

(*ii*)For $\ell = 0$, Eq. (10) defines generalized moments of the form $E_{g_{GBG}^{II}}\{(F(X))^k\}$, which are quite analogous to the classic moments $E(X^k)$ of any distribution, for k a positive integer. Given that the first four moments $E(X^k)$, $k = 1, 2, 3, 4$, are used to define Pearson's coefficients of skewness and kurtosis of a distribution, it would be of interest to be investigated the role of the analogous generalized moments $M_k = E_{g_{GBG}^{II}}\{(F(X))^k\}$ in studying the skewness and kurtosis behavior of the respective GBG-II distributions.

3.3 Representation of the GBG-II Distribution

When a new distribution is proposed it should includes a theoretical representation result to connect the new model with existing distributions. It is easy to see that the GBG-II distribution is represented by means of the classic beta distribution as follows.

Let Y be a random variable which is distributed according to a classic beta distribution $Beta(p,q)$, with density $\frac{1}{B(p,q)}t^{p-1}(1-t)^{q-1}$, $0 < t < 1$. Consider also a parent distribution function F with respective density f. Then it can be easily seen that the random variable

$$X = F^{-1}\left\{b\left(\frac{Y}{1-cY}\right)^{1/\alpha}\right\}, \text{ with } 0 \le c \le 1 \text{ and } \alpha, b > 0,$$

is distributed according to a GBG-II distribution with density (5). This representation can be used to generate random numbers from the new distribution. In this direction, someone has to generate random numbers y from a *Beta*(p,q) distribution. Let $t = b\left(\frac{y}{1-cy}\right)^{1/\alpha}$ and compute $F^{-1}(t)$. If $x = F^{-1}(t)$, then x is a random number from the GBG-II distribution with density (5).

3.4 Shannon Entropy

The role of Shannon entropy $\mathcal{H}_{Sh}$ is seminal in many fields in science and engineering. In probability and statistics Shannon entropy can be also considered as a descriptive quantity which provides with useful information about the shape of a distribution. The applications of this universal quantity are extended from the formulation of stochastic dependence between two or more random variables to the development of goodness of fit tests. The maximum entropy method introduced by Jaynes [8] in statistical physics has become a popular procedure for the estimation of the unknown distribution in finance and econometrics. Due to its importance all the recent contributions on new distributions include explicit expressions of their Shannon entropies.

Shannon entropy of the GBG-II distribution is defined by

$$\mathcal{H}_{Sh}(g_{GBG}^{II}) = -\int g_{GBG}^{II}(x) \ln g_{GBG}^{II}(x) dx$$

and it is analyzed as follows

$$\mathcal{H}_{Sh}(g_{GBG}^{II}) = -\ln \frac{\alpha}{b^{\alpha p} B(p,q)} - I_1 - (\alpha p - 1) I_2 - (q-1) I_3 + (p+q) I_4,$$

where I_m, $m = 1,2,3,4$, are the integrals

$$I_1 = \int (\ln f(x)) g_{GBG}^{II}(x) dx,$$

$$I_2 = \int (\ln F(x)) g_{GBG}^{II}(x) dx,$$

$$I_3 = \int \{\ln[1-(1-c)(F(x)/b)^{\alpha}]\} \, g_{GBG}^{II}(x) dx,$$

and

$$I_4 = \int \{\ln(1 + c(F(x)/b)^{\alpha})\} \, g_{GBG}^{II}(x) dx.$$

If we consider the density function

$$h(\omega) = \frac{1}{B(p,q)(1-c)^p}\omega^{p-1}(1-\omega)^{q-1}\left(1+\frac{c}{1-c}\omega\right)^{-(p+q)}, \qquad (12)$$

with $0 < \omega < 1$, $p > 0$, $q > 0$ and $0 \le c < 1$, then it can be seen, after heavy algebraic manipulations, that

$$I_1 = (1-c)E_h\left\{\ln\left[f\left(F^{-1}\left(\frac{bW^{1/\alpha}}{(1-c)^{1/\alpha}}\right)\right)\right]\right\},$$

$$I_2 = \frac{1}{\alpha}E_h\{\ln W\} - \left(\frac{1}{\alpha}\ln(1-c) - \ln b\right),$$

$$I_3 = E_h\{\ln(1-W)\},$$

and

$$I_4 = E_h\left\{\ln\left(1+\frac{c}{1-c}W\right)\right\},$$

where W denotes a random variable with density $h(\omega)$, given by (12). Based on the above expressions for the integrals I_m, $m = 1, 2, 3, 4$, it can be easily seen, after some manipulations, that the Shannon entropy $H_{Sh}(g^{II}_{GBG})$ of the GBG-II distribution is given by

$$\begin{aligned}\mathcal{H}_{Sh}(g^{II}_{GBG}) &= -\ln\tfrac{\alpha}{b^{\alpha p}B(p,q)} + \tfrac{\alpha p-1}{\alpha}\ln\tfrac{1-c}{b^{\alpha}} \\ &\quad -E_h\left\{\ln\left[f\left(F^{-1}\left(\tfrac{bW^{1/\alpha}}{(1-c)^{1/\alpha}}\right)\right)\right]\right\} - \tfrac{\alpha-1}{\alpha}E_h\{\ln W\} \\ &\quad +\mathcal{H}_{Sh}(h) - \ln\left(B(p,q)(1-c)^p\right),\end{aligned} \qquad (13)$$

where W denotes a random variable with density $h(\omega)$, defined by (12).

Remark 4.
(i) The mean value $E_h\{\ln W\}$ can be expressed in a series form as follows:

$$E_h\{\ln W\} = -\frac{1}{B(p,q)(1-c)^p}\sum_{\nu=1}^{\infty}\frac{1}{\nu}B(p,\nu+q)\ {}_2F_1(p+q,p;p+q+\nu;c/(1-c)).$$

For $c = 0$, it is easily obtained that $E_h\{\ln W\} = (d/dp)\ln B(p,q)$.

(ii) Equation (12) defines the density function of a generalized beta distribution which is not included, to the best of our knowledge, in the literature and it needs therefore an independent study. For $c = 0$ it reduces to the classic beta distribution.

(iii) The Shannon entropy of the GBG-1 distribution with density given by (6), can be obtained by using (13) for $c = 0$. After some algebraic manipulation it can be shown that

$$\begin{aligned}\mathcal{H}_{Sh}(g^{(1)}_{GBG}) &= -\ln\tfrac{\alpha}{b^{\alpha p}B(p,q)} - \tfrac{\alpha-1}{\alpha}[\Psi(p) - \Psi(p+q)] + (p+q-2)\Psi(p+q) \\ &\quad -(p-1)\Psi(p) - (q-1)\Psi(q) - E_h\left\{\ln\left[f\left(F^{-1}\left(bW^{1/\alpha}\right)\right)\right]\right\},\end{aligned}$$

where $W \sim Beta(p,q)$. For $F(x) = x$, $0 < x < 1$ and $b = 1$ the above formula coincides with the formula of Proposition 1 of Alexander and Sarabia [2].

It is clear that Shannon entropy of the GBG-II distribution is not obtained in a closed form. It is also observed that it is factorized into two parts. The first is related to the parameters of the model and the other one is related to the parent distribution F. Moreover, all the members of the family of GBG-II distributions are discriminated between each other by means of the term $E_h\left\{\ln\left[f\left(F^{-1}\left(bW^{1/\alpha}/(1-c)^{1/\alpha}\right)\right)\right]\right\}$, which depends on the parent distribution F and the parent density f. This term plays a key role to introduce a test for discriminating between the members of the GBG-II distribution, in a manner similar to that which is developed in Zografos and Balakrishnan [24].

4 Examples

It is quite clear that the GBG-II distribution, with density (5), is the basis for the definition of a multitude of univariate distributions for special cases of the parent model F and special choices of the parameters of g_{GBG}^{II}. In this section two broad models and their relationship with existing distributions will be defined.

4.1 Generalized Beta Generated by Weibull (GBW)

Let the parent distribution F is the two parameter Weibull distribution with scale parameter $\lambda > 0$ and shape parameter $\gamma > 0$. The parent density is $f(x) = \gamma\lambda^\gamma x^{\gamma-1}\exp\left(-(\lambda x)^\gamma\right)$ and $F(x) = 1-\exp\left(-(\lambda x)^\gamma\right)$, for $x > 0$. Based on (5), the density of the GBW distribution is given by

$$\begin{aligned} g_{GBW}^{(II)}(x) &= \tfrac{\alpha}{b^{\alpha p}B(p,q)}\gamma\lambda^\gamma x^{\gamma-1}\exp\left(-(\lambda x)^\gamma\right)\left[1-\exp\left(-(\lambda x)^\gamma\right)\right]^{\alpha p-1} \\ &\times\left\{1-(1-c)\left[\left(1-\exp\left(-(\lambda x)^\gamma\right)\right)/b\right]^\alpha\right\}^{q-1} \\ &\times\left\{1+c\left[\left(1-\exp\left(-(\lambda x)^\gamma\right)\right)/b\right]^\alpha\right\}^{-(p+q)}, \end{aligned}$$

for $0 < x < \frac{1}{\lambda}\left\{-\ln\left(1-\frac{b}{(1-c)^{1/\alpha}}\right)^{1/\gamma}\right\}$ and $0 \leq c < 1$, with α, b, p, q positive. This is a broad family leading to several models. For $\gamma = 1$, it is obtained the generalized beta exponential distribution (GBED) with density

$$\begin{aligned} g_{GBEG}^{(II)}(x) &= \tfrac{\alpha}{b^{\alpha p}B(p,q)}\lambda\exp\left(-\lambda x\right)\left[1-\exp\left(-(\lambda x)\right)\right]^{\alpha p-1} \\ &\times\left\{1-(1-c)\left[\left(1-\exp\left(-(\lambda x)\right)\right)/b\right]^\alpha\right\}^{q-1} \\ &\times\left\{1+c\left[\left(1-\exp\left(-(\lambda x)\right)\right)/b\right]^\alpha\right\}^{-(p+q)}, \end{aligned}$$

for $0 < x < \frac{1}{\lambda}\left\{-\ln\left(1-\frac{b}{(1-c)^{1/\alpha}}\right)\right\}$ and $0 \leq c < 1$, with α, b, p, q positive. This model is not appeared in the literature, to the best of our knowledge. For $c = 0$ and $b = 1$ this last distribution leads to the beta generalized exponential distribution which has been introduced and studied recently by Barreto-Souza *et al.* [4]. If, in addition, $p = q = 1$, then the GBED is reduced to the exponentiated exponential family of Gupta and Kundu [7].

4.2 Generalized Beta Generated by Kumaraswamy (GBK)

Let the parent distribution is the Kumaraswamy distribution (cf. Jones [10]) with positive shape parameters α_1 and β_1 and density and distribution functions, given by

$$f(x) = \alpha_1\beta_1 x^{\alpha_1-1}(1-x^{\alpha_1})^{\beta_1-1},\ 0<x<1,$$
$$F(x) = 1-(1-x^{\alpha_1})^{\beta_1},\ 0<x<1.$$

The Generalized Beta Generated by Kumaraswamy (GBK) distribution has density

$$\begin{aligned} g_{GBK}^{(II)}(x) = &\tfrac{\alpha}{b^{\alpha p}B(p,q)}\alpha_1\beta_1 x^{\alpha_1-1}(1-x^{\alpha_1})^{\beta_1-1}\left[1-(1-x^{\alpha_1})^{\beta_1}\right]^{\alpha p-1}\\ &\times\left\{1-(1-c)\left[\left(1-(1-x^{\alpha_1})^{\beta_1}\right)/b\right]^{\alpha}\right\}^{q-1}\\ &\times\left\{1+c\left[\left(1-(1-x^{\alpha_1})^{\beta_1}\right)/b\right]^{\alpha}\right\}^{-(p+q)}, \end{aligned}$$

for $0 < x < \left[1-\left(1-\frac{b}{(1-c)^{1/\alpha}}\right)^{1/(\beta_1-1)}\right]^{1/\alpha_1}$ and $0 \le c < 1$, with α, b, p, q positive. For $c = 0$ and $b = 1$ it is reduced to the density

$$\begin{aligned} g_{GBK}^{(II)}(x) = &\frac{\alpha\alpha_1\beta_1}{B(p,q)}x^{\alpha_1-1}(1-x^{\alpha_1})^{\beta_1-1}\left[1-(1-x^{\alpha_1})^{\beta_1}\right]^{\alpha p-1}\\ &\times\left\{1-\left[1-(1-x^{\alpha_1})^{\beta_1}\right]^{\alpha}\right\}^{q-1}, \end{aligned} \tag{14}$$

for $0 < x < 1$. It is a quite general model which leads to Kumaraswamy distribution, the beta distribution, the beta Burr XII distribution, introduced recently by Paranaiba *et al.* [21], etc. Plots of the above density for specific parameter values are given in the next figures.

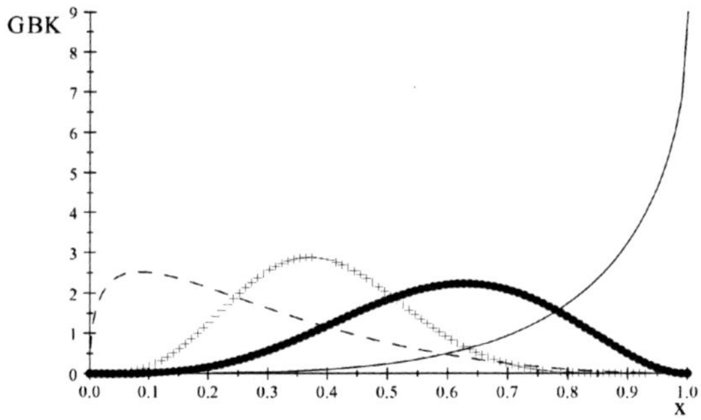

Fig. 1. Plot of the density of GBK for selected values of α_1 and β_1

Figure 1 corresponds to the density (14) for $\alpha = 1, p = 3$ and $q = 3$, while $\alpha_1 = 3/2, \beta_1 = 1/2$ (solid line), $\alpha_1 = 1/2, \beta_1 = 3/2$ (dash), $\alpha_1 = 3/2, \beta_1 = 7/2$ (cross) and $\alpha_1 = 3/2, \beta_1 = 3/2$ (box). It is clear that the parameters α_1 and β_1 are related to the skewness of the GBK distribution with density (14).

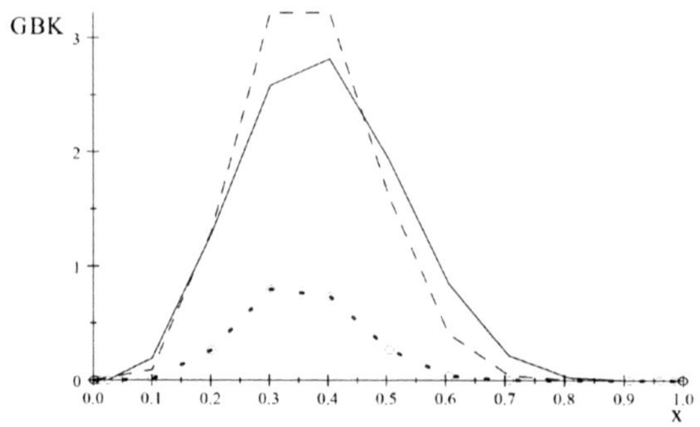

Fig. 2. Plot of the density of GBK for selected values of p and q

Figure 2 is the plot of (14) for $\alpha = 1$, $\alpha_1 = 3/2$ and $\beta_1 = 7/2$, while $p = 3, q = 2$ (solid line), $p = 4, q = 3$ (dash) and $p = 5, q = 4$ (dash and point). It is clear that the parameters p and q are related to the kurtosis of the GBK distribution with density (14).

5 Conclusions

A broad family of univariate distributions is defined in this paper. This family is created on the basis of a generalized beta distribution which was proposed by McDonald and Xu [15]. The introduced model includes several univariate distributions, as particular or limiting cases, with different shapes. Given that the proposed family is supported on a bounded domain it would be probably a useful model to formulate size phenomena, such as the distribution of the income. Some properties of the introduced model have been studied in this paper. However, much more work is in order, related to the study of other properties and the development of statistical inference on the parameters of the GBG-II distribution. The investigation of the usefulness of the proposed model to formulate and analyze real data is also an appealing problem.

References

1. Abramowitz, M., Stegun, I.A.: Handbook of Mathematical Functions With Formulas, Graphs, and Mathematical Tables. Dover, New York (1965)
2. Alexander, C., Sarabia, J.M.: Generalized beta generated distributions. ICMA Centre Discussion Papers in Finance DP2010-09 (2010)
3. Akinsete, A., Famoye, F., Lee, C.: The beta-Pareto distribution. Statistics 42, 547–563 (2008)

4. Barreto-Souza, W., Santos, A.H.S., Cordeiro, G.M.: The beta generalized exponential distribution. J. Statist. Comp. Simul. 80, 159–172 (2010)
5. Brown, B.W., Spears, F.M., Levy, L.B.: The log F: a distribution for all seasons. Comput. Statist. 17, 47–58 (2002)
6. Eugene, N., Lee, C., Famoye, F.: Beta-normal distribution and its applications. Comm. Statist.-Theor. Meth. 31, 497–512 (2002)
7. Gupta, R.D., Kundu, D.: Exponentiated exponential family: an alternative to gamma and Weibull distributions. Biom. J. 43, 117–130 (2001)
8. Jaynes, E.T.: Information theory and statistical mechanics. Phys. Rev. 106, 620–630 (1957)
9. Jones, M.C.: Families of distributions arising from distributions of order statistics. TEST 13, 1–43 (2004)
10. Jones, M.C.: Kumaraswamy's distribution: A beta-type distribution with some tractability advantages. Statist. Meth. 6, 70–81 (2009)
11. Kleiber, C., Kotz, S.: Statistical Size Distributions in Economics and Actuarial Sciences. Wiley, New York (2003)
12. Kong, L., Lee, C., Sepanski, J.H.: On the properties of beta-gamma distribution. J. Modern Appl. Statist. Meth. 6, 187–211 (2007)
13. Lee, C., Famoye, F., Olumolade, O.: Beta-Weibull distribution: some properties and applications to censored data. J. Modern Appl. Statist. Meth. 6, 173–186 (2007)
14. McDonald, J.B.: Some generalized functions for the size distribution of income. Econometrica 52, 647–663 (1984)
15. McDonald, J.B., Xu, Y.J.: A generalization of the beta distribution with applications. J. Economet. 66, 133–152 (1995)
16. McDonald, J.B., Ransom, M.: The geralized beta distribution as a model for the distribution of income: Estimation of related measures of inequality. In: Chotipakanich, D. (ed.) Economic Studies in Inequality: Social Exclusion and Well-Being: Modeling Income Distributions and Lorenz Curves, vol. 5, pp. 147–166. Springer, Heidelberg (2008)
17. Nadarajah, S., Gupta, A.K.: The beta Fréchet distribution. Far. East J. Theor. Stat. 14, 15–24 (2004)
18. Nadarajah, S., Kotz, S.: The beta Gumbel distribution. Math. Probl. Eng. 4, 323–332 (2004)
19. Nadarajah, S., Kotz, S.: The beta exponential distribution. Reliab. Eng. Syst. Safety 91, 689–697 (2006)
20. Olapade, A.K.: On extended type I generalized logistic distribution. Int. J. Math. Math. Sci. 57, 3069–3074 (2004)
21. Paranaiba, P.F., Ortega, E.M.M., Cordeiro, G.M., Pescim, R.R.: The beta burr XII distribution with application to lifetime data. Comp. Statist. Data Anal. 55, 1118–1136 (2010)
22. Pescim, R.R., Demetrio, C.G.B., Cordeiro, G.M., Ortega, E.M.M., Urbano, M.R.: The beta generalized half-normal distribution. Comp. Statist. Data Anal. 54, 945–957 (2010)
23. Zografos, K.: On some beta generated distributions and their maximum entropy characterization: The beta-Weibull distribution. In: Barnett, N.S., Dragomir, S.S. (eds.) Advances in Inequalities from Probability Theory & Statistics, pp. 237–260. Nova Science Publishers, New Jersey (2008)
24. Zografos, K., Balakrishnan, N.: On families of Beta- and generalized gamma- generated distributions and associated inference. Statist. Meth. 6, 344–362 (2009)

Part IV

Divergence Measures and Statistical Applications

Using Power-Divergence Statistics to Test for Homogeneity in Product-Multinomial Distributions*

Noel Cressie[1] and Frederick M. Medak[2]

[1] Department of Statistics, The Ohio State University, Columbus OH 43210, USA
ncressie@stat.osu.edu
[2] Deceased

Summary. Testing for homogeneity in the product-multinomial distribution, where the hypotheses are hierarchical, uses maximum likelihood estimation and the loglikelihood ratio statistic G^2. We extend these ideas to the power-divergence family of test statistics, which is a one-parameter family of goodness-of-fit statistics that includes the loglikelihood ratio statistic G^2, Pearson's X^2, the Freeman-Tukey statistic, the modified loglikelihood ratio statistic, and the Neyman-modified chi-squared statistic. Explicit minimum-divergence estimators can be obtained for all members of the one-parameter family, which allows a straightforward analysis of divergence. An analysis of fourteen retrospective studies on the association between smoking and lung cancer demonstrates the ease of interpretation of the resulting analysis of divergence.

Keywords: analysis of divergence, loglikelihood ratio statistic, Pearson's X^2, power-divergence family.

1 Introduction

The search for similarities among independent groups of experimental units underlies many scientific investigations. Assume that the experimental units within each group are distributed among a set of categories according to a discrete probability distribution. For example, suppose there are r independent groups (or strata) and that each experimental unit in a group falls into one of c categories according to some discrete probability distribution. For instance, such assumptions are often appropriate when categorical data are collected at several locations. Our objective in this paper is to present statistical methods for determining which groups are homogeneous, that is, which groups have similar distributions of experimental units among the categories.

* I would like to acknowledge the fruitful collaboration in this area with my friend and colleague, Leandro Pardo. His wife, Marisa, was my friend too, and she will be greatly missed. This article was prepared in her memory.

Suppose that a random sample of experimental units of size n_i is drawn from the probability distribution of the c categories in the i-th group. Then, the number of experimental units belonging to each of the categories, divided by n_i, yields an estimate of the true probability distribution ($i = 1, \dots, r$).

To discover which, if any, subsets of the r groups exhibit homogeneity, we proceed as follows. We assume that the product-multinomial distribution provides an appropriate probability model for the sampling scheme; that is, the r multinomial distributions of order c, corresponding to the r different groups, are assumed independent.

We hypothesize that there is homogeneity among the distributions corresponding to a specified subset of the r groups. For each group in the specified subset, we could obtain, via maximum likelihood, an estimate (under the corresponding hypothesis) of the distribution. Finally, to test the fit of the estimated distributions, we could compare the appropriate loglikelihood ratio statistic to a chi-squared distribution with a suitable number of degrees of freedom.

Cressie and Read [6] have shown that the loglikelihood ratio statistic G^2, Pearson's chi-squared statistic X^2, the Freeman-Tukey statistic, the modified loglikelihood ratio statistic, and the Neyman-modified chi-squared statistic are members of a one-parameter family of goodness-of-fit statistics called the power-divergence family. Cressie and Pardo [3] extend this further to the family of ϕ-divergence goodness-of-fit statistics, of which the power-divergence family is an important subfamily.

Predicated on the equivalence of maximizing likelihood and minimizing the loglikelihood ratio statistic for the Poisson, multinomial, and product-multinomial distributions, [6], [4], and [5] consider estimation procedures based on minimizing members of the power-divergence family. In particular, for the product-multinomial, [13] obtain closed-form expressions for the minimum-power-divergence estimators. Using such expressions, we show how to test hierarchical hypotheses of homogeneity in the product-multinomial distribution. Further, we show that under mild assumptions the corresponding test statistics have limiting central chi-squared distributions and are asymptotically independent.

In Section 2, we define and give properties of the power-divergence family. Section 3 is devoted to developing hierarchical testing procedures for homogeneity in the product-multinomial distribution. Then, in Section 4, we give an example with data to illustrate these ideas. Conclusions are given in Section 5.

2 The Power-Divergence Family

Define the positive orthant of k-dimensional Euclidean space as,

$$\mathbb{R}_+^k \equiv \{\mathbf{z} \in \mathbb{R}^k \colon z_i > 0\,, \quad i = 1, \dots, k\}\,, \tag{1}$$

and the power-divergence as in[13] as,

$$2I^{\lambda}(\mathbf{x}\colon \mathbf{y}) \equiv (2/\{\lambda(\lambda+1)\}) \sum_{i=1}^{k} \{x_i[(x_i/y_i)^{\lambda} - 1] + \lambda(y_i - x_i)\}\,;$$

$$-\infty < \lambda < \infty\,, \qquad (2)$$

where $\mathbf{x} \in \overline{\mathbb{R}}_{+}^{k}$, $\mathbf{y} \in \mathbb{R}_{+}^{k}$ for $\lambda \geq 0$, $\mathbf{x} \in \mathbb{R}_{+}^{k}$, $\mathbf{y} \in \overline{\mathbb{R}}_{+}^{k}$ for $\lambda < 0$, and $\overline{\mathbb{R}}_{+}^{k} \equiv \{\mathbf{z} \in \mathbb{R}^k \colon z_i \geq 0, \quad i = 1, \ldots k\}$ is the closure of $\mathbb{R}_{+}^{k}$. The cases $\lambda = 0$ and -1 are defined by the limits, $\lambda \to 0$ and $\lambda \to -1$, respectively, yielding:

$$2I^{0}(\mathbf{x}\colon \mathbf{y}) \equiv 2 \sum_{i=1}^{k} \{x_i \log(x_i/y_i) + (y_i - x_i)\}\,, \qquad (3)$$

and

$$2I^{-1}(\mathbf{x}\colon \mathbf{y}) \equiv 2 \sum_{i=1}^{k} \{y_i \log(y_i/x_i) + (x_i - y_i)\}\,, \qquad (4)$$

where $\omega \cdot \log(\omega) \equiv 0$, for $\omega = 0$.

Next, define the $(k-1)$-dimensional simplex,

$$\Delta_k \equiv \left\{ \boldsymbol{\gamma} \in \mathbb{R}^k \colon \sum_{i=1}^{k} \gamma_i = 1 \ \text{ and } \ \gamma_i > 0\,,\ i = 1, \ldots, k \right\}. \qquad (5)$$

Consequently,

$$2I^{\lambda}(\mathbf{p}\colon \mathbf{q}) \equiv (2/\{\lambda(\lambda+1)\}) \sum_{i=1}^{k} p_i[(p_i/q_i)^{\lambda} - 1]\,; \quad -\infty < \lambda < \infty\,, \qquad (6)$$

where $\mathbf{p} \in \overline{\Delta}_k$, $\mathbf{q} \in \Delta_k$ for $\lambda \geq 0$, $\mathbf{p} \in \Delta_k$, $\mathbf{q} \in \overline{\Delta}_k$ for $\lambda < 0$, and $\overline{\Delta}_k \equiv \{\boldsymbol{\gamma} \in \mathbb{R}^k \colon \sum_{i=1}^{k} \gamma_i = 1, \gamma_i \geq 0, i = 1, \ldots, k\}$ is the closure of Δ_k. Cressie and Read [6] call the set of divergences, $\{2I^{\lambda}(\mathbf{p}\colon \mathbf{q}) \colon -\infty < \lambda < \infty\}$, the power-divergence family (with index λ).

Some properties of the power-divergence family ([6], [13]) include:

$$2I^{\lambda}(\mathbf{x}\colon \mathbf{y}) \geq 0\,, \quad \text{with equality if and only if } \ \mathbf{x} = \mathbf{y}\,; \qquad (7)$$

$$2I^{\lambda}(\mathbf{x}\colon \mathbf{y}) = 2I^{-(\lambda+1)}(\mathbf{y}\colon \mathbf{x})\,, \ \text{ for } \ \mathbf{x}, \mathbf{y} \in \mathbb{R}_{+}^{k}\,; \qquad (8)$$

$$2I^{\lambda}(\mathbf{x}\colon \mathbf{y}) \quad \text{is strictly convex in both } \mathbf{x} \text{ and } \mathbf{y}\,; \qquad (9)$$

$$2I^{\lambda}(\mathbf{x}\colon \mathbf{y}) \quad \text{is continuous in both } \mathbf{x} \text{ and } \mathbf{y}\,; \qquad (10)$$

$$2I^{\lambda}(\mathbf{x}\colon \mathbf{y}) \equiv \sum_{i=1}^{k} h^{\lambda}(x_i, y_i)\,, \qquad (11)$$

where

$$h^\lambda(x_i, y_i) \equiv (2/\{\lambda(\lambda+1)\})\{x_i[(x_i/y_i)^\lambda - 1] + \lambda(y_i - x_i)\}\,, \tag{12}$$

$$h^\lambda(x_i, y_i) \geq 0\,, \tag{13}$$

and

$$h^\lambda(x_i, y_i) = 0 \quad \text{if and only if} \quad x_i = y_i\,. \tag{14}$$

Property (7) indicates that the members of the power-divergence family resemble distance functions, while (8) shows that the symmetry property of a distance function holds only for the $\lambda = -1/2$ member of the power-divergence family. (In fact, the $\lambda = -1/2$ member is the square of a distance function, sometimes called Matusita's distance or the Hellinger distance.)

An additional property, which generalizes the recursivity and the strong nonadditivity properties (e.g., [13], p. 111) and is especially useful when considering the product-multinomial distribution, is the following. Let $\{\mathbf{e}_1, \ldots, \mathbf{e}_k\}$ represent the standard basis for $\mathbb{R}^k$; that is, $\mathbf{e}_i$ $(i = 1, \ldots, k)$ is the i-th column of the $k \times k$ identity matrix I_k. Further, for $i = 1, \ldots, p \leq k$, define $\mathbf{h}_i \equiv \sum_{j \in \mathcal{S}_i} \mathbf{e}_j$, where $\mathcal{S}_i \subset \{1, \ldots, k\}$, $\mathcal{S}_i \cap \mathcal{S}_{i'} = \emptyset$ for $i \neq i'$, and $\bigcup_{i=1}^{p} \mathcal{S}_i = \{1, \ldots, k\}$. Then, letting $H \equiv [\mathbf{h}_1, \ldots, \mathbf{h}_p]$, which is a $k \times p$ matrix, one obtains,

$$\begin{aligned} 2I^\lambda(\mathbf{x}\colon \mathbf{y}) &= 2I^\lambda(H^T\mathbf{x}\colon H^T\mathbf{y}) \\ &\quad + \sum_{i=1}^{p} (\mathbf{h}_i^T\mathbf{x})^{\lambda+1}(\mathbf{h}_i^T\mathbf{y})^{-\lambda} \cdot 2I^\lambda(\mathbf{x}_i/\mathbf{h}_i^T\mathbf{x}\colon \mathbf{y}_i/\mathbf{h}_i^T\mathbf{y})\,, \end{aligned} \tag{15}$$

where $\mathbf{x}_i \equiv B_i^T\mathbf{x}$, $\mathbf{y}_i \equiv B_i^T\mathbf{y}$, and the columns of B_i are those members of the standard basis indexed by $\mathcal{S}_i$. Thus, B_i is a $k \times |\mathcal{S}_i|$ matrix, where $|\mathcal{S}_i|$ is the cardinality of $\mathcal{S}_i$ $(i = 1, \ldots, p)$.

The one-parameter family of divergences given in (6) is now extended to a one-parameter family of test statistics. Let $\mathbf{X}$ be a k-dimensional random vector with nonnegative integer components where $\sum_{i=1}^{k} X_i = n$, and n is fixed. Define, $\boldsymbol{\pi} \equiv E(\mathbf{X}/n)$; then $\boldsymbol{\pi}$ is a discrete probability distribution. The power-divergence family of test statistics is defined as,

$$2nI^\lambda(\mathbf{X}/n\colon \hat{\boldsymbol{\pi}}) = (2/\{\lambda(\lambda+1)\}) \sum_{i=1}^{k} X_i[(X_i/n\hat{\pi}_i)^\lambda - 1]\,; \quad -\infty < \lambda < \infty\,, \tag{16}$$

where $\widehat{\boldsymbol{\pi}}$ is an estimate of $\boldsymbol{\pi}$ based on the data $\mathbf{X}$. The family defined in (16) contains many of the most common goodness-of-fit statistics. Upon letting $\lambda = 1, 0, -1/2, -1, -2$, we obtain, respectively,

$$\begin{aligned}
X^2 &= \sum_{i=1}^{k} (X_i - n\widehat{\pi}_i)^2/n\widehat{\pi}_i && \text{(Pearson's chi-squared, } \lambda = 1\text{)}\,, \\
CR^2 &= (9/5) \sum_{i=1}^{k} X_i((X_i/n\widehat{\pi}_i)^{2/3} - 1) && \text{(Cressie-Read, } \lambda = 2/3\text{)}\,, \\
G^2 &= 2 \sum_{i=1}^{k} X_i \log(X_i/n\widehat{\pi}_i) && \text{(loglikelihood ratio, } \lambda = 0\text{)}, \\
FT^2 &= 4 \sum_{i=1}^{k} (X_1^{1/2} - (n\widehat{\pi}_i)^{1/2})^2 && \text{(Freeman-Tukey, } \lambda = -1/2\text{)}, \\
GM^2 &= 2 \sum_{i=1}^{k} n\widehat{\pi}_i \log(n\widehat{\pi}_i/X_i) && \text{(modified loglikelihood ratio, } \lambda = -1\text{)}, \\
NM^2 &= \sum_{i=1}^{k} (X_i - n\widehat{\pi}_i)^2/X_i && \text{(Neyman-modified chi-squared, } \lambda = -2\text{)}.
\end{aligned}$$

An extension, particularly useful in the product-multinomial case, is obtained as follows. Define,

$$\mathbf{X} \equiv \begin{bmatrix} \mathbf{X}_1 \\ \vdots \\ \mathbf{X}_r \end{bmatrix}, \tag{17}$$

where, for $i = 1, \ldots, r$, $\mathbf{X}_i$ is a c-dimensional random vector with nonnegative components, $n_i = \mathbf{1}_c^T \mathbf{X}_i$, n_i is fixed, and $\mathbf{1}_c$ is a c-dimensional vector of ones. Further, define $\boldsymbol{\pi}_i \equiv E(\mathbf{X}_i/n_i)$, for $i = 1, \ldots, r$, so that $\{\boldsymbol{\pi}_i : i = 1, \ldots, r\}$ is a discrete probability distribution. Then, by (15), we obtain the following equality between power-divergence test statistics:

$$2I^{\lambda}(\mathbf{X} : \widehat{\boldsymbol{\mu}}) = \sum_{i=1}^{r} 2n_i I^{\lambda}(\mathbf{X}_i/n_i : \widehat{\boldsymbol{\pi}}_i)\,, \tag{18}$$

where $\widehat{\boldsymbol{\pi}}_i$ is an estimate of $\boldsymbol{\pi}_i$ based on the data $\mathbf{X}$ for $i = 1, \ldots, r$, and $\widehat{\boldsymbol{\mu}} \equiv (n_1 \widehat{\boldsymbol{\pi}}_i^T, \ldots, n_r \widehat{\boldsymbol{\pi}}_r^T)^T$.

3 Minimum Power-Divergence Estimation and Testing of Hierarchical Hypotheses

We now consider estimation procedures for the product-multinomial distribution. We shall apply these procedures to hierarchical testing for homogeneity among the independent multinomials comprising the product-multinomial.

Define,

$$\mathbf{X} \equiv \begin{bmatrix} \mathbf{X}_1 \\ \vdots \\ \mathbf{X}_r \end{bmatrix}, \tag{19}$$

to be an rc-dimensional vector of nonnegative integers, where the c-dimensional random vectors $\mathbf{X}_i$, $i = 1, \ldots, r$, are independent. Further, suppose $\mathbf{X}$ follows the r-dimensional product-multinomial distribution of order c with parameters,

$$\mathbf{n} \equiv \begin{bmatrix} n_1 \\ \vdots \\ n_r \end{bmatrix}, \tag{20}$$

and

$$\boldsymbol{\pi} \equiv \begin{bmatrix} \boldsymbol{\pi}_1 \\ \vdots \\ \boldsymbol{\pi}_r \end{bmatrix}, \tag{21}$$

where $\mathbf{n}$ is r-dimensional and $\boldsymbol{\pi}$ is rc-dimensional. We denote,

$$\mathbf{X} \sim \mathrm{Mult}_c^r(\mathbf{n}, \boldsymbol{\pi})\,; \tag{22}$$

observe that,

$$\Pr(\mathbf{X} = \mathbf{x}) = \prod_{i=1}^{r} \Pr(\mathbf{X}_i = \mathbf{x}_i)\,, \tag{23}$$

where, for each $i = 1, \ldots, r$,

$$\Pr(\mathbf{X}_i = \mathbf{x}_i) \equiv (n_i!/x_{i1}! \cdots x_{ic}!) \prod_{j=1}^{c} \pi_{ij}^{x_{ij}}\,; \tag{24}$$

$x_{ij} \geq 0$ for $j = 1, \ldots, c$; $\sum_{j=1}^{c} x_{ij} = n_i$; $\pi_{ij} > 0$ for $j = 1, \ldots, c$; and $\sum_{j=1}^{c} \pi_{ij} = 1$.

Define the general null model for $\boldsymbol{\pi}$ to be,

$$H_0 \colon \boldsymbol{\pi} \in \Pi_0\,, \tag{25}$$

where $\Pi_0 \subseteq \Delta_c^r$ and

$$\Delta_c^r \equiv \{\boldsymbol{\gamma} \in \mathbb{R}^{rc} \colon \boldsymbol{\gamma} = (\boldsymbol{\gamma}_1^T, \ldots, \boldsymbol{\gamma}_r^T)^T\,,\ \boldsymbol{\gamma}_i \in \Delta_c\,,\ i = 1, \ldots, r\}\,. \tag{26}$$

Now the method of maximum likelihood is equivalent to minimizing G^2 (i.e., $\lambda = 0$) over $\boldsymbol{\pi} \in \Pi_0$ [2]. That is, find $\widehat{\boldsymbol{\pi}}$ such that

$$G^2(\mathbf{X}\colon \widehat{\boldsymbol{\mu}}) \equiv 2I^0(\mathbf{X}\colon \widehat{\boldsymbol{\mu}}) = \inf_{\boldsymbol{\pi}\in\Pi_0} 2I^0(\mathbf{X}\colon \boldsymbol{\mu}(\boldsymbol{\pi}))\,, \tag{27}$$

where $\widehat{\boldsymbol{\pi}} \equiv (\widehat{\boldsymbol{\pi}}_1^T, \ldots, \widehat{\boldsymbol{\pi}}_r^T)^T$, $\boldsymbol{\mu}(\boldsymbol{\pi}) \equiv (n_1\boldsymbol{\pi}_1^T, \ldots, n_r\boldsymbol{\pi}_r^T)^T$, $\widehat{\boldsymbol{\mu}} \equiv \boldsymbol{\mu}(\widehat{\boldsymbol{\pi}})$ and $2I^0(\mathbf{X}\colon \boldsymbol{\mu}(\boldsymbol{\pi})) = \sum_{i=1}^{r} 2n_i I^0(\mathbf{X}_i/n_i\colon \boldsymbol{\pi}_i)$.

This leads to the natural generalization, namely, find $\widehat{\boldsymbol{\pi}}^{(\lambda)}$ such that,

$$2I^\lambda(\mathbf{X}\colon \widehat{\boldsymbol{\mu}}^{(\lambda)}) \equiv \inf_{\boldsymbol{\pi}\in\Pi_0} 2I^\lambda(\mathbf{X}\colon \boldsymbol{\mu}(\boldsymbol{\pi}))\,; \quad -\infty < \lambda < \infty\,, \tag{28}$$

where $\widehat{\boldsymbol{\pi}}^{(\lambda)} \equiv (\widehat{\boldsymbol{\pi}}_1^{(\lambda)^T}, \ldots, \widehat{\boldsymbol{\pi}}_r^{(\lambda)^T})^T$, $\boldsymbol{\mu}(\boldsymbol{\pi}) \equiv (n_1\boldsymbol{\pi}_1^T, \ldots, n_r\boldsymbol{\pi}_r^T)^T$, $\widehat{\boldsymbol{\mu}}^{(\lambda)} \equiv \boldsymbol{\mu}(\widehat{\boldsymbol{\pi}}^{(\lambda)})$, and $2I^\lambda(\mathbf{X}\colon \boldsymbol{\mu}(\boldsymbol{\pi})) = \sum_{i=1}^{r} 2n_i I^\lambda(\mathbf{X}_i/n_i\colon \boldsymbol{\pi}_i)$.

Then $\widehat{\boldsymbol{\pi}}^{(\lambda)}$ (if it exists) is called the minimum power-divergence estimator (MPE) based on the data $\mathbf{X}$. If it exists, such an estimator is unique by the *strict* convexity of I^λ.

Now consider testing for homogeneity among subsets of the r groups. That is, consider hypotheses of the form:

$$H_0\colon \boldsymbol{\pi}_{r_0+1} = \cdots = \boldsymbol{\pi}_{r_1} = \boldsymbol{\pi}^{(1)}, \ldots, \boldsymbol{\pi}_{r_{k-1}+1} = \cdots = \boldsymbol{\pi}_{r_k} = \boldsymbol{\pi}^{(k)}\,, \tag{29}$$

where $r_0 = 0$, $r_k = r$, and $\boldsymbol{\pi}^{(i)} \in \Delta_c$ for $i = 1, \ldots, k \le r$ are unspecified. That is, in (25), $\Pi_0 = \{\boldsymbol{\pi} \in \Delta_c^r\colon \boldsymbol{\pi}_{r_{j-1}+1} = \cdots = \boldsymbol{\pi}_{r_j},\ j = 1, \ldots, k\}$. In (29), we have assumed, without loss of generality, that the subscripts are consecutive integers. Read and Cressie [13] show that the MPEs, $\widehat{\boldsymbol{\pi}}^{(1;\lambda)}, \ldots, \widehat{\boldsymbol{\pi}}^{(k;\lambda)}$ of $\boldsymbol{\pi}^{(1)}, \ldots, \boldsymbol{\pi}^{(k)}$, respectively, are for $t = 1, \ldots, k$ given by

$$\widehat{\pi}_s^{(t;\lambda)} = z_s(t;\lambda) \Big/ \sum_{j=1}^{c} z_j(t;\lambda)\,; \quad s = 1, \ldots, c\,, \tag{30}$$

where $z_s(t;\lambda) \equiv \left[\sum_{i=r_{t-1}+1}^{r_t} n_i (x_{is}/n_i)^{(\lambda+1)}\right]^{1/(\lambda+1)}$; $s = 1, \ldots, c$.

The following theorem establishes the asymptotic behavior of the *minimized* power-divergence statistic under the model specified in (29).

Theorem 1. *Suppose* $\mathbf{X} \sim Mult_c^r(\mathbf{n}, \boldsymbol{\pi})$ *where* $\boldsymbol{\pi}$ *satisfies the null hypothesis specified in (29); that is,* $\boldsymbol{\pi} \in \Pi_0$. *Define* $n \equiv \sum_{i=1}^{r} n_i$, *and suppose that* $n_i/n \to \gamma_i$, *as* $n \to \infty$, *where* $0 < \gamma_i < \infty$, *for* $i = 1, \ldots, r$. *Then,*

$$2I^{\lambda}(\mathbf{X}\colon \widehat{\boldsymbol{\mu}}^{(\lambda)}) \xrightarrow{\mathcal{D}} \chi^2_{(r-k)(c-1)}\,, \tag{31}$$

as $n \to \infty$, *where* "$\xrightarrow{\mathcal{D}}$" *denotes "convergence in distribution" and* $\widehat{\boldsymbol{\mu}}^{(\lambda)} \equiv (n_1\widehat{\boldsymbol{\pi}}^{(1;\lambda)^T}, \ldots, n_{r_1}\widehat{\boldsymbol{\pi}}^{(1;\lambda)^T}, \ldots, n_{r_{k-1}+1}\widehat{\boldsymbol{\pi}}^{(k;\lambda)^T}, \ldots, n_r\widehat{\boldsymbol{\pi}}^{(k;\lambda)^T})^T$.

Proof. See the Appendix. □

Next, consider a sequence of hypotheses $H_{0,1}, \ldots, H_{0,m}$ of the form,

$$\begin{aligned} H_{0,i}\colon \boldsymbol{\pi}_{r_{0,i}+1} = \cdots = \boldsymbol{\pi}_{r_{1,i}} = \boldsymbol{\pi}^{i(1)}, \ldots, \\ \boldsymbol{\pi}_{r_{k_i-1,i}+1} = \cdots = \boldsymbol{\pi}_{r_{k_i,i}} = \boldsymbol{\pi}^{i(k_i)}\,, \end{aligned} \tag{32}$$

for $i = 1, \ldots, m$, where the number of groups of equalities satisfies $1 \le k_m < \cdots < k_1 \le r$, $r_{0,i} = 0$, and $r_{k_i,i} = r$ $(i = 1, \ldots, m)$. Further, the groups of equalities in $H_{0,i}$ are formed from those in $H_{0,i-1}$ by combining groups of equalities; hence, the sequence of hypotheses in (32) is *hierarchical.* Now, (32) is of the form,

$$H_{0,i}\colon \boldsymbol{\pi} \in \Pi_{0,i}\,, \tag{33}$$

where

$$\Pi_{0,i} \equiv \{\boldsymbol{\pi} \in \Delta_c^r\colon \boldsymbol{\pi}_{r_{j-1,i}+1} = \cdots = \boldsymbol{\pi}_{r_{j,i}},\ j = 1, \ldots, k_i\}\,; \quad i = 1, \ldots, m\,, \tag{34}$$

and $\Pi_{0,m} \subset \cdots \subset \Pi_{0,1}$.

Next, define an *analysis of divergence* for the hierarchy of hypotheses given in (33) as,

$$\begin{aligned} 2I^{\lambda}(\mathbf{X}\colon \widehat{\boldsymbol{\mu}}^{m(\lambda)}) &= 2I^{\lambda}(\mathbf{X}\colon \widehat{\boldsymbol{\mu}}^{1(\lambda)}) \\ &\quad + \sum_{j=2}^{m}\{2I^{\lambda}(\mathbf{X}\colon \widehat{\boldsymbol{\mu}}^{j(\lambda)}) - 2I^{\lambda}(\mathbf{X}\colon \widehat{\boldsymbol{\mu}}^{j-1(\lambda)})\}\,, \end{aligned} \tag{35}$$

where $\widehat{\boldsymbol{\pi}}^{i(\lambda)} \equiv (\widehat{\boldsymbol{\pi}}_1^{i(\lambda)^T}, \ldots, \widehat{\boldsymbol{\pi}}_r^{i(\lambda)^T})^T$ denotes the MPE (index λ) of $\boldsymbol{\pi}$ under $H_{0,i}$, and $\widehat{\boldsymbol{\mu}}^{i(\lambda)} \equiv (n_1\widehat{\boldsymbol{\pi}}_1^{i(\lambda)^T}, \ldots, n_r\widehat{\boldsymbol{\pi}}_r^{i(\lambda)^T})^T$. Thus,

$$2I^{\lambda}(\mathbf{X}\colon \widehat{\boldsymbol{\mu}}^{j(\lambda)}) - 2I^{\lambda}(\mathbf{X}\colon \widehat{\boldsymbol{\mu}}^{j-1(\lambda)}) \ge 0\,,$$

since $\Pi_{0,j} \subset \Pi_{0,j-1}$.

The following theorem establishes the asymptotic distributional and independence properties of the terms in the analysis of divergence given in (35).

Theorem 2. *Assume* $\mathbf{X} \sim Mult_c^r(\mathbf{n}, \boldsymbol{\pi})$. *Define* $n \equiv \sum_{i=1}^{r} n_i$, *and suppose that* $n_i/n \to \gamma_i$, *as* $n \to \infty$, *where* $0 < \gamma_i < \infty$, *for* $i = 1, \ldots, r$. *Let* $H_{0,1}, \ldots, H_{0,m}$

be the sequence of hierarchical hypotheses given in (35), and assume $H_{0,i'}$ is true (i.e., $\boldsymbol{\pi} \in \Pi_{0,i'}$). Then,

$$2I^{\lambda}(\mathbf{X} \colon \widehat{\boldsymbol{\mu}}^{i(\lambda)}) - 2I^{\lambda}(\mathbf{X} \colon \widehat{\boldsymbol{\mu}}^{j(\lambda)}) \xrightarrow{\mathcal{D}} \chi^2_{(k_j - k_i)(c-1)}\,, \tag{36}$$

as $n \to \infty$, where $1 \le j < i \le i'$ and k_t is the number of sets of equalities under $H_{0,t}$ (see (32)). Additionally, the differences $2I^{\lambda}(\mathbf{X} \colon \widehat{\boldsymbol{\mu}}^{i_2(\lambda)}) - 2I^{\lambda}(\mathbf{X} \colon \widehat{\boldsymbol{\mu}}^{i_1(\lambda)})$ and $2I^{\lambda}(\mathbf{X} \colon \widehat{\boldsymbol{\mu}}^{j_2(\lambda)}) - 2I^{\lambda}(\mathbf{X} \colon \widehat{\boldsymbol{\mu}}^{j_1(\lambda)})$ are asymptotically independent central chi-squared random variables for $1 \le j_1 < j_2 \le i_1 < i_2 \le i'$.

Proof. See the Appendix. □

4 Using an Analysis of Divergence for Testing Homogeneity

Table 1 displays data obtained from fourteen independent studies assessing the association between smoking and lung cancer ([7], [9]). These studies were conducted in the United States and Northwestern Europe (i.e., England, Finland, Germany, and the Netherlands). The sample proportions for the fourteen studies are displayed in Table 2.

Table 1. Fourteen independent retrospective studies on the association between smoking and lung cancer. [Source: [9], p. 167]

	Control Patients		Lung Cancer Patients			
Study	Non-Smokers	Smokers	Non-Smokers	Smokers	Total	Country
1	14	72	3	83	172	Germany
2	43	227	3	90	363	Germany
3	19	81	7	129	236	Netherlands
4	54	246	4	724	1028	Finland
5	12	174	5	88	279	England
6	61	1296	7	1350	2714	England
7	114	666	8	597	1385	USA
8	81	534	18	459	1092	USA
9	27	106	3	60	196	USA
10	131	299	32	412	874	USA
11	636	1729	39	451	2855	USA
12	28	259	5	260	552	USA
13	125	397	12	70	604	USA
14	56	462	19	499	1036	USA

Table 2. Sample proportions for the fourteen independent retrospective studies on the association between smoking and lung cancer in Table 1

	Control Patients		Lung Cancer Patients	
Study	Non-Smokers	Smokers	Non-Smokers	Smokers
1	0.081	0.419	0.017	0.482
2	0.118	0.625	0.008	0.248
3	0.081	0.343	0.029	0.547
4	0.052	0.239	0.004	0.704
5	0.043	0.624	0.018	0.315
6	0.022	0.477	0.002	0.497
7	0.082	0.481	0.006	0.431
8	0.074	0.489	0.016	0.420
9	0.138	0.541	0.015	0.306
10	0.150	0.342	0.037	0.471
11	0.223	0.606	0.014	0.158
12	0.051	0.469	0.009	0.471
13	0.207	0.657	0.020	0.116
14	0.054	0.446	0.018	0.482

We wish to determine which, if any, of the fourteen studies exhibit homogeneity. A common goal is to find a parsimonious aggregation, that is, a small value for k in (29), of the independent multinomials. A general strategy is to start by testing for homogeneity among all the independent multinomials and then to separate them sequentially (in a hierarchical way). Consider the following sequence of hierarchical hypotheses:

$$H_{0,i}\colon \boldsymbol{\pi} \in \Pi_{0,i}\,; \qquad i = 1, \dots, 5\,, \tag{37}$$

where

$$\begin{aligned}
\Pi_{0,1} &\equiv \{\boldsymbol{\pi} \in \Delta_4^{14}\colon \boldsymbol{\pi}_7 = \boldsymbol{\pi}_8\}\,,\\
\Pi_{0,2} &\equiv \{\boldsymbol{\pi} \in \Delta_4^{14}\colon \boldsymbol{\pi}_1 = \boldsymbol{\pi}_2, \boldsymbol{\pi}_5 = \boldsymbol{\pi}_6, \boldsymbol{\pi}_7 = \boldsymbol{\pi}_8 = \boldsymbol{\pi}_9\}\\
\Pi_{0,3} &\equiv \{\boldsymbol{\pi} \in \delta_4^{14}\colon \boldsymbol{\pi}_1 = \boldsymbol{\pi}_2, \boldsymbol{\pi}_5 = \boldsymbol{\pi}_6, \boldsymbol{\pi}_7 = \cdots = \boldsymbol{\pi}_{14}\}\,,\\
\Pi_{0,4} &\equiv \{\boldsymbol{\pi} \in \Delta_4^{14}\colon \boldsymbol{\pi}_1 = \cdots = \boldsymbol{\pi}_6, \boldsymbol{\pi}_7 = \cdots = \boldsymbol{\pi}_{14}\}\,,
\end{aligned} \tag{38}$$

and

$$\Pi_{0,5} \equiv \{\boldsymbol{\pi} \in \Delta_4^{14}\colon \boldsymbol{\pi}_1 = \cdots = \boldsymbol{\pi}_{14}\}\,.$$

Notice that $\Pi_{0,5} \subset \cdots \subset \Pi_{0,1}$, where $H_{0,5}$ is the hypothesis of homogeneity among all studies; $H_{0,4}$ specifies homogeneity among the European studies and among the US studies; $H_{0,3}$ specifies homogeneity within countries; $H_{0,2}$ specifies homogeneity within countries for the European studies and homogeneity for the nationwise US studies; and $H_{0,1}$ specifies homogeneity for two nationwide US studies.

An analysis of divergence for $H_{0,1}, \ldots, H_{0,5}$ is presented in Table 3. We assume that the ratios of the independent multinomial sample sizes to the total sample size have limiting values between 0 and 1, as the total sample size goes to infinity; hence, from Theorem 2, we know that under the more restrictive hypothesis $H_{0,i}$, the statistic corresponding to testing the model $H_{0,i}$ against the model $H_{0,i-1}$, in the analysis of divergence, has an asymptotic central chi-squared distribution with degrees of freedom equal to the degrees of freedom for the model $H_{0,i}$ minus the degrees of freedom for the model $H_{0,i-1}$. Notice in Table 3 that, among $\lambda = -1/2, 0, 1/2, 2/3$, and 1, the parameters $\lambda = 2/3$ and $\lambda = 1/2$ produce intermediate values.

Based on Table 3, we would reject $H_{0,i}$ in favor of $H_{0,i-1}$ for $i = 5, 4, 3$, and 2. That is, among $H_{0,1}, \ldots, H_{0,5}$, the only viable candidate is $H_{0,1}$. Observe that the minimized power-divergence statistics corresponding to $H_{0,1}$ (from the 5th line of Table 3) have values in the range of 7.05 to 7.47. These values are less than 7.81, the approximate 95% critical value obtained from the χ_3^2-distribution. Thus, applying Theorem 3.1, we may conclude that the model $H_{0,1}: \boldsymbol{\pi} \in \Pi_{0,1}$ adequately describes the data; that is, US nationwide studies 7 and 8 have similar distributions of smokers and non-smokers with and without lung cancer, but the other studies exhibit inhomogeneity.

That hypotheses $H_{0,5}, \ldots, H_{0,2}$ were rejected may seem somewhat surprising since, in view of Table 2, many of the studies compared appear to give similar proportions of subjects in the four categories. But, with sample sizes $\{n_i\}$ often of the order of 10^3, slight differences in proportions become important (in relation to the standard deviation of their differences).

Table 3. Analysis of divergence for $\lambda = -1/2$, 0, 1/2, 2/3, 1, corresponding to the sequence of hypotheses specified in (37) and (38). The value in the last column is the 0.95 quantile from the χ^2 distribution based on the indicated degrees of freedom (df).

		λ					
Source	df	-1/2	0	1/2	2/3	1	χ^2
$H_{0,5}$ vs. $H_{0,4}$	3	726.45	652.77	600.51	587.60	568.12	7.81
$H_{0,4}$ vs. $H_{0,3}$	9	363.88	355.36	349.77	348.43	346.49	16.9
$H_{0,3}$ vs. $H_{0,2}$	15	1169.15	1111.61	1061.21	1046.22	1019.10	25.0
$H_{0,2}$ vs. $H_{0,1}$	9	86.10	87.55	89.32	89.89	90.92	16.9
$2I^\lambda$ for $H_{0,1}$	3	7.47	7.45	7.31	7.23	7.05	7.81
Total	39	2353.05	2214.74	2108.12	2079.37	2031.68	

5 Conclusions

We have developed a method for hierarchical testing of homogeneity of probabilities within the product-multinomial distribution. The concept of an analysis of divergence and its use in testing a sequence of hierarchical homogeneity hypotheses were outlined. Further, under mild conditions, the terms in the analysis of divergence were shown to have independent limiting chi-squared distributions.

The example in Section 4 illustrates our methods. A recommendation has been given in [6], [13] to use $\lambda = 2/3$, based on power calculations, comparisons of exact versus nominal levels for various hypothesis testing situations, and the rate of convergence of moment approximations. Further, [1] recommends $\lambda \in (2/3, 5/4)$ based on the small-sample coverage properties of confidence intervals for the ratio of two binomial proportions. In addition, [14] compared the choices $\lambda = 0$, $\lambda = 2/3$, and $\lambda = 1$, and recommended either $\lambda = 2/3$ or $\lambda = 1$ based on small-sample comparisons for loglinear models fitted to two- and three-dimensional contingency tables. Moreover, [10] recommends $\lambda = 2/3$ or $\lambda = 1$ (with $\lambda = 1$ having a slight edge), based on small-sample comparisons of loglinear models fitted to three-dimensional contingency tables. However, closer scrutiny of Table IV in [10] reveals that just the opposite is true ([13], p. 79). That is, $\lambda = 2/3$ has a slight edge over $\lambda = 1$. Finally, [11] recommends $\lambda = 2/3$ or $\lambda = 1/2$ based on nominal versus exact levels of confidence regions under the trinomial distribution, the rate at which the exact levels approach the nominal levels for the confidence regions, and the area of the confidence regions. In conclusion, we prefer to use the analysis of divergence corresponding to $\lambda = 2/3$, rather than that associated with G^2 (i.e., $\lambda = 0$) or X^2 (i.e., $\lambda = 1$), when testing hierarchical hypotheses using an analysis of divergence such as in Table 3; see also [3].

References

1. Bedrick, E.J.: A family of confidence intervals for the ratio of two binomial populations. Biometrics 43, 993–998 (1987)
2. Bishop, Y.M.M., Fienberg, S.E., Holland, P.W.: Discrete Multivariate Analysis: Theory and Practice. MIT Press, Cambridge (1975)
3. Cressie, N., Pardo, L.: Minimum ϕ-divergence estimator and hierarchical testing in loglinear models. Statistica Sinica 10, 867–884 (2000)
4. Cressie, N., Pardo, L.: Model checking in loglinear models using ϕ-divergences and MLEs. J. Statist. Plan. Infer. 103, 437–453 (2002)
5. Cressie, N., Pardo, L., Pardo, M.C.: Size and power considerations for testing loglinear models using ϕ-divergence test statistics. Statistica Sinica 13, 555–570 (2003)
6. Cressie, N., Read, T.R.C.: Multinomial goodness-of-fit tests. J. Royal Statist. Soc. Ser. B 46, 440–464 (1984)
7. Dorn, H.F.: The relationship of cancer of the lung and the use of tobacco. The American Statistician 8, 7–13 (1954)
8. Fienberg, S.E.: The Analysis of Cross-Classified Categorical Data. MIT Press, Cambridge (1980)
9. Gokhale, D.V., Kullback, S.: The Information in Contingency Tables. Marcel Dekker, New York (1978)
10. Hosmane, B.: An empirical investigation of chi-squared tests for the hypothesis of no three-factor interaction in $I \times J \times K$ contingency tables. J. Statist. Comp. Simul. 28, 167–178 (1987)

11. Medak, F., Cressie, N.: Confidence regions in ternary diagrams based on the power-divergence statistics. Math. Geol. 23, 1045–1057 (1991)
12. Read, T.R.C.: Choosing a Goodness-of-Fit Test. PhD Thesis, School of Mathematical Sciences, The Flinders University of South Australia, Adelaide, South Australia (1982)
13. Read, T.R.C., Cressie, N.A.C.: Goodness-of-Fit Statistics for Discrete Multivariate Data. Springer, New York (1988)
14. Rudas, T.: A Monte Carlo comparison of the small sample behavior of the Pearson, the likelihood ratio, and the Cressie-Read statistics. J. Statist. Comp. Simul. 17, 107–120 (1986)
15. Searle, S.R.: Linear Models. Wiley, New York (1971)

Appendix

In this section, the proofs of Theorems 1 and 2 are given. In order to facilitate these proofs, we introduce some useful notation.

First, notice that H_0 in (30) partitions $\mathcal{I} \equiv \{1,\ldots,r\}$ into k sets $\mathcal{I}_1,\ldots,\mathcal{I}_k$; that is,

$$\mathcal{I}_i \equiv \{r_{i-1}+1,\ldots,r_i\}\,; \quad i=1,\ldots,k\,, \tag{39}$$

where $\mathcal{I}_i \cap \mathcal{I}_j = \emptyset$, for $i \neq j$; $\bigcup_{j=1}^{k} \mathcal{I}_j = \mathcal{I}$, $1 \leq k \leq r$; $r_0 = 0$; and $r_k = r$.

Without loss of generality, (39) denotes the members of $\mathcal{I}_i$ $(i=1,\ldots,k)$ as consecutive integers.

Now, define

$$\boldsymbol{\psi}_j \equiv \sum_{t\in\mathcal{I}_j} (n_t)^{1/2}\mathbf{e}_t\,; \quad j=1,\ldots,k, \tag{40}$$

and

$$\boldsymbol{\phi}_j \equiv \sum_{t\in\mathcal{I}_j} \mathbf{e}_t\,; \quad j=1,\ldots,k\,, \tag{41}$$

where $\{\mathbf{e}_1,\ldots,\mathbf{e}_r\}$ is the standard basis for $\mathbb{R}^r$. Then,

$$D_{\boldsymbol{\phi}_j}\boldsymbol{\psi}_i = \begin{cases} \boldsymbol{\psi}_j\,, & \text{if } i=j \\ \mathbf{0}\,, & \text{if } i\neq j\,, \end{cases} \tag{42}$$

and

$$D_{\boldsymbol{\phi}_j}D_{\boldsymbol{\phi}_i} = \begin{cases} D_{\boldsymbol{\phi}_j}\,, & \text{if } i=j \\ 0\,, & \text{if } i\neq j\,, \end{cases} \tag{43}$$

where $D_{\mathbf{x}}$ is the $r\times r$ diagonal matrix whose diagonal elements are given by $\mathbf{x}\in\mathbb{R}^r$.

Further, define the $r\times r$ matrix P_j as,

$$P_j \equiv \boldsymbol{\psi}_j(\boldsymbol{\psi}_j^T\boldsymbol{\psi}_j)^{-1}\boldsymbol{\psi}_j^T\,; \quad j=1,\ldots,k\,. \tag{44}$$

Then, for $j=1,\ldots,k$, P_j satisfies,

$$P_j^2 = P_j\,, \tag{45}$$

$$P_j^T = P_j\,, \tag{46}$$

$$P_jP_{j'} = 0\,, \quad \text{if } j\neq j'\,, \tag{47}$$

and

$$D_{\boldsymbol{\phi}_j} P_{j'} = \begin{cases} P_j \,, \text{ if } j = j' \\ 0 \,, \quad \text{if } j \neq j' \,. \end{cases} \tag{48}$$

Recall, any matrix P_j satisfying (45) and (46) is a projection matrix. Now, as $n \to \infty$,

$$(\boldsymbol{\psi}_j^T \boldsymbol{\psi}_j)^{-1/2} \boldsymbol{\psi}_j \to \boldsymbol{\beta}_j \,; \quad j = 1, \ldots, k \,, \tag{49}$$

where $\boldsymbol{\beta}_j \equiv \sum_{t \in I_j} (\gamma_t / \sum_{s \in I_j} \gamma_s)^{1/2} \mathbf{e}_t$. Hence, as $n \to \infty$,

$$P_j \to \boldsymbol{\beta}_j \boldsymbol{\beta}_j^T \equiv P_j^* \,; \quad j = 1, \ldots, k \,. \tag{50}$$

Note, P_j^*, $j = 1, \ldots, k$, satisfies (45) through (48).

Proof of Theorem 1:

Denote the true value of $\boldsymbol{\pi} \in \Pi_0$ by,

$$(\widetilde{\boldsymbol{\pi}}^{(1)^T}, \ldots, \widetilde{\boldsymbol{\pi}}^{(1)^T}, \ldots, \widetilde{\boldsymbol{\pi}}^{(k)^T}, \ldots, \widetilde{\boldsymbol{\pi}}^{(k)^T})^T \,, \tag{51}$$

where $\widetilde{\boldsymbol{\pi}}^{(i)} \in \Delta_c$, for $i = 1, \ldots, k$. Then, under H_0, the first-order Taylor series expansion of the MPE $\widehat{\boldsymbol{\pi}}^{(i;\lambda)}$ is,

$$\widehat{\boldsymbol{\pi}}^{(i;\lambda)} = \widetilde{\boldsymbol{\pi}}^{(i)} + [\{[\boldsymbol{\psi}_j^T \boldsymbol{\psi}_j)^{-1/2} \boldsymbol{\psi}_j^T\} \otimes H] \boldsymbol{\omega}_n + o_p(1/n) \,; \quad i = 1, \ldots, k \,, \tag{52}$$

where $\otimes$ denotes the Kronecker (or tensor) product; H is the $c \times c$ matrix,

$$H \equiv \begin{bmatrix} I_{c-1} & 0 \\ -\mathbf{1}_{c-1}^T & 0 \end{bmatrix} \,; \tag{53}$$

I_{c-1} is the $(c-1) \times (c-1)$ identity matrix; $\mathbf{1}_{c-1}$ is a $(c-1)$-dimensional of ones; and

$$\boldsymbol{\omega}_n \equiv \begin{bmatrix} n_1^{1/2} (\mathbf{X}_1 / n_1 - \widetilde{\boldsymbol{\pi}}^{(1)}) \\ \vdots \\ n_r^{1/2} (\mathbf{X}_r / n_r - \widetilde{\boldsymbol{\pi}}^{(r)}) \end{bmatrix} \,. \tag{54}$$

Under H_0, as $n \to \infty$,

$$\boldsymbol{\omega}_n \xrightarrow{\mathcal{D}} \boldsymbol{\omega} \sim N_{rc}(\mathbf{0}, \Sigma) \,, \tag{55}$$

where $\Sigma \equiv \sum_{j=1}^{k} [D_{\boldsymbol{\phi}_j} \otimes \{D_{\widetilde{\boldsymbol{\pi}}^{(j)}} - \widetilde{\boldsymbol{\pi}}^{(j)} \widetilde{\boldsymbol{\pi}}^{(j)^T}\}]$.

Now, from [3] and (52),

$$2I^{\lambda}(\mathbf{X}\colon \widehat{\boldsymbol{\mu}}^{(\lambda)}) = \sum_{j=1}^{k}\sum_{i\in I_j} 2n_i I^1(\mathbf{X}_i/n_i\colon \widehat{\boldsymbol{\pi}}^{(j;\lambda)}) + o_p(1)$$

$$= \boldsymbol{\omega}_n^T[\sum_{j=1}^{k}(\{D_{\boldsymbol{\phi}_j} - P_j\}\otimes D^{-1}_{\widetilde{\boldsymbol{\pi}}^{(j)}})]\boldsymbol{\omega}_n + o_p(1)\,,$$

where $\widehat{\boldsymbol{\mu}}^{\lambda} \equiv (n_1\widehat{\boldsymbol{\pi}}^{(1;\lambda)^T},\ldots,n_r\widehat{\boldsymbol{\pi}}^{(k;\lambda)^T})^T$. Thus, as $n\to\infty$,

$$2I^{\lambda}(\mathbf{X}\colon \widehat{\boldsymbol{\mu}}^{(\lambda)}) \xrightarrow{\mathcal{D}} \boldsymbol{\omega}^T Q\boldsymbol{\omega} \sim \chi^2_{(r-k)(c-1)}\,, \tag{56}$$

where $Q \equiv \sum_{j=1}^{k}(\{D_{\boldsymbol{\phi}_j} - P_j^*\}\otimes D^{-1}_{\widetilde{\boldsymbol{\pi}}^{(j)}})$, P_j^* is given by (50), and $\boldsymbol{\omega}$ is given by (55). Since $Q\Sigma Q\Sigma Q = Q\Sigma Q$ and $\mathrm{trace}(Q\Sigma) = (r-k)(c-1)$, (56) follows from standard results on quadratic forms (e.g., [15], p. 69).

Proof of Theorem 2:

Without loss of generality, consider

$$2I^{\lambda}(\mathbf{X}\colon \widehat{\boldsymbol{\mu}}^{2(\lambda)}) - 2I^{\lambda}(\mathbf{X}\colon \widehat{\boldsymbol{\mu}}^{1(\lambda)})\,, \tag{57}$$

where

$$2I^{\lambda}(\mathbf{X}\colon \widehat{\boldsymbol{\mu}}^{i(\lambda)}) \equiv \inf_{\boldsymbol{\pi}\in\Pi_{0,i}} 2I^{\lambda}(\mathbf{X}\colon \boldsymbol{\mu}(\boldsymbol{\pi}))\,; \qquad i=1,2\,; \tag{58}$$

$$\boldsymbol{\mu}(\boldsymbol{\pi}) \equiv (n_1\pi_1^T,\ldots,n_r\pi_r^T)^T\,; \tag{59}$$

$$\Pi_{0,1} \equiv \{\boldsymbol{\pi}\in\Delta_c^r\colon \boldsymbol{\pi}_{r_{i-1}+1} = \cdots = \boldsymbol{\pi}_{r_i},\ i=1,\ldots,k,\ r_0=0\,,\ r_k\equiv r\}\,; \tag{60}$$

$$\Pi_{0,2} \equiv \{\boldsymbol{\pi}\in\Delta_c^r\colon \boldsymbol{\pi}_1 = \cdots = \boldsymbol{\pi}_r\}\,; \tag{61}$$

$$\widehat{\boldsymbol{\mu}}^{1(\lambda)} \equiv (n_1\widehat{\boldsymbol{\pi}}^{1(1;\lambda)^T},\ldots,n_{r_1}\widehat{\boldsymbol{\pi}}^{1(1;\lambda)^T},\ldots,n_{r_{k-1}+1}\widehat{\boldsymbol{\pi}}^{1(k;\lambda)^T},\ldots,n_r\widehat{\boldsymbol{\pi}}^{1(k;\lambda)^T})^T\,;$$

and

$$\widehat{\boldsymbol{\mu}}^{2(\lambda)} \equiv (n_1\widehat{\boldsymbol{\pi}}^{2(1;\lambda)^T},\ldots,n_r\widehat{\boldsymbol{\pi}}^{2(1;\lambda)^T})^T\,. \tag{62}$$

In addition, denote the partitions of $\mathcal{I}$ induced by $H_{0,1}$ and $H_{0,2}$, respectively, as

$$\mathcal{P}^{(1)} \equiv \{\mathcal{I}_1^{(1)},\ldots,\mathcal{I}_k^{(1)}\}\,,\ \text{and}\ \mathcal{P}^{(2)} \equiv \{\mathcal{I}_1^{(2)}\}\,, \tag{63}$$

where $\mathcal{I}_1^{(2)} \equiv \mathcal{I}$.

Now, suppose $H_{0,2}$ is true and denote the true value of $\boldsymbol{\pi} \in \Pi_{0,2}$ by

$$(\widetilde{\boldsymbol{\pi}}^{(2)^T}, \ldots, \widetilde{\boldsymbol{\pi}}^{(2)^T})^T, \tag{64}$$

where $\widetilde{\boldsymbol{\pi}}^{(2)} \in \Delta_c$. Then, the first-order Taylor series expansions of the MPEs $\widehat{\boldsymbol{\pi}}^{1(i;\lambda)}$, $i = 1, \ldots, k$, and $\widehat{\boldsymbol{\pi}}^{2(1;\lambda)}$ are given, respectively, by

$$\widehat{\boldsymbol{\pi}}^{1(i;\lambda)} = \widetilde{\boldsymbol{\pi}}^{(2)} + [\{(\boldsymbol{\psi}_i^{(1)^T}\boldsymbol{\psi}_i^{(1)})^{-1/2}\boldsymbol{\psi}_i^{(1)^T}\} \otimes H]\boldsymbol{\omega}_n + o_p(1/n)\,; \quad i = 1, \ldots, k\,,$$

and

$$\widehat{\boldsymbol{\pi}}^{2(1;\lambda)} = \widehat{\boldsymbol{\pi}}^{(2)} + [\{(\boldsymbol{\psi}_1^{(2)^T}\boldsymbol{\psi}_1^{(2)})^{-1/2}\boldsymbol{\psi}_1^{(2)^T}\} \otimes H]\boldsymbol{\omega}_n + o_p(1/n)\,, \tag{65}$$

where $\boldsymbol{\psi}_i^{(1)}$, $i = 1, \ldots, k$, and $\boldsymbol{\psi}_1^{(2)}$ are defined analogously to (40) and (41), H is given by (53), and

$$\boldsymbol{\omega}_n \equiv \begin{bmatrix} n_1^{1/2}(\mathbf{X}_1/n_1 - \widetilde{\boldsymbol{\pi}}^{(2)}) \\ \vdots \\ n_r^{1/2}(\mathbf{X}_r/n_r - \widetilde{\boldsymbol{\pi}}^{(2)}) \end{bmatrix}. \tag{66}$$

Under $H_{0,2}$, as $n \to \infty$,

$$\boldsymbol{\omega}_n \xrightarrow{\mathcal{D}} \boldsymbol{\omega} \sim N_{rc}(\mathbf{0}, \Sigma)\,, \tag{67}$$

where $\Sigma \equiv \sum_{j=1}^{k}(D_{\boldsymbol{\phi}_j^{(1)}} \otimes \{D_{\widetilde{\boldsymbol{\pi}}^{(2)}} - \widetilde{\boldsymbol{\pi}}^{(2)}\widetilde{\boldsymbol{\pi}}^{(2)^T}\})$ and $\boldsymbol{\phi}_j^{(1)}$ is defined analogously to (41), for $j = 1, \ldots, k$.

Then under $H_{0,2}$,

$$\begin{aligned}
&2I^\lambda(\mathbf{X}\colon \widehat{\boldsymbol{\mu}}^{2(\lambda)}) - 2I^\lambda(\mathbf{X}\colon \widehat{\boldsymbol{\mu}}^{1(\lambda)}) \\
&\quad = \sum_{i \in I} 2n_i I^1(\mathbf{X}_i/n_i\colon \widehat{\boldsymbol{\pi}}^{2(1;\lambda)}) - \sum_{j=1}^{k}\sum_{i \in \mathcal{I}_j^{(1)}} 2n_i I^1(\mathbf{X}_i/n_i\colon \widehat{\boldsymbol{\pi}}^{1(j;\lambda)}) + o_p(1) \\
&\quad = \boldsymbol{\omega}_n^T\left[\left\{\sum_{j=1}^{k} P_j^{(1)} - P_1^{(2)}\right\} \otimes D^{-1}_{\widetilde{\boldsymbol{\pi}}^{(2)}}\right]\boldsymbol{\omega}_n + o_p(1)\,,
\end{aligned}$$

where $P_i^{(1)}$, $i = 1, \ldots, k$, and $P_1^{(2)}$ are defined analogously to (44). Hence, as $n \to \infty$,

$$2I^\lambda(\mathbf{X}\colon \widehat{\boldsymbol{\mu}}^{2(\lambda)}) - 2I^\lambda(\mathbf{X}\colon \widehat{\boldsymbol{\mu}}^{1(\lambda)}) \xrightarrow{\mathcal{D}} \boldsymbol{\omega}^T Q \boldsymbol{\omega} \sim \chi^2_{(k-1)(c-1)}\,, \tag{68}$$

where

$$Q \equiv \left\{\sum_{j=1}^{k} P_j^{(1)^*} - P_1^{(2)^*}\right\} \otimes D^{-1}_{\widetilde{\boldsymbol{\pi}}^{(2)}}\,, \tag{69}$$

and $P_i^{(1)^*}$, $i = 1, \ldots, k$, and $P_1^{(2)^*}$ are defined analogously to (50). The distributional result in (67) is a consequence of $Q\Sigma Q\Sigma Q = Q\Sigma Q$, trace$(Q\Sigma) = (k-1)(c-1)$, and standard results on quadratic forms (e.g., [15], p. 69).

Now, we prove the second part of Theorem 2. Suppose, $H_{0,i'}$, $i' > 1$, is true. Let,

$$\mathcal{P}^{(j_t)} \equiv \{\mathcal{I}_1^{(j_t)}, \ldots, \mathcal{I}_{k_{j_t}}^{(j_t)}\}; \qquad t = 1, 2, \tag{70}$$

and

$$\mathcal{P}^{(i_t)} \equiv \{\mathcal{I}_1^{(i_t)}, \ldots, \mathcal{I}_{k_{i_t}}^{(i_t)}\}; \qquad t = 1, 2, \tag{71}$$

be the partitions of $\mathcal{I}$ induced by $1 \le j_1 < j_2 \le i_1 < i_2 \le i'$; so, $k_{j_1} > k_{j_2} \ge k_{i_1} > k_{i_2}$. From the hierarchical-hypotheses restriction imposed on (33), we have, for $1 \le s < t \le m$,

$$\mathcal{I}_q^{(t)} = \bigcup_{d \in \mathcal{S}_q^{(s,t)}} \mathcal{I}_d^{(s)}; \quad q = 1, \ldots, k_t, \tag{72}$$

and define

$$P_q^{(t)} \equiv \sum_{d \in \mathcal{S}_q^{(s,t)}} P_d^{(s)}; \quad q = 1, \ldots, k_t, \tag{73}$$

where $\mathcal{S}_q^{(s,t)}$ is the set of indices d such that $\mathcal{I}_d^{(s)} \subset \mathcal{I}_q^{(t)}$.

Thus, as $n \to \infty$,

$$2I^\lambda(\mathbf{X} \colon \widehat{\boldsymbol{\mu}}^{j_2(\lambda)}) - 2I^\lambda(\mathbf{X} \colon \widehat{\boldsymbol{\mu}}^{j_1(\lambda)}) \xrightarrow{\mathcal{D}} \boldsymbol{\omega}^T Q_1 \boldsymbol{\omega}, \tag{74}$$

and

$$2I^\lambda(\mathbf{X} \colon \widehat{\boldsymbol{\mu}}^{i_2(\lambda)}) - 2I^\lambda(\mathbf{X} \colon \widehat{\boldsymbol{\mu}}^{i_1(\lambda)}) \xrightarrow{\mathcal{D}} \boldsymbol{\omega}^T Q_2 \boldsymbol{\omega}, \tag{75}$$

where

$$Q_1 \equiv \sum_{j=1}^{j_1} \left[\left\{ \sum_{d \in \mathcal{S}_j^{(j_1, j_2)}} P_d^{(j_2)^*} - P_j^{(j_1)^*} \right\} \otimes D^{-1}_{\widetilde{\boldsymbol{\pi}}_j^{(i')}} \right], \tag{76}$$

$$Q_2 \equiv \sum_{i=1}^{i_1} \left[\left\{ \sum_{d \in \mathcal{S}_i^{(i_1, i_2)}} P_d^{(i_2)^*} - P_i^{(i_1)^*} \right\} \otimes D^{-1}_{\widetilde{\boldsymbol{\pi}}_i^{(i')}} \right], \tag{77}$$

and

$$(\widetilde{\boldsymbol{\pi}}^{i'(1)^T}, \ldots, \widetilde{\boldsymbol{\pi}}^{i'(1)^T}, \ldots, \widetilde{\boldsymbol{\pi}}^{i'(k_{i'})^T}, \ldots, \widetilde{\boldsymbol{\pi}}^{i'(k_{i'})^T})^T, \tag{78}$$

denotes the true value of $\boldsymbol{\pi} \in \Pi_{0,i'}$. Further,

$$\begin{bmatrix} n_1^{1/2}(\mathbf{X}_1/n_1 - \widetilde{\boldsymbol{\pi}}_1^{(i')}) \\ \vdots \\ n_r^{1/2}(\mathbf{X}_r/n_r - \widetilde{\boldsymbol{\pi}}_{k_{i'}}^{(i')}) \end{bmatrix} \xrightarrow{\mathcal{D}} \boldsymbol{\omega} \sim N_{rc}(\mathbf{0}, \Sigma)\,, \tag{79}$$

as $n \to \infty$, where $\Sigma \equiv \sum_{j=1}^{k_{i'}} [D_{\boldsymbol{\phi}_j^{(i')}} \otimes \{D_{\boldsymbol{\pi}^{i'(j)}} - \widetilde{\boldsymbol{\pi}}^{i'(j)}\widetilde{\boldsymbol{\pi}}^{i'(j)^T}\}]$.

Since $\Sigma Q_1 \Sigma Q_2 \Sigma = 0$, $\boldsymbol{\omega}^T Q_1 \boldsymbol{\omega}$ and $\boldsymbol{\omega}^T Q_2 \boldsymbol{\omega}$ are independent chi-squared random variables by standard results on quadratic forms (e.g., [15], p. 71).

Statistical Information Tools for Multivariate Discrete Data

Ove Frank

Department of Statistics, Stockholm University,
SE-10691 Stockholm, Sweden
Ove.Frank@stat.su.se

Summary. An exposition is given of various information theoretic measures appropriate for general statistical analysis of multivariate discrete data. Such measures ought to be better known and used in descriptive and exploratory statistics, and they can also be beneficial as test statistics of probabilistic models.

Keywords: information divergence, entropy, flatness, spread, dependence, redundancy, variable selection, data editing.

1 Introduction

Discrete multivariate data on categorical or numerical variables are common in all parts of statistics. Less well known in general statistics are the measures of spread, flatness, association and dependence that are based on entropy and developed in information theory. This exposition illustrates how a systematic use of such measures can be beneficial in both exploratory and confirmatory statistical analyses involving data on nominal and ordinal as well as numerical scales.

The next section introduces some basic notation and discusses a few examples of situations in which it might be useful with a systematic approach to variable selection and preparation of data for further statistical analysis. Sections 3 - 5 introduce the information theoretic measures and explain how they can be used to examine and compare distributions of values on one or several variables. The statistics used as descriptive or exploratory tools can also be used in confirmatory analyses of testing and estimating probabilistic models. Section 6 illustrates such tests. Section 7 gives some remarks on literature and further research. Connections with graphical modeling and statistical lattice theory are mentioned.

2 Variables and Data

To fix ideas, consider an opinion poll using a questionnaire with 15 items that can be answered by yes or no or don't know. Consequently there are 3^{15} or more than 14 millions possible response patterns, and only a few of these can be represented among the questionnaires collected from participants in the investigation. The distribution of response patterns is usually not considered as a single 15-dimensional distribution. Normally data are separated into several sets of two- or three-dimensional distributions of items that are related and considered to be of interest to be seen presented together in plots or tables. Some items might be found to be redundant since no respondent answered them by yes or no, and some items might appear to be unreliable or not sufficiently discriminating between opinions. There might also be other good reasons to explore the quality of data and prepare interesting subsets of data for the final analysis.

Another example of a situation that requires analysis of multivariate discrete data is given by economic, social or medical exploratory investigations in which there is an abundance of available variables on different types of scales. It might seem convenient to transform some variables to different scales, to simplify some variables to fewer outcomes by aggregation of similar outcomes, to combine some variables to an index, etcetera. Such data editing work could benefit from the tools described in this exposition.

Generally, consider m discrete variables $X_1, X_2, \ldots, X_m$ having finitely many categorical or numerical outcomes. For variable X_i the possible outcomes are labeled by integers $1, \ldots, r_i$ for $i = 1, \ldots, m$. Thus the m-variate variable $(X_1, \ldots, X_m)$ has at most $r = r_1 \cdot r_2 \cdot \ldots \cdot r_m$ possible outcomes. If data consist of n observations on $(X_1, \ldots, X_m)$, the distribution of observations on the possible outcomes is given by the frequencies $n(i_1, \ldots, i_m)$ of observations equal to $(i_1, \ldots, i_m)$ for $i_1 = 1, \ldots, r_1$, ..., $i_m = 1, \ldots, r_m$. If the variables are considered as random variables, the relative frequency $n(i_1, \ldots, i_m)/n$ estimates the probability

$$P(X_1 = i_1, \ldots, X_m = i_m) = p_{1,\ldots,m}(i_1, \ldots, i_m).$$

If the variables should not a priori be considered as random variables, a probability distribution can be identified with the empirical distribution, and probabilistic terms still be used.

It is convenient to use X, Y, Z for arbitrary subsets of variables among $X_1, \ldots, X_m$. The notation X, Y or Z can be used for single variables as well as for overlapping or disjoint subsets of variables. The numbers of outcomes of X, Y, Z are denoted by r, s, t. The trivariate probabilities are given by

$$P(X = i, Y = j, Z = k) = p_{X,Y,Z}(i, j, k)$$

for $i = 1, \ldots, r, j = 1, \ldots, s, k = 1, \ldots, t$. The trivariate distribution can be factorized according to

$$p_{X,Y,Z} = p_X \cdot p_{Y|X} \cdot p_{Z|X,Y}$$

where the conditional distributions are defined when the univariate and bivariate conditions have positive probabilities.

3 Univariate and Bivariate Distributions

Consider a discrete variable X with finitely many outcomes labeled by integers $1, \ldots, r$ and a probability distribution p_X with probabilities $p_X(i) = p_i > 0$ for $i = 1, \ldots, r$ satisfying $p_1 + \ldots + p_r = 1$. The entropy of X is defined as

$$H_X = \sum_{i=1}^{r} p_i \log \frac{1}{p_i}.$$

It is a weighted mean of the logarithms of the inverted probabilities. Since $1/p_i \geq 1$, the entropy is non-negative and it is equal to 0 only if $r = 1$. In this case there is a one-point distribution p_X and X is a constant. The entropy is also the logarithm of the weighted geometric mean of the inverted probabilities. Since the weighted geometric mean is at most equal to the weighted arithmetic mean, it follows that $H_X \leq \log r$. Similarly, the weighted harmonic mean is a lower bound, which yields that

$$H_X \geq \log \frac{1}{\sum_{i=1}^{r} p_i^2}$$

with equality only for the uniform distribution. Consequently,

$$0 \leq -\log\left(\sum_{i=1}^{r} p_i^2\right) \leq H_X \leq \log r.$$

Thus, the entropy is 0 if and only if there is only one outcome, and the entropy is $\log r$ if and only if there are r outcomes with a flat distribution. In intermediate cases when the entropy is strictly between 0 and $\log r$, it can be interpreted as the logarithm of the approximate number of outcomes that would correspond to a flat distribution. The integer closest to $\exp H_X$ is this approximate number. The relative entropy $H_X / \log r$ can be considered as a measure of flatness. If the mode is taken as a measure of centrality of a probability distribution, then a natural measure of spread is the relative entropy, which can be considered to give the size of the efficient or operative part of the outcome space.

The bivariate entropy of (X, Y) is given by

$$H_{X,Y} = \sum_{i=1}^{r} \sum_{j=1}^{s} \Phi(p_{i,j}) \quad \text{where } \Phi(p) = p \log \frac{1}{p} \quad \text{if } p > 0 \text{ and } \Phi(0) = 0.$$

Here $p_{X,Y}(i,j) = p_{i,j} \geq 0$, $p_X(i) = p_{i\cdot} > 0$ for $i = 1, \ldots, r$ and $p_Y(j) = p_{\cdot j} > 0$ for $j = 1, \ldots, s$. Using the factorization $p_{X,Y} = p_X \cdot p_{Y|X}$ it follows that

$$H_{X,Y} = H_X + H_{Y|X},$$

where

$$H_{Y|X} = \sum_{i=1}^{r} p_{i\cdot} \sum_{j=1}^{s} \Phi\left(\frac{p_{i,j}}{p_{i\cdot}}\right).$$

Due to the convexity from above of Φ, it follows that

$$H_{Y|X} \leq \sum_{j=1}^{s} \Phi(p_{\cdot j}) = H_Y$$

with equality if and only if X and Y are independent, which is denoted by $X \perp Y$. The conditional entropy of Y for $X = i$ is given by $\sum_{j=1}^{s} \Phi(p_{i,j}/p_{i\cdot})$, and it takes values between 0 and $\log s_i$, where s_i is the number of outcomes of Y of positive conditional probability for $X = i$. Now $\sum_{j=1}^{s} \Phi(p_{i,j}/p_{i\cdot}) = 0$ only if $s_i = 1$, which means that there is a unique j with $p_{i,j} = p_{i\cdot}$, say $j = f(i)$. If this holds true for all $i = 1, \ldots, r$, then $H_{Y|X} = 0$ and there is a (univalent) function f such that $Y = f(X)$. In this case, variable X uniquely determines variable Y, which is written $X \gg Y$. Thus

$$H_{X,Y} = H_X + H_{Y|X} \leq H_X + H_Y$$

with equality only if $X \perp Y$. Furthermore, $H_X \leq H_{X,Y}$ with equality only if $X \gg Y$. Similarly, $H_Y \leq H_{X,Y}$ with equality only if $Y \gg X$. If $X \gg Y$, then $H_{Y|X} = 0$ and

$$H_X = H_{X,Y} = H_Y + H_{X|Y} \geq H_Y.$$

This can be expressed in terms of the joint entropy

$$J_{X,Y} = H_X + H_Y - H_{X,Y} = H_Y - H_{Y|X} = H_X - H_{X|Y}$$

according to

$$0 \leq J_{X,Y} \leq \min(H_X, H_Y)$$

with equality to the left only for independence $X \perp Y$ and equality to the right only for functional dependence $X \gg Y$ or $Y \gg X$. The joint entropy $J_{X,Y}$ measures the amount of entropy shared by variables X and Y. It can also be interpreted as a measure of divergence of the distribution $p_{X,Y}$ from the distribution $p_X \cdot p_Y$. The influence of X on Y can be measured by relative joint entropy $J_{X,Y}/H_Y$. The ratio $J_{X,Y}/H_Y$ is a measure between 0 and 1 indicating the degree of dependence from X to Y. It is 0 for independence $X \perp Y$, and it is 1 for functional dependence $X \gg Y$.

4 Trivariate Distributions

Consider three variables X, Y, Z having r, s, t outcomes and simultaneous probabilities

$$P(X = i, Y = j, Z = k) = p_{X,Y,Z}(i, j, k) = p_{i,j,k} \geq 0$$

satisfying

$$p_X(i) = p_{i\cdot\cdot} > 0,\ p_Y(j) = p_{\cdot j\cdot} > 0,\ p_Z(k) = p_{\cdot\cdot k} > 0,$$

$$\sum_{i=1}^{r}\sum_{j=1}^{s}\sum_{k=1}^{t} p_{ijk} = p_{\dots} = 1$$

for all $i = 1, \dots, r$, $j = 1, \dots, s$, $k = 1, \dots, t$. From the factorization of the trivariate distribution according to

$$p_{X,Y,Z} = p_X \cdot p_{Y|X} \cdot p_{Z|X,Y}$$

follows the additivity of entropy in the sense

$$H_{X,Y,Z} = H_X + H_{Y|X} + H_{Z|X,Y}.$$

The bivariate entropy inequality

$$\max(H_X, H_Y) \leq H_{X,Y} \leq H_X + H_Y$$

yields corresponding trivariate inequalities

$$\max(H_{X,Y}, H_Z) \leq H_{X,Y,Z} \leq H_{X,Y} + H_Z$$

$$\max(H_{X,Z}, H_Y) \leq H_{X,Y,Z} \leq H_{X,Z} + H_Y$$

$$\max(H_{Y,Z}, H_X) \leq H_{X,Y,Z} \leq H_{Y,Z} + H_X$$

that can be combined to

$$\max(H_{X,Y}, H_{X,Z}, H_{Y,Z}) \leq H_{X,Y,Z} \leq \min(H_{X,Y} + H_Z, H_{X,Z} + H_Y, H_{Y,Z} + H_X)$$

with equality to the left only if one variable is a function of the other two and equality to the right only if one variable is independent of the pair of the other two. Note that $(X, Y) \perp Z$ implies $X \perp Z$ and $Y \perp Z$, but the converse is not necessarily true.

By expanding $H_{X,Y,Z} = H_X + H_{Y,Z|X}$ and using $H_{Y,Z|X} \leq H_{Y|X} + H_{Z|X}$ where $H_{Y|X} = H_{X,Y} - H_X$ and $H_{Z|X} = H_{X,Z} - H_X$ it follows that

$$H_{X,Y,Z} \leq H_{X,Y} + H_{X,Z} - H_X$$

with equality only if $Y \perp Z|X$. Interchanging the variables yields the companion inequalities

$$H_{X,Y,Z} \leq H_{X,Y} + H_{Y,Z} - H_Y,$$

$$H_{X,Y,Z} \leq H_{Y,Z} + H_{X,Z} - H_Z.$$

Combining these three inequalities leads to the following upper bound to trivariate entropy

$$H_{X,Y,Z} \leq \min(H_{X,Y} + H_{X,Z} - H_X, H_{X,Y} + H_{Y,Z} - H_Y, H_{Y,Z} + H_{X,Z} - H_Z)$$

with equality only if conditional on one variable, the other two are independent. If the sum of the three univariate entropies is denoted by $S_1 = H_X + H_Y + H_Z$, and the sum of the three bivariate entropies is denoted by $S_2 = H_{X,Y} + H_{X,Z} + H_{Y,Z}$, it is possible to express the present upper bound to $H_{X,Y,Z}$ as

$$S_2 - \max(H_{X,Y} + H_Z, H_{X,Z} + H_Y, H_{Y,Z} + H_X),$$

and this expression can be seen to be smaller than or equal to the previous upper bound given as

$$\min(H_{X,Y} + H_Z, H_{X,Z} + H_Y, H_{Y,Z} + H_X).$$

This follows from the observation that the difference between the bounds equals

$$\begin{aligned} &\min(H_{X,Y} + H_Z, H_{X,Z} + H_Y, H_{Y,Z} + H_X) - S_2 \\ &+ \max(H_{X,Y} + H_Z, H_{X,Z} + H_Y, H_{Y,Z} + H_X) \\ &= S_1 - \text{med}(H_{X,Y} + H_Z, H_{X,Z} + H_Y, H_{Y,Z} + H_X) \end{aligned}$$

where med stands for median value. This difference is non-negative, and it is equal to zero only if there is independence between the two variables of the remaining bivariate entropy. Hence the sharpest bounds to trivariate entropy are given by

$$\begin{aligned} &\max(H_{X,Y}, H_{X,Z}, H_{Y,Z}) \leq H_{X,Y,Z} \\ &\leq \min(H_{X,Y} + H_{X,Z} - H_X, H_{X,Y} + H_{Y,Z} - H_Y, H_{Y,Z} + H_{X,Z} - H_Z). \end{aligned}$$

The lower bound is larger than or equal to $S_2/3$ and the upper bound is smaller than or equal to $(2S_2 - S_1)/3$.

5 Multivariate Distributions

For a discrete m-variate distribution of variables $(X_1, \ldots, X_m)$, let variable X_k have r_k outcomes denoted by integers $1, \ldots, r_k$ for $k = 1, \ldots, m$. Probabilities are given as

$$P(X_1 = i_1, \ldots, X_m = i_m) = p_{1,\ldots,m}(i_1, \ldots, i_m)$$

for $i_k = 1, \ldots, r_k$ and $k = 1, \ldots, m$. The marginal distribution of $(X_k : k \in K)$ where K is a subset of $\{1, \ldots, m\}$ is denoted p_K. The conditional distribution of $(X_k : k \in K)$ given the outcomes of $(X_k : k \in C)$ where K and C are disjoint subsets of $\{1, \ldots, m\}$ is given by $p_{K|C}$. In particular, the factorization

$$p_{1,\ldots,m} = p_1 \cdot p_{2|1} \cdot p_{3|1,2} \cdot \cdots \cdot p_{m|1,\ldots,m-1}$$

implies the additivity of entropy according to

$$H_{1,\ldots,m} = H_1 + H_{2|1} + H_{3|1,2} + \ldots + H_{m|1,\ldots,m-1}.$$

There are m univariate entropies $H_1, \ldots, H_m$ with sum $S_1 = \sum_{i=1}^{m} H_i$. There are $m(m-1)/2$ bivariate entropies $H_{i,j}$ for $1 \leq i < j \leq m$ with sum S_2. Generally, there are $m!/k!(m-k)!$ k-variate entropies with sum S_k for $k = 1, \ldots, m$. The average of the k-variate entropies is denoted by $M_k = k!(m-k)!S_k/m!$ for $k = 1, \ldots, m$ and the average entropy per variable is M_k/k among the k-variate distributions for $k = 1, \ldots, m$. Generally, M_k is non-decreasing and M_k/k is non-increasing with increasing k for $k = 1, \ldots, m$. A way to prove that is to consider the inequalities

$$H_{K\setminus\{i\}} \leq H_K \leq H_{K\setminus\{i\}} + H_{K\setminus\{j\}} - H_{K\setminus\{i,j\}}$$

where K is a subset of $k \geq 2$ elements from $\{1, \ldots, m\}$, and i and j are two distinct elements chosen from K. There are $m!/k!(m-k)!$ choices for K, and for each K there are k choices for i and $k(k-1)/2$ choices for i and j. The $k[m!/k!(m-k)!]$ left inequalities sum to

$$M_{k-1} \leq M_k$$

and the $[k(k-1)/2][m!/k!(m-k)!]$ right inequalities sum to

$$M_k \leq 2M_{k-1} - M_{k-2}$$

for $k = 2, \ldots, m$, where $S_0 = M_0 = 0$. Hence M_k is convex from above, and it follows by iteration that

$$M_k \leq k \cdot \frac{M_{k-1}}{k-1}$$

so that $M_{k-1} \leq M_k$ and $M_{k-1}/(k-1) \geq M_k/k$ for $k = 2, \ldots, m$. In particular, for $k = 2$ it holds that $M_2/2 \leq M_1 \leq M_2$ which is $S_2 \leq (m-1)S_1 \leq 2S_2$ and expands to $H_{1,2} \leq H_1 + H_2 \leq 2H_{1,2}$ for $m = 2$ and expands to $H_{1,2} + H_{1,3} + H_{2,3} \leq 2(H_1 + H_2 + H_3) \leq 2(H_{1,2} + H_{1,3} + H_{2,3})$ for $m = 3$. Similarly, for $k = 3$ it holds that $M_3/3 \leq M_2/2 \leq M_2 \leq M_3$ which is $2S_3 \leq (m-2)S_2 \leq 2(m-2)S_2 \leq 6S_3$ and expands to $2H_{1,2,3} \leq H_{1,2}+H_{1,3}+H_{2,3} \leq 2(H_{1,2}+H_{1,3}+H_{2,3}) \leq 6H_{1,2,3}$ for $m = 3$. The bounds of the bivariate and trivariate entropies given in Sections 3 and 4 are somewhat tighter, and such tighter bounds can be obtained from the general formula

$$\max_i H_{K\setminus\{i\}} \leq H_K \leq \min_{i,j}(H_{K\setminus\{i\}} + H_{K\setminus\{j\}} - H_{K\setminus\{i,j\}}),$$

where K, i and j are defined as above.

The variables in a multivariate distribution can be checked for dependencies and successively simplified by using entropy screening. First the univariate entropies are used to eliminate variables with no or almost no variation revealed by entropies close to zero. Next the bivariate entropies $H_{i,j}$ are compared to H_i and H_j to find out whether some variables are redundant and can be omitted because they are determined as functions of another variable. If $H_{i,j}$ is close to its maximal value $H_i + H_j$ the two variables are almost independent. Functional dependence and independence can conveniently be investigated using appropriate bounds for the relative joint entropy $J_{i,j}/H_j$. When no remaining variable is functionally dependent on another variable, there might still be redundancy because some variable might be functionally dependent on two other variables. Such redundancies can be detected by checking trivariate entropies for similarity with bivariate entropies. Functional dependencies involving more variables can be found by similar comparisons between higher order multivariate entropies. A simple example will be used to demonstrate convenient checking procedures.

Consider 15 observations on six binary variables X, Y, Z, U, V, W. Data are given as a binary 15 by 6 matrix in Table 1. Some rows are equal and the distinct rows can be reported with their frequencies.

Table 1. Values on six variables representing answers by 15 respondents to six items in a questionnaire

	X	Y	Z	U	V	W
1	0	0	1	0	0	0
2	0	0	1	0	0	0
3	0	1	0	1	1	0
4	0	1	1	1	0	1
5	0	1	1	1	0	1
6	1	0	0	1	1	0
7	1	0	0	1	1	0
8	1	0	0	1	1	0
9	1	0	0	1	1	0
10	1	0	0	1	1	0
11	1	0	1	1	0	1
12	1	0	1	1	0	1
13	1	1	0	1	1	1
14	1	1	0	1	1	1
15	1	1	0	1	1	1
$\sum$	10	6	6	13	9	7

Table 2 is a frequency table for the simultaneous outcomes of the six variables. Since all variables are binary their entropies are given by $\Phi(p)+\Phi(1-p)$ where p and $1-p$ are the relative frequencies of the two outcomes of the variable and $\Phi(p) = p\log(1/p)$ for $0 < p \leq 1$ and $\Phi(0) = 0$. The entropy is convex from above with maximum $\log 2$ for $p = 1/2$, and the entropies of the variables can be ordered according to $\min(p, 1-p)$. In general, the entropy of a distribution $p_1, \ldots, p_r$ with sum 1 is given by $\sum_i \Phi(p_i)$. When $p_i = n_i/n$ are relative frequencies it is convenient to abbreviate the entropy as $h(n_1, \ldots, n_r) = \sum_i \Phi(n_i/n)$. The column frequencies in Table 1 yield the univariate entropies: $H_X = h(5, 10)$, $H_Y = h(6, 9)$, $H_Z = h(6, 9)$, $H_U = h(2, 13)$, $H_V = h(6, 9)$, $H_W = h(7, 8)$. Consequently, $\log 2 > H_W > H_Y = H_Z = H_V > H_X > H_U > 0$.

Table 2. Frequencies of different outcome patterns

X	Y	Z	U	V	W	Frequency
0	0	1	0	0	0	2
0	1	0	1	1	0	1
0	1	1	1	0	1	2
1	0	0	1	1	0	5
1	0	1	1	0	1	2
1	1	0	1	1	1	3

Table 3 shows an entropy matrix for the bivariate entropies with the univariate entropies in the diagonal. It reveals that $H_{Z,V} = H_Z = H_V$, so that Z and V are equivalent. In fact, $V = 1 - Z$, and V can be omitted. There are no further functional dependencies between single variables.

Table 3. Entropy matrix

	X	Y	Z	U	V	W
X	$h(5,10)$ $= \mathbf{.636}$	$h(2,3,3,7)$ $= \mathbf{1.269}$	$h(1,2,4,8)$ $= \mathbf{1.137}$	$h(2,3,10)$ $= \mathbf{.861}$	$h(1,2,4,8)$ $= \mathbf{1.137}$	$h(2,3,5,5)$ $= \mathbf{1.323}$
Y		$h(6,9)$ $= \mathbf{.672}$	$h(2,4,4,5)$ $= \mathbf{1.339}$	$h(2,6,7)$ $= \mathbf{.991}$	$h(2,4,4,5)$ $= \mathbf{1.339}$	$h(1,2,5,7)$ $= \mathbf{1.171}$
Z			$h(6,9)$ $= \mathbf{.672}$	$h(2,4,9)$ $= \mathbf{.928}$	$h(6,9)$ $= \mathbf{.672}$	$h(2,3,4,6)$ $= \mathbf{1.310}$
U				$h(2,13)$ $= \mathbf{.393}$	$h(2,4,9)$ $= \mathbf{.928}$	$h(2,6,7)$ $= \mathbf{.991}$
V					$h(6,9)$ $= \mathbf{.672}$	$h(2,3,4,6)$ $= \mathbf{1.310}$
W						$h(7,8)$ $= \mathbf{.691}$

The remaining five variables have their bivariate and trivariate entropies given in Table 4. Table 4 reveals that $H_{X,Y} = H_{X,Y,U}$ and $H_{Z,W} = H_{Z,U,W}$, so that $(X,Y) \gg U$ and $(Z,W) \gg U$. Variable U can be omitted. Table 4 also reveals that four of the trivariate entropies are equal, and in fact equal to the entropy of the initial frequency distribution in Table 2. The remaining variables X, Y, Z, W have the same simultaneous entropy $h(1,2,2,2,3,5)$ as any of its triples (X,Y,Z), (X,Y,W), (X,Z,W), (Y,Z,W). Any remaining variable is functionally determined by the other three variables. Thus W could be removed leaving X, Y, Z as the only remaining three variables.

Table 4. Bivariate and trivariate entropies

X,Y	$h(2,3,3,7) = 1.269$	X,Y,Z	$h(1,2,2,2,3,5) = 1.676$
X,Z	$h(1,2,4,8) = 1.137$	X,Y,U	$h(2,3,3,7) = 1.269$
X,U	$h(2,3,10) = .861$	X,Y,W	$h(1,2,2,2,3,5) = 1.676$
X,W	$h(2,3,5,5) = 1.323$	X,Z,U	$h(1,2,2,2,8) = 1.322$
Y,Z	$h(2,4,4,5) = 1.339$	X,Z,W	$h(1,2,2,2,3,5) = 1.676$
Y,U	$h(2,6,7) = .991$	X,U,W	$h(1,2,2,5,5) = 1.450$
Y,W	$h(1,2,5,7) = 1.171$	Y,Z,U	$h(2,2,2,4,5) = 1.525$
Z,U	$h(2,4,9) = .928$	Y,Z,W	$h(1,2,2,2,3,5) = 1.676$
Z,W	$h(2,3,4,6) = 1.310$	Y,U,W	$h(1,2,2,5,5) = 1.450$
U,W	$h(2,6,7) = .991$	Z,U,W	$h(2,3,4,6) = 1.310$

Table 5. Matrix of joint entropy and relative entropy for the column variable

	X	Y	Z	U	V	W
	.636	.039	.171	.168	.171	.004
X	**1.00**	**.06**	**.25**	**.43**	**.25**	**.01**
	.039	.672	.005	.074	.005	.192
Y	**.06**	**1.00**	**.01**	**.19**	**.01**	**.28**
	.171	.005	.672	.137	.672	.053
Z	**.27**	**.01**	**1.00**	**.35**	**1.00**	**.08**
	.168	.074	.137	.393	.137	.093
U	**.26**	**.11**	**.20**	**1.00**	**.20**	**.14**
	.171	.005	.672	.137	.672	.053
V	**.27**	**.01**	**1.00**	**.35**	**1.00**	**.08**
	.004	.192	.053	.093	.053	.691
W	**.01**	**.29**	**.08**	**.24**	**.08**	**1.00**

When redundancies and functional dependencies have been identified, the structure and strength of the joint entropies can be examined. Table 5 gives a matrix of the joint entropies and their relative contributions to the entropies of the column variables. By convenient rounding a structural pattern might appear as in the plot shown in Figure 1. Functional dependencies from variables or groups of variables surrounded by a border line are marked by thick arrows to the variable determined by them. Degrees of dependencies between pairs of variables are shown by association arrows and weak association arrows. According to the independence tests given in the next section, the arrows of association and weak association in Figure 1 represent joint entropy values in Table 5 that correspond to rejection of independence with confidence levels higher than 98% and 80%, respectively.

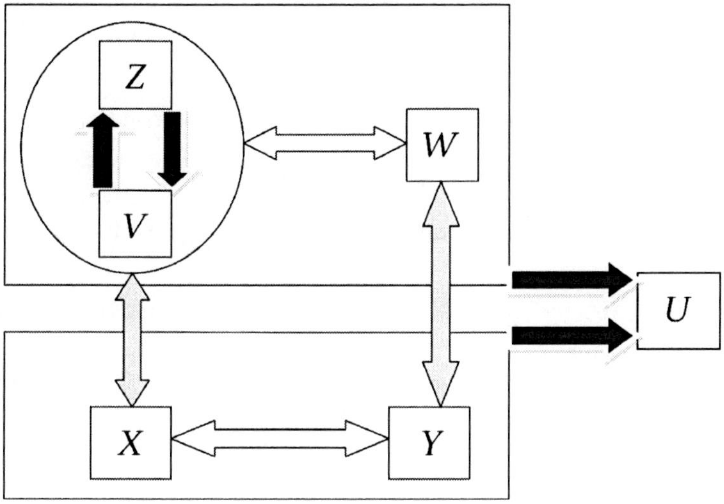

Fig. 1. Functional dependencies and associations between variables

6 Statistical Tests of Goodness of Fit

Exploratory statistical analyses of empirical distributions can be purely descriptive as described above. Dependence measures were conveniently rounded to distinguish between strong and weak associations. If probabilistic models should be used, a confirmatory statistical analysis is required, and the dependence measures need to be compared to critical limits determined by statistical significance. This section is devoted to some goodness of fit tests of hypothetical multivariate discrete distributions by using entropy measures. In particular, significance interpretations are given for the critical limits of dependence measures.

Let X be a random variable with r outcomes $1, \ldots, r$ of positive probabilities $p_X(i)$ for $i = 1, \ldots, r$. Let $p_i = n_i/n \geq 0$ denote the relative frequency of outcome i among n independent observations of X. The probability distribution p_X is estimated by the empirical distribution p, and the likelihood-function is estimated by $L(p) = \exp(-nH)$ where $H = H(p)$ is the entropy of

p. If there is a hypothetical specification of the distribution p_X, the likelihood-ratio test of p_X has a rejection region defined by large values of $L(p)/L(q)$. Here $L(p)$ is the estimated likelihood-function with no restrictions on p_X, and $L(q)$ is the estimated likelihood-function when p_X is restricted by the hypothesis and estimated by q. Twice the estimated log-likelihood-ratio is known to be approximately distributed as chi-square with d degrees of freedom for large n. Here $d = d(p) - d(q)$ where $d(p)$ and $d(q)$ are the numbers of parameters estimated to get p and q, respectively. The $\chi^2(d)$-statistic can be given as $2n$ times the divergence from p to q according to

$$\chi^2(d) = 2 \log \frac{L(p)}{L(q)} = 2n \sum_{i=1}^{r} p_i \log \frac{p_i}{q_i} = 2n\, D(p, q).$$

A convenient form of the critical region that will be used throughout this section is

$$\chi^2(d) > d + 2\sqrt{2d} = d + \sqrt{8d}.$$

It has a confidence level approximately equal to 95%. The divergence is equal to

$$D(p, q) = -\sum_{i=1}^{r} q_i \Phi(pi/qi) \text{ where } \Phi(x) = x \log(1/x),$$

and due to convexity from above of Φ it follows that $D(p, q) \geq 0$ with equality if and only if $p = q$. The empirical distribution has $d(p) = r - 1$. The hypothesis that X is uniformly distributed on r outcomes specifies $p_X(i) = q_i = 1/r$ with $d(q) = 0$. Hence uniformity is tested by

$$\chi^2(r - 1) = 2n\, D(p, q) = 2n[\log r - H],$$

and it follows that uniformity is rejected if H deviates from its maximum value $\log r$ by more than $[r - 1 + \sqrt{8(r - 1)}]/2n$.

As another example, consider a bivariate random variable (X, Y) with empirical distribution $p_{i,j} = n_{i,j}/n$ for $i = 1, \ldots, r$ and $j = 1, \ldots, s$ with $d(p) = rs - 1$. Under the hypothesis that X and Y are independent, the distribution of (X, Y) is estimated by $q_{i,j} = n_{i\cdot} n_{\cdot j}/n^2$ with $d(q) = r-1+s-1$. It follows that independence is tested by

$$\chi^2((r - 1)(s - 1)) = 2n\, D(p, q) = 2n[H_X + H_Y - H_{X,Y}] = 2n\, J_{X,Y}$$

where the entropies and joint entropy are understood to be the empirical versions $H_X + H_Y = H(q)$ and $H_{X,Y} = H(p)$ calculated with relative frequencies. Hence independence is rejected when the empirical joint entropy is larger than $[(r - 1)(s - 1) + \sqrt{8(r - 1)(s - 1)}]/2n$.

As a third example consider a trivariate random variable (X, Y, Z) which has empirical distribution p with probabilities $p_{i,j,k} = n_{i,j,k}/n$ for $i = 1, \ldots, r$, $j = 1, \ldots, s$, $k = 1, \ldots, t$ and $d(p) = rst - 1$. Under the hypothesis that conditional on Z, the other two variables X and Y are independent, the distribution $p_{X,Y,Z} = p_Z \cdot p_{X|Z} \cdot p_{Y|Z}$ is estimated by empirical distribution q with relative frequencies $q_{i,j,k} = p_{\cdot\cdot k} \cdot (p_{i \cdot k}/p_{\cdot\cdot k}) \cdot (p_{\cdot j,k}/p_{\cdot\cdot k}) = p_{i \cdot k} \cdot p_{\cdot jk}/p_{\cdot\cdot k}$, and it follows that $d(q) = t - 1 + t(r - 1 + s - 1)$. Here $d = d(p) - d(q) = (r-1)(s-1)t$ and conditional independence $X \perp Y|Z$ is tested by

$$\chi^2((r-1)(s-1)t) = 2n\, D(p,q) = 2n\,(H_{X,Z} + H_{Y,Z} - H_Z - H_{X,Y,Z}),$$

where the entropies are empirical versions $H_{X,Z} + H_{Y,Z} - H_Z = H(q)$ and $H_{X,Y,Z} = H(p)$. Hence the conditional independence is rejected when the empirical trivariate entropies have a difference $H(q) - H(p)$ that is larger than $[(r-1)(s-1)t + \sqrt{8(r-1)(s-1)t}\,]/2n$.

7 Comments on Literature and Related Topics

The classic paper by Shannon [16] that introduced entropy as a measure of information in long sequences of letters, digits or signals in communicated messages can be considered as a source of impact not only to the theory of information and communication but also to broader fields of data security and statistics. Kullback [12], Theil [17], Gokhale and Kullback [6], Hamming [8], Ellis [2], Krippendorff [11], Cover and Thomas [1], Hankerson, Harris and Johnson [9] are examples of text books that illustrate the rich development of information concepts in various directions.

The divergence measure used in statistics is often referred to as Kullback-Leibler information, and it is sometimes considered as an alternative to Fisher information. Fisher information is developed from likelihood theory for parametric families of probability distributions that satisfy certain regularity conditions, and it doesn't have such general applicability as Kullback-Leibler information. The divergence measure has also been generalized in different ways, and in this context it seems appropriate to especially mention the work by Frank, Menéndez and Pardo [5].

Modern advanced texts in probability and statistics like Kallenberg [10] and Schervish [15] have sections on information and use information measures as theoretical tools to evaluate and compare distributions and to derive asymptotic results. Less well spread are information measures as tools in applied statistics. It is likely that continued research on the development of information-based screening methods for data editing and variable selection combined with development of convenient computer software will make these methods more accessible and common in applied statistics. Some first steps towards making entropy methods more accessible are taken by Frank and Lorenc [4].

Goodman and Kruskal [7] include information measures in their thorough exposition of association measures. Association can be considered as being intermediate between functional dependence and complete independence. All these concepts are central to the variable screening illustrated in the previous section. For continued research on variable screening, comparisons with related graphical models and implication lattices should be of interest. Graphical models plot conditional dependencies between variables and implication lattices plot incomplete functional relations between variables. Whittaker [18] and Lauritzen [13] are texts on graphical models. Luksch, Skorsky and Wille [14] and Frank [3] describe plots of implications between variables.

References

1. Cover, T., Thomas, J.: Elements of Information Theory. Wiley, New York (1991)
2. Ellis, R.S.: Entropy, Large Deviations, and Statistical Mechanics. Springer, New York (1985)
3. Frank, O.: Structural plots of multivariate binary data. J. Soc. Struct. 1(4), 1–19 (2000)
4. Frank, O., Lorenc, B.: Entropy combinations that detect conditional, pair-wise and mutual independence. Statistics Department, Stockholm University (2002)
5. Frank, O., Menéndez, M.L., Pardo, L.: Asymptotic distributions of weighted divergence between discrete distributions. Comm. Statist.-Theor. Meth. 27(4), 867–885 (1998)
6. Gokhale, D., Kullback, S.: The Information in Contingency Tables. Marcel-Decker, New York (1978)
7. Goodman, L.A., Kruskal, W.H.: Measures of Association for Cross-Classifications. Springer, New York (1979)
8. Hamming, R.W.: Coding and Information Theory. Prentice-Hall, Englewood Cliffs (1980)
9. Hankerson, D., Harris, G., Johnson Jr., P.: Introduction to Information Theory and Data Compression. CRC Press, Boca Raton (1998)
10. Kallenberg, O.: Foundations of Modern Probability, 2nd edn. Springer, New York (2002)
11. Krippendorff, K.: Information Theory; Structural Models for Qualitative Data. Sage, Newbury Park (1986)
12. Kullback, S.: Information Theory and Statistics. Wiley, New York (1959)
13. Lauritzen, S.L.: Graphical Models. Oxford University Press, Oxford (1996)
14. Luksch, P., Skorsky, M., Wille, R.: Drawing concept lattices with a computer. In: Gaul, W., Schader, M. (eds.) Classification as a Tool of Research, pp. 269–274. Elsevier Science, Amsterdam (1986)
15. Schervish, M.J.: Theory of Statistics. Springer, New York (1995)
16. Shannon, C.E.: A mathematical theory of communication. Bell Syst. Tech. J. 27, 379–423, 623–656 (1948)
17. Theil, H.: Economics and Information Theory. North Holland, Amsterdam (1967)
18. Whittaker, J.: Graphical Models in Applied Multivariate Statistics. Wiley, New York (1990)

Minimum Phi-Divergence Estimators of a Set of Binomial Probabilities*

María Luisa Menéndez[1] and Leandro Pardo[2]

[1] Department of Applied Mathematics, E.T.S.A.M., Technical University of Madrid, Spain
[2] Department of Statistics and O.R. I, Complutense University of Madrid, Spain
lpardo@mat.ucm.es

Summary. In this paper we consider the product Bernoulli model with different probability success and the related problem of estimation of the probabilities when it is suspected that they are equal. If we are completely sure that the probabilities are equal we must use a restricted estimator but in many situations it is not clear if the probabilities are equal or not and then a better procedure will be to use a "preliminary test estimator". Based on minimum phi-divergence estimator (MϕE) we study, in this paper, some estimators for the parameters of the product Bernoulli model: Unrestricted MϕE, Restricted MϕE, Preliminary MϕE, Shrinkage MϕE, Shrinkage preliminary MϕE, James-Stein MϕE, Positive-part of Stein-Rule MϕE and Modified preliminary MϕE. Asymptotic quadratic bias as well as asymptotic quadratic risk are studied under contiguous alternative hypotheses.

1 Introduction

We consider v Binomial populations of parameters 1 and θ_i, $0 < \theta_i < 1$, $i = 1, \ldots, v$. Let $y_{i1}, \ldots, y_{in_i}$ be a random sample of size n_i from the ith Binomial population. We denote $y_i = \sum_{j=1}^{n_i} y_{ij}$, $i = 1, \ldots, v$. The maximum likelihood estimator (MLE) of $\boldsymbol{\theta} = (\theta_1, \ldots, \theta_v)^T$ maximizes the expression

$$l(\boldsymbol{\theta}) = \sum_{i=1}^{v} \log\left(\theta_i^{y_i}\left(1 - \theta_i^{n_i - y_i}\right)\right), \tag{1}$$

i.e., $\widehat{\boldsymbol{\theta}} = \arg\max_{\boldsymbol{\theta} \in \Theta} l(\boldsymbol{\theta})$, being

$$\Theta = \left\{\boldsymbol{\theta} : \boldsymbol{\theta} = (\theta_1, \ldots, \theta_v)^T,\ 0 < \theta_i < 1,\ i = 1, \ldots, v\right\}. \tag{2}$$

* A draft for this paper was written in the summer 2007 in Soto del Real. We planned to undertake later a meticulous proofreading together along Christmas 2007 and submit it to a journal. Marisa's health problems prevented us to do it. Unfortunately I have had to take up again our former planning three years later in the summer 2010 in Soto del Real, but this time alone. Nevertheless, I'm really certain Marisa has been supporting me strongly from wherever she is.

It is well-known that $\widehat{\boldsymbol{\theta}} = (\widehat{\theta}_1, \ldots, \widehat{\theta}_\upsilon)^T$, being

$$\widehat{\theta}_i = y_i/n_i, \qquad i = 1, \ldots, \upsilon. \tag{3}$$

Now if we assume that non-sample prior information on the values $\theta_1, \ldots, \theta_\upsilon$ is available (either from previous studies or from practical experience of the researches or experts) and this non-sample prior information can be expressed by the hypothesis

$$H_0 : \theta_1 = \ldots = \theta_\upsilon = \theta_0, \ (\theta_0 \text{ unknown}). \tag{4}$$

The maximum likelihood estimator, under (3), can be expressed by

$$\widetilde{\theta} = \arg \max_{\boldsymbol{\theta} \in \boldsymbol{\Theta}_0} l(\boldsymbol{\theta}),$$

where $\Theta_0 = \{\boldsymbol{\theta} \in \Theta : \theta_1 = \ldots = \theta_\upsilon = \theta_0\}$. It is well-known that $\widetilde{\theta} = \frac{1}{n}\sum_{i=1}^{\upsilon} y_i$; $(n = n_1 + \ldots + n_\upsilon)$.

We define,

$$\widetilde{\boldsymbol{\theta}} = (\widetilde{\theta}, \ldots, \widetilde{\theta})^T = (\tfrac{1}{n}\textstyle\sum_{i=1}^{\upsilon} y_i, \ldots, \tfrac{1}{n}\sum_{i=1}^{\upsilon} y_i) = \widetilde{\theta} \boldsymbol{J}_\upsilon$$

where $\boldsymbol{J}_\upsilon = \left(1, \overset{(\upsilon}{\ldots}, 1\right)^T$.

In the following we refer to $\widehat{\boldsymbol{\theta}}$ as the unrestricted maximum likelihood estimator of $\boldsymbol{\theta}$ and $\widetilde{\boldsymbol{\theta}}$ as the restricted maximum likelihood estimator (RMLE) of $\boldsymbol{\theta}$. When H_0 holds, $\widetilde{\boldsymbol{\theta}}$ has a smaller risk (under quadratic loss) than $\widehat{\boldsymbol{\theta}}$. On the other hand when H_0 does not hold, $\widehat{\boldsymbol{\theta}}$ may perform better than $\widetilde{\boldsymbol{\theta}}$. When the prior information on H_0 is rather uncertain it may be desirable to have a preliminary test estimator (PTE),

$$\widehat{\boldsymbol{\theta}}^{PTE} = \widetilde{\boldsymbol{\theta}} + (1 - I_{(0,\chi^2_{\upsilon-1})}(LR))(\widehat{\boldsymbol{\theta}} - \widetilde{\boldsymbol{\theta}}) \tag{5}$$

where LR is the likelihood ratio test or a shrinkage estimator (SE),

$$\widehat{\boldsymbol{\theta}}^S = \widetilde{\boldsymbol{\theta}} + \left(1 - (\nu - 3)(LR)^{-1}\right)(\widehat{\boldsymbol{\theta}} - \widetilde{\boldsymbol{\theta}}), \ (\upsilon > 3). \tag{6}$$

Ali and Saleh [1] introduced these estimators and compared them with the restricted and unrestricted maximum likelihood estimators. This comparison was carried out under quadratic loss.

A first generalization of the results obtained by Ali and Saleh [1] can be obtained if we consider the family of estimators

$$\widehat{\boldsymbol{\theta}}_h = \widetilde{\boldsymbol{\theta}} + (1 - h(LR))(\widehat{\boldsymbol{\theta}} - \widetilde{\boldsymbol{\theta}}), \tag{7}$$

where h is a real function.

Estimators (4) and (5) can be considered elements of the family (7). For $h(x) = I_{(0,\chi^2_{v-1},\alpha)}(x)$ we get the PTE given in (4) and for $h(x) = (v-3)\,x^{-1}$ the SE given in (5).

In this paper we consider the family given in (7) but instead of considering the restricted maximum likelihood estimator of $\boldsymbol{\theta}$ we shall consider the restricted minimum phi-divergence estimator (RMϕE), $\widetilde{\boldsymbol{\theta}}_\phi$, of $\boldsymbol{\theta}$ and instead of the LR we consider the family of phi-divergence test statistics, $T_n^{\phi_1,\phi_2}$. RMLE of $\boldsymbol{\theta}$, $\widehat{\boldsymbol{\theta}}$, can be considered as a particular case of the RMϕE (see definition in (12)) with $\phi_2(x) = x\log x - x + 1$ and at the same time the LR can be considered an element of the family of phi-divergence test statistics, $T_n^{\phi_1,\phi_2}$, (see definition in (15)) with $\phi_1(x) = \phi_2(x) = x\log x - x + 1$. For more details about MϕE and phi-divergence test statistics see Pardo [13]. The RMϕE where considered for the first time in Pardo *et al.* [12].

Section 2 is devoted to introduce a new family of estimators based on (7) as well as in RMϕE and phi-divergence test statistics. Some asymptotic distributional results are presented in Section 3. The asymptotic bias as well as the asymptotic risk of the new family of estimators are given in Section 4 and 5, respectively.

2 The Proposed Family of Estimators

It is not difficult to see that $l(\boldsymbol{\theta})$, given in (1), can be written as

$$l(\boldsymbol{\theta}) = k - \sum_{i=1}^{v} n_i D_{Kull}(\widehat{\boldsymbol{p}}_i, \boldsymbol{p}(\theta_i))$$

where $D_{Kull}(\widehat{\boldsymbol{p}}_i, \boldsymbol{p}(\theta_i))$ is the Kullback-Leibler divergence measure between the probability vectors $\widehat{\boldsymbol{p}}_i = (y_i/n_i, (n_i - y_i)/n_i)^T$ and $\boldsymbol{p}(\theta_i) = (\theta_i, 1-\theta_i)$, $i = 1, \ldots, v$.

Therefore the unrestricted and restricted maximum likelihood estimators of $\boldsymbol{\theta}$ can be defined by

$$\widehat{\boldsymbol{\theta}} = \arg\min_{\boldsymbol{\theta}\in\boldsymbol{\Theta}} \sum_{i=1}^{v} n_i D_{Kull}(\widehat{\boldsymbol{p}}_i, \boldsymbol{p}(\theta_i)) \tag{8}$$

and

$$\widetilde{\theta} = \arg\min_{\boldsymbol{\theta}\in\boldsymbol{\Theta}_0} \sum_{i=1}^{v} n_i D_{Kull}(\widehat{\boldsymbol{p}}_i, \boldsymbol{p}(\theta_i)), \tag{9}$$

respectively.

If instead of considering in (8) and (9) the Kullback-Leibler divergence measure we consider a more general family of divergences we could get a family of restricted and unrestricted estimators. In this paper we consider the family of phi-divergence measures introduced by Csiszár [4] and Ali and Silvey [2], simultaneously. The phi-divergence measure between the probability vectors $\widehat{\boldsymbol{p}}_i$ and $\boldsymbol{p}(\theta_i)$ is given by

$$D_\phi\left(\widehat{\boldsymbol{p}}_i, \boldsymbol{p}(\theta_i)\right) = \theta_i \phi\left(\frac{y_i}{n_i\theta_i}\right) + (1-\theta_i)\,\phi\left(\frac{n_i - y_i}{(1-\theta_i)\,n_i}\right) \tag{10}$$

where ϕ is a convex function defined for $x > 0$, such that at $x = 1$, $\phi(1) = 0$, $\phi''(1) > 0$. In the following we shall assume the conventions $0\phi(0/0) = 0$ and $o\phi(p/0) = p\lim_{u\to\infty}\phi(u)/u$. For a systematic study of phi-divergence measures see Pardo [13].

As a natural extension of the restricted and unrestricted maximum likelihood estimators it is possible to consider the MϕE and RMϕE defined by

$$\widehat{\boldsymbol{\theta}}_\phi = \arg\min_{\boldsymbol{\theta}\in\boldsymbol{\Theta}} \textstyle\sum_{i=1}^{v} n_i D_\phi\left(\widehat{\boldsymbol{p}}_i, \boldsymbol{p}(\theta_i)\right) \tag{11}$$

and

$$\widetilde{\theta}_\phi = \arg\min_{\boldsymbol{\theta}\in\boldsymbol{\Theta}_0} \textstyle\sum_{i=1}^{v} n_i D_\phi\left(\widehat{\boldsymbol{p}}_i, \boldsymbol{p}(\theta_i\right), \tag{12}$$

respectively. We shall denote $\widetilde{\boldsymbol{\theta}}_\phi = \left(\widetilde{\theta}_\phi, \ldots, \widetilde{\theta}_\phi\right) = \widetilde{\theta}_\phi \boldsymbol{J}_v$.

We can observe the following: Minimizing (11) over Θ is equivalent to finding the minimum in θ_i of the function $D_\phi\left(\widehat{\boldsymbol{p}}_i, \boldsymbol{p}(\theta_i)\right)$. We have,

$$\frac{\partial D_\phi\left(\widehat{\boldsymbol{p}}_i, \mathbf{p}(\theta_i)\right)}{\partial\theta_i} = \phi\left(\frac{y_i}{n_i\theta_i}\right) + \theta_i\phi'\left(\frac{y_i}{n_i\theta_i}\right)\frac{y_i}{n_i}\left(-\frac{1}{\theta_i^2}\right) - \phi\left(\frac{n_i - y_i}{n_i(1-\theta_i)}\right)$$
$$+ (1-\theta_i)\,\phi'\left(\frac{n_i - y_i}{n_i(1-\theta_i)}\right)\frac{n_i - y_i}{n_i(1-\theta_i)^2} = 0$$

and $\widehat{\theta}_i^\phi = y_i/n_i$, $i = 1, \ldots, v$, is a solution. Therefore the MϕE , $\widehat{\boldsymbol{\theta}}^\phi$, is independent of the function ϕ and therefore it coincides with the unrestricted maximum likelihood estimator.

This situation changes when we consider the RMϕE obtained by minimizing the expression (12). For a general function ϕ it is not possible to have explicit expression for the RMϕE. It is necessary to have explicit expression of the function ϕ to get explicit expression of the RMϕE. For instance if we consider the family

$$\phi_\lambda(x) = \begin{cases} \frac{1}{\lambda(\lambda+1)}\left(x^{\lambda+1} - x - \lambda(x-1)\right) & \lambda \neq 0, -1 \\ x\log x - x + 1 & \lambda = 0 \\ \log x + x - 1 & \lambda = -1 \end{cases} \tag{13}$$

we get for $\lambda \neq 0$ or -1,

$$\widetilde{\theta}_\lambda(y_1, \ldots, y_n) = \frac{\left(\sum_{i=1}^{v}\frac{y_i^{\lambda+1}}{n_i^\lambda}\right)^{\frac{1}{\lambda+1}}}{\left(\sum_{i=1}^{v}\frac{y_i^{\lambda+1}}{n_i^\lambda}\right)^{\frac{1}{\lambda+1}} + \left(\sum_{i=1}^{v}\frac{(n_i-y_i)^{\lambda+1}}{n_i^\lambda}\right)^{\frac{1}{\lambda+1}}}$$

for $\lambda = 0$ (maximum likelihood estimator),

$$\widetilde{\theta}_0\left(y_1,\ldots,y_n\right)=\frac{\sum_{i=1}^{v}y_i}{\sum_{i=1}^{v}y_i+\sum_{i=1}^{v}\left(n_i-y_i\right)}=\frac{1}{n}\sum_{i=1}^{v}y_i$$

and for $\lambda=-1$,

$$\widetilde{\theta}_{-1}\left(y_1,\ldots,y_n\right)=\frac{1}{1+\left[\prod_{i=1}^{v}\left(\frac{n_i-y_i}{y_i}\right)^{n_i}\right]^{\frac{1}{n}}}.$$

The family of divergence measures obtained from (13) is called power-divergence measure and it was introduced by Cressie and Read [3]. Based on $\widehat{\boldsymbol{\theta}}$ and $\widetilde{\boldsymbol{\theta}}_{\phi_2}$ we consider the following family of estimators

$$\widehat{\boldsymbol{\theta}}^{h}_{\phi_1,\phi_2}=\widetilde{\boldsymbol{\theta}}_{\phi_2}+\left(1-h\left(T_n^{\phi_1,\phi_2}\right)\right)\left(\widehat{\boldsymbol{\theta}}-\widetilde{\boldsymbol{\theta}}_{\phi_2}\right) \tag{14}$$

where $T_n^{\phi_1,\phi_2}$ is the family of phi-divergence test statistics for testing (3) and whose expression is given by

$$T_n^{\phi_1,\phi_2}=\frac{2n}{\phi_1''\left(1\right)}\sum_{i=1}^{v}n_iD_\phi\left(\widehat{\boldsymbol{p}}_i,p(\widetilde{\boldsymbol{\theta}}_{\phi_2})\right). \tag{15}$$

In Theorem 1 we shall present the asymptotic distribution of $T_n^{\phi_1,\phi_2}$ under H_0 as well as under contiguous alternative hypotheses to H_0. In Menéndez *et al.* ([6], [7], [8], [9], [10], [11]), Pardo and Menéndez [14] and Pardo *et al.* [15] can be seen some estimators of type (14) for different statistical problems.

The election of different functions h gives some well-known estimators. If we choose $h(x)=0$ for all x we get the unrestricted maximum likelihood estimator, i.e., $\widehat{\boldsymbol{\theta}}^{h}_{\phi_1,\phi_2}=\widehat{\boldsymbol{\theta}}$. For $h(x)=1$ for all x we get the RMϕ_2E, $\widehat{\boldsymbol{\theta}}^{h}_{\phi_1,\phi_2}=\widetilde{\boldsymbol{\theta}}_{\phi_2}$. For $h(x)=1-a$ for all x $a\in(0,1)$, the shrinkage Mϕ_2E, $\widehat{\boldsymbol{\theta}}^{h}_{\phi_1,\phi_2}=\widehat{\boldsymbol{\theta}}^{SRE}_{\phi_1,\phi_2}$. For $h(x)=I_{\left(0,\chi^2_{v-1},\alpha\right)}(x)$ the preliminary Mϕ_2E, $\widehat{\boldsymbol{\theta}}^{h}_{\phi_1,\phi_2}=\widehat{\boldsymbol{\theta}}^{PTE}_{\phi_1,\phi_2}$. For $h(x)=aI_{\left(0,\chi^2_{v-1},\alpha\right)}(x)$ the shrinkage preliminary Mϕ_2E, $\widehat{\boldsymbol{\theta}}^{h}_{\phi_1,\phi_2}=\widehat{\boldsymbol{\theta}}^{SPT}_{\phi_1,\phi_2}$. For $h(x)=(v-3)\,x^{-1}$ with $(v>3)$ the James-Stein Mϕ_2E, $\widehat{\boldsymbol{\theta}}^{h}_{\phi_1,\phi_2}=\widehat{\boldsymbol{\theta}}^{S}_{\phi_1,\phi_2}$. For $h(x)=1-\left(1-(v-3)\,x^{-1}\right)I_{(v-3,\infty)}(x)$, $(v>3)$, the positive part of Stein-Rule Mϕ_2E, $\widehat{\boldsymbol{\theta}}^{h}_{\phi_1,\phi_2}=\widehat{\boldsymbol{\theta}}^{S_+}_{\phi_1,\phi_2}$. For $h(x)=1-(1-(v-2)\,x^{-1})I_{\left[\chi^2_{v-1},\alpha,\infty\right)}(x)$, $(v>3)$ the modified preliminary Mϕ_2E, $\widehat{\boldsymbol{\theta}}^{h}_{\phi_1,\phi_2}=\widehat{\boldsymbol{\theta}}^{PTE_+}_{\phi_1,\phi_2}$.

3 Some Asymptotic Distributional Results

We define the two following probability vectors,

$$\boldsymbol{p}_n\left(\boldsymbol{\theta}\right)=\left(n_1\theta_1/n\ ,\left(1-\theta_1\right)n_1/n,\ldots,n_v\theta_v/n,n_v\left(1-\theta_v\right)/n\right)^T$$

and $\widehat{\boldsymbol{p}} = \left(y_1/n,\ \left(n_1 - y_1\right)/n, \ldots, y_v/n,\ \left(n_v - y_v\right)/n\right)^T$. We denote

$$\boldsymbol{D}_{\theta_0} = \operatorname{diag}\left(\theta_0^{-1/2}, \left(1-\theta_0\right)^{-1/2}\right).$$

We have,

$$\operatorname{diag}\left(\boldsymbol{p}_n\left(\theta_0\right)^{-1/2}\right) = \operatorname{diag}\left(\left(\frac{n_1}{n}\right)^{-1/2}, \ldots, \left(\frac{n_v}{n}\right)^{-1/2}\right) \otimes \boldsymbol{D}_{\theta_0}.$$

and

$$\left(\frac{\partial \boldsymbol{p}_n\left(\boldsymbol{\theta}\right)}{\partial \boldsymbol{\theta}}\right)_{\boldsymbol{\theta}=\boldsymbol{\theta}_0} = \operatorname{diag}\left(\frac{n_1}{n}, \ldots, \frac{n_v}{n}\right) \otimes \left(1,-1\right)^T \qquad \boldsymbol{\theta}_0 = \left(\theta_0, ..^{(v}., \theta_0\right).$$

By $\otimes$ we are denoting the Kronecker product between the respective matrices. It is well-known (see p.225 in Pardo [13]) that Fisher information matrix, for this model, is defined by $\boldsymbol{I}_n\left(\theta_0\right) = \boldsymbol{A}_n\left(\theta_0\right)^T \boldsymbol{A}_n\left(\theta_0\right) = \boldsymbol{\Lambda}_n \theta_0^{-1}\left(1-\theta_0\right)^{-1}$, where

$$\boldsymbol{A}_n\left(\theta_0\right) = \boldsymbol{\Lambda}_n^{-1/2} \otimes \left(\theta_0^{-1/2}, -\left(\left(1-\theta_0\right)\right)^{-1/2}\right)^T$$

being $\boldsymbol{\Lambda}_n = \operatorname{diag}\left(n_1/n, \ldots, n_v/n\right)$.

The unrestricted maximum likelihood, $\widehat{\boldsymbol{\theta}}$, of $\boldsymbol{\theta}_0 = \left(\theta_0, ..^{(v}., \theta_0\right)$ admits the following BAN decomposition

$$\widehat{\boldsymbol{\theta}} = \boldsymbol{\theta}_0 + \boldsymbol{I}_n\left(\theta_0\right)^{-1} \boldsymbol{A}_n\left(\theta_0\right) \operatorname{diag}\left(\boldsymbol{p}_n\left(\theta_0\right)^{-1/2}\right)\left(\widehat{\boldsymbol{p}} - \boldsymbol{p}_n\left(\theta_0\right)\right) + o_p\left(n^{-1/2}\right) \tag{16}$$

and its asymptotic distribution is $\sqrt{n}\left(\widehat{\boldsymbol{\theta}} - \boldsymbol{\theta}_0\right) \underset{n\to\infty}{\xrightarrow{L}} \mathcal{N}\left(\mathbf{0}, \boldsymbol{I}\left(\theta_0\right)^{-1}\right)$ where $\boldsymbol{I}\left(\theta_0\right) = \lim_{n\to\infty} \boldsymbol{I}_n\left(\theta_0\right) = \boldsymbol{\Lambda}\theta_0^{-1}\left(1-\theta_0\right)^{-1}$, $\boldsymbol{\Lambda} = \operatorname{diag}\left(\lambda_1, \ldots, \lambda_v\right)$ and $\lambda_j = \lim_{n\to\infty} n_j/n$, $j = 1, \ldots, v$.

The null hypothesis given in (3) can be written by $h_i\left(\boldsymbol{\theta}\right) = \theta_i - \theta_v = 0$, $i = 1, \ldots, v-1$. If we denote by

$$\boldsymbol{B} = \left(\frac{\partial h_i\left(\boldsymbol{\theta}\right)}{\partial \theta_j}\right)_{\substack{i=1,\ldots,v-1\\ j=1,\ldots,v}} = \left(\boldsymbol{I}_{(v-1)\times(v-1)}, -\boldsymbol{J}_{v-1}\right)_{(v-1)\times v}$$

the hypothesis (3) can be given by $H_0 : \boldsymbol{B\theta} = \mathbf{0}_{(v-1)\times 1}$ or equivalently, $g(\boldsymbol{\theta}) = \mathbf{0}$ where $g(\boldsymbol{x}) = \boldsymbol{Bx}$.

The BAN decomposition of the RMϕE, $\widetilde{\boldsymbol{\theta}}_\phi$, of $\boldsymbol{\theta}_0 = \left(\theta_0, ..^{(v}., \theta_0\right)$ admits the expansion

$$\widetilde{\boldsymbol{\theta}}_\phi = \boldsymbol{\theta}_0 + \boldsymbol{H}_n(\theta_0)\boldsymbol{I}_n\left(\theta_0\right)^{-1} \boldsymbol{A}_n\left(\theta_0\right)^T \operatorname{diag}\left(\boldsymbol{p}_n\left(\theta_0\right)^{-1/2}\right)\left(\widehat{\boldsymbol{p}} - \boldsymbol{p}_n\left(\theta_0\right)\right) + o_p\left(n^{-1/2}\right) \tag{17}$$

being $\boldsymbol{H}_n(\theta_0) = \boldsymbol{I}_{v\times v} - \boldsymbol{I}_n\left(\theta_0\right)^{-1} \boldsymbol{B}^T\left(\boldsymbol{B}\boldsymbol{I}_n\left(\theta_0\right)^{-1}\boldsymbol{B}^T\right)^{-1}\boldsymbol{B}$.

For more details see Pardo [12]. Based on (16) and (17) we have

$$\sqrt{n}\left(\widetilde{\boldsymbol{\theta}}_{\phi}-\boldsymbol{\theta}_0\right)=\boldsymbol{H}_n(\theta_0)\sqrt{n}\left(\widehat{\boldsymbol{\theta}}-\boldsymbol{\theta}_0\right)+o_p(1). \tag{18}$$

It is not difficult to see that

$$\boldsymbol{H}(\theta_0)=\lim_{n\to\infty}\boldsymbol{H}_n(\theta_0)=\boldsymbol{J}_{\upsilon}\boldsymbol{J}_{\upsilon}^T\mathrm{diag}(\boldsymbol{\lambda}), \qquad \boldsymbol{\lambda}=(\boldsymbol{\lambda}_1,\ldots,\lambda_{\upsilon})^T. \tag{19}$$

Let $\boldsymbol{\theta}_n\in\Theta-\Theta_0$ be a given alternative and let $\boldsymbol{\theta}_0$ be the element in Θ_0 closest to $\boldsymbol{\theta}_n$ in the Euclidean distance sense. A first possibility for introducing contiguous alternative hypothesis is to consider a fixed $\Delta\in\mathbb{R}^{\upsilon}$ and allowing $\boldsymbol{\theta}_n$ to move toward $\boldsymbol{\theta}_0$ as n increases in the following way

$$H_{1,n}:\boldsymbol{\theta}_n=\boldsymbol{\theta}_0+n^{-1/2}\boldsymbol{\Delta}, \qquad \boldsymbol{\Delta}=(\Delta_1,\ldots,\Delta_n)^T. \tag{20}$$

A second approach is to relax the condition $g(\boldsymbol{\theta})=\mathbf{0}$ defining Θ_0. Let $\boldsymbol{\delta}\in\mathbb{R}^{\upsilon-1}$ and consider the following sequence, $\boldsymbol{\theta}_n$, of parameters approaching Θ_0 according to $H_{1,n}^*: g(\boldsymbol{\theta})=n^{-1/2}\boldsymbol{\delta}$. Note that a Taylor series expansion of $g(\boldsymbol{\theta}_n)$ around $\boldsymbol{\theta}_0\in\Theta_0$ yields

$$g(\boldsymbol{\theta}_n)=g(\boldsymbol{\theta})+\boldsymbol{B}\left(\boldsymbol{\theta}_n-\boldsymbol{\theta}_0\right)+o\left(\|\boldsymbol{\theta}_n-\boldsymbol{\theta}_0\|\right). \tag{21}$$

By substituting $\boldsymbol{\theta}_n=\boldsymbol{\theta}_0+n^{-1/2}\boldsymbol{\Delta}$ in (21) and taking into account that $g(\boldsymbol{\theta}_0)=\mathbf{0}$, we get

$$g(\boldsymbol{\theta}_n)=n^{-1/2}\,\boldsymbol{B\Delta}+o\left(\|\boldsymbol{\theta}_n-\boldsymbol{\theta}_0\|\right) \tag{22}$$

so that the equivalence in the limit is obtained for $\boldsymbol{\delta}=\boldsymbol{B\Delta}$.

In the following we shall denote $\boldsymbol{Q_B}=\left(\boldsymbol{BI}(\theta_0)^{-1}\boldsymbol{B}^T\right)^{-1/2}$. It is not difficult to see, under $H_{1,n}$, that $\sqrt{n}g(\widehat{\boldsymbol{\theta}})\xrightarrow[n\to\infty]{L}\mathcal{N}\left(\boldsymbol{B\Delta},\boldsymbol{Q}_{\boldsymbol{B}}^{\mathbf{2}}\right)$.

Now we are going to give, without proof, some asymptotic distributional results under $H_{1,n}$.

Theorem 1. *Under $H_{1,n}$ the following results follows,*

a) $\boldsymbol{X}_n\equiv\sqrt{n}\left(\widehat{\boldsymbol{\theta}}-\boldsymbol{\theta}_0\right)\xrightarrow[n\to\infty]{L}\mathcal{N}\left(\Delta,\boldsymbol{I}(\theta_0)^{-1}\right)$, $\boldsymbol{I}(\theta_0)=\boldsymbol{\Lambda}\theta_0^{-1}(1-\theta_0)^{-1}$.

b) $\boldsymbol{Y}_n\equiv\sqrt{n}\left(\widetilde{\boldsymbol{\theta}}_{\phi}-\boldsymbol{\theta}_0\right)\xrightarrow[n\to\infty]{L}\mathcal{N}\left(\boldsymbol{J}_{\upsilon}\boldsymbol{J}_{\upsilon}^T\mathrm{diag}(\boldsymbol{\lambda})\boldsymbol{\Delta},\boldsymbol{\Sigma_Y}\right)$, *where*
$\boldsymbol{\Sigma_Y}=\boldsymbol{I}(\theta_0)^{-1}-\boldsymbol{I}(\theta_0)^{-1}\boldsymbol{B}^T\boldsymbol{Q}_{\boldsymbol{B}}^2\boldsymbol{BI}(\theta_0)^{-1}=\theta_0(1-\theta_0)\boldsymbol{J}_{\upsilon}\boldsymbol{J}_{\upsilon}^T$.
If we assume $\boldsymbol{\lambda}^T\boldsymbol{\Delta}=\mathbf{0}$, *then* $\boldsymbol{J}_{\upsilon}\boldsymbol{J}_{\upsilon}^T\mathrm{diag}(\boldsymbol{\lambda})\boldsymbol{\Delta}=\mathbf{0}$.

c) $\boldsymbol{Z}_n\equiv\sqrt{n}\left(\widehat{\boldsymbol{\theta}}-\widetilde{\boldsymbol{\theta}}_{\phi}\right)\xrightarrow[n\to\infty]{L}\mathcal{N}\left(\boldsymbol{\delta}^*,\boldsymbol{\Sigma_Z}\right)$ *where* $\boldsymbol{\delta}^*=\left(\boldsymbol{I}_{\upsilon\times\upsilon}-\mathrm{diag}(\boldsymbol{\lambda})\boldsymbol{J}_{\upsilon}\boldsymbol{J}_{\upsilon}^T\right)\Delta$ *and*
$\boldsymbol{\Sigma_Z}=\boldsymbol{I}(\theta_0)^{-1}\boldsymbol{B}^T\boldsymbol{Q}_{\boldsymbol{B}}^{\mathbf{2}}\boldsymbol{BI}(\theta_0)^{-1}=\left(\boldsymbol{I}_{\upsilon\times\upsilon}-\boldsymbol{H}(\theta_0)\right)\boldsymbol{I}(\theta_0)^{-1}$.

d) We have $\begin{pmatrix}\boldsymbol{X}_n\\ \boldsymbol{Z}_n\end{pmatrix}\xrightarrow[n\to\infty]{L}\mathcal{N}\left(\begin{pmatrix}\boldsymbol{\Delta}\\ \boldsymbol{\delta}^*\end{pmatrix},\begin{pmatrix}\boldsymbol{I}(\theta_0)^{-1} & \boldsymbol{\Sigma_Z}\\ \boldsymbol{\Sigma_Z}^T & \boldsymbol{\Sigma_Z}\end{pmatrix}\right)$, *where* $\boldsymbol{\delta}^*$ *and* $\boldsymbol{\Sigma_Z}$ *where defined in c).*

e) We have, $\begin{pmatrix} \boldsymbol{Y}_n \\ \boldsymbol{Z}_n \end{pmatrix} \xrightarrow[n\to\infty]{L} \mathcal{N}\left(\begin{pmatrix} \mathbf{0} \\ \boldsymbol{\delta}^* \end{pmatrix}, \begin{pmatrix} \boldsymbol{\Sigma}_{\boldsymbol{Y}} & \mathbf{0} \\ \mathbf{0} & \boldsymbol{\Sigma}_{\boldsymbol{Z}} \end{pmatrix} \right).$

f) The asymptotic distribution of $T_n^{\phi_1,\phi_2}$*, under* $H_{1,n}$*, is a noncentral chi-square distribution with* $\upsilon-1$ *degrees of freedom and noncentrality parameter*

$$\mu = \theta_0^{-1}\left(1-\theta_0\right)^{-1} \operatorname{diag}\left(\boldsymbol{\lambda}\right) \boldsymbol{J}_\upsilon \boldsymbol{J}_\upsilon^T = \boldsymbol{\delta}^T \boldsymbol{Q}_{\boldsymbol{B}}^{\mathbf{2}} \boldsymbol{\delta} \qquad (\boldsymbol{\delta} = \boldsymbol{B}\boldsymbol{\Delta}). \tag{23}$$

4 Asymptotic Quadratic Bias of $\widehat{\boldsymbol{\theta}}^{\boldsymbol{h}}_{\boldsymbol{\phi_1},\boldsymbol{\phi_2}}$ under $\boldsymbol{H_{1,n}}$

Let $\widehat{\boldsymbol{\theta}}_*$ be a suitable estimator of $\boldsymbol{\theta}$ and we denote by $F_{\widehat{\boldsymbol{\theta}}_*}$ the asymptotic distribution of $\sqrt{n}(\widehat{\boldsymbol{\theta}}_* - \boldsymbol{\theta}_n)$. The asymptotic bias of $\widehat{\boldsymbol{\theta}}_*$ is defined by $B(\widehat{\boldsymbol{\theta}}_*) = \int \boldsymbol{x} dF_{\widehat{\boldsymbol{\theta}}_*}(\boldsymbol{x})$. $B(\widehat{\boldsymbol{\theta}}_*)$ is not in a scalar form and in order to be able to do comparisons we consider the asymptotic quadratic bias of it, $B^*(\widehat{\boldsymbol{\theta}}_*) = B(\widehat{\boldsymbol{\theta}}_*)^T \boldsymbol{I}(\theta_0) B(\widehat{\boldsymbol{\theta}}_*)$. The following theorem gives the expression of $B^*(\widehat{\boldsymbol{\theta}}_*)$.

Theorem 2. *The asymptotic quadratic bias,* $B^*\left(\widehat{\boldsymbol{\theta}}^h_{\phi_1,\phi_2}\right)$*, of*

$$\widehat{\boldsymbol{\theta}}^h_{\phi_1,\phi_2} = \widetilde{\boldsymbol{\theta}}_{\phi_2} + \left(1 - h\left(T_n^{\phi_1,\phi_2}\right)\right)(\widehat{\boldsymbol{\theta}} - \widetilde{\boldsymbol{\theta}}_{\phi_2}), \tag{24}$$

under $H_{1,n}$*, is given by* $B^*\left(\widehat{\boldsymbol{\theta}}^h_{\phi_1,\phi_2}\right) = E\left[h\left(\chi^2_{\upsilon+1}\left(\mu\right)\right)\right]^2 \mu$*, where* μ *was defined in (23).*

Proof. We denote by $\boldsymbol{\eta}_n$ the $(\upsilon-1)$ $-$dimensional random vector defined by

$$\boldsymbol{\eta}_n = \left(\boldsymbol{B}\boldsymbol{I}_n\left(\theta_0\right)^{-1}\boldsymbol{B}^T\right)^{-1/2} \sqrt{n} g\left(\widehat{\boldsymbol{\theta}}\right) \tag{25}$$

whose asymptotic distribution is normal with vector mean $\boldsymbol{Q}_{\boldsymbol{B}}\boldsymbol{B}\boldsymbol{\Delta}$ and variance-covariance matrix the identity matrix $\boldsymbol{I}_{(\upsilon-1)\times(\upsilon-1)}$.

A second order Taylor expansion gives

$$T_n^{\phi_1,\phi_2} = \sqrt{n}(\widehat{\boldsymbol{\theta}} - \widetilde{\boldsymbol{\theta}}_{\phi_2})^T \boldsymbol{I}_n\left(\theta_0\right) \sqrt{n}(\widehat{\boldsymbol{\theta}} - \widetilde{\boldsymbol{\theta}}_{\phi_2}) + o_p\left(1\right).$$

Now using (17) and (19) we get

$$T_n^{\phi_1,\phi_2} = \sqrt{n} g(\widehat{\boldsymbol{\theta}})^T \left(\boldsymbol{B}\boldsymbol{I}_n\left(\theta_0\right)^{-1}\boldsymbol{B}^T\right)^{-1} \sqrt{n} g(\widehat{\boldsymbol{\theta}}) + o_p\left(1\right) = \boldsymbol{\eta}_n^T \boldsymbol{\eta}_n + o_p\left(1\right),$$

with $\boldsymbol{\eta}_n$ defined in (25).

We have,

$$\begin{aligned} \sqrt{n}(\widehat{\boldsymbol{\theta}}^h_{\phi_1,\phi_2} - \boldsymbol{\theta}_n) &= \sqrt{n}\left(\widetilde{\boldsymbol{\theta}}_{\phi_2} - \boldsymbol{\theta}_n\right) + \left(1 - h\left(T_n^{\phi_1,\phi_2}\right)\right)\sqrt{n}(\widehat{\boldsymbol{\theta}} - \widetilde{\boldsymbol{\theta}}_{\phi_2}) + o_p\left(1\right) \\ &= \sqrt{n}(\widehat{\boldsymbol{\theta}} - \boldsymbol{\theta}_n) - \sqrt{n}(\widehat{\boldsymbol{\theta}} - \widetilde{\boldsymbol{\theta}}_{\phi_2}) h(T_n^{\phi_1,\phi_2}) + o_p\left(1\right) \\ &= \sqrt{n}(\widehat{\boldsymbol{\theta}} - \boldsymbol{\theta}_n) \\ &\quad - \boldsymbol{I}_n\left(\theta_0\right)^{-1}\boldsymbol{B}^T\left(\boldsymbol{B}\boldsymbol{I}_n\left(\theta_0\right)^{-1}\boldsymbol{B}^T\right)^{-1/2} \boldsymbol{\eta}_n h\left(\boldsymbol{\eta}_n^T\boldsymbol{\eta}_n\right) + o_p(1). \end{aligned}$$

Therefore, applying 2.2.9 of page 32 in Saleh [16] we get,

$$B(\widehat{\boldsymbol{\theta}}^{h}_{\phi_1,\phi_2}) = -\boldsymbol{I}(\theta_0)^{-1}\boldsymbol{B}^T\boldsymbol{Q}_{\boldsymbol{B}}\boldsymbol{\delta}E\left[h\left(\chi^2_{v+1}(\mu)\right)\right]$$

and therefore

$$B^*(\widehat{\boldsymbol{\theta}}^{h}_{\phi_1,\phi_2}) = E\left[h\left(\chi^2_{v+1}(\mu)\right)\right]^2\boldsymbol{\delta}^T\boldsymbol{Q}^2_{\boldsymbol{B}}\boldsymbol{\delta}^T = \mu E\left[h\left(\chi^2_{v+1}(\mu)\right)\right]^2.$$

In the following by $G_r(x;\lambda)$ we shall denote the distribution function of a non-central chi-square with r degrees of freedom and noncentrality parameter λ evaluated at x.

In the following theorem we are going to give some relations among $\widehat{\boldsymbol{\theta}}$, $\widetilde{\boldsymbol{\theta}}_{\phi_2}$, $\widehat{\boldsymbol{\theta}}^{SRE}_{\phi_2}$, $\widehat{\boldsymbol{\theta}}^{PTE}_{\phi_1,\phi_2}$, $\widehat{\boldsymbol{\theta}}^{SPT}_{\phi_1,\phi_2}$, $\widehat{\boldsymbol{\theta}}^{S}_{\phi_1,\phi_2}$, $\widehat{\boldsymbol{\theta}}^{S_+}_{\phi_1,\phi_2}$ and $\widehat{\boldsymbol{\theta}}^{PTE_+}_{\phi_1,\phi_2}$ on the basis of the asymptotic quadratic bias.

Theorem 3. *The estimators $\widehat{\boldsymbol{\theta}}$, $\widetilde{\boldsymbol{\theta}}_{\phi_2}$, $\widehat{\boldsymbol{\theta}}^{SRE}_{\phi_2}$, $\widehat{\boldsymbol{\theta}}^{PTE}_{\phi_1,\phi_2}$, $\widehat{\boldsymbol{\theta}}^{SPT}_{\phi_1,\phi_2}$, $\widehat{\boldsymbol{\theta}}^{S}_{\phi_1,\phi_2}$, $\widehat{\boldsymbol{\theta}}^{S_+}_{\phi_1,\phi_2}$ and $\widehat{\boldsymbol{\theta}}^{PTE_+}_{\phi_1,\phi_2}$ can be ordered in according to the asymptotic quadratic bias in the following way:*

a) $B^*(\widehat{\boldsymbol{\theta}}) \leq B^*(\widehat{\boldsymbol{\theta}}^{SRE}_{\phi_2}) \leq B^*(\widetilde{\boldsymbol{\theta}}_{\phi_2})$; $B^*(\widehat{\boldsymbol{\theta}}) \leq B^*(\widehat{\boldsymbol{\theta}}^{SPT}_{\phi_2}) \leq B^*(\widehat{\boldsymbol{\theta}}^{PTE}_{\phi_1,\phi_2})$.

b) $B^*(\widehat{\boldsymbol{\theta}}^{S}_{\phi_1,\phi_2}) \leq B^*(\widehat{\boldsymbol{\theta}}^{PTE}_{\phi_1,\phi_2})$ *iff* $G_{v+1}\left(\chi^2_{v-1,\alpha};\mu\right) \geq (v-3)E\left[\chi^{-2}_{v+1}(\mu)\right]$ *for all α and μ.*

c) $B^*(\widehat{\boldsymbol{\theta}}^{S_+}_{\phi_1,\phi_2}) \leq B^*(\widehat{\boldsymbol{\theta}}^{PTE}_{\phi_1,\phi_2})$ *iff* $G_{v+1}\left(\chi^2_{v-1,\alpha};\mu\right) \geq E\left[1-\left(1-(v-3)\chi^{-2}_{v+1}(\mu)\right)\right.$ $\left.\times I_{(v-3,\infty)}\left(\chi^2_{v+1}(\mu)\right)\right]$ *for all α and μ.*

d) $B^*(\widehat{\boldsymbol{\theta}}^{S_+}_{\phi_1,\phi_2}) \leq B^*(\widehat{\boldsymbol{\theta}}^{S}_{\phi_1,\phi_2})$ *and* $B^*(\widehat{\boldsymbol{\theta}}^{PTE_+}_{\phi_1,\phi_2}) \leq B^*(\widehat{\boldsymbol{\theta}}^{PTE}_{\phi_1,\phi_2})$.

Proof. Based on the expressions of

$$B^*(\widehat{\boldsymbol{\theta}}), B^*(\widetilde{\boldsymbol{\theta}}_{\phi_2}), B^*(\widehat{\boldsymbol{\theta}}^{SRE}_{\phi_2}), B^*(\widehat{\boldsymbol{\theta}}^{PTE}_{\phi_1,\phi_2}),$$

$$B^*(\widehat{\boldsymbol{\theta}}^{SPT}_{\phi_2}), B^*(\widehat{\boldsymbol{\theta}}^{S}_{\phi_1,\phi_2}), B^*(\widehat{\boldsymbol{\theta}}^{S_+}_{\phi_1,\phi_2}), B^*(\widehat{\boldsymbol{\theta}}^{PTE_+}_{\phi_1,\phi_2})$$

parts *a)*, *b)* and *c)* are immediate. We are going to establish d). We denote $s = B^*(\widehat{\boldsymbol{\theta}}^{S}_{\phi_1,\phi_2}) - B^*(\widehat{\boldsymbol{\theta}}^{S_+}_{\phi_1,\phi_2})$ and we have,

$$\begin{aligned}
s &= (v-3)^2\mu E\left[\chi^{-2}_{v+1}(\mu)\right]^2 - \mu\left\{G_{v+1}(v-3;\mu) + (v-3)E\left[\chi^{-2}_{v+1}(\mu)\right]\right.\\
&\quad \left.-(v-3)E\left[\chi^{-2}_{v+1}(\mu)I_{(0,v-3)}\left(\chi^2_{v+1}(\mu)\right)\right]\right\}^2\\
&= (v-3)^2\mu\left\{E\left[\chi^{-2}_{v+1}(\mu)I_{(0,v-3)}\left(\chi^2_{v+1}(\mu)\right)\right] - \frac{1}{v-3}G_{v+1}(v-3;\mu)\right\}\\
&\quad \left\{2E\left[\chi^{-2}_{v+1}(\mu)\right] + \frac{1}{v-3}G_{v+1}(v-3;\mu) - E\left[\chi^{-2}_{v+1}(\mu)I_{(0,v-3)}\left(\chi^2_{v+1}(\mu)\right)\right]\right\}.
\end{aligned}$$

Using the probability density function of a noncentral chi-square random variable as well as relations 2.2.13a and 2.2.13g in pages 32 and 33 in Saleh [16], it is not difficult to establish that

$$\frac{1}{\upsilon-3}G_{\upsilon+1}\left(\upsilon-3;\mu\right)\leq E\left[\chi_{\upsilon+1}^{-2}\left(\mu\right)I_{(0,\upsilon-3)}\left(\chi_{\upsilon+1}^{2}\left(\mu\right)\right)\right]$$

and

$$2E\left[\chi_{\upsilon+1}^{-2}\left(\mu\right)\right]+\frac{1}{\upsilon-3}G_{\upsilon+1}\left(\upsilon-3;\mu\right)\geq E\left[\chi_{\upsilon+1}^{-2}\left(\mu\right)I_{(0,\upsilon-3)}\left(\chi_{\upsilon+1}^{2}\left(\mu\right)\right)\right],$$

therefore, $B^{*}\left(\widehat{\boldsymbol{\theta}}_{\phi_1,\phi_2}^{S}\right)>B^{*}\left(\widehat{\boldsymbol{\theta}}_{\phi_1,\phi_2}^{S_+}\right)$.

It is clear that

$$\begin{aligned}E\left[\chi_{\upsilon+1}^{-2}\left(\mu\right)\right]&=\int_0^\infty x^{-1}dG_{\upsilon+1}\left(x;\mu\right)\geq\int_0^{\upsilon-3}x^{-1}dG_{\upsilon+1}\left(x;\mu\right)\\&=E\left[\chi_{\upsilon+1}^{-2}\left(\mu\right)I_{(0,\upsilon-3)}\left(\chi_{\upsilon+1}^{2}\left(\mu\right)\right)\right].\end{aligned}$$

Therefore, $B^{*}\left(\widehat{\boldsymbol{\theta}}_{\phi_1,\phi_2}^{PTE_+}\right)\leq B^{*}\left(\widehat{\boldsymbol{\theta}}_{\phi_1,\phi_2}^{PTE}\right)$.

5 Asymptotic Quadratic Risk under Null and Contiguous Alternative Hypotheses

Let $\widehat{\boldsymbol{\theta}}_*$ be a suitable estimator of $\boldsymbol{\theta}$ and we denote by $F_{\widehat{\boldsymbol{\theta}}_*}$ the asymptotic distribution of $\sqrt{n}(\widehat{\boldsymbol{\theta}}_*-\boldsymbol{\theta}_n)$ and $\boldsymbol{W}$ a positive semidefinite matrix. The asymptotic distributional quadratic risk (ADQR) of $\widehat{\boldsymbol{\theta}}_*$ is given by $R(\widehat{\boldsymbol{\theta}}_*;\boldsymbol{W})=\int\boldsymbol{x}^T\boldsymbol{W}\boldsymbol{x}d F_{\widehat{\boldsymbol{\theta}}_*}(\boldsymbol{x})$. The following theorem gives the ADQR of $\widehat{\boldsymbol{\theta}}_{\phi_1,\phi_2}^{h}$ defined in (14).

Theorem 4. *The ADQR of* $\widehat{\boldsymbol{\theta}}_{\phi_1,\phi_2}^{h}$ *is given by*

$$\begin{aligned}R\left(\widehat{\boldsymbol{\theta}}_{\phi_1,\phi_2}^{h};\boldsymbol{W}\right)=&\ \boldsymbol{\delta}^T\boldsymbol{L}^T\boldsymbol{W}\boldsymbol{L}\boldsymbol{\delta}\left\{E\left[h\left(\chi_{\upsilon+3}^{2}\left(\mu\right)\right)^2\right]-2E\left[h\left(\chi_{\upsilon+3}^{2}\left(\mu\right)\right)\right]\right.\\&\left.+2E\left[h\left(\chi_{\upsilon+1}^{2}\left(\mu\right)\right)\right]\right\}+\operatorname{trace}\left(\boldsymbol{I}\left(\theta_0\right)^{-1}\boldsymbol{W}\right)\\&-\operatorname{trace}\left(\boldsymbol{\Sigma}_{\boldsymbol{Z}}\boldsymbol{W}\right)\left\{2E\left[h\left(\chi_{\upsilon+1}^{2}\left(\mu\right)\right)\right]-E\left[h\left(\chi_{\upsilon+1}^{2}\left(\mu\right)\right)^2\right]\right\},\end{aligned}\qquad(26)$$

where $\mu=\theta_0^{-1}\left(1-\theta_0\right)^{-1}\operatorname{diag}\left(\boldsymbol{\lambda}\right)\boldsymbol{J}_\upsilon\boldsymbol{J}_\upsilon^T$.

Proof. We know,

$$\sqrt{n}(\widehat{\boldsymbol{\theta}}_{\phi_1,\phi_2}^{h}-\boldsymbol{\theta}_n)=\sqrt{n}(\widehat{\boldsymbol{\theta}}-\boldsymbol{\theta}_n)-\sqrt{n}(\widehat{\boldsymbol{\theta}}-\widetilde{\boldsymbol{\theta}}_{\phi_2})h(T_n^{\phi_1,\phi_2})+o_p\left(1\right).$$

Then,

$$R\left(\widehat{\boldsymbol{\theta}}^{h}_{\phi_1,\phi_2};\boldsymbol{W}\right)=\lim_{n\longrightarrow\infty}E\left[\sqrt{n}\left(\widehat{\boldsymbol{\theta}}^{h}_{\phi_1,\phi_2}-\boldsymbol{\theta}_n\right)^T\boldsymbol{W}\sqrt{n}\left(\widehat{\boldsymbol{\theta}}^{h}_{\phi_1,\phi_2}-\boldsymbol{\theta}_n\right)\right]$$

$$\begin{aligned}
&=\lim_{n\to\infty}E\left[\left\{\sqrt{n}\left(\widehat{\boldsymbol{\theta}}-\boldsymbol{\theta}_n\right)-\sqrt{n}\left(\widehat{\boldsymbol{\theta}}-\widetilde{\boldsymbol{\theta}}_{\phi_2}\right)h\left(T_n^{\phi_1,\phi_2}\right)\right\}^T\boldsymbol{W}\right.\\
&\times\left.\left\{\sqrt{n}\left(\widehat{\boldsymbol{\theta}}-\boldsymbol{\theta}_n\right)-\sqrt{n}\left(\widehat{\boldsymbol{\theta}}-\widetilde{\boldsymbol{\theta}}_{\phi_2}\right)h\left(T_n^{\phi_1,\phi_2}\right)\right\}\right]\\
&=\lim_{n\to\infty}E\left[\sqrt{n}\left(\widehat{\boldsymbol{\theta}}-\boldsymbol{\theta}_n\right)^T\boldsymbol{W}\sqrt{n}\left(\widehat{\boldsymbol{\theta}}-\boldsymbol{\theta}_n\right)\right]\\
&-\lim_{n\to\infty}E\left[\sqrt{n}\left(\widehat{\boldsymbol{\theta}}-\boldsymbol{\theta}_n\right)^T\boldsymbol{W}\sqrt{n}\left(\widehat{\boldsymbol{\theta}}-\widetilde{\boldsymbol{\theta}}_{\phi_2}\right)h\left(T_n^{\phi_1,\phi_2}\right)\right]\\
&-\lim_{n\to\infty}E\left[\sqrt{n}\left(\widehat{\boldsymbol{\theta}}-\widetilde{\boldsymbol{\theta}}_{\phi_2}\right)^T h\left(T_n^{\phi_1,\phi_2}\right)\boldsymbol{W}\sqrt{n}\left(\widehat{\boldsymbol{\theta}}-\boldsymbol{\theta}_n\right)\right]\\
&+\lim_{n\to\infty}E\left[\sqrt{n}\left(\widehat{\boldsymbol{\theta}}-\widetilde{\boldsymbol{\theta}}_{\phi_2}\right)h\left(T_n^{\phi_1,\phi_2}\right)^2\boldsymbol{W}\sqrt{n}\left(\widehat{\boldsymbol{\theta}}-\widetilde{\boldsymbol{\theta}}_{\phi_2}\right)\right]\\
&=L_1-L_2-L_3+L_4.
\end{aligned}$$

If $\boldsymbol{X}$ is a random vector with vector mean $\boldsymbol{a}$ and variance-covariance matrix $\boldsymbol{\Sigma}$, then $E\left[\boldsymbol{X}^T\boldsymbol{A}\boldsymbol{X}\right]=\text{trace}(\boldsymbol{A}\boldsymbol{\Sigma})+\boldsymbol{a}^T\boldsymbol{A}\boldsymbol{a}$. In our case $\sqrt{n}(\widehat{\boldsymbol{\theta}}-\boldsymbol{\theta}_n)\underset{n\to\infty}{\overset{L}{\longrightarrow}}$ $\mathcal{N}(\boldsymbol{0},\boldsymbol{I}\left(\theta_0\right)^{-1})$, therefore

$$L_1=\lim_{n\to\infty}E\left[\sqrt{n}\left(\widehat{\boldsymbol{\theta}}-\boldsymbol{\theta}_n\right)^T\boldsymbol{W}\sqrt{n}\left(\widehat{\boldsymbol{\theta}}-\boldsymbol{\theta}_n\right)\right]=\text{trace}\left(\boldsymbol{I}\left(\theta_0\right)^{-1}\boldsymbol{W}\right).$$

We know

$$\begin{aligned}
\sqrt{n}\left(\widehat{\boldsymbol{\theta}}-\widetilde{\boldsymbol{\theta}}_{\phi_2}\right)&=\boldsymbol{I}_n\left(\theta_0\right)^{-1}\boldsymbol{B}^T\left(\boldsymbol{B}\boldsymbol{I}_n\left(\theta_0\right)^{-1}\boldsymbol{B}^T\right)^{-1}\sqrt{n}g\left(\widehat{\boldsymbol{\theta}}\right)+o_p\left(1\right)\\
&=\boldsymbol{I}_n\left(\theta_0\right)^{-1}\boldsymbol{B}^T\left(\boldsymbol{B}\boldsymbol{I}_n\left(\theta_0\right)^{-1}\boldsymbol{B}^T\right)^{-1/2}\boldsymbol{\eta}_n+o_p\left(1\right),
\end{aligned}$$

where $\boldsymbol{\eta}_n$ is an asymptotic $(\upsilon-1)$-dimensional normal random vector with vector mean $\boldsymbol{Q}_{\boldsymbol{B}}\boldsymbol{\delta}$ and variance-covariance matrix the identity.

Then,

$$L_4=E\left[h\left(\boldsymbol{\eta}^T\boldsymbol{\eta}\right)^2\boldsymbol{\eta}^T\boldsymbol{Q}_{\boldsymbol{B}}\boldsymbol{B}\boldsymbol{I}\left(\theta_0\right)^{-1}\boldsymbol{W}\boldsymbol{I}\left(\theta_0\right)^{-1}\boldsymbol{B}^T\boldsymbol{Q}_{\boldsymbol{B}}\boldsymbol{\eta}\right].$$

Now we apply the following result: "Let $\boldsymbol{Z}$ be a p-dimensional normal random vector distributed as a $\mathcal{N}\left(\boldsymbol{a},\boldsymbol{I}\right)$. Then for a measurable function φ and a positive definite matrix $\boldsymbol{A}$, we have

$$E\left[\boldsymbol{Z}^T\boldsymbol{A}\boldsymbol{Z}\varphi\left(\boldsymbol{Z}^T\boldsymbol{Z}\right)\right]=\text{trace}\left(\boldsymbol{A}\right)E\left[\varphi\left(\chi^2_{p+2}\left(\mu\right)\right)\right]+E\left[\varphi\left(\chi^2_{p+4}\left(\mu\right)\right)\right],$$

$\mu = \boldsymbol{a^T A a}$" (see Judge and Bock [5]), and we get

$$L_4 = \text{trace}\left(\boldsymbol{B}^T\boldsymbol{Q_B}\boldsymbol{B}\boldsymbol{I}\left(\boldsymbol{\theta_0}\right)^{-1}\boldsymbol{W}\boldsymbol{I}\left(\boldsymbol{\theta_0}\right)^{-1}\boldsymbol{B}^T\boldsymbol{Q_B}\right)E\left[h\left(\chi^2_{\upsilon+1}\left(\mu\right)\right)^2\right]$$
$$+\boldsymbol{\delta}^T\boldsymbol{Q}^2_{\boldsymbol{B}}\boldsymbol{B}\boldsymbol{I}\left(\theta_0\right)^{-1}\boldsymbol{B}^{T-1}\boldsymbol{B}\boldsymbol{I}\left(\theta_0\right)^{-1}\boldsymbol{W}\boldsymbol{I}\left(\boldsymbol{\theta_0}\right)^{-1}\boldsymbol{B}^T\boldsymbol{Q}^2_{\boldsymbol{B}}\boldsymbol{\delta}E\left[h\left(\chi^2_{\upsilon+3}\left(\mu\right)\right)^2\right]$$
$$=\text{trace}\left(\boldsymbol{\Sigma_Z}\boldsymbol{W}\right)E\left[h\left(\chi^2_{\upsilon+1}\left(\mu\right)\right)^2\right]+E\left[h\left(\chi^2_{\upsilon+3}\left(\mu\right)\right)^2\right]\boldsymbol{\delta}^T\boldsymbol{L}^T\boldsymbol{W}\boldsymbol{L}\boldsymbol{\delta},$$

being $\boldsymbol{L} = \boldsymbol{I}\left(\boldsymbol{\theta_0}\right)^{-1}\boldsymbol{B}^T\boldsymbol{Q}^2_{\boldsymbol{B}}$.

We denote by $\boldsymbol{S}$ the random vector obtained by $\boldsymbol{S}_n \equiv \sqrt{n}(\widehat{\boldsymbol{\theta}}-\boldsymbol{\theta}_n) \underset{n\to\infty}{\overset{L}{\longrightarrow}} \boldsymbol{S}$, where $\boldsymbol{S}$ is a υ-dimensional normal vector with vector mean $\mathbf{0}$ and variance-covariance matrix $\boldsymbol{I}(\boldsymbol{\theta}_0)^{-1}$. Now we are going to obtain L_2,

$$L_2 = \lim_{n\to\infty} E\left[\sqrt{n}\left(\widehat{\boldsymbol{\theta}}-\boldsymbol{\theta}_n\right)^T\boldsymbol{W}\sqrt{n}\left(\widehat{\boldsymbol{\theta}}-\widetilde{\boldsymbol{\theta}}_{\phi_2}\right)h\left(T_n^{\phi_1,\phi_2}\right)\right]$$
$$=E\left[\boldsymbol{S}^T\boldsymbol{W}h\left(\boldsymbol{\eta}^T\boldsymbol{\eta}\right)\boldsymbol{I}\left(\boldsymbol{\theta_0}\right)^{-1}\boldsymbol{B}^T\left(\boldsymbol{B}\boldsymbol{I}\left(\theta_0\right)^{-1}\boldsymbol{B}^T\right)^{-1/2}\boldsymbol{\eta}\right]$$
$$=E\left[E\left[\boldsymbol{S}^T/\boldsymbol{\eta}\right]\boldsymbol{W}h\left(\boldsymbol{\eta}^T\boldsymbol{\eta}\right)\boldsymbol{I}\left(\boldsymbol{\theta_0}\right)^{-1}\boldsymbol{B}^T\boldsymbol{Q_B}\boldsymbol{\eta}\right].$$

In relation to $E[\boldsymbol{S}^T/\boldsymbol{\eta}]$, we have $\boldsymbol{S}_n \equiv \sqrt{n}(\widehat{\boldsymbol{\theta}}-\boldsymbol{\theta}_n) \underset{n\to\infty}{\overset{L}{\longrightarrow}} \boldsymbol{S}$ and $\boldsymbol{\eta}_n \equiv \boldsymbol{Q_B}\sqrt{n}\boldsymbol{B}$ $(\widehat{\boldsymbol{\theta}}-\boldsymbol{\theta}_n)+\boldsymbol{k}$, therefore,

$$\begin{pmatrix}\boldsymbol{S}_n\\ \boldsymbol{\eta}_n\end{pmatrix} = \begin{pmatrix}\boldsymbol{I}\\ \left(\boldsymbol{B}\boldsymbol{I}_n\left(\theta_0\right)^{-1}\boldsymbol{B}^T\right)^{-1/2}\boldsymbol{B}\end{pmatrix}\sqrt{n}\left(\widehat{\boldsymbol{\theta}}-\boldsymbol{\theta}_n\right)+\begin{pmatrix}\mathbf{0}\\ \boldsymbol{k}\end{pmatrix}$$

and

$$\begin{pmatrix}\boldsymbol{S}_n\\ \boldsymbol{\eta}_n\end{pmatrix}\underset{n\to\infty}{\overset{L}{\longrightarrow}}\mathcal{N}\left(\begin{pmatrix}\mathbf{0}\\ \boldsymbol{Q_B}\boldsymbol{\delta}\end{pmatrix},\begin{pmatrix}\boldsymbol{I}\left(\theta_0\right)^{-1} & \boldsymbol{M}\\ \boldsymbol{M}^T & \boldsymbol{I}\end{pmatrix}\right),$$

being $\boldsymbol{M} = \boldsymbol{I}\left(\theta_0\right)^{-1}\boldsymbol{B}^T\boldsymbol{Q}^{-2}_{\boldsymbol{B}}$.

Finally,
$E\left[\boldsymbol{S}^T/\boldsymbol{\eta}=\boldsymbol{y}\right] = \boldsymbol{I}\left(\theta_0\right)^{-1}\boldsymbol{B}^T\boldsymbol{Q_B}\left(\boldsymbol{y}-\boldsymbol{Q_B}\boldsymbol{\delta}\right) = \boldsymbol{I}\left(\theta_0\right)^{-1}\boldsymbol{B}^T\boldsymbol{Q_B}-\boldsymbol{L}\boldsymbol{\delta}.$
Then,

$$L_2 = E\left[h\left(\boldsymbol{\eta}^T\boldsymbol{\eta}\right)\left[\boldsymbol{\eta}^T\boldsymbol{Q_B}\boldsymbol{B}\boldsymbol{I}\left(\theta_0\right)^{-1}-\boldsymbol{\delta}^T\boldsymbol{L}^T\right]\boldsymbol{W}\boldsymbol{I}\left(\theta_0\right)^{-1}\boldsymbol{B}^T\boldsymbol{Q_B}\boldsymbol{\eta}\right]$$
$$=E\left[h\left(\boldsymbol{\eta}^T\boldsymbol{\eta}\right)\boldsymbol{\eta}^T\boldsymbol{Q_B}\boldsymbol{B}\boldsymbol{I}\left(\theta_0\right)^{-1}\boldsymbol{W}\boldsymbol{I}\left(\theta_0\right)^{-1}\boldsymbol{B}^T\boldsymbol{Q_B}\boldsymbol{\eta}\right]$$
$$-E\left[h\left(\boldsymbol{\eta}^T\boldsymbol{\eta}\right)\boldsymbol{\delta}^T\boldsymbol{L}^T\boldsymbol{W}\boldsymbol{I}\left(\theta_0\right)^{-1}\boldsymbol{B}^T\boldsymbol{Q_B}\boldsymbol{\eta}\right]=a_1-a_2.$$

Now we get a_1 and a_2,

$$\begin{aligned} a_1 &= \operatorname{trace}\left(\boldsymbol{Q_B}\boldsymbol{B}\boldsymbol{I}(\theta_0)^{-1}\boldsymbol{W}\boldsymbol{I}(\theta_0)^{-1}\boldsymbol{B}^T\boldsymbol{Q_B}\right)E\left[h\left(\chi^2_{v+1}(\mu)\right)\right] \\ &+ \boldsymbol{\delta}^T\boldsymbol{Q}^2_{\boldsymbol{B}}\boldsymbol{B}\boldsymbol{I}(\theta_0)^{-1}\boldsymbol{W}\boldsymbol{I}(\theta_0)^{-1}\boldsymbol{B}^T\boldsymbol{Q}^2_{\boldsymbol{B}}\boldsymbol{\delta}E\left[h\left(\chi^2_{v+3}(\mu)\right)\right] \\ &= \operatorname{trace}\left(\boldsymbol{I}(\theta_0)^{-1}\boldsymbol{B}^T\boldsymbol{Q}^2_{\boldsymbol{B}}\boldsymbol{B}\boldsymbol{I}(\theta_0)^{-1}\boldsymbol{W}\right)E\left[h\left(\chi^2_{v+3}(\mu)\right)\right] \\ &+ \boldsymbol{\delta}^T\boldsymbol{L}^T\boldsymbol{W}\boldsymbol{L}\boldsymbol{\delta}E\left[h\left(\chi^2_{v+3}(\mu)\right)\right] \end{aligned}$$

and $a_2 = \boldsymbol{\delta}^T\boldsymbol{L}^T\boldsymbol{W}\boldsymbol{L}\boldsymbol{\delta}E\left[h\left(\chi^2_{v+1}(\mu)\right)\right]$. Then,

$$\begin{aligned} L_2 &= \operatorname{trace}\left(\boldsymbol{\Sigma_Z}\boldsymbol{W}\right)E\left[h\left(\chi^2_{v+3}(\mu)\right)\right] \\ &\quad + \boldsymbol{\delta}^T\boldsymbol{L}^T\boldsymbol{W}\boldsymbol{L}\boldsymbol{\delta}\ \left(E\left[h\left(\chi^2_{v+3}(\mu)\right)\right] - E\left[h\left(\chi^2_{v+1}(\mu)\right)\right]\right). \end{aligned}$$

Remark 1. Under H_0 we have $\boldsymbol{\delta} = \boldsymbol{0}$ and $\mu = 0$, therefore, taking into account that trace($\boldsymbol{I}(\theta_0)^{-1}\boldsymbol{W}$) = trace $(\boldsymbol{\Sigma_Y}\boldsymbol{W})$ + trace $(\boldsymbol{\Sigma_Z}\boldsymbol{W})$, we get

$$R\left(\widehat{\boldsymbol{\theta}}^h_{\phi_1,\phi_2};\boldsymbol{W}\right) = \operatorname{trace}\left(\boldsymbol{\Sigma_Y}\boldsymbol{W}\right) + \left(\boldsymbol{\Sigma_Z}\boldsymbol{W}\right)E\left[\left(1-h\left(\chi^2_{v+1}(0)\right)\right)^2\right]$$

and $R\left(\widehat{\boldsymbol{\theta}}^h_{\phi_1,\phi_2};\boldsymbol{W}\right)$ is an increasing function of $E\left[\left(1-h\left(\chi^2_{v+1}(0)\right)\right)^2\right]$. Based on this result it is not difficult to establish, under H_0, the following relations,

i) $R\left(\widetilde{\boldsymbol{\theta}}_{\phi_2};\boldsymbol{W}\right) \leq R\left(\widehat{\boldsymbol{\theta}}^{PTE}_{\phi_1,\phi_2};\boldsymbol{W}\right) \leq R\left(\widehat{\boldsymbol{\theta}}^{SPT}_{\phi_1,\phi_2};\boldsymbol{W}\right) \leq R\left(\widehat{\boldsymbol{\theta}};\boldsymbol{W}\right)$.

ii) Assuming $v > 3$,
$R\left(\widetilde{\boldsymbol{\theta}}_{\phi_2};\boldsymbol{W}\right) \leq R\left(\widehat{\boldsymbol{\theta}}^{S_+}_{\phi_1,\phi_2};\boldsymbol{W}\right) \leq R\left(\widehat{\boldsymbol{\theta}}^{S}_{\phi_1,\phi_2};\boldsymbol{W}\right) \leq R\left(\widehat{\boldsymbol{\theta}};\boldsymbol{W}\right)$ and
$R\left(\widetilde{\boldsymbol{\theta}}_{\phi_2};\boldsymbol{W}\right) \leq R\left(\widehat{\boldsymbol{\theta}}^{PTE_+}_{\phi_1,\phi_2};\boldsymbol{W}\right) \leq R\left(\widehat{\boldsymbol{\theta}}^{PTE}_{\phi_1,\phi_2};\boldsymbol{W}\right) \leq R\left(\widehat{\boldsymbol{\theta}};\boldsymbol{W}\right)$.

iii) $R\left(\widetilde{\boldsymbol{\theta}}_{\phi_2};\boldsymbol{W}\right) \leq R\left(\widehat{\boldsymbol{\theta}}^{SRE}_{\phi_1,\phi_2};\boldsymbol{W}\right) \leq R\left(\widehat{\boldsymbol{\theta}};\boldsymbol{W}\right)$.

Theorems 5 and 6 presents some results, without proof, under contiguous alternative hypotheses.

Theorem 5. *We assume $v \geq 4$. Under $H_{1,n}$ we have*

$$R\left(\widehat{\boldsymbol{\theta}}^{S_+}_{\phi_1,\phi_2};\boldsymbol{W}\right) \leq R\left(\widehat{\boldsymbol{\theta}}^{S}_{\phi_1,\phi_2};\boldsymbol{W}\right).$$

If in addition trace $(\boldsymbol{\Sigma_Z}\boldsymbol{W}) \geq \frac{v+1}{2}Ch_{\max}\left(\boldsymbol{I}(\theta_0)^{-1}\boldsymbol{W}\right)$, *where* $Ch_{\max}(\boldsymbol{G})$ *represents the largest eigenvalue of the matrix* $\boldsymbol{G}$, *we get*

$$R\left(\widehat{\boldsymbol{\theta}}^{S}_{\phi_1,\phi_2};\boldsymbol{W}\right) \leq R\left(\widehat{\boldsymbol{\theta}};\boldsymbol{W}\right).$$

Theorem 6. *The following relations are verified:*

i) $R\left(\widetilde{\boldsymbol{\theta}}_{\phi_1,\phi_2};\boldsymbol{W}\right) \leq R\left(\widehat{\boldsymbol{\theta}};\boldsymbol{W}\right)$ *iff* $\mu \leq \operatorname{trace}\left(\boldsymbol{\Sigma}_{\boldsymbol{Z}}\boldsymbol{W}\right).Ch_{\max}(\boldsymbol{I}\left(\theta_0\right)^{-1}\boldsymbol{W})^{-1}$ *and* $R\left(\widehat{\boldsymbol{\theta}};\boldsymbol{W}\right) \leq R\left(\widetilde{\boldsymbol{\theta}}_{\phi_1,\phi_2};\boldsymbol{W}\right)$ *iff* $\mu \geq \operatorname{trace}\left(\boldsymbol{\Sigma}_{\boldsymbol{Z}}\boldsymbol{W}\right)Ch_{\min}(\boldsymbol{I}\left(\theta_0\right)^{-1}\boldsymbol{W})^{-1}$, *being* $Ch_{\min}(\boldsymbol{G})$ *the smallest eigenvalue of the matrix* $\boldsymbol{G}$.

ii) $R\left(\widehat{\boldsymbol{\theta}};\boldsymbol{W}\right) \leq R\left(\widehat{\boldsymbol{\theta}}^{PTE}_{\phi_1,\phi_2};\boldsymbol{W}\right)$ *iff*

$$\mu \geq \operatorname{trace}\left(\boldsymbol{\Sigma}_{\boldsymbol{Z}}\boldsymbol{W}\right)G_{v+1}\left(\chi^2_{v-1,\alpha};\mu\right)\left(Ch_{\min}\left(\boldsymbol{I}\left(\theta_0\right)^{-1}\boldsymbol{W}\right)\right)^{-1} \times \left[2G_{v+1}\left(\chi^2_{v-1,\alpha};\mu\right) - G_{v+3}\left(\chi^2_{v-1,\alpha};\mu\right)\right]^{-1}$$

and $R\left(\widehat{\boldsymbol{\theta}}^{PTE}_{\phi_1,\phi_2};\boldsymbol{W}\right) \leq R\left(\widehat{\boldsymbol{\theta}};\boldsymbol{W}\right)$ *iff*

$$\mu \leq \operatorname{trace}\left(\boldsymbol{\Sigma}_{\boldsymbol{Z}}\boldsymbol{W}\right)G_{v+1}\left(\chi^2_{v-1,\alpha};\mu\right)\left(Ch_{\max}\left(\boldsymbol{I}\left(\theta_0\right)^{-1}\boldsymbol{W}\right)\right)^{-1} \times \left[2G_{v+1}\left(\chi^2_{v-1,\alpha};\mu\right) - G_{v+3}\left(\chi^2_{v-1,\alpha};\mu\right)\right]^{-1}.$$

If $\mu \to \infty$ *or* $\alpha \to 1$, $R\left(\widehat{\boldsymbol{\theta}}^{PTE}_{\phi_1,\phi_2};\boldsymbol{W}\right) \to R\left(\widehat{\boldsymbol{\theta}};\boldsymbol{W}\right)$.

iii) $R\left(\widetilde{\boldsymbol{\theta}}_{\phi_1,\phi_2};\boldsymbol{W}\right) \leq R\left(\widehat{\boldsymbol{\theta}}^{PTE}_{\phi_1,\phi_2};\boldsymbol{W}\right)$ *iff*

$$\mu \leq \left(1 - G_{v+1}\left(\chi^2_{v-1,\alpha};\mu\right)\right)\operatorname{trace}\left(\boldsymbol{\Sigma}_{\boldsymbol{Z}}\boldsymbol{W}\right)\left(Ch_{\max}\left(\boldsymbol{\Sigma}_{\boldsymbol{Z}}\boldsymbol{W}\right)\right)^{-1} \times \left[1 - 2G_{v+1}\left(\chi^2_{v-1,\alpha};\mu\right) + G_{v+3}\left(\chi^2_{v-1,\alpha};\mu\right)\right]^{-1}$$

and $R\left(\widehat{\boldsymbol{\theta}}^{PTE}_{\phi_1,\phi_2};\boldsymbol{W}\right) \leq R\left(\widetilde{\boldsymbol{\theta}}_{\phi_1,\phi_2};\boldsymbol{W}\right)$ *iff*

$$\mu \geq \left(1 - G_{v+1}\left(\chi^2_{v-1,\alpha};\mu\right)\right)\operatorname{trace}\left(\boldsymbol{\Sigma}_{\boldsymbol{Z}}\boldsymbol{W}\right)\left(Ch_{\min}\left(\boldsymbol{\Sigma}_{\boldsymbol{Z}}\boldsymbol{W}\right)\right)^{-1} \times \left[1 - 2G_{v+1}\left(\chi^2_{v-1,\alpha};\mu\right) + G_{v+3}\left(\chi^2_{v-1,\alpha};\mu\right)\right]^{-1}.$$

If $\alpha \to 0$, $R\left(\widehat{\boldsymbol{\theta}}^{PTE}_{\phi_1,\phi_2};\boldsymbol{W}\right) \to R\left(\widetilde{\boldsymbol{\theta}}_{\phi_1,\phi_2};\boldsymbol{W}\right)$.

iv) $R\left(\widehat{\boldsymbol{\theta}}^{S_+}_{\phi_1,\phi_2};\boldsymbol{W}\right) \leq R\left(\widehat{\boldsymbol{\theta}}^{PTE_+}_{\phi_1,\phi_2};\boldsymbol{W}\right)$ *if* $\chi^2_{v-1,\alpha} < v - 3$.

v) $R\left(\widehat{\boldsymbol{\theta}}^{SRE}_{\phi_1,\phi_2};\boldsymbol{W}\right) \leq R\left(\widehat{\boldsymbol{\theta}};\boldsymbol{W}\right)$ *iff*

$$\mu \leq \left(1 - a^2\right)\operatorname{trace}\left(\boldsymbol{\Sigma}_{\boldsymbol{Z}}\boldsymbol{W}\right)\left(1 - a\right)^{-2}\left(Ch_{\max}\left(\boldsymbol{I}\left(\boldsymbol{\theta}_0\right)^{-1}\boldsymbol{W}\right)\right)^{-1}$$

and $R\left(\widehat{\boldsymbol{\theta}};\boldsymbol{W}\right) \leq R\left(\widehat{\boldsymbol{\theta}}^{SRE}_{\phi_1,\phi_2};\boldsymbol{W}\right)$ *iff*

$$\lambda \geq \left(1 - a^2\right)\operatorname{trace}\left(\boldsymbol{\Sigma}_{\boldsymbol{Z}}\boldsymbol{W}\right)\left(1 - a\right)^{-2}\left(Ch_{\max}\left(\boldsymbol{I}\left(\boldsymbol{\theta}_0\right)^{-1}\boldsymbol{W}\right)\right)^{-1}.$$

Acknowledgement This work was partially supported by Grant MTM 2009-10072.

References

1. Ali, A.M., Ehsanes Saleh, A.K.: Asymptotic theory for simultaneous. Estimation of binomial means. Statistica Sinica 1, 271–294 (1991)
2. Ali, S.M., Silvey, S.D.: A general class of coefficient of divergence of one distribution from another. J. Royal Statist. Soc. 28(1), 131–142 (1966)
3. Cressie, N., Read, T.R.C.: Multinomial goodness-of-fit tests. J. Royal Statist. Soc. 46, 440–464 (1984)
4. Csiszár, I.: Eine Informationstheoretische Ungleichung und ihre Anwendung auf den Bewis der Ergodizität on Markhoffschen Ketten. Pub. Math. Inst. Hungarian Acad. Sci. 8, 84–108 (1963)
5. Judge, G.G., Bock, M.E.: The Statistical Implications of Pre-test and Stein-Rule Estimators in Econometrics. North-Holland, Amsterdam (1978)
6. Menéndez, M.L., Pardo, L., Pardo, M.C.: Preliminary test estimators and phi-divergence measures in generalized linear models with binary data. J. Multivar. Anal. 99(10), 2265–2284 (2008)
7. Menéndez, M.L., Pardo, L., Pardo, M.C.: Stein-type estimation in logistic regression models based on minimum phi-divergence estimators. J. Korean Statist. Soc. 38(1), 73–86 (2009)
8. Menéndez, M.L., Pardo, L., Pardo, M.C.: Confidence sets and coverage probabilities based on preliminary estimators in logistic regression models. J. Comp. Appl. Math. 224(1), 193–203 (2009)
9. Menéndez, M.L., Pardo, L., Pardo, M.C.: Preliminary Phi-divergence Test Estimators for Linear Restrictions in a Logistic Regression Model. Statist. Pap. 50, 277–300 (2009)
10. Menéndez, M.L., Pardo, L., Zografos, K.: Preliminary test and Stein-type estimation of location parameter for elliptically contoured distributions. Adv. Appl. Statist. 11(1), 101–136 (2009)
11. Menéndez, M.L., Pardo, L., Zografos, K.: Preliminary Test Estimators in Intraclass Correlation Model under Unequal Familial Sizes. Math. Meth. Statist. 19(1), 73–87 (2010)
12. Pardo, J.A., Pardo, L., Zografos, K.: Minimum phi-divergence estimators with constraints in multinomial populations. J. Statist. Plan. Infer. 104(1), 221–237 (2002)
13. Pardo, L.: Statistical Inference Based on Divergence Measures. Chapman & Hall/CRC, New York (2006)
14. Pardo, L., Menéndez, M.L.: On some pre-test and Stein-rule phi-divergence test estimators in the independence model of categorical data. Journal of Statistical Planning Inference 138(7), 2163–2179 (2008)
15. Pardo, L., Menéndez, M.L., Martín, N.: Pretesting in Polytomous Logistic Regression Models Based on Phi-divergence Measures. In: Arnold, B.C., Balakrishnan, N., Sarabia, J.M., Mínguez, R. (eds.) Advances in Mathematical and Statistical Modeling Series: Statistics for Industry and Technology. Birkhäuser, Basel (2008)
16. Ehsanes Saleh, A.K.: Theory of Preliminary Test and Stein-Type Estimation with Applications. Wiley, New York (2006)

Detection of Outlying Points in Ordered Polytomous Regression*

María Carmen Pardo and Julio A. Pardo

Department of Statistics and O.R (I),
Complutense University of Madrid,
28040-Madrid, Spain
mcapardo@mat.ucm.es, japardo@mat.ucm.es

Summary. Ordered polytomous logistic models are used to model relationships between a polytomous response variable and a set of regressor variables when the response of an individual unit is restricted to one of a finite number of ordinal values. A method of checking the model's goodness of fit is described as well as a procedure to find outlying points. Finally, the paper concludes with an analysis of real data.

Keywords: ordered polytomous logistic regression, minimum $\phi-$divergence estimation, goodness-of-fit tests, outlying points.

1 Introduction

The application of ordered polytomous logistic regression in medical research has greatly increased in recent years. Ordered polytomous logistic regression is an useful technique for relating a dependent ordered categorical variable to categorical independent variables. Ordered polytomous regression has been discussed by McCullagh [11], Anderson and Philips [4], Anderson [3], Agresti [1], Liu and Agresti [10].

In an ordered model, the response $\boldsymbol{Y}$ of an individual unit is restricted to one of J ordered values. For example, the severity of a medical condition may be: none, mild, and severe. The ordered polytomous logistic model or cumulative logit model assumes that the ordinal nature of the observed response is due to methodological limitations in collecting the data that results in lumping together values of an otherwise continuous response variable (McKelvey and Zavoina [13]). Suppose $\boldsymbol{Y}$ takes values $y_1, ..., y_J$ on some scale, where $y_1 < ... < y_J$. It is assume that the observable variable is a categorized version of a continuous latent variable such that $\boldsymbol{Y} = y_r \Longleftrightarrow \alpha_{r-1} < U \leq \alpha_r$,

* Memories bring back to our minds and cannot be easily expressed. Our scientific collaboration with Marisa has been wide and fruitful, but it seems rather small in contrast to our personal relationship. Thank you, sister, for all the moments we have shared with you and for your invaluable and unselfish support and affection. You will be always in our hearts.

$r = 1, ..., J$, where $-\infty = \alpha_0 < \alpha_1 < .... < \alpha_J = \infty$. It is further assumed that the latent variable U is determined by the explanatory variable vector $\boldsymbol{x}^T = (x_1, ..., x_m) \in \mathbb{R}^m$ in the linear form $U = -\boldsymbol{x}^T\beta + \epsilon$, where $\boldsymbol{\beta}$ is a vector of regression coefficients and ϵ is a random variable with a logistic distribution function. It follows that

$$\Pr\left(\boldsymbol{Y} \leq y_r/\boldsymbol{x}^T\right) = \Pr(U \leq \alpha_r) = \frac{\exp\left(\alpha_r + \boldsymbol{x}^T\boldsymbol{\beta}\right)}{1 + \exp\left(\alpha_r + \boldsymbol{x}^T\boldsymbol{\beta}\right)}, \qquad r = 1, ..., J. \tag{1}$$

Given $\boldsymbol{x}$, $\boldsymbol{Y}$ is a multinomial with probability vector $\boldsymbol{\pi}^T = (\pi_1, ..., \pi_J)$ and $\pi_r = P(\boldsymbol{Y} = y_r \mid \boldsymbol{x}^T)$, $r = 1, ..., J$. Suppose we observe the sample $\boldsymbol{Y}_1 = \boldsymbol{y}_1, ..., \boldsymbol{Y}_N = \boldsymbol{y}_N$ jointly with the explanatory variables $\boldsymbol{x}_1, ..., \boldsymbol{x}_N$. The model (1) can be expressed as

$$\eta_i = \log\left(\frac{\Pr\left(\boldsymbol{Y} \leq y_r/\boldsymbol{x}^T\right)}{\Pr\left(\boldsymbol{Y} > y_r/\boldsymbol{x}^T\right)}\right) = \boldsymbol{Z}_i^T\boldsymbol{\gamma} \tag{2}$$

where

$$\boldsymbol{Z}_i^T = \begin{pmatrix} 1 & & & & \boldsymbol{x}_i^T \\ & 1 & & & \boldsymbol{x}_i^T \\ & & \cdot & & \cdot \\ & & & \cdot & \cdot \\ & & & \cdot & \cdot \\ & & & 1 & \boldsymbol{x}_i^T \end{pmatrix} \tag{3}$$

and $\boldsymbol{\gamma}^T = \left(\alpha_1, ..., \alpha_{J-1}, \boldsymbol{\beta}^T\right)$. The usual way to estimate the vector $\boldsymbol{\gamma}$ of unknown parameters is using the maximum likelihood estimator (MLE). As a natural extension of the unrestricted MLE, Pardo [17] defined the unrestricted minimum ϕ-divergence estimator as follows.

$$\widehat{\boldsymbol{\gamma}}_\phi = \arg\min_{\boldsymbol{\gamma}\in\mathbb{R}^p} D_\phi\left(\widehat{\boldsymbol{p}}, \boldsymbol{p}\left(\boldsymbol{\gamma}\right)\right) \tag{4}$$

with $p = J - 1 + m$,

$$\widehat{\boldsymbol{p}} = \left(\frac{y_{11}}{n}, ..., \frac{y_{J1}}{n}, \frac{y_{12}}{n}, ..., \frac{y_{J2}}{n}, ..., \frac{y_{1N}}{n}, ..., \frac{y_{JN}}{n}\right)^T,$$

with $y_{Ji} = n\left(\boldsymbol{x}_i\right) - \sum_{s=1}^{J-1} y_{si}$, $i = 1, ..., N$, $n\left(\boldsymbol{x}_i\right)$ is the number of observations with explicative variable $\boldsymbol{x}_i$, $n = \sum_{i=1}^N n\left(\boldsymbol{x}_i\right)$,

$$\boldsymbol{p}\left(\boldsymbol{\gamma}\right) = \left(\frac{n\left(\boldsymbol{x}_1\right)}{n}\widetilde{\boldsymbol{\pi}}_1^T, ..., \frac{n\left(\boldsymbol{x}_N\right)}{n}\widetilde{\boldsymbol{\pi}}_N^T\right)^T$$

being $\widetilde{\boldsymbol{\pi}}_i^T = (\pi_{i,1}, ..., \pi_{i,J})$ and

$$D_\phi\left(\widehat{\boldsymbol{p}}, \boldsymbol{p}\left(\boldsymbol{\gamma}\right)\right) = \sum_{l=1}^{J}\sum_{i=1}^{N} \pi_l(\boldsymbol{Z}_i^T\boldsymbol{\gamma})\frac{n\left(\boldsymbol{x}_i\right)}{n}\phi\left(\frac{y_{li}/n}{\pi_l\left(\boldsymbol{Z}_i^T\boldsymbol{\gamma}\right)n\left(\boldsymbol{x}_i\right)/n}\right) \tag{5}$$

is the ϕ–divergence measure defined by Ali and Silvey [2], being $\phi \in \Phi$ and Φ is the class of all convex functions $\phi(x)$, $x > 0$, such that at $x = 1, \phi(1) = \phi'(1) = 0, \phi''(1) > 0$, and at $x = 0$, $0\phi(0/0) = 0$ and $0\phi(p/0) = p\lim_{u\to\infty}\phi(u)/u$. For more details about ϕ-divergences see Vajda [19] and Pardo [15]. For $\phi(x) = x\log x - x + 1$, we obtain as particular case the MLE in (4).

Under mild regularity conditions, Pardo [17] established the asymptotic expansion of the unrestricted minimum ϕ-divergence estimator which is given by

$$\widehat{\boldsymbol{\gamma}}_{\phi} = \boldsymbol{\gamma}^0 + \boldsymbol{I}_{F,n}\left(\boldsymbol{\gamma}^0\right)^{-1} \boldsymbol{Z}\,Diag\left(\left(\boldsymbol{C}_{n,i}^0\right)_{i=1,..,N}\right) Diag\left(\boldsymbol{p}\left(\boldsymbol{\gamma}^0\right)^{-1/2}\right)\left(\widehat{\boldsymbol{p}} - \boldsymbol{p}\left(\boldsymbol{\gamma}^0\right)\right)$$
$$+ \left\|\widehat{\boldsymbol{p}} - \boldsymbol{p}\left(\boldsymbol{\gamma}^0\right)\right\| \boldsymbol{\alpha}_1\left(\widehat{\boldsymbol{p}}; \widehat{\boldsymbol{p}} - \boldsymbol{p}\left(\boldsymbol{\gamma}^0\right)\right) \tag{6}$$

where $\boldsymbol{\gamma}^0$ is the true parameter value of the parameter $\boldsymbol{\gamma}$,

$$\boldsymbol{C}_{n,i}^0 = \left(\boldsymbol{C}_{n,i}\right)_{\boldsymbol{\gamma}=\boldsymbol{\gamma}}^0 = \left[\left(\frac{n\left(\boldsymbol{x}_i\right)}{n}\right)^{1/2} \frac{\partial \widetilde{\boldsymbol{\pi}}\left(\boldsymbol{\eta}_i\right)}{\partial \boldsymbol{\eta}_i^T} Diag\left(\widetilde{\boldsymbol{\pi}}(\boldsymbol{\eta}_i)^{-1/2}\right)\right]_{\boldsymbol{\gamma}=\boldsymbol{\gamma}}^0, \quad i = 1, ..., N,$$

$$\boldsymbol{I}_{F,n}\left(\boldsymbol{\gamma}\right) = \boldsymbol{Z}\mathbf{V}_n\left(\boldsymbol{\gamma}\right)\boldsymbol{Z}^T \tag{7}$$

with

$$\boldsymbol{Z} = \left(\boldsymbol{Z}_1, ..., \boldsymbol{Z}_N\right),$$
$$\boldsymbol{V}_n\left(\boldsymbol{\gamma}\right) = Diag\left(\boldsymbol{V}_{n,1}\left(\boldsymbol{\gamma}\right), ..., \boldsymbol{V}_{n,N}\left(\boldsymbol{\gamma}\right)\right)$$

being

$$\boldsymbol{V}_{n,i}\left(\boldsymbol{\gamma}\right) = \frac{n\left(\boldsymbol{x}_i\right)}{n} \frac{\partial \boldsymbol{\pi}\left(\boldsymbol{\eta}_i\right)}{\partial \boldsymbol{\eta}_i} \boldsymbol{\Sigma}_i^{-1}\left(\boldsymbol{\gamma}\right) \frac{\partial \boldsymbol{\pi}\left(\boldsymbol{\eta}_i\right)}{\partial \boldsymbol{\eta}_i^T}, \tag{8}$$

$\boldsymbol{\Sigma}_i\left(\boldsymbol{\gamma}\right)$ is the inverse of the covariance matrix of $\boldsymbol{Y}_i$ and

$$\boldsymbol{\alpha}_1 : \mathbb{R}^{JN \times JN} \to \mathbb{R}^p \text{ verifies that } \boldsymbol{\alpha}_1\left(\boldsymbol{p}; \boldsymbol{p} - \boldsymbol{p}\left(\boldsymbol{\gamma}^0\right)\right) \to \mathbf{0} \text{ as } \boldsymbol{p} \to \boldsymbol{p}\left(\boldsymbol{\gamma}^0\right). \tag{9}$$

Suppose that in addition to the linear predictor (2) there are r linearly independent restrictions on the parameter vector $\boldsymbol{\gamma}$, $\boldsymbol{K}^T\boldsymbol{\gamma} = \boldsymbol{\xi}$ where $\boldsymbol{K}^T$ is any matrix of r rows and p columns and $\boldsymbol{\xi}$ is a vector, of order r of specified constants. There is only the limitation on $\boldsymbol{K}^T$ in the sense that it must have full row rank, i.e., rank$(\boldsymbol{K}^T) = r$. To fit the unknown parameters, Pardo [17] defined the restricted minimum ϕ-divergence estimator given by

$$\widehat{\boldsymbol{\gamma}}_{\phi}^{\Theta_0} = \arg\min_{\boldsymbol{\gamma}\in\boldsymbol{\Theta}_0} D_{\phi}\left(\widehat{\boldsymbol{p}}, \boldsymbol{p}\left(\boldsymbol{\gamma}\right)\right).$$

where $\Theta_0 = \left\{\gamma \in \mathbb{R}^p \;/\; \boldsymbol{K}^T\gamma = \boldsymbol{\xi}\right\}$.

An important step of the modeling process is assessing how well the data are described by the model. The tests that are used to evaluate fit in this manner are referred to as "goodness-of-fit tests". The most common goodness-of-fit tests for ordered polytomous logistic models include the Pearson χ^2-test and the deviance test. In Section 2, we propose a new family of test statistics based on the $\phi-$divergence that contains the above test statistics. Furthermore, we estimate the unknown parameters of the model using the above extension of the maximum likelihood estimator, the minimum $\phi-$divergence estimator.

In practice, however, the model building process can be highly influenced by peculiarities in the data. For univariate generalized models, Cook and Weisberg [5], McCullagh and Nelder [12], Pregibon [18], and Williams [20] have discussed diagnostic tools for detecting outlyings. Lesaffre and Albert [9] have extended Pregibon's regression diagnostics to the case where several groups are envisaged. A wider extension was made by Fahrmeir and Tutz [7] for multivariate extensions of generalized linear models. Pardo [16] shows that maximum likelihood and deviance-based diagnostics for multivariate extensions of generalized linear models extend naturally to the $\phi-$divergences family. In this paper, new measures based on divergences for detecting outlying points in ordered polytomous logistic models are presented in Section 3. For illustration, the procedure is applied to a data set in Section 4.

2 Goodness-of-Fit

After estimating the coefficients $\boldsymbol{\gamma}^T = \left(\alpha_1, ..., \alpha_{J-1}, \boldsymbol{\beta}^T\right)$, we begin with the summary measures of goodness-of-fit, as they give an overall indication of the fit of the model. Because these are summary statistics, they may not be very specific about the individual components. A small value for one of these statistics does not rule out the possibility of some substantial and thus interesting deviation from fit for a few subjects. On the other hand, a large value for one of these statistics is a clear indication of a substantial problem with the model.

The two most common measures for fitting the ordered polytomous logistic regression are the Pearson chi-square statistic defined by

$$X^2 = \sum_{i=1}^{N} X_i^2, \tag{10}$$

where

$$X_i^2 = \sum_{l=1}^{J} \frac{\left(y_{li} - n\left(\boldsymbol{x}_i\right)\pi_l\left(\boldsymbol{Z}_i^T\widehat{\boldsymbol{\gamma}}\right)\right)^2}{n\left(\boldsymbol{x}_i\right)\pi_l\left(\boldsymbol{Z}_i^T\widehat{\boldsymbol{\gamma}}\right)} \tag{11}$$

with $\widehat{\gamma}$ the MLE of γ and the deviance statistic defined by

$$D = 2\sum_{i=1}^{N} d_i, \tag{12}$$

where

$$d_i = \sum_{l=1}^{J} y_{li} \log \frac{y_{li}}{n(\boldsymbol{x}_i)\,\pi_l\left(\boldsymbol{Z}_i^T \widehat{\gamma}\right)}. \tag{13}$$

It is known that the asymptotic distribution of the test statistics given in (10) and (12) is a chi-square with $(J-1)\,N - p$ degrees of freedom.

The statistics (10) and (12) are used for testing the goodness-of-fit

$$H_0 : \mathbf{p} = \mathbf{p}(\gamma) \tag{14}$$

which are a particular case of the family of statistics

$$B_n^{\phi_1,\phi_2} = \frac{2n}{\phi_1''(1)} D_{\phi_1}\left(\widehat{\boldsymbol{p}}, \boldsymbol{p}\left(\widehat{\gamma}_{\phi_2}\right)\right). \tag{15}$$

Note that the above test statistics uses one measure of divergence (ϕ_2) for estimation and another different measure of divergence (ϕ_1) for testing. In particular if we consider $\phi_1(x) = \phi_2(x) = x\log x - x - 1$ and for $\phi_1(x) = \frac{1}{2}(x-1)^2$ and $\phi_2(x) = x\log x - x - 1$ we obtain (12) and (10), respectively.

Theorem 1. *Let $\boldsymbol{Y}_i$, $i = 1, ..., N$, independent random variables with multinomial distributions of parameters $\left(n(\boldsymbol{x}_i), \boldsymbol{\pi}\left(\boldsymbol{Z}_i^T \gamma\right)\right)$. Assume conditions of Theorem 2 of Pardo [17] and that $\phi_1 \in \Phi$. Then under the null hypothesis (14), the family of test statistics $B_n^{\phi_1,\phi_2}$, given in (15), is asymptotically distributed as a chi-square with $(J-1)\,N - p$ degrees of freedom.*

Proof. Firstly, we obtain the asymptotic distribution of the random vector

$$\boldsymbol{W} = Diag\left(\boldsymbol{p}\left(\gamma^0\right)^{-1/2}\right)\sqrt{n}\left(\widehat{\boldsymbol{p}} - \boldsymbol{p}\left(\widehat{\gamma}_{\phi_2}\right)\right).$$

Using a Taylor expansion of order one we have

$$\boldsymbol{p}\left(\widehat{\gamma}_{\phi_2}\right) - \boldsymbol{p}\left(\gamma^0\right) = \left(\frac{\partial \boldsymbol{p}(\gamma)}{\partial \gamma}\right)_{\gamma=\gamma^0}\left(\widehat{\gamma}_{\phi_2} - \gamma^0\right) + \left\|\widehat{\gamma}_{\phi_2} - \gamma^0\right\| \boldsymbol{\alpha}_2\left(\widehat{\gamma}_{\phi_2}; \widehat{\gamma}_{\phi_2} - \gamma^0\right)$$

where

$$\left(\frac{\partial \boldsymbol{p}(\gamma)}{\partial \gamma}\right)_{\gamma=\gamma^0} = \boldsymbol{S}_n\left(\gamma^0\right)\boldsymbol{Z}^T$$

with $\boldsymbol{S}_n\left(\gamma^0\right) = Diag\left(\frac{n\left(\boldsymbol{x}_1\right)}{n}\frac{\partial\widetilde{\boldsymbol{\pi}}\left(\boldsymbol{\eta}_1\right)}{\partial\boldsymbol{\eta}_1},...,\frac{n\left(\boldsymbol{x}_N\right)}{n}\frac{\partial\widetilde{\boldsymbol{\pi}}\left(\boldsymbol{\eta}_N\right)}{\partial\boldsymbol{\eta}_N}\right)$ and evaluating in $\boldsymbol{\gamma}^0$ and

$$\boldsymbol{\alpha}_2 : \mathbb{R}^p \rightarrow \mathbb{R}^{JN\times JN} \text{ verifies that } \boldsymbol{\alpha}_2\left(\boldsymbol{\gamma};\boldsymbol{\gamma}-\boldsymbol{\gamma}^0\right) \rightarrow \mathbf{0} \text{ as } \boldsymbol{\gamma}\rightarrow\boldsymbol{\gamma}^0. \tag{16}$$

Denoting by

$$\begin{aligned}\boldsymbol{L}_n(\boldsymbol{\gamma}^0) = \boldsymbol{S}_n\left(\boldsymbol{\gamma}^0\right)\boldsymbol{Z}^T\boldsymbol{I}_{F,n}(\boldsymbol{\gamma}^0)^{-1}\boldsymbol{Z}Diag\left(\left(\boldsymbol{C}^0_{n,i}\right)_{i=1,...,N}\right)\\ \times Diag\left(\boldsymbol{p}\left(\boldsymbol{\gamma}^0\right)^{-1/2}\right)\end{aligned} \tag{17}$$

we have by (6)

$$\begin{aligned}\mathbf{p}(\widehat{\gamma}_{\phi_2}) - \mathbf{p}(\boldsymbol{\gamma}^0) = \mathbf{L}_n(\boldsymbol{\gamma}^0)\left(\widehat{\mathbf{p}}-\mathbf{p}\left(\boldsymbol{\gamma}^0\right)\right) + \mathbf{S}_n\left(\boldsymbol{\gamma}^0\right)\mathbf{Z}^T\left\|\widehat{\mathbf{p}}-\mathbf{p}(\boldsymbol{\gamma}^0)\right\|\boldsymbol{\alpha}_1(\widehat{\mathbf{p}};\widehat{\mathbf{p}}-\mathbf{p}(\boldsymbol{\gamma}^0))\\ + \left\|\widehat{\gamma}_{\phi_2}-\boldsymbol{\gamma}^0\right\|\boldsymbol{\alpha}_2(\widehat{\gamma}_{\phi_2};\widehat{\gamma}_{\phi_2}-\boldsymbol{\gamma}^0).\end{aligned}$$

Therefore,

$$\begin{pmatrix}\widehat{\boldsymbol{p}}-\boldsymbol{p}\left(\boldsymbol{\gamma}^0\right)\\ \boldsymbol{p}\left(\widehat{\gamma}_{\phi_2}\right)-\boldsymbol{p}\left(\boldsymbol{\gamma}^0\right)\end{pmatrix} = \begin{pmatrix}\boldsymbol{I}\\ \boldsymbol{L}_n(\boldsymbol{\gamma}^0)\end{pmatrix}\left(\widehat{\boldsymbol{p}}-\boldsymbol{p}\left(\boldsymbol{\gamma}^0\right)\right)$$

$$+\begin{pmatrix}\mathbf{0}\\ \boldsymbol{S}_n(\boldsymbol{\gamma}^0)\boldsymbol{Z}^T\left\|\widehat{\boldsymbol{p}}-\boldsymbol{p}\left(\boldsymbol{\gamma}^0\right)\right\|\boldsymbol{\alpha}_1\left(\widehat{\boldsymbol{p}};\widehat{\boldsymbol{p}}-\boldsymbol{p}\left(\boldsymbol{\gamma}^0\right)\right)+\left\|\widehat{\gamma}_{\phi_2}-\boldsymbol{\gamma}^0\right\|\boldsymbol{\alpha}_2\left(\widehat{\gamma}_{\phi_2};\widehat{\gamma}_{\phi_2}-\boldsymbol{\gamma}^0\right)\end{pmatrix}.$$

Applying the Central Limit Theorem we have

$$\sqrt{n}\left(\widehat{\boldsymbol{p}}-\boldsymbol{p}\left(\boldsymbol{\gamma}^0\right)\right) \underset{n\rightarrow\infty}{\overset{L}{\rightarrow}} \mathcal{N}\left(\mathbf{0},\boldsymbol{\Sigma}_{\boldsymbol{p}_\lambda(\boldsymbol{\gamma}^0)}\right),$$

where $\boldsymbol{\Sigma}_{\boldsymbol{p}_\lambda(\boldsymbol{\gamma}^0)} = Diag\left(\boldsymbol{p}_\lambda\left(\boldsymbol{\gamma}^0\right)\right)\left(\boldsymbol{I}-\boldsymbol{X}\left(\boldsymbol{\gamma}^0\right)\right)$ and

$$\boldsymbol{X}\left(\boldsymbol{\gamma}^0\right) = \boldsymbol{X}_0\left(\boldsymbol{X}_0^T Diag\left(\boldsymbol{p}_\lambda\left(\boldsymbol{\gamma}^0\right)\right)\boldsymbol{X}_0\right)^{-1}\boldsymbol{X}_0^T Diag\left(\boldsymbol{p}_\lambda\left(\boldsymbol{\gamma}^0\right)\right) \tag{18}$$

with

$$\boldsymbol{X}_0 = \begin{pmatrix}\mathbf{1}_{\boldsymbol{J}} & \mathbf{0} & ... & \mathbf{0}\\ \mathbf{0} & \mathbf{1}_{\boldsymbol{J}} & ... & \mathbf{0}\\ . & . & ... & .\\ \mathbf{0} & \mathbf{0} & ... & \mathbf{1}_{\boldsymbol{J}}\end{pmatrix}_{JN\times N}, \tag{19}$$

being $\mathbf{1}_{\boldsymbol{J}}$ the unit vector $J\times 1-$dimensional, then $\sqrt{n}\left\|\widehat{\boldsymbol{p}}-\boldsymbol{p}\left(\boldsymbol{\gamma}^0\right)\right\|$ is bounded in probability. Also $\sqrt{n}\left\|\widehat{\gamma}_{\phi_2}-\boldsymbol{\gamma}^0\right\|$ is bounded in probability since

$$\sqrt{n}\left(\widehat{\gamma}_{\phi_2}-\boldsymbol{\gamma}^0\right) \underset{n\rightarrow\infty}{\overset{L}{\longrightarrow}} \mathcal{N}\left(0,\boldsymbol{I}_{F,\lambda}\left(\boldsymbol{\gamma}^0\right)^{-1}\right).$$

So,

$$\sqrt{n}\left(\mathbf{S}_n(\gamma^0)\mathbf{Z}^T||\widehat{\mathbf{p}}-\mathbf{p}(\gamma^0)||\alpha_1(\widehat{\mathbf{p}};\widehat{\mathbf{p}}-\mathbf{p}(\gamma^0)) + ||\widehat{\gamma}_{\phi_2}-\gamma^0||\alpha_2(\widehat{\gamma}_{\phi_2};\widehat{\gamma}_{\phi_2}-\gamma^0)\right)$$

converges to 0. Then

$$\sqrt{n}\begin{pmatrix} \widehat{\boldsymbol{p}}-\boldsymbol{p}\left(\boldsymbol{\gamma}^{0}\right) \\ \boldsymbol{p}\left(\widehat{\boldsymbol{\gamma}}_{\phi_2}\right)-\boldsymbol{p}\left(\boldsymbol{\gamma}^{0}\right) \end{pmatrix} \xrightarrow[n\to\infty]{L} \mathcal{N}\left(\mathbf{0},\boldsymbol{\Sigma}_3\left(\boldsymbol{\gamma}^{0}\right)\right)$$

where

$$\boldsymbol{\Sigma}_3\left(\boldsymbol{\gamma}^{0}\right)=\begin{pmatrix} \boldsymbol{A}\left(\boldsymbol{\gamma}^{0}\right) & \boldsymbol{A}\left(\boldsymbol{\gamma}^{0}\right)\boldsymbol{L}_\lambda\left(\boldsymbol{\gamma}^{0}\right)^T \\ \boldsymbol{L}_\lambda\left(\boldsymbol{\gamma}^{0}\right)\boldsymbol{A}\left(\boldsymbol{\gamma}^{0}\right) & \boldsymbol{L}_\lambda\left(\boldsymbol{\gamma}^{0}\right)\boldsymbol{A}\left(\boldsymbol{\gamma}^{0}\right)\boldsymbol{L}_\lambda\left(\boldsymbol{\gamma}^{0}\right)^T \end{pmatrix}$$

with $\boldsymbol{A}\left(\boldsymbol{\gamma}^{0}\right)=Diag\left(\boldsymbol{p}_\lambda\left(\boldsymbol{\gamma}^{0}\right)\right)\left(\boldsymbol{I}-\boldsymbol{X}\left(\boldsymbol{\gamma}^{0}\right)\right)$ and $\boldsymbol{L}_\lambda\left(\boldsymbol{\gamma}^{0}\right)=\lim_{n\to\infty}\boldsymbol{L}_n\left(\boldsymbol{\gamma}^{0}\right)$.

From above

$$\sqrt{n}\left(\widehat{\boldsymbol{p}}-\boldsymbol{p}\left(\widehat{\boldsymbol{\gamma}}_{\phi_2}\right)\right) \xrightarrow[n\to\infty]{L} \mathcal{N}\left(\mathbf{0},\boldsymbol{\Sigma}_4\left(\boldsymbol{\gamma}^{0}\right)\right)$$

where

$$\boldsymbol{\Sigma}_4\left(\boldsymbol{\gamma}^{0}\right)=\boldsymbol{A}\left(\boldsymbol{\gamma}^{0}\right)-\boldsymbol{A}\left(\boldsymbol{\gamma}^{0}\right)\boldsymbol{L}_\lambda\left(\boldsymbol{\gamma}^{0}\right)^T-\boldsymbol{L}_\lambda\left(\boldsymbol{\gamma}^{0}\right)\boldsymbol{A}\left(\boldsymbol{\gamma}^{0}\right)+\boldsymbol{L}_\lambda\left(\boldsymbol{\gamma}^{0}\right)\boldsymbol{A}\left(\boldsymbol{\gamma}^{0}\right)\boldsymbol{L}_\lambda\left(\boldsymbol{\gamma}^{0}\right)^T$$

or equivalently

$$\sqrt{n}Diag\left(\boldsymbol{p}(\boldsymbol{\gamma}^{0})^{-1/2}\right)\left(\widehat{\boldsymbol{p}}-\boldsymbol{p}\left(\widehat{\boldsymbol{\gamma}}_{\phi_2}\right)\right) \xrightarrow[n\to\infty]{L} \mathcal{N}(\mathbf{0},\boldsymbol{\Sigma}_5\left(\boldsymbol{\gamma}^{0}\right)), \tag{20}$$

where

$$\boldsymbol{\Sigma}_5\left(\boldsymbol{\gamma}^{0}\right)=Diag\left(\boldsymbol{p}_\lambda(\boldsymbol{\gamma}^{0})^{-1/2}\right)\boldsymbol{\Sigma}_4\left(\boldsymbol{\gamma}^{0}\right)Diag\left(\boldsymbol{p}_\lambda(\boldsymbol{\gamma}^{0})^{-1/2}\right). \tag{21}$$

Taking into account that

a) As

$$\mathbf{D}_\lambda Diag\left(\boldsymbol{p}_\lambda\left(\boldsymbol{\gamma}^{0}\right)^{1/2}\right)\boldsymbol{X}_0=\mathbf{0}, \tag{22}$$

where $\mathbf{D}_\lambda=Diag\left(\left(\boldsymbol{C}^0_{\lambda,i}\right)_{i=1,\ldots,N}\right)$ with $\mathbf{C}^0_{\lambda,i}=\lim_{n\to\infty}\mathbf{C}^0_{n,i}$, then

$$\begin{aligned} &Diag\left(\boldsymbol{p}_\lambda(\boldsymbol{\gamma}^{0})^{-1/2}\right)\boldsymbol{L}_\lambda\left(\boldsymbol{\gamma}^{0}\right)Diag\left(\boldsymbol{p}_\lambda\left(\boldsymbol{\gamma}^{0}\right)\right)\boldsymbol{X}\left(\boldsymbol{\gamma}^{0}\right)Diag\left(\boldsymbol{p}_\lambda(\boldsymbol{\gamma}^{0})^{-1/2}\right)\\ &\quad=Diag\left(\boldsymbol{p}_\lambda(\boldsymbol{\gamma}^{0})^{-1/2}\right)\boldsymbol{S}_\lambda\left(\boldsymbol{\gamma}^{0}\right)\boldsymbol{Z}^T\left(\boldsymbol{Z}\boldsymbol{V}_\lambda\left(\boldsymbol{\gamma}^{0}\right)\boldsymbol{Z}^T\right)^{-1}\boldsymbol{Z}\mathbf{D}_\lambda\\ &\quad\quad\times Diag\left(\boldsymbol{p}_\lambda\left(\boldsymbol{\gamma}^{0}\right)^{1/2}\right)\boldsymbol{X}\left(\boldsymbol{\gamma}^{0}\right)Diag\left(\boldsymbol{p}_\lambda(\boldsymbol{\gamma}^{0})^{-1/2}\right)=\mathbf{0} \end{aligned}$$

with $\mathbf{S}_\lambda(\boldsymbol{\gamma}^{0})=\lim_{n\to\infty}\mathbf{S}_n(\boldsymbol{\gamma}^{0})$ and $\boldsymbol{V}_\lambda(\boldsymbol{\gamma}^{0})=\lim_{n\to\infty}\boldsymbol{V}_n(\boldsymbol{\gamma}^{0})$.

b) By (18) and taking into account (22),

$$\begin{aligned} &Diag\left(\boldsymbol{p}_\lambda(\boldsymbol{\gamma}^{0})^{1/2}\right)\boldsymbol{X}\left(\boldsymbol{\gamma}^{0}\right)\boldsymbol{L}_\lambda\left(\boldsymbol{\gamma}^{0}\right)^T Diag\left(\boldsymbol{p}_\lambda(\boldsymbol{\gamma}^{0})^{-1/2}\right)\\ &\quad=Diag\left(\boldsymbol{p}_\lambda(\boldsymbol{\gamma}^{0})^{1/2}\right)\boldsymbol{X}_0\left(\boldsymbol{X}_0^T Diag\left(\boldsymbol{p}_\lambda\left(\boldsymbol{\gamma}^{0}\right)\right)\boldsymbol{X}_0\right)^{-1}\\ &\quad\quad\times\boldsymbol{X}_0^T Diag\left(\boldsymbol{p}_\lambda\left(\boldsymbol{\gamma}^{0}\right)\right)\boldsymbol{L}_\lambda\left(\boldsymbol{\gamma}^{0}\right)^T Diag\left(\boldsymbol{p}_\lambda(\boldsymbol{\gamma}^{0})^{-1/2}\right)\\ &=Diag\left(\boldsymbol{p}_\lambda(\boldsymbol{\gamma}^{0})^{1/2}\right)\boldsymbol{X}_0\left(\boldsymbol{X}_0^T Diag\left(\boldsymbol{p}_\lambda\left(\boldsymbol{\gamma}^{0}\right)\right)\boldsymbol{X}_0\right)^{-1}\boldsymbol{X}_0^T Diag\left(\boldsymbol{p}_\lambda\left(\boldsymbol{\gamma}^{0}\right)^{1/2}\right)\\ &\quad\times\mathbf{D}_\lambda^T\mathbf{Z}^T\left(\mathbf{Z}\mathbf{V}_\lambda(\boldsymbol{\gamma}^{0})\mathbf{Z}^T\right)^{-1}\boldsymbol{Z}\boldsymbol{S}_\lambda(\boldsymbol{\gamma}^{0})^T Diag\left(\boldsymbol{p}_\lambda(\boldsymbol{\gamma}^{0})^{-1/2}\right)=\mathbf{0}. \end{aligned}$$

In a similar way

$$Diag\left(\mathbf{p}_\lambda(\gamma^0)^{\frac{1}{2}}\right)\mathbf{L}_\lambda\left(\gamma^0\right)^T\mathbf{X}\left(\gamma^0\right)Diag\left(\boldsymbol{p}_\lambda(\gamma^0)^{-1/2}\right)=\mathbf{0}.$$

As

$$\begin{aligned}Diag\left(\boldsymbol{p}_\lambda(\gamma^0)^{-1/2}\right)\boldsymbol{L}_\lambda\left(\gamma^0\right)Diag\left(\boldsymbol{p}_\lambda\left(\gamma^0\right)\right)\boldsymbol{L}_\lambda\left(\gamma^0\right)_\lambda^T Diag\left(\boldsymbol{p}_\lambda(\gamma^0)^{-1/2}\right)\\ = Diag\left(\boldsymbol{p}_\lambda(\gamma^0)^{-1/2}\right)\boldsymbol{L}_\lambda\left(\gamma^0\right)Diag\left(\boldsymbol{p}_\lambda(\gamma^0)^{1/2}\right)\end{aligned} \tag{23}$$

then

$$\begin{aligned}\boldsymbol{\Sigma}_5(\gamma^0) = \mathbf{I} - Diag\left(\mathbf{p}_\lambda(\gamma^0)^{\frac{1}{2}}\right)\mathbf{X}\left(\gamma^0\right)Diag\left(\mathbf{p}_\lambda(\gamma^0)^{-\frac{1}{2}}\right)\\ -Diag\left(\mathbf{p}_\lambda(\gamma^0)^{\frac{1}{2}}\right)\mathbf{L}_\lambda\left(\gamma^0\right)^T Diag\left(\mathbf{p}_\lambda(\gamma^0)^{-\frac{1}{2}}\right).\end{aligned}$$

Furthermore, the matrix $\boldsymbol{\Sigma}_5\left(\gamma^0\right)$ is idempotent since

$$\begin{aligned}\boldsymbol{\Sigma}_5\left(\gamma^0\right)^2 &= \boldsymbol{\Sigma}_5\left(\gamma^0\right) - Diag\left(\boldsymbol{p}_\lambda(\gamma^0)^{1/2}\right)\boldsymbol{X}\left(\gamma^0\right)Diag\left(\boldsymbol{p}_\lambda(\gamma^0)^{-1/2}\right)\\ &+Diag\left(\boldsymbol{p}_\lambda(\gamma^0)^{1/2}\right)\boldsymbol{X}\left(\gamma^0\right)\boldsymbol{X}\left(\gamma^0\right)Diag\left(\boldsymbol{p}_\lambda(\gamma^0)^{-1/2}\right)\\ &+Diag\left(\boldsymbol{p}_\lambda(\gamma^0)^{1/2}\right)\boldsymbol{X}\left(\gamma^0\right)\boldsymbol{L}_\lambda\left(\gamma^0\right)^T Diag\left(\boldsymbol{p}_\lambda(\gamma^0)^{-1/2}\right)\\ &-Diag\left(\boldsymbol{p}_\lambda(\gamma^0)^{1/2}\right)\boldsymbol{L}_\lambda\left(\gamma^0\right)^T Diag\left(\boldsymbol{p}_\lambda(\gamma^0)^{-1/2}\right)\\ &+Diag\left(\boldsymbol{p}_\lambda(\gamma^0)^{1/2}\right)\boldsymbol{L}_\lambda\left(\gamma^0\right)^T\boldsymbol{X}\left(\gamma^0\right)Diag\left(\boldsymbol{p}_\lambda(\gamma^0)^{-1/2}\right)\\ &+Diag\left(\boldsymbol{p}_\lambda(\gamma^0)^{1/2}\right)\boldsymbol{L}_\lambda\left(\gamma^0\right)^T\boldsymbol{L}_\lambda\left(\gamma^0\right)^T Diag\left(\boldsymbol{p}_\lambda(\gamma^0)^{-1/2}\right)\\ &= \boldsymbol{\Sigma}_5\left(\gamma^0\right)\end{aligned}$$

where the last equality is obtained by b), that $\boldsymbol{X}\left(\gamma^0\right)\boldsymbol{X}\left(\gamma^0\right)=\boldsymbol{X}\left(\gamma^0\right)$ and that $\boldsymbol{L}_\lambda\left(\gamma^0\right)^T\boldsymbol{L}_\lambda\left(\gamma^0\right)^T=\boldsymbol{L}_\lambda\left(\gamma^0\right)^T$.

Such as $\boldsymbol{W}^T\boldsymbol{W}$ (Ferguson [8]) has a chi-square distribution with r degrees of freedom if $\boldsymbol{\Sigma}_5\left(\gamma^0\right)$ is a projection of rank r. The rank of this matrix, to be idempotent, coincides with its trace. As

i)

$$Traza(\boldsymbol{I}) = JN$$

ii)

$$\begin{aligned}&Traza\left(Diag\left(\boldsymbol{p}_\lambda(\gamma^0)^{1/2}\right)\boldsymbol{L}_\lambda\left(\gamma^0\right)^T Diag\left(\boldsymbol{p}_\lambda(\gamma^0)^{-1/2}\right)\right)\\ &= Traza\left(\mathbf{D}_\lambda\mathbf{Z}^T\left(\boldsymbol{Z}\boldsymbol{V}_\lambda(\gamma^0)\,\boldsymbol{Z}^T\right)^{-1}\boldsymbol{Z}\mathbf{D}_\lambda^T\right)\\ &= Traza\left(\left(\mathbf{Z}\mathbf{V}_\lambda(\gamma^0)\mathbf{Z}^T\right)^{-1}\mathbf{Z}\mathbf{D}_\lambda^T\mathbf{D}_\lambda\mathbf{Z}^T\right)\\ &= Traza(\boldsymbol{I}_{p\times p}) = p\end{aligned}$$

iii)

$$Traza\left(Diag\left(\boldsymbol{p}_{\lambda}(\gamma^0)^{1/2}\right)\boldsymbol{X}\left(\gamma^0\right)Diag\left(\boldsymbol{p}_{\lambda}(\gamma^0)^{-1/2}\right)\right) = Traza(\boldsymbol{I}_{N\times N}) = N.$$

We have that

$$Traza(\boldsymbol{\Sigma}_5\left(\boldsymbol{\gamma}^0\right)) = JN - N - p = N\left(J-1\right) - p.$$

Taking a second order expansion of $B_n^{\phi_1,\phi_2}$ around $\mathbf{p} = \mathbf{p}\left(\boldsymbol{\gamma}^0\right)$,

$$B_n^{\phi_1,\phi_2} = \boldsymbol{W}^T\boldsymbol{W} + o_p(1),$$

the result follows. □

3 Detection of Outlying Points

In Section 2 we introduced statistics for checking model fit in a global sense. The disadvantage of these single overall test statistics of goodness-of-fit is that it will not usually give constructive guidance on how to deal with any failure of the original model and is likely to be insensitive in detecting specific types of departure. Observations that are badly predicted (or allocated) are termed outlying. Sometimes, it is possible to identify outlying through a visual inspection of the data. However, it may not be possible to do it and an outlying observations identification procedure is needed. Residuals are the most common diagnostic method to identify observations that are not well explained by the model. An observation that leads to an abnormally large residual is an outlying.

In this section, an alternative way for identifying an "outlying" at a designated case, say i, consists in considering the model,

$$\boldsymbol{\eta}_j = \begin{cases} \boldsymbol{Z}_i^T\boldsymbol{\gamma}+\boldsymbol{\lambda}_i & j = i \\ \boldsymbol{Z}_j^T\boldsymbol{\gamma} & j \neq i \end{cases} \quad j = 1, ..., N. \tag{24}$$

This model is analogue of the mean slippage model commonly used for outlying detection in linear regression (Cook and Weisberg [5], p.20). To test that observation $\mathbf{x}_i$ is an "outlying" is equivalent to

$$H_0 : \boldsymbol{\lambda}_i = \mathbf{0} \quad \text{versus} \quad H_{1,i} : \boldsymbol{\lambda}_i \neq \mathbf{0}. \tag{25}$$

If the null hypothesis is rejected, the ith observation will be an "outlying"or if, we express the model given in (24) as

$$\boldsymbol{\eta}_j = \left(\boldsymbol{Z}_j^*\right)^T\boldsymbol{\gamma}^*, \qquad j = 1, ..., N$$

where $\left(\boldsymbol{Z}_i^*\right)^T = \left(\boldsymbol{Z}_i^T, \mathbf{1}\right)$ and $\left(\boldsymbol{Z}_j^*\right)^T = \left(\boldsymbol{Z}_i^T, \mathbf{0}\right)$ $j \neq i$ with $\boldsymbol{Z}_i^T$ defined in (3), $\mathbf{1}$ is a vector of ones and $\mathbf{0}$ is a vector of ceros of dimension $J-1$ both and

$(\gamma^*)^T = \left(\alpha_1, ..., \alpha_{J-1}, \boldsymbol{\beta}^T, \boldsymbol{\lambda}_i\right)$. Then, to test the hypothesis given in (25) is equivalent to test

$$H_0 : \boldsymbol{K}^T \boldsymbol{\gamma}^* = 0 \qquad \text{versus} \qquad H_{1,i} : \boldsymbol{K}^T \boldsymbol{\gamma}^* \neq 0,$$

where $\boldsymbol{K}^T = (\mathbf{0}_{m+J-1}, 1)$ and $\text{rank}(\boldsymbol{K}^T) = 1$. For the above test, Pardo [17] proposed the following test statistic

$$T_{n,i}^{\phi_1,\phi_2} = \frac{2n}{\phi_1''(1)} D_{\phi_1}\left(\boldsymbol{p}\left(\widehat{\boldsymbol{\beta}}_{\phi_2}^{H_{1,i}}\right), \boldsymbol{p}\left(\widehat{\boldsymbol{\beta}}_{\phi_2}^{H_0}\right)\right)$$

where $\widehat{\beta}_{\phi_2}^{H_{1,i}}$ and $\widehat{\beta}_{\phi_2}^{H_0}$ are the restricted minimum ϕ_2-divergence estimators under the alternative hypothesis and under the null hypothesis, respectively. The asymptotic distribution of this statistic is a chi-square with $r = 1$ degrees of freedom. When the candidate case for an "outlying" is unknown, a multiple testing procedure, such as one based on the first Bonferroni inequality (Miller [14]), must be used to find significance levels. In our case, we will say that $\mathbf{x}_i$ is an "outlying" if

$$T_{n,i}^{\phi_1,\phi_2} \geq \chi^2_{1,1-\alpha/N},$$

and the probability to reject incorrectly an observation is given by

$$\Pr\left(\bigcup_{i=1}^{N}\left(T_{n,i}^{\phi_1,\phi_2} \geq \chi^2_{1,1-\alpha/N}\right)\right) \leq \sum_{i=1}^{N} \Pr\left(T_{n,i}^{\phi_1,\phi_2} \geq \chi^2_{1,1-\alpha/N}\right) = \alpha.$$

It is straighforward to extend this procedure to study if a set of observations $I = \{\boldsymbol{x}_{i_1}..., \boldsymbol{x}_{i_k}\}$ are outlying.

4 Numerical Example

As an illustration of the new tools for diagnostic presented in previous sections we consider data on the perspectives of students, psychology students at the University of Regensburg were asked if they expected to find adequate employment after getting their degree. The response categories were ordered with respect to their expectation. The responses were 'don't expect adequate employment' (category 1), 'not sure' (category 2), and 'immediately after the degree' (category 3). The data are given in Fahrmeir and Tutz [7]. Table 1 shows the data for different ages of the students.

To fit the model we use the minimum $\phi-$divergence estimator with $\phi = \phi_{(a)}$ being $\phi_{(a)}$ a parametric family introduced by Cressie and Read [6] that is defined as

$$\begin{aligned}
&\phi_{(a)}(x) = (a(a+1))^{-1}\left(x^{a+1} - x - a(x-1)\right);\ a \neq 0, a \neq -1,\\
&\phi_{(0)}(x) = \lim_{a\to 0}\phi_{(a)}(x) = x\log x - x + 1, \qquad (27)\\
&\phi_{(-1)}(x) = \lim_{a\to -1}\phi_{(a)}(x) = x - \log x - 1.
\end{aligned}$$

Fitting the ordered polytomous logistic regression

$$\Pr(\boldsymbol{Y} \leq r/Age) = (1+\exp(-(\alpha_r + \beta \log(Age))))^{-1}, \qquad r = 1, 2,$$

we obtain the following minimum $\phi_{(a_2)}$−divergence estimations for $a_2 = 0, 2/3$ and 1 shown in Table 2.

First of all, the goodness-of-fit statistics $B_n^{\phi_{(a_1)},\phi_{(a_2)}}$ are calculated to value the fit of the model and they are shown in Tables 3, 4 and 5 for $a_2 = 0, 2/3$ and 1, respectively as well as their p-values.

Table 1. Grouped data for job expectations of psychology students in Regensburg

Number of obs.	Age	Y 1 2 3	$n(\boldsymbol{x}_i)$
1	19	1 2 0	3
2	20	5 18 2	25
3	21	6 19 2	27
4	22	1 6 3	10
5	23	2 7 3	12
6	24	1 7 5	13
7	25	0 0 3	3
8	26	0 1 0	1
9	27	0 2 1	3
10	29	1 0 0	1
11	30	0 0 2	2
12	31	0 1 0	1
13	34	0 1 0	1

Table 2. Minimum $\phi_{(a_2)}$-divergence estimators

a_2	α_1	α_2	β
0	14.9884	18.1497	-5.4027
2/3	8.4044	11.2404	-3.2143
1	5.8553	8.526	-2.3661

Table 3. $B_n^{\phi_{(a_1)},\phi_{(0)}}$ for different values of a_1 and their p-values

a_1	-1/2	0	2/3	1	2
$B_n^{\phi_{(a_1)},\phi_{(0)}}$	36.2456	26.7334	31.5669	42.4664	240.5135
p-values	0.0389	0.2675	0.1095	0.0080	0.0000

Table 4. $B_n^{\phi(a_1),\phi(2/3)}$ for different values of a_1 and their p-values

a_1	-1/2	0	2/3	1	2
$B_n^{\phi(a_1),\phi(2/3)}$	39.9762	28.5346	27.9965	31.3110	72.9037
p-values	0.0154	0.1962	0.2159	0.1153	0.0000

Table 5. $B_n^{\phi(a_1),\phi(1)}$ for different values of a_1 and their p-values

a_1	-1/2	0	2/3	1	2
$B_n^{\phi(a_1),\phi(1)}$	42.9673	30.5972	28.5852	30.4439	53.7457
p-values	0.0070	0.1330	0.1945	0.1371	0.0003

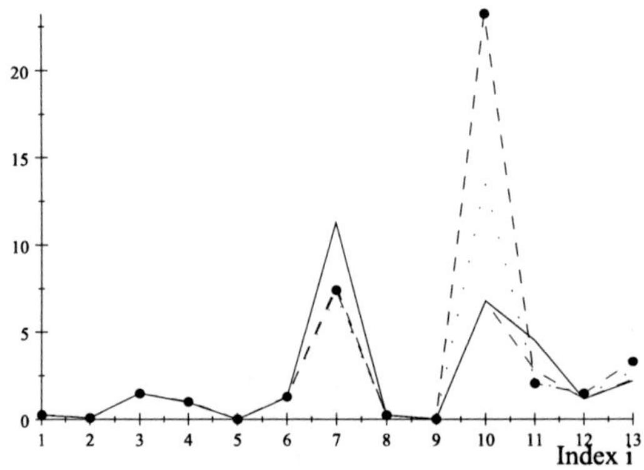

Fig. 1. $T_{n,i}^{\phi(a_1),\phi(0)}$ for $a_1 = -\frac{1}{2}, 0, \frac{2}{3}$ and 1. Solid line $a_1 = -\frac{1}{2}$, dashed line $a_1 = 0$, dotted line $a_1 = \frac{2}{3}$ and dash-dotted line $a_1 = 1$.

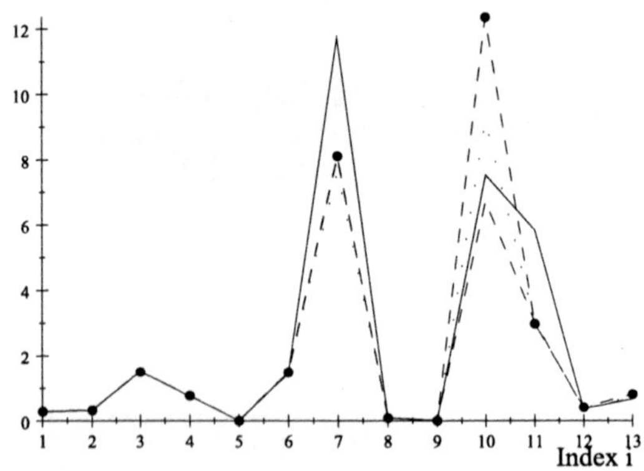

Fig. 2. $T_{n,i}^{\phi(a_1),\phi(\frac{2}{3})}$ for $a_1 = -\frac{1}{2}, 0, \frac{2}{3}$ and 1. Solid line $a_1 = -\frac{1}{2}$, dashed line $a_1 = 0$, dotted line $a_1 = \frac{2}{3}$ and dash-dotted line $a_1 = 1$.

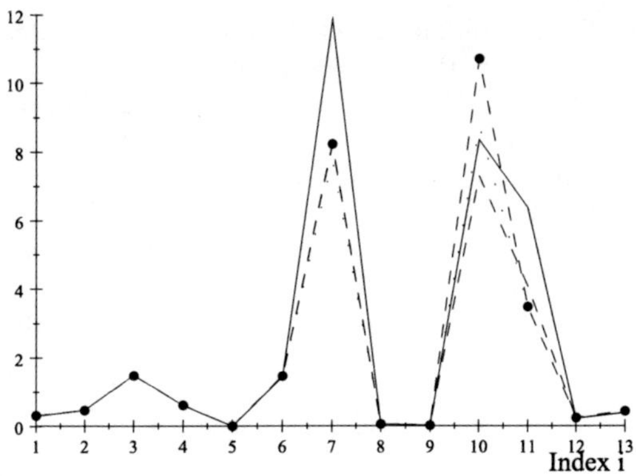

Fig. 3. $T_{n,i}^{\phi_{(a_1)},\phi_{(1)}}$ for $a_1 = -\frac{1}{2}, 0, \frac{2}{3}$ and 1. Solid line $a_1 = -\frac{1}{2}$, dashed line $a_1 = 0$, dotted line $a_1 = \frac{2}{3}$ and dash-dotted line $a_1 = 1$.

The small value of Pearson statistic $\left(B_n^{\phi_{(1)},\phi_{(0)}}\right)$ p-value shows a rather bad fit. The strong difference between Pearson test statistic and the other statistics may be, such as Fahrmeir and Tutz [7] pointed out in their book, ought to the assumptions for asymptotics may be violated. Note that the number of observations for some explicative variable is 1. Anyway, we go on with the example as Fahrmeir and Tutz [7] do since, on one hand, the bad fit suggests further investigation and on the other hand, a global measure as $B_n^{\phi_{(a_1)},\phi_{(a_2)}}$ may not detect some types of deviations of the model.

To detect 'outlying' using procedure given in Section 3 we draw the values of the statistics $T_{n,i}^{\phi_{(a_1)},\phi_{(a_2)}}$, $i = 1, ..., N = 13$ in Figures 1, 2 and 3 for $a_2 = 0$ (MLE), $a_2 = 2/3$ (minimum Cressie-Read estimator) and $a_2 = 1$ (minimum chi-square estimator), respectively.

The three figures identify observations 7 and 10 as outlying as Fahrmeir and Tutz [7]. Note that if we consider $\alpha = 0.123$, this means that at most this is the probability to reject incorrectly at least one observation. Furthermore, this means that $\boldsymbol{x}_i$ is an outlying if

$$T_{n,i}^{\phi_{(a_1)},\phi_{(a_2)}} \geq \chi^2_{1,\alpha/N} = 6.7335,$$

that is to say, only observations 7 and 10 will be considered outlying. This conclusion matches with that given in Pardo [16].

Acknowledgement. This work was supported by Grant MTM2009-06997.

References

1. Agresti, A.: Analysis of Ordinal Categorical Data. Wiley, New York (1984)
2. Ali, S.M., Silvey, S.D.: A general class of coefficients of divergence of one distribution from another. J. Royal Statist. Soc. Ser. B 26, 131–142 (1966)

3. Anderson, J.A.: Regression and ordered categorical variables (with discussion). J. Royal Statist. Soc. Ser. B 46, 1–30 (1984)
4. Anderson, J.A., Philips, P.R.: Regression, discrimination and measurement models for ordered categorical variables. Appl. Statist. 30, 22–31 (1981)
5. Cook, R.D., Weisberg, S.: Residuals and Influence in Regression. Chapman & Hall, New York (1982)
6. Cressie, N.A.C., Read, T.R.C.: Multinomial goodness-of-fit tests. J. Royal Statist. Soc. Ser. B 46, 440–464 (1984)
7. Fahrmeir, L., Tutz, G.: Multivariate Statistical Modelling Based on Generalized Linear Models. Springer, New York (2001)
8. Ferguson, T.S.: A Course in Large Sample Theory. Wiley, New York (1996)
9. Lesaffre, E., Albert, A.: Multiple-group Logistic Regression Diagnostics. Appl. Statist. 38, 425–440 (1989)
10. Liu, I., Agresti, A.: The analysis of ordered categorical data: an overview and a survey of recent developments (with discussion). TEST 14(1), 1–73 (2005)
11. McCullagh, P.: Regression models for ordinal data (with discussion). J. Royal Statist. Soc. Ser. B 42, 109–142 (1980)
12. McCullagh, P., Nelder, J.A.: Generalized Linear Models, 2nd edn. Chapman & Hall, New York (1989)
13. McKelvey, R., Zavoina, W.: A statistical model for the analysis of ordinal level dependent variables. J. Math. Sociol. 4, 103–120 (1975)
14. Miller, R.: Simultaneous Inference. McGraw Hill, New York (1966)
15. Pardo, L.: Statistical Inference Based on Divergence Measures. Chapman & Hall, New York (2006)
16. Pardo, M.C.: High leverage points and outliers in Generalized Linear Models for ordinal data. In: Skiadas, C.H. (ed.) Advances in Data Analysis, Statistics for Industry and Technology, Part 2, pp. 67–80. Birkhauser, Boston (2010)
17. Pardo, M.C.: Testing equality restrictions in Generalized Linear Models. Metrika (2010), doi:10.1007/s00184-009-0275-y
18. Pregibon, D.: Logistic regression diagnostics. Ann. Statist. 9, 705–724 (1981)
19. Vajda, I.: Theory of Statistical Inference and Information. Kluwer Acad. Pub., Dordrecht (1989)
20. Williams, D.A.: Generalized Linear Model diagnostics using the deviance and single case deletions. Appl. Statist. 36, 181–191 (1987)

Part V

Modelling in Engineering Problems

Finite Element Numerical Solution for Modelling Ground Deformation in Volcanic Areas*

María Charco[1] and Pedro Galán del Sastre[2]

[1] Instituto de Astronomía y Geodesia (CSIC-UCM),
Ciudad Universitaria, Pza. de Ciencias, 3, 28040 Madrid, Spain
mcharco@iag.csic.es

[2] Departamento de Matemática Aplicada al Urbanismo,
a la Edificación y al Medio Ambiente, E.T.S.A.
Universidad Politécnica de Madrid, Avda. Juan de Herrera 4,
28040 Madrid, Spain
pedro.galan@upm.es

Summary. Current understanding of critical stages prior to a volcanic eruption is generally based on elastostatic analysis. We investigate the elastic response of the Earth to an internal load that simulates the effect of a pressurized magma reservoir. Firstly, equations describing the Earth's deformation for elastic models are introduced and the corresponding boundary value problem is formulated in a weak sense. Then, a numerical tool to compute the displacement and stress fields produced by pressurized sources in volcanic areas is described. In doing so, we propose the Finite Element Method for simulating the deformation that Teide volcano (Tenerife, Canary Islands) would undergo, if a hypothetical magma intrusion would take place in a shallow magma reservoir beneath its summit. Furthermore, the numerical approach can be used to estimate the influence of parameters such as size, depth and shape of a pressurized reservoir, the topography and the medium heterogeneities over ground deformation modelling. Therefore, such numerical approaches can be useful to design and/or improve the geodetic monitoring system in volcanic areas.

1 Introduction

It is expected that changes within magma system leading to eruption in volcanic areas will result in precursory deformation measurable by geodetic techniques. Consequently, geodetic techniques are being used extensively to monitor ground deformation at active volcanoes. Furthermore, as part of the geodetic monitoring system, various computational methods have been proposed for modelling ground deformation since the analysis of such effects

* This work is dedicated to our friend and colleague Maria Luisa Menéndez.

is one of the most important tools for understanding the volcano processes within the Earth.

The inflation and deflation cycles that geodetic techniques support make that the Earth can be treated as an elastic solid. Within this elastic frame a variety of models have been proposed to account for the observed deformation, such as models that include spherical and ellipsoidal sources, vertical and horizontal magma migration, collapse structures, fluid migration and structural topography (e.g., [22, 24, 12, 20, 5, 13, 9]). The most commonly used is the Mogi point dilatation model [22]. It represents the simplest analytical solution for an inflating/deflating source in a homogeneous elastic half-space with free surface. However, in the case of fully 3-D rheology and/or complicated geometrical structures, a numerical method is needed.

The aim of this study is to provide a numerical tool to compute the displacement and stress fields produced by pressurized sources for studying the deformation that Tenerife (Canary Islands) would undergo, if a hypothetical overpressurization of a shallow magmatic system beneath Teide volcano would take place. In doing so, we propose the Finite Element Method (FEM). The FEM is robust and accurate when dealing with problems for which a complicated geometry requires an irregular or unstructured mesh. The description of the problem including equations and formulations is presented in Section 2. In Section 3, the numerical procedure joint with a validation test are introduced. Finally, we present the application of the methodology to Tenerife in Section 4.

2 Problem Statement

Deformation of Earth's surface reflects tectonic, magmatic, and hydrothermal processes at depth that result in strain that is transmitted to the surface through the mechanical properties of the crust. This is a key assumption behind geodetic effects modelling. Nowadays, ground deformation understanding is generally based on elastostatic analysis. In this section, we present a brief description of the equations to compute static deformation of an elastic heterogeneous Earth in response to an internal load.

2.1 Equation of Motion

Consider a solid $\Omega \subset \mathbb{R}^3$ with boundary Γ, the conservation of linear momentum states that:

$$\frac{D}{Dt}\int_{\Omega} \rho \mathbf{v} = \int_{\Gamma} \mathbf{T} + \int_{\Omega} \rho \mathbf{f}, \tag{1}$$

where $\mathbf{T}$ is the traction acting on the boundary Γ, that is related to stress, σ, via $\mathbf{T} = \sigma.\mathbf{n}$ ($\mathbf{n}$ being the unit outward normal vector), $\rho\mathbf{f}$ are body forces per unit mass (ρ being the density), $\mathbf{v}$ is the instantaneous particle velocity and

$\frac{D}{Dt}$ denotes the material time derivative operator. For an arbitrary domain Ω, the momentum balance takes the form:

$$\int_{\Omega} \left[\nabla \cdot \sigma + \rho \mathbf{f} - \rho \frac{D\mathbf{v}}{Dt} \right] = 0, \tag{2}$$

after application of the divergence and Reynolds transport theorems [19]. This must be held for all volume elements so that, at each point, we have the Cauchy's Equations of motion:

$$\nabla \cdot \sigma + \rho \mathbf{f} = \rho \frac{D\mathbf{v}}{Dt}. \tag{3}$$

We limit our discussion to slow static changes that occur over long time and to permanent offsets associated with volcanic events. Geosciences researchers do not like to label any change as static, so they often refer to these very low frequency ground movements as quasi-static ground deformation. In quasi-static processes, the stress equilibrium exists at every point at each instant of time. This join with the fact that, for small deformation, $\frac{D\mathbf{v}}{Dt} = \frac{\partial^2 \mathbf{u}}{\partial t^2}$, where $\mathbf{u}$ is the displacement field, the equation of motion (3) reduce to the quasi-static equilibrium equation:

$$\nabla \cdot \sigma + \rho \mathbf{f} = 0. \tag{4}$$

Note that this is an Eulerian description where we are neglecting the initial stress field at depth within the Earth [18].

2.2 Constitutive Law

The Equation (4) together with the constitutive relation between stress and strain is enough to describe the physics of a solid material. Considering a purely elastic Earth model,

$$\sigma_{ij} = \sum_{k,l=1}^{3} C_{ijkl} \varepsilon_{kl}, \tag{5}$$

where $1 \leq i, j \leq 3$, ε denotes the strain tensor and $\mathbf{C}$ is a positive-definite fourth order tensor of elastic coefficients satisfying the symmetries $C_{ijkl} = C_{jikl}$, $C_{ijkl} = C_{ijlk}$ and $C_{ijkl} = C_{klij}$. The first two of these are implied by the symmetry of σ and ε while the third is followed from energy considerations [19].

The fact that most materials have some internal organization helps to simplify the stress-strain relationship (5). The mathematical models we discuss represents the Earth as an ideal elastic body that is mechanically isotropic. The constitutive relation (5) for an isotropic, linearly elastic solid has the form (Hooke's law):

$$\sigma_{ij} = \lambda \sum_{k=1}^{3} \varepsilon_{kk} \delta_{ij} + 2\mu \varepsilon_{ij}, \tag{6}$$

where δ_{ij} is the Kronecker delta, μ represents the shear modulus, also called the rigidity modulus or the second Lamé coefficient, that relates shear stress to strain providing a material rigidity or stiffness under shear, and λ is the first Lamé coefficient (no physical meaning).

The constitutive relation between stress and strain help us to completely formulate the Equation (4) in terms of displacements. We assume small deformations, i.e., strain and displacement field, $\mathbf{u}$, are related as,

$$\varepsilon_{ij} = \frac{1}{2}\left(\frac{\partial u_i}{\partial x_j} + \frac{\partial u_j}{\partial x_i}\right). \tag{7}$$

Then, Equation (6) may be expressed alternatively as,

$$\sigma_{ij} = \lambda \delta_{ij} \nabla \cdot \mathbf{u} + 2\mu \left(\frac{\partial u_i}{\partial x_j} + \frac{\partial u_j}{\partial x_i}\right). \tag{8}$$

Substitution in the equilibrium equation (4) yields the Navier equation of motion in terms of displacement field, $\mathbf{u}$:

$$\mu \nabla^2 \mathbf{u} + (\lambda + \mu) \nabla (\nabla \cdot \mathbf{u}) + \rho \mathbf{f} = 0. \tag{9}$$

A complication when interpreting ground deformations in volcanic areas is the choice of the rheology of the medium. Although for many purposes it is useful to consider a purely elastic, homogeneous medium, there are cases in which such assumption may have led to misleading results (e.g. [4, 6]). In general, volcano structure involves sequences of deposition and emplacement of various materials, magma intrusion, crystallization and alteration, fracture and shallow hydrothermal systems. Thus, conceptual models of volcanic structure (based on field observations, seismic tomography and geochemical data) do not occur with the assumption of homogeneous material properties at a volcano-wide scale, i.e., the rheological behavior of the rocks can have lateral and depth variations, so that $\lambda = \lambda(\mathbf{x})$, $\mu = \mu(\mathbf{x})$, $\rho = \rho(\mathbf{x})$. In the absence of material property data for a particular volcano of interest, reasonable material property specifications can be extracted from laboratory experiments [11, 10].

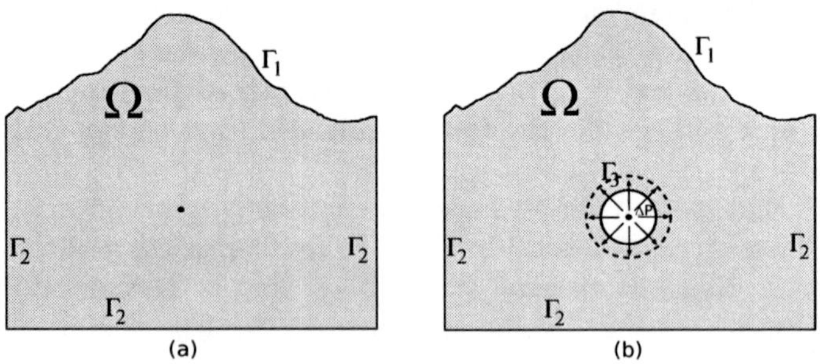

Fig. 1. 2-D section of the domain Ω: (a) for the problem (4)-(10); (b) for a model considering a spherical cavity that expands with uniform pressure ΔP

2.3 Boundary Conditions

The complete problem statement requires appropriate boundary conditions for Equation (4). We assume that the domain Ω is bounded and has a Lipschitz boundary Γ. As it is shown in Figure 1a, the boundary of the domain Ω is divided into parts Γ_1 and Γ_2. The Equation (4) is then followed by the boundary conditions:

$$\left.\begin{array}{ll}\sigma \cdot \mathbf{n} = 0 & \text{on } \Gamma_1 \\ \mathbf{u} = 0 & \text{on } \Gamma_2\end{array}\right\} \tag{10}$$

where $\mathbf{n}$ is the outward normal vector to Ω whereas $\overline{\Gamma_1 \cup \Gamma_2} = \Gamma$, $\Gamma_1 \cap \Gamma_2 = \emptyset$ and are non-empty. The first condition describes a free surface, the second corresponds to the fact that, for sufficiently large computational domain Ω, the displacement field is very small on the subterranean boundaries of Ω. These assumptions lead to a well-posed problem.

2.4 Body Force

The inflation/deflation of magma reservoirs has usually been modeled by a spherical pressurized cavity with radius a inside the medium (Figure 1b). In such a case, we should add to the boundary condition (10) the following condition on the reservoir walls:

$$\sigma \cdot \mathbf{n} = -\Delta P \text{ on } \Gamma_3, \tag{11}$$

that specifies that the normal stresses at the cavity walls are equal to a uniform pressure increment ΔP. When interpreting ground deformation at the Earth surface, these a priory constraints specify in a unique way the stress-strain distribution at depth. However, it is well known that the solution for the spherical cavity can also be obtained assuming three orthogonal force dipoles or center of dilatation (e.g.,[21]), or three orthogonal tensile dislocations (e.g., [27]). These three conceptually different source models yield the same displacement outside the source provided that the source strength is suitable chosen.

In this study we consider the inflation/deflation of a pressurized spherical cavity applying equivalent body forces. In such a case, the displacement and stress fields due to a center of dilatation are equivalent to the fields obtained by the superposition of three mutually orthogonal dipoles of identical strength, f_0, i.e.,

$$\mathbf{f} = f_0 \nabla \delta_{\mathbf{x}=\mathbf{x}'} \tag{12}$$

where $\delta_{\mathbf{x}=\mathbf{x}'}$ is the Dirac delta distribution that represents a point force at $\mathbf{x}'$ and $f_0 = a^3 \Delta P \frac{\lambda+2\mu}{\mu} \pi$ is the source strength, (a being the radius of the source). Since the Dirac delta is the limit of a sequence of Gaussian functions when their variance tends to zero, we use the body force

$$\mathbf{f} = f_0 \frac{1}{\sigma_{x_1}\sigma_{x_2}\sigma_{x_3}\pi^{3/2}} \nabla \left(\mathrm{e}^{-\left(\frac{(x_1 - x_1')^2}{\sigma_{x_1}^2} + \frac{(x_2 - x_2')^2}{\sigma_{x_2}^2} + \frac{(x_3 - x_3')^2}{\sigma_{x_3}^2} \right)} \right) \tag{13}$$

for solving (9)-(10), where $\sigma_{x_i}^2$ is the variance of the Gaussian function in x_i direction. This ensures that the body force function is sufficiently smooth. Moreover, the Gaussian function can be used to model sources of different shapes. In fact when $\sigma_{x_i} = a$ for all $i = 1, 2, 3$, (13) represents a spherical source whereas (13), for $\sigma_{x_1} = a$, $\sigma_{x_2} = b$, $\sigma_{x_3} = c$ and $f_0 = abc\Delta P\frac{\lambda+2\mu}{\mu}\pi$, can be used to describe an elliptical source with semiaxis a, b and c.

2.5 Weak Formulation

We use the notation $\mathbf{L^2}(\Omega) = \left[L^2(\Omega)\right]^3$, $\mathbf{H^1}(\Omega) = \left[H^1(\Omega)\right]^3$ throughout this work. The problem is to find the tensor $\sigma = ((\sigma_{ij}))_{1\le i,j\le 3}$ and the displacement $\mathbf{u} = (u_1, u_2, u_3) \in \mathbf{L}^2(\Omega)$ that satisfies Equations (4)-(10) for a prescribed $\mathbf{f} \in \mathbf{L}^2(\Omega)$ and $\rho \in L^\infty(\Omega)$. For this task, the boundary value problem is reformulated in a weak sense (e.g., [8]) in order to obtain reliable numerical solutions. Let us define the space $\mathbf{V}$ of the test functions, a bilinear form $a : \mathbf{V} \times \mathbf{V} \to \mathbb{R}$ and a linear functional $L : \mathbf{V} \to \mathbb{R}$ by

$$\mathbf{V} = \left\{\mathbf{v} \in \mathbf{H^1}(\Omega) : \mathbf{v}_{|\Gamma_2} = \mathbf{0}\right\}, \tag{14}$$

$$a(\mathbf{u}, \mathbf{v}) = \int_\Omega \left[\lambda(\nabla\cdot\mathbf{u})(\nabla\cdot\mathbf{v}) + 2\mu\sum_{i,j=1}^{3}\varepsilon_{ij}(\mathbf{u})\,\varepsilon_{ij}(\mathbf{v})\right], \tag{15}$$

$$L(\mathbf{v}) = \int_\Omega \rho\mathbf{f}\cdot\mathbf{v}, \tag{16}$$

respectively. The problem may now be formulated in a weak sense as follows: find $\mathbf{u} \in \mathbf{V}$ that satisfies the following variational equality,

$$a(\mathbf{u}, \mathbf{v}) = L(\mathbf{v}) \ \forall \mathbf{v} \in \mathbf{V}. \tag{17}$$

Moreover we assume $\rho, \lambda, \mu \in L^\infty(\Omega)$ are positive real-valued functions. It is straightforward to prove that problem (17) is equivalent to (4) with boundary conditions (10) applying the Green formulae.

Theorem 1. *Assume that* $\mathbf{f} \in \mathbf{L}^2(\Omega)$ *and* $\rho, \lambda, \mu \in L^\infty(\Omega)$ *are positive real-valued functions. Then, the variational problem (17) has a unique solution.*

Proof. The proof is straightforward using the Lax-Milgram theorem and Korn inequality [8].

Note that the term $a(\mathbf{u}, \mathbf{v})$ can be interpreted as the work of the internal elastic forces and $L(\mathbf{v})$ as the work of external (body and surface) forces. Thus the expression (17) is a reformulation of the virtual work theorem.

3 Numerical Solution: The Finite Element Method

The goal of this work is to develop a numerical tool to perform static calculations in order to compute the horizontal and vertical displacement and stress fields induced by pressurized sources that may be used to represent magma reservoirs. In doing so we propose the Finite Element Method.

Suppose that $\Omega \subset \mathbb{R}^3$ is an open bounded subset with Lipschitz boundary Γ and D_h is a partition of $\bar{\Omega}$ such that $D_h = \{R_j\}_{j=1}^{N_e} \subset \bar{\Omega}$, where R_j denotes a quadrilateral prism and N_e is the number of finite elements in the partition. As usual, we assume that $\bar{\Omega} = \bigcup_{j=1}^{N_e} R_j$ and the elements R_j satisfy the regularity conditions: (i) any face of R_j is either a subset of Γ or any other face of any R_i, with $i \neq j$; (ii) there exists $\alpha > 0$ such that $h_j/\varepsilon_j < \alpha$ for all $1 \leq j \leq N_e$, where $h_j = \text{diam}R_j$ and $\varepsilon_j = \sup\{\text{diam}S : S \text{ a ball contained in } R_j\}$.

Let $\hat{R} = [-1,1]^3 \subset \mathbb{R}^3$ be the reference element, then we define the set of polynomials of degree$\leq m$, with m an integer, $P_m\left(\hat{R}\right) = P_m([-1,1]) \otimes P_m([-1,1]) \otimes P_m([-1,1])$. Thus, $P_m(R_j) = \left\{\hat{p} \circ T_j^{-1} \in C(R_j) : \hat{p} \in P_m\left(\hat{R}\right)\right\}$ where we have used the continuous bijective transformation $T_j : \hat{R} \to R_j$.

Then, the finite element subspaces V_h and V_{h0} associated to the partition D_h are defined as

$$\begin{aligned} V_h &= \left\{v_h \in C\left(\bar{\Omega}\right) : v_{h|R_j} \in P_m(R_j) \text{ for all } 1 \leq j \leq N_e\right\} \\ V_{h0} &= \left\{v_h \in V_h : v_{h|\Gamma_2} = 0\right\} \end{aligned}$$

so that $\mathbf{V}_h = V_h \times V_h \times V_h$ and $\mathbf{V}_{h0} = V_{h0} \times V_{h0} \times V_{h0}$.

Let $N = \dim V_{h0}$ and $\{\varphi_i\}_{i=1}^{N}$ a basis of V_{h0}, then we shall denote

$$\boldsymbol{\psi}_i = \begin{cases} (\varphi_i, 0, 0), & \text{if } 1 \leq i \leq N \\ (0, \varphi_{i-N}, 0), & \text{if } N < i \leq 2N \\ (0, 0, \varphi_{i-2N}), & \text{if } 2N < i \leq 3N \end{cases}$$

so that $\{\boldsymbol{\psi}_i\}_{i=1}^{3N}$ is a basis of $\mathbf{V}_{h0}$. Thus, the finite element solution of (4) with the boundary conditions (10) is computed by solving the variational formulation: find $\mathbf{u}_h \in \mathbf{V}_{h0}$ such that

$$a(\mathbf{u}_h, \mathbf{v}_h) = L(\mathbf{v}_h) \ \forall \mathbf{v}_h \in \mathbf{V}_{h0},$$

or equivalently: find $\mathbf{u}_h \in \mathbf{V}_{h0}$ such that

$$a(\mathbf{u}_h, \boldsymbol{\psi}_k) = L(\boldsymbol{\psi}_k) \ \forall k = 1, 2, \ldots 3N \tag{18}$$

We define the stiffness matrix $K \in M_{3N\times 3N}(\mathbb{R})$ and the real vector $\mathbf{b} \in \mathbb{R}^{3N}$ such that $k_{ij} = a(\boldsymbol{\psi}_i, \boldsymbol{\psi}_j)$ and $b_i = L(\boldsymbol{\psi}_i)$. Since $\mathbf{u}_h = \sum_{i=1}^{3N} u_i \boldsymbol{\psi}_i$, the problem (18) becomes: find $\mathbf{u} = (u_i)_{i=1}^{3N} \subset \mathbb{R}^{3N}$ such that $K\mathbf{u} = \mathbf{b}$.

Note that the stiffness matrix K is symmetric and positive definite since the bilinear form a is symmetric and coercive [8], thus we can use the conjugate gradient algorithm to compute the numerical solution.

3.1 Validation of the Numerical Procedure: The Mogi Model

Validation is performed on a problem with a known analytical solution which is the problem of a small pressurized spherical cavity embedded in an elastic half-space [22]. In such a model all the perturbation quantities tends to zero as $|\mathbf{x}| \to \infty$. The most useful solution is an approximate one given by [20] that holds when the source is small compare to its depth.

Displacements calculated analytically are compared to numerical results found using FEM. Note that the FEM solution is computed in a $200 \times 200 \times 50$ km^3 domain that approximate the conditions that the displacement field satisfies in the boundary of the Mogi model. The displacements are caused by a center of dilatation of 50 MPa km^3 strength located at 4 km depth in a homogeneous medium with 50 GPa for Young modulus and 0.21 for Poisson ratio. These corresponds to average for basalts values given by [16]. The variance of the Gaussian function (13) that represents the center of dilatation is chosen according to the mesh element size. In this case, $\sigma_{x_i} = 50$ m for $i = 1, 2, 3$. Figure 2 shows that the homogeneous FEM solution agrees with the displacement predicted by Mogi analytical solution confirming the reliability of the chosen mesh and boundary conditions. Since the domain for obtaining the solution of the Mogi model is different, we cannot compare both solutions at the exact source location $(0, 0, 4)$ km. In fact, since the source is assumed to be small (point-like in the ideal situation), the deformation of the interior is typically ignored with the Mogi model. Nevertheless, even in the proximity of the source location there is a good agreement between both solutions (Figure 2c).

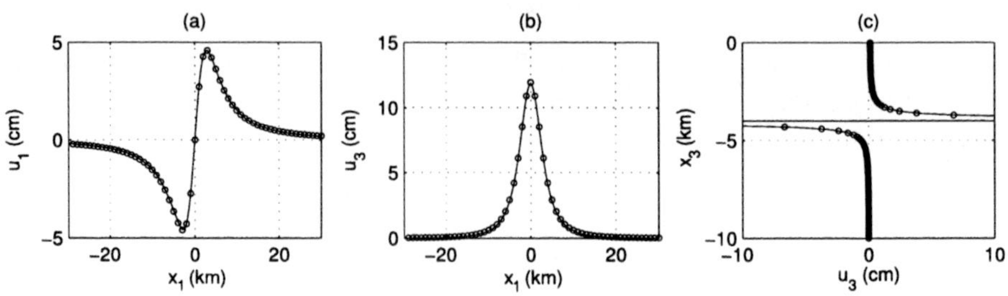

Fig. 2. Model comparison of (a) surface horizontal displacements, $u_1(x_1, 0, 0)$, (b) surface vertical displacements, $u_3(x_1, 0, 0)$, and (c) vertical displacements, $u_3(0, 0, x_3)$ between Mogi's analytical solution (circles) and FEM solution (solid line). The displacements are caused by a center of dilatation of 50 MPa km^3 strength located at 4 km depth in a homogeneous medium.

4 Application: Tenerife (Canary Islands)

Tenerife is the largest island of the Canary archipelago where volcanism has continued over the last 30 Myr. Teide-Pico Viejo complex, which remains active nowadays, dominates the eruptive system of the island [1, 2]. Teide and Pico Viejo are two large stratovolcanoes that overlap to form an elongated doubled edifice. The highest altitude corresponds to the youngest summit of Teide (3718 m). The last explosive eruption in the island occurred in Montaña Blanca, located on the Teide flanks, 2000 years ago. Therefore, due to the hazard of this kind of volcanic activity, we consider the possibility of eruption in the area of this emission center. To show the application possibilities of the methodology described above to simulate ground deformation in volcanic areas, we present here a 3-D finite element numerical solution for the volcanic island of Tenerife. The goal of the simulation is to estimate the deformation that would undergo the island, if a hypothetical overpressurization of a shallow magmatic system beneath Teide would take place.

In this application the complexities taken into account are the topography and the medium heterogeneities. A system of cartesian coordinates with the origin located at the sea level, just below the Teide summit, is assumed. In this case x_1 and x_2 axis are orientated along WE and SN directions respectively, and x_3 axis points up out of the medium. The 3-D computational domain is a volume extending $200 \times 200\ \text{km}^2$ in the x_1 and x_2 direction and from 100 km below sea level to the Earth surface in the x_3 direction. The whole island of Tenerife is covered. The boundary at the ground surface is constructed from a Digital Elevation Model provided by the Instituto Geográfico Nacional (IGN), and a bathymetry model from the 1-minute global elevation database [26]. The uppermost part of the domain is considered as

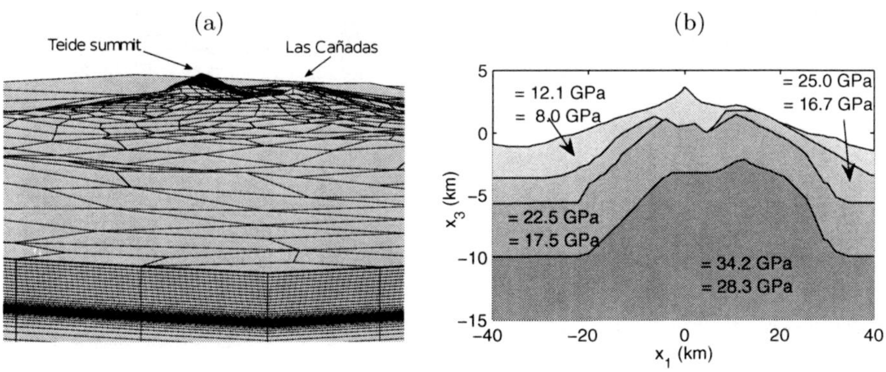

Fig. 3. (a) Detail of the mesh of the computational domain with a spatial resolution of 50 m in the submit area and around the source location and coarse at greater distances; (b) a vertical profile at $x_2 = 0$ of the 3-D elastic parameter model for Tenerife island

a free surface. The displacements of the outermost lateral boundaries and on the bottom are fixed to zero representing the tendency of the displacement field at infinity. The computational domain was meshed into 88055 quadrilateral prism elements and 91000 nodes, see Figure 3a. This mesh is locally refined through the vicinity of the chamber and near the Earth's surface and takes into account most important morphologic features of the island. The source of 50 MPa km^3 is located at 4 km depth below Teide volcano summit, i.e., approximately 300 m below sea level. An important aspect concerns to the crustal properties. Figure 3b shows a vertical WE profile of the 3-D model of elastic parameters for this study. The elastic parameters were estimated using density structures determined by means of seismic and gravity data [7, 28, 17]. We assume an empirical Nafe and Drake relationships between the compressional velocity and density [23] and the Poisson ratio is interpolated from PREM [15]. Since there is no analytical solution for validation, convergence test were performed in order to include medium heterogeneities. In this application ground displacements at nodes are calculated as primary solutions. However, strain and stress can also be obtained as derived solutions.

The results of this application are illustrated in Figure 4a,b,c. Most of the deformation is restricted to the vicinity of the volcano since at distances

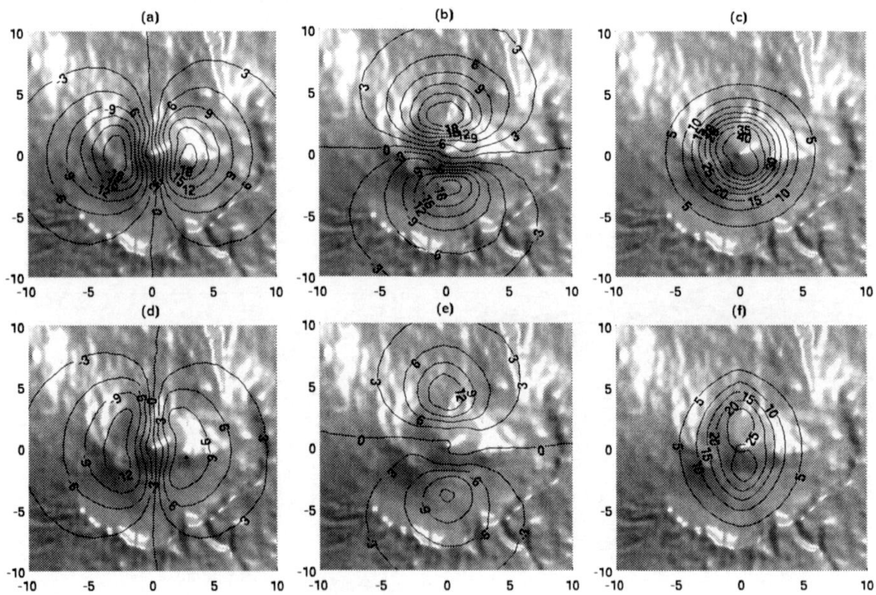

Fig. 4. Surface displacement field (cm) caused by sources of 50 MPa km^3 strength located at 4 km depth below Teide volcano summit considering medium heterogeneities: (a) u_{x_1}, (b) u_{x_2} and (c) u_{x_3}, caused by a spherical source; (d) u_{x_1}, (e) u_{x_2} and (f) u_{x_3}, caused by a elliptical source

greater than 10 km from the volcano summit, the related deformation might be indiscernible from background noise. In particular, most of the predicted displacements lie inside the caldera walls. The largest component of the displacement field is the vertical one that takes a maximum value of 43 cm at the volcano summit area. Given the precision attainable nowadays by permanent GPS (Global Positioning System) networks ($0.2 - 0.6$ cm and $0.5 - 1$ cm for horizontal and vertical coordinates respectively [14, 3]) and by GPS campaigns ($1 - 3$ cm [25]) such a kind of monitoring systems would be able to detect the magnitude of the displacement field due to the shallow source described above. In the elastic case the displacement field is proportional to the pressure change of the source. Therefore, considering the precision of that geodetic techniques, even a pressure change of 5 MPa at 4 km depth below Teide volcano summit would be detected.

Comparing analytical solutions in a half-space, as the one showed in Figure 2, with the numerical solution considering the topography of the island results in changes in the displacement field pattern. In fact, the vertical displacement in $x_1 = x_2 = 0$ at the surface of the island (corresponding to the top of the edifice) is actually found to be a local minimum (Figure 4c), while in the half-space example u_3 reaches its maximum value (Figure 2b). This result is in agreement with the computations of the Indirect Boundary Element Method by [9].

The displacement field depends on several parameters such as size, depth and shape of the overpressurized reservoir, reservoir pressure change, the topography and the heterogeneities of the medium. Different tests have been performed in order to evaluate the influence of some of these parameters on the elastic solutions. First, we perform a test, to estimate the influence of the shape of the chamber, considering an ellipsoidal reservoir with $a = 1/3$ km (x_1-axis), $b = 3$ km (x_2-axis) and $c = 1$ km (x_3-axis) . This choice ensures that both, the spherical source that we have used in the other test and the ellipsoidal source, have the same volume. Important discrepancies are observed when considering the ellipsoidal source (Figure 4d,e,f). The magnitude of both, horizontal and vertical displacements, is reduced with respect to the results showed at Figure 4a,b,c. Comparing both Figure 4a,b,c and Figure 4d,e,f, we can see that the uplift pattern is elongated through NS and consequently quasi-elliptical in shape. Such effects are due to the redistribution of ΔP according to the source shape.

In order to test the influence of the medium heterogeneities, a numerical solution is obtained considering a homogeneous medium with the same elastic parameters as the domain area where the source is embedded in the heterogeneous model (Figure 5a,b,c). The magnitude of the vertical and horizontal displacements is reduced with respect to the one provided by the heterogeneous model. In the heterogeneous model the upper part of the domain is softer than the lower one where the source is located (Figure 3b). Surface displacements are affected mostly by the elastic properties of the domain area between the source and the surface. Therefore, this test shows that

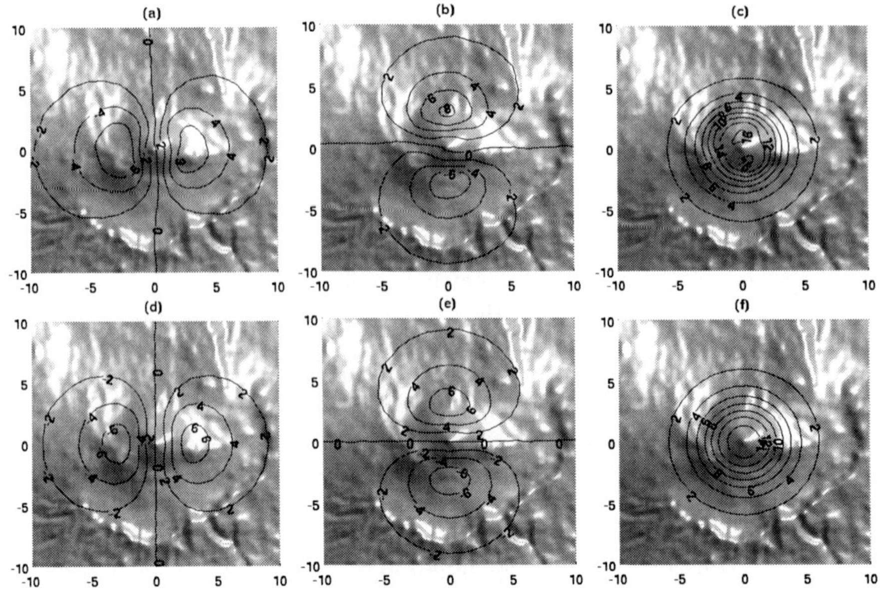

Fig. 5. Surface displacement field (cm) caused by sources of 50 MPa km^3 strength located at 4 km depth below Teide volcano summit considering a homogeneous medium: (a) u_{x_1}, (b) u_{x_2} and (c) u_{x_3}, assuming the real topography of Tenerife island; (d) u_{x_1}, (e) u_{x_2} and (f) u_{x_3}, assuming the island as a cone with the same height of Teide volcano and 16.5° slope

the cumulative effect of heterogeneities above the source are important when interpreting ground deformation.

A 3-D axisymmetrical cone with an approximate topography has also been studied (Figure 5d,e,f). The island is assumed to be an axisymmetrical cone with height equal to that of Teide and with average slope of the flanks of 16.5°. In this case, we consider again the homogeneous medium described above. The comparison between Figure 5d,e,f and the results obtained by the 3-D realistic topography (Figure 5a,b,c) illustrates how both are similar in magnitude, comparing with the results obtained by considering medium heterogeneities. However, the real topography alters the axisymmetrical pattern of the displacement field caused by a spherical source located under axisymmetrical volcanoes.

Since surface deformation can be ascribed to a wide variety of tectonic, magmatic, hydrothermal and shallow processes, numerical modelling of ground deformation at volcanic areas provides further insights into the mechanism as well as valuable information on the dependence of the results on rheology and structural features of the medium. Thus the proposed approaches may be used as preliminary pictures to design or improve the geodetic monitoring system in Tenerife (Canary Islands).

5 Concluding Remarks

We have developed a 3-D numerical model by using the Finite Element Method to simulate the displacement and stress fields predicted by pressurized reservoirs at depth. We have focused on volcanic context because volcanic activity produces deformation that can be precursors of future eruptions. A theoretical study is provided in order to simulate deformation at Teide volcano (Tenerife, Canary Islands). The study assumes that the displacement field is caused by the presence of a shallow magmatic system. In view of the results we expect that changes within the magmatic system leading to eruption will result in precursory deformation measurable by GPS networks.

The 3-D FEM model implemented for modelling elastic deformation has the next properties:

- Real surface topographies can be taken into account,
- The perturbation due to inflation/deflation of reservoirs of any shape can be modeled,
- The model can include structural features of the medium such as faults or fractures,
- The vertical and lateral heterogeneities of the medium can be considered through a density model.

In summary, the proposed model shows that the medium and source features affect the magnitude and the pattern of the deformation field. Therefore we point out that structural features and heterogeneities of the medium as well as complex configurations can influence the estimation of source parameters in volcanic areas, which is crucial for both the correct interpretation of geodetic data and the correct evaluation of volcanic crisis.

Acknowledgement. The research of the first author has been support by 200930I053 MICINN-CSIC Grant. The second author work has been partially funded by Grant CGL2007-66440-C04-01 from Ministerio de Educación y Ciencia de España. The numerical codes used in this work are freely available through authors' request.

References

1. Ablay, G., Ernst, G., Martí, J., Sparks, R.S.J.: The 2ka subplinian eruption of Montaña Blanca, Tenerife. Bull Volcanol. 57, 337–355 (1995)
2. Ablay, G., Carroll, M.R., Palmer, M.R., Martí, J., Sparks, R.S.J.: Basanite-phonolite lineages of Teide-Pico Viejo volcanic complex, Tenerife, Canary Islands. J. Petrol 39, 905–936 (1998)
3. Bartel, B.A., Hamburger, M.W., Mertens, C.M., Lowry, A.R., Corpuz, E.: Dynamics of active magmatic and hydrothermal systems at Taal Volcano, Philippines, from continuous GPS measurements. J. Geophys. Res. 108(10), ECV4-1–ECV4-15 (2003)

4. Berrino, G., Corrado, G., Luongo, G., Toro, B.: Ground deformation and gravity changes accompanying the 1982 Pozzuoli uplift. Bull. Volcanol. 44, 187–200 (1984)
5. Bonafede, M.: Axi-symmetric deformation of thermo-poro-elastic halfspace: Inflation of a magma chamber. Geophys. J. Int. 103, 289–299 (1990)
6. Bonafede, M., Dragoni, M., Quareni, F.: Displacement and stress field produced by a centre of dilatation and by a pressure source in a viscoelastic half-space: application to the study of ground deformation and seismic activity at Campi Flegrei, Italy. Geophys. J. R. Astr. Soc. 87, 455–485 (1986)
7. Boshard, E., MacFarlane, D.J.: Crustal structure of the western Canary Island from seismic refraction and gravity data. J. Geophys. Res. 75, 4901–4918 (1970)
8. Brenner, S.C., Scott, R.: The Mathematical Theory of Finite Element Methods. Springer, New York (1994)
9. Charco, M., Luzón, F., Fernández, J., Tiampo, K.F., Sánchez-Sesma, F.J.: Three dimensional indirect boundary element method for deformation and gravity changes in volcanic areas: Application to Teide volcano (Tenerife, Canary Islands). J. Geophys. Res. 112(B8), B08409.1–B08409.17 (2007)
10. Christensen, N.I.: Poisson's ratio and crustal seismology. J. Geophys. Res. 101, 3139–3156 (1996)
11. Christensen, N.I., Mooney, W.: Seismic velocity structure and composition of continental crust: A global view. J. Geophys. Res. 100, 9761–9788 (1995)
12. Davis, P.M.: Surface deformation due to inflation of an arbitrary oriented triaxial ellipsoidal cavity in an elastic halfspace, within reference to Kilauea volcano, Hawaii. J. Geophys. Res. 91, 7429–7438 (1986)
13. De Natale, G., Pingue, F.: Ground deformation modeling in volcanic areas. In: Scarpa, R., Tilling, R. (eds.) Monitoring and Mitigation of Volcanic Hazards, pp. 365–388. Springer, Heidelberg (1996)
14. Dixon, T., Mao, A., Bursik, M., Heflin, M., Langbein, J., Stein, R., Webb, F.: Continuous monitoring of surface deformation at Long Valley Caldera, California, with GPS. J. Geophys. Res. 102, 12017–12034 (1997)
15. Dziewonski, A.M., Anderson, D.L.: Preliminary Reference Earth Model. Phys. Earth Planet. Int. 25, 297–356 (1981)
16. Goodman, R.E.: Introduction to Rock Mechanics. Willey, New York (1989)
17. Gottsmann, J., Camacho, A.G., Martí, J., Wooller, L., Fernández, J., García, A., Rymer, H.: Shallow structure beneath the Central Volcanic Complex of Tenerife from new gravity data: Implications for its evolution and recent reactivation. Phys. Earth Planet Int. 168, 212–230 (2008)
18. Love, A.E.H.: Some problems in Geodynamics. Cambridge Univ. Press, New York (1911)
19. Malvern, L.E.: Introduction to the Mechanics of a Continuous Medium. Prentice-Hall, Englewood Cliffs (1969)
20. McTigue, D.F.: Elastic stress and deformation near a finite spherical magma body: resolution of a point source paradox. J. Geophys. Res. 92, 12931–12940 (1987)
21. Mindlin, R.D.: Force at a point in the interior of a semi-infinite solid. Physics 7, 195–202 (1936)
22. Mogi, K.: Relations of eruptions of various volcanoes and deformation around them. Bull Earthquake Res.Inst. Univ. Tokyo 36, 99–134 (1958)

23. Nafe, J.E., Drake, C.L.: Physical properties of marine sediments. In: Hill, N. (ed.) The sea, vol. 3, pp. 794–815. Interscience, New York (1963)
24. Rundle, J.B.: Static elastic-gravitational deformation of a layered half-space by point couple-sources. J. Geophys. Res. 85, 5355–5363 (1980)
25. Segall, P., Davis, J.: GPS applications for geodynamics and earthquake studies. Annu. Rev. Earth Planet Sci. 25, 301–336 (1997)
26. Smith, W.H.F., Sandwell, D.T.: Global seafloor topography from satellite altimetry and ship depth soundings. Science 277, 1956–1962 (1997)
27. Steketee, J.A.: On Volterra's dislocations in a semi-infinite elastic medium. Can J. Phys. 36, 192–205 (1958)
28. Watts, A.B., Peire, C., Collier, J., Dalwood, R., Canales, J.P., Henstock, T.J.: A seismic study of lithosferic flexure in the vicinity of Tenerife, Canary Islands. Earth Planet Sci. Lett. 146(3-4), 431–447 (1997)

A Finite Volume Scheme for Simulating the Coupling between Deep Ocean and an Atmospheric Energy Balance Model*

Arturo Hidalgo[1] and Lourdes Tello[2]

[1] Dpto. Matemática Aplicada y Métodos Informáticos, E.T.S.I. Minas, Universidad Politécnica de Madrid, Spain
arturo.hidalgo@upm.es

[2] Dpto. Matemática Aplicada, E.T.S. Arquitectura, Universidad Politécnica de Madrid, Spain
l.tello@upm.es

Summary. In this work we consider a model including the coupling surface/deep ocean first proposed in [20]. It is a diagnostic model which can be used to understand the long-term climate evolution. The unknown is the temperature over each parallel and the effect of the deep ocean on the Earth surface temperature is considered. One of the difficulties of this problem is the dynamic and diffusive boundary condition. The purpose of this work is to approximate the solutions by a finite volume scheme. We also compare the solution of the studied model with the solution of an energy balance model without deep ocean effect.

1 The Model

A great attention is being paid to problems related to climate change because of its socio-economic and ecological implications. Climatology is a source of research problems in many fields ranging from Geology to Mathematics. Climate system is very complex and involves many components and complicate mechanisms. Different climate models consider only a few of these components to understand only some of the mechanisms. In this section we describe a mathematical model which describes the coupling surface/deep ocean.

The model represents the evolution of the temperature inside an ocean of depth H. The spatial domain considered is $\Omega = (-1, 1) \times (-H, 0)$ where the spatial variables (x, z) represent $x = sin(latitude)$ and $-z = depth$. The boundary of Ω is denoted by $\Gamma_H \cup \Gamma_0 \cup \Gamma_1 \cup \Gamma_{-1}$, where $\Gamma_H = \{(x,z) \in \overline{\Omega} : \ z = -H\}$, $\Gamma_0 = \{(x,z) \in \overline{\Omega} : \ z = 0\}$, $\Gamma_1 = \{(x,z) \in \overline{\Omega} : x = 1\}$, $\Gamma_{-1} = \{(x,z) \in \overline{\Omega} : x = -1\}$ and $\overline{\Omega}$ denotes the closure of Ω.

* This work is dedicated to our colleague and friend María Luisa Menéndez.

The governing equation for the ocean interior is given by

$$U_t - \left(\frac{K_H}{R^2}(1-x^2)U_x\right)_x - K_V U_{zz} + wU_z = 0 \quad \text{in } \Omega \times (0,T), \tag{1}$$

where the subscript x, z or t denotes partial derivation with respect to the variable x, z or t, $U = U(x,z,t)$ represents the temperature, K_H and K_V are thermal conductivities in x and z direction respectively, w is the velocity, assumed vertical, R is the radius of the Earth and T is the final simulation time. In this model the temperature is assumed to be constant over each parallel, therefore it only depends on latitude and depth. In the rest of the paper we shall denominate the equation (1) as DOM (Deep Ocean Model).

Concerning the boundary conditions for the ocean bottom, Γ_H, we have

$$wx\frac{\partial U}{\partial x} + K_V\frac{\partial U}{\partial z} = 0 \qquad \text{on } \Gamma_H \times (0,T). \tag{2}$$

Budyko [3] and Sellers [14] formulated one layer thermodynamic models of the Earth's zonally averaged, mean annual surface temperature field as a balance. Both models include one important nonlinear mechanism: ice albedo feedback. The boundary condition in Γ_0 is based on such a balance:

$$Du_t - \frac{DK_{H_0}}{R^2}\left((1-x^2)^{\frac{p}{2}}|u_x|^{p-2}u_x\right)_x + Bu + C + K_V\frac{\partial U}{\partial n} + wxu_x \in \frac{1}{\rho c}QS(x)\beta(u). \tag{3}$$

According to Budyko's model, $Bu + C$ represents the emitted energy by cooling (that is, the Newton's cooling law) with B and C representing cooling parameters (assumed constant in time and space), D is the depth of the mixed layer, K_{H_0} is the horizontal thermal diffusivity in the mixed layer, Q is a solar parameter, $S(x)$ is an insolation function and β is the co-albedo. The coalbedo represents the fraction of the incoming radiation flux which is absorbed by the surface. Coalbedo is eventually discontinuous in u and it is introduced in the equation as a continuous graph in order to get a well posed problem.

In the rest of the paper we shall denominate the equation (3) as EBM (Energy Balance Model). Let us remark that upper-case letter U is used for the deep ocean temperature while lower-case u represents surface temperature, this means that $U|_{\Gamma_0} = u$.

Regarding the diffusive part in the equation (3) we have followed the ideas proposed in [16] where nonlinear diffusion is considered and it is given by the term $\operatorname{div}(|\nabla u|\nabla u)$, which after being changed into spherical coordinates and assuming u is constant over each parallel give raise to

$$\left((1-x^2)^{\frac{3}{2}}|u_x|\,u_x\right)_x,$$

where x is the sine of the latitude. We notice that this is a particular case of the p-Laplacian operator, $\Delta_p u := \operatorname{div}(|\nabla u|^{p-2}\nabla u)$, with $p = 3$. From (1), (2) and (3) and using $p = 3$ we get the final expression of the model:

$$
\begin{aligned}
&U_t - \left(\frac{K_H}{R^2}(1-x^2)U_x\right)_x - K_V U_{zz} + wU_z = 0 \quad \text{in } \Omega \times (0,T),\\
&wxU_x + K_V U_z = 0 \quad \text{in } \Gamma_H \times (0,T),\\
&Du_t - \frac{DK_{H_0}}{R^2}\left((1-x^2)^{3/2}|u_x|\,u_x\right)_x + K_V\frac{\partial U}{\partial n}\\
&+ wxu_x + Bu + C \in \frac{1}{\rho c} QS(x)\beta(u) \quad \text{in } \Gamma_0 \times (0,T),\\
&(1-x^2)^{3/2}|U_x|\,U_x = 0 \quad \text{in } (\Gamma_{-1}\times(0,T)) \cup (\Gamma_1 \times (0,T)),\\
&U(0,x,z) = U_0(x,z) \quad \text{in } \Omega,\\
&u(0,x,0) = u_0(x) \quad \text{in } \Gamma_0,
\end{aligned}
\tag{4}
$$

where we have added boundary conditions also in vertical borders.

Structural hypotheses

(H_1) β is a bounded maximal monotone graph, i.e. $|v| \le M \quad \forall v \in \beta(s)$, $s \in \mathbb{R}$.
(H_2) $S : \Gamma_0 \to \mathbb{R}$, $s_1 \ge S(x) \ge s_0 > 0$ a.e. $x \in \Gamma_0$.
(H_3) $w \in C^1(\Omega)$.
(H_4) The constants B, C, R, Q, ρ, c, K_V, K_H and K_{H_0} are positive.

The existence of solutions of this problem under the previous hypotheses is proved in [7] by fixed point arguments and extended to higher dimension in [8]. One of the model main features is its high sensitivity to variation of parameters. Multiplicity of steady states depending on the parameter Q was studied in [9].

Remark 1. The term $K_V \frac{\partial U}{\partial n}$ stands for the coupling atmosphere-ocean in the sense of analyzing the influence of the ocean temperature in the atmosphere. In this work we shall show results with and without this term.

Many works are dedicated to the mathematical treatment of global climate energy balance models (one layer), among them, we mention [5] and the references there in, [12], [13], [21] and [10]. In [2] a finite element approach is given to a 2D climate energy balance model without deep ocean effect.

2 Numerical Approximation

In this section we are interested in computing a numerical solution for the problem (4). To start with we rewrite this problem as advection-reaction-diffusion equations, both for the upper boundary EBM and for the DOM. In particular, for the EBM equation we have

$$u_t - \left(f\left(x, u(x,t), u_x(x,t)\right)\right)_x = \sigma\left(x, u(x,t), \frac{\partial U}{\partial n}(x,0,t)\right), \tag{5}$$

where we have used the flux

$$f\left(x, u(x,t), u_x(x,t)\right) := \frac{K_{H_0}}{R^2}(1-x^2)^{3/2}\left|u_x(x,t)\right| u_x(x,t) - \frac{w}{D} x u(x,t), \tag{6}$$

and the source

$$\sigma\left(x, u, \frac{\partial U}{\partial n}\right) := \frac{1}{D}\left(-C + \frac{Q}{\rho c} S(x)\beta(u) + (\omega + x\omega_x - B)u(x,t) - K_V \frac{\partial U}{\partial n}\right). \tag{7}$$

DOM reads

$$U_t(x,z,t) - (F(x, U_x(x,z,t)))_x - (G(U(x,z,t), U_z(x,z,t)))_z = \gamma(x, U(x,z,t)), \tag{8}$$

where

$$F\left(x, U_x(x,z,t)\right) := \frac{K_H}{R^2}\left(1-x^2\right) U_x(x,z,t), \tag{9}$$

$$G(U(x,z,t), U_z(x,z,t)) := K_V U_z(x,z,t) - wU(x,z,t), \tag{10}$$

$$\gamma(x, U(x,z,t)) := \omega_z U(x,z,t). \tag{11}$$

In this work we have obtained a numerical solution of this problem using the finite volume method with Weighted Essentially Non-Oscillatory (WENO) reconstruction in space and third-order Runge-Kutta TVD for time integration. For each time step, a numerical solution of the EBM model equation, (3), is computed and then used as a Dirichlet boundary condition for the DOM, given by (1). In the following subsections we briefly describe the numerical scheme developed.

2.1 The Finite Volume Framework

In this section we construct a semi-discrete finite-volume scheme for both the 1D part of the problem as formulated in equation (5), and the 2D part, which is formulated in (8). To start with, we consider the upper boundary condition, equation (5). Let us discretize the 1D domain $[-1,1]$ in N_x intervals, denominated as control volumes. Then, we integrate equation (5) over each control volume and divide by its length. That is, given one general control volume $S_i = \left[x_{i-\frac{1}{2}}, x_{i+\frac{1}{2}}\right]$ of dimension $\Delta x_i = x_{i+\frac{1}{2}} - x_{i-\frac{1}{2}}$ we integrate equation (5) in S_i and divide it by its length Δx_i to obtain the following ordinary differential equation (ODE)

$$\frac{du_i(t)}{dt} = \frac{1}{\Delta x_i}\left(f_{i+\frac{1}{2}} - f_{i-\frac{1}{2}}\right) + \sigma_i(t) \equiv l_i(u(t)), \tag{12}$$

where

$$u_i(t) = \frac{1}{\Delta x_i}\int_{x_{i-\frac{1}{2}}}^{x_{i+\frac{1}{2}}} u(x,t)dx\ , \tag{13}$$

is the spatial cell average of the solution $u(x,t)$ in the control volume S_i at time t,

$$f_{i+\frac{1}{2}} = f\left(x_{i+\frac{1}{2}}, u\left(x_{i+\frac{1}{2}}, t\right), u_x\left(x_{i+\frac{1}{2}}, t\right)\right), \tag{14}$$

is the right interface numerical flux at time t, and

$$\sigma_i(t) = \frac{1}{\Delta x_i}\int_{x_{i-\frac{1}{2}}}^{x_{i+\frac{1}{2}}} \sigma\left(x, u, \frac{\partial U}{\partial n}\right) dx\ , \tag{15}$$

is the spatial average of the source term $\sigma(u(x,t))$ in the control volume S_i at time t.

In a similar way we apply a finite volume scheme for the DOM. We discretize the 2D domain $[-1,1]\times[0,-H]$ in $N_x \cdot N_z$ rectangular cells, also denominated as control volumes. Let $V_{i,j}$ be one of these 2D control volumes of dimensions $\Delta x_i \times \Delta z_j$ where $\Delta x_i = x_{i+\frac{1}{2}} - x_{i-\frac{1}{2}}$ and $\Delta z_j = z_{j+\frac{1}{2}} - z_{j-\frac{1}{2}}$. We integrate the equation in this control volume to yield

$$\frac{dU_{i,j}(t)}{dt} = \frac{1}{\Delta x_i}\left(F_{i+\frac{1}{2},j} - F_{i-\frac{1}{2},j}\right) + \frac{1}{\Delta z_j}\left(G_{i,j+\frac{1}{2}} - G_{i,j-\frac{1}{2}}\right) + \Gamma_{ij} \equiv L_{ij}(t), \tag{16}$$

where

$$U_{i,j}(t) = \frac{1}{\Delta x_i \Delta z_j}\int_{x_{i-\frac{1}{2}}}^{x_{i+\frac{1}{2}}}\int_{z_{j-\frac{1}{2}}}^{z_{j+\frac{1}{2}}} U(x,z,t)dzdx\ , \tag{17}$$

is the cell average of the unknown inside the cell V_{ij}, while the value $F_{i+\frac{1}{2},j}$ is the right intercell numerical flux in $x-$direction and $G_{i,j+\frac{1}{2}}$ is the upper intercell numerical flux in $z-$direction at time t, and

$$F_{i+\frac{1}{2},j} = \frac{1}{\Delta z_j}\int_{z_{j-\frac{1}{2}}}^{z_{j+\frac{1}{2}}} F\left(x_{i+\frac{1}{2}}, U_x\left(x_{i+\frac{1}{2}}, z, t\right)\right) dz, \tag{18}$$

$$G_{i,j+\frac{1}{2}} = \frac{1}{\Delta x_i}\int_{x_{i-\frac{1}{2}}}^{x_{i+\frac{1}{2}}} G\left(U\left(x, z_{j+\frac{1}{2}}, t\right), U_z\left(x, z_{j+\frac{1}{2}}, t\right)\right) dx, \tag{19}$$

are the spatial average of physical fluxes over cell faces at time t and

$$\Gamma_{ij} = \frac{1}{\Delta x_i \Delta z_j} \int_{x_{i-\frac{1}{2}}}^{x_{i+\frac{1}{2}}} \int_{z_{j-\frac{1}{2}}}^{z_{j+\frac{1}{2}}} \gamma(x, U(x,z,t)) dz dx, \tag{20}$$

is the spatial average of the source term $\gamma(x, U(x,z,t))$ over the control volume V_{ij}.

The numerical solution of the EBM given by (12) and the DOM given by (16) may be advanced in time by means of, for instance, a TVD Runge-Kutta method. The one we have used in this paper is the third-order method, as described in [15, 17], whose expressions are

$$\eta^{k,1} = \eta^n + \Delta t \Lambda(\eta^n), \qquad \eta^{k,2} = \frac{3}{4}\eta^n + \frac{1}{4}\eta^{k,1} + \frac{1}{4}\Delta t \Lambda(\eta^{k,1}),$$

$$\eta^{k+1} = \frac{1}{3}\eta^n + \frac{2}{3}\eta^{k,2} + \frac{2}{3}\Delta t \Lambda(\eta^{k,2}), \tag{21}$$

where $\eta^n := u(x_{i+\frac{1}{2}}, t^n)$ for the EBM and $\eta^n := U(x_{i+\frac{1}{2}}, t^n)$ for the DOM. Moreover the operator $\Lambda(\cdot)$ is the operator $l(\cdot)$ for the EBM and the operator $L(\cdot)$ for the DOM part.

The process we have applied to solve the problem can be summarized as follows:

1. Compute the initial cell averages of the solution, both for the EBM and the DOM. Therefore we have the values

$$u_i^0 = \frac{1}{\Delta x_i} \int_{x_{i-\frac{1}{2}}}^{x_{i+\frac{1}{2}}} u(x,0) dx, \qquad U_i^0 = \frac{1}{\Delta x_i \Delta z_j} \int_{x_{i-\frac{1}{2}}}^{x_{i+\frac{1}{2}}} \int_{z_{j-\frac{1}{2}}}^{z_{j+\frac{1}{2}}} U(x,0) dz dx. \tag{22}$$

2. For each time step:
   a) Solve the EBM, according to the following steps:
      i) Compute the intercell numerical fluxes using WENO technique, which will be briefly explained in the next subsection.
      ii) Use the intercell numerical fluxes (14) to obtain the operator $l(\cdot)$ for the EBM using (12).
      iii) Solve the ODE (12) using third-order TVD Runge-Kutta scheme (21) to obtain the cell averages of the numerical solution of the EBM, u_i^n.
   b) Solve the DOM using the solution of the EBM, u_i^n, as Dirichlet boundary condition at the upper boundary. The steps to follow are:
      i) Compute the intercell numerical fluxes using WENO technique.
      ii) Use the numerical fluxes (18), (19), to obtain the operator $L(\cdot)$ for the DOM using (16).
      iii) Solve the ODE (16) using third-order TVD Runge-Kutta scheme (21) to obtain the cell averages of the numerical solution of the DOM, U_i^n.

It can be seen that we need to know intercell values and intercell spatial derivatives from cell averages of the solution in order to be able to calculate the numerical fluxes to be used in (12) and (16). This is achieved via the procedure usually called *reconstruction*. The one used in this paper is described in the next subsection. Details about WENO procedure can be found in many references such as [4, 1, 17, 18].

2.2 WENO Reconstruction

We distinguish two different parts: reconstruction for the Energy Balance Model and reconstruction for the Deep Ocean Model.

WENO reconstruction for Energy Balance Model

The numerical solution of (12) requires the computation of the intercell numerical fluxes $f_{i\pm\frac{1}{2}}$. If we consider the expression (14) we notice that we need to obtain the solution and the spatial derivative of the solution at each cell interface from the spatial cell averages that are given by

$$u_i(t) = \frac{1}{\Delta x_i}\int_{x_{i-\frac{1}{2}}}^{x_{i+\frac{1}{2}}} u(x,t)dx. \qquad (23)$$

Since we are working in one space dimension, for an order of accuracy r we have r candidate stencils each one of them with r cells. We can denote the r stencils as $\{S_{i-r+1}, S_{i-r+2}, \cdots, S_i\}$, $\{S_{i-r+2}, S_{i-r+3}, \cdots, S_{i+1}\}, \cdots$, $\{S_i, S_{i+1}, \cdots, S_{i+r-1}\}$. For each one of those stencils we can consider a $(r-1)-th$ degree interpolating polynomial $p_l(x), \quad l = 0, \cdots, r-1$. The WENO procedure defines the reconstructed values: $u(x_{i+\frac{1}{2}}, t)$, $u_x(x_{i+\frac{1}{2}}, t)$ as a convex combination of the rth- order accurate values of all polynomials taken with positive nonlinear weights.

Each one of the polynomials considered must be conservative, in the sense that the integral average of the polynomial is equal to the integral average of the solution within each cell in the stencil

$$\frac{1}{\Delta x_k}\int_{S_k} p_l(x)dx = u_k(t), \qquad 0 \le l \le r-1, \quad 0 \le k \le r-1 \qquad (24)$$

where S_k is each one of the r cells in the stencil used to construct the polynomial p_l.

Let us now denote as $u^{(k,0)}_{i+\frac{1}{2}}, (1 \le k \le r)$ the r values taken by the r polynomials at the cell interface $x_{i+\frac{1}{2}}$ and we denote as $u^{(k,1)}_{i+\frac{1}{2}}, (1 \le k \le r)$ the r values taken by the first spatial derivative of the r polynomials at $x_{i+\frac{1}{2}}$. We define the values $u(x_{i+\frac{1}{2}}, t)$ and $u_x(x_{i+\frac{1}{2}}, t)$ as

$$u(x_{i+\frac{1}{2}},t)=\sum_{k=0}^{r-1}\omega_k u_{i+\frac{1}{2}}^{(k,0)},\quad u_x(x_{i+\frac{1}{2}},t)=\sum_{k=0}^{r-1}\omega_k u_{i+\frac{1}{2}}^{(k,1)} \tag{25}$$

where ω_r are the so-called nonlinear weights. They are calculated using

$$\omega_j=\frac{\alpha_j}{\sum_{k=0}^{r-1}\alpha_k}\quad \text{where}\quad \alpha_j=\frac{d_j}{(\varepsilon+\beta_j)^p}\quad (0\le j\le r-1). \tag{26}$$

where we use $p=2$ and $\varepsilon=10^{-6}$ which is introduced to avoid division by zero. In the expression (26) we have used the so-called smoothness indicators that are obtained from

$$\beta_k=\sum_{m=0}^{r-1}\int_{x_{i-\frac{1}{2}}}^{x_{i+\frac{1}{2}}}\frac{d^m}{dx^m}(p_k(x))^2\Delta x^{2m-1}dx\qquad(0\le k\le r-1). \tag{27}$$

Apart from the numerical fluxes we also need to compute the source term integral given in (15). One possible option is to use a Gaussian quadrature formula, as the following two-point one, which for the reference interval $\hat{S}=[-1,1]$ reads

$$\int_{-1}^{1}\phi(\xi)d\xi\approx\phi\left(-\frac{1}{\sqrt{3}}\right)+\phi\left(\frac{1}{\sqrt{3}}\right). \tag{28}$$

We can easily map the interval $S_i=[x_{i-\frac{1}{2}},x_{i+\frac{1}{2}}]$ onto the reference interval $\hat{S}$ by means of a linear transformation and obtain the gaussian quadrature points, denoted as $x_{\alpha,i}$ and $x_{\beta,i}$, and weights, denoted as $w_{\alpha,i}$ and $w_{\beta,i}$, for S_i which are expressed as $x_{\alpha,i}=x_i-\frac{\Delta x_i}{2\sqrt{3}}$ and $x_{\beta,i}=x_i+\frac{\Delta x_i}{2\sqrt{3}}$, where $x_i=(x_{i+\frac{1}{2}}+x_{i-\frac{1}{2}})/2$. The weights are $w_{\alpha,i}=w_{\beta,i}=\Delta x_i$.

Therefore we can approximate the integral of the source term as

$$\sigma_i(t)=\frac{1}{\Delta x_i}\int_{x_{i-\frac{1}{2}}}^{x_{i+\frac{1}{2}}}\sigma(u(x,t))dx\approx\frac{1}{2}\left(\sigma\left(u\left(x_{\alpha,i},t\right)\right)+\sigma\left(u\left(x_{\beta,i},t\right)\right)\right). \tag{29}$$

The expression (29) requires to compute the values $u\left(x_{\alpha,i},t\right)$ and $u\left(x_{\beta,i},t\right)$, which can be achieved using the WENO polynomials, p_l, previously used. We obtain the value of the unknown for each Gaussian point as

$$u\left(x_\alpha,t\right)=\sum_{k=0}^{r-1}\Omega_k u^{(k,\alpha)},\qquad u\left(x_\beta,t\right)=\sum_{k=0}^{r-1}\Omega_k u^{(k,\beta)}, \tag{30}$$

where $u^{(k,\alpha)}$ and $u^{(k,\beta)}$ are the values of each polynomial at the gaussian points x_α,x_β and Ω_k are the nonlinear weights for WENO interpolation.

In this work we have used piecewise-cubic reconstruction which means that we have taken the value $r=4$. Therefore, we use the four following stencils:

$$\{S_{i-3}, S_{i-2}, S_{i-1}, S_i\}, \qquad \{S_{i-2}, S_{i-1}, S_i, S_{i+1}\}$$
$$\{S_{i-1}, S_i, S_{i+1}, S_{i+2}\}, \qquad \{S_i, S_{i+1}, S_{i+2}, S_{i+3}\}.$$

The extrapolated values of the solution for $r = 4$ at cell interface $x_{i+\frac{1}{2}}$ are given by the general expression:

$$u_{i+\frac{1}{2}}^{(k,0)} = \sum_{j=-3}^{3} C_j u_{i+j}(t), \tag{31}$$

where the coefficients C_j are given in Table 1.

The extrapolated derivatives of the solution for $r = 4$ at cell interface $x_{i+\frac{1}{2}}$ are given by the general expression:

$$u_{i+\frac{1}{2}}^{(k,1)} = \frac{1}{\Delta x_i} \sum_{j=-3}^{3} D_j u_{i+j}(t), \tag{32}$$

where the coefficients D_j are given in Table 2.

Table 1. WENO coefficients for intercell extrapolated values

k	C_{-3}	C_{-2}	C_{-1}	C_0	C_1	C_2	C_3
0	0	0	0	1/4	13/12	-5/12	1/12
1	0	0	-1/12	7/12	7/12	-1/12	0
2	0	1/12	-5/12	13/12	1/4	0	0
3	-1/4	13/12	-23/12	25/12	0	0	0

Table 2. WENO coefficients for derivative extrapolated values

k	D_{-3}	D_{-2}	D_{-1}	D_0	D_1	D_2	D_3
0	0	0	0	-11/12	9/12	3/12	-1/12
1	0	0	1/12	-15/12	15/12	-1/12	0
2	0	1/12	-3/12	11/12	9/12	0	0
3	-11/12	45/12	-69/12	35/12	0	0	0

As optimal weights for intercell values computation we have followed the idea first proposed in [11] in which a much higher weight is assigned to the central stencils. The values of this optimal weights we propose are: $d_0 = d_3 = 1$, $d_1 = d_2 = 10^{10}$. The smoothness indicators, optimal weights in Gaussian calculation and extrapolated values at Gaussian points can be found in [1, 17].

Influence of the ocean temperature on the interface

In formula (7) we have the normal derivative $\frac{K_V}{D}\frac{\partial U}{\partial n}$ which represents the influence of the temperature in the deep ocean on the interface with the atmosphere. In order to consider this term in the formulation we first approximate the normal derivative and integrate this term in the term in the 1D control volume $[x_{i-\frac{1}{2}}, x_{i+\frac{1}{2}}]$ to yield: $\frac{\partial U}{\partial n} \approx \frac{u_i(t)-U_{i,N_z}(t)}{\Delta z/2}$ where $u_i(t)$, where we have considered integral averages of the solucion of the EBM and DOM, as defined in (13) and (17). In the numerical examples the effect of the consideration of this term will be studied.

WENO reconstruction for Deep Ocean Model

In the 2D part of our problem, that is the DOM, we need to obtain the numerical solution of the scheme (16) which requires to compute the intercell numerical fluxes $F_{i\pm\frac{1}{2},j}$ and $G_{i,j\pm\frac{1}{2}}$. Following the ideas put forward in the previous subsection we shall produce a reconstruction polynomial which allows us to obtain point-wise values and spatial derivatives wherever they are needed (in particular at cell interfaces) from cell averages of the solution calculated in the previous time step

$$U_{ij}(t) = \frac{1}{\Delta x_i \Delta z_j} \int_{x_{i-\frac{1}{2}}}^{x_{i+\frac{1}{2}}} \int_{z_{j-\frac{1}{2}}}^{z_{j+\frac{1}{2}}} U(x,z,t)dzdx. \tag{33}$$

Following a similar procedure to the one used in the EBM case we apply WENO reconstruction. In order to proceed we can choose between using a fully 2D reconstruction polynomial or a dimension-by-dimension reconstruction, which consists of obtaining two one 1D polynomial for each cartesian direction. The dimension-by-dimension option is the one we have chosen in this work, as it is easier to implement, it is less computationally expensive and it gives good results. Furthermore it can be extended straightforward to the three-dimensional case. See for example [4, 17, 18] for details on applications of this kind of reconstruction.

Therefore we are using two 1D reconstructions for each 2D control volume and applying WENO procedure for both of them. This means that, for each reconstruction we can use the same process as described previously in this section.

3 Numerical Treatment of Boundary Conditions

In order to obtain a numerical solution of the EBM using the numerical method explained so far, we need to consider boundary conditions. The ones we have applied are $\frac{\partial u}{\partial x}(t,-1) = \frac{\partial u}{\partial x}(t,1) = 0$. In order to carry out WENO reconstruction we can consider the usually called *ghost cells* on the left and on the right side of the domain as depicted in Figure 1.

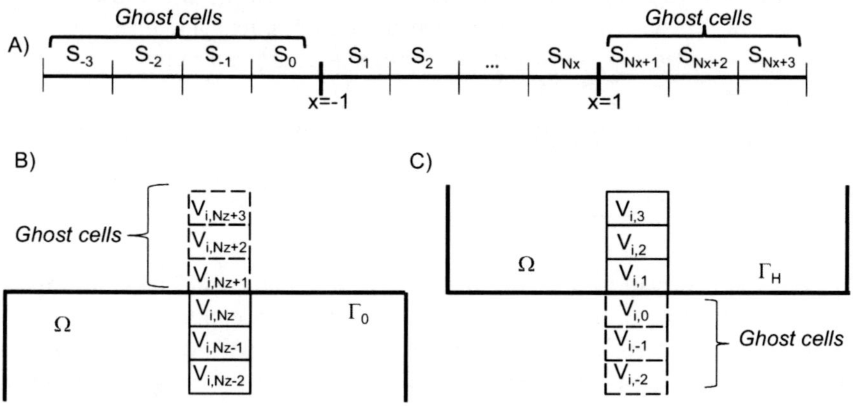

Fig. 1. Ghost cells for boundary conditions. A) 1D EBM, B) 2D DOM Upper boundary, C) 2D DOM Lower boundary.

The usual manner to apply the no flow boundary conditions in 1D is, for each time step, to set $u_0^n = u_1^n$, $u_{-1}^n = u_2^n$, $u_{-2}^n = u_3^n$, $u_{-3}^n = u_4^n$ and $u_{N_z+1}^n = u_N^n$,$u_{N+2}^n = u_{N-1}^n$,$u_{N+3}^n = u_{N-2}^n$, where u_i^n stands for the solution in cell S_i at time step t^n.

Concerning the boundary conditions for the DOM we need to use ghost cells for each boundary. In the boundaries Γ_{-1} and Γ_1 we shall consider no flow boundary conditions so their treatment will be identical to that mentioned for the EBM. Let us consider now the upper boundary Γ_0. We must impose the numerical solution of the EBM as Dirichlet boundary condition. Let us denote by u_i^n the EBM solution for interval S_i at time t^n. In order to use this boundary condition we consider three rows of ghost cells on top of the domain. See Figure 1B.

Using simple linear interpolation we can assign one value to each ghost cell: $U_{i,N_z+1}^n = 2u_i^n - U_{i,N_z}$, $U_{i,N_z+2}^n = 2u_i^n - U_{i,N_z-1}$ and $U_{i,N_z+3}^n = 2u_i^n - U_{i,N_z-2}$ where $U_{i,j}^n$ is the numerical solution for the 2D cell $V_{i,j}$ at time t^n.

Finally we consider the bottom boundary condition. One way to proceed consists of approximating the spatial derivatives appearing in (2) by the following centred formulae

$$\frac{\partial U}{\partial x}(x_i, -H, t^n) \approx \frac{1}{2\Delta x}(U_{i+1,1}^n - U_{i-1,1}^n),$$

$$\frac{\partial U}{\partial z}(x_i, -H, t^n) \approx \frac{1}{(2j-1)\Delta z}(U_{i,j}^n - U_{i,1-j}^n), \quad j = 1, 2, 3, \qquad (34)$$

where we have assumed that all the finite volumes in the mesh have the same size $\Delta x \times \Delta z$. Now we introduce the expressions (34) into the bottom boundary condition given in (2) to obtain the ghost-cell values

$$U^n_{i,1-j} = \omega x_i \frac{2j-1}{2\Delta x}(U^n_{i+1,1} - U^n_{i-1,1}) + K_V U^n_{i,j}, \quad j = 1, 2, 3. \tag{35}$$

4 Numerical Example

We consider the system deep ocean-atmosphere. The main processes involved are incoming solar radiation onto the ocean surface, cooling in the interface ocean-atmosphere sinking cold water owing to ice melting in the poles. We consider that melting water sinks all the way down to the bottom, spreads throughout the bottom of the ocean and, in certain latitudes, rises up towards the surface.

The mathematical model is the one proposed in (4) using the initial conditions $U(x, z, 0) = 18e^{-x^2-z^2}$ for the ocean interior and $u(x, 0) = 80e^{-x^2} - 60$. The physical data used in this example are depicted in Table 3. The first column represents the physical parameters, most of them taken from [20] while the data in the second column have been obtained applying the formula: $100 \cdot Data/L^2$ where L is a length representing either the radius of the Earth ($R \approx 6.378 \times 10^6 m$), when referring to K_H and K_{H0}, or the ocean depth ($H \approx 4000m$), when referring to K_V. The multiplication by 100 is due to the conversion of time units into centuries.

We also need to define, to be used in (3), the function $S(x)$. It represents how the incoming solar radiation is distributed throughout the surface of the

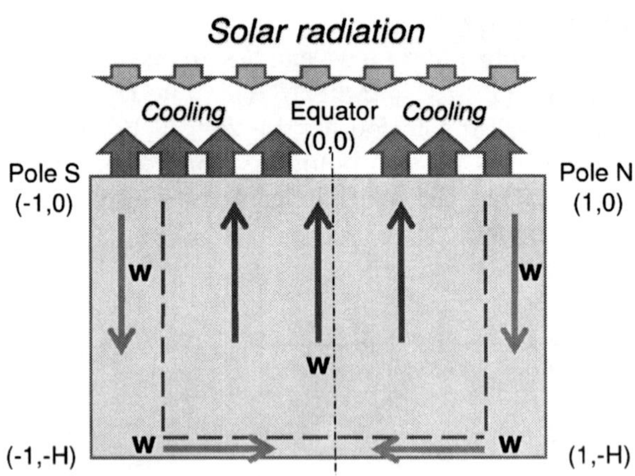

Fig. 2. Physical process involved in Experience 1

Table 3. Physical data used in the model

Parameter	*Data*	*Scaled data*
K_H	$2 \times 10^{10} m^2 yr^{-1}$	0.049
K_{H_0}	$2.26 \times 10^8 m^2 yr^{-1}$	0.555×10^{-3}
K_V	$2000 m^2 yr^{-1}$	0.0125
C, B	$190 W m^{-2}$, $2 W m^{-2 \circ} C^{-1}$	190, 2
c, ρ	1, 1	1, 1
Q	340	340
D	60	60

ocean such that most of the heat goes to the Equator and a little amount of heat goes to the poles. Moreover this function must be non-negative everywhere. To simulate this effect we give the value $S(x) = 1 - \frac{1}{2}P_2(x)$ where $P_2(x) = \frac{1}{2}(3x^2 - 1)$ is the second Legendre polynomial in the interval $[-1, 1]$. The coalbedo $\beta(u)$ is given by

$$\beta(u) = \begin{cases} m \text{ if } u < -10, \\ [m, M] \text{ if } u = -10, \\ M \text{ if } u > -10, \text{ with } 0 < m < M, \end{cases} \tag{36}$$

where $m = 0.4$ and $M = 0.69$. The considered velocity depends only on x and is defined as

$$\omega(x, z) = W(x) = \frac{10(x + 0.75)(x - 0.75)}{(0.1 + 10\,|x + 0.75|)(0.1 + 10\,|x - 0.75|)}, \tag{37}$$

which is displayed in Figure 3.

The spatial domain is the rectangle $[-1, 1] \times [0, -1]$. We have discretized the domain using 40 cells in $x-$direction and also 40 cells in $z-$direction. Regarding the discretization in time we have taken the time step

$$\Delta t = \min(\alpha \Delta x^2 (K_H)^{-1}, \alpha \Delta z^2 (K_H)^{-1}, \alpha \Delta x^2 (K_{H_0} |\frac{du}{dx}|)^{-1}),$$

where $\alpha = 0.3$ is a diffusion parameter which controls the stability of the numerical scheme. Numerical experiments have shown that higher values of the parameter α yield an unstable numerical solution. In Figure 4 we display the contour plot of the temperature inside the deep ocean.

Figure 4 shows that if we consider the effect of deep ocean on the atmosphere the range of temperatures is more narrow than if we do not consider it, which is due to the thermostatic behaviour of the ocean. More precisely, if we call $\tilde{v}(x, t)$ the approximate temperature at the upper boundary without the deep ocean effect (without the term $K_V \frac{\partial U}{\partial n}$) and if we call $v(x, t)$ the approximate temperature at the upper boundary with the deep ocean effect (obtained by this numerical aproximation) then we get that, for every t,

$$\min_{\Gamma_0}(\tilde{v}(x, t)) \le \min_{\Gamma_0}(v(x, t)) \le \max_{\Gamma_0}(v(x, t)) \le \max_{\Gamma_0}(\tilde{v}(x, t)).$$

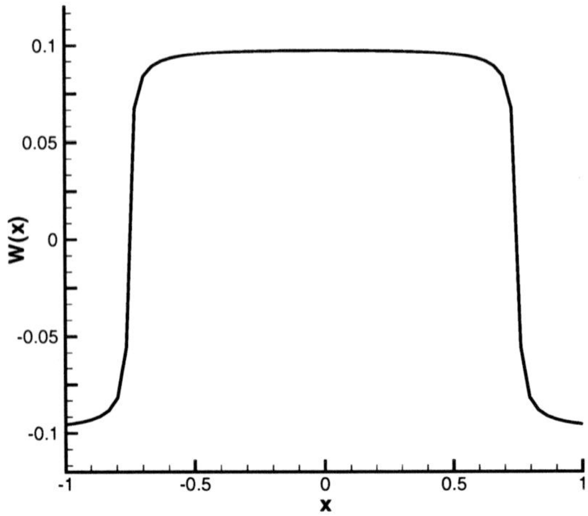

Fig. 3. Velocity profile $W(x)$ used in the example

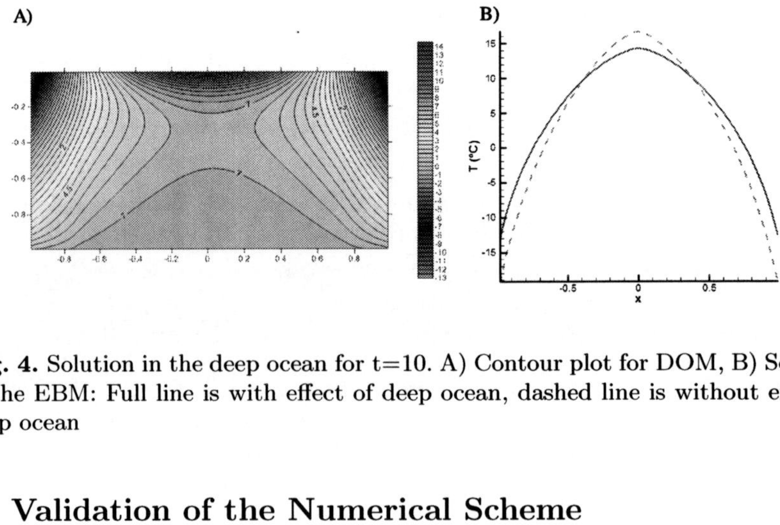

Fig. 4. Solution in the deep ocean for t=10. A) Contour plot for DOM, B) Solution of the EBM: Full line is with effect of deep ocean, dashed line is without effect of deep ocean

5 Validation of the Numerical Scheme

In order to carry out the validation of our numerical scheme we consider the function

$$U(x,z,t) = \frac{1}{1+t}(x^2-1)^2(1+z)^2, \tag{38}$$

and construct the auxiliary test problem

$$U_t - (\frac{K_H}{R^2}(1-x^2)U_x)_x - K_V U_{zz} + wU_z = \Phi(x,z,t) \quad \text{in } \Omega \times (0,T),$$

$$wxU_x + K_V U_z = 0 \quad \text{in } \Gamma_H \times (0,T),$$

$$Du_t - \frac{DK_{H_0}}{R^2}\left((1-x^2)^{3/2}\,|u_x|\,u_x\right)_x + K_V \frac{\partial U}{\partial n}$$

$$+wxU_x + C + Bu = \frac{1}{\rho c} QS(x)\beta(x,U) + \psi(x,t) \quad \text{in } \Gamma_0 \times (0,T),$$

$$(1-x^2)^{3/2}\,|U_x|\,U_x = 0 \quad \text{in } \Gamma_{-1} \times (0,T) \cup \Gamma_1 \times (0,T),$$

$$U(x,z,0) = (1+x^2)(1+z^2) \quad \text{in } \Omega,$$

$$u(x,0,0) = 1 + x^2 \quad \text{in } \Gamma_0, \tag{39}$$

where the source terms $\Phi(x,z,t)$ and $\psi(x,t)$ have been added to the original model such that (38) be the exact solution of the problem (39). Their expressions are not displayed here but they can be easily obtained using a symbolic computation tool. The physical and discretization parameters are the same as those used in the previous example.

In the first two plots in Figure 5 it is displayed the contour plot of the numerical solution and the exact one. The third plot represents the comparison between the numerical solution and the exact one in the upper boundary, which means the 1D EBM, and in the DOM along the lines $z = -0.25$ and $z = -0.5$. The results show a good agreement between the numerical solution and the exact one.

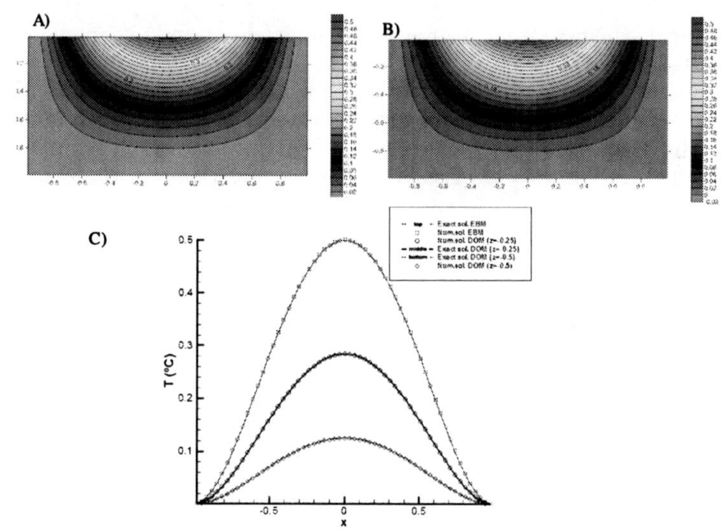

Fig. 5. A) Contour plot of numerical solution of (39), B) Contour plot of exact solution of (39), C) Comparison exact solution (full line) versus numerical solution (symbols) for: Upper boundary (1D EBM) and DOM, along the lines $z = -0.25$ and $z = -0.5$ (with $t = 1$).

6 Conclusions

We have obtained a numerical solution of the coupled model atmosphere-deep ocean, by means of a finite volume approach with WENO−7 reconstruction and third order TVD Runge-Kutta for time discretization. The evolution of the temperature in the deep ocean is due to the combined effect of water going down from the Earth poles combined with heating-cooling processes taking place in the interface atmosphere-ocean. We have checked the numerical solution using a test problem in which we have added a source term in order to have an analytical solution, obtaining good results. The results obtained also show that temperatures are smoothed by the effect of the ocean, in the sense that the maximum decreases and the minimum increases.

Acknowledgement. This work was partially supported by the projects of the research groups "Modelos Matemáticos No Lineales (MOMAT)" and "Técnicas Numéricas Avanzadas en Ciencias de la Tierra y la Energía (TNTE)" of Universidad Politécnica de Madrid (Spain) and CAM and project MTM2008 - 06208 of DGISGP (Spain). The authors want to thank Prof. Carlos Conde and Prof. Ildefonso Díaz for their useful suggestions.

References

1. Balsara, D.S., Shu, C.W.: Monotonicity preserving weighted essentially non-oscillatory schemes with increasingly high order of accuracy. J. Comp. Phys. 160, 405–452 (2000)
2. Bermejo, R., Carpio, J., Díaz, J.I., Tello, L.: Mathematical and Numerical Analysis of a Nonlinear Diffusive Climate Energy Balance Model. Math. Comp. Model 49, 1180–1210 (2009)
3. Budyko, M.I.: The effects of solar radiation variations on the climate of the Earth. Tellus 21, 611–619 (1969)
4. Casper, J., Atkins, H.: A finite-volume high order ENO scheme for two-dimensional hyperbolic systems. J. Comp. Phys. 106, 62–76 (1993)
5. Díaz, J.I. (ed.): The Mathematics of Models in Climatology and Environment. ASI NATO Global Change Series I, vol. 48. Springer, Heidelberg (1996)
6. Díaz, J.I., Hetzer, G., Tello, L.: An energy balance climate model with hysteresis. Nonlin Anal. 64, 2053–2074 (2006)
7. Díaz, J.I., Tello, L.: Sobre un modelo climático de balance de energía superficial acoplado con un océano profundo. Actas XVII CEDYA/ VI CMA. Univ. Salamanca (2001)
8. Díaz, J.I., Tello, L.: A 2D climate energy balance model coupled with a 3D deep ocean model. Elect. J. Differ. Equat. Conf. 16, 129–135 (2007)
9. Díaz, J.I., Tello, L.: On a climate model with a dynamic nonlinear diffusivity boundary condition. Discr. Contin. Dynam. Syst. Ser. S 1(2), 253–262 (2008)
10. Díaz, J.I., Hetzer, G., Tello, L.: An energy balance climate model with hysteresis. Nonlinear analysis, Series S 1(2), 2053–2074 (2006)
11. Dumbser, M., Enaux, C., Toro, E.F.: Finite volume schemes of very high order of accuracy for stiff hyperbolic balance laws. J. Comp. Phys. 228(5), 1573–1590 (2009)

12. Hetzer, G.: The structure of the principal component for semilinear diffusion equations from energy balance climate models. Houston Journal of Math. 16, 203–216 (1990)
13. North, G.R.: Multiple solutions in energy balance climate models. Paleogeography, Paleoclimatology, Paleoecology 82, 225–235 (1990)
14. Sellers, W.D.: A global climatic model based on the energy balance of the earth-atmosphere system. J. Appl. Meteorol 8, 392–400 (1969)
15. Shu, C.W.: Total-variation-diminishing time discretizations. SIAM J. Sci. Stat. Comput. 9, 1073–1084 (1988)
16. Stone, P.H.: A simplified radiative - dynamical model for the static stability of rotating atmospheres. Journal of the Atmospheric Sciences 29(3), 405–418 (1972)
17. Titarev, V.A., Toro, E.F.: Finite-volume WENO schemes for three-dimensional conservation laws. J. Comp. Phys. 201, 238–260 (2004)
18. Titarev, V.A., Toro, E.F.: ADER schemes for three-dimensional non-linear hyperbolic systems. J. Comp. Phys. 204(2), 715–736 (2005)
19. Toro, E.F., Hidalgo, A.: ADER finite volume schemes for nonlinear reaction–diffusion equations. Appl. Num. Math. 59(1), 73–100 (2009)
20. Watts, R.G., Morantine, M.: Rapid climatic change and the deep ocean. Climate Change 16, 83–97 (1990)
21. Xu, X.: Existence and regularity theorems for a free boundary problem governing a simple climate model. Applicable Anal. 42, 33–59 (1991)

On the Existence and Location of the Free–Boundary for an Equilibrium Problem in Nuclear Fusion*

J. Francisco Padial

Dept. Matemática Aplicada, E.T.S. de Arquitectura,
Univ. Politécnica de Madrid,
Avda. Juan de Herrera 4, 28040 Madrid, Spain
jf.padial@upm.es

Summary. We consider a non linear elliptic problem arising in the mathematical modeling of the nuclear fusion by magnetic confinement of a plasma in Stellarator devices. We prove some results about the existence, location and size of plasma region and vacuum region, and thus about the existence and location of the boundary of the plasma (the free–boundary for the mathematical model). From the point of view of its mathematical approach, we introduce a suitable auxiliary problem in order to obtain a subsolution of this new problem and for which our original solution is a supersolution. Finally, by applying a comparison principle of solutions, we obtain some appropriate estimates on the location and size of the plasma region and vacuum region.

1 Introduction

We consider the following nonlinear elliptic problem. Let Ω be a bounded and regular set of $\mathbb{R}^2$. Given the parameters $F_v > 0$, $\gamma < 0$, functions $a, b \in L^\infty(\Omega)$, $a \not\equiv 0, b > 0$ a.e. in Ω. The problem is to find two functions (u, F) such that $u : \Omega \to \mathbb{R}$, $F : \mathbb{R} \to \mathbb{R}$ with $F(s) = F_v$ for all $s \leq 0$, satisfying

$$(\mathcal{P})\begin{cases} -\Delta u(x) = a(x) F(u(x)) + \frac{1}{2}\left(F^2(u(x))\right)' + b(x) p'(u(x)) \text{ in } \Omega, \\ u(x) = \gamma, \quad \text{in } \partial\Omega, \\ \displaystyle\int_{\{x\in\Omega: u(x)>s\}} \frac{1}{2}\left(F^2(u(x))\right)' + b(x) p'(u(x))\, dx = j(s_+, \|u_+\|_{L^\infty(\Omega)}) \\ \qquad \text{for all } s \in [\operatorname{ess\,inf}_\Omega u, \operatorname{ess\,sup}_\Omega u]. \end{cases}$$

Here, and in what follows, we shall assume that $p \in C^1(\mathbb{R})$ such that $p(0) = 0$, $0 \leq p'(t) \leq \lambda t_+$, and $|p'(t) - p'(s)| \leq L|t - s|^\alpha$ for some $\lambda > 0$, $L > 0$

* The author would like to thank L. Pardo, N. Balakrishnan and M.Á. Gil for their kind invitation to present a contribution in this book in tribute to Professor María Luisa Menéndez.

and $\alpha \in]0,1[$. For function j, we assume $j \in \mathcal{C}(\mathbb{R} \times \mathbb{R}^+)$, $j(\sigma,\sigma) = 0$ for all $\sigma \geq 0$, $j_1' \in \mathcal{C}(\mathbb{R}^+ \times \mathbb{R}^+)$ and $\eta := \sup\{|j_1'(s,\sigma)| : (s,\sigma) \in \mathbb{R}^+ \times \mathbb{R}^+\} < +\infty$. We shall always use j_1' to denote the derivative of j with respect to the first component (i.e. $j_1' := \frac{\partial}{\partial s} j\,(s,\sigma)$) and by p' we denote the derivative of p.

Problem $(\mathcal{P})$ appears in the mathematical treatment of a bidimensional model arising in the magnetically confined plasma in a Stellarator device. One of the difficulties of the magnetically controlled plasma fusion, is to determinate the conditions on the magnetic field and on the current density in order to maintain the plasma far from the camera wall. This variables will be the unknowns of magnetic confinement problem. In the case of Stellarator devices, the plasma is assumed to be an ideal fluid and a perfect conductor. This problem can be modeled by using the ideal incompressible MHD system (magnetohydrodynamic system). The nonlinear elliptic partial differential equation is obtained in [4, Appendix A], [5] and [12] from ideal 3D MHD system, introducing a set of special toroidal coordinates (the *Boozer vacuum coordinates system* [2]) and the averaging arguments of [7]. We just remark that $(\mathcal{P})$ can be viewed as a free–boundary problem since the sets $\Omega_p := \{x \in \Omega : u(x) \geq 0\}$ and $\Omega_v := \{x \in \Omega : u(x) < 0\}$ are a priori unknowns and then their interface. Physically, these two domains correspond with the *plasma region* (Ω_p), i.e. the region in the Stellarator device where the plasma is confined, and the *vacuum region* (Ω_v), i.e., where is present no plasma. The unknown function u is called the *flux function* and its gradient represents two of the components of the averaged magnetic field that is confining the plasma. The unknown scalar function F mapping on u, gives the third components of the averaged magnetic field. Finally, a characteristic of an ideal Stellarator is that it has zero net current within each flux magnetic surface, but in practice, however, this ideal condition does not hold, and a known current arises in the interior of each magnetic surface (see [3] for a physical modeling). Using the change of variables introduced in [4], the condition of a nonzero current inside each magnetic surface can be expressed in terms of the family of integrals that appears in $(\mathcal{P})$, involving a given function j. This family of integral identities is known as the *current–carrying Stellarator condition.*

In the case of ideal Stellarator, the net current within each flux magnetic surface is zero and this can be expressed in terms of the integral condition of problem $(\mathcal{P})$ assuming $j \equiv 0$. The existence and regularity of a weak solution (u, F) of this problem has been established in [5] and the existence and location of the plasma region has been studied in [8]. For the case of non ideal Stellarator, the condition of non–zero current inside each magnetic surface can be expressed in terms of the *current–carrying Stellarator condition* of problem $(\mathcal{P})$ assuming $j \not\equiv 0$. The fact that $j \not\equiv 0$ makes appear a new nonlinear terms in its mathematical treatment with respect to the case $j \equiv 0$. The existence and regularity of a weak solution (u, F) for this problem, when $j \not\equiv 0$, have been established in [4] and [12]. The existence and location of the plasma region for the corresponding evolution problem was studied in [13].

The main goal of this paper is to obtain the existence and the estimate on the location and size of plasma region $\Omega_p = \{x \in \Omega : u(x) \geq 0\}$ and the vacuum region $\Omega_v = \{x \in \Omega : u(x) < 0\}$ for the non ideal Stellarator device, that is, when $j \not\equiv 0$. That is stated as follow:

Theorem 1. *Let Ω be a bounded open regular subset of $\mathbb{R}^2$ (with C^1 boundary $\partial\Omega$) and such that*

$$\exists\, x_0 \in \Omega \textit{ verifying } R_p := \left(\frac{-4\gamma}{F_v \operatorname*{essinf}_{x\in\Omega} a(x)}\right)^{\frac{1}{2}} < d(x, \partial\Omega).$$

Assume that $\operatorname*{essinf}_{\Omega} a > 0$. *Let (u, F) be a weak solution of $(\mathcal{P})$ such that u has not flat region and $F \in W^{1,\infty}(\operatorname{ess\,inf}_{x\in\Omega} u, \operatorname{ess\,sup}_{x\in\Omega} u)$. Then, if λ and η are small enough, we have that*

$$\Omega_{R_p} := \{x \in \Omega : d(x, \partial\Omega) \geq R_p\} \subset \Omega_p = \{x \in \Omega : u(x) \geq 0\}.$$

In particular meas $\{x \in \Omega\colon d(x, \partial\Omega) \geq R_p\} \leq |\Omega_p|$ *(d denotes the Euclidean distance).*

Analogously, we find a similar estimate for the location and size of the vacuum region $\Omega_v := \{x \in \Omega : u(x) < 0\}$:

Theorem 2. *Let Ω be an open bounded regular (with C^1 boundary $\partial\Omega$) subset of $\mathbb{R}^2$ and such that*

$$\exists\, x_0 \in \Omega \textit{ verifying } R_p := \left(\frac{-4\gamma}{F_v \operatorname*{essinf}_{x\in\Omega} a(x)}\right)^{\frac{1}{2}} < d(x, \partial\Omega)$$

and let $\hat{R}$ be a positive number such that

$$0 < \rho < \hat{R} \leq R_p + \rho$$

for a $\rho > 0$. Assume that for any $\bar{x} \in \partial\Omega$ the segment $\bar{x} + r\mathbf{n}, 0 < r \leq \hat{R}$ belongs to Ω where $\mathbf{n}$ is the inward normal unit vector to $\partial\Omega$. Let (u, F) be a weak solution of $(\mathcal{P})$ such that u has not flat region and $F \in W^{1,\infty}(\operatorname{ess\,inf}_{x\in\Omega} u, \operatorname{ess\,sup}_{x\in\Omega} u)$. Then, if λ and η are small enough we have that

$$\left\{x \in \Omega : d(x, \partial\Omega) \leq \hat{R} - \rho\right\} \subset \Omega_v = \{x \in \Omega : u(x) < 0\}.$$

In particular meas $\left\{x \in \Omega\colon d(x, \partial\Omega) \leq \hat{R} - \rho\right\} \leq |\Omega_p|$.

The structure of the rest of the paper is as follows. In Section 2 we introduce the notion of decreasing and relative rearrangement and we define the non–local problem $(\mathcal{P}_*)$ which is equivalent to our problem $(\mathcal{P})$, but now with only one unknown. We give some results proved in previous works concerning to the existence of solutions for the problem $(\mathcal{P})$ and some a priori estimate of solutions. Finally we prove some properties for the weak solutions. In Section 3 we prove the main results concerning to the existence and estimate the location and size of the plasma region and the vacuum region.

2 Previous Results on the Existence and a Priori Estimates for the Solution of Problem $(\mathcal{P})$

We start recalling the notion of the decreasing and relative rearrangement. Let Ω be a bounded and connected open measurable set of $\mathbb{R}^2$ (we assume a 2d–setting motivated by the physical modeling but the definitions and results that follows hold for any dimension $N > 1$). Given a measurable function $u : \Omega \to \mathbb{R}$, the *distribution function of* u is given by $m_u(\sigma) :=\text{meas}\{x \in \Omega : u(x) > \sigma\}$ (the Lebesgue measure of the set $\{x \in \Omega : u(x) > \sigma\}$ will be denoted by $|u > \sigma|$). It is well–know that the function $m_u(\cdot)$ is decreasing and right semicontinuous. We shall say that u has a flat region at the level σ if $\text{meas}\{x \in \Omega : u(x) = \sigma\}$ (denoting by $|u = \sigma|$) is strictly positive. The generalized inverse of m_u is called the *decreasing rearrangement* of u with respect to x and it is defined as the function $u_* : [0, |\Omega|] \to \bar{\mathbb{R}}$ such that $u_*(s) := \inf\{\sigma \in \mathbb{R} : m_u(\sigma) \le s\}$ for all $s \in \Omega_*$, where $\Omega_* :=]0, |\Omega|[$ (see e.g. [10], [11] for more details about its definition and properties). We recall some properties: u_* is decreasing, $u_*(0) = ||u_+||_{L^\infty(\Omega)} = \text{ess}\sup_{x\in\Omega} u(x)$, u_* and u are equimeasurable, and the mapping $u \in L^p(\Omega)$ to $u_* \in L^p(\Omega_*)$ is a contraction for $1 \le p \le +\infty$. Moreover, if u has not flat region, then m_u and u_* are continuous and $u_*(m_u(\sigma)) = \sigma$ (that is, $u_*^{-1} = m_u$). On the other hand, if $u \in W^{1,p}(\Omega)$, $1 \le p \le +\infty$, then $u_* \in W^{1,p}_{loc}(\Omega_*)$.

Given a measurable function $u : \Omega \to \mathbb{R}$, and $b \in L^p(\Omega)$ with $1 \le p \le \infty$, we define the function $w : \Omega_* \to \mathbb{R}$ as

$$w(s) = \int\limits_{\{x\in\Omega:u(x)>u_*(s)\}} b(x)\,dx \qquad + \int\limits_0^{s-|u(\cdot)>u_*(s)|} \left(b|_{\{x\in\Omega:u(x)=u_*(s)\}}\right)_*(\sigma)\,d\sigma.$$

The *relative rearrangement of* b *with respect to* u is the functions $b_{*u} \in L^p(\Omega_*)$ defined by $b_{*u}(s) := \frac{dw(t)(s)}{ds} = \lim\limits_{\sigma\to 0} \frac{(u+\sigma b)_*(s)-u_*(s)}{\sigma}$ for all $s \in \Omega_*$. Notice that by this definition, if u has not flat region (that implies that $s-|u > u_*(s)| = 0$) then $b_{*u}(s) := \frac{d}{ds}\int_{\{x\in\Omega:u(x)>u_*(s)\}} b(x)\,dx$ for all $s \in \Omega_*$. Also, for any measurable function u, the mapping $b \in L^p(\Omega)$ to $b_{*u} \in L^p(\Omega_*)$ is a contraction for $1 \le p \le +\infty$ and in particular $||b_{*u}||_{L^\infty(\Omega_*)} \le ||b||_{L^\infty(\Omega)}$ (further details on the decreasing and relative rearrangement can be found, for instance, in [4], [9], [10], [11], [14], [15], [16] and their references).

The notion of decreasing and relative rearrangement and their properties will be the key to reformulate the problem $(\mathcal{P})$ as a new equivalent non–local problem $(\mathcal{P}_*)$ defined later with only one unknown.

Before starting the results concerning to the existence and to the a priori estimates of the solution of the problem $(\mathcal{P})$, we introduce the following useful convex cone $V(\Omega) := \{v \in H^1(\Omega) : \Delta v \in L^\infty(\Omega), v_{|\partial\Omega} \leq 0\}$. We recalling the existence result and some a priori estimates given in [4] and [12] about the solution of problem $(\mathcal{P})$.

Theorem 3. *Suppose that* $\gamma \leq 0$ *and* $\inf_\Omega |a| > 0$. *Then there exist* $\Lambda_1, \Lambda_2 > 0$ *such that if*

$$\lambda \|b\|_{L^\infty(\Omega)} + \eta < \Lambda_1 \quad and \quad \Lambda_2 < \inf_\Omega |a| F_v$$

there is a couple (u, F), $u \in V(\Omega)$ *and* $F \in W^{1,\infty}(]\operatorname{ess\,inf}_\Omega u, \operatorname{ess\,sup}_\Omega u[)$ *solution of* $(\mathcal{P})$ *satisfying also that* $\operatorname{meas}\{x \in \Omega : \nabla u(x) = 0\} = 0$.

Following [4], [5] and [12], we can determinate the unknown function F in terms of the function u by derivating the *integral condition* of $(\mathcal{P})$ with respect to the level s. So, for any pair of functions (u, F) that verify the *integral condition* of $(\mathcal{P})$, with $u \in W^{2,p}(\Omega)$, $p \geq 1$, $F \in W^{1,\infty}(]\operatorname{ess\,inf}_\Omega u, \operatorname{ess\,sup}_\Omega u[)$, $F(s) = F_v$, if $s \leq 0$ and such that $\operatorname{meas}\{x \in \Omega : \nabla u(x) = 0\} = 0$ (that is means that u has *not flat region* and thus we can use the fact that m_u and u_* are continuous functions, $u_*(m_u(\sigma)) = \sigma$ and $b_{*u}(s) := \frac{d}{ds}\int_{\{x\in\Omega : u(x) > u_*(s)\}} b(x)\, dx$), we can obtain that

Proposition 1. *Given* (u, F) *solution of* $(\mathcal{P})$, *with* $\operatorname{meas}\{x \in \Omega : \nabla u(x) = 0\} = 0$, *then* $F(s) = G_u(s)$ *for all* $s \in [\operatorname{ess\,inf}_{x\in\Omega} u, \operatorname{ess\,sup}_{x\in\Omega} u]$ *and* $F(u(x)) = G_u(u(x))$ *a.e.* $x \in \bar{\Omega}$ *where the function* G_u *is defined as*

$$G_u(s) := \left[F_v^2 - 2\int_0^{s_+} p'(r)\, b_{*u}(|u > \sigma|)\, d\sigma \right. \tag{1}$$
$$\left. + 2\int_0^{s_+} j_1'(\sigma, u_{+*}(0)) u_{+*}'(|u > \sigma|) dr \right]_+^{\frac{1}{2}}$$

(See [4] and [12] for its proof). In order to simplify the notation, we set

$$G_u(u(x)) = \left[F_v^2 - F_1(x, u(x), b_{*u}) + F_2(x, u(x))\right]_+^{\frac{1}{2}} \tag{2}$$

where

$$F_1(x, u(x), b_{*u}) := 2\int_{|u>0|}^{|u>u_+(x)|} [p(u_*)]'(\sigma)\, b_{*u}(\sigma)\, d\sigma, \quad \text{and} \tag{3}$$
$$F_2(x, u(x)) := 2\int_{|u>0|}^{|u>u_+(x)|} j_1'(u_{+*}(\sigma), u_{+*}(0))(u_{+*}'(\sigma))^2 d\sigma.$$

This proposition gives us the possibility to define a new non local problem $(\mathcal{P}_*)$ for which u (solution of $(\mathcal{P})$) will be also a solution.

Theorem 4. *Let (u, F) be a solution of $(\mathcal{P})$ given by Theorem 3, Then u is a weak solution of the non local problem*

$$(\mathcal{P}_*)\begin{cases} -\Delta u(x) = a(x)\, G_u(u(x)) + H(u(x), b_{*u}) + J(u(x)) \text{ in } \Omega, \\ u(x) - \gamma \in H_0^1(\Omega) \quad \text{on } x \in \partial\Omega \end{cases} \tag{4}$$

where the functions H and J are given by

$$H(u(x), b_{*u}) := p'(u(x))\,[b(x) - b_{*u}(|u > u(x)|)] \tag{5}$$

and

$$J(u(x)) := j_1'(u_+(x), u_{+*}(0))u'_{+*}(|u > u(x)|)\,. \tag{6}$$

Reciprocally, if u is a weak solution of $(\mathcal{P}_*)$ such that u has not flat region (meas$\{x \in \Omega : \nabla u(x) = 0\} = 0$) *and such that $G_u > 0$ in* $[\operatorname{ess\,inf}_{x\in\Omega} u, \operatorname{ess\,sup}_{x\in\Omega} u]$, *then (u, G_u) is a solution of $(\mathcal{P})$ where G_u is defined as in (1).*

Remark 1. Notice that u is the only one unknown of the problem $(\mathcal{P}_*)$.

Remark 2. We can verify that if $s \leq 0$ then $G_u(s) = F_v > 0$ (it comes from (1)). If $u(x) \leq 0$ then $G_u(u(x)) = F_v$ (from (2) and (3)), $H(u(x), b_{*u}) = 0$ and $J(u(x)) = 0$ from the definition of H, J and the hypotheses on p and j_1'.

Remark 3. This theorem allows us to work with u as weak solution of $(\mathcal{P}_*)$ or (u, G_u) as a weak solution of $(\mathcal{P})$ indistinctly. Thus we only consider the above mentioned regularity with λ and μ small enough in what follows (see [4] and [5] for more details).

Notice that, in the existence Theorem 3, assuming Λ_1 small enough (from λ and η small enough) we can define the positive number ν such that

$$\nu := \frac{1}{4\pi}\left[2^{1/2}\eta^{1/2}|\Omega|^{1/2}\|a\|_{L^\infty(\Omega)} + \lambda|\Omega|\mathrm{osc}_\Omega b + \eta\right] < 1 \tag{7}$$

(with $\mathrm{osc}_\Omega b = \operatorname{ess\,sup}_{x\in\Omega} b - \operatorname{ess\,inf}_{x\in\Omega} b$). The existence of solution was proved in [4] by using a Galerkin type methods. The solution found in this way, it is such that verifies the following

Proposition 2. *For $\Lambda_1, \Lambda_2 > 0$ small enough, there exists a solution (u, F) of problem $(\mathcal{P})$ given by Theorem 3 such that*

$$\|u_+\|_{L^\infty(\Omega)} \leq \frac{|\Omega|}{4\pi}\frac{\|a\|_{L^\infty(\Omega)}F_v}{(1-\nu)} := S, \tag{8}$$

$$\|\Delta u\|_{L^\infty(\Omega)} \leq \frac{\|a\|_{L^\infty(\Omega)}F_v}{1-\nu} = \frac{4\pi}{|\Omega|}S, \tag{9}$$

$$\left\|\frac{du_{+*}}{ds}\right\|_{L^\infty(\Omega_*)} \leq \frac{1}{4\pi}\frac{\|a\|_{L^\infty(\Omega)}F_v}{1-\nu} = \frac{1}{|\Omega|}S$$

where ν is a positive number given by (7).

In particular, we can obtain the following a priori estimates:

Corollary 1. *Given a solution (u, F) of problem $(\mathcal{P})$ as in Proposition 2 we have that*

$$0 \le F_1(x, u(x), b_{*u}) \le 2\lambda \|b\|_{L^\infty(\Omega)} S, \quad |F_2(x,u)| \le \frac{2\eta}{|\Omega|} S^2,$$
$$|H(u, b_{*u})| \le \lambda \mathrm{osc}_\Omega b S, \quad |J(u(x))| \le \frac{\eta}{|\Omega|} S.$$

Proof. From Proposition 2 and the definition of F_1, F_2, H and J, we can prove that

$$0 \le F_1(x, u, b_{*u}) \le 2\frac{\lambda|\Omega|}{4\pi} \|b\|_{L^\infty(\Omega)} \|\Delta u\|_{L^\infty(\Omega)} \le 2\lambda \|b\|_{L^\infty(\Omega)} S,$$
$$|F_2(x, u(x))| \le 2\eta|\Omega| \frac{\|\Delta u\|^2_{L^\infty(\Omega)}}{(4\pi)^2} \le \frac{2\eta}{|\Omega|}\left(\frac{|\Omega|}{4\pi}\frac{\|a\|_{L^\infty(\Omega)} F_v}{1-\nu}\right)^2 = \frac{2\eta}{|\Omega|} S^2,$$
$$|H(u(x), b_{*u})| \le \lambda \mathrm{osc}_\Omega b \|u_+\|_{L^\infty(\Omega)} \le \lambda \mathrm{osc}_\Omega b S,$$
$$|J(u(x))| \le \frac{\eta}{4\pi}\|\Delta u\|_{L^\infty(\Omega)} \le \frac{\eta}{4\pi}\frac{\|a\|_{L^\infty(\Omega)} F_v}{1-\nu} = \frac{\eta}{|\Omega|} S. \qquad \square$$

From the above corollary we can obtain the following estimates for $F(u(x))$ for a.e. $x \in \Omega$ and $F(s)$ for all $s \in [\mathrm{ess\,inf}_\Omega u, \mathrm{ess\,sup}_\Omega u]$:

Corollary 2. *Given a solution (u, F) of problem $(\mathcal{P})$ as in Proposition 2 we have that for a.e. $x \in \Omega$*

$$0 \le F(u(x)) \le F_v + S\sqrt{\frac{2\eta}{|\Omega|}}, \quad F^2(u(x)) \ge \left[F_v^2 - 2\lambda \|b\|_{L^\infty(\Omega)} S - \frac{2\eta}{|\Omega|} S^2\right]_+$$

and

$$\left|\frac{1}{2}\left(F^2(u(x))\right)'\right| \le \left(\eta \frac{1}{|\Omega|} + \lambda \|b\|_{L^\infty(\Omega)}\right) S.$$

The same estimate holds for the function $F(s)$ and $s \in [\mathrm{ess\,inf}_\Omega u, \mathrm{ess\,sup}_\Omega u]$, that is

$$0 \le F(s) \le F_v + S\sqrt{\frac{2\eta}{|\Omega|}}, \quad F^2(s) \ge \left[F_v^2 - 2\lambda \|b\|_{L^\infty(\Omega)} S - \frac{2\eta}{|\Omega|} S^2\right]_+$$

and

$$\left|\frac{1}{2}\left(F^2(s)\right)'\right| \le \frac{\eta}{|\Omega|} S + \lambda \|b\|_{L^\infty(\Omega)} s_+ \le \left(\eta \frac{1}{|\Omega|} + \lambda \|b\|_{L^\infty(\Omega)}\right) S.$$

Moreover, for all $x \in \Omega$,

$$|a(x) F(u(x)) + \tfrac{1}{2}\left(F^2(u(x))\right)' + b(x) p'(u(x))| \le K$$

with $K := ||a||_{L^\infty(\Omega)} F_v + S\left(||a||_{L^\infty(\Omega)}\sqrt{\frac{2\eta}{|\Omega|}} + \frac{\eta}{|\Omega|} + 2\lambda ||b||_{L^\infty(\Omega)}\right)$.

Proof. Considering that estimates given in Corollary 1, we get to

$$0 \leq F(u(x)) = G_u(u(x)) = \left[F_v^2 - F_1(x,u,b_{*u}) + F_2(x,u)\right]_+^{\frac{1}{2}}$$
$$\leq \left[F_v^2 + \frac{2\eta}{|\Omega|}S^2\right]^{\frac{1}{2}} \leq F_v + S\sqrt{\frac{2\eta}{|\Omega|}},$$

$$F^2(u(x)) = \left[F_v^2 - F_1(x,u,b_{*u}) + F_2(x,u)\right]_+$$
$$\geq \left[F_v^2 - 2\lambda\|b\|_{L^\infty(\Omega)} S - \frac{2\eta}{|\Omega|}S^2\right]_+$$

and

$$\left|\frac{1}{2}\left(F^2(u(x))\right)'\right| \leq \eta\frac{1}{|\Omega|}S + \lambda\|b\|_{L^\infty(\Omega)} u_+(x)$$
$$\leq \left(\eta\frac{1}{|\Omega|} + \lambda\|b\|_{L^\infty(\Omega)}\right) S.$$

Analogously, we obtain the estimates for $F(s)$, $F^2(s)$ and $\left|\frac{1}{2}\left(F^2(s)\right)'\right|$ and using these, we obtain the estimate about the right hand side of problem $(\mathcal{P})$. □

Corollary 3. *For this solution (u,F), assuming λ and η small enough in order to have*

$$F_v^2 - 2\lambda\|b\|_{L^\infty(\Omega)} S - \frac{2\eta}{|\Omega|}S^2 > 0, \tag{10}$$

we have that $F(s) > 0$ for all $s \in [\inf_\Omega u, \sup_\Omega u]$ and $F(s)$ is Lipschitz, *i.e. there exists a positive number $C(\lambda,\eta)$ only dependents of λ and η, such that*

$$|F(s) - F(\sigma)| \leq C(\lambda,\eta)|\sigma - s| \qquad \forall s,\sigma \in \left[\inf_\Omega u, \sup_\Omega u\right].$$

Moreover $C(\lambda,\eta)$ goes to zero when λ and η go to zero.

Proof. From the assumption (10) and the last two corollaries, we have that

$$F(s) > 0 \text{ for all } s \in [\operatorname{ess\,inf}_\Omega u, \operatorname{ess\,sup}_\Omega u].$$

Thus,

$$F^2(s) = \left[F_v^2 - 2\int_0^{s_+} p'(r)\, b_{*u}(|u > r|)\right.$$
$$\left. + 2\int_0^{s_+} j_1'(r_+, u_{+*}(0))\, u_{+*}'(|u > r|)\, dr\right]_+$$
$$= F_v^2 - 2\int_0^{s_+} p'(r)\, b_{*u}(|u > r|)$$
$$+ 2\int_0^{s_+} j_1'(r_+, u_{+*}(0))\, u_{+*}'(|u > r|)\, dr.$$

By the last identity (we can assume $\sigma \geq s \geq 0$ without lost of generality)

$$F^2(s) - F^2(\sigma) = 2\int_{s_+}^{\sigma_+} p'(r)\, b_{*u}(|u>r|)\, dr - 2\int_{s_+}^{\sigma_+} j_1'(r_+, u_{+*}(0))\, u_{+*}'(|u>r|)\, dr$$

and from the estimates of p', b_{*u}, j_1' and u_{+*}' we have that

$$|F^2(s) - F^2(\sigma)| \leq \lambda \|b\|_{L^\infty(\Omega)} |\sigma_+^2 - s_+^2| + 2\eta \frac{1}{|\Omega|} S|\sigma_+ - s_+|.$$

On the other hand

$$F(s) + F(\sigma) \geq 2\left[F_v^2 - 2\lambda \|b\|_{L^\infty(\Omega)} S - \frac{2\eta}{|\Omega|} S^2\right]^{1/2}$$

from Corollary 2. Notice that $F_v^2 - 2\lambda \|b\|_{L^\infty(\Omega)} S - \frac{2\eta}{|\Omega|} S^2 > 0$ form assumption (10) of corollary. Since

$$|F(s) - F(\sigma)| = \frac{|F^2(s) - F^2(\sigma)|}{F(s) + F(\sigma)},$$

by replacing the last two inequalities in this identity and considering that σ_+ and s_+ are less or equal than $\|u_+\|_{L^\infty(\Omega)} \leq S$, we get that for all $s, \sigma \in [\operatorname{ess\,inf}_{x\in\Omega} u, \operatorname{ess\,sup}_{x\in\Omega} u]$

$$|F(s) - F(\sigma)| = \frac{\left[\lambda \|b\|_{L^\infty(\Omega)} |\sigma_+ + s_+| + 2\eta \frac{1}{|\Omega|} S\right]}{2\left[F_v^2 - 2\lambda \|b\|_{L^\infty(\Omega)} S - \frac{2\eta}{|\Omega|} S^2\right]^{1/2}} |\sigma - s|,$$

$$|F(s) - F(\sigma)| \leq \frac{\left(\lambda \|b\|_{L^\infty(\Omega)} + \eta \frac{1}{|\Omega|}\right) S}{\left[F_v^2 - 2\lambda \|b\|_{L^\infty(\Omega)} + 2\frac{\eta}{|\Omega|} S^2\right]^{1/2}} |\sigma - s|.$$

Let $C(\lambda, \eta) := \frac{\left(\lambda \|b\|_{L^\infty(\Omega)} + \eta \frac{1}{|\Omega|}\right) S}{\left[F_v^2 - 2\lambda \|b\|_{L^\infty(\Omega)} + 2\frac{\eta}{|\Omega|} S^2\right]^{1/2}}$ be. $C(\lambda, \eta)$ is a positive constant. The last assertion of the corollary comes form the definition of C, S and ν. □

Lemma 1. *Let (u, F) be the given solution of $(\mathcal{P})$, then*

$$u(x) \geq \gamma \quad a.e.\ x \in \Omega.$$

Proof. Since $\gamma < 0$, multiplying the elliptic equation of problem $(\mathcal{P})$ by $(\gamma - u)_+ := \max(0, (\gamma - u))$ we obtain

$$\begin{aligned}
0 &\geq -\int_{\Omega} -\Delta\left(\gamma - u\left(x\right)\right)\left(\gamma - u\left(x\right)\right)_{+} dx \\
&= -\int_{\Omega} |\nabla\left(\gamma - u\left(x\right)\right)_{+}|^{2} dx \\
&= \int_{\Omega} \left(a\left(x\right) F\left(u\left(x\right)\right) + \frac{1}{2}\left(F^{2}\left(u\left(x\right)\right)\right)' + b\left(x\right) p'\left(u\left(x\right)\right)\right)\left(\gamma - u\left(x\right)\right)_{+} \\
&= \int_{\{y\in\Omega : u(y)<\gamma\}} a\left(x\right) F_{v} dx \geq 0.
\end{aligned}$$

Thus

$$\int_{\Omega} |\nabla\left(\gamma - u\left(x\right)\right)_{+}|^{2} dx = 0$$

and then $(\gamma - u)_{+}$ is constant for a.e. $x \in \bar{\Omega}$. Since $u(x) = \gamma$ on $\partial\Omega$, we obtain that $(\gamma - u)_{+} \equiv 0$ in Ω and so $u \geq \gamma$ in Ω. □

3 Estimate on the Location and Size of the Plasma Region and the Vacuum Region

We consider the following approach: *(i)* to give a condition for the existence of the free–boundary (i.e. $\Omega_p \neq \emptyset$), *(ii)* to verify that the solution u is a supersolution for an auxiliary problem in a test balls in Ω, *(iii)* to give a suitable local subsolution $\underline{u}$ for this auxiliary problem satisfying the hypotheses of the comparison principle in the sense of Hopf [6] and finally *(iv)* to compare u with $\underline{u}$. This way to work will be the same for to study the plasma region and the vacuum region.

Let φ_1 be a normalized eigenfunction associated to the first eigenvalue λ_1 of the operator $-\Delta$ on Ω with Dirichlet boundary condition, i.e., $\varphi_1 \in H_0^1(\Omega)$ and $-\Delta\varphi_1 = \lambda_1\varphi_1$ on Ω. We know that $\varphi_1 > 0$ on Ω. Besides, we can renormalize it such that $\lambda_1 \int_{\Omega} \varphi_1 dx = 1$.

The following proposition guaranties the existence of free–boundary and thus $\text{meas}\{x \in \Omega : u(x) > 0\} = |\Omega_p| > 0$.

Proposition 3. *Assume that*

$$-\gamma < F_v \int_{\Omega} a(x)\varphi_1(x)dx,$$

then any solution u of $(\mathcal{P}_*)$ satisfies $u_+ \not\equiv 0$.

Proof. Arguing by contradiction (see [4]). □

We will use the following general result

Lemma 2. *Let $B \subset R^2$ and open ball of radius R centred at the origin and assume $\hat{a} \in L^{\infty}(B)$ be radially symmetric (i.e. $\hat{a}(x) = \tilde{a}(|x|)$ a.e. $x \in B$). Then, unique solution $u \in W^{2,p}(B)$, $p \in [1, \infty)$, to problem*

$$(PB)\begin{cases} -\Delta \underline{u} = \hat{a} & in\ B \\ \underline{u} \ \ = \gamma & on\ \partial B \end{cases}$$

satisfies:

$$\begin{aligned} &if \int_0^R \left(\frac{1}{r}\int_0^r s\tilde{a}(s)\,ds\right)dr + \gamma = 0 \ then\ \underline{u}(0) = 0, \\ &if \int_0^R \left(\frac{1}{r}\int_0^r s\tilde{a}(s)\,ds\right)dr + \gamma < 0 \ then\ \underline{u}(0) < 0, \\ &if \int_0^R \left(\frac{1}{r}\int_0^r s\tilde{a}(s)\,ds\right)dr + \gamma > 0 \ then\ \underline{u}(0) > 0. \end{aligned}$$

Moreover, if $\int_0^r s\tilde{a}(s)\,ds \geq 0 \quad \forall r \in (0,R]$ *then* u *decreases along the radius* $r = |x|$.

Proof. The existence, regularity and uniqueness of $\underline{u}$, solution of (PB) is a well-known result (see for instance [1]). Moreover, $\underline{u}$ is a radial symmetric function in B (i.e. , so, $\underline{u}(x) = \tilde{\underline{u}}(|x|)\ x \in B$) and is the unique solution to ordinary differential equation

$$\begin{cases} -\dfrac{1}{r}\dfrac{\partial}{\partial r}\left(r\dfrac{\partial}{\partial r}\tilde{\underline{u}}\right) = \tilde{a}(r) \ \text{ in } 0 < r < R, \\ \tilde{\underline{u}}(R) \qquad\qquad = \gamma \qquad \tilde{\underline{u}}(0) = 0. \end{cases}$$

By integration, we obtain the exact solution for previous ordinary differential equation

$$\tilde{\underline{u}}(r) = \int_0^R \left(\frac{1}{r}\int_0^r s\tilde{a}(s)\,ds\right)dr + \gamma \quad r \in [0,R]$$

and from this and the fact that $\underline{u}(x) = \tilde{\underline{u}}(|x|)\ x \in B$, we prove the lemma. □

Proof (Theorem 1). Let $x_0 \in \Omega$ such that $d(x_0,\partial\Omega) \geq R_p$ with $R_p = \sqrt{\frac{-4\gamma}{F_v\,\mathrm{essinf}_{x\in\Omega}\,a}}$ and $B_0 := B_{R_p}(x_0) = \{x \in \Omega : d(x,x_0) < R_p\}$. Since (u,F) is a solution of problem $(\mathcal{P})$,

$$0 = -\Delta u(x) - a(x)F(u(x)) - \frac{1}{2}\left(F^2(u(x))\right)' - b(x)p'(u(x)) \quad \text{in } \Omega.$$

Now, by the properties of b, p', we have that $b(x)p'(u(x)) \geq 0$ a.e. $x \in \Omega$ and by estimates on $\left(F^2(u(x))\right)'$ and u (see Corollary 2) and the fact that u has not flat region, we have that

$$0 \leq -\Delta u(x) - aF(u(x)) + 2\lambda\|b\|_{L^\infty(\Omega)}u_+(x) + \eta\frac{1}{|\Omega|}S \quad \text{in } \Omega.$$

Then

$$-\Delta u + f(x,u) \geq 0 \text{ in } B_0 \tag{11}$$

where $f : \Omega \times \mathbb{R} \to \mathbb{R}^+$ is defined by

$$f(x,\tau) = -aF(\tau) + 2\lambda \|b\|_{L^\infty(\Omega)} \tau_+ + \eta \frac{1}{|\Omega|} S.$$

Notice that $f(x,\cdot)$ is non–decreasing in τ since F is a non–increasing function. So we can apply the comparison principle for quasi-linear problems (see e.g. [6]). Now, we consider the solution $\underline{u}$ given in Lemma 2 with $B = B_0 = B_{R_p}(x_0)$ and $\hat{a} := F_v \operatorname*{essinf}_{x\in\Omega} a(x)$. So, $\underline{u}$ is verifies

$$(PB) \begin{cases} -\Delta \underline{u}(x) = F_v \operatorname*{essinf}_{x\in\Omega} a(x) & \text{in } B_0, \\ \underline{u}(x) \quad = \gamma & \text{on } \partial B_0 \end{cases}$$

and by property of R_p, $\underline{u}(x) < 0$ for all $x \in B_0 \setminus \{x_0\}$. Notice that, by integration, the exact solution $\underline{u}$ is given by

$$\underline{u}(x) = \gamma + F_v \left(R_p - |x - x_0|^2\right) \operatorname*{essinf}_{x\in\Omega} a(x) \quad \text{for all } x \in B_0$$

and

$$\underline{u}(x_0) = \gamma + F_v R_p \operatorname*{essinf}_{x\in\Omega} a(x) = 0$$

from the definition of R_p. On the other hand

$$\begin{aligned} f(x,\underline{u}) &= -a(x) F_v + \eta \frac{1}{|\Omega|} S \\ &\le -F_v \operatorname*{essinf}_{\Omega} a + \eta \frac{1}{|\Omega|} S < 0 \quad \text{a.e. in } B_0 \end{aligned} \tag{12}$$

from the assumption of theorem. So, we get that

$$\begin{aligned} -\Delta \underline{u} + f(x,\underline{u}) < 0 \le -\Delta u + f(x,u) \quad &\text{in } B_0, \\ \underline{u}(x) = \gamma \le u(x) \quad &\text{on } \partial\Omega \end{aligned}$$

form (11), (12) and the property $u \ge \gamma$ in Ω (see Lemma 1). By the comparison principle, we conclude that

$$u \ge \underline{u} \;\text{ in } \bar{B}_0 \;\text{ then }\; u(x_0) \ge \underline{u}(x_0) = 0.$$

Since the solution u has not flat region, we can deduce that

$$u > 0 \text{ a.e. in } \{x \in \Omega : d(x,\partial\Omega) \ge R_p\}$$

and thus

$$\{x \in \Omega : d(x,\partial\Omega) \ge R_p\} \subset \Omega_p := \{x \in \Omega : u(x) > 0\}. \qquad \square$$

Remark 4. Theorem 1 gives a sufficient condition on the size and the geometry of Ω for the existence of free–boundary and, hence, for the size and geometry of plasma region Ω_p.

Analogously, we prove an estimate for the vacuum region $\Omega_v := \{x \in \Omega : u(x) < 0\}$:

Proof (Theorem 2). Let $\rho = \left(\frac{2S}{K}\right)^{1/2}$ be the positive constant introduced in Theorem 2 with K the bound obtained in Corollary 2. Then $R_v := \hat{R} - \left(\frac{2S}{K}\right)^{1/2}$. We take a point $x_1 \in \Omega$ such that $d(x, \partial\Omega) = R_v$ and a point $\bar{x}_1 \in \partial\Omega$ such that $d(x_1, \partial\Omega) = d(x_1, \bar{x}_1)$. Then $x_1 = \bar{x}_1 + R_v\mathbf{n}$ and $\{x : x = \bar{x}_1 + r\mathbf{n},\ 0 \leq r \leq \hat{R}\} \subset \bar{\Omega}$. On this segment, (u, F) satisfies the nonlinear equation

$$-u''(r) = a(r)\,F(u(r)) + \frac{1}{2}\left(F^2(u(r))\right)' + b(r)\,p'(u(r))$$

for $0 < r < \hat{R}$ (here, for a given function $h : \Omega \to \mathbb{R}$, we use the notation $h(r) := h(\bar{x}_1 + r\mathbf{n})$). From Corollary 2, $-u''(r) \leq K$. Moreover

$$u(0) = u(\bar{x}_1) = \gamma \quad \text{and} \quad u(\hat{R}) = u(\bar{x}_1 + \hat{R}\mathbf{n}) \leq \|u_+\|_{L^\infty(\Omega)}\,.$$

As in the proof of Theorem 1, we can to find a bound for the right hand side and u verifies the equations

$$\begin{cases} -u''(r) \leq K \ \text{ in } (0, \hat{R}), \\ \quad u(0) = \gamma,\ u(\hat{R}) \leq \|u_+\|_{L^\infty(\Omega)} \leq S\,. \end{cases}$$

Notice that $u \geq \gamma$ and $\gamma < 0$, thus if $u(\hat{R}) \leq 0$ then $u(\hat{R}) \leq \|u_+\|_{L^\infty(\Omega)}$; and if $u(\hat{R}) > 0$ then $u(\hat{R}) = u_+(\hat{R}) \leq \|u_+\|_{L^\infty(\Omega)}$. Now, we consider the real function $v(r) := S - \frac{1}{2}K(\hat{R} - r)^2$ for $r \in [0, \hat{R}]$. Then, by definition of v, on has that $v(\hat{R}) = S$, $v(R_v) = S - \frac{1}{2}K(\hat{R} - R_v)^2 = 0$ and since v is increasing in $(0, \hat{R})$ then $v(r) \leq 0$ in $(0, R_v)$. One the other hand, v is the unique solution to the linear boundary problem

$$(BP)\begin{cases} -v''(r) = K & \text{in } (0, \hat{R}), \\ \quad v(0) \ \ = \gamma, & v(\hat{R}) = S. \end{cases}$$

Thus, $u(r) \leq v(r)$ in $(0, \hat{R})$. Moreover, by construction v

$$v(R_v) = 0 > v(r) \geq u(r) \ \text{ for any } r \in [0, R_v).$$

In particular for all $0 < r < R_v$, one has that $u(\bar{x}_1 + r\mathbf{n}) < 0$ and then the segment $\{x = \bar{x}_1 + r\mathbf{n}, 0 < r < R_v\} \subset \Omega_v = \{x \in \Omega : u(x) < 0\}$. Thus $\{x \in \Omega : d(x, \partial\Omega) \leq \hat{R} - \rho\} \subset \Omega_v = \{x \in \Omega : u(x) < 0\}$. □

Remark 5. It is known that if Ω is a disk, then the level sets of u are concentric circles. From the last two theorems we can deduce how is the shape of free–boundary for a more general domain Ω. In Theorem 2 we have assumed that

$$0 < \left(\frac{2S}{K}\right)^{1/2} < \hat{R} \leq R_p + \left(\frac{2S}{K}\right)^{1/2}.$$

If we choose $\hat{R} = R_p + \left(\frac{2S}{K}\right)^{1/2}$, then $R_v = \hat{R} - \left(\frac{2S}{K}\right)^{1/2} = R_p$ (R_p is the constant given in Theorem 1). Since $\Omega = \{x \in \Omega : d(x, \partial\Omega) \leq R_p\} \cup \{x \in \Omega: d(x, \partial\Omega) \geq R_p\} = \Omega_p \cup \Omega_v$ and from Theorem 1 and Theorem 2 we have that $\{x \in \Omega : d(x, \partial\Omega) \leq R_p\} \subset \Omega_p$, $\{x \in \Omega: d(x, \partial\Omega) \geq R_p\} \subset \Omega_v$ we can obtain exactly the free–boundary

$$\partial\Omega_p = \{x : u(x) = 0\} = \{x : d(x, \partial\Omega) = R_p\}.$$

The hypothesis of Theorem 1 and Theorem 2 about Ω, that is, Ω be an open bounded regular (with C^1 boundary $\partial\Omega$) subset of $\mathbb{R}^2$ and such that

$$\exists\ x_0 \in \Omega \text{ verifying } R_p := \left(\frac{-4\gamma}{F_v \operatorname*{essinf}_{x\in\Omega} a(x)}\right)^{\frac{1}{2}} < d(x, \partial\Omega)$$

defines a type of domains Ω more general that a disk for which we can obtain the shape of free–boundary for the solution of problem $(\mathcal{P})$.

Acknowledgement. The author was partially supported by the projects No. CCG07 - UPM/000-3199 of Universidad Politécnica de Madrid and CAM; and by the projects No. MTM2008-06208 of DGISGP (Spain).

References

1. Brezis, H.: Analyse Fonctionelle. North–Holland, Amsterdam (1983)
2. Boozer, A.H.: Establishment of magnetic coordinates for a given magnetic field. Phys. Fluids 25, 520–521 (1982)
3. Cooper, W.A.: Global External Ideal Magnetohydrodynamic Instabilities in Three-dimensional Plasmas, Theory of Fusion Plasmas. In: Proc. of the Joint Varenna–Laussane Workshop, Compositori, Bologna (1990)
4. Díaz, J.I., Padial, J.F., Rakotoson, J.M.: Mathematical treatment of the magnetic confinement in a current carrying Stellarator. Nonlin. Anal. 34, 857–887 (1998)
5. Díaz, J.I., Rakotoson, J.M.: On a nonlocal stationary free boundary problem arising in the confinement of a plasma in a Stellarator geometry. Arch. Rat. Mech. Anal. 134, 53–95 (1996)
6. Gilbarg, D., Trudinger, N.: Elliptic Partial Differential Equations of Second Order. Springer, Berlin (1983)
7. Hender, T.C., Carreras, B.A.: Equilibrium calculations for helical axis Stellarators. Phys. Fluids 27, 2101–2120 (1984)
8. Lerena, M.B.: On the existence of the free-boundary for some problems arising in plasma physics. Nonlin. Anal. 55, 419–439 (2003)
9. Mossino, J.: Inégalités isoperimétriques. Collection "Travaux en cours", Herman Paris (1984)

10. Mossino, J., Rakotoson, J.M.: Isoperimetric inequalities in parabolic equations. Ann. Scuola Normale Sup. Pisa - Classe Sci. Sér 4 13(1), 51–73 (1986)
11. Mossino, J., Temam, R.: Directional derivative of the increasing rearrangement mapping and application to a queer differential equation in plasma physics. Duke Math. J. 48, 475–495 (1981)
12. Padial, J.F.: EDP's no lineales originadas en plasmas de fusión y filtración en medios porosos. PhD Thesis, Universidad Complutense de Madrid, Spain (1995)
13. Padial, J.F.: Existence and estimate of the location of the free-boundary for a non local inverse elliptic-parabolic problem arising in nuclear fusion. AIMS Journal (to appear, 2011)
14. Rakotoson, J.M.: Some properties of the relative rearrangement. J. Math. Anal. Appl. 135, 488–500 (1988)
15. Rakotoson, J.M.: Strong continuity of the relative rearrangement maps and application to a Galerkin approach for nonlocal problems. Appl. Math. Lett. 8, 61–63 (1995)
16. Rakotoson, J.M.: Réarrangement Relatif: Un instrument d'estimations dans les problèmes aux limites. Mathématiques et Applications, SMAI. Springer, Paris (2008)

Part VI

Theory of Games

A New Power Index for Spatial Games

José María Alonso-Meijide[1], María Gloria Fiestras-Janeiro[2], and Ignacio García-Jurado[3]

[1] Departamento de Estatística e Investigacíın Operativa, Universidade de Santiago de Compostela, 27002 Lugo, Spain
josemaria.alonso@usc.es

[2] Departamento de Estatística e Investigación Operativa, Universidade de Vigo, 36271 Vigo, Spain
fiestras@uvigo.es

[3] Departamento de Matemáticas, Universidade da Coruña, 15071 A Coruña, Spain
igjurado@udc.es

Summary. In this paper we present a new power index for spatial games. We study some of its properties and indicate its advantages in comparison with the other existing indexes. Finally we illustrate our results with an example taken from a Spanish regional Parliament.

1 Introduction

One interesting application of cooperative game theory is its use in the analysis of voting institutions. In a voting institution a finite group of agents N makes decisions by voting. The decisions considered here are of the kind "we take a particular action" or "we do not take the action." A voting institution is characterized when we give N and the class of subsets of N which have enough power to win a voting, i.e. the class of winning coalitions. This is what we call a *simple game*.

Spatial games are a more sophisticated model which incorporates ideological considerations to voting institutions. A *spatial game* is a simple game together with a collection of points, one for each voter, whose coordinates describe the ideological positions of the voters with respect to certain selected variables. The voters vote for taking or not actions which might also be located in the ideological space. A winning coalition can enforce any decision on which its members agree. Moreover, we assume that the closer an action is to the ideological position of a voter the more likely is that the voter votes for it.

The problem we tackle in this paper is the following: we would like to define an index which measures the power of each voter in one of these spatial games, if we mean by power a voter's capacity for influencing the group decisions. This problem has already been treated in [3], [5] and [4]. In those papers two indexes for spatial games are proposed, both being modifications of the Shapley-Shubik index, originally introduced for simple games (see [6]).

The organization of the paper is as follows. In Section 2 we formally introduce spatial games and two power indexes for them. In Section 3 we propose a new power index and study some of its properties. Finally, in Section 4 we apply this index to analyse a real Parliament: the Parliament which ruled Catalonia (a Spanish region) in the period 2003-2006.

2 Spatial Games

We start this section giving the definition of a simple game, a mathematical model which formalizes voting institutions. N denotes the finite set of players $\{1, 2, \ldots, n\}$. Any subset $S \subseteq N$ is said to be a coalition. A winning coalition is a subset of N which has the power to enforce a decision when it has been unanimously adopted by its members.

Definition 1. *A simple game $(N, \mathcal{W})$ consists of a finite set of players N and a family of winning coalitions $\mathcal{W}$, which is a collection of subsets of N with the following properties:*

- $\emptyset \notin \mathcal{W}$*,*
- $N \in \mathcal{W}$*,*
- *if $S \in \mathcal{W}$ and $S \subset T$, then $T \in \mathcal{W}$ (monotonicity).*

A minimal winning coalition of $(N, \mathcal{W})$ is a winning coalition which does not contain any other winning coalition as a proper subset, i.e., $S \subseteq N$ is a minimal winning coalition if $S \in \mathcal{W}$ and $T \notin \mathcal{W}$ for every $T \subset S$. The set of minimal winning coalitions is denoted by $\mathcal{W}^m$. By the monotonicity property, it is clear that a simple game is characterized if we give its set of minimal winning coalitions.

A power index is a function g which assigns to each simple game $(N, \mathcal{W})$ a vector $g(N, \mathcal{W}) \in \mathbb{R}^N$ where each component $g_i(N, \mathcal{W})$ provides a measure of the power of player i in the simple game $(N, \mathcal{W})$.

The most well-known power index is the Shapley-Shubik index, introduced in [6]. It is the restriction of the Shapley value, defined for a wider class of cooperative games, to the class of simple games (see, for instance, [2] for details on the Shapley value). The Shapley-Shubik index can be easily introduced using the notion of pivot, that we define below.

Definition 2. *Let $(N, \mathcal{W})$ be a simple game and let π be a permutation of N. For every $i \in N$, let $S(i, \pi)$ be the set of players that precede i in the order defined by π. Then, i is the pivot of π if and only if $S(i, \pi) \notin \mathcal{W}$, while $S(i, \pi) \cup \{i\} \in \mathcal{W}$. Notice that the properties in the definition of simple game imply that each permutation π has a unique pivot. We denote by $\Pi_i(N, \mathcal{W})$ the set of permutations of N for which i is the pivot in $(N, \mathcal{W})$.*

Definition 3. *The Shapley-Shubik index provides for every simple game* $(N, \mathcal{W})$ *the vector* $g(N, \mathcal{W})$ *such that, for each* $i \in N$,

$$g_i(N, \mathcal{W}) = \frac{|\Pi_i(N, \mathcal{W})|}{n!},$$

where $|\Pi_i(N, \mathcal{W})|$ *denotes the cardinal of the set* $\Pi_i(N, \mathcal{W})$.

Notice that the Shapley-Shubik index of a simple game can be interpreted as the vector assigning to each player the probability that this player is a pivot, assuming that all permutations of N are equally probable. This assumption is quite reasonable when dealing with simple games, where only the set of winning coalitions is known. However, it is not so reasonable when dealing with more sophisticated models. One of those models are spatial games, introduced in [3]. A spatial game is a simple game together with a collection of points, one for each player, whose coordinates describe the ideological positions of the players with respect to certain selected variables.

Definition 4. *An m-dimensional spatial game is a triplet* $(N, \mathcal{W}, \{Q^j\}_{j \in N})$ *such that:*

- $(N, \mathcal{W})$ *is a simple game, and*
- $Q^j \in \mathbb{R}^m$ *for all* $j \in N$. Q^j *is said to be j's* ideal point *and represents j's ideological position in the ideological space* $\mathbb{R}^m$. *We assume that* $Q^j \neq Q^k$ *for all different* $j, k \in N$.

In spatial games the players vote for taking or not actions which can be located in the ideological space. It is reasonable to assume that the closer an action is to the ideological position of a player the more likely that the player votes for it. In this paper we define an index which measures the power of each voter in one of these spatial games, if we mean by power a voter's capacity for influencing the group decisions. Moreover, our index is a modification of the Shapley-Shubik index for spatial games.

There exist two other modifications of the Shapley-Shubik index for spatial games. All such modifications are based on the same idea: when the ideological positions of the players are added to a simple game, the assumption that all permutations are equally probable is not reasonable any more.

2.1 The Owen Index

The first modification is the Owen index introduced in [3]. The Owen index is defined as follows. Take a spatial game $(N, \mathcal{W}, \{Q^j\}_{j \in N})$. Now consider the sphere $\mathbb{S}$ with the lowest dimension which contains all the ideal points of the players. If we choose a point $z \in \mathbb{S}$, the distances between z and the ideal points of the players can be computed. Then z determines a permutation of N: the one which gives rise to the order of increasing distances from the ideal points $\{Q^j\}_{j \in N}$ to z (notice that the set of all points of the sphere which produce ties in the collection of distances has measure zero). Now, for each $i \in N$, define A_i as the set of points $z \in \mathbb{S}$ such that i is the pivot in the permutation determined by z. The Owen index is defined as follows.

Definition 5. *The Owen index provides for every spatial game* $(N, \mathcal{W}, \{Q^j\}_{j \in N})$ *the vector* $f^O(N, \mathcal{W}, \{Q^j\}_{j \in N})$ *such that, for each* $i \in N$,

$$f_i^O(N, \mathcal{W}, \{Q^j\}_{j \in N}) = \frac{\lambda(A_i)}{\lambda(\mathbb{S})},$$

where λ *denotes the Lebesgue measure.*

A fundamental drawback of the Owen index is that its computation can be extremely hard even for small problems. Hence, Shapley proposed another index for spatial games in [5], which was further analysed by Owen and Shapley in [4]. We call it the Owen-Shapley index.

2.2 The Owen-Shapley Index

In order to define the Owen-Shapley index take an m-dimensional spatial game $(N, \mathcal{W}, \{Q^j\}_{j \in N})$. The unit sphere in $\mathbb{R}^m$ is the set $\mathbb{S}_m^1 = \{u \in \mathbb{R}^m \mid \sum_{i=1}^m u_i^2 = 1\}$. Now, every $u \in \mathbb{S}_m^1$ determines the permutation of N which gives rise to the following order: given $i, j \in N$, i precedes j if and only if $\sum_{k=1}^m u_k Q_k^i < \sum_{k=1}^m u_k Q_k^j$ (notice that $\sum_{k=1}^m u_k Q_k^i$ is the scalar product of the vectors u and Q^i). Owen and Shapley proved that the set of points in $\mathbb{S}_m^1$ which produce ties when comparing these scalar products has measure zero. Now, for each $i \in N$, define B_i as the set of vectors $u \in \mathbb{S}_m^1$ such that i is the pivot in the permutation determined by u. The Owen-Shapley index is defined as follows.

Definition 6. *The Owen-Shapley index provides for every* m*-dimensional spatial game* $(N, \mathcal{W}, \{Q^j\}_{j \in N})$ *the vector* $f^{OS}(N, \mathcal{W}, \{Q^j\}_{j \in N})$ *such that, for each* $i \in N$,

$$f_i^{OS}(N, \mathcal{W}, \{Q^j\}_{j \in N}) = \frac{\lambda(B_i)}{\lambda(\mathbb{S}_m^1)},$$

where λ *denotes the Lebesgue measure.*

The computation of the Owen-Shapley index is also quite hard, but it is feasible if the dimensions of the problem are small. We provide now an example.

Example 1. In this example we illustrate the computation of the Owen-Shapley index in the following two-dimensional spatial game. The simple game is given by $N = \{1, 2, 3, 4\}$ and $\mathcal{W}^m = \{\{1, 4\}, \{2, 4\}, \{3, 4\}, \{1, 2, 3\}\}$. The ideal points of the players are displayed in Figure 1. Take for instance the vector $u = (\frac{1}{\sqrt{5}}, \frac{2}{\sqrt{5}}) \in \mathbb{S}_2^1$. The permutation determined by u is $\pi(1) = 1$, $\pi(2) = 3$, $\pi(3) = 4$, $\pi(4) = 2$. After some algebra, it can be obtained that the realization of the Owen-Shapley index in this game is $(0, 0.1982, 0.3137, 0.4881)$.

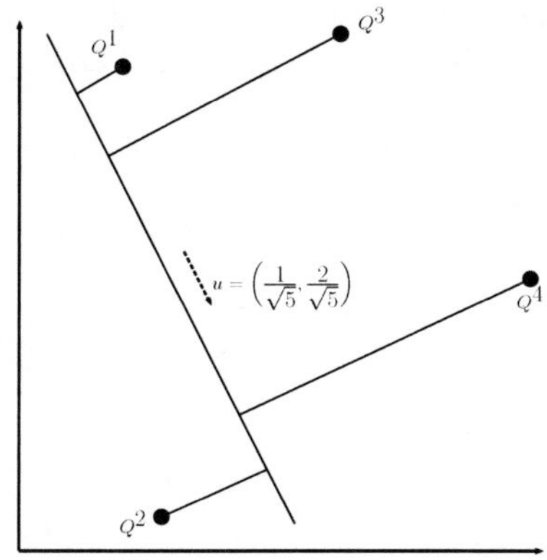

Fig. 1. Order determined by the vector u

3 A New Power Index for Spatial Games

In this section we propose a new variation of the Shapley-Shubik index for spatial games. In our variation, all permutations are possible and the probability of each permutation depends on the distance among the ideal points of the players taking into account the induced order. This new index that we call *distance index* have good computational properties and other advantages with respect to the two existing indexes presented before. Let us formally introduce the distance index.

To start with we give a notion which is in the basis of our index: the *length* of a permutation. Let $(N, \mathcal{W}, \{Q^j\}_{j \in N})$ be an m-dimensional spatial game. Given a permutation $\pi : N \to N$ of the set of players, we consider the derived polygonal given by the sequence of segments $[Q^{\pi(1)}, Q^{\pi(2)}], [Q^{\pi(2)}, Q^{\pi(3)}], \ldots, [Q^{\pi(n-1)}, Q^{\pi(n)}]$. The length of π is the sum of the lengths of these segments. The formal definition is given below.

Definition 7. *Let $(N, \mathcal{W}, \{Q^j\}_{j \in N})$ be an m-dimensional spatial game and let π be a permutation of N. The length of π is defined as*

$$l(\pi) = \sum_{i=1}^{n-1} \sqrt{\sum_{j=1}^{m} (Q_j^{\pi(i)} - Q_j^{\pi(i+1)})^2}.$$

The length of a permutation can be seen as a measure of its internal instability. The more lengthy a permutation is, the less likely that it describes a cooperation route. The distance index is a modification of the Shapley-Shubik index for spatial games which considers that each permutation has a probability inversely proportional to its length. Remember that $\Pi_i(N, \mathcal{W})$ denotes the set of permutations of N for which i is the pivot in $(N, \mathcal{W})$. Denote by $\Pi(N)$ the set of all permutations of N.

Definition 8. *The distance index provides for every m-dimensional spatial game $(N, \mathcal{W}, \{Q^j\}_{j\in N})$ the vector $f^D(N, \mathcal{W}, \{Q^j\}_{j\in N})$ such that, for each $i \in N$,*

$$f_i^D(N, \mathcal{W}, \{Q^j\}_{j\in N}) = \frac{\sum_{\pi\in\Pi_i(N,\mathcal{W})} \frac{1}{l(\pi)}}{\sum_{\pi\in\Pi(N)} \frac{1}{l(\pi)}}.$$

Example 2. We compute the distance index for the spatial game given in Example 1. Figure 2 displays the segments needed for the computation of $l(\pi)$ for permutation $\pi(i) = i$, for all $i \in N$. After some algebra, it can be obtained that the realization of the distance index in this game is (0.1507, 0.1718, 0.1765, 0.5010).

Let us see now a couple of properties of the distance index. The first one has to do with the so-called null players. A null player of a simple game $(N, \mathcal{W})$ is an $i \in N$ such that $i \notin S$ for all $S \in \mathcal{W}^m$; so, a null player is one which is not necessary for the formation of a winning coalition. Intuitively, it is clear that a null player has no power in the voting institution described by the simple game. On the contrary, every player which is not null has some power in that voting institution. If we add a set of ideal points to the simple game, null players still have no power, and the other players still have some power (maybe modified by the new elements of the model). These intuitive considerations are well reflected by the distance index as the following proposition shows.

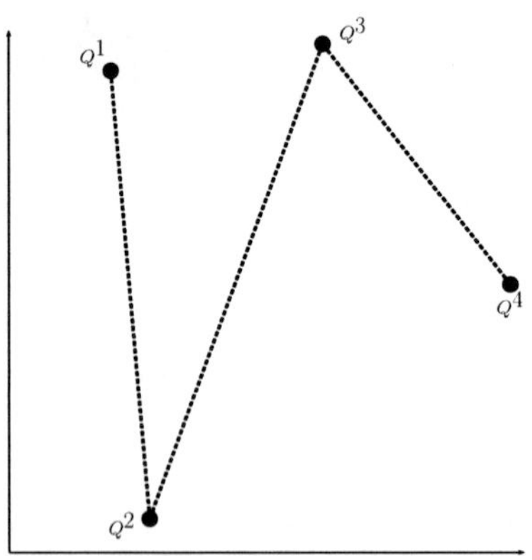

Fig. 2. Length of π

Proposition 1. *Let $(N, \mathcal{W}, \{Q^j\}_{j\in N})$ be an m-dimensional spatial game and take $i \in N$. Then $f_i^D(N, \mathcal{W}, \{Q^j\}_{j\in N}) = 0$ if and only if i is a null player of $(N, \mathcal{W})$.*

Proof. $f_i^D(N, \mathcal{W}, \{Q^j\}_{j\in N}) = 0$ if and only if $\Pi_i(N, \mathcal{W})$ is an empty set, and $\Pi_i(N, \mathcal{W})$ is an empty set if and only if i is a null player of $(N, \mathcal{W})$.

Notice that the Owen-Shapley index does not satisfy the property in Proposition 1 as Example 1 shows. In that example, player 1 is not a null player but it has power zero according to the Owen-Shapley index. The interest of this property will emerge again in the example discussed in Section 4.

The second property has to do with symmetry taking into account the spatial features of the model. First we introduce some notation. Let $(N, \mathcal{W}, \{Q^j\}_{j\in N})$ be an m-dimensional spatial game and take $i \in N$. We build now the partition of $\Pi_i(N, \mathcal{W})$ given by the lengths of its elements:

- $\Pi_i^0(N, \mathcal{W}) = \emptyset$.
- For every integer $r \geq 1$ with $\Pi_i(N, \mathcal{W}) \setminus \cup_{k=0}^{r-1} \Pi_i^k(N, \mathcal{W}) \neq \emptyset$, define $\Pi_i^r(N, \mathcal{W})$ recursively as the following set:

$$\{\pi \in \Pi_i(N, \mathcal{W}) \setminus \cup_{k=0}^{r-1} \Pi_i^k(N, \mathcal{W}) \mid l(\pi) \leq l(\tilde{\pi})\ \forall \tilde{\pi} \in \Pi_i(N, \mathcal{W}) \setminus \cup_{k=0}^{r-1} \Pi_i^k(N, \mathcal{W})\}.$$

It is clear that there exists $r_i \in \mathbb{N}$ such that $\Pi_i(N, \mathcal{W}) = \cup_{r=1}^{r_i} \Pi_i^r(N, \mathcal{W})$. The set $\Pi_i^1(N, \mathcal{W})$ contains all the permutations with the smallest length among those for which i is the pivot, and $\Pi_i^{r_i}(N, \mathcal{W})$ contains all permutations with the largest length among those for which i is the pivot. For each $i \in N$ and each $r \in \{1, \ldots, r_i\}$, $\pi, \tilde{\pi} \in \Pi_i^r(N, \mathcal{W})$ if and only if $l(\pi) = l(\tilde{\pi})$. Next we give the notion of *spatially symmetric* players in a spatial game and illustrate this definition with an example.

Definition 9. *Let $(N, \mathcal{W}, \{Q^j\}_{j\in N})$ be an m-dimensional spatial game and take $i, k \in N$. We say that i and k are spatially symmetric players if and only if $r_i = r_k$ and, moreover, for every $r \in \{1, \ldots, r_i\}$ it holds that:*

- *$|\Pi_i^r(N, \mathcal{W})| = |\Pi_k^r(N, \mathcal{W})|$, and*
- *$l(\pi) = l(\tilde{\pi})$ for every $\pi \in \Pi_i^r(N, \mathcal{W})$ and $\tilde{\pi} \in \Pi_k^r(N, \mathcal{W})$.*

Example 3. Let $(N, \mathcal{W}, \{Q^i\}_{i\in N})$ be a two-dimensional spatial game where N is the set $\{1, 2, 3, 4, 5\}$, the collection of minimal winning coalitions is

$$\mathcal{W}^m = \{\{1, 2\}, \{1, 3, 4\}, \{1, 3, 5\}, \{1, 4, 5\}, \{2, 3, 4\}, \{2, 3, 5\}, \{2, 4, 5\}\},$$

and the ideal points are those displayed in Figure 3. It is not difficult to check that 1 and 2 are spatially symmetric. It can also be seen that, even though players 3 and 5 satisfy that $|\Pi_3(N, \mathcal{W})| = |\Pi_5(N, \mathcal{W})|$, they are not spatially symmetric. To verify that 3 and 5 are not spatially symmetric take for instance the permutation $\tilde{\pi}$ given by $\tilde{\pi}(1) = 3$, $\tilde{\pi}(2) = 2$, $\tilde{\pi}(3) = 5$, $\tilde{\pi}(4) = 4$, and $\tilde{\pi}(5) = 1$. Clearly $\tilde{\pi} \in \Pi_5(N, \mathcal{W})$; however there is no $\pi \in \Pi_3(N, \mathcal{W})$ such that $l(\pi) = l(\tilde{\pi})$.

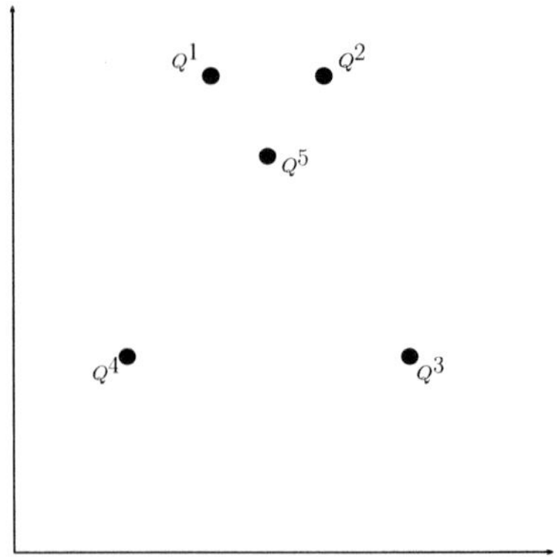

Fig. 3. Ideal points of the spatial game in Example 3

Clearly, spatially symmetric players in a spatial game are in some sense interchangeable and have the same power. Thus a sensible power index for spatial games should treat spatially symmetric players accordingly. The distance index does have this property as the following proposition shows.

Proposition 2. *Let* $(N, \mathcal{W}, \{Q^j\}_{j \in N})$ *be an* m*-dimensional spatial game with a pair of spatially symmetric players* $i, k \in N$*. Then*

$$f_i^D(N, \mathcal{W}, \{Q^j\}_{j \in N}) = f_k^D(N, \mathcal{W}, \{Q^j\}_{j \in N}).$$

Proof. For each $r \in \{1, \ldots, r_i\}$, take $\pi^r \in \Pi_i^r(N, \mathcal{W})$ and $\tilde{\pi}^r \in \Pi_k^r(N, \mathcal{W})$. Then,

$$f_i^D(N, \mathcal{W}, \{Q^j\}_{j \in N}) = \frac{\sum_{r=1}^{r_i} \frac{|\Pi_i^r(N,\mathcal{W})|}{l(\pi^r)}}{\sum_{\pi \in \Pi(N)} \frac{1}{l(\pi)}} = \frac{\sum_{r=1}^{r_i} \frac{|\Pi_k^r(N,\mathcal{W})|}{l(\tilde{\pi}^r)}}{\sum_{\pi \in \Pi(N)} \frac{1}{l(\pi)}} = f_k^D(N, \mathcal{W}, \{Q^j\}_{j \in N}).$$

The distance index satisfies the two properties above but there are other power indexes for spatial games satisfying them. An open and interesting question for future research is to find a collection of properties (including the two treated in this paper) which characterize the distance value.

4 An Example

In this section we illustrate with an example the behaviour of the distance index and we compare it with the Owen-Shapley index.

The example deals with the Catalonia Parliament resulting after the elections held on November 16, 2003, Catalonia being one of the seventeen Spanish regions. This Parliament has also been analysed in other papers, like for instance in [1]. The parties that obtained some of the 135 seats of the Parliament in those elections were:

CiU	Catalan nationalist middle-of-the-road federation of two parties; 46 seats.
PSC	Moderate left-wing socialist party. It is a Catalan party federated to the Spanish socialist party; 42 seats.
ERC	Catalan nationalist left-wing party; 23 seats.
PPC	Moderate right-wing party. It is the Catalan delegation of the main Spanish right-wing party; 15 seats.
ICV	Coalition of Catalan left-wing parties (federated to a Spanish left-wing federation) and Catalan ecological groups; 9 seats.

The analysis performed in [1] does not use spatial indexes. However, the authors implicitly build a spatial game. They write that "in Catalonia, politics is based on two main axes: the classical left-to-right axis and a cross one going from Spanish centralism to Catalanism (Catalan nationalism)." Moreover, they give the coordinates of the five parties above in this two-dimensional ideological space. We display those coordinates in Figure 4. Spatial Game 1 (in short SG1) is the spatial game given by the simple game associated to the Catalan Parliament (when a coalition is winning if it has 68 or more seats) plus the collection of points in Figure 4.

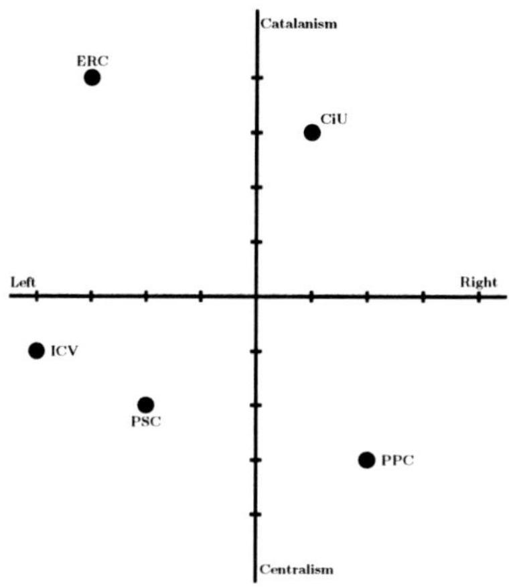

Fig. 4. Positions of the parties at the beginning of the Legislature

The coordinates in [1] are possibly appropriate to describe the ideological position of the parties in 2003, at the beginning of the Legislature 2003-2006 (prematurely finished). However, at its end, PSC and ICV had moved in the centralism-Catalanism axis in the direction of Catalanism. Thus, the coordinates in Figure 5 seem to be more appropriate to describe the ideological positions of the parties in 2005 and 2006. Spatial Game 2 (in short SG2) is the spatial game given by the simple game associated to the Catalan Parliament plus the collection of points in Figure 5.

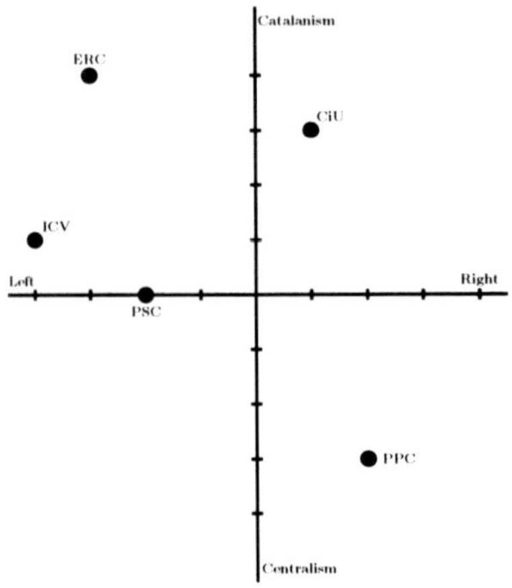

Fig. 5. Positions of the parties at the end of the Legislature

Now we compute the Owen-Shapley index and the distance index for the spatial games SG1 and SG2. The results are displayed in Table 1.

Table 1. Spatial Indexes for SG1 and SG2

Parties	SG1 Owen-Shapley	SG1 Distance	SG2 Owen-Shapley	SG2 Distance
CiU	0.4304	0.4017	0.3280	0.3943
PSC	0.2246	0.2249	0.3280	0.2348
ERC	0.3450	0.2502	0.3440	0.2442
PPC	0	0.0580	0	0.0578
ICV	0	0.0652	0	0.0688

Now let us make some comments on the results in Table 1. First a minor comment: although CiU and PSC appear to have the same power in SG2 according to the Owen-Shapley index, it must be noted that this is a rounding effect. In fact, the power of CiU is slightly larger according to this index.

Observe that PPC and ICV have no power according to the Owen-Shapley index. It is clear that, because of their number of seats and of their ideological positions, their influence is not large; but they do have some influence. In fact ICV was represented in the Government of Catalonia in that period. This feature is better reflected by the distance index, which gives small but non-zero power to both parties.

It is interesting to notice that, according to the distance index, the two parties which moved their positions throughout the Legislature 2003-2006, PSC and ICV, increased their power. This effect is not perceived by the

Owen-Shapley index, since it considers that PPC and ICV have no power in both scenarios. An especially relevant application of the spatial indexes is that they can help a party to find positions in the ideological space that, while being compatible with its most important ideological characteristics, enforce its influence and power. This example suggests that the distance index is more sensitive and, then, more useful in this respect.

Acknowledgement. The authors acknowledge the financial support of *Ministerio de Ciencia e Innovación* and FEDER through project ECO2008-03484-C02-02, and the financial support of *Xunta de Galicia* through project INCITE09-207-064-PR.

References

1. Alonso-Meijide, J.M., Carreras, F., Puente, A.: Axiomatic characterizations of the symmetric coalitional binomial semivalues. Discrete Appl. Math. 155, 2282–2293 (2007)
2. Moretti, S., Patrone, F.: Transversality of the Shapley value. Top 16, 1–41 (2008)
3. Owen, G.: Political Games. Naval Res. Logist. Quart. 18, 345–355 (1972)
4. Owen, G., Shapley, L.S.: Optimal location of candidates in ideological space. Int. J. Game Theor. 18, 339–356 (1989)
5. Shapley, L.S.: A comparison of power indices and a nonsymmetric generalization. Document P-5872. Rand Collection (1977)
6. Shapley, L.S., Shubik, M.: A method for evaluating the distribution of power in a committee system. Amer. Polit Sci. Rev. 48, 787–792 (1954)

International Environmental Agreements and Game Theory

Emilio Cerdá-Tena

Dpto. de Fundamentos del Análisis Económico I
Facultad de Ciencias Económicas y Empresariales,
Universidad Complutense de Madrid, Somosaguas, 28223 Madrid, Spain
ecerdate@ccee.ucm.es

Summary. Since there is no supranational institution which can enforce International Environmental Agreements (IEAs), international cooperation proves difficult in practice. Global emissions exhibit negative externalities in countries other than their country of origin and hence there is a high interdependence between countries and strategic considerations play an important role. Game Theory analyzes the interaction between agents and formulates hypotheses about their behavior and the final outcomes in games. Hence, international environmental problems are particularly suited for analysis by this method. The purpose of this chapter is to present an introduction to the main aspects of the formation and stability of IEAs using Game Theory.

1 Introduction

When we study the interlinkages between the economy and the environment, we see that environmental resources provide four functions: first, the production sector extracts energy resources (such as oil) and material resources (such as copper) from the environment and these are transformed into outputs, second, the environment is a sink or receptacle for waste products obtained directly from production or from consumption, third, the environment acts as a supplier of amenities which are "consumed" directly by individuals in the form of clean air, water for household uses or recreational services in natural areas, and fourth, the environment provides global life-support services such as maintenance of an atmospheric composition suitable for life, maintenance of temperature and climate or recycling of water and nutrients.

Along with other economic goods and services, environmental resources contribute positively or negatively, directly or indirectly to individual well-being. The environmental functions, and therefore the assets which provide them, are in fact also economic goods or services because in modern society they are not free; their provision, maintenance and conservation entails the sacrifice of other goods or services. They are distinguished, however, from conventional economic goods and services in that their use does not always involve market transactions. Consequently, explicit market-determined valuation, that is prices, usually does not exist for them. That is why they are often referred to as "non-market" goods (Shechter [17]).

Economic agents through consumption or production impose external costs on society in the form of air, soil and water pollution, landscape degradation, health risks and loss of biodiversity. An important reason for the absence of prices for environmental commodities is that they are typically public goods. A *pure public good* has the following two properties: *non-rivalry* (consumption of the public good by one consumer does not reduce the quantity available for consumption by any other) and *non-excludability* (if the public good is supplied no consumer can be excluded from consuming it). This means that once an environmental good, for example clean air, is provided, consumption by one economic agent does not interfere with another agent's consumption, that is, the additional costs of another economic agent consuming the environmental good is zero. Moreover, if clean air is provided, no consumer can be excluded from consuming that good. Most public goods eventually suffer from congestion when too many consumers try to use them simultaneously. For example, parks and roads are public goods that can become congested. The effect of congestion is to reduce the benefit the public good yields to each user. Public goods that are excludable, but at a cost, or suffer from congestion beyond some level of use are called *impure.*

Public goods do not conform to the assumptions required for a competitive economy to be efficient. Their characteristics of non-rivalry and non-excludability lead to the wrong incentives for consumers. Since they can share in consumption, each consumer has an incentive to rely on others to make purchases of the public good. This reliance on others to purchase is called *free-riding,* and it is this that leads to inefficiency. Efficiency in consumption for private goods is guaranteed by each consumer equating their marginal rate of substitution to the price ratio. The strategic interaction inherent with public goods does not ensure such equality. The efficient condition in the case of public goods involves the sum of marginal rates of substitution and is termed the *Samuelson rule* (Hindriks and Myles [12]).

The government assumes a role in the provision of environmental commodities by means of environmental protection. A wide array of the so-called "policy instruments" have been designed to implement and enforce environmental policies, including command and control regulations, environmental taxes, subsidies, marketable pollution permits, deposit-refund systems, bonding systems, liability systems or voluntary agreements. In the absence of markets the government decides about the optimal level of provision of environmental protection on the basis of cost-benefit analysis. At first sight this framework of domestic environmental policy is also applicable to environmental problems that occur internationally. However, in the case of an international environmental problem, the impacts of the externality are not confined to the country of origin. The absence of an international institution or government with the jurisdiction to enforce environmental policy internationally means that the only way to tackle an international environmental problem such as ozone layer depletion, climate change or acid rain is through self-enforcing International Environmental Agreements (IEA). This requires that countries which harm the environment as a consequence of pursuing their self-interest should be given enough incentives to make it beneficial for them to join an agreement by outweighing the possible increased costs of complying with it.

Game Theory has been a fundamental tool in the design and study of IEAs since the pioneering papers by Tulkens [18] and fundamentally Mäler [15]. The purpose of this chapter is to present an introduction to the main aspects of the formation and stability of IEAs using Game Theory.

2 International Environmental Agreements

In accordance with Folmer and De Zeeuw [10], we start this section by considering the three most important reasons for countries involved in an international environmental problem to cooperate, that is, to conclude an agreement with respect to an overall emissions reduction programme including an abatement specification per country: 1) Effectiveness. Unilateral actions or actions by a small proportion of the countries involved in an international environmental problem are usually futile. 2) Efficiency. In many instances there are substantial differences in abatement costs among the various countries. Efficiency requires that abatement takes place where the least-cost option exists. 3) Welfare. The foregoing implies that cooperation leads to higher welfare or total net benefits for all countries involved together in comparison with the non-cooperation outcome. An individual country's welfare, however, may suffer from cooperation.

As there is no international institution or government that can establish binding agreements, cooperation faces three fundamental constraints (Finus [8]): 1) IEAs have to be profitable for all potential participants. Profitability implies that countries must find it beneficial to participate in the IEA. A country must have a higher welfare from being a signatory country than a non-signatory one. 2) The countries must agree on the particular design of an IEA by consensus. Critical issues are the level of abatement, the allocation of abatement burdens, and the level, kind, as well as the net donors and recipients of compensation payments. Generally, it seems relatively easy for countries to agree on "framework conventions", which are mainly declarations of intentions, but far more difficult to agree on "protocols" with explicit and serious emission reductions. 3) IEAs must be self-enforcing. No country can be forced to sign an IEA, and signatories to an IEA can always withdraw from the agreement. Two types of free-riding can be distinguished: the first type implies that a country is either not a member of an IEA or is a member of an agreement that contributes less to the improvement of environmental quality than members of other agreements. The second type of free-riding implies that a country is a member of an IEA but does not comply with the terms of the agreement.

In Barrett [2] information on more than 300 international environmental treaties is provided (from the International Convention for the Prevention of Pollution of the sea by oil, adopted in 1954, to the Persistent Organic Pollutants Treaty, adopted in 2000) and a number of case studies is analysed in detail. The essential lesson of that book is that treaties should not just tell countries what to do . Treaties must make it in the interest of countries to behave differently. That is, they must restructure the underlying game. Most importantly, they must create incentives for states to participate in a treaty and for parties to comply.

In the process of treaty-making the following five stages can be distinguished: pre-negotiation, negotiation, ratification, implementation and renegotiation.

3 Game Theory

Game Theory provides general mathematical techniques for analyzing situations in which two or more individuals (or groups of individuals) make decisions that will influence one another's welfare. An important objective of Game Theory consists of, given whatever game, to deduce through analysis which are the reasonable ways of playing and, as a consequence, to predict the results of the game.

Game Theory was born officially in 1944, with the book *Theory of Games and Economic Behaviour,* by John von Neumann (a mathematician) and Oskar Morgenstern (an economist). In 1994 Nash, Selten and Harsanyi received the Nobel Prize in Economics, which can be considered as a recognition of the importance of Game Theory in Economics. In 2005 Aumann and Schelling, other important specialists in Game Theory, also received the Nobel Prize in Economics. Also, Vickrey in 1996 (with Mirlees), Akerlof in 2001 (with Spence and Stiglitz) and Maskin and Myerson in 2007 (with Hurwicz), all with important contributions to Game Theory, received the Nobel Prize in Economics. Nowadays it is a fundamental tool in Economic Analysis. For example, in whatever advanced book of Microeconomics some chapters on Game Theory are found. A lot of different research problems appearing in Economics are studied and analysed using Game Theory.

As Finus [9] writes, since the early papers by Hoel [13], Chander and Tulkens [6], Carraro and Siniscalco [3] and Barrett [1], there is a sharply increasing number of publications that analyze the formation and stability of IEAs using Game Theory. This is not surprising for at least two reasons. Firstly, Game Theory is a mathematical method that studies the interaction between agents based on behavioral assumptions about the preference of agents and makes prediction about the outcome of these interactions by applying various equilibrium concepts. Thus, Game Theory seems to be an ideal tool to study IEAs as they provide a public good with transboundary externalities from which nobody can be excluded. Secondly, environmental problems with an international dimension become more and more threatening and receive an increasing coverage in politics, the media and the public. However, despite good intentions and many public statements by politicians, progress is often slow: free-riding is still the most important obstacle for successful IEAs. In order to mitigate free-riding, it is important to bear in mind the strategic considerations of the actors causing transboundary environmental externalities for which Game Theory is an ideal method. In that interesting paper, Finus also analyses and comments on some critiques about game theoretic analysis of these problems.

4 International Environmental Problems and Game Theory

In this section we will present the basic framework, some simple illustrations and how the basic framework can be extended using issue linkage (another important extension of the basic framework is through side payments). Some general books on Environmental Economics such as Kolstad [14], Perman, Ma, McGilvray and Common [16] or Hanley, Shogren and White [11] contain some introduction about this topic. Finus [7] is a specialized book.

4.1 Basic Framework

Consider N countries denoted by subscripts $i = 1, 2, \ldots, N$. Let e_i be the emissions of country i. Since production and consumption are not possible without emissions, let us consider a gross benefit function, for country i, defined as

$$B_i = B_i(e_i), \quad i = 1, 2, \ldots, N. \tag{1}$$

Country i benefits from its own emissions e_i. It is usually assumed that benefits increase ($B_i' > 0$) at a decreasing rate ($B_i'' \leq 0$). Emissions can be viewed as an input in the production and consumption of goods.

Pollution also causes damage in the form of environmental degradation. Country i suffers damages from its own (e_i) and foreign (e_j, $j \neq i$) emissions. The transportation coefficient T_{ij}, $0 \leq T_{ij} \leq 1$, indicates the proportion of pollution generated in country j, which is deposited in country i. For local pollutants, $T_{ii} = 1$ and $T_{ij} = T_{ji} = 0$, for $j \neq i$. For global pollutants, such as greenhouse gases, $T_{ij} = 1$, $\forall i, j$. Let us define the damage function, for country i, defined as

$$D_i = D_i\left(\sum_{j=1}^{N} T_{ij} e_j\right), \quad i = 1, 2, \ldots, N. \tag{2}$$

It is usually assumed that damages increase in depositions ($D_i' > 0$) at an increasing rate ($D_i'' \geq 0$).

Combining 1 and 2 gives the net benefit function, defined as

$$W_i = B_i(e_i) - D_i\left(\sum_{j=1}^{N} T_{ij}e_j\right), \qquad i = 1, 2, \ldots, N. \tag{3}$$

Now we consider three approaches that countries can adopt with respect to an international environmental problem: the market, non-cooperative and fully cooperative approach.

Under the **market approach**, the gross benefit function (or the net benefit function, ignoring the damage component) is maximized. Therefore a total absence of pollution regulation in each country is assumed. That is, for country i,the market outcome is given by

$$\max_{e_i} \ B_i(e_i), \quad i = 1, 2, \ldots, N. \tag{4}$$

The optimal solution has to satisfy the first order conditions

$$B_i'(e_i) = 0, \quad i = 1, 2, \ldots, N. \tag{5}$$

Let e_i^M the optimal level of emissions of country i under the market approach.

The market approach is quite rare in industrialized countries nowadays because of the growing environmental concern.

In the **non-cooperative approach** each of the N countries pursues its own interest, that is each country $i = 1, 2, \ldots, N$, solves the following problem

$$\max_{e_i} \ W_i = B_i(e_i) - D_i\left(\sum_{j=1}^{N} T_{ij}e_j\right), \quad \text{taking } e_j,\ \forall j \neq i \text{ as given} \tag{6}$$

The optimal solution of this problem has to satisfy the first order conditions

$$B_i'(e_i) = T_{ii}D_i'\left(\sum_{j=1}^{N} T_{ij}e_j\right), \text{ taking } e_j,\ \forall j \neq i \text{ as given}, \quad i = 1, 2, \ldots, N. \tag{7}$$

In this approach, country i will continue increasing its pollution as long as the benefits of each additional unit of pollution exceed the damage to country i itself. The optimal level of emissions e_i^{NC}, under the non-cooperative approach, given the emissions of the other countries, is determined by the equality between the marginal benefits and marginal damage in the home country.

In the case of **fully cooperative approach**, country i not only takes its own benefits and damages into account but also the damages of its emissions in other countries. Therefore the corresponding problem of country i (for $i = 1, 2, \dots, N$) is

$$\max_{e_i} \; U_i = B_i(e_i) - \sum_{k=1}^{N} D_k \left(\sum_{j=1}^{N} T_{kj} e_j \right), \quad \text{taking } e_j, \forall j \neq i \text{ as given.} \quad (8)$$

The first order conditions for optimality are, in this case

$$B_i'(e_i) = \sum_{k=1}^{N} T_{ki} D_k' \left(\sum_{j=1}^{N} T_{kj} e_j \right), \text{ taking } e_j, \forall j \neq i \text{ as given, } i = 1, 2, \dots, N. \quad (9)$$

Let e_i^{FC} be the optimal level of emissions of country i, under the fully cooperative approach, given the emissions of the other countries.

A comparison of 5, 7 and 9 shows that under the market approach there are no restrictions on emissions whereas under the non-cooperative approach emissions in the home country are restricted by the damage they cause in the country itself, and under the fully cooperative approach they are even further restricted because the damages in other countries are taken into account as well. In Figure 1 the result is illustrated for the case in which both marginal benefit and cost functions are linear.

In Figure 1, the amount of emissions of country i is represented on the horizontal axis, while marginal benefits (MB) and marginal costs (MC) of pollution are represented on the vertical axis. Since the marginal costs curve

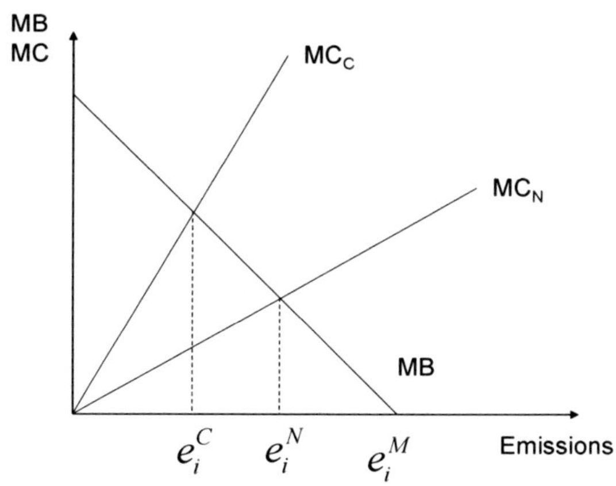

Fig. 1. The market, non-cooperative and fully cooperative outcome

under the full cooperative approach comprises both the marginal costs in the home country and those in other countries, MC_{FC} coincides with or lies above MC_{NC}. It follows that emissions under the fully cooperative approach (e_i^{FC}) are equal to or lower than emissions under the non-cooperative approach (e_i^{NC}). Emissions under the market outcome are equal to or exceed emissions under the non-cooperative approach (Folmer and De Zeeuw [10]).

The simultaneous solution of the N first order conditions in 7 delivers the non-cooperative Nash equilibrium emission vector

$$e^{NC} = (e_1^{NC}, e_2^{NC}, \ldots, e_N^{NC}). \tag{10}$$

Since this equilibrium implies that countries form singleton coalitions, it is usually assumed that it represents the *status quo* before an IEA is adopted. If all the countries were to pursue the common interest, that is, in the full cooperative approach, they would maximize the aggregate payoff over all countries. Then the mathematical problem would be

$$\max_{e_1,\ldots,e_N} \sum_{k=1}^{N} W_k, \tag{11}$$

then the first order corresponds to 9 and the simultaneous solution of that N first-order conditions delivers the full cooperative (also called globally or socially optimal) emission vector $e^{FC} = (e_1^{FC}, e_2^{FC}, \ldots, e_N^{FC})$. This may be interpreted as if all countries form a grand coallition and jointly maximize the aggregate welfare of their coalition. Since $e^{FC} \neq e^{NC}$ as long as there is some transboundary pollution ($T_{ij} \neq 0$, for some $i \neq j$), global welfare could be raised through cooperation, that is, $\sum_{i=1}^{N} W_i(e^{NC}) < \sum_{i=1}^{N} W_i(e^{FC})$ (Finus [8]).

The main impediments to the fully cooperative approach are the following: 1) The fully cooperative approach may imply net welfare gains for some countries and at the same time, net welfare losses for others. Those countries that incur negative benefits would have an incentive not to cooperate or to default from a concluded agreement. 2) Even if the net benefits of cooperation are positive , a country has an incentive to free-ride. The reason is that by staying out of an agreement or by defaulting from a concluded one it may be possible for a country to reap virtually the same benefits of pollution control as by joining it, without incurring abatement costs. Hence it will be better off because the net benefits will be larger than when it cooperates fully. Free-riding is an especially attractive option in the case of global environmental problems because under these circumstances each country's contribution is a relatively small proportion of total pollution. This implies that each country's loss of environmental quality is small or even negligible (Folmer and De Zeeuw [10]).

An equivalent way to tackle the problem consists of taking abatements (from the optimal market level of emissions) instead of emissions as strategies. Then the payoff function for a country i is represented as net benefits from abatement, that is benefits from abatements minus abatement costs. Benefits from emissions reduction (or abatement) are a function of the overall level of abatement of the N countries because of the public good nature of the problem. Abatement costs only affects country i. The payoff function of country i is then

$$\pi_i = b_i\left(\sum_{j=1}^{N}\sigma_{ij}q_j\right) - c_i(q_i), \quad i = 1,\ldots,n \tag{12}$$

where q_i is the abatement level of country i, and σ_{ij} is the part of the abatement of country j that benefits i.

4.2 Some Simple Illustrations

Let us consider the following two player pollution abatement game: There are two countries, 1 and 2. Each of the countries has to choose between two strategies: pollute or abate. The unit of pollution abatement comes at a cost of 9 to the abater, but confers a benefit of 7 to each country (if both abate, each country has a benefit of 14). The pay-offs from the four possible outcomes are indicated in Table1.

Table 1. A two player abatement game

Country 2 Country 1	Abate	Pollute
Abate	5,5	-2,<u>7</u>
Pollute	<u>7</u>,-2	<u>0</u>,<u>0</u>

If Country 2 chooses Abate, then the best choice of Country 1 is Pollute, because $7 > 5$. If Country 2 chooses Pollute, then the best choice of country 1 is Pollute, because $0 > -2$. Therefore, for whatever decision of Country 2, the best reply of Country 1 is Pollute, in accordance with the pay-offs of the game. Pollute is a dominant strategy for Country 1. As the game is symmetric, the same applies to Country 2. Then (Pollute, Pollute) is an equilibrium in dominant strategies, and the corresponding pay-offs are (0,0). It is the non-cooperative equilibrium, which is not Pareto efficient, because if the countries play (Abate, Abate), then they both obtain higher pay-offs (5,5). (Abate, Abate) is the full cooperative solution. However this solution is not an equilibrium, because each country has incentives to deviate unilaterally from abatement because then they can obtain 7, which is higher than 5. This is a Prisoner's Dilemma game.

Let us assume now that a fine of 6 is imposed if a country pollutes. Then the previous game changes to the game of Table 2. Now (Abate, Abate) is an equilibrium in dominant strategies, with pay-offs (5,5), and that equilibrium is Pareto efficient. The non-cooperative and the cooperative solutions are the same. However in practice, even if countries have previously negotiated an agreement with built-in penalty clauses for default, countries have incentives to renege on promises and not to pay the fine.The problem again is the absence of an international institution or government with the jurisdiction to enforce environmental policy internationally.

Let us now assume a situation in which the pay-offs of the game are as in Table 1, except that doing nothing exposes both countries to serious pollution damage, at a cost of 4 to both countries. This situation corresponds to what in the Game Theory literature is called a "chicken" game. We have two Nash equilibria (Abate, Pollute) with pay-offs $(-2, 7)$ and (Pollute, Abate) with pay-offs $(7, -2)$. The first is better for Country 2 and the second is better for Country 1.

Table 2. A fine of 6 if a country pollutes

Country 2 Country 1	Abate	Pollute
Abate	5,5	−2,1
Pollute	1,−2	−6, −6

Table 3. A "chicken" game

Country 2 Country 1	Abate	Pollute
Abate	5, 5	−2,7
Pollute	7,−2	−4, −4

Game Theory predicts that a Nash equilibrium will be played. Here there are two Nash equilibria (bottom left and top right cells), so there is some indeterminacy in this game. How can this indeterminacy be removed? One possibility arises from commitment or reputation: Suppose that Country 2 commits itself to pollute, and that Country 1 regards this commitment as credible. Then the left column of the matrix becomes irrelevant, and country 1, faced with pay-offs of either -2 or -4, will choose Abate. Another possibility arises if the game is played sequentially rather than simultaneously. Now the question is: can it be advantageous to move first? The answer is yes. Suppose that country 1 chooses first. In Figure 2 we have the extensive form of the game.

This game can be solved by backward induction. Assume that when player 2 has to choose, the game has arrived to the top part (because previously player 1 chose Abate), in that case player 2 has to choose between a pay-off of 5 and a pay-off of 7, he will prefer 7 and therefore will take the strategy

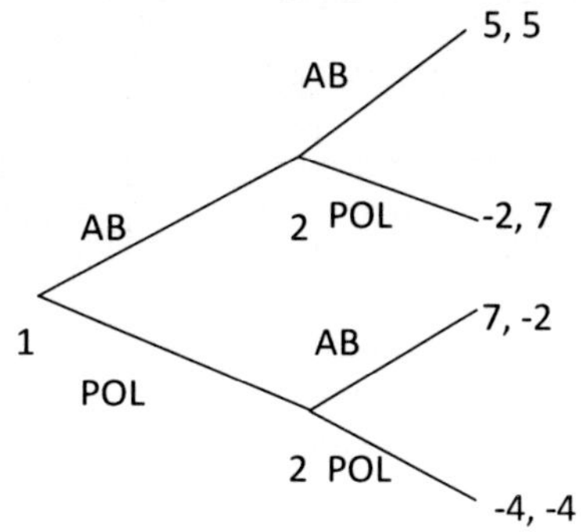

Fig. 2. Extensive form of the "chicken" game

Pollute. If player 2 has to choose in the bottom part (because previously player 1 chose Pollute), he will take the strategy Abate, because a pay-off of -2 is better than a pay-off of -4. Then, player 1 knows that if he chooses Abate, as player 2 will choose Pollute, he will win -2, and that if he chooses Pollute, as player 2 will choose Abate, he will win 7. As $7 > -2$, player 1 will play Pollute and player 2 will play Abate, and the pay-offs will be $(7, -2)$, and the player having the first move will obtain a better result.

4.3 Issue Linkage

Countries are usually simultaneously engaged in several areas of negotiation. The basic framework previously introduced can be enlarged by coupling the negotiations on the environmental problem with negotiations on other issues. Such issue linkage was introduced in the economic literature on international environmental cooperation to solve the problem of asymmetries among countries. Simultaneous involvement in negotiations on several issues of interest opens up the possibilities for exchanging concessions in fields of relative strength. A prerequisite for such an exchange or interconnection is that the net benefits of cooperation are reversed in the problems in which the countries are involved. The following example taken from Folmer and De Zeeuw [10] illustrates this point very well.

Consider the environmental problem in which the costs associated with abatement of a unit of pollution are 6 for Country 1 and 5 for Country 2. The benefits of reduction of a unit of pollution are 5 in Country 1 and 2 in Country 2, regardless of which country makes the reduction. We can see this game in Table 4. Again (Pollute, Pollute) is an equilibrium in dominant strategies, with pay-offs (0,0). The cooperative solution is (Abate, Abate) with pay-offs $(4, -1)$, but that solution is good for Country 1 and bad for Country 2. Country 1 could compensate Country 2 through side payments.

Table 4. The international environmental game

Country 2 / Country 1	Abate (A)	Pollute (B)
Abate (A)	$4, -1$	$-1,\underline{2}$
Pollute (B)	$\underline{5}, -3$	$\underline{0},\underline{0}$

Suppose that the same two countries are also involved in a trade dispute with net benefits that are an exact mirror image of the net benefits of the environmental game. The international trade game is represented in Table 5.

Table 5. The international trade game

Country 2 / Country 1	Cooperate (C)	Not to cooperate (N)
Cooperate (C)	$1, 4$	$-3,\underline{5}$
Not to cooperate (N)	$\underline{2}, -1$	$\underline{0},\underline{0}$

This game has an equilibrium in dominant strategies (Not to cooperate, Not to cooperate), with pay-offs (0,0). The cooperative solution is (Cooperate, Cooperate), with pay-offs (1,4), but that solution is better for Country 2 than for Country 1.

The trade and environmental game can be interconnected in the sense that the interconnected game comprises simultaneous play of the two games such that each country plays a strategy containing actions of both games rather than one strategy for each game separately. Pay-offs of the interconnected game are unweighted sums of the pay-off from the constituting isolated games. Interconnection yields the game represented in Table 6. That game has an equilibrium in dominant strategies $((P, N), (P, N))$ with pay-offs (0,0). The cooperative solution is $((A, C), (A, C))$, with pay-offs (5,3). However in this case a cooperative solution can be played because that is better for both players, and then in the first game country 2 chooses Abate, but in the second game player 1 plays Cooperate. Each country contributes in the game in which the other obtains better results, and globally are better. Each player knows that if he defaults in one game, then the other player will also default in the other game.

Table 6. The interconnected game

Country 2 / Country 1	(A,C)	(A,N)	(P,C)	(P,N)
(A, C)	$5, 3$	$1, 4$	$0, 6$	$-4,\underline{7}$
(A, N)	$6, -2$	$4, -1$	$1, 1$	$-1,\underline{2}$
(P, C)	$6, 1$	$2, 2$	$1, 4$	$-3,\underline{5}$
(P, N)	$\underline{7}, -4$	$\underline{5}, -3$	$\underline{2}, -1$	$\underline{0},\underline{0}$

5 Concluding Remarks

As has been written in Section 3, there is a sharply increasing number of publications that analyze the formation and stability of IEAs using Game Theory. This chapter is introductory. With the references given at the end of the chapter it is possible to learn much more about this important research area. In the literature two approaches can be distinguished: cooperative and non-cooperative.The main results of each of the two approaches can be found in Finus [8]. The advantage of the non-cooperative over the cooperative approach is that it captures externalities between players and coalitions better. The non-cooperative approach has developed strongly in the last 20 years, creating a place for the Coalition Theory Network.

A lot of works using Game Theory to study International Climate Change Negotiations have been published recently. Abstract models used to obtain analytical results have to capture the fundamental elements of the problem to be studied but usually have to leave aside other elements which can have some influence on the results. In non-cooperative coalition theory many empirical papers have been published, using models of the global economy in which some of the assumptions of the theoretical models are relaxed and where important interactions among countries are taken into account. A good survey is Carraro and Massetti [4], with comments in Cerdá [5].

References

1. Barrett, S.: Self-enforcing International Environmental Agreements. Oxford Econ. Pap. 46, 878–894 (1994)
2. Barrett, S.: Environment & Statecraft. The Strategy of Environmental Treaty-Making. Oxford University Press, New York (2003)
3. Carraro, C., Siniscalco, D.: Strategies for the International Protection of the Environment. J. Pub. Econ. 52, 309–328 (1993)
4. Carraro, C., Massetti, E.: International climate change negotiations: lessons from theory. In: Cerdá, E., Labandeira, X. (eds.) Climate Change Policies. Global Challenges and Future Prospects. Edward Elgar, Cheltenham (2010)
5. Cerdá, E.: A discussion of International climate change negotiations: lessons from theory, by Carlo Carraro and Emanuele Massetti. In: Cerdá, E., Labandeira, X. (eds.) Climate Change Policies. Global Challenges and Future Prospects, Edward Elgar, Cheltenham (2010)
6. Chander, P., Tulkens, H.: Theoretical Foundations of Negotiations and Cost-Sharing in Transfrontier Pollution Problems. European Econ. Rev. 36, 288–299 (1992)
7. Finus, M.: Game Theory and International Environmental Cooperation. Edward Elgar, Cheltenham (2001)
8. Finus, M.: Stability and design of international environmental agreements: the case of global and transboundary pollution. In: Folmer, H., Tietenberg, T. (eds.) International Yearbook of Environmental and Resource Economics 2003/4, pp. 82–158. Edward Elgar, Cheltenham (2003)

9. Finus, M.: Game theoretic research on the design of international environmental agreements: insights, critical remarks and future challenges. Int. Rev. Environ. Res. Econ. 2, 1–39 (2008)
10. Folmer, H., De Zeeuw, A.: International environmental problems and policy. In: Folmer, H., Landis Gabel, H. (eds.) Principles of Environmental and Resource Economics. A Guide for Students and Decision-Makers, pp. 447–478. Edward Elgar, Cheltenham (2000)
11. Hanley, N., Shogren, J., White, B.: Environmental Economics in Theory and Practice, 2nd edn. Palgrave Macmillan, New York (2007)
12. Hindriks, J., Myles, G.D.: Intermediate Public Economics. The MIT Press, Massachusetts (2006)
13. Hoel, M.: International Environment Conventions: The Case of Uniform Reductions of Emissions. Environ. Res. Econ. 2, 141–159 (1992)
14. Kolstad, C.D.: Environmental Economics. Oxford University Press, Oxford (2000)
15. Mäler, K.G.: The acid rain game. In: Folmer, H., Van Ierland, E. (eds.) Valuation Methods and Policy Making in Environmental Economics, pp. 231–252. Elsevier, Amsterdam (1989)
16. Perman, R., Ma, Y., McGilvray, J., Common, M.: Natural Resource and Environmental Economics. Pearson Edu. Lim., Harlow (2003)
17. Shechter, M.: Valuing the environment. In: Folmer, H., Landis Gabel, H. (eds.) Principles of Environmental and Resource Economics, pp. 72–103. Edward Elgar, Cheltenham (2000)
18. Tulkens, H.: An economic model of international negotiations relating to transfrontier pollution. In: Krippendorf, K. (ed.) Communication and Control in Society, pp. 199–212. Gordon and Breach, New York (1979)

Part VII

Model-Based Methods for Survey Sampling

An Area-Level Model with Fixed or Random Domain Effects in Small Area Estimation Problems*

María Dolores Esteban[1], Montserrat Herrador[2], Tomáš Hobza[3], and Domingo Morales[1]

[1] Operations Research Center,
Miguel Hernández University of Elche, Elche, Spain
md.esteban@umh.es, d.morales@umh.es

[2] Instituto Nacional de Estadística, Madrid, Spain
herrador@ine.es

[3] Department of Mathematics,
Czech Technical University in Prague, Czech Republic
tomas.hobza@fjfi.cvut.cz

Summary. A Fay-Herriot model having both fixed and random effects is introduced to estimate linear parameters of small areas. The model is applicable to data having a small subset of domains where direct estimates of the variable of interest cannot be described in the same way as in its complementary subset of domains. Algorithms and formulas to fit the model, to calculate EBLUPs and to estimate mean squared errors are given. A Monte Carlo simulation experiment is carried out to investigate the gain of precision obtained by using the proposed model. An application to Spanish Labour Force Survey data is also given.

Keywords: small area estimation, linear mixed models, Fay-Herriot regression model, fixed effects, random effects, EBLUP, Labour Force Survey.

1 Introduction

An area level linear mixed model was first proposed by Fay and Herriot [2] to estimate average per-capita income in small places of the United States. Since then, Empirical Best Linear Predictors (EBLUP) are commonly used to estimate domain linear parameters. These models typically assume that the regression parameter is constant but the intercept is random with realizations on the domains. Searle *et al.* [9] provide a detailed description of linear mixed models and Ghosh and Rao [3], and more recently Rao [7] and Jiang and Lahiri [4], discuss their applications to small area estimation.

* The research in this paper was done in memory of our beloved friend María Luisa Menéndez. It was a great honor to coauthor papers and to live beautiful moments with her.

When estimating totals or means we may often find that there exist a small proportion of domains where direct estimates behave in a different manner; for example they can be much higher than the rest. In these cases traditional random intercept models do not fit well to data because some domains may be responsible for producing an overestimated intercept variance affecting negatively to EBLUP estimates. We may also consider the case, quite common in practice, that some few direct estimates have been obtained with large sample sizes and therefore they are reliable. For this last case it should be interesting to use models with the property that direct estimates coincide with EBLUP estimates in some selected domains. It is thus necessary to introduce EBLUP estimators of linear parameters based on an area-level linear mixed model with one factor having both fixed and random levels.

A general theory for a case where a factor has both fixed and random effects was developed under a one-way ANOVA model by Njuho and Milliken [5]. In this paper their model is extended to an area level linear regression model with an intercept being fixed in part of the domains and being random in the rest of them. Estimation procedures for the fixed effects, variance components and regression parameters are considered and EBLUP estimators of domain parameters are derived. The approximation given by Prasad and Rao [6] and extended to a general class of linear mixed models by Das *et al.* [1] is applied to obtain estimators of the mean squared errors of the EBLUP estimates.

The paper is organized as follows. In Sections 2-4 we introduce the proposed model, we give a Fisher-scoring algorithm to calculate the maximum likelihood estimators of model parameters, we derive the expression of the EBLUP estimator of a domain linear parameter and we give an estimator of its mean squared error (MSE). In Section 5 we present a simulation experiment to investigate the behavior of the EBLUP estimates under some proposed setups. In Section 6 we illustrate the use of the proposed model with data from the Spanish Labour Force Survey (SLFS) and from some administrative registers. Finally, in Section 7 we give some conclusions.

2 The Proposed Model

We suppose a model having both fixed and random levels which can be written in terms of fixed effect (F) part and random effect (R) part in the following way

$$(F)\ \ y_d = x_d\gamma + \mu_d + e_d, \quad d = 1, \dots, D_F,$$

$$(R)\ \ y_d = x_d\gamma + u_d + e_d, \quad d = D_F + 1, \dots, D,$$

where y_d is a direct estimate of a linear parameter (typically, a mean or a total) of area d, x_d is a row vector of auxiliary variables, γ is a column vector of unknown parameters, $\mu_1, \dots, \mu_{D_F}$ are the unknown parameters corresponding to the fixed effect levels and $u_{D_F+1}, \dots, u_D$ are i.i.d. random variables independent of the random errors e_d.

Using matrix notation parts (F) and (R) of the model can be written in the form

$$\boldsymbol{y}_F = \boldsymbol{X}_F\gamma + \boldsymbol{\mu} + \boldsymbol{e}_F = [\boldsymbol{X}_F\, \boldsymbol{I}_{D_F}] \begin{pmatrix} \gamma \\ \boldsymbol{\mu} \end{pmatrix} + \boldsymbol{e}_F\,,$$

where the dimensions of $\boldsymbol{y}_F$, $\boldsymbol{X}_F$, $\boldsymbol{\mu}$ and γ are $D_F \times 1, D_F \times p, D_F \times 1$ and $p \times 1$ respectively,

$$\boldsymbol{y}_R = \boldsymbol{X}_R\gamma + \boldsymbol{u} + \boldsymbol{e}_R = [\boldsymbol{X}_R\, \boldsymbol{I}_{D_R}] \begin{bmatrix} \gamma \\ \boldsymbol{u} \end{bmatrix} + \boldsymbol{e}_R\,,$$

where the dimensions of $\boldsymbol{y}_R$, $\boldsymbol{X}_R$, $\boldsymbol{u}$ and γ are $D_R \times 1, D_R \times p, D_R \times 1$ and $p \times 1$ respectively, and $D_R = D - D_F$. We can express the complete model in the form

$$\boldsymbol{y} = \begin{pmatrix} \boldsymbol{y}_F \\ \boldsymbol{y}_R \end{pmatrix} = \begin{bmatrix} \boldsymbol{X}_F & \boldsymbol{I}_{D_F} \\ \boldsymbol{X}_R & \boldsymbol{0}_{D_R\times D_F} \end{bmatrix} \begin{pmatrix} \gamma \\ \boldsymbol{\mu} \end{pmatrix} + \begin{bmatrix} \boldsymbol{0}_{D_F\times D_R} \\ \boldsymbol{I}_{D_R} \end{bmatrix} \boldsymbol{u} + \begin{pmatrix} \boldsymbol{e}_F \\ \boldsymbol{e}_R \end{pmatrix} \qquad (1)$$

or more simply

$$\boldsymbol{y} = \boldsymbol{X\beta} + \boldsymbol{Zu} + \boldsymbol{e}\,, \qquad (2)$$

where $\boldsymbol{y} = \boldsymbol{y}_{D\times 1}, \boldsymbol{X} = \boldsymbol{X}_{D\times(p+D_F)}, \boldsymbol{\beta} = \boldsymbol{\beta}_{(p+D_F)\times 1}, \boldsymbol{Z} = \boldsymbol{Z}_{D\times D_R}, \boldsymbol{u} = \boldsymbol{u}_{D_R\times 1}$ and $\boldsymbol{e} = \boldsymbol{e}_{D\times 1}$. We further assume that $\text{rank}(\boldsymbol{X}) = p + D_F$, $\boldsymbol{u} \sim \mathcal{N}(0, \boldsymbol{\Sigma}_u)$ and $\boldsymbol{e} \sim \mathcal{N}(0, \boldsymbol{\Sigma}_e)$ are independent, $\boldsymbol{\Sigma}_u = \sigma_u^2 \boldsymbol{I}_{D_R}$ and $\boldsymbol{\Sigma}_e = \text{diag}\{\sigma_1^2, \ldots, \sigma_D^2\}$, where $\sigma_1^2, \ldots, \sigma_D^2$ are known. Thus $\boldsymbol{y} \sim \mathcal{N}(\boldsymbol{X\beta}, \boldsymbol{V})$ with $\boldsymbol{V} = \boldsymbol{Z\Sigma}_u\boldsymbol{Z}^t + \boldsymbol{\Sigma}_e = \text{diag}(v_1^2, \ldots, v_D^2)$, where

$$v_d^2 = \begin{cases} \sigma_d^2 & \text{for } d = 1, \ldots, D_F, \\ \sigma_d^2 + \sigma_u^2 & \text{for } d = D_F + 1, \ldots, D. \end{cases} \qquad (3)$$

If $\sigma_u^2 > 0$ is known, the best linear unbiased estimator (BLUE) of $\boldsymbol{\beta} = (\beta_1, \ldots, \beta_{p+D_F})^t$ and the best linear unbiased predictor (BLUP) of $\boldsymbol{u} = (u_1, \ldots, u_{D_R})^t$ are

$$\widehat{\boldsymbol{\beta}} = (\widehat{\gamma}^t, \widehat{\boldsymbol{\mu}}^t)^t = (\boldsymbol{X}^t\boldsymbol{V}^{-1}\boldsymbol{X})^{-1}\boldsymbol{X}^t\boldsymbol{V}^{-1}\boldsymbol{y} \quad \text{and} \quad \widehat{\boldsymbol{u}} = \boldsymbol{\Sigma}_u\boldsymbol{Z}^t\boldsymbol{V}^{-1}\left(\boldsymbol{y} - \boldsymbol{X}\widehat{\boldsymbol{\beta}}\right).$$

Components of $\widehat{\boldsymbol{u}}$ are

$$\widehat{u}_d = \frac{\sigma_u^2}{\sigma_u^2 + \sigma_d^2}\left(y_d - x_d\widehat{\gamma}\right), \qquad d = D_F + 1, \ldots, D.$$

BLUP of the components of the linear parameter $\boldsymbol{\tau} = \boldsymbol{X\beta} + \boldsymbol{Zu}$ are

$$\widehat{\tau}_d^{blup} = \boldsymbol{x}_d\widehat{\boldsymbol{\beta}} + \boldsymbol{z}_d\widehat{\boldsymbol{u}} = \begin{cases} x_d\widehat{\gamma} + \widehat{\mu}_d, & d = 1, \ldots, D_F, \\ x_d\widehat{\gamma} + \widehat{u}_d = \frac{\sigma_u^2}{\sigma_u^2+\sigma_d^2}\, y_d + \frac{\sigma_d^2}{\sigma_u^2+\sigma_d^2}\, x_d\widehat{\gamma}, & d = D_F + 1, \ldots, D, \end{cases} \qquad (4)$$

where $\boldsymbol{x}_d$ ($\boldsymbol{z}_d$) is the row d of matrix $\boldsymbol{X}$ ($\boldsymbol{Z}$). EBLUP of the components of $\boldsymbol{\tau}$ are obtained by substituting σ_u^2 by an estimator $\widehat{\sigma}_u^2$ in (4).

3 Maximum Likelihood Estimates

The parameter space of the supposed model is

$$\Theta = \{\boldsymbol{\theta}^t = (\boldsymbol{\beta}^t, \sigma_u^2) : \boldsymbol{\beta} \in R^{p+D_F}, \sigma_u^2 \geq 0\} \tag{5}$$

and the corresponding log-likelihood function is

$$\ell(\boldsymbol{\beta}, \sigma_u^2; \boldsymbol{y}) = -\frac{D}{2}\ln 2\pi - \frac{1}{2}\ln|\boldsymbol{V}| - \frac{1}{2}(\boldsymbol{y} - \boldsymbol{X\beta})^t \boldsymbol{V}^{-1}(\boldsymbol{y} - \boldsymbol{X\beta}).$$

The derivatives of the log-likelihood function with respect to parameters are

$$\begin{aligned}
\boldsymbol{S}_\beta &= \boldsymbol{X}^t\boldsymbol{V}^{-1}(\boldsymbol{y} - \boldsymbol{X\beta}) = \sum_{d=1}^{D} \boldsymbol{x}_d^t \frac{y_d - \boldsymbol{x}_d\boldsymbol{\beta}}{v_d^2}, \\
S_{\sigma_u^2} &= -\frac{1}{2}\mathrm{tr}(\boldsymbol{V}^{-1}\boldsymbol{V}_u) + \frac{1}{2}(\boldsymbol{y} - \boldsymbol{X\beta})^t\boldsymbol{V}^{-1}\boldsymbol{V}_u\boldsymbol{V}^{-1}(\boldsymbol{y} - \boldsymbol{X\beta}) \\
&= -\frac{1}{2}\sum_{d=D_F+1}^{D}\frac{1}{v_d^2} + \frac{1}{2}\sum_{d=D_F+1}^{D}\frac{(y_d - \boldsymbol{x}_d\boldsymbol{\beta})^2}{v_d^4},
\end{aligned}$$

where $\boldsymbol{V}_u = \frac{\partial \boldsymbol{V}}{\partial \sigma_u^2} = \mathrm{diag}(b_1, \ldots, b_D)$ with $b_j = 0$ if $1 \leq j \leq D_F$ and $b_j = 1$ if $D_F + 1 \leq j \leq D$. The second order derivatives of the log-likelihood function with respect to parameters are

$$\begin{aligned}
\boldsymbol{H}_{\beta\beta} &= -\boldsymbol{X}^t\boldsymbol{V}^{-1}\boldsymbol{X}, \quad \boldsymbol{H}_{\beta\sigma_u^2} = -\boldsymbol{X}^t\boldsymbol{V}^{-1}\boldsymbol{V}_u\boldsymbol{V}^{-1}(\boldsymbol{y} - \boldsymbol{X\beta}), \\
H_{\sigma_u^2\sigma_u^2} &= \frac{1}{2}\mathrm{tr}(\boldsymbol{V}^{-1}\boldsymbol{V}_u\boldsymbol{V}^{-1}\boldsymbol{V}_u) - (\boldsymbol{y} - \boldsymbol{X\beta})^t\boldsymbol{V}^{-1}\boldsymbol{V}_u\boldsymbol{V}^{-1}\boldsymbol{V}_u\boldsymbol{V}^{-1}(\boldsymbol{y} - \boldsymbol{X\beta}).
\end{aligned}$$

The components of the Fisher information matrix are

$$\boldsymbol{F}_{\beta\beta} = \boldsymbol{X}^t\boldsymbol{V}^{-1}\boldsymbol{X} = \sum_{d=1}^{D} v_d^{-2}\boldsymbol{x}_d^t\boldsymbol{x}_d, \quad \boldsymbol{F}_{\beta\sigma_u^2} = \boldsymbol{F}_{\sigma_u^2\beta} = \boldsymbol{0},$$

$$F_{\sigma_u^2\sigma_u^2} = \frac{1}{2}\mathrm{tr}(\boldsymbol{V}^{-1}\boldsymbol{V}_u\boldsymbol{V}^{-1}\boldsymbol{V}_u) = \frac{1}{2}\sum_{d=D_F+1}^{D} v_d^{-4}.$$

Updating equations of the Fisher-scoring algorithm are

$$\sigma_u^{2(k+1)} = \sigma_u^{2(k)} + F^{-1}_{\sigma_u^{2(k)}\sigma_u^{2(k)}} S_{\sigma_u^{2(k)}}, \quad \boldsymbol{\beta}^{(k+1)} = \boldsymbol{\beta}^{(k)} + F^{-1}_{\beta^{(k)}\beta^{(k)}} S_{\beta^{(k)}}. \tag{6}$$

4 MSE of EBLUP

Prasad and Rao [6] give an approximation to the mean squared error of the EBLUP in Fay-Herriot models. In our case the approximation is

$$MSE(\widehat{\tau}_d^{eblup}) \approx g_1(\sigma_u^2) + g_2(\sigma_u^2) + g_3(\sigma_u^2),$$

where

$$g_1(\sigma_u^2) = \begin{cases} 0 & \text{if } 1 \le d \le D_F \\ \frac{\sigma_u^2 \sigma_d^2}{\sigma_u^2 + \sigma_d^2} & \text{if } D_F + 1 \le d \le D \end{cases}$$

$$g_2(\sigma_u^2) = \begin{cases} \boldsymbol{x}_d \boldsymbol{F}_{\beta\beta}^{-1} \boldsymbol{x}_d^t & \text{if } 1 \le d \le D_F \\ \frac{\sigma_d^4}{(\sigma_u^2 + \sigma_d^2)^2} \boldsymbol{x}_d \boldsymbol{F}_{\beta\beta}^{-1} \boldsymbol{x}_d^t & \text{if } D_F + 1 \le d \le D \end{cases}$$

$$g_3(\sigma_u^2) = \begin{cases} 0 & \text{if } 1 \le d \le D_F \\ \frac{\sigma_d^4}{(\sigma_u^2 + \sigma_d^2)^3} var(\widehat{\sigma}_u^2) & \text{if } D_F + 1 \le d \le D \end{cases}$$

where $var(\widehat{\sigma}_u^2) \approx F_{\sigma_u^2, \sigma_u^2}^{-1}$ for MLE. Mean squared error is estimated by

$$mse(\widehat{\tau}_d^{eblup}) = g_1(\widehat{\sigma}_u^2) + g_2(\widehat{\sigma}_u^2) + 2g_3(\widehat{\sigma}_u^2). \tag{7}$$

Remark 1. An interesting property of the model with fixed and random effects is that $\widehat{\tau}_d^{eblup} = y_d$ and $mse(\widehat{\tau}_d^{eblup}) = \sigma_d^2$ for every $d = 1, \ldots, D_F$. These properties can be proved by straightforward calculations and are very useful from the applied point of view. For example, imagine that for several domains we have direct estimators based on large sample sizes, which are thus reliable. Then we may be interested in using model-based estimators having the property that they take the same values as the direct ones for the mentioned domains. Note also that their corresponding estimates of the model-based mean squared errors, $mse(\widehat{\tau}_d^{eblup})$, will be equal to the design-based variance estimates, σ_d^2, of the direct estimators.

5 Model-Based Simulation Experiment

We consider the model (1) with D ($D = 30$) small areas and $D_F = D/10$ small areas with fixed effect. The algorithm of the simulation experiment is described by the following steps:

1. Sample generation

Model parameters are $\sigma_u^2 = 1$, $\gamma = 1$ and $\sigma_d^2 = 1$, $d = 1, \ldots, D$. Auxiliary variable is

$$x_d = \frac{d}{D}, \quad d = 1, \ldots, D.$$

Target variable is

$$y_d = \begin{cases} x_d\gamma + \mu_d + e_d & \text{if } d = 1, \ldots D_F, \\ x_d\gamma + u_d + e_d & \text{if } d = D_F + 1, \ldots, D, \end{cases}$$

where $u_d \sim \mathcal{N}(0, \sigma_u^2)$ and $e_d \sim \mathcal{N}(0, \sigma_d^2)$ are independent and

$$\mu_d = 2 + \frac{d}{D_F}, \quad d = 1, \dots D_F.$$

2. Parameter estimation and prediction

For each area d, the parameter of interest is

$$\tau_d = \begin{cases} x_d\gamma + \mu_d \text{ if } d = 1, \dots D_F, \\ x_d\gamma + u_d \text{ if } d = D_F + 1, \dots, D. \end{cases}$$

We calculate:

1) the maximum likelihood estimates $\widehat{\boldsymbol{\beta}}$, $\widehat{\sigma}_u^2$ of the parameters $\boldsymbol{\beta}$, σ_u^2 using the Fisher-Scoring algorithm (6) with the corresponding formulas for the Fisher information matrix $\boldsymbol{F}$ and for the vector of scores $\boldsymbol{S}$ from model (1),
2) the EBLUP $\widehat{\tau}_d^{eblup}$ of τ_d using the formula (4),
3) the MSE estimator $mse(\widehat{\tau}_d^{eblup})$ using formula (7),
4) the maximum likelihood estimates $\widehat{\boldsymbol{\beta}}^*$, $\widehat{\sigma}_u^{2*}$ of the parameters $\boldsymbol{\beta}$, σ_u^2 using the Fisher-Scoring algorithm (6) under the assumption that the model does not include fixed effects, i.e. $D_F = 0$,
5) the corresponding EBLUP $\widehat{\tau}_d^{eblup*}$ of τ_d^* using the formulas (4) under the assumption $D_F = 0$, and
6) the MSE estimator $mse(\widehat{\tau}_d^{eblup*})$ using formula (7) under the assumption $D_F = 0$.

3. Repetition and performance measures

Steps 1-2 are repeated $K = 10^4$ times obtaining thus in each iteration $\tau_d^{(k)}$, $\widehat{\tau}_d^{eblup(k)}$ and $mse(\widehat{\tau}_d^{eblup(k)})$. The following performance measures are calculated.

$$MEAN_d = \frac{1}{K}\sum_{k=1}^{K} \tau_d^{(k)}, \quad mean_d = \frac{1}{K}\sum_{k=1}^{K} \widehat{\tau}_d^{eblup(k)},$$

$$BIAS_d = mean_d - MEAN_d,$$

$$MSE_d = \frac{1}{K}\sum_{k=1}^{K} \left(\widehat{\tau}_d^{eblup(k)} - \tau_d^{(k)}\right)^2, \quad mse_d = \frac{1}{K}\sum_{k=1}^{K} mse(\widehat{\tau}_d^{eblup(k)}),$$

$$E_d = \frac{1}{K}\sum_{k=1}^{K} \left(mse(\widehat{\tau}_d^{eblup(k)}) - MSE_d\right)^2.$$

and also, in the same way, $mean_d^*$, $BIAS_d^*$, MSE_d^*, mse_d^* and E_d^*.

Concerning the estimation of τ_d, performance measures are plotted in Figure 1. Concerning the estimation of the mean squared error derived from the estimation of τ_d, performance measures are plotted in Figures 2 and 3. In all the figures the areas with fixed effect were shifted to the right end, i.e. they correspond to the values of $d = 28, 29, 30$.

In Figure 1 (left) we observe that if the EBLUP estimator is derived under the true model with $D_F = 3$, then it is basically unbiased. However, if it is derived under incorrect model with $D_F = 0$, then the unbiasedness is preserved in the domains of the random effect part but a high negative bias appears in the domains of the fixed effect part. In Figure 1 (right) we observe that MSEs of EBLUPs derived under the incorrect model with $D_F = 0$ are slightly greater than the ones of EBLUPs derived under the true model in the domains of the random effect part and much greater in some of the remaining domains.

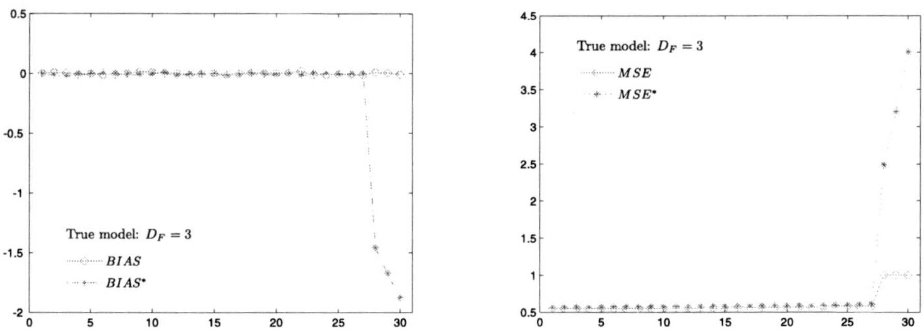

Fig. 1. $BIAS_d$, $BIAS^*_d$ (left) and MSE_d, MSE^*_d (right) for $D_F = 3$

Taking into account the different scales, in Figure 2 we observe the same pattern. If the EBLUP and its MSE estimator are derived under the true model with $D_F = 3$, then the MSE estimator is basically unbiased. However if they are derived under the incorrect model with $D_F = 0$, then a moderate negative bias appears in the domains of the random effect part and a high negative bias in the domains of the fixed effect part.

In Figure 3 we observe that the estimation of the MSE of the EBLUP is less precise for the estimator derived under the incorrect model in the domains of the fixed effect part. In the domains of the random effect part E_d is slightly smaller that E^*_d. This is not appreciated in Figure 3 because of the scale of the vertical axis. It is worth to note that for the domains of the fixed part it holds $MSE_d = \frac{1}{K}\sum_{k=1}^{K} e_d^2 \approx 1$ and $mse_d = 1$ $(K = 10^4)$.

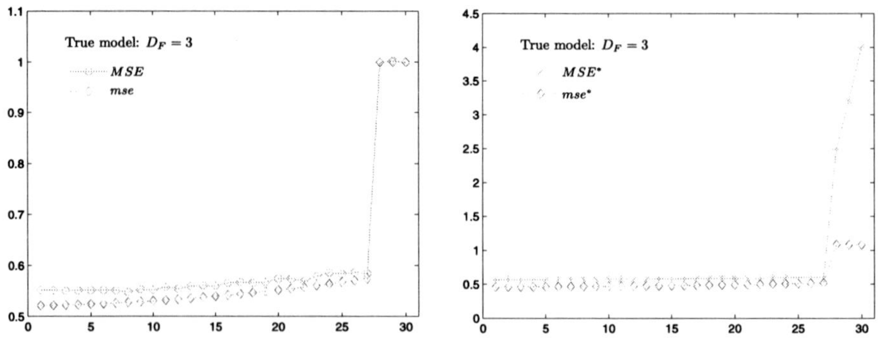

Fig. 2. MSE_d, mse_d (left) and MSE_d^*, mse_d^* (right) for $D_F = 3$

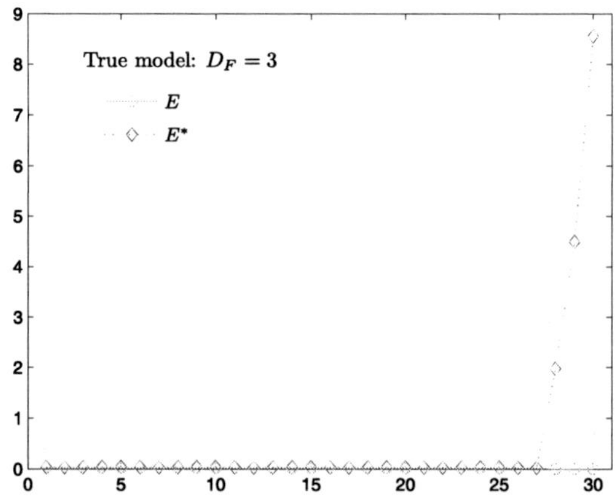

Fig. 3. E_d and E_d^* for $D_F = 3$

6 An Application to the Spanish Labour Force Survey

Data set was elaborated by the Spanish Instituto Nacional de Estadística (INE) and contains aggregated data from the Canary Islands in the second trimester of 2003. Statistical sources were the Spanish Labour Force Survey (SLFS) and the Spanish administrative register of unemployed people. In the data set there are $D = 50$ records corresponding to 25 areas and $D = 50$ domains (areas crossed with sex). Target population contains all the individuals aged 16 or more with legal residence in the Canary Islands during the studied period. Target variable is the direct estimate of the domain mean of ILO (International Labour Office) unemployed people. Auxiliary variables are the population means ($\overline{\mathbf{X}}_d$) of the 12 SEX*AGE*WORK categories described in Table 1. All the domains are assigned to the random part of the model, except the two domains with larger sample size (domains (1,1) and (2,1)).

Let P_d, N_d and s_d be the domain population, size and sample respectively. Means of variables y and $\boldsymbol{x}$ in domain d are

$$\overline{Y}_d = \frac{1}{N_d} \sum_{j \in P_d} y_{dj}, \quad \overline{\boldsymbol{X}}_d = \frac{1}{N_d} \sum_{j \in P_d} \boldsymbol{x}_{dj},$$

and direct estimates of N_d, $\overline{Y}_d$ and of the design-based variance of $\overline{Y}_d$ are

$$\widehat{N}_d = \sum_{j \in s_d} w_{dj}, \quad \widehat{\overline{Y}}_d^{dir} = \frac{1}{\widehat{N}_d} \sum_{j \in s_d} w_{dj} y_{dj},$$

$$\widehat{V}_\pi(\widehat{\overline{Y}}_d^{dir}) = \frac{1}{\widehat{N}_d^2} \sum_{j \in s_d} w_{dj}(w_{dj} - 1)(y_{dj} - \widehat{\overline{Y}}_d^{dir})^2.$$

where the w_{dj} are the sampling weights. In the Spanish LFS w_{dj} are calibrated inverses of inclusion probabilities. Formula of $\widehat{V}_\pi(\widehat{\overline{Y}}_d^{dir})$ is obtained from Särndal *et al.* [9] under the assumptions that sampling weights are the inverses of the first order inclusion probabilities, $w_{dj} = 1/\pi_{dj}$, and that equalities $\pi_{dii} = \pi_{di}$ and $\pi_{dij} = \pi_{di}\pi_{dj}$, if $i \neq j$, hold for the second order inclusion probabilities.

Table 1. Description of the variables in the data file

Variable	Description
AREA	small territories of Canary Islands: 1-25
SEX	sex categories: 1 if man, 2 if woman
AGE	age categories: 1 for 16-24, 2 for 25-54, 3 for ≥ 55
WORK	registered unemployment in public office: 1 if YES, 2 if NO
DOMAIN (d)	sex-area categories: 1-50 for (1,1),...,(1,25),(2,1),...,(2,25)
UNEMPLOYED (y)	ILO unemployment status: 1 if YES, 0 if NO
SEXAGEWORK ($\boldsymbol{x}$)	SEX∗AGE∗WORK categories: 1-12 for (1,1,1), (1,1,2),(1,2,1),...,(2,3,2)

By taking $\sigma_d^2 = \widehat{V}_\pi(\widehat{\overline{Y}}_d^{dir})$ we formulate the area-level linear mixed model

$$\widehat{\overline{Y}}_d^{dir} = \begin{cases} \overline{\boldsymbol{X}}_d\gamma + \mu_d + e_d \; d = 1, \ldots, D_F, \\ \overline{\boldsymbol{X}}_d\gamma + u_d + e_d \; d = D_F + 1, \ldots, D, \end{cases} \tag{8}$$

where $u_d \sim N(0, \sigma_u^2)$ and $e_d \sim N(0, \sigma_d^2)$ are independent. We consider the cases $D_F = 2$ and $D_F = 0$ to obtain EBLUP estimates labeled eb2 and eb0 respectively. The two domains with larger sample sizes, (1,1) and (1,2) with 1149 and 1247 observations respectively, have been assigned to the fixed effect part of the model. For area 1, with the largest sample size for men and women, eb2 estimates have the desirable property of being equal to the direct estimates. For the model with $D_F = 2$, the number of parameters (γ, μ_d and σ_u^2) to be estimated is 12+2+1=15, which is quite high with respect to the number of domains $D = 50$. For this reason we do not recommend to use models with $D_F > 2$ in this practical case. Coefficients of variation (CV) are calculated with the formulas

$$cv(dir) = \frac{[\widehat{V}_\pi(\widehat{\overline{Y}}_d^{dir})]^{1/2}}{\widehat{\overline{Y}}_d^{dir}}, \quad cv(ebi) = \frac{[mse(\widehat{\overline{Y}}_d^{ebi})]^{1/2}}{\widehat{\overline{Y}}_d^{ebi}}, \quad i = 2, 0.$$

Tables 2 and 3 present the estimates of the means (proportions) of unemployed men and women and the corresponding estimates of their CV's (multiplied by 100), in the Canary Islands during the second trimester of 2003. The two EBLUP estimates decrease in general the coefficient of variation in comparison with the direct estimator which can be seen particularly from

Table 2. Domain means and CV's (×100) for men

area	n	dir	eb2	eb0	cv(dir)	cv(eb2)	cv(eb0)
1	1149	0.0682	0.0682	0.0688	11.55	11.55	10.88
2	726	0.0644	0.0639	0.0635	14.78	13.87	13.95
20	193	0.1217	0.0829	0.0820	21.13	21.29	21.50
13	167	0.0630	0.0714	0.0695	29.59	21.03	21.35
3	144	0.0164	0.0205	0.0202	72.14	51.72	52.39
12	143	0.0800	0.0576	0.0568	31.80	30.50	30.90
15	126	0.0819	0.0652	0.0638	29.38	26.19	26.66
10	115	0.0618	0.0692	0.0673	37.12	25.38	25.86
14	99	0.0148	0.0160	0.0158	71.07	60.93	61.70
18	97	0.0204	0.0158	0.0156	70.27	77.86	78.63
17	86	0.0930	0.0903	0.0870	34.45	23.27	23.72
8	85	0.0304	0.0223	0.0220	57.63	63.33	64.23
7	80	0.0772	0.0422	0.0420	40.18	45.97	46.07
24	75	0.0877	0.0661	0.0636	36.58	29.32	30.14
9	73	0.0250	0.0268	0.0257	70.05	53.14	55.11
22	73	0.0489	0.0653	0.0627	56.44	28.46	29.21
4	60	0.0134	0.0096	0.0095	99.17	122.25	123.11
16	44	0.0497	0.0518	0.0541	68.72	51.47	49.07
5	41	0.0746	0.0195	0.0179	55.50	111.64	120.84
19	37	0.1099	0.0557	0.0541	47.42	39.39	40.38
11	35	0.0640	0.0535	0.0507	68.17	42.20	44.06
21	20	0.0434	0.0077	0.0096	98.05	294.02	234.36
23	19	0.0458	0.0513	0.0459	97.93	50.78	55.04
25	13	0.0000	0.0035	0.0033		0.00	0.00
6	12	0.0000	0.0040	0.0039		0.00	0.00

Table 3. Domain means and CV's (×100) for women

area	n	dir	eb2	eb0	cv(dir)	cv(eb2)	cv(eb0)
1	1247	0.0777	0.0777	0.0774	10.14	10.14	9.68
2	859	0.0742	0.0739	0.0740	12.54	11.76	11.73
20	214	0.1074	0.0934	0.0936	20.53	17.76	17.68
13	160	0.0570	0.0559	0.0565	35.34	29.40	28.93
12	156	0.0332	0.0378	0.0381	41.86	31.92	31.62
3	152	0.0963	0.0869	0.0878	30.03	22.84	22.45
10	143	0.0530	0.0536	0.0546	34.91	28.85	27.92
15	132	0.0580	0.0547	0.0546	37.83	30.03	30.09
18	111	0.0487	0.0472	0.0478	40.14	33.21	32.63
17	95	0.1207	0.0878	0.0885	27.44	23.78	23.53
24	79	0.0244	0.0357	0.0358	70.32	39.61	39.48
8	78	0.0478	0.0486	0.0485	56.06	38.58	38.63
14	77	0.0469	0.0472	0.0485	57.50	39.64	38.02
9	76	0.0231	0.0199	0.0197	69.83	69.00	69.68
7	74	0.0362	0.0407	0.0405	69.71	44.34	44.49
22	71	0.0993	0.0813	0.0828	39.37	26.10	25.26
4	61	0.0175	0.0154	0.0153	98.83	94.61	94.90
19	43	0.0430	0.0465	0.0471	69.34	40.86	40.11
5	41	0.0190	0.0218	0.0210	98.95	70.95	73.12
16	41	0.0738	0.0887	0.0910	55.83	33.57	32.07
11	34	0.0584	0.0683	0.0679	68.53	37.29	37.42
21	20	0.1027	0.0312	0.0308	66.90	78.51	79.37
23	19	0.0478	0.0846	0.0859	97.73	33.21	32.39
6	18	0.0936	0.0785	0.0799	69.00	33.34	32.48
25	15	0.1684	0.0746	0.0761	63.59	37.89	36.82

the Table 3 for women. EBLUP under model (8) with $D_F = 2$ ($D_F = 0$) is denoted by eb2 (eb0). As the two considered models are not significatively different one can not observe remarkable differences between the corresponding eblup estimates. Nevertheless, the possibility of obtaining model-based estimates that coincide with the direct ones in selected domains makes the use of model (8) with $D_F = 2$ very attractive.

7 Conclusions

In this paper an area-level mixed model having both fixed and random intercepts has been introduced in order to estimate linear parameters of small areas when some of the areas needs a separated treatment. Algorithms and formulas to fit the model, to calculate EBLUP and to estimate mean squared errors are given. An appealing property of the EBLUP based on the proposed model is that it coincides with the modeled direct estimate in the areas with fixed effects. So it is recommended to put domains with reliable direct estimates in the fixed part of the model.

In the presented simulation experiment it is shown that if the proposed model is true and the standard linear mixed model is used, then a severe lack of precision is achieved. An application to real data from the Spanish Labour Force survey shows that the introduced new EBLUP give, without any loss of efficiency with respect to the standard EBLUP, the same estimates as the direct one in selected desired domains with large sample size. This is an interesting property from the point of view of modelers and official statisticians.

Acknowledgement. The authors are grateful to the Czech and the Spanish Government for their economical support under Grants MSMTV 1M0572, MSM6840770039 and MTM2009-09473. The authors also thank the Instituto Nacional de Estadística for providing the Spanish Labour Force survey data.

References

1. Das, K., Jiang, J., Rao, J.N.K.: Mean squared error of empirical predictor. Ann. Statist. 32, 818–840 (2004)
2. Fay, R.E., Herriot, R.A.: Estimates of income for small places: an application of James-Stein procedures to census data. J. Amer. Statist. Assoc. 74(366), 269–277 (1979)
3. Ghosh, M., Rao, J.N.K.: Small area estimation: An appraisal. Statist. Sci. 9., 55–93 (1994)
4. Jiang, J., Lahiri, P.: Mixed model prediction and small area estimation. TEST 15, 1–96 (2006)
5. Njuho, P.M., Milliken, G.A.: Analysis of linear models with one factor having both fixed and random effects. Comm. Statist.-Theor. Meth. 34, 1979–1989 (2005)
6. Prasad, N.G.N., Rao, J.N.K.: The estimation of the mean squared error of small-area estimators. J. Amer. Statist. Assoc. 85, 163–171 (1990)
7. Rao, J.N.K.: Small Area Estimation. Wiley, New York (2003)
8. Särndal, C.E., Swensson, B., Wretman, J.: Model Assisted Survey Sampling. Springer, Berlin (1992)
9. Searle, S.R., Casella, G., McCullogh, C.E.: Variance Components. Wiley, New York (1982)

Small Area Estimation of Poverty Proportions under Random Regression Coefficient Models*

Tomáš Hobza[1] and Domingo Morales[2]

[1] Department of Mathematics,
Czech Technical University in Prague, Czech Republic
hobza@fjfi.cvut.cz

[2] Centro de Investigación Operativa,
Universidad Miguel Hernández de Elche, Spain
d.motrales@umh.es

Summary. In this paper a random regression coefficient model is used to provide estimates of small area poverty proportions. As poverty variable is dichotomic at the individual level, the sample data from Spanish Living Conditions Survey is previously aggregated to the level of census sections. EBLUP estimates based on the proposed model are obtained. A closed-formula procedure to estimate the mean squared error of the EBLUP estimators is given and empirically studied. Results of several simulations studies are reported as well as an application to real data.

1 Introduction

In small area estimation samples are drawn from a finite population, but estimations are required for subsets (called small areas or domains) where the effective sample sizes are too small to produce reliable direct estimates. An estimator of a small area parameter is called direct if it is calculated just with the sample data coming from the corresponding small area. Thus, the lack of sample data from the target small area affects seriously the accuracy of the direct estimators, and this fact has given rise to the development of new tools for obtaining more precise estimates. See a description of this theory in the monograph of Rao [8], or in the reviews of Ghosh and Rao [2], Rao [7], Pfeffermann [5] and more recently Jiang and Lahiri [3]. Mixed models increase the effective information used in the estimation process by linking all observations of the sample, and at the same time they can allow for between-area variation. Further flexibility is obtained by using random coefficient regression models, which allows the coefficient of auxiliary variables to vary across sampling units or domains. Moura and Holt [4] suggested the application of random coefficient models in small area estimation. This paper

* The research in the paper is dedicated to our beloved friend María Luisa Menéndez. Her professional honesty and kindness will inspire us in our day-to-day work.

follows their recommendation and presents and application to the estimation of poverty proportions by using data from the Spanish Living Conditions Survey.

The paper is organized as follows. Section 2 introduces the considered random coefficient model and Section 3 derives the corresponding EBLUP estimates. Section 4 deals with the problem of estimating mean squared errors. Section 5 presents several simulation experiments designed to investigate some practical issues. Section 6 is devoted to the application to real data. Finally, Section 7 gives some conclusions.

2 A Random Regression Coefficient Model

We consider two models. The first one, which will be called *Model B* in the sequel, is the random regression coefficient model

$$y_{dj} = \sum_{k=0}^{p} \beta_k x_{kdj} + \sum_{k=0}^{p} u_{kd} x_{kdj} + e_{dj}, \quad d = 1, \ldots, D, \ j = 1, \ldots, n_d, \tag{1}$$

where y_{dj} is the jth observation from area d, x_{kdj} are auxiliary variables and β_k are unknown regression parameters. Further, random regression coefficients $u_{kd} \overset{iid}{\sim} N(0, \sigma_k^2)$ and random errors $e_{dj} \sim N(0, w_{dj}^{-1}\sigma_e^2)$ are independent, $d = 1, \ldots, D$, $j = 1 \ldots, n_d$, $k = 0, \ldots, p$. If $x_{0dj} = 1$ for any d and j then model (1) contains a random intercept of the form $\beta_0 + u_{0d}$ for area d. The model variance and covariance parameters are σ_e^2, σ_k^2, $k = 0, \ldots, p$, $(2 + p$ parameters).

In this paper we will compare model (1) with the standard nested regression model (denoted as *Model A*)

$$y_{dj} = \sum_{k=0}^{p} \beta_k x_{kdj} + u_{0d} + e_{dj}, \quad d = 1, \ldots, D, \ j = 1, \ldots, n_d, \tag{2}$$

where $u_{0d} \overset{iid}{\sim} N(0, \sigma_0^2)$ and $e_{dj} \overset{iid}{\sim} N(0, w_{dj}^{-1}\sigma_e^2)$ are independent, $d = 1, \ldots, D$, $j = 1 \ldots, n_d$. In this section we briefly describe some basic facts for the application of Model B to small area estimation. The corresponding derivations for Model A are straightforward.

In matrix notation model (1) can be written in the form

$$\boldsymbol{y} = \boldsymbol{X\beta} + \sum_{k=0}^{p} \boldsymbol{Z}_k \boldsymbol{u}_k + \boldsymbol{e}, \tag{3}$$

where $n = \sum_{d=1}^{D} n_d$, $\boldsymbol{\beta} = \boldsymbol{\beta}_{(p+1)\times 1}$, $\boldsymbol{y} = \underset{1\leq d\leq D}{\text{col}}(\boldsymbol{y}_d)$, $\boldsymbol{y}_d = \underset{1\leq j\leq n_d}{\text{col}}(y_{dj})$, $\boldsymbol{e} = \underset{1\leq d\leq D}{\text{col}}(\boldsymbol{e}_d)$, $\boldsymbol{e}_d = \underset{1\leq j\leq n_d}{\text{col}}(e_{dj})$, $\boldsymbol{u}_k = \underset{1\leq d\leq D}{\text{col}}(u_{kd})$, $\boldsymbol{X} = \underset{1\leq d\leq D}{\text{col}}(\boldsymbol{X}_d)$, $\boldsymbol{X}_d = \underset{0\leq k\leq p}{\text{col}^t}(\boldsymbol{x}_{k,n_d})$, $\boldsymbol{x}_{k,n_d} = \underset{1\leq j\leq n_d}{\text{col}}(x_{kdj})$, $\boldsymbol{Z}_k = \underset{1\leq d\leq D}{\text{diag}}(\boldsymbol{x}_{k,n_d})$, $\boldsymbol{I}_a = \underset{1\leq j\leq a}{\text{diag}}(1)$, $\boldsymbol{W} = \underset{1\leq d\leq D}{\text{diag}}(\boldsymbol{W}_d)$, $\boldsymbol{W}_d = \underset{1\leq j\leq n_d}{\text{diag}}(w_{dj})$, with $w_{dj} > 0$ known, $d = 1, \ldots, D$,

$j = 1, \ldots, n_d$. Note that model (1) is a multilevel model that can alternatively be written as in Moura and Holt [4], i.e.

$$\boldsymbol{y}_d = \boldsymbol{X}_d \boldsymbol{\gamma}_d + \boldsymbol{e}_d, \quad \boldsymbol{\gamma}_d = \boldsymbol{\beta} + \boldsymbol{u}_{.d}, \quad d = 1, \ldots, D, \tag{4}$$

where $\boldsymbol{u}_{.d} = \underset{0 \le k \le p}{\text{col}} (u_{kd})$.

Variance matrices of Model B are $\boldsymbol{V}_e = \text{var}(\boldsymbol{e}) = \sigma_e^2 \boldsymbol{W}^{-1}$, $\boldsymbol{V}_{u_k} = \text{var}(\boldsymbol{u}_k) = \sigma_k^2 \boldsymbol{I}_D$, $k = 0, 1, \ldots, p$, and

$$\boldsymbol{V} = \text{var}(\boldsymbol{y}) = \boldsymbol{V}_e + \sum_{k=0}^{p} \boldsymbol{Z}_k \boldsymbol{V}_{u_k} \boldsymbol{Z}_k^t = \underset{1 \le d \le D}{\text{diag}} (\boldsymbol{V}_d),$$

where

$$\boldsymbol{V}_d = \sigma_e^2 \boldsymbol{W}_d^{-1} + \sum_{k=0}^{p} \sigma_k^2 \boldsymbol{x}_{k,n_d} \boldsymbol{x}_{k,n_d}^t, \quad d = 1, \ldots, D.$$

For model fitting it is worthwhile to consider the alternative parameters $\sigma^2 = \sigma_e^2$, $\varphi_k = \sigma_k^2 / \sigma_e^2$, $k = 0, 1, \ldots, p$, in such a way that $\boldsymbol{V} = \sigma^2 \boldsymbol{\Sigma}$ and $\boldsymbol{V}_d = \sigma^2 \boldsymbol{\Sigma}_d$, where $\boldsymbol{\Sigma} = \underset{1 \le d \le D}{\text{diag}} (\boldsymbol{\Sigma}_d)$ and

$$\boldsymbol{\Sigma}_d = \boldsymbol{W}_d^{-1} + \sum_{k=0}^{p} \varphi_k \boldsymbol{x}_{k,n_d} \boldsymbol{x}_{k,n_d}^t, \quad d = 1, \ldots, D. \tag{5}$$

Let $\boldsymbol{\varphi} = (\sigma^2, \varphi_0, \varphi_1, \ldots, \varphi_p)$ be the vector of variance components, with $\sigma^2 > 0$, $\varphi_0 > 0$, $\varphi_1 > 0, \ldots, \varphi_p > 0$. Let $\boldsymbol{u} = \underset{0 \le k \le p}{\text{col}} (\boldsymbol{u}_k)$ with variance $\boldsymbol{V}_u = \text{var}(\boldsymbol{u}) = \underset{0 \le k \le p}{\text{diag}} (\boldsymbol{V}_{u_k})$ and $\boldsymbol{Z} = \underset{0 \le k \le p}{\text{col}^t} (\boldsymbol{Z}_k)$. Using this notation the model (3) can be written in the general form

$$\boldsymbol{y} = \boldsymbol{X}\boldsymbol{\beta} + \boldsymbol{Z}\boldsymbol{u} + \boldsymbol{e}.$$

If $\boldsymbol{\varphi}$ is known, then the BLUE of $\boldsymbol{\beta} = (\beta_0, \beta_1, \ldots, \beta_p)^t$ is

$$\hat{\boldsymbol{\beta}} = (\boldsymbol{X}^t \boldsymbol{V}^{-1} \boldsymbol{X})^{-1} \boldsymbol{X}^t \boldsymbol{V}^{-1} \boldsymbol{y} = \left(\sum_{d=1}^{D} \boldsymbol{X}_d^t \boldsymbol{\Sigma}_d^{-1} \boldsymbol{X}_d \right)^{-1} \left(\sum_{d=1}^{D} \boldsymbol{X}_d^t \boldsymbol{\Sigma}_d^{-1} \boldsymbol{y}_d \right)$$

and the BLUP of $\boldsymbol{u}$ is $\hat{\boldsymbol{u}} = \boldsymbol{V}_u \boldsymbol{Z}^t \boldsymbol{V}^{-1} (\boldsymbol{y} - \boldsymbol{X}\hat{\boldsymbol{\beta}})$, i.e.

$$\hat{\boldsymbol{u}} = \underset{0 \le k \le p}{\text{diag}} (\boldsymbol{V}_{u_k}) \underset{0 \le k \le p}{\text{col}} (\boldsymbol{Z}_k^t) \underset{1 \le d \le D}{\text{diag}} (\boldsymbol{V}_d^{-1}) \underset{1 \le d \le D}{\text{col}} (\boldsymbol{y}_d - \boldsymbol{X}_d \hat{\boldsymbol{\beta}}).$$

The empirical BLUE and BLUP (EBLUE and EBLUP) are obtained by substituting the variance parameters by convenient estimates. We will now describe the Fisher-scoring algorithm to calculate the residual maximum likelihood estimates of the variance components.

The REML log-likelihood is

$$l_{REML}(\boldsymbol{\sigma}) = -\frac{1}{2}(n-p)\log 2\pi - \frac{1}{2}(n-p)\log\sigma^2 - \frac{1}{2}\log|\boldsymbol{K}^t\boldsymbol{\Sigma}\boldsymbol{K}| - \frac{1}{2\sigma^2}\boldsymbol{y}^t\boldsymbol{P}\boldsymbol{y},$$

where

$$\begin{aligned}\boldsymbol{P} &= \boldsymbol{K}(\boldsymbol{K}^t\boldsymbol{\Sigma}\boldsymbol{K})^{-1}\boldsymbol{K}^t = \boldsymbol{\Sigma}^{-1} - \boldsymbol{\Sigma}^{-1}\boldsymbol{X}(\boldsymbol{X}^t\boldsymbol{\Sigma}^{-1}\boldsymbol{X})^{-1}\boldsymbol{X}^t\boldsymbol{\Sigma}^{-1},\\ \boldsymbol{K} &= \boldsymbol{W} - \boldsymbol{W}\boldsymbol{X}(\boldsymbol{X}^t\boldsymbol{W}\boldsymbol{X})^{-1}\boldsymbol{X}^t\boldsymbol{W}\end{aligned}$$

are such that $\boldsymbol{P}\boldsymbol{X} = \boldsymbol{0}$ and $\boldsymbol{P}\boldsymbol{\Sigma}\boldsymbol{P} = \boldsymbol{P}$. From (5) it follows that $\boldsymbol{\Sigma}$ can be written in the form

$$\boldsymbol{\Sigma} = \boldsymbol{W}^{-1} + \sum_{k=0}^{p}\varphi_k\boldsymbol{A}_k,$$

where $\boldsymbol{A}_k = \boldsymbol{Z}_k\boldsymbol{Z}_k^t = \underset{1\le d\le D}{\text{diag}}\,(\boldsymbol{x}_{k,n_d}\boldsymbol{x}_{k,n_d}^t)$, $k = 0,1,\ldots,p$. As $\frac{\partial\boldsymbol{P}}{\partial\varphi_k} = -\boldsymbol{P}\boldsymbol{A}_k\boldsymbol{P}$, by taking partial derivatives with respect to σ^2 and φ_k, $k = 0,1,\ldots,p$, one gets

$$S_{\sigma^2} = -\frac{n-p}{2\sigma^2} + \frac{1}{2\sigma^4}\boldsymbol{y}^t\boldsymbol{P}\boldsymbol{y}, \quad S_{\varphi_k} = -\frac{1}{2}\,\text{tr}\{\boldsymbol{P}\boldsymbol{A}_k\} + \frac{1}{2\sigma^2}\boldsymbol{y}^t\boldsymbol{P}\boldsymbol{A}_k\boldsymbol{P}\boldsymbol{y},$$

$k = 0,1,\ldots,p$. The second partial derivatives are

$$\begin{aligned}H_{\sigma^2\sigma^2} &= \frac{n-p}{2\sigma^4} - \frac{1}{\sigma^6}\boldsymbol{y}^t\boldsymbol{P}\boldsymbol{y}, \quad H_{\sigma^2\varphi_i} = -\frac{1}{2\sigma^4}\boldsymbol{y}^t\boldsymbol{P}\boldsymbol{A}_i\boldsymbol{P}\boldsymbol{y},\\ H_{\varphi_i\varphi_j} &= \frac{1}{2}\,\text{tr}\{\boldsymbol{P}\boldsymbol{A}_i\boldsymbol{P}\boldsymbol{A}_j\} - \frac{1}{\sigma^2}\boldsymbol{y}^t\boldsymbol{P}\boldsymbol{A}_i\boldsymbol{P}\boldsymbol{A}_j\boldsymbol{P}\boldsymbol{y}, \quad i,j = 0,1,\ldots,p.\end{aligned}$$

By taking expectations and multiplying by -1, we obtain the components of the Fisher information matrix $(i,j = 0,1,\ldots,p)$

$$F_{\sigma^2\sigma^2} = \frac{n-p}{2\sigma^4}, \quad F_{\sigma^2\varphi_j} = \frac{1}{2\sigma^2}\,\text{tr}\{\boldsymbol{P}\boldsymbol{A}_j\}, \quad F_{\varphi_i\varphi_j} = \frac{1}{2}\,\text{tr}\{\boldsymbol{P}\boldsymbol{A}_i\boldsymbol{P}\boldsymbol{A}_j\}.$$

To calculate the REML estimates, the Fisher-scoring updating formula is

$$\boldsymbol{\varphi}^{k+1} = \boldsymbol{\varphi}^k + \boldsymbol{F}^{-1}(\boldsymbol{\varphi}^k)\boldsymbol{S}(\boldsymbol{\varphi}^k).$$

The following seeds can be used as starting values in the Fisher-scoring algorithm

$$\sigma^{2(0)} = \theta_0^{(0)} = \varphi_1^{(0)} = \ldots = \varphi_p^{(0)} = S^2/(p+2),$$

where $S^2 = \frac{1}{n-p}(\boldsymbol{y} - \boldsymbol{X}\tilde{\boldsymbol{\beta}})^t\boldsymbol{W}(\boldsymbol{y} - \boldsymbol{X}\tilde{\boldsymbol{\beta}})$ and $\tilde{\boldsymbol{\beta}} = (\boldsymbol{X}^t\boldsymbol{W}\boldsymbol{X})^{-1}\boldsymbol{X}^t\boldsymbol{W}\boldsymbol{y}$.

The asymptotic distributions of the REML estimators of $\boldsymbol{\varphi}$ and $\boldsymbol{\beta}$ are

$$\hat{\boldsymbol{\varphi}} \sim N_{p+2}(\boldsymbol{\varphi}, \boldsymbol{F}^{-1}(\boldsymbol{\varphi})), \quad \hat{\boldsymbol{\beta}} \sim N_{p+1}(\boldsymbol{\beta}, (\boldsymbol{X}'\boldsymbol{V}^{-1}\boldsymbol{X})^{-1}),$$

so that the $1-\alpha$ asymptotic confidence intervals for φ_k and β_k are

$$\hat{\varphi}_k \pm z_{\alpha/2}\,\nu_{kk}^{1/2}, \quad \text{and} \quad \hat{\beta}_k \pm z_{\alpha/2}\,q_{kk}^{1/2}, \; k=0,1,\ldots,p,$$

where $\hat{\varphi} = \varphi^{\kappa}$, κ is the last iteration in the Fisher-scoring algorithm, $\boldsymbol{F}^{-1}(\varphi^{\kappa}) = (\nu_{k\ell})_{k,\ell=-1,0,\ldots,p}$, $(\boldsymbol{X}'\boldsymbol{V}^{-1}(\varphi^{\kappa})\boldsymbol{X})^{-1} = (q_{k\ell})_{k,\ell=0,1,\ldots,p}$ and z_α is the α-quantile of the $N(0,1)$ distribution. The confidence interval for σ^2 is obtained in the same way by using the corresponding diagonal element of the matrix $\boldsymbol{F}^{-1}$.

3 EBLUP of the Domain Mean

In this section we consider a finite population of N elements following the model introduced in (1) with population sizes N_d in the place of sample sizes n_d. From the population a sample of size n with n_d elements in area d, $n = \sum_{d=1}^{D} n_d$, is selected. Without loss of generality we can reorder the population so that $\boldsymbol{y} = (\boldsymbol{y}_s^t, \boldsymbol{y}_r^t)^t$, where $\boldsymbol{y}_s$ is the vector of n observed elements and $\boldsymbol{y}_r$ is the vector of $N-n$ unobserved elements. In the following, the index s for the sample and the index r for the rest of the population will be used when appropriate. In this notation and taking into account the reordering we can write

$$\boldsymbol{V} = \text{var}[\boldsymbol{y}] = \begin{pmatrix} \boldsymbol{V}_{ss} & \boldsymbol{V}_{sr} \\ \boldsymbol{V}_{rs} & \boldsymbol{V}_{rr} \end{pmatrix}.$$

We are interested in the estimation of the mean of the small area d, i.e.

$$\overline{Y}_d = \frac{1}{N_d}\sum_{j=1}^{N_d} y_{dj} = \boldsymbol{a}^t\boldsymbol{y} = \boldsymbol{a}_s^t\boldsymbol{y}_s + \boldsymbol{a}_r^t\boldsymbol{y}_r,$$

where $\boldsymbol{a}^t = \frac{1}{N_d}\left(\boldsymbol{0}_{N_1}^t,\ldots,\boldsymbol{0}_{N_{d-1}}^t, \boldsymbol{1}_{N_d}^t, \boldsymbol{0}_{N_d+1}^t,\ldots,\boldsymbol{0}_{N_D}^t\right)$ and $\boldsymbol{0}_m^t = (0,\ldots,0)_{1\times m}$. From the general theorem of prediction it follows that the BLU predictor of $\overline{Y}_d$, under Model B, is

$$\widehat{\overline{Y}}_d^{blupB} = \boldsymbol{a}_s^t\boldsymbol{y}_s + \boldsymbol{a}_r^t\left[\boldsymbol{X}_r\widehat{\boldsymbol{\beta}} + \widehat{\boldsymbol{V}}_{rs}\widehat{\boldsymbol{V}}_{ss}^{-1}(\boldsymbol{y}_s - \boldsymbol{X}_s\widehat{\boldsymbol{\beta}})\right]. \tag{6}$$

In our case it holds $\boldsymbol{V}_{e,rs} = \boldsymbol{0}$, $\boldsymbol{V}_{rs} = \boldsymbol{Z}_r\boldsymbol{V}_u\boldsymbol{Z}_s^t + \boldsymbol{V}_{e,rs} = \boldsymbol{Z}_r\boldsymbol{V}_u\boldsymbol{Z}_s^t$ and $\hat{\boldsymbol{u}} = \hat{\boldsymbol{V}}_u\boldsymbol{Z}_s^t\hat{\boldsymbol{V}}_{ss}^{-1}(\boldsymbol{y}_s - \boldsymbol{X}_s\hat{\boldsymbol{\beta}})$, so

$$\begin{aligned}\widehat{\overline{Y}}_d^{blupB} &= \boldsymbol{a}_s^t\boldsymbol{y}_s + \boldsymbol{a}_r^t\left[\boldsymbol{X}_r\hat{\boldsymbol{\beta}} + \boldsymbol{Z}_r\hat{\boldsymbol{V}}_u\boldsymbol{Z}_s^t\hat{\boldsymbol{V}}_{ss}^{-1}(\boldsymbol{y}_s - \boldsymbol{X}_s\hat{\boldsymbol{\beta}})\right] \\ &= \boldsymbol{a}^t\left[\boldsymbol{X}\hat{\boldsymbol{\beta}} + \sum_{k=0}^{p}\boldsymbol{Z}_k\hat{\boldsymbol{u}}_k\right] + \boldsymbol{a}_s^t\left[\boldsymbol{y}_s - \boldsymbol{X}_s\hat{\boldsymbol{\beta}} - \sum_{k=0}^{p}\boldsymbol{Z}_{k,s}\hat{\boldsymbol{u}}_k\right].\end{aligned}$$

Since $\boldsymbol{a}^t$ can be written in the form $\boldsymbol{a}^t = \frac{1}{N_d}\underset{1\le\ell\le D}{\text{col}^t}\{\delta_{d\ell}\boldsymbol{1}_{N_\ell}^t\}$, where $\delta_{ab}=1$ if $a=b$ and $\delta_{ab}=0$ if $a\neq b$, it holds that $\boldsymbol{a}^t\boldsymbol{X}\hat{\boldsymbol{\beta}} = \sum_{k=0}^{p}\overline{X}_{kd}\hat{\beta}_k$ and

$$\boldsymbol{a}^t \boldsymbol{Z}_k \hat{\boldsymbol{u}}_k = \frac{1}{N_d} \underset{1 \le \ell \le D}{\mathrm{col}^t} \{\delta_{d\ell} \mathbf{1}_{N_\ell}^t\} \underset{1 \le \ell \le D}{\mathrm{diag}} (\boldsymbol{x}_{k,N_\ell}) \hat{\boldsymbol{u}}_k = \overline{X}_{kd} \hat{u}_{kd},$$

where $\overline{X}_{kd} = \frac{1}{N_d} \sum_{j=1}^{N_d} x_{kdj}$. Thus the EBLUP B of $\overline{Y}_d$ is

$$\hat{\overline{Y}}_d^{eblupB} = \sum_{k=0}^{p} \overline{X}_{kd} \hat{\beta}_k + \sum_{k=0}^{p} \overline{X}_{kd} \hat{u}_{kd} + f_d \left[\overline{y}_{d,s} - \sum_{k=0}^{p} \overline{X}_{kd,s} \hat{\beta}_k - \sum_{k=0}^{p} \overline{X}_{kd,s} \hat{u}_{kd} \right],$$

where $\overline{y}_{d,s} = \frac{1}{n_d} \sum_{j=1}^{n_d} y_{dj}$, $\overline{X}_{kd,s} = \frac{1}{n_d} \sum_{j=1}^{n_d} x_{kdj}$ and $f_d = \frac{n_d}{N_d}$. EBLUP under Model A is similarly introduced and it is denoted by EBLUP A in the sequel. The mean squared error (MSE) of the EBLUP and its proposed estimator are given in the next section.

4 MSE of EBLUP

Following Prasad and Rao [6] and Das, Jiang and Rao [1], the mean squared error (MSE) of the EBLUP of $\overline{Y}_d$, under Model B, is

$$MSE(\hat{\overline{Y}}_d^{eblupB}) = g_1(\boldsymbol{\varphi}) + g_2(\boldsymbol{\varphi}) + g_3(\boldsymbol{\varphi}) + g_4(\boldsymbol{\varphi}),$$

where

$$\begin{aligned}
g_1(\boldsymbol{\varphi}) &= \boldsymbol{a}_r^t \boldsymbol{Z}_r \boldsymbol{T}_s \boldsymbol{Z}_r^t \boldsymbol{a}_r, \\
g_2(\boldsymbol{\varphi}) &= [\boldsymbol{a}_r^t \boldsymbol{X}_r - \boldsymbol{a}_r^t \boldsymbol{Z}_r \boldsymbol{T}_s \boldsymbol{Z}_s^t \boldsymbol{V}_{e,s}^{-1} \boldsymbol{X}_s] \boldsymbol{Q}_s [\boldsymbol{X}_r^t \boldsymbol{a}_r - \boldsymbol{X}_s^t \boldsymbol{V}_{e,s}^{-1} \boldsymbol{Z}_s \boldsymbol{T}_s \boldsymbol{Z}_r^t \boldsymbol{a}_r], \\
g_3(\boldsymbol{\varphi}) &\approx \mathrm{tr}\left\{ (\nabla \boldsymbol{b}^t) \boldsymbol{V}_s (\nabla \boldsymbol{b}^t)^t E\left[(\hat{\boldsymbol{\varphi}} - \boldsymbol{\varphi})(\hat{\boldsymbol{\varphi}} - \boldsymbol{\varphi})^t \right] \right\}, \\
g_4(\boldsymbol{\varphi}) &= \boldsymbol{a}_r^t \boldsymbol{V}_{e,r} \boldsymbol{a}_r,
\end{aligned}$$

and $\boldsymbol{T}_s = \boldsymbol{V}_u - \boldsymbol{V}_u \boldsymbol{Z}_s^t \boldsymbol{V}_s^{-1} \boldsymbol{Z}_s \boldsymbol{V}_u$, $\boldsymbol{Q}_s = (\boldsymbol{X}_s^t \boldsymbol{V}^{-1} \boldsymbol{X}_s)^{-1}$, $\boldsymbol{b}^t = \boldsymbol{a}_r^t \boldsymbol{Z}_r \boldsymbol{V}_u \boldsymbol{Z}_s^t \boldsymbol{V}_s^{-1}$. The Prasad-Rao (PR) estimator of $MSE(\hat{\overline{Y}}_d^{eblupB})$ is

$$mse_d^B = mse(\hat{\overline{Y}}_d^{eblup\,B}) = g_1(\hat{\boldsymbol{\varphi}}) + g_2(\hat{\boldsymbol{\varphi}}) + 2g_3(\hat{\boldsymbol{\varphi}}) + g_4(\hat{\boldsymbol{\varphi}}),$$

where $\hat{\boldsymbol{\varphi}}$ is REML estimator of $\boldsymbol{\varphi}$. In what follows we present the calculation of $g_1 - g_4$ for Model B. The derivations under Model A are straightforward. We employ the notation $mse_d^\ell = mse(\hat{\overline{Y}}_d^{eblup\ell})$, $\ell = A, B$, under Models A and B.

4.1 Calculation of $g_1(\boldsymbol{\varphi})$ under Model B

To calculate $g_1(\boldsymbol{\varphi}) = \boldsymbol{a}_r^t \boldsymbol{Z}_r \boldsymbol{T}_s \boldsymbol{Z}_r^t \boldsymbol{a}_r$, basic elements are

$$\boldsymbol{a}_r^t = \frac{1}{N_d} \underset{1 \le \ell \le D}{\mathrm{col}^t} (\delta_{d\ell} \mathbf{1}_{N_\ell - n_\ell}^t), \quad \boldsymbol{Z}_r = \underset{0 \le k \le p}{\mathrm{col}^t} (\boldsymbol{Z}_{k,r}), \quad \boldsymbol{V}_u = \sigma^2 \underset{0 \le k \le p}{\mathrm{diag}} (\varphi_k \boldsymbol{I}_D)$$

and

$$\boldsymbol{T}_s = \boldsymbol{V}_u - \boldsymbol{V}_u\boldsymbol{Z}_s^t\boldsymbol{V}_s^{-1}\boldsymbol{Z}_s\boldsymbol{V}_u = \sigma^2 \underset{0\le k\le p}{\text{diag}}(\varphi_k\boldsymbol{I}_D)$$
$$- \sigma^2 \underset{0\le k\le p}{\text{col}}(\varphi_k\boldsymbol{Z}_{k,s}^t)\underset{1\le\ell\le D}{\text{diag}}(\boldsymbol{\Sigma}_{\ell,s}^{-1})\underset{0\le k\le p}{\text{col}^t}(\varphi_k\boldsymbol{Z}_{k,s}) = (\boldsymbol{T}_{k_1k_2})_{k_1,k_2=0,1,\ldots,p}\,.$$

where $\delta_{k_1k_2}=0$ if $k_1\neq k_2$, $\delta_{k_1k_2}=1$ if $k_1=k_2$ and

$$\boldsymbol{T}_{k_1k_2} = \sigma^2\varphi_{k_1}\delta_{k_1k_2}\boldsymbol{I}_D - \sigma^2\varphi_{k_1}\varphi_{k_2}\boldsymbol{Z}_{k_1,s}^t\underset{1\le\ell\le D}{\text{diag}}(\boldsymbol{\Sigma}_{\ell,s}^{-1})\boldsymbol{Z}_{k_2,s}.$$

Therefore

$$g_1(\theta) = \frac{1}{N_d^2}\underset{1\le\ell\le D}{\text{col}^t}(\delta_{d\ell}\boldsymbol{1}_{N_\ell-n_\ell}^t)\underset{0\le k\le p}{\text{col}^t}(\boldsymbol{Z}_{k,r})\boldsymbol{T}_s\underset{0\le k\le p}{\text{col}}(\boldsymbol{Z}_{k,r}^t)\underset{1\le\ell\le D}{\text{col}}(\delta_{d\ell}\boldsymbol{1}_{N_\ell-n_\ell})$$
$$= (1-f_d)^2\sigma^2\left\{\sum_{k=0}^{p}\varphi_k\overline{X}_{kd}^{*2} - \sum_{k_1=0}^{p}\sum_{k_2=0}^{p}\varphi_{k_1}\varphi_{k_2}\overline{X}_{k_1d}^{*}\boldsymbol{x}_{k_1,n_d}^t\boldsymbol{\Sigma}_{d,s}^{-1}\boldsymbol{x}_{k_2,n_d}\overline{X}_{k_2d}^{*}\right\},$$

where $f_d = n_d/N_d$ and $\overline{X}_{kd}^* = \frac{1}{N_d-n_d}\sum_{j\in r}x_{kdj} = (1-f_d)^{-1}(\overline{X}_{kd} - f_d\overline{X}_{kd,s})$.

4.2 Calculation of $g_2(\boldsymbol{\varphi})$ under Model B

From the definition of $g_2(\boldsymbol{\varphi})$ it follows that it can be written in the form

$$g_2(\boldsymbol{\varphi}) = [\boldsymbol{a}_1^t - \boldsymbol{a}_2^t]\boldsymbol{Q}_s[\boldsymbol{a}_1 - \boldsymbol{a}_2],$$

where $\boldsymbol{Q}_s$ is defined on page 320. The first vector from the difference $[\boldsymbol{a}_1^t - \boldsymbol{a}_2^t]$ is

$$\boldsymbol{a}_1^t = \boldsymbol{a}_r^t\boldsymbol{X}_r = \frac{1}{N_d}\boldsymbol{1}_{N_d-n_d}^t\boldsymbol{X}_{rd} = (1-f_d)\overline{\boldsymbol{X}}_d^*,$$

where $\overline{\boldsymbol{X}}_d^* = (\overline{X}_{0d}^*, \overline{X}_{1d}^*, \ldots, \overline{X}_{pd}^*)$. The second vector can be written as

$$\boldsymbol{a}_2^t = \boldsymbol{a}_r^t\underset{0\le k\le p}{\text{col}^t}(\boldsymbol{Z}_{k,r})\boldsymbol{T}_s\underset{0\le k\le p}{\text{col}}(\boldsymbol{Z}_{k,s}^t)\sigma^{-2}\boldsymbol{W}_s\boldsymbol{X}_s$$

and after some straightforward algebra it takes the form

$$\boldsymbol{a}_2^t = (1-f_d)\left\{\sum_{k=0}^{p}\varphi_k\overline{X}_{kd}^*\boldsymbol{x}_{k,n_d}^t\right.$$
$$\left. - \sum_{k_1=0}^{p}\sum_{k_2=0}^{p}\varphi_{k_1}\varphi_{k_2}\overline{X}_{k_1d}^*\boldsymbol{x}_{k_1,n_d}^t\boldsymbol{\Sigma}_{d,s}^{-1}\boldsymbol{x}_{k_2,n_d}\boldsymbol{x}_{k_2,n_d}^t\right\}\boldsymbol{W}_{d,s}\boldsymbol{X}_{d,s}.$$

4.3 Calculation of $g_3(\varphi)$ under Model B

We recall that $g_3(\varphi) \approx \operatorname{tr}\left\{(\nabla \boldsymbol{b}^t)\boldsymbol{V}_s(\nabla \boldsymbol{b}^t)^t E\left[(\hat{\varphi}-\varphi)(\hat{\varphi}-\varphi)^t\right]\right\}$, where

$$\boldsymbol{b}^t = \boldsymbol{a}_r^t \boldsymbol{Z}_r \boldsymbol{V}_u \boldsymbol{Z}_s^t \boldsymbol{V}_s^{-1} = \boldsymbol{a}_r^t \sum_{k=0}^{p} \varphi_k \boldsymbol{Z}_{k,r} \boldsymbol{Z}_{k,s}^t \underset{1\le \ell \le D}{\operatorname{diag}} (\boldsymbol{\Sigma}_{\ell,s}^{-1}).$$

As $\frac{\partial \boldsymbol{\Sigma}_{\ell,s}}{\partial \sigma^2} = 0$ and $\frac{\partial \boldsymbol{\Sigma}_{\ell,s}}{\partial \varphi_k} = \boldsymbol{x}_{k,n_\ell}\boldsymbol{x}_{k,n_\ell}^t$ $(k=0,\ldots,p)$, the derivative with respect to σ^2 is $\frac{\partial \boldsymbol{b}^t}{\partial \sigma^2} = 0$ and the remaining derivatives are

$$\begin{aligned}\frac{\partial \boldsymbol{b}^t}{\partial \varphi_k} &= \boldsymbol{a}_r^t \boldsymbol{Z}_{k,r}\boldsymbol{Z}_{k,s}^t \underset{1\le \ell \le D}{\operatorname{diag}} (\boldsymbol{\Sigma}_{\ell,s}^{-1}) \\ &- \boldsymbol{a}_r^t \left(\sum_{i=0}^{p} \varphi_i \boldsymbol{Z}_{i,r}\boldsymbol{Z}_{i,s}^t\right) \underset{1\le \ell \le D}{\operatorname{diag}} (\boldsymbol{\Sigma}_{\ell,s}^{-1}\boldsymbol{x}_{k,n_\ell}\boldsymbol{x}_{k,n_\ell}^t\boldsymbol{\Sigma}_{\ell,s}^{-1}), \quad k=0,1,\ldots,p.\end{aligned}$$

As $\boldsymbol{Z}_{k,r} = \underset{1\le \ell \le D}{\operatorname{diag}} (\boldsymbol{x}_{k,N_\ell - n_\ell})$, we obtain for $k=0,1,\ldots,p$

$$\begin{aligned}\frac{\partial \boldsymbol{b}^t}{\partial \varphi_k} &= (1-f_d)\left[\underset{1\le \ell \le D}{\operatorname{col}^t} (\delta_{d\ell}\overline{X}_{k\ell}^{*}\,\boldsymbol{x}_{k,n_\ell}^t \boldsymbol{\Sigma}_{\ell,s}^{-1}) \right.\\ &\left. - \underset{1\le \ell \le D}{\operatorname{col}^t}\left(\delta_{d\ell}\left(\sum_{i=0}^{p}\varphi_i \overline{X}_{i\ell}^{*}\,\boldsymbol{x}_{i,n_\ell}^t\right)\boldsymbol{\Sigma}_{\ell,s}^{-1}\boldsymbol{x}_{k,n_\ell}\boldsymbol{x}_{k,n_\ell}^t\boldsymbol{\Sigma}_{\ell,s}^{-1}\right)\right].\end{aligned}$$

Let us define $\boldsymbol{H}(\varphi) = (h_{k_1,k_2})_{k_1,k_2=-1,0,1,\ldots,p}$, where $h_{-1,k} = h_{k,-1} = 0$, $k=-1,0,1,\ldots,p$ and

$$\begin{aligned}h_{k_1,k_2} &= \frac{\partial \boldsymbol{b}^t}{\partial \varphi_{k_1}}\boldsymbol{V}_s\left(\frac{\partial \boldsymbol{b}^t}{\partial \varphi_{k_2}}\right)^t = \sigma^2(1-f_d)^2\left\{\overline{X}_{k_1 d}^{*}\boldsymbol{x}_{k_1,n_d}^t\boldsymbol{\Sigma}_{d,s}^{-1}\boldsymbol{x}_{k_2,n_d}\overline{X}_{k_2 d}^{*}\right.\\ &- \overline{X}_{k_1 d}^{*}\boldsymbol{x}_{k_1,n_d}^t\boldsymbol{\Sigma}_{d,s}^{-1}\boldsymbol{x}_{k_2,n_d}\boldsymbol{x}_{k_2,n_d}^t\boldsymbol{\Sigma}_{d,s}^{-1}\sum_{i=0}^{p}\varphi_i\boldsymbol{x}_{i,n_d}\overline{X}_{id}^{*}\\ &- \left(\sum_{i=0}^{p}\varphi_i\overline{X}_{id}^{*}\boldsymbol{x}_{i,n_d}^t\right)\boldsymbol{\Sigma}_{d,s}^{-1}\boldsymbol{x}_{k_1,n_d}\boldsymbol{x}_{k_1,n_d}^t\boldsymbol{\Sigma}_{d,s}^{-1}\boldsymbol{x}_{k_2,n_d}\\ &\left.\cdot\left[\overline{X}_{k_2 d}^{*} - \boldsymbol{x}_{k_2,n_d}^t\boldsymbol{\Sigma}_{d,s}^{-1}\sum_{i=0}^{p}\varphi_i\boldsymbol{x}_{i,n_d}\overline{X}_{id}^{*}\right]\right\}\end{aligned}$$

for any $k_1, k_2 = 0,1,\ldots,p$. Then

$$g_3(\varphi) \approx \operatorname{tr}\left\{\boldsymbol{H}(\varphi)\boldsymbol{F}^{-1}(\varphi)\right\},$$

where $\boldsymbol{F}(\varphi)$ is the REML Fisher information matrix which approximates the covariance matrix $E\left[(\hat{\varphi}-\varphi)(\hat{\varphi}-\varphi)^t\right]$.

4.4 Calculation of $g_4(\varphi)$ under Model B

We recall that $g_4(\varphi) = \boldsymbol{a}_r^t \boldsymbol{V}_{e,r} \boldsymbol{a}_r$, where

$$\boldsymbol{a}_r^t = \frac{1}{N_d} \underset{1\leq \ell \leq D}{\text{col}^t} (\delta_{d\ell} \mathbf{1}_{N_\ell - n_\ell}^t), \quad \boldsymbol{V}_{e,r}^{-1} = \sigma^{-2} \boldsymbol{W}_r = \sigma^{-2} \underset{1\leq d \leq D}{\text{diag}} \{\boldsymbol{W}_{d,r}\}.$$

Therefore

$$g_4(\varphi) = \frac{\sigma^2}{N_d^2} \mathbf{1}_{N_d - n_d}^t \underset{j\in r_d}{\text{diag}} \{w_{dj}^{-1}\} \mathbf{1}_{N_d - n_d} = \frac{\sigma^2}{N_d^2} \sum_{j \in r_d} \frac{1}{w_{dj}}.$$

5 Simulation Experiments

In this section we present several simulation experiments. The first one is designed to check the behavior of the REML estimates under Model B. The second simulation experiment is planned to study the behavior of EBLUP a, $a = A, B$, under Models A and B. Finally, the third simulation experiment is carried out to analyze the behavior of the MSE estimates.

In all the simulations, samples are generated as follows.

- Explanatory variable: Take $a_d = 1$, $b_d = 2 + \frac{8d}{D}$, $d = 1, \ldots, D$. For $d = 1, \ldots, D$, $j = 1, \ldots, n_d$, generate
$$x_{1dj} = (b_d - a_d) U_{dj} + a_d \quad \text{with } U_{dj} = \frac{j}{n_d + 1}, \quad j = 1, \ldots, n_d.$$
- Random effects and errors: For $d = 1, \ldots, D$, $j = 1, \ldots, n_d$, generate
$$u_{0d} \sim N(0, \sigma^2 \varphi_0), \quad u_{1d} \sim N\left(0, \sigma^2 \varphi_1\right), \quad e_{dj} \sim N(0, \sigma^2),$$
with $\sigma^2 = \varphi_0 = 1$ and $\varphi_1 = 2$.
- Target variable: For $d = 1, \ldots, D$, $j = 1, \ldots, n_d$, generate
$$y_{dj} = \beta_0 + \beta_1 x_{dj} + u_{1d} x_{dj} + u_{0d} + w_{dij}^{-1/2} e_{dj}, \quad \text{with } \beta_0 = 2, \ \beta_1 = 1.$$
(Just skipping the term $u_{1d} x_{dj}$ in the case of Model A.)

5.1 Simulation 1

The steps of the simulation experiment are:

1. Repeat $K = 10^4$ times $(k = 1, \ldots, K)$
   1.1. Generate a sample of size $n = \sum_{d=1}^D n_d$ and calculate the REML estimates $\gamma_{(k)} \in \{\hat{\beta}_{0(k)}, \hat{\beta}_{(1k)}, \hat{\sigma}_{(k)}^2, \hat{\varphi}_{0(k)}, \hat{\varphi}_{1(k)}\}$.
2. Output:

$$EMSE(\hat{\gamma}) = \frac{1}{K} \sum_{k=1}^K (\hat{\gamma}_{(k)} - \gamma)^2, \quad BIAS(\hat{\gamma}) = \frac{1}{K} \sum_{k=1}^K (\hat{\gamma}_{(k)} - \gamma).$$

Table 1 presents the obtained performance measures. In all the presented cases we observe that EMSE decreases as sample size increases. The conclusion is that the implemented Fisher-scoring algorithm is running properly and thus the obtained REML parameter estimates are reliable.

Table 1. BIAS and EMSE for $K = 10^4$ under Model B

n	300		600		1200		2400	
n_d	5		10		20		40	
$D = 60$	BIAS	EMSE	BIAS	EMSE	BIAS	EMSE	BIAS	EMSE
$\beta_0 = 2$	-0.001	0.052	-0.001	0.032	-0.002	0.024	0.000	0.020
$\beta_1 = 1$	-0.001	0.020	0.000	0.018	-0.001	0.018	0.000	0.017
$\sigma^2 = 1$	0.006	0.010	0.002	0.004	0.001	0.002	0.001	0.001
$\varphi_0 = 1$	-0.050	0.335	-0.007	0.129	-0.001	0.070	0.002	0.050
$\varphi_1 = 1$	-0.020	0.055	-0.005	0.043	-0.002	0.038	-0.002	0.037

5.2 Simulation 2

The second simulation experiment is designed to investigate the behavior of EBLUPa, $a = A, B$, under Models A and B. The steps of simulation experiment are:

1. Generate deterministically $N = \sum_{d=1}^{D} N_d$ x-values with $N_d = 100$, $D = 60$ as described at the beginning of this section and calculate $\overline{X}_d$, $d = 1, \ldots, D$.
2. Repeat $K = 10^4$ times ($k = 1, \ldots, K$)
   2.1. Generate a population of size N and extract a sample of size $n = \sum_{d=1}^{D} n_d$ ($n_d = 10$) under Model B (Model A).
   2.2 Calculate the REML estimates under Models A and B.
   2.3 Calculate the true value $\overline{Y}_d^{(k)}$ and its estimates $\widehat{\overline{Y}}_d^{eblup\,a(k)}$ for $a = A, B$.
3. For any $a = A, B$ the output is:

$$mean_d^a = \frac{1}{K}\sum_{k=1}^{K}\widehat{\overline{Y}}_d^{eblup\,a(k)}, \quad MEAN_d = \frac{1}{K}\sum_{k=1}^{K}\overline{Y}_d^{(k)},$$

$$EMSE_d^a = \frac{1}{K}\sum_{k=1}^{K}(\widehat{\overline{Y}}_d^{eblup\,a(k)} - \overline{Y}_d^{(k)})^2, \quad EMSE^a = \frac{1}{D}\sum_{d=1}^{D}EMSE_d^a,$$

and

$$BIAS_d^a = \frac{1}{K}\sum_{k=1}^{K}(\widehat{\overline{Y}}_d^{eblup\,a(k)} - \overline{Y}_d^{(k)}), \quad BIAS^a = \frac{1}{D}\sum_{d=1}^{D}BIAS_d^a.$$

Table 2 presents the basic performance measures of simulations 2. $DIFr$, $r = A, B$, is used to denote the differences $EMSE^a - EMSE^r$, $a = A, B$, $a \neq r$. Figure 1 plot $EMSE_d^a$ of estimators $\widehat{\overline{Y}}_d^{eblup\,a}$, $a = A, B$ under Model A and B, respectively.

We observe that if Model B is true, EBLUP estimate may lose a significative amount of precision by assuming the wrong Model A. However, the loss of efficiency is negligible in the reciprocal case.

Table 2. BIAS and EMSE for $D = 60$ and $K = 10^4$

	Model B		Model A	
$N_d = 100,\ n_d = 10$	eblupA	eblupB	eblupA	eblupB
$10^2 BIAS$	0.0046	0.0065	0.0992	0.0993
$10^2 EMSE$	10.7272	8.513	8.2237	8.2253
$10^2 DIFr$	2.2142			0.0016

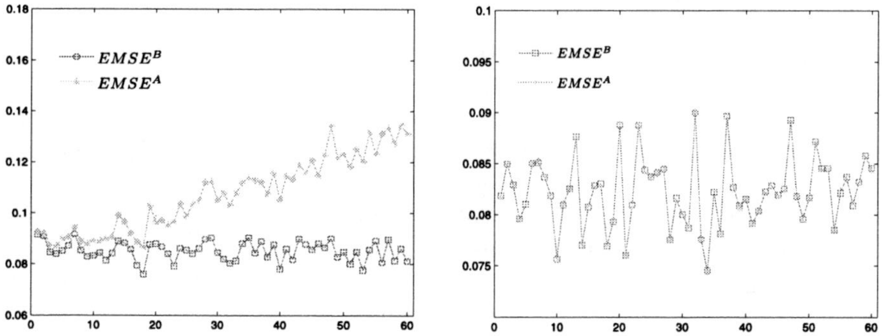

Fig. 1. $EMSE_d$ values under the true Model B (left) and Model A (right)

5.3 Simulation 3

The third simulation experiment is designed to analyze the behavior of the MSE estimates. The steps of the simulation experiment under Model B (Model A) are:

1-2. Do steps 1-2.3 as in Simulation 2. Do new step 2.4 as follows.
  2.4. Calculate the MSE estimates $mse_d^{A(k)}$ and $mse_d^{B(k)}$.
3. For $a = A, B$ the output is:

$$E_d^a = \frac{1}{K}\sum_{k=1}^{K}(mse_d^{a(k)} - EMSE_d^a)^2, \quad B_d^a = \frac{1}{K}\sum_{k=1}^{K}(mse_d^{a(k)} - EMSE_d^a).$$

$$E^a = \frac{1}{D}\sum_{d=1}^{D} E_d^a, \quad B^a = \frac{1}{D}\sum_{d=1}^{D} B_d^a,$$

where the values $EMSE_d^a$ are taken from the results of Simulation 2.

Table 3 presents basic performance measures of Simulation 3.

From the table it can be seen that the two estimators mse_d^B and mse_d^A have basically the same behavior under the true Model A. However, under the true Model B mse_d^A has a very poor behavior when it is used to estimate $MSE(\widehat{\overline{Y}}_d^{eblupA})$.

Table 3. B^a and E^a values for $K = 10^4$

$N_d = 100$	Model B		Model A	
$n_d = 10$	mse_d^A	mse_d^B	mse_d^A	mse_d^B
$10^2 B$	46.5629	0.0098	-0.0146	0.0054
$10^2 E$	23.0612	0.0029	0.0025	0.0026

6 Estimation of Poverty Proportions

In this section we use data from the 2006 Spanish Living Conditions Survey (SLCS) with global sample size 34694. The SLCS is the Spanish version of the European Statistics on Income and Living Conditions (EU-SILC), which is one of the statistical operations that have been harmonized for EU countries. Its main goal is to provide a reference source on comparative statistics on the distribution of income and social exclusion in the European environment. The sample includes 16000 dwellings distributed in 2000 census sections.

We consider $D = 52$ domains (provinces) and we are interested in studying the household normalized net annual incomes at the domain level. The aim of normalizing the household income is to adjust for the varying size and composition of households. The definition of the total number of normalized household members is the modified OECD scale used by EUROSTAT, where OECD is the acronym for the Organization for Economic Cooperation and Development. This scale gives a weight of 1.0 to the first adult, 0.5 to the second and each subsequent person aged 14 and over and 0.3 to each child aged under 14 in the household. The *normalized size* of a household is the sum of the weights assigned to each person. So the total number of normalized household members is

$$H_{di} = 1 + 0.5(H_{di\geq 14} - 1) + 0.3H_{di<14}$$

where $H_{di\geq 14}$ is the number of people aged 14 and over and $H_{di<14}$ is the number of children aged under 14. The normalized net annual income of a household (z) is obtained by dividing its net annual income by its normalized size. Following the standards of the Spanish Statistical Office, the Poverty Threshold is fixed as the 60% of the median of the normalized incomes in Spanish households. The Spanish poverty thresholds (in euros) in 06 is $z_{2006} =$ 6556.60. This is z_0-value used in the calculation of the direct estimates of the poverty proportion

$$\overline{Y}_d = \frac{1}{N_d}\sum_{j=1}^{N_d} y_{dj}, \quad y_{dj} = I(z_{dj} < z_0),$$

where $I(z_{dj} < z_0) = 1$ if $z_{dj} < z_0$ and $I(z_{dj} < z_0) = 0$ otherwise.

The considered auxiliary variables are *nationality* (x_0) and *employed* (x_1), both with values 0-1 at the individual level (1 for Spanish citizenship and employed). In the SLCS the target variable y is measured at the household level and the auxiliary variables x_1 and x_2 at the individual level. For this reason a data file has been built containing the survey data aggregated at the level of census sections (territories with around 2000 people). In the census section file the y variable and the x-variables are calculated by taking weighted averages on the territory.

Table 4 presents the REML estimates of model parameters and the corresponding 90% confidence intervals. We observe that confidence intervals for parameters φ_0 and φ_1 are strictly positive, suggesting that Model B fits better to data than Model A.

Figure 2 presents the domain mean estimates and their estimated mean squared error. It shows that EBLUP B has slightly different behavior from EBLUP A estimates. Figure 2 also shows that the EBLUP estimates behave more smoothly than the direct ones, which are calculated by means of the formula

$$\hat{\overline{Y}}_d^{dir} = \frac{1}{\hat{N}_d}\sum_{j=1}^{n_d} \omega_{dj} y_{dj}, \quad \hat{N}_d = \sum_{j=1}^{n_d} \omega_{dj},$$

where the ω_{dj}'s are SLCS calibrated sampling weights.

Concerning mean squared errors, EBLUP B is the estimator giving the best results. EBLUP estimators produce some gain of efficiency with respect to the direct ones. For comparison purposes, design-based mean squared errors of direct estimators where approximated by

$$mse(\hat{\overline{Y}}_d^{dir}) = \frac{1}{\hat{N}_d^2}\sum_{j=1}^{n_d} \omega_{dj}(\omega_{dj}-1)\big(y_{dj} - \hat{\overline{Y}}_d^{dir}\big)^2. \tag{7}$$

The last formula is taken from Särndal *et al.* [9], pp. 43, 185 and 391, with the simplifications $\omega_{dj} = 1/\pi_{dj}$, $\pi_{dj,dj} = \pi_{dj}$ and $\pi_{di,dj} = \pi_{di}\pi_{dj}$, $i \neq j$ in the second order inclusion probabilities.

By observing the signs of the regression parameters, we interpret that poverty proportion tends to be smaller in those domains with larger proportion of people with non Spanish citizenship (may be because immigrants tends to go to regions with greater richness where it is easier to find job) and larger proportion of employed people.

Table 4. Parameter estimates and 90% confidence intervals for models B and A.

	Model A		**Model B**	
	Estim.	90% CI	Estim.	90% CI
β_0	**0.2942**	(0.1722 , 0.4162)	**0.3336**	(0.2197 , 0.4475)
β_1	**-0.2900**	(-0.4946 , -0.0854)	**-0.2958**	(-0.5294 , -0.0621)
σ^2	**0.0453**	(0.0430 , 0.0477)	**0.0457**	(0.0433 , 0.0480)
φ_0	**0.1481**	(0.0865 , 0.2098)	**0.0689**	(0.0061 , 0.1317)
φ_1			**0.1382**	(0.0478 , 0.2287)

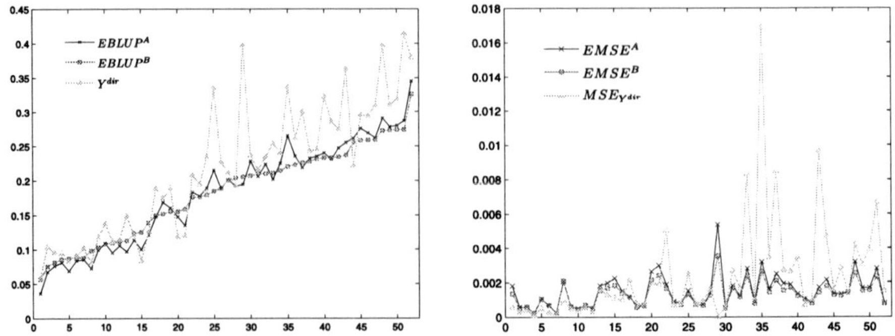

Fig. 2. Direct estimates and EBLUP estimates (left) and its estimated mean square error (right)

7 Conclusions

This paper investigate the use of EBLUPs, based on random regression coefficient models, in small area estimation. By looking at the presented simulations and application to real data, we may conclude that fixed regression coefficients are sometimes too rigid for modeling real data. Some extra variability, and better performance of EBLUP estimates, might be obtained by allowing some variability on the regression (beta) parameters.

Acknowledgement. The authors are grateful to the Czech and Spanish Governments for their economical support under Grants MSMTV 1M0572, MSM6840770039 and MTM2009-09473. The authors also thanks the Instituto Nacional de Estadística for providing the Spanish EU-SILC data.

References

1. Das, K., Jiang, J., Rao, J.N.K.: Mean squared error of empirical predictor. Ann. Statist. 32, 818–840 (2004)
2. Ghosh, M., Rao, J.N.K.: Small area estimation: An appraisal. Statist. Sci. 9, 55–93 (1994)
3. Jiang, J., Lahiri, P.: Mixed model prediction and small area estimation. TEST 15, 1–96 (2006)
4. Moura, F.A.S., Holt, D.: Small area estimation using multilevel models. Surv. Meth. 25(N.1), 73–80 (1999)
5. Pfeffermann, D.: Small Area Estimation. New Developments and Directions. Int. Statist. Rev. 70, 125–143 (2002)
6. Prasad, N.G.N., Rao, J.N.K.: The estimation of the mean squared error of small-area estimators. J. Amer. Statist. Assoc. 85, 163–171 (1990)
7. Rao, J.N.K.: Some recent advances in model-based small area estimation. Surv. Meth. 25, 175–186 (1999)
8. Rao, J.N.K.: Small Area Estimation. Wiley, New York (2003)
9. Särndal, C.E., Swensson, B., Wretman, J.: Model assisted survey sampling. Springer, Berlin (1992)

Robust Henderson III Estimators of Variance Components in the Nested Error Model

Betsabé Pérez, Daniel Peña, and Isabel Molina

Universidad Carlos III de Madrid, Department of Statistics,
Postal address: C/Madrid 126, 28903 Getafe, Spain
{betsabe.perez,daniel.pena,isabel.molina}@uc3m.es

Summary. This work deals with robust estimation of variance components under a nested error model. Traditional estimation methods include Maximum Likelihood (ML), Restricted Maximum Likelihood (REML) and Henderson method III (H3). However, when outliers are present, these methods deliver estimators with poor properties. Robust modifications of the ML and REML have been proposed in the literature (see for example, Fellner [3], Richardson and Welsh [14] and Richardson [13]). In this work we explore some robust alternatives based on the idea of Henderson method III. The work is organized as follows. In section 2, we introduce the nested error model. In section 3, we describe the traditional methods for estimating variance components. In section 4, several robustified versions of the H3 estimators of the variance components are presented. In section 5, we present some results on diagnostics methods. In section 6 we perform a Monte Carlo study to compare the new robust estimation methods with the non-robust alternatives.

Keywords: Henderson method III, linear mixed models, nested error model, outliers, robust estimation, variance components

1 Introduction

In the last decades, linear mixed models (Laird and Ware [8]) have received considerable attention in the literature from a practical and theoretical point of view (e.g. McCulloch and Searle [10], Verbeke and Molenberghs [17] and Jiang [7]). These models are frequently used in small area estimation or to analyze repeated measures data, because they model flexibly the within-subject correlation often present in these type of data. However, there are many other fields of application of these models, such as clinical trials (Vangeneugden *et al.* [16]), air pollution studies (Wellenius *et al.* [18]), etc. Despite the many applications in which these models are used, only few works have been done on model diagnostics, an important step to validate the model. Christensen et al. [2] studied case deletion diagnostics. Banerjee and Frees [1] developed influence diagnostics. Galpin and Zewotir [5] extended some results of the ordinary linear regression influence diagnostics to the linear mixed models context such as residuals, leverages and outliers when the variance components are known. However, in practice the variance components need to be estimated from sample data. If sample data are contaminated, then the estimation might be affected and this will in turn affect all diagnostic tools.

Here we focus on a particular linear mixed model with only one random factor, called nested error model. For this model, we propose several robust alternatives to the H3 estimators of variance components. Section 2 describes the data structure and the model. Section 3 summarizes the most common methods for estimation. Section 4 introduces our proposed robust estimators. Section 5 describes diagnostic tools for these models and finally, Section 6 reports the results of a simulation study that compares the robustness properties of the proposed estimators with those of the traditional non-robust ones. Finally, Section 7 gives some concluding remarks.

2 The Model

In this section we introduce the nested error model and describe some of its properties. Consider that the sample observations come from D different populations groups, with n_d observations coming from d-th group, $d = 1, \ldots, D$ and $n = \sum_{d=1}^{D} n_d$ being the total sample size. Let us denote y_{dj} the value of the study variable for j-th sample unit from d-th group and $\mathbf{x}_{dj}$ a (column) vector containing the values of p auxiliary variables for the same unit. We consider the model

$$y_{dj} = \mathbf{x}_{dj}^T \boldsymbol{\beta} + u_d + e_{dj} \quad j = 1, \ldots, n_d \quad d = 1, \ldots, D, \tag{1}$$

where $\boldsymbol{\beta}$ is the $p \times 1$ vector of fixed parameters, u_d is the random effect of d-th group and e_{dj} is the model error. Random effects and errors are supposed to be independent with distributions

$$u_d \overset{iid}{\sim} N(0, \sigma_u^2) \quad \text{and} \quad e_{dj} \overset{iid}{\sim} N(0, \sigma_e^2).$$

Stacking the model elements y_{dj}, $\mathbf{x}_{dj}^T$ and e_{dj} in columns, we can express the model in matrix notation as

$$\mathbf{y} = \mathbf{X}\boldsymbol{\beta} + \mathbf{Z}\mathbf{u} + \mathbf{e}, \quad \mathbf{u} \sim N(\mathbf{0}, \sigma_u^2 \mathbf{I}_D), \quad \mathbf{e} \sim N(\mathbf{0}, \sigma_e^2 \mathbf{I}_n). \tag{2}$$

where $\mathbf{u} = (u_1, \ldots, u_D)^T$ and $\mathbf{Z}$ is the $n \times D$ design matrix associated with $\mathbf{u}$, containing in its columns the indicators of the groups $d = 1, \ldots, D$.

The expectation and covariance matrix of $\mathbf{y}$ are given by

$$E(\mathbf{y}) = \mathbf{X}\boldsymbol{\beta} \quad \text{and} \quad Var(\mathbf{y}) = \sigma_u^2 \mathbf{Z}\mathbf{Z}^T + \sigma_e^2 \mathbf{I}_n := \mathbf{V}.$$

Let us define the vector of variance components $\boldsymbol{\theta} = (\sigma_u^2, \sigma_e^2)^T$. When $\boldsymbol{\theta}$ is known, Henderson [6] obtained the Best Linear Unbiased Estimator (BLUE) of $\boldsymbol{\beta}$ and the Best Linear Unbiased Predictor (BLUP) of $\mathbf{u}$, which are given respectively by

$$\tilde{\boldsymbol{\beta}} = (\mathbf{X}^T\mathbf{V}^{-1}\mathbf{X})^{-1}\mathbf{X}^T\mathbf{V}^{-1}\mathbf{y}, \tag{3}$$

$$\tilde{\mathbf{u}} = \sigma_u^2\mathbf{Z}^T\mathbf{V}^{-1}(\mathbf{y} - \mathbf{X}\tilde{\boldsymbol{\beta}}). \tag{4}$$

3 Estimation of Variance Components

The estimator of $\boldsymbol{\beta}$ and the predictor of $\mathbf{u}$ in (3) and (4) respectively depend on $\boldsymbol{\theta}$, which in practice is unknown and needs to be estimated from sample data. The empirical versions of (3) and (4), (EBLUE and EBLUP respectively) are obtained by replacing a suitable estimator $\hat{\boldsymbol{\theta}} = (\hat{\sigma}_u^2, \hat{\sigma}_e^2)^T$ for $\boldsymbol{\theta}$ in (3) and (4) and are given by

$$\hat{\boldsymbol{\beta}} = (\mathbf{X}^T\hat{\mathbf{V}}^{-1}\mathbf{X})^{-1}\mathbf{X}^T\hat{\mathbf{V}}^{-1}\mathbf{y}, \tag{5}$$

$$\hat{\mathbf{u}} = \hat{\sigma}_u^2\mathbf{Z}^T\hat{\mathbf{V}}^{-1}(\mathbf{y} - \mathbf{X}\hat{\boldsymbol{\beta}}), \tag{6}$$

where $\hat{\mathbf{V}} = \hat{\sigma}_u^2\mathbf{Z}\mathbf{Z}^T + \hat{\sigma}_e^2\mathbf{I}_n$.

Next we describe the ML, REML and H3 methods to estimate variance components.

Maximum likelihood

Maximum likelihood estimation is usually done by assumming that $\mathbf{y}$ has a multivariate normal distribution. Under this assumption, the likelihood is given by

$$f(\boldsymbol{\theta}|\mathbf{y}) = (2\pi)^{-\frac{n}{2}}|\mathbf{V}|^{-1/2}exp\left\{-\frac{1}{2}(\mathbf{y} - \mathbf{X}\boldsymbol{\beta})^T\mathbf{V}^{-1}(\mathbf{y} - \mathbf{X}\boldsymbol{\beta})\right\}.$$

The log-likelihood is

$$\ell(\boldsymbol{\theta}|\mathbf{y}) = \ln(f(\boldsymbol{\theta}|\mathbf{y})) = c - \frac{1}{2}[\ln|\mathbf{V}| + (\mathbf{y} - \mathbf{X}\boldsymbol{\beta})^T\mathbf{V}^{-1}(\mathbf{y} - \mathbf{X}\boldsymbol{\beta})],$$

where c is denotes a constant. Using the relations

$$\frac{\partial \ln|\mathbf{V}|}{\partial\boldsymbol{\theta}} = tr\left\{\mathbf{V}^{-1}\frac{\partial\mathbf{V}}{\partial\boldsymbol{\theta}}\right\} \quad \text{and} \quad \frac{\partial\mathbf{V}^{-1}}{\partial\boldsymbol{\theta}} = -\mathbf{V}^{-1}\frac{\partial\mathbf{V}}{\partial\boldsymbol{\theta}}\mathbf{V}^{-1},$$

the first order partial derivatives of ℓ with respect to $\boldsymbol{\beta}$, σ_u^2 and σ_e^2 are

$$\begin{aligned}
\frac{\partial\ell(\boldsymbol{\theta}|\mathbf{y})}{\partial\boldsymbol{\beta}} &= \mathbf{X}^T\mathbf{V}^{-1}(\mathbf{y} - \mathbf{X}\boldsymbol{\beta}),\\
\frac{\partial\ell(\boldsymbol{\theta}|\mathbf{y})}{\partial\sigma_u^2} &= -\frac{1}{2}tr\left\{\mathbf{V}^{-1}\mathbf{Z}\mathbf{Z}^T\right\} + \frac{1}{2}(\mathbf{y} - \mathbf{X}\boldsymbol{\beta})^T\mathbf{V}^{-1}\mathbf{Z}\mathbf{Z}^T\mathbf{V}^{-1}(\mathbf{y} - \mathbf{X}\boldsymbol{\beta}),\\
\frac{\partial\ell(\boldsymbol{\theta}|\mathbf{y})}{\partial\sigma_e^2} &= -\frac{1}{2}tr\{\mathbf{V}^{-1}\} + \frac{1}{2}(\mathbf{y} - \mathbf{X}\boldsymbol{\beta})^T\mathbf{V}^{-1}\mathbf{V}^{-1}(\mathbf{y} - \mathbf{X}\boldsymbol{\beta}),
\end{aligned}$$

and equating to zero we obtain the equations

$$\mathbf{X}^T\mathbf{V}^{-1}\mathbf{y} = \mathbf{X}\mathbf{V}^{-1}\mathbf{X}\boldsymbol{\beta}, \quad (7)$$

$$tr\{\mathbf{V}^{-1}\mathbf{Z}\mathbf{Z}^T\} = (\mathbf{y} - \mathbf{X}\boldsymbol{\beta})^T\mathbf{V}^{-1}\mathbf{Z}\mathbf{Z}^T\mathbf{V}^{-1}(\mathbf{y} - \mathbf{X}\boldsymbol{\beta}), \quad (8)$$

$$tr\{\mathbf{V}^{-1}\} = (\mathbf{y} - \mathbf{X}\boldsymbol{\beta})^T\mathbf{V}^{-1}\mathbf{V}^{-1}(\mathbf{y} - \mathbf{X}\boldsymbol{\beta}). \quad (9)$$

From (7), the ML estimating equation for $\boldsymbol{\beta}$ is

$$\hat{\boldsymbol{\beta}} = (\mathbf{X}^T\hat{\mathbf{V}}^{-1}\mathbf{X})^{-1}\mathbf{X}^T\hat{\mathbf{V}}^{-1}\mathbf{y}$$

Equations (8) and (9) do not have analytic solution and need to be solved using numerical methods such as Newton-Raphson or Fisher-Scoring.

Restricted maximum likelihood

REML approach starts by transforming $\mathbf{y}$ into two independent vectors, $\mathbf{y}_1 = K_1\mathbf{y}$ and $\mathbf{y}_2 = K_2\mathbf{y}$. The distribution of $\mathbf{y}_1$ does not depend on $\boldsymbol{\beta}$ and satisfies $E(\mathbf{y}_1) = \mathbf{0}$, which means that $K_1\mathbf{X} = \mathbf{0}$. On the other hand, $\mathbf{y}_2$ is independent of $\mathbf{y}_1$, which means that $K_1\mathbf{V}K_2^T = \mathbf{0}$. The matrix K_1 is chosen to have maximum rank, i.e. $n - p$, so the rank of K_2 is p. The likelihood function of $\mathbf{y}$ is the product of the likelihoods of $\mathbf{y}_1$ and $\mathbf{y}_2$. The variance components coming from the REML approach are the ML estimators of these parameters based on $\mathbf{y}_1$, see Patterson and Thompson [11]. Similarly to the ML case, the obtained equations do not have analytic solutions and need to be solved using iterative techniques. This method takes into account the degrees of freedom of estimation of $\boldsymbol{\beta}$ when estimating the variance components and for this reason it gives less biased estimators.

Henderson method III

ML and REML estimators of $\boldsymbol{\theta}$ are typically obtained under the assumption that the vector $\mathbf{y}$ has a multivariate normal distribution. In many circumstances, however, this assumption does not hold. An alternative method which does not rely on the normality assumption and provides explicit solutions to the variance components estimators is the Henderson method III. This method works as follows. First, consider a general linear mixed model $\mathbf{y} = \mathbf{X}\boldsymbol{\beta} + \mathbf{e}$, where $\boldsymbol{\beta}$ might contain fixed and random effects. Let us split $\boldsymbol{\beta}$ into two subvectors $\boldsymbol{\beta} = (\boldsymbol{\beta}_1, \boldsymbol{\beta}_2)$ and rewrite the model as

$$\mathbf{y} = \mathbf{X}_1\boldsymbol{\beta}_1 + \mathbf{X}_2\boldsymbol{\beta}_2 + \mathbf{e}. \quad (10)$$

If we treat $\boldsymbol{\beta}_1$ and $\boldsymbol{\beta}_2$ as fixed, the total and residual sum of squares are respectively

$$SST = \mathbf{y}^T\mathbf{y} \quad \text{and} \quad SSE(\boldsymbol{\beta}_1, \boldsymbol{\beta}_2) = \mathbf{y}^T\mathbf{P}\mathbf{y},$$

where $\mathbf{P} = \mathbf{I}_n - \mathbf{X}(\mathbf{X}^T\mathbf{X})^{-1}\mathbf{X}^T$. The sum of squares of the regression is

$$SSR(\boldsymbol{\beta}_1, \boldsymbol{\beta}_2) = SST - SSE(\boldsymbol{\beta}_1, \boldsymbol{\beta}_2) = \mathbf{y}^T\mathbf{Q}\mathbf{y}, \tag{11}$$

where $\mathbf{Q} = \mathbf{X}(\mathbf{X}^T\mathbf{X})^{-1}\mathbf{X}^T$. Secondly, consider the reduced model

$$\mathbf{y} = \mathbf{X}_1\boldsymbol{\beta}_1 + \varepsilon, \tag{12}$$

considering $\boldsymbol{\beta}_1$ as fixed. The residual sum of squares is given by

$$SSE(\boldsymbol{\beta}_1) = \mathbf{y}^T\mathbf{P}_1\mathbf{y},$$

where $\mathbf{P}_1 = \mathbf{I}_n - \mathbf{X}_1(\mathbf{X}_1^T\mathbf{X}_1)^{-1}\mathbf{X}_1^T$. The sum of squares of the regression is

$$SSR(\boldsymbol{\beta}_1) = SST - SSE(\boldsymbol{\beta}_1) = \mathbf{y}^T\mathbf{Q}_1\mathbf{y},$$

where $\mathbf{Q}_1 = \mathbf{X}_1(\mathbf{X}_1^T\mathbf{X}_1)^{-1}\mathbf{X}_1^T$. The reduction in sum of squares due to introducing $\mathbf{X}_2$ in the model with only $\mathbf{X}_1$ is

$$SSR(\boldsymbol{\beta}_2|\boldsymbol{\beta}_1) = SSR(\boldsymbol{\beta}_1, \boldsymbol{\beta}_2) - SSR(\boldsymbol{\beta}_1). \tag{13}$$

Now consider model (2) and rewrite it as (10) taking $\boldsymbol{\beta}_1 = \boldsymbol{\beta}$, $\boldsymbol{\beta}_2 = \mathbf{u}$, $\mathbf{X}_1 = \mathbf{X}$ and $\mathbf{X}_2 = \mathbf{Z}$. This method calculates the expectations of (11) and (13) and equates the sum of squares to their expectations obtaining two equations. Solving for σ_e^2 and σ_u^2 in the resulting equations, we obtain unbiased estimators for σ_e^2 and σ_u^2 (for more details see Searle *et al.* [15], chapter 5). Let $\hat{\mathbf{e}}$ and $\hat{\boldsymbol{\varepsilon}}$ be the vectors of residuals of the submodels (10) and (12), respectively. If rank($\mathbf{X}$) $= p$ and rank($\mathbf{X}|\mathbf{Z}$) $= p + D$, then the Henderson III estimators of the variance components are given by

$$\hat{\sigma}_e^2 = \frac{\sum_{d=1}^{D}\sum_{j=1}^{n_d}\hat{e}_{dj}^2}{n-p-D}, \quad \hat{\sigma}_u^2 = \frac{\sum_{d=1}^{D}\sum_{j=1}^{n_d}\hat{\varepsilon}_{dj}^2 - \hat{\sigma}_e^2(n-p)}{tr\left\{\mathbf{Z}^T[\mathbf{I} - \mathbf{X}(\mathbf{X}^T\mathbf{X})^{-1}\mathbf{X}^T]\mathbf{Z}\right\}}, \tag{14}$$

where $\hat{e}_{dj}$ is the residual corresponding to observation $(\mathbf{x}_{dj}^T, y_{dj})$ in model (10) and $\hat{\varepsilon}_{dj}$ is the residual for the same observation but obtained from model (12).

4 Robust Estimation of Variance Components

In this section we introduce some new robust estimators of variance components based on Henderson method III. We have chosen this method for three reasons; first, because it provides explicit formulas of the estimators, fact which will help to decrease the computational time; second, it does not need the normality assumption; third, the estimation procedure consists simply of solving two standard regression problems. Let us rewrite the estimators as

$$\hat{\sigma}_e^2 = \frac{n[\sum_{d=1}^{D}\sum_{j=1}^{n_d} \hat{e}_{dj}^2/n]}{n-p-D}, \quad \hat{\sigma}_u^2 = \frac{n[\sum_{d=1}^{D}\sum_{j=1}^{n_d} \hat{\varepsilon}_{dj}^2/n] - \hat{\sigma}_e^2(n-p)}{tr\,\{\mathbf{Z}^T[\mathbf{I}-\mathbf{X}(\mathbf{X}^T\mathbf{X})^{-1}\mathbf{X}^T]\mathbf{Z}\}}, \quad (15)$$

These two estimators contain in the numerator sample means of squared residuals obtained from models (10) and (12) respectively. Then a small fraction of outliers, even a single observation, might seriously affect these estimators. To avoid this problem, we propose to use robust methods to fit the two models (10) and (12) and then, replacing the means of squared residuals in (15) by other more robust functions of residuals. Model (12) is a standard linear regression model, which can be robustly fitted using any method available in the literature such as L_1 estimation, M estimation or the fast method of Peña and Yohai [12]. Model (10) is a model with a categorical variable that distributes the observations into groups, which can be robustly fitted using an adaptation of the principal sensibility components method of Peña and Yohai [12] to the grouped data structure, or by the M-S estimation of Maronna and Yohai [9]. These fitting methods would then provide better residuals e_{dj} and ε_{dj}, which are in turn used to find robust estimators of the variance components similar to (15). Below we describe different estimators obtained using robust functions of these new residuals obtained using the robust fit of models (10) and (12).

MADH3 estimators

In (15), we substitute the mean of the squared residuals by the square of the normalized median of absolute deviations (MAD), given by

$$MAD = 1.481 \cdot \text{Med}(|\hat{\xi}_{dj}|, \hat{\xi}_{dj} \neq 0),$$

where $\hat{\xi}_{dj}$ is the residual of observation $(\mathbf{x}_{dj}^T, y_{dj})$ under the corresponding fitted model, either (10) or (12). Then, our first robust proposal for the estimation of the variance components is given by

$$\hat{\sigma}_{e,MADH3}^2 = \frac{n[1.481 \cdot \text{Med}_i(|\hat{e}_{dj}|, \hat{e}_{dj} \neq 0)]^2}{n-p-D} \quad (16)$$

$$\hat{\sigma}_{u,MADH3}^2 = \frac{n[1.481 \cdot \text{Med}_i(|\hat{\varepsilon}_{dj}|, \hat{\varepsilon}_{dj} \neq 0)]^2 - \hat{\sigma}_{e,MADH3}^2(n-p)}{tr\,\{\mathbf{Z}^T[\mathbf{I}-\mathbf{X}(\mathbf{X}^T\mathbf{X})^{-1}\mathbf{X}^T]\mathbf{Z}\}} \quad (17)$$

TH3 estimators

Trimming consists of giving zero weight to a percentage of extreme cases. In this case, instead of this, in the two equations given in (15) we trim the residuals that are outside the interval (b_1, b_2) with

$$b_1 = q_1 - k(q_3 - q_1) \quad \text{and} \quad b_2 = q_3 + k(q_3 - q_1). \quad (18)$$

Here, q_1 and q_2 are the first and third sample quartiles of residuals and k is a constant. Based on results obtained from different simulation studies, we propose to use the constant $k = 2$, just slightly smaller than that one used as outer frontier in the box-plot for detecting outliers.

RH3 estimators

Instead of replacing extreme residuals by zero as in the previous proposal, we can smooth the residuals appearing in (15) according to an appropriate smoothing function. Here we consider the Tukey's biweight function, given by

$$\Psi(x) = x[1 - (x/k)^2]^2, \quad \text{if } |x| \le k. \tag{19}$$

5 Model Diagnostics

In this section we describe some diagnostics tools for the nested error model. Considering that $\boldsymbol{\theta}$ is known, the vector of predicted values is defined as

$$\tilde{\mathbf{y}} = (\mathbf{I} - \mathbf{R})\mathbf{y},$$

where

$$\mathbf{R} = \mathbf{V}^{-1} - \mathbf{V}^{-1}\mathbf{X}(\mathbf{X}^T\mathbf{V}^{-1}\mathbf{X})^{-1}\mathbf{X}^T\mathbf{V}^{-1}. \tag{20}$$

This relation evokes the definition of the hat matrix as

$$\mathbf{H} = \mathbf{I} - \mathbf{R}.$$

The diagonal elements $(1 - r_{dj})$ of matrix $\mathbf{H}$ are measures of the leverage effect of the observations and are called *leverages*. Thus, Galpin and Zewotir [5] proposed the use of the r_{dj}s to identify influential observations. If r_{dj} approaches zero, this indicates that the corresponding observation $(\mathbf{x}_{dj}^T, y_{dj})$ has a large leverage effect.

Due to the data structure in nested error models, it seems more relevant to study the leverage effect of full groups instead of isolated observations. Here we define the leverage effect of group d as

$$h_d = \overline{\mathbf{x}}_d^T(\mathbf{X}^T\mathbf{V}^{-1}\mathbf{X})^{-1}\overline{\mathbf{x}}_d, \quad d = 1, \ldots, D, \tag{21}$$

where $\overline{\mathbf{x}}_d = n_d^{-1}\sum_{j=1}^{n_d}\mathbf{x}_{dj}$.

Concerning residuals, the i-th internally studentized residual is defined as

$$t_{dj} = \frac{e_{dj}}{\sqrt{\text{var}(e_{dj})}} = \frac{e_{dj}}{\sigma_e\sqrt{r_{dj}}} \tag{22}$$

In practice, the variance components involved in (20), (21) and (22) need to be estimated. When there are outliers, these might affect the estimators of variance components, and these in turn will change the distribution of standarized residuals. Better versions of these diagnostic tools can be obtained using the robust variance components estimators introduced in Section 4.

6 Monte Carlo Simulation

In this section we describe a Monte Carlo simulation study that compares the robust estimators of the variance components with the traditional ones. For this, we generated data coming from $D = 10$ groups. The group sample sizes n_d, $d = 1, \ldots, D$ were respectively 20, 20, 30, 30, 40, 40, 50, 50, 60 and 60, with a total sample size of $n = 400$. We considered $p = 4$ auxiliary variables, and they were generated from normal distributions with means and standard deviations coming from a real data set from the Australian Agricultural and Grazing Industries Survey. Thus, the values of the four auxiliary variables were generated respectively as $X_1 \sim N(3.3, 0.6)$, $X_2 \sim N(1.7, 1.2)$, $X_3 \sim N(1.7, 1.6)$ and $X_4 \sim N(2.4, 2.6)$.

The simulation study is based on $L = 500$ iterations. In each iteration, we generated group effects as $u_d \overset{iid}{\sim} N(0, \sigma_u^2)$ with $\sigma_u^2 = 0.25$. Similarly, we generated errors as $e_{dj} \overset{iid}{\sim} N(0, \sigma_e^2)$ with $\sigma_e^2 = 0.25$. Then we generated the model responses y_{dj}, $j = 1, \ldots, n_d$, $d = 1, \ldots, D$, from model (1). Observe that in principle there is no contamination. Finally, we introduced contamination according to three different scenarios:

A. *No contamination.*
B. *Groups with a mean shift:* A subset $\mathcal{D}_c \subseteq \{1, 2, \ldots, D\}$ of groups was selected for contamination. For each selected group $d \in \mathcal{D}_c$, half of the observations were replaced by $c_{d1} = \bar{y}_d + k\, s_{Y,d}$ and the other half by $c_{d2} = \bar{y}_d - k\, s_{Y,d}$ with $k = 5$, where $\bar{y}_d$ and $s_{Y,d}$ are respectively the mean and the standard deviation of the outcome for the clean data in d-th group. This increases the between group variability σ_u^2.
C. *Groups with high variability:* A small percentage of contaminated observations was introduced in each selected group $d \in \mathcal{D}_c$, similarly as described in Scenario B. This increases the within group variability σ_e^2.

After each iteration, we fitted the two models (10) and (12) using the procedure of Peña and Yohai [12], using in the first model an adaptation of this method for grouped data. Then, we calculated the traditional estimators H3, ML and REML, and the proposed robust estimators, MADH3, TH3 and RH3. After the $L = 500$ iterations, we computed their empirical bias and mean squared error (MSE). Table 1 reports the resulting empirical bias and percent MSE of each estimator under Scenario A, without contamination. Observe in that table that in absence of outlying observations, the traditional non-robust estimators, H3, ML and REML, provide the minimum MSE, but the robust alternatives TH3 and RH3 are not too far away from them. However, under Scenario B with full groups contaminated with a mean shift (Tables 2 and 3), the estimators ML, REML and H3 of σ_u^2 increase considerably their MSE. The estimator TH3 achieves the minimum MSE, followed by RH3. Under Scenario C with contamination introduced to make the within cluster variability increase (Tables 4 and 5), now the estimators ML, REML, and H3 of σ_e^2 increase considerably their MSE whereas the robust estimator TH3 resists quite well.

Table 1. Theoretical values $\sigma_u^2 = \sigma_e^2 = 0.25$. Scenario 0: No contamination

Method	Estimators		Bias		MSE $\times 10^2$	
	$\hat{\sigma}_u^2$	$\hat{\sigma}_e^2$	$\hat{\sigma}_u^2$	$\hat{\sigma}_e^2$	$\hat{\sigma}_u^2$	$\hat{\sigma}_e^2$
H3	0.24	0.25	-0.0081	0.0014	1.43	0.03
ML	0.22	0.25	-0.0298	-0.0011	1.16	0.03
REML	0.25	0.25	-0.0046	0.0014	1.32	0.03
MADH3	0.25	0.25	0.0041	0.0018	2.33	0.09
TH3	0.23	0.25	-0.0189	-0.0019	1.04	0.04
RH3	0.24	0.23	-0.0136	-0.0179	1.25	0.06

Table 2. Theoretical values $\sigma_u^2 = \sigma_e^2 = 0.25$. Scenario B: One outlying group

Method	Estimators		Bias		MSE $\times 10^2$	
	$\hat{\sigma}_u^2$	$\hat{\sigma}_e^2$	$\hat{\sigma}_u^2$	$\hat{\sigma}_e^2$	$\hat{\sigma}_u^2$	$\hat{\sigma}_e^2$
H3	1.28	0.24	1.0286	-0.0095	123.73	0.04
ML	1.15	0.24	0.9000	-0.0120	123.27	0.04
REML	1.28	0.24	1.0285	-0.0096	123.38	0.04
MADH3	0.44	0.23	0.1884	-0.0169	7.84	0.10
TH3	0.24	0.24	-0.0089	-0.0142	1.25	0.05
RH3	0.46	0.22	0.2106	-0.0277	6.04	0.10

Table 3. Theoretical values $\sigma_u^2 = \sigma_e^2 = 0.25$. Scenario B: Two outlying groups

Method	Estimators		Bias		MSE $\times 10^2$	
	$\hat{\sigma}_u^2$	$\hat{\sigma}_e^2$	$\hat{\sigma}_u^2$	$\hat{\sigma}_e^2$	$\hat{\sigma}_u^2$	$\hat{\sigma}_e^2$
H3	2.79	0.23	2.5375	-0.0242	715.98	0.08
ML	2.13	0.22	1.8807	-0.0266	495.49	0.10
REML	2.37	0.23	2.1179	-0.0242	500.14	0.08
MADH3	1.10	0.21	0.8529	-0.0437	91.67	0.25
TH3	0.27	0.22	0.0227	-0.0319	2.13	0.13
RH3	0.76	0.21	0.5088	-0.0412	31.52	0.19

Table 4. Theoretical values $\sigma_u^2 = \sigma_e^2 = 0.25$. Scenario C: 10% of atypical observations shared among groups

Method	Estimators		Bias		MSE $\times 10^2$	
	$\hat{\sigma}_u^2$	$\hat{\sigma}_e^2$	$\hat{\sigma}_u^2$	$\hat{\sigma}_e^2$	$\hat{\sigma}_u^2$	$\hat{\sigma}_e^2$
H3	0.23	0.60	-0.0175	0.3512	1.47	12.58
ML	0.21	0.60	-0.0397	0.3450	1.23	12.15
REML	0.24	0.60	-0.0144	0.3512	1.35	12.58
MADH3	0.28	0.27	0.0253	0.0198	2.78	0.14
TH3	0.24	0.25	-0.0073	-0.0012	1.17	0.04
RH3	0.22	0.30	-0.0266	0.0487	1.22	0.26

Table 5. Theoretical values $\sigma_u^2 = \sigma_e^2 = 0.25$. Scenario C: 20% of atypical observations shared among groups

Method	Estimators		Bias		MSE $\times 10^2$	
	$\hat{\sigma}_u^2$	$\hat{\sigma}_e^2$	$\hat{\sigma}_u^2$	$\hat{\sigma}_e^2$	$\hat{\sigma}_u^2$	$\hat{\sigma}_e^2$
H3	0.22	0.93	-0.0268	0.6814	1.50	47.19
ML	0.20	0.92	-0.0489	0.6719	1.32	45.89
REML	0.23	0.93	-0.0236	0.6814	1.39	47.19
MADH3	0.30	0.29	0.0473	0.0406	3.48	0.29
TH3	0.25	0.25	0.0045	0.0003	1.27	0.04
RH3	0.21	0.37	-0.0400	0.1151	1.18	1.35

7 Discussion

In this work we present three robust versions of H3 estimators called MADH3, TH3 and RH3 estimators. These robust estimators are obtained by first, fitting in a robust way the two models (10) and (12), and then replacing the means of squared residuals in H3 estimators by other robust functions of the residuals coming from those robust fittings. In simulations we have analyzed the robustness of our proposed estimators against two different kind of contamination scenarios: when the between groups variability is increased by including a mean shift in some groups, and when the within group variability is increased by introducing given percentages of outliers shared among the clusters. The new robust estimator TH3 gets the best results in these simulations, achieving great efficiency under both types of contamination but preserving at the same time good efficiency when there is not contamination.

Acknowledgement. This work is supported by the grant SEJ2007-64500 from the Spanish Ministerio de Educación y Ciencia and the Collaborative Project 217565, Call identifier FP7-SSH-2007-1, from the European Commision.

References

1. Banerjee, M., Frees, E.W.: Influence diagnostics for linear longitudinal models. J. Amer. Statist. Assoc. 92, 999–1005 (1997)
2. Christensen, R., Pearson, L.M., Johnson, W.: Case-deletion diagnostics for mixed models. Technometrics 34, 38–45 (1992)
3. Fellner, W.H.: Robust estimation of variance components. Technometrics 28, 51–60 (1986)
4. Galpin, J.S., Zewotir, T.: Influence diagnostics for linear mixed models. J. Data Sci. 3, 153–177 (2005)
5. Galpin, J.S., Zewotir, T.: A unified approach on residuals, leverages and outliers in the linear mixed models. TEST 16, 58–75 (2007)
6. Henderson, C.R.: Best linear unbiased estimation and prediction under a selection model. Biometrics 31, 423–447 (1975)
7. Jiang, J.: Linear and Generalized Linear Models and their Applications. Springer Series in Statistics. Springer, Heidelberg (2007)
8. Laird, N.M., Ware, J.H.: Random-effects models for longitudinal data. Biometrics 38, 963–974 (1982)
9. Maronna, R.A., Yohai, V.J.: Robust regression with both continuous and categorical predictors. J. Statist. Plan Infer. 89, 197–214 (2000)
10. McCulloch, C., Searle, S.: Generalized, Linear and Mixed Models. Wiley, New York (2001)
11. Patterson, H.D., Thompson, R.: Recovery of inter-block information when block sizes are unequal. Biometrika 58, 545–554 (1971)
12. Peña, D., Yohai, V.J.: A fast procedure for outlier diagnostics in large regression problems. J. Amer. Statist. Assoc. 94, 434–445 (1999)
13. Richardson, A.M.: Bounded influence estimation in the mixed linear model. J. Amer. Statist. Assoc. 92, 154–161 (1997)
14. Richardson, A.M., Welsh, A.H.: Robust restricted maximum likelihood in mixed linear models. Biometrics 51, 1429–1439 (1995)
15. Searle, S.R., Casella, G., McCulloch, C.E.: Variance Components. Wiley Series in Probability and Mathematical Statistics. Wiley, New York (1992)
16. Vangeneugden, T., Laenen, A., Geys, H., Renard, D., Molenberghs, G.: Applying linear mixed models to estimate reliability in clinical trial data with repeated measurements. Contr. Clin. Trials 25, 13–30 (2004)
17. Verbeke, G., Molenberghs, G.: Linear Mixed Models for Longitudinal Data. Springer, Heidelberg (2009)
18. Wellenius, G.A., Yeh, G.Y., Coull, B.A., Suh, H.H., Phillips, R.S., Mittlemann, M.A.: Effects of ambient air pollution on functional status in patients with chronic congestive heart failure: a repeated-measures study. Environ. Health 6(26) (2007)

Imputation and Inference with Multivariate Adaptive Regression Splines*

Ismael Sánchez-Borrego[1], María del Mar Rueda[1], and Juan F. Muñoz[2]

[1] Department of Statistics and Operational Research,
University of Granada, Spain
ismasb@ugr.es, mrueda@ugr.es

[2] Department of Quantitative Methods for the Economy and the Business,
University of Granada, Spain
jfmunoz@ugr.es

Summary. The problem of missing data is often addressed with imputation. Traditional single imputation methods, such as the ratio imputation, multiple regression imputation, nearest neighbor imputation, respondent mean imputation or hot deck imputation, have been widely used to compensate for non-response. Nonparametric regression methods have been recently applied to the estimation of the regression function in a wide range of settings and areas of research. The focus of this work is on replacing missing observations on a variable of interest by imputed values obtained from a new algorithm based on Multivariate Adaptive Regression Splines. Some imputation methods can lead to serious underestimation for measures of population distributions. This bias can be reduced by adding to the imputed values a small disturbance drawn from a given distribution. Two different methods of adding the random disturbance are also described. Numerical examples are presented to illustrate the theoretical results and analyze the precision of the proposed method.

1 Introduction

Information plays a very important role in our life. Scientists from different research areas have developed methods to analyze huge amounts of data and to extract useful information. Unfortunately, traditional methods usually cannot deal directly with real-world data because of missing values.

Incomplete data sets are a commonplace problem in several areas of research and in many applications. In fact, they can be encountered in a wide range of fields, including social and behavioral sciences, biological systems and computer vision. More examples can be found in clinical studies, in engineering applications, in industrial and research databases, among others.

* The authors would like to thank the editors for this opportunity to contribute to this volume in honour to María Luisa Menéndez.

Incomplete data sets are very common in statistical situations. They can be a serious problem as standard statistical methods generally work with complete data sets. For example, applying complete-data based methods with this loss of scientific information may result in a loss of efficiency and in an incorrect conclusion.

There exist many techniques to manage data with missing values, but no one is absolutely better than the others. Different situations require different solutions. Incomplete-data problems are often addressed with imputation, which consists on replacing a missing value by a specific value obtained from an imputation method. By treating these imputed values as true observations, traditional analysis may be carried out using the standard procedures developed for data without any missing observations.

Single imputation methods have become one of the most popular tools utilized to deal with non-response, which can be classified into two categories: random and deterministic. Traditional single imputation methods include the ratio imputation, multiple regression imputation, nearest neighbor imputation, respondent mean imputation and hot deck imputation (see e.g. Särndal and Lundström [29], Schafer [28], Little and Rubin [17]). Comparisons of several imputation methods are given by Montaquila and Ponikowski [19], Hu *et al.* [15] and Nitter [21]. Iacus and Porro [16] describe applications of regression tree to hot-deck missing data imputation. Ding and Simonoff [9] consider popular missing data methods for classification tree algorithms applied to binary response. Other important related works in this area are given by D'Ambrosio *et al.* [8] and Conversano and Siciliano [7].

The hot deck imputation is one of the most popular random imputation methods in practice. The nearest neighbor imputation method consists on imputing a non-respondent by a respondent from the same variable identified by distance minimization. The nearest neighbor imputation method is used in many survey agencies and it has a long history of application in surveys conducted by Statistic Canada, the US Bureau of Labor Statistic and the US Census Bureau. The nearest neighbor method does not assume a parametric regression model, which implies that this method is more robust against model violations than the methods based upon a linear regression model. Chen and Shao [4] studied some theoretical results regarding the validity of the nearest neighbor method.

The mean imputation method distorts substantially the distribution of the data, and the concentration of all imputed values at the mean creates spikes in the distribution. Also, the true variance can be seriously underestimated when the mean imputation is used. An alternative is to use the ratio or regression imputation methods, which have some relevant advantages in comparison with the mean imputation method. For example, the ratio and regression imputation methods provide imputed values with a larger variability, which may reduce the aforementioned problems of spikes in the distribution and the underestimation of the true variance.

Multiple regression estimation is one of the most commonly used imputation methods. Nevertheless, this method is based on a parametric regression model, and it states strong assumptions about the functional form of the underlying regression function. When this assumption does not hold, nonparametric methods are more appropriate, as only smoothing assumptions over the regression function are made.

Nonparametric regression methods have been recently applied to the estimation of the regression function in a wide range of settings and areas of research. In particular, the local linear kernel smoother (Ruppert and Wand [27] and Fan and Gijbels [11], among others) is well known for its good theoretical and practical properties. Nevertheless, some nonparametric regression methods like the above mentioned one do not perform well when the number of independent variables in the model is large. This problem is usually referred to as the 'curse of dimensionality'.

Penalized splines regression or P-splines is a nonparametric regression method developed by Eilers and Marx [10]. They are regression splines fit by least squares with a more general roughness penalty than the original smoothing spline. Hastie [14] and Marx and Eilers [18] illustrated their flexibility for additive models and Breidt *et al.* [1] applied them to the survey sampling setting. An overview of different applications of P-splines can be found in Wand [31].

Multivariate Adaptive Regression Splines (MARS) were introduced by Friedman [13] in the general context of multivariate nonparametric regression. It is a flexible tool that relies on an adaptive and recursive construction of the system of basis functions. MARS can easily automate variable selection and can handle a large number of independent variables. They are an attractive smoothing method, because of their flexibility and their potential for application to different settings.

2 An Imputation Method Using MARS

Let $U = \{1, \ldots, N\}$ be the population of N units from which a random sample s of fixed size n is drawn according to a specified sampling design d. Let y_j be the value of the response variable y for the unit j, and $x_{1j}, \ldots, x_{kj}$ the corresponding values of the auxiliary variables $x_1, \ldots, x_k$.

Let n_2 be the number of missing values in a given sample s of size n. We denote s_r the set of observed data in s, of size $n_1 = n - n_2$. If $j \in s_r$, y_j is an observed value of the study variable y. If $j \in s - s_r$, the observation y_j is missing and will be estimated and denoted by $\widehat{y}_j$.

We assume the following model:

$$y_j = m(x_{1j}, \ldots, x_{kj}) + e_j, \quad j = 1, \ldots, N, \tag{1}$$

where the e_j, $j = 1, \ldots, N$, are independent and identically distributed with $E(e_j) = 0$ and $Var(e_j) = \sigma^2$, for $j = 1, ..., N$. The unknown regression function $m(\mathbf{x})$ is defined over $D \subseteq \mathbb{R}^k$. We use the MARS technique to estimate this unknown regression function.

Assuming the recursive partitioning regression methodology (Morgan and Sunquist [20] and Breiman *et al.* [2]), the estimation of $m(\mathbf{x})$, say $\widehat{m}(\mathbf{x})$, can be expressed as

$$\widehat{m}(\mathbf{x}) = \sum_{l=1}^{L} a_l B_l(\mathbf{x}), \qquad (2)$$

where L is the number of basis functions, B_l takes the form $B_l(\mathbf{x}) = I[\mathbf{x} \in R_l]$, where $\{R_l\}_1^L$ are disjoint regions and I is the indicator function having the value 1 if its argument is true and 0 otherwise. The aim of this method is to adjust the coefficient values $\{a_l\}$ $l = 1, \ldots, L$ to best fit the data and to derive a data-driven set of basis functions. The partitioning is performed through recursive splitting of the subregions. At each stage of the partitioning all existing subregions (called "parent" subregions) are split into two subregions, so called "daughter" subregions. The basis functions produced by this method have the form

$$B_l(\mathbf{x}) = \prod_{k=1}^{K_l} H\left[s_{kl}(x_{v(k,l)} - t_{kl})\right], \qquad (3)$$

where t_{kl} are the knot locations, K_l is the number of factors, $s_{kl} = \pm 1$, $v(k,l)$ label the predictor variables. H is a step function defined by $H(x) = I(x \geq 0)$.

Recursive partitioning regression can be seen as a forward/backward regression stepwise regression procedure giving rise to a local variable selection method. It tends to overfit the data with a large model and then reduce it with an backward stepwise strategy. Nevertheless, each basis function represents disjoint regions, so removing a basis implies leaving a hole in the predictor space and the model will predict a zero response. Hence, this stepwise deletion basis strategy does not work, as it can not remove a basis without seriously affecting the quality of the fit. Another limitation of the method is that recursive partitioning regression models (2) are piecewise constant and discontinuous at region boundaries, which has an effect on the accuracy of the approximation.

MARS can be regarded as a series of generalization to recursive partitioning regression. MARS produce continuous models by replacing the step function with a continuous one. The continuous functions used are the two-sided truncated splines, which are a mixture of functions of the form

$$b_q^{\pm}(x - t) = [\pm(x - t)]_+^q, \qquad (4)$$

where t is the knot location, q is the order of the spline and the subscript indicates the positive part of the argument. Using the two-sided truncated power basis, these multivariate spline basis functions are defined by

$$B_l^{(q)}(\mathbf{x}) = \prod_{k=1}^{K_l} \left[s_{kl}(x_{v(k,l)} - t_{kl})\right]_+^q,$$

where t_{kl} are the knot locations, K_l is the number of factors and $s_{kl} = \pm 1$. Note that spline basis (3) are a subset of these basis ($q = 0$).

To improve the backward stepwise procedure of recursive partitioning regression, MARS enlarges the eligible set to include each basis of the complete tensor product. The simple constant basis function $B_1(x) = 1$ is never removed and unlike recursive partitioning regression, every basis functions (parent and daughters) are eligible for the next stage of splitting. Hence, removing a basis function does not produce a hole in the predictor space as regions are not disjoint and overlap.

The resulting MARS estimator after these two algorithms is a model of the form

$$\widehat{m}(\mathbf{x}) = a_0 + \sum_{l=1}^{L} a_l \prod_{k=1}^{K_l} \left[s_{kl}(x_{v(k,l)} - t_{kl}) \right]_+ , \tag{5}$$

where a_l are the fitted coefficients of the basis function.

Now, we can calculate the "predicted" value $\widehat{y}_j = \widehat{m}(x_j)$ for each missing value y_j, with $j \in s - s_r$.

The missing values can be replaced by the predicted values. However, if we use $\widehat{y}_j$ as the imputed value, this method artificially reduces the variance of the variable of interest. To overcome the underestimated variance issue, we may add a small disturbance d_j. Most often a normal distribution is used to draw the random disturbance. Thus, we propose to generate d_{1j} from a normal distribution with a zero mean and a variance $\widehat{\sigma}^2$ obtained from the observed data, and the following complete set of observations of size n is obtained

$$y_{1j}^* = \begin{cases} y_j & j \in s_r \\ \widehat{y}_j + d_{1j} & j \in s - s_r, \end{cases}$$

being $\widehat{y}_j = \widehat{m}(\mathbf{x}_j)$ the MARS estimated value of y_j.

A second method of calculating the disturbance is to consider the normalized residuals, i.e.,

$$d_{2j} = \frac{(y_j - \widehat{m}(\mathbf{x}_j))^2}{\widehat{\sigma}^2} \left(1 - \frac{k}{n_1}\right)^{-1/2}, \quad j = 1, \ldots, n_1,$$

and the complete set of observations is now given by

$$y_{2j}^* = \begin{cases} y_j & j \in s_r \\ \widehat{y}_j + d_{2j} & j \in s - s_r, \end{cases}$$

Once the missing observations have been estimated, we have a complete set of values, and standard complete-data methods of analysis can be therefore used. In addition, parameters such as mean, variance and the distribution function can be estimated by using the values of any of the variables y_1^* and y_2^*.

We describe the algorithm including every required operation.

Step 1. Determine if there are missing values of the variable of interest.
Step 2. If Step 1 is true, then a selection of the auxiliary variables with completed observations is performed. Then, go to Step 4.
Step 3. If Step 1 is false, standard complete data methods of analysis can be used. Then, Halt.
Step 4. MARS estimation procedure (Friedman [13]), consisting of modifications to the recursive partitioning regression procedure:

- Use of truncated power basis functions ($q = 1$) instead of a step function.
- Not removing the parent basis function B_1 and making each basis function eligible for further splitting.

Step 5. Generate disturbance d_{1j} or d_{2j} to obtain a complete set of observations y^*_{1j} or y^*_{2j} respectively, then Halt.

3 Some Numerical Examples

In this section, simulations studies are carried out to illustrate the performance of the proposed imputation method under different scenarios.

We consider the random sample

$$(y_j, x_j, \delta_j)\,, \quad j = 1, \ldots, N, \tag{6}$$

where all the x_j's are observed and $\delta_j = 0$ if y_j is missing and $\delta_j = 1$ otherwise. By a purely nonparametric approach to (6), Chu and Cheng [6] and Nitter [21] among others, assume that the data are missing completely at random (MCAR). A relaxed version of MCAR is missing at random (MAR) (Cheng [5] and Little and Rubin [17]). MAR assumes that there is a chance mechanism $p(x)$, such that

$$P(\delta = 1 \mid X = x, Y = y) = P(\delta = 1 \mid X = x) = p(x).$$

MCAR considers δ independent of both X and Y, for example, $p(x)$ being a constant between 0 and 1.

First, we consider a simulated population, where x is a standard normal random variable, and y is obtained by assuming the model (1) with $m(x) = x$. Following Cheng [5], the piecewise linear missing function

$$p(x) = \begin{cases} 0.9 - 0.2|x| & \text{if } |x| \leq 4.5 \\ 0.1 & \text{otherwise} \end{cases}$$

is considered.

Simulation studies are based on $R = 1000$ samples drawn under simple random sampling without replacement. Missing values, with percentages $p = \{10\%, 30\%, 50\%\}$, are generated missing at random (MAR) according to the probabilities given by $p(x)$.

Various imputation methods are evaluated in the problem of the estimation of the population mean $\overline{Y} = N^{-1} \sum_{i \in U} y_i$. The proposed estimators $\widehat{\theta}_{prop1} = n^{-1} \sum_{i \in s} y^*_{1j}$ and $\widehat{\theta}_{prop2} = n^{-1} \sum_{i \in s} y^*_{2j}$ based on the complete data sets are empirically compared to the following estimators based on different imputation methods. First, we consider the local linear kernel regression estimator ($\widehat{\theta}_{LL}$) (Fan and Gijbels [11], among others), which incorporates the disturbance used by $\widehat{\theta}_{prop1}$. We also consider the hot-deck and the nearest-neighbor imputation methods to obtain imputed values, and they are used to obtain, respectively, the estimators ($\widehat{\theta}_{HD}$) and ($\widehat{\theta}_{NN}$), which are calculated as $\widehat{\theta}_{prop1}$ after considering the corresponding imputed values. Finally, imputed values are also obtained by using the regression imputation method, which assumes the functional form of the underlying regression function to be linear. The estimator based on regression imputation method is named as $\widehat{\theta}_{reg}$.

The various imputation methods are compared to the estimator based on the observed units, $\widehat{\theta}_r = n_1^{-1} \sum_{i \in s_r} y_i$, in terms of relative bias RB and relative efficiency RE, where

$$RB(\widehat{\theta}) = 100 \times \frac{1}{R} \sum_{i=1}^{R} \frac{\widehat{\theta}(s_i) - \overline{Y}}{\overline{Y}},$$

$$RE(\widehat{\theta}) = \frac{MSE(\widehat{\theta})}{MSE(\widehat{\theta}_r)} = \frac{\sum_{i=1}^{R} \left(\widehat{\theta}(s_i) - \overline{Y}\right)^2}{\sum_{i=1}^{R} \left(\widehat{\theta}_r(s_i) - \overline{Y}\right)^2},$$

being R the number of replications, MSE is the mean square error, and $\widehat{\theta}$ is the population mean estimator considered in the comparison.

The calculations, imputes values and estimators were obtained using the *R* program and the mda package. Programming details are available from the authors.

Tables 1 and 2 show the RE and the RB for the estimators obtained from the different imputation methods. We observe that estimators based on the proposed imputation method have a good performance in comparison to theirs competitors.

The regression estimator ($\widehat{\theta}_{reg}$) performs better than the local linear kernel regression smoother, since the regression model is well-specified. However, a superior efficiency can be gained by the nonparametric regression estimators when the model is misspecified.

Table 1. Relative efficiency (RE) for the simulated population. The sample size is $n = 100$.

p	$\widehat{\theta}_r$	$\widehat{\theta}_{prop1}$	$\widehat{\theta}_{prop2}$	$\widehat{\theta}_{LL}$	$\widehat{\theta}_{reg}$	$\widehat{\theta}_{HD}$	$\widehat{\theta}_{NN}$
10%	1.00	0.75	0.66	0.81	0.69	0.78	0.70
30%	1.00	0.77	0.61	0.84	0.69	0.71	1.14
50%	1.00	0.46	0.36	0.50	0.44	0.42	1.57

Non-parametric estimators ($\widehat{\theta}_{prop1}$, $\widehat{\theta}_{prop2}$, $\widehat{\theta}_{LL}$) are clearly more efficient than the estimator $\widehat{\theta}_r$, whereas estimators based on MARS ($\widehat{\theta}_{prop1}$, $\widehat{\theta}_{prop2}$) are more efficient than the estimator based on local linear kernel regression ($\widehat{\theta}_{LL}$). Proposed estimators are also more efficient than estimators $\widehat{\theta}_{HD}$ and $\widehat{\theta}_{NN}$. We also observe that $\widehat{\theta}_{prop2}$ is always the most efficient estimator.

Estimators are generally more efficient than $\widehat{\theta}_r$ as the proportion of missing data, p, increases. An exception is the estimator $\widehat{\theta}_{NN}$, which is even less efficient than $\widehat{\theta}_r$ when the p is larger than 30%.

Estimators give small values of RB, which indicates that they have a good performance in term of bias. However, estimators $\widehat{\theta}_{prop1}$, $\widehat{\theta}_{prop2}$ and $\widehat{\theta}_{reg}$ give values of RB slightly smaller than the other estimators.

Table 2. Relative bias (RB) for the simulated population. The sample size is $n = 100$.

p	$\widehat{\theta}_r$	$\widehat{\theta}_{prop1}$	$\widehat{\theta}_{prop2}$	$\widehat{\theta}_{LL}$	$\widehat{\theta}_{reg}$	$\widehat{\theta}_{HD}$	$\widehat{\theta}_{NN}$
10%	-0.28	-0.07	-0.05	-0.27	0.09	-0.48	-0.06
30%	-0.70	-0.01	-0.09	-0.25	0.05	-0.80	-0.33
50%	0.56	0.04	-0.03	-0.29	0.05	-0.22	-0.37

The various imputations methods are now evaluated under three real-life populations, which are described as follows. First, we considered the Canadian Prestige Occupational population (Fox [12]). The variable of interest is the Prestige score for occupation (Pineo *et al.* [22]), whereas the auxiliary variables are the average education and the average income of 102 incumbents in years. A scatter plot of this population is given by Figure 1.

The second population consists of 338 Sugar cane farms. The variable of interest is income from cane and the auxiliary variables are the area assigned for growing cane and the costs. This population was used by Chambers and Dunstan [3] and by Rao, Kovar and Mantel [23]. This population is plotted on Figure 2.

The third population comprises 281 Swedish municipalities and it was used by Särndal *et al.* [30]. This population is named as MU284, and it involves three auxiliary variables. The variable of interest is revenues from the 1985

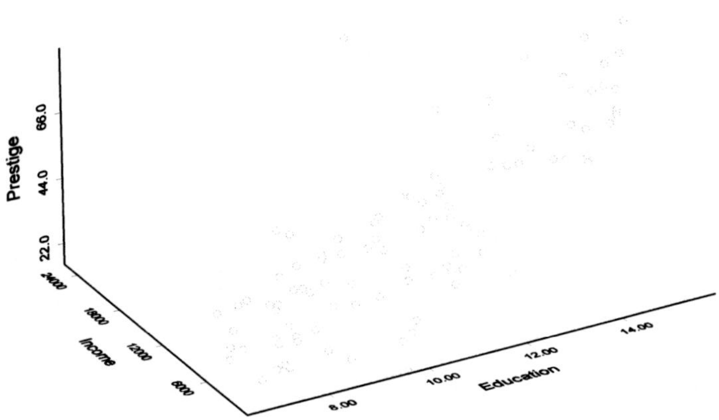

Fig. 1. Scatter plot for the Prestige population

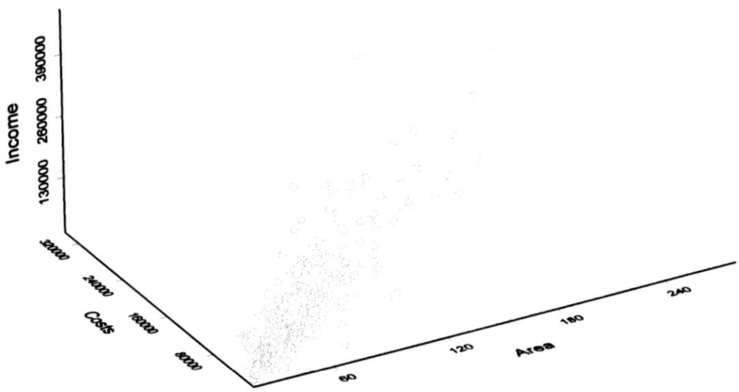

Fig. 2. Scatter plot for the Sugar Cane population

municipal taxation and the auxiliary variables are the number of conservative seats, the number of socialist seats in municipal councils in 1982 and the number of municipal employees in 1984.

Imputation methods are again evaluated in the problem of the estimation of the population mean $\overline{Y}$. The aforementioned estimators are considered. However, $\widehat{\theta}_{prop2}$ had the best performance among the non-parametric estimators, and $\widehat{\theta}_{prop1}$ and $\widehat{\theta}_{LL}$ are thus omitted. Missing data are selected for each sample under a missing completely at random (MCAR) mechanism and three proportions of missing values (constant values of $p(x)$ in MCAR: $p = 0.1$, $p = 0.3$ and $p = 0.5$) are considered.

Tables 3, 4 and 5 show the values of RE obtained from the real populations. We observe that estimators have negligible biases, with values of RB smaller than 1%, and they are thus omitted.

Table 3. Relative efficiency (RE) for Prestige population. The sample size is $n = 40$.

p	$\widehat{\theta}_r$	$\widehat{\theta}_{prop2}$	$\widehat{\theta}_{reg}$	$\widehat{\theta}_{HD}$	$\widehat{\theta}_{NN}$
10%	1	0.83	0.83	1.10	0.94
30%	1	0.62	0.67	1.16	0.93
50%	1	0.38	0.51	1	0.84

Table 4. Relative efficiency (RE) for Sugar cane population. The sample size is $n = 75$.

p	$\widehat{\theta}_r$	$\widehat{\theta}_{prop}$	$\widehat{\theta}_{reg}$	$\widehat{\theta}_{HD}$	$\widehat{\theta}_{NN}$
10%	1	0.92	0.93	1.10	0.96
30%	1	0.70	0.74	1.20	0.78
50%	1	0.46	0.54	1.01	0.60

Table 5. Relative efficiency (RE) for MU284 population. The sample size is $n = 75$.

p	$\widehat{\theta}_r$	$\widehat{\theta}_{prop}$	$\widehat{\theta}_{reg}$	$\widehat{\theta}_{HD}$	$\widehat{\theta}_{NN}$
10%	1	0.87	0.88	1.15	0.87
30%	1	0.59	0.62	1.19	0.59
50%	1	0.42	0.44	1.01	0.43

Results derived from this simulation study indicate that the estimator $\widehat{\theta}_{prop2}$ based on the proposed imputation method is more efficient than alternative estimators for every population under study. As p increases, the proposed estimator continues showing a satisfactory performance in comparison to alternatives estimators.

4 Conclusion and Comments

Among modern strategies used to cope with missing data, a major problem faced by survey statisticians, imputation is one of the most common. Our goal in this chapter is to study the application of nonparametric methods to the problem of imputation.

Multiple regression estimator ($\widehat{\theta}_{reg}$) performs well when the regression model is well-specified. However, a superior efficiency can be gained by the nonparametric based estimator when the model is misspecified. As the superpopulation model is typically unknown, the proposed estimator is likely to be a desirable choice for finite population parameter estimation.

The proposed method has good flexibility, since it can handle large data sets and is convenient for computer realization. The proposed method is based on MARS, which is well-known for its easy applicability and advantageous properties. We may conclude that the proposed method may be a good alternative to other classical imputation estimators. Moreover, it can be applied to other areas of research and applications where incomplete data exists.

An alternative to single imputation is to use Multiple imputation (MI) (Rubin [24], [25] and [26]). There are plenty of MI computational tools available for creating multiple imputations, such as NORM (which performs MI under a multivariate normal model and is free at http://www.stat.edu.psu/˜jls), S-plus missing data library, Solas v3.0 (a commercial programme for imputing binary and categorical variables), MICE (which incorporates many conditional distributions and regression models for S-plus or R), among others.

MI is an attractive method, since it can be highly efficient even for small values of M, the number of multiple imputations (only 3-5 imputations are needed). Nevertheless, the first step of MI involves building an imputation model, and the choice of a correct imputation model is one of the uncertainties of the MI method. Although MI tends to correct itself from choosing an imperfect one, the use of an appropriate model may improve the efficiency of the imputation. Other limitations can be also observed by the MI. For example, MI may not lead to consistent variance estimators for stratified multistage surveys. Moreover, several statistical agencies seem to prefer single imputation, mainly due to operational difficulties in maintaining multiple complete data sets, especially in large-scale surveys.

Nonparametric regression methods are more flexible as they do not place restrictions on the functional form of the regression function and they are more efficient than parametric regression estimators when the parametric model is incorrectly specified.

Acknowledgement. This work is partially supported by Ministerio de Educación y Ciencia (contract No. MTM2009-10055).

References

1. Breidt, F.J., Claeskens, G., Opsomer, J.D.: Model-assisted estimation for complex surveys using penalized splines. Biometrika 92, 831–846 (2005)
2. Breiman, L., Friedman, J.H., Olhsen, R.A., Stone, C.J.: Classification and Regression Trees. Wadsworth, Belmont (1984)
3. Chambers, R.L., Dunstan, R.: Estimating distribution functions from survey data. Biometrika 73, 597–604 (1986)
4. Chen, J., Shao, J.: Nearest neighbor imputation for survey data. J. Offic. Statist. 16, 113–131 (2000)
5. Cheng, P.E.: Nonparametric Estimation of Mean Functionals with Data Missing ar Random. J. Amer. Statist. Assoc. 89, 81–87 (1994)

6. Chu, C.K., Cheng, P.E.: Nonparametric regression estimation with missing data. J. Statist. Plan Infer. 48, 85–99 (1995)
7. Conversano, C., Siciliano, R.: Incremental Tree-Based Missing Data Imputation with Lexicographic Ordering. J. Classif. 26(3), 361–379 (2009)
8. D'Ambrosio, A., Aria, M., Siciliano, R.: Robust tree-based incremental imputation method for data fusion. In: Berthold, M., Shawe-Taylor, J., Lavrač, N. (eds.) IDA 2007. LNCS, vol. 4723, pp. 174–183. Springer, Heidelberg (2007)
9. Ding, Y., Simonoff, J.S.: An investigation of missing data methods for classification trees applied to binary response data. J. Mach. Learn Res. 11, 131–170 (2010)
10. Eilers, P.H.C., Marx, B.D.: Flexible smoothing with B-splines and penalties. Statist. Sci. 11(2), 86–121 (1996)
11. Fan, J., Gijbels, I.: Local Polynomial Modelling and Its Applications. Chapman & Hall, London (1996)
12. Fox, J.: Applied Regression Analysis, Linear Models and Related Methods. Sage Publications, Hamilton (1997)
13. Friedman, J.H.: Multivariate Adaptive Regression Splines. Ann. Statist. 19(1), 1–67 (1991)
14. Hastie, T.: Pseudosplines. J. Royal Statist. Soc. Ser. B 58, 376–396 (1996)
15. Hu, M., Salvucci, S., Lee, R.: A Study of Imputation Algorithms. Working Paper No. 2001-17. Washington DC: U.S. Department of Education, National Center for Education Statistics, 27 Stata Statistical Software (2001)
16. Iacus, S.M., Porro, G.: Missing data imputation, matching and other applications of random recursive partitioning. Comp. Statist. Data Anal. 52(2), 773–789 (2007)
17. Little, R.J.A., Rubin, D.: Statistical Analysis with missing data. Wiley, New York (2002)
18. Marx, B.D., Eilers, P.H.C.: Direct generalized additive modelling with penalized likelihood. Comp. Statist. Data Anal. 28, 193–209 (1998)
19. Montaquila, J.M., Ponikowski, C.H.: An evaluation of alternative imputation methods. Proc. Section on Surv. Res. Meth. Amer. Statist. Assoc., 281–286 (1995)
20. Morgan, J.N., Sonquist, J.A.: Problems in the analysis of survey data, and a proposal. J. Amer. Statist. Assoc. 58, 415–434 (1963)
21. Nitter, T.: The additive model affected by missing completely at random in the covariate. Comput. Statist. 19(2), 261–282 (2004)
22. Pineo, P.C., Porter, J., McRoberts, H.A.: The 1971 census and the socioeconomic misclassification of occupations. Can Rev. Sociol. Anthropol. 14, 147–157 (1977)
23. Rao, J.N.K., Kovar, J.G., Mantel, H.J.: On estimating distribution functions and quantiles from survey data using auxiliary information. Biometrika 77, 365–375 (1990)
24. Rubin, D.B.: Formalizing subjective notions about the effect of nonrespondents in sample surveys. J. Amer. Statist. Assoc. 72, 53–543 (1977)
25. Rubin, D.B.: Multiple Imputation for Nonresponse in Surveys. Wiley, New York (1987)
26. Rubin, D.B.: Multiple imputations in sample surveys. Proc. Section on Surv. Res. Meth. Amer. Statist. Assoc., 20–34 (1978)

27. Ruppert, D., Wand, M.P.: Multivariate locally weighted least squares regression. Ann. Statist. 22(3), 1346–1370 (1994)
28. Schafer, J.: Analysis of Incomplete Multivariate Data. Chapman & Hall, London (1997)
29. Särndal, C.E., Lunström, S.: Estimation in Surveys with Nonresponse. Wiley Series in Survey Methodology. Wiley, New York (2005)
30. Särndal, C.E., Swensson, B., Wretman: Model Assisted Survey Sampling. Springer, New York (1992)
31. Wand, M.: Smoothing and mixed models. Comput. Statist. 18, 223–249 (2003)

Part VIII

Probability Theory

Multifractional Random Systems on Fractal Domains

José Miguel Angulo and María Dolores Ruiz-Medina

Faculty of Sciences, Campus Fuente Nueva s/n,
University of Granada, 18071 Granada, Spain
{jmangulo,mruiz}@ugr.es

Summary. In this paper, the effect of the domain geometry on the local regularity/singularity properties of the solution to the trace of a multifractional pseudodifferential equation on a fractal domain is studied. The singularity spectrum of the Gaussian solution to this type of models is trivial due to regularity assumptions on the variable order of its fractional derivatives. The theory of reproducing kernel Hilbert spaces (RKHSs) and generalized random fields is applied in this study. Specifically, the associated family of RKHSs is isomorphically identified with the trace on a compact fractal domain of a multifractional Sobolev space. The fractal defect modifies the variable order of weak-sense factional derivatives of the functions in these spaces. In the Gaussian case, random fields on fractal domains having sample paths with variable local Hölder exponent are introduced in this framework.

Keywords: fractal geometry, multifractional local Hölder exponent, multifractional pseudodifferential operator, reproducing kernel Hilbert space.

1 Introduction

Heterogeneous fractal models were originally introduced to describe complex, non-stationary, physical and enginering systems (see, for example, [22], [21]). Special attention has been paid to modeling the spatial distribution of the kinetic energy dissipation rate in fully developed turbulence ([10], [17], [35]). We also highlight the characterization of pore systems in rocks ([39]); spatial variability of bioactive marine sediments ([29]); spatial variability of soil properties ([16], [19], [28], [42]); hydraulic conductivity ([55], [34]); scaling of intrinsic permeability ([11]); fluctuations of geophysical fields ([36], [41], [44]). Topography, earthquake activity and surface gravity over various scale ranges also provide empirical evidence of multifractality ([31], [37], [58], and especially [20]). Some recent developments on multifractal random measures and processes include [8], [7], [2], [3], [4], and references therein.

In this paper, we consider the case where the heterogeneous fractality of the physical law is altered by the fractal geometry of the disordered domain of definition. This issue has aroused great interest (see, for instance, [18], [46], [50], [48], [54]), with a number of problems still remaining open. The motivation of this paper lies on the need to characterize the influence of a disordered (fractal) medium on the structure of a chaotic system governed by a multifractional pseudodifferential equation. The heterogeneous fractality displayed by the solution to this equation is then affected by the fractal dimension of the domain. We consider the theory of trace operators on functional spaces (see [57]) for characterization of the properties of the solution, which is introduced in a generalized random field framework. This differs from the approaches where warped processes are considered (for example, warped fractional Brownian motion or warped fractional Lévy motion; see [43]). Specifically, we introduce a class of generalized random fields whose reproducing kernel Hilbert space (RKHS) is isomorphic to the trace of a multifractional Sobolev space (see [23], [27]) on a fractal domain. Factorization of the covariance is then established, allowing their characterization as solutions of multifractional pseudodifferential equations on fractal domains (see [51], for the fractal case, and [50], for the variable regularity order case on $\mathbb{R}^n$). Embeddings between fractional Besov spaces allow the strong-sense interpretation of the results derived in the weak sense ([32], [57], [52]). Thus, the weak-sense fractional variable order of differentiation of the solution is interpreted in terms of its Hölder spectrum. The definition of a class of random fields on fractal domains with increments having heterogeneous local mean quadratic variation is then obtained. In particular, the multifractional exponent of the associated variogram depends on the second-order variable regularity order of the model and on the local dimension of the fractal domain.

In the Gaussian case, the multifractional exponent of the variogram defines the local regularity properties of the sample paths of the solution. The modulus of continuity of the sample paths then depends on the exponent of the mean quadratic local variation of the increments (see [1]). The methodological proposal made in this paper thus opens a new research line in relation to the introduction of continuous multifractional Gaussian processes extending fractional Brownian motion, e.g., multifractional Brownian motion (see [33], [9]), and generalized multifractional Brownian motion (see [6], [5]).

The outline of the paper is as follows: In Section 2, the pseudoduality condition introduced in [50] is reformulated in the context of fractal domains. This condition is used to derive the isomorphic identification of the RKHS of the generalized random field solution with the trace of a multifractional Sobolev space on a fractal domain. Its characterization as the solution to a multifractional psedodifferential equation on a fractal domain then follows in Section 3. The spectral properties of the solution are given in Section 4. In Section 5, the Hölder continuity and regularity properties of the solution, in the second-order moment sense, and in the sample-path sense, for the

Gaussian case, are derived. Some examples are considered in Section 6. Section 7 provides a final synthesis and concluding remarks. Auxiliary definitions and results on multifractional Sobolev spaces, multifractional pseudodifferential operators, and trace operators on fractals are provided in the Appendix.

2 Reproducing Kernel Hilbert Spaces of Variable Order on Fractal Domains

The trace theorems on compact fractal domains formulated in [57] (see also Appendix) lead to the identification of the multifractional Sobolev space $H^{s(\cdot)}(\mathcal{M})$ as the trace of $H^{s(\cdot)+\frac{n-\alpha}{2}}(\mathbb{R}^n)$ on the fractal domain $\mathcal{M}$, with $\frac{n-\alpha}{2}$ denoting the fractal deffect of the domain $\mathcal{M}$, and $s \in \mathbb{R}$ representing the weak-sense variable order of differentiation of the functions belonging to $H^{s(\cdot)}(\mathcal{M})$. Our present objective, in this section, is to identify isomorphically the RKHS of the random solution to a multifractional pseudodifferential equation on a compact fractal domain (a special case of $d-$set, with $d = \alpha$; see Appendix) with the trace space $H^{s(\cdot)}(\mathcal{M})$, for $s \in \mathbb{R}$. This identification allows to establish the continuous extension of the solution to $\mathbb{R}^n$, in the second-order moment sense, that is, its extension is defined as a second-order random field with RKHS isomorphic to the multifractional Sobolev space $H^{s(\cdot)+\frac{n-\alpha}{2}}(\mathbb{R}^n)$. In the Gaussian case, the extension operator can also be applied to the sample paths for their characterization over $\mathbb{R}^n$.

Let us consider for the base complete probability space $(\Omega, \mathcal{A}, P)$, the space $\mathcal{L}^2(\Omega, \mathcal{A}, P)$ defined as the Hilbert space of real-valued zero-mean random variables defined on $(\Omega, \mathcal{A}, P)$ with finite second-order moment and with the inner product

$$< X, Y >_{\mathcal{L}^2(\Omega)} = E[XY], \quad X, Y \in \mathcal{L}^2(\Omega, \mathcal{A}, P). \tag{1}$$

Definition 1. *(See [50]) Let $\beta(\cdot)$ be a real-valued function in $\mathcal{B}^{\infty}(\mathbb{R}^n)$, and let $X_{\beta(\cdot)}$ be defined from $H^{-\beta(\cdot)}(\mathbb{R}^n)$ into $\mathcal{L}^2(\Omega, \mathcal{A}, P)$. We say that $X_{\beta(\cdot)}$ is a* fractional generalized random field of variable order *(FGRFVO) $\beta(\cdot)$ if it is linear and continuous, in the mean-square sense, with respect to the norm defined on $H^{-\beta(\cdot)}(\mathbb{R}^n)$.*

We consider the Hilbert space $H\left(X_{\beta(\cdot)}\right)$, which is defined as the closed span in the $\mathcal{L}^2(\Omega, \mathcal{A}, P)-$topology of the random components of $X_{\beta(\cdot)}$. The covariance function $B_{\beta(\cdot)}$ of $X_{\beta(\cdot)}$ defines a positive, symmetric and continuous operator of variable order $R_{\beta(\cdot)} : H^{-\beta(\cdot)}(\mathbb{R}^n) \longrightarrow H^{\beta(\cdot)}(\mathbb{R}^n)$ by the identity

$$\begin{aligned} B_{\beta(\cdot)}(f, g) &= E\left[X_{\beta(\cdot)}(f) X_{\beta(\cdot)}(g)\right] = R_{\beta(\cdot)}(f)(g) \\ &= \langle \left[R_{\beta(\cdot)}(f)\right]^*, g \rangle_{H^{-\beta(\cdot)}(\mathbb{R}^n)}, \end{aligned} \tag{2}$$

for all $f, g \in H^{-\beta(\cdot)}(\mathbb{R}^n)$. We refer to $R_{\beta(\cdot)}$ as the covariance operator of $X_{\beta(\cdot)}$, which generates the RKHS $\mathcal{H}(X_{\beta(\cdot)})$ of $X_{\beta(\cdot)}$, constituted by the functions of $H^{\beta(\cdot)}(\mathbb{R}^n)$ defined from the elements of $H(X_{\beta(\cdot)})$ as follows:

$$\phi(f) = E\left[X X_{\beta(\cdot)}(f)\right], \ \forall f \in H^{-\beta(\cdot)}(\mathbb{R}^n), \text{ for a certain } X \in H(X_{\beta(\cdot)}). \quad (3)$$

The RKHS $\mathcal{H}(X_{\beta(\cdot)})$ is isometric to the dual of the Hilbert space $H(X_{\beta(\cdot)})$; thus, each function in this space defines an element of the dual of $H(X_{\beta(\cdot)})$.

Definition 2. *(See [50]) Let $\beta(\cdot)$ be as given in Definition 1. We say that $\widetilde{X}_{\beta(\cdot)} : H^{\beta(\cdot)}(\mathbb{R}^n) \to \mathcal{L}^2(\Omega, \mathcal{A}, P)$ is a* pseudodual generalized random field of variable order *for the FGRFVO $X_{\beta(\cdot)}$ if the following conditions hold:*

i) $\widetilde{X}_{\beta(\cdot)}$ is linear and continuous, in the mean-square sense, with respect to the norm defined on $H^{\beta(\cdot)}(\mathbb{R}^n)$;
ii) the space $H\left(\widetilde{X}_{\beta(\cdot)}\right)$ coincides with the space $H\left(X_{\beta(\cdot)}\right)$, and
iii) for all $\phi \in H^{\beta(\cdot)}(\mathbb{R}^n)$ and $g \in H^{-\beta(\cdot)}(\mathbb{R}^n)$, the inner product $\left\langle X_{\beta(\cdot)}(g), \widetilde{X}_{\beta(\cdot)}(\phi)\right\rangle_{H\left(X_{\beta(\cdot)}\right)}$ is given by

$$\begin{aligned}\left\langle X_{\beta(\cdot)}(g), \widetilde{X}_{\beta(\cdot)}(\phi)\right\rangle_{H\left(X_{\beta(\cdot)}\right)} &= \widetilde{X}_{\beta(\cdot)}(\phi)\left[X_{\beta(\cdot)}(g)\right] = \left[(I+R)g\right](\phi) \\ &= \left[(I+R)^*\phi\right](g), \end{aligned} \quad (4)$$

where $R \in \boldsymbol{\mathcal{S}}^{-\infty}_{\rho,\delta} = \bigcap_{m\in\mathbb{R}} \boldsymbol{\mathcal{S}}^{m}_{\rho,\delta}$, for certain δ and ρ with $0 \leq \delta < \rho \leq 1$. Here, A^* denotes the formal adjoint of the operator A.

The following Hilbert spaces are also considered:

$$H\left(\widetilde{X}_{\beta(\cdot)}\right) = \overline{\mathrm{sp}}^{\mathcal{L}^2(\Omega,\mathcal{A},P)}\left\{\widetilde{X}_{\beta(\cdot)}(\phi) : \ \phi \in H^{\beta(\cdot)}(\mathbb{R}^n)\right\}, \quad (5)$$

and the associated RKHS $\mathcal{H}(\widetilde{X}_{\beta(\cdot)})$, isometric to the dual space of $H(\widetilde{X}_{\beta(\cdot)})$, constituted by the functions $f \in H^{-\beta(\cdot)}(\mathbb{R}^n)$ satisfying

$$f(\phi) = E\left[Y \widetilde{X}_{\beta(\cdot)}(\phi)\right], \ \forall \phi \in H^{\beta(\cdot)}(\mathbb{R}^n), \text{ for a certain } Y \in H(\widetilde{X}_{\beta(\cdot)}). \quad (6)$$

The space of distributions $H^{-\beta(\cdot),\mathcal{M}}(\mathbb{R}^n)$ of variable weak-sense order of differentiation $\beta(\cdot)$ with compact support contained in $\mathcal{M}$ is now introduced as the space of test functions needed for the definition of the trace on a fractal domain $\mathcal{M}$ of a FGRFVO $X_{\beta(\cdot)}$ defined on $H^{-\beta(\cdot)}(\mathbb{R}^n)$.

Definition 3. *Let $\mathcal{M}$ be a compact fractal domain with local dimension given by the singularity exponent α of $\mu_{\mathcal{M}}$, the fractal measure with support $\mathcal{M}$ (see Appendix), and let $X_{\beta(\cdot)}$ be a FGRFVO with variable mean-square regularity order $\beta(\cdot)$. For $\beta(\cdot) = s(\cdot) + \frac{n-\alpha}{2}$, the restriction $X^{\mathcal{M}}_{s(\cdot)}$ of $X_{\beta(\cdot)}$ to $\mathcal{M}$ is defined as*

$$X_{s(\cdot)}^{\mathcal{M}}(f) = X_{\beta(\cdot)}(f), \quad \forall f \in H^{-\beta(\cdot),\mathcal{M}}(\mathbb{R}^n).$$

We refer to $X_{s(\cdot)}^{\mathcal{M}}$ *as the trace on the compact fractal domain* $\mathcal{M}$ *of* $X_{\beta(\cdot)}$*, in the second-order moment sense.*

The existence of a pseudodual generalized random field with variable weak-sense second-order of differentiation, and with support contained in the fractal domain $\mathcal{M}$, allows the derivation of a covariance factorization and white-noise multifractional filter representation on fractal domains.

Definition 4. *For* $\beta(\cdot) = s(\cdot) + \frac{n-\alpha}{2}$*, we say that* $\widetilde{X}_{s(\cdot)}^{\mathcal{M}} : H^{s(\cdot)}(\mathcal{M}) \longrightarrow \mathcal{L}^2(\Omega, \mathcal{A}, P)$*, with support contained in the compact fractal set* $\mathcal{M}$*, is the pseudodual of* $X_{s(\cdot)}^{\mathcal{M}}$ *if it satisfies the following conditions:*

i) $\widetilde{X}_{s(\cdot)}^{\mathcal{M}}$ *is continuous, in the mean-square sense, with respect to the norm defined on* $H^{s(\cdot)}(\mathcal{M})$*;*

ii) $H\left(X_{s(\cdot)}^{\mathcal{M}}\right) = H\left(\widetilde{X}_{s(\cdot)}^{\mathcal{M}}\right)$*, and*

iii) $< X_{s(\cdot)}^{\mathcal{M}}(f), \widetilde{X}_{s(\cdot)}^{\mathcal{M}}(\phi) >_{H\left(X_{s(\cdot)}^{\mathcal{M}}\right)} = \int_{\mathcal{M}}(I+R)(f)(\mathbf{z})\phi(\mathbf{z})\mu_{\mathcal{M}}(d\mathbf{z})$*, for* $\phi \in H^{s(\cdot)}(\mathcal{M})$ *and* $f \in H^{-\beta(\cdot),\mathcal{M}}(\mathbb{R}^n)$*, where* $\mu_{\mathcal{M}}$ *is the fractal measure with singularity exponent* α *defining the local dimension of the fractal compact set* $\mathcal{M}$*, and* R *is as given in Definition 2.*

Note that the spaces $H\left(X_{s(\cdot)}^{\mathcal{M}}\right)$ and $\mathcal{H}\left(X_{s(\cdot)}^{\mathcal{M}}\right)$, as well as the spaces $H\left(\widetilde{X}_{s(\cdot)}^{\mathcal{M}}\right)$ and $\mathcal{H}\left(\widetilde{X}_{s(\cdot)}^{\mathcal{M}}\right)$, are defined as before, for a FGRFVO and its pseudodual, considering the spaces $H^{-\beta(\cdot),\mathcal{M}}(\mathbb{R}^n)$ and $H^{s(\cdot)}(\mathcal{M})$ instead of the spaces $H^{-\beta(\cdot)}(\mathbb{R}^n)$ and $H^{\beta(\cdot)}(\mathbb{R}^n) = [H^{-\beta(\cdot)}(\mathbb{R}^n)]^*$, respectively.

The pseudoduality condition on fractal sets, introduced in Definition 4, allows the definition of a bounded parametrix on $\mathcal{M}$ for the following operators:

$$J_{\mathcal{M}} : H\left(X_{s(\cdot)}^{\mathcal{M}}\right) \longrightarrow \mathcal{H}\left(X_{s(\cdot)}^{\mathcal{M}}\right) \subseteq H^{s(\cdot)}(\mathcal{M}), \quad \text{and} \tag{7}$$

$$J'_{\mathcal{M}} : H\left(\widetilde{X}_{s(\cdot)}^{\mathcal{M}}\right) \longrightarrow \mathcal{H}\left(\widetilde{X}_{s(\cdot)}^{\mathcal{M}}\right) \subseteq H^{-\beta(\cdot),\mathcal{M}}(\mathbb{R}^n), \tag{8}$$

respectively defined as

$$X \longrightarrow J_{\mathcal{M}}[X] = \varphi_X \text{ with } \varphi_X(f) = EXX_{s(\cdot)}^{\mathcal{M}}(f), \ f \in H^{-\beta(\cdot),\mathcal{M}}(\mathbb{R}^n), \text{ and}$$
$$Y \longrightarrow J'_{\mathcal{M}}[Y] = g_Y \text{ with } g_Y(\phi) = EY\widetilde{X}_{s(\cdot)}^{\mathcal{M}}(\phi), \ \phi \in H^{s(\cdot)}(\mathcal{M}).$$

Specifically, the pseudoduality condition means that

$$J_{\mathcal{M}}\left(\widetilde{X}_{s(\cdot)}^{\mathcal{M}}(\phi)\right) = (I+R)^*(\phi), \quad \forall \phi \in H^{s(\cdot)}(\mathcal{M}), \quad \text{and}$$
$$J'_{\mathcal{M}}\left(X_{s(\cdot)}^{\mathcal{M}}(f)\right) = (I+R)(f), \quad \forall f \in H^{-\beta(\cdot),\mathcal{M}}(\mathbb{R}^n), \tag{9}$$

where, as before, $*$ stands for the formal adjoint. Furthermore, from the definition of RKHS,

$$J_{\mathcal{M}}^{-1} R_{X_{s(\cdot)}^{\mathcal{M}}}(f) = X_{s(\cdot)}^{\mathcal{M}}(f), \quad \forall f \in H^{-\beta(\cdot),\mathcal{M}}(\mathbb{R}^n), \quad \text{and}$$
$$(J'_{\mathcal{M}})^{-1} R_{\widetilde{X}_{s(\cdot)}^{\mathcal{M}}}(\phi) = \widetilde{X}_{s(\cdot)}^{\mathcal{M}}(\phi), \quad \forall \phi \in H^{s(\cdot)}(\mathcal{M}). \tag{10}$$

Equations (9) and (10) lead to the identities

$$J'_{\mathcal{M}} J_{\mathcal{M}}^{-1} R_{X_{s(\cdot)}^{\mathcal{M}}}(f) = (I+R)(f), \quad \forall f \in H^{-\beta(\cdot),\mathcal{M}}(\mathbb{R}^n), \quad \text{and}$$
$$J_{\mathcal{M}}(J'_{\mathcal{M}})^{-1} R_{\widetilde{X}_{s(\cdot)}^{\mathcal{M}}}(\phi) = (I+R)^*(\phi), \quad \forall \phi \in H^{s(\cdot)}(\mathcal{M}). \tag{11}$$

From the pseudoduality condition, we also have

$$J_{\mathcal{M}}^{-1}(I+R)^*(\phi) = \left[J'_{\mathcal{M}}\right]^*(\phi), \quad \forall \phi \in H^{s(\cdot)}(\mathcal{M}), \quad \text{and}$$

$$J_{\mathcal{M}}\left[J'_{\mathcal{M}}\right]^* = (I+R)^*. \tag{12}$$

Similarly, we have

$$J'_{\mathcal{M}} J_{\mathcal{M}}^* = (I+R). \tag{13}$$

From equations (11), (12) and (13), we obtain

$$R_{X_{s(\cdot)}^{\mathcal{M}}} = J_{\mathcal{M}} J_{\mathcal{M}}^*, \quad \text{and}$$
$$R_{\widetilde{X}_{s(\cdot)}^{\mathcal{M}}} = J'_{\mathcal{M}}[J'_{\mathcal{M}}]^*. \tag{14}$$

The RKHS of $X_{s(\cdot)}^{\mathcal{M}}$ coincides, as a set of functions, with the space $H^{s(\cdot)}(\mathcal{M})$, since by definition $\mathcal{H}\left(X_{s(\cdot)}^{\mathcal{M}}\right) \subseteq H^{s(\cdot)}(\mathcal{M}) \subseteq (I+R)^*(H^{s(\cdot)}(\mathcal{M}))$, and from the pseudoduality condition,

$$J_{\mathcal{M}}\left(\widetilde{X}_{s(\cdot)}^{\mathcal{M}}(\phi)\right) = (I+R)^*(\phi), \quad \forall \phi \in H^{s(\cdot)}(\mathcal{M}). \tag{15}$$

Thus, $(I+R)^*(H^{s(\cdot)}(\mathcal{M})) \subseteq \mathcal{H}\left(X_{s(\cdot)}^{\mathcal{M}}\right)$. The spaces $\mathcal{H}\left(X_{s(\cdot)}^{\mathcal{M}}\right)$ and $H^{s(\cdot)}(\mathcal{M})$ are also isomorphically related by the identity operator (see Proposition 1 below). Similarly, an isomorphic relationship can also be established between $\mathcal{H}\left(\widetilde{X}_{s(\cdot)}^{\mathcal{M}}\right)$ and $H^{-\beta(\cdot),\mathcal{M}}(\mathbb{R}^n)$ from the identity

$$J'_{\mathcal{M}}\left(X_{s(\cdot)}^{\mathcal{M}}(f)\right) = (I+R)(f), \quad \forall f \in H^{-\beta(\cdot),\mathcal{M}}(\mathbb{R}^n),$$

and the ellipticity of $J_{\mathcal{M}}^*$ is obtained from the mean-square continuity of $\widetilde{X}_{s(\cdot)}^{\mathcal{M}}$.

3 Multifractional White-Noise Filter Representation on Fractal Domains

The FGRFVOs $X_{s(\cdot)}^{\mathcal{M}}$ and $\widetilde{X}_{s(\cdot)}^{\mathcal{M}}$ admit a linear multifractional pseudodifferential representation in terms of generalized white noise and the multifractional pseudodifferential operators $J'_{\mathcal{M}}$ and $J_{\mathcal{M}}$, respectively, involved in the covariance factorizations obtained in the previous section.

The space $L^2_{\mu_{\mathcal{M}}}(\mathcal{M})$ of square integrable functions with respect to the fractal measure $\mu_{\mathcal{M}}$ defines the RKHS of a white-noise process $\varepsilon_{\mu_{\mathcal{M}}}$ on a compact fractal domain $\mathcal{M}$. The following relationships then hold between the Hilbert spaces of random variables and the RKHSs associated with $\varepsilon_{\mu_{\mathcal{M}}}$, $X_{s(\cdot)}^{\mathcal{M}}$ and $\widetilde{X}_{s(\cdot)}^{\mathcal{M}}$:

$$H\left(X_{s(\cdot)}^{\mathcal{M}}\right) \underset{J_{\mathcal{M}}}{\equiv} \mathcal{H}\left(X_{s(\cdot)}^{\mathcal{M}}\right) \underset{i}{\simeq} H^{s(\cdot)}(\mathcal{M}) \underset{\left[\langle D_{\mathbf{x}}\rangle^{s(\cdot)}\right]_{\mu_{\mathcal{M}}}}{\equiv} L^2_{\mu_{\mathcal{M}}}(\mathcal{M}) \underset{J_0}{\equiv} H(\varepsilon_{\mu_{\mathcal{M}}}), \text{ and}$$

$$H\left(\widetilde{X}_{s(\cdot)}^{\mathcal{M}}\right) \underset{J'_{\mathcal{M}}}{\equiv} \mathcal{H}\left(\widetilde{X}_{s(\cdot)}^{\mathcal{M}}\right) \underset{\widetilde{i}}{\simeq} H^{-\beta(\cdot),\mathcal{M}}(\mathbb{R}^n) \underset{\left[\langle D_{\mathbf{x}}\rangle^{-\beta(\cdot)}\right]_{\mu_{\mathcal{M}}}}{\equiv} L^2_{\mu_{\mathcal{M}}}(\mathcal{M}) \underset{J_0}{\equiv} H(\varepsilon_{\mu_{\mathcal{M}}}), \tag{16}$$

where

$$J_0X(g) = E\left[X\varepsilon_{\mu_{\mathcal{M}}}(g)\right], \quad \forall g \in L^2_{\mu_{\mathcal{M}}}(\mathcal{M}) \text{ and } X \in H\left(\varepsilon_{\mu_{\mathcal{M}}}\right).$$

The following result provides the linear filters relating the FGRFVOs $X_{s(\cdot)}$ and $\widetilde{X}_{s(\cdot)}$ with white noise.

Proposition 1. *Under the pseudoduality condition given in Definition 4, the following white-noise linear filter representations on the fractal domain $\mathcal{M}$ hold:*

$$X_{s(\cdot)}^{\mathcal{M}} L_{\mathcal{M}} f \underset{\text{m.s.}}{=} \varepsilon_{\mu_{\mathcal{M}}}\left((I+R)f\right), \quad \forall f \in L^2_{\mu_{\mathcal{M}}}(\mathcal{M}), \quad \textit{and} \tag{17}$$

$$\widetilde{X}_{s(\cdot)}^{\mathcal{M}} \widetilde{L}_{\mathcal{M}} g \underset{\text{m.s.}}{=} \varepsilon_{\mu_{\mathcal{M}}}\left((I+R)^*g\right), \quad \forall g \in L^2_{\mu_{\mathcal{M}}}(\mathcal{M}), \tag{18}$$

where

$$L_{\mathcal{M}} = J'_{\mathcal{M}} J_0^{-1}, \quad \textit{and} \tag{19}$$

$$\widetilde{L}_{\mathcal{M}} = J_{\mathcal{M}} J_0^{-1}, \tag{20}$$

with J_0 representing the isometric isomorphism between the spaces $H(\varepsilon_{\mu_{\mathcal{M}}})$ and $\mathcal{H}(\varepsilon_{\mu_{\mathcal{M}}}) = L^2_{\mu_{\mathcal{M}}}(\mathcal{M})$.

Proof. Operators $J'_{\mathcal{M}}$ and $J_{\mathcal{M}}$ can be composed with the operator J_0^{-1} from the identifications given in equation (16).

From condition (iii) of Definition 4, for all $f, g \in L^2_{\mu_{\mathcal{M}}}(\mathcal{M})$, the following identities hold:

$$\left\langle X^{\mathcal{M}}_{s(\cdot)}\left([(I+R)]^{-1} J'_{\mathcal{M}} J_0^{-1} f\right), X^{\mathcal{M}}_{s(\cdot)}\left([(I+R)]^{-1} J'_{\mathcal{M}} J_0^{-1} g\right)\right\rangle_{H\left(X_{s(\cdot)}\right)}$$
$$= \langle J_0^{-1} f, J_0^{-1} g\rangle_{H(\varepsilon_{\mu_{\mathcal{M}}})} = \langle \varepsilon_{\mu_{\mathcal{M}}}(f), \varepsilon_{\mu_{\mathcal{M}}}(g)\rangle_{H(\varepsilon_{\mu_{\mathcal{M}}})}. \tag{21}$$

From equation (21),

$$X^{\mathcal{M}}_{s(\cdot)}\left([(I+R)]^{-1} J'_{\mathcal{M}} J_0^{-1} f\right) \underset{\text{m.s.}}{=} \varepsilon_{\mu_{\mathcal{M}}}(f), \quad \forall f \in L^2_{\mu_{\mathcal{M}}}(\mathcal{M}).$$

That is,

$$X^{\mathcal{M}}_{s(\cdot)}(g) \underset{\text{m.s.}}{=} \varepsilon_{\mu_{\mathcal{M}}}\left(J_0 \left[J'_{\mathcal{M}}\right]^{-1} (I+R)(g)\right), \quad \forall g \in H^{-\beta(\cdot),\mathcal{M}}(\mathbb{R}^n).$$

Hence,

$$X^{\mathcal{M}}_{s(\cdot)}\left(J'_{\mathcal{M}} J_0^{-1} f\right) \underset{\text{m.s.}}{=} \varepsilon_{\mu_{\mathcal{M}}}\left((I+R)f\right), \quad \forall f \in L^2_{\mu_{\mathcal{M}}}(\mathcal{M}).$$

Equation (18) can be obtained in a similar way from condition (iii) of Definition 4.

From the above result, the factorization of the bicontinuous covariance operator $R_{X^{\mathcal{M}}_{s(\cdot)}}$ can also be rewritten as

$$R_{X^{\mathcal{M}}_{s(\cdot)}} = \widetilde{L}_{\mathcal{M}} \widetilde{L}^*_{\mathcal{M}};$$

similarly,

$$R_{\widetilde{X}^{\mathcal{M}}_{s(\cdot)}} = L_{\mathcal{M}} L^*_{\mathcal{M}}.$$

4 Second-Order Spectral Properties

Examples of multifractional pseudodifferential models on compact fractal domains can be constructed as restrictions to such domains of elliptic pseudodifferential operators of variable order (see [23] for the definition and properties of these operators on $\mathbb{R}^n$), and of equivalent versions of such operators in the class described in [50]. The compactness of the restrictions comes from the compactness of the fractal domain. Their continuous spectra are then empty and they have pure point spectra.

The covariance operators $R_{X^{\mathcal{M}}_{s(\cdot)}}$ and $R_{\widetilde{X}^{\mathcal{M}}_{s(\cdot)}}$ generate closed bilinear forms which are respectively equivalent to the ones defining the inner products in the spaces $H^{-\beta(\cdot),\mathcal{M}}(\mathbb{R}^n)$ and $H^{s(\cdot)}(\mathcal{M})$. Hence, their spectral properties are equivalent to the spectral properties of the pseudodifferential operators of variable order generating the inner products in such spaces. Specifically, the eigenvalues $\left\{\lambda_k\left(R_{X^{\mathcal{M}}_{s(\cdot)}}\right) : k \in \mathbb{N}\right\}$ of $R_{X^{\mathcal{M}}_{s(\cdot)}}$ then satisfy

$$k^{-2\frac{\overline{s}}{\alpha}} \leq \lambda_k\left(R_{X^{\mathcal{M}}_{s(\cdot)}}\right) \leq k^{-2\frac{\underline{s}}{\alpha}}, \quad k \in \mathbb{N},$$

and similarly, the eigenvalues of $R_{\widetilde{X}^{\mathcal{M}}_{s(\cdot)}}$ satisfy

$$k^{2\frac{\underline{s}}{\alpha}} \leq \lambda_k\left(R_{\widetilde{X}^{\mathcal{M}}_{s(\cdot)}}\right) \leq k^{2\frac{\overline{s}}{\alpha}}, \quad k \in \mathbb{N},$$

where

$$\overline{s} = \sup_{\mathbf{x}\in\mathcal{M}} s(\mathbf{x}) \quad \underline{s} = \inf_{\mathbf{x}\in\mathcal{M}} s(\mathbf{x}).$$

In the case where $\underline{s} > \alpha/2$, with $0 < \alpha < n$, we have

$$\sum_{k\in\mathbb{N}} \left|\lambda_k\left(R_{X^{\mathcal{M}}_{s(\cdot)}}\right)\right| \leq \sum_{k\in\mathbb{N}} k^{-2\underline{s}/\alpha} < \infty, \tag{22}$$

which means that $R_{X^{\mathcal{M}}_{s(\cdot)}}$ is in the trace class.

5 Hölder Continuity of the Solution to Multifractional Equations on Fractals

In this section we consider the case where the functions of the RKHS $\mathcal{H}\left(X^{\mathcal{M}}_{s(\cdot)}\right)$ are Hölder continuous. Specifically, we consider the case where

$$\underline{s} = \inf_{\mathbf{x}\in\mathcal{M}} s(\mathbf{x}) > \alpha/2. \tag{23}$$

In this case, the multifractional pseudodifferential white-noise filter (17) defines a mean-square Hölder continuous ordinary solution with global Hölder exponent $\underline{s} - \alpha/2$. The local Hölder exponent, in the second-order moment sense, that is, the local Hölder exponent of the functions in the RKHS, is given by the function $s(\cdot) - \alpha/2$, which can be computed from the multifractional exponent defining the local behaviour of the variogram (see Proposition 2 below).

In the Gaussian case, the local exponent of mean quadratic variation of the increments can be related to the sample-path Hölder continuity (see [1]). Hence, Gaussian multifractional models on fractals can be introduced in the sample-path sense within this framework.

Proposition 2. *Assume that the pseudoduality condition and condition (23) hold. Then, there exists a unique mean-square Hölder continuous ordinary random field $\mathcal{X}^{\mathcal{M}}_{s(\cdot)}$ satisfying*

$$L^{*}_{\mathcal{M}}\mathcal{X}^{\mathcal{M}}_{s(\cdot)}(\mathbf{x}) \underset{\text{m.s.}}{=} \varepsilon_{\mu_{\mathcal{M}}}(\mathbf{x}), \quad \forall \mathbf{x} \in \mathcal{M}. \tag{24}$$

The local mean quadratic variation of its increments is defined from the function $2s(\cdot) - \alpha$. That is,

$$E\left[\mathcal{X}^{\mathcal{M}}_{s(\mathbf{x}+\mathbf{h})}(\mathbf{x}+\mathbf{h}) - \mathcal{X}^{\mathcal{M}}_{s(\mathbf{x})}(\mathbf{x})\right]^2 \leq C\|\mathbf{h}\|^{2\underline{s}_{\mathbf{h}}(\mathbf{x})-\alpha}, \quad C > 0,$$

where

$$\underline{s}_{\mathbf{h}}(\mathbf{x}) = \inf_{\mathbf{y}\in\mathcal{M}\cap\Lambda_{\mathbf{h}}(\mathbf{x})} s(\mathbf{y}),$$

with $\Lambda_{\mathbf{h}}(\mathbf{x})$ representing a neighborhood of radius $|\mathbf{h}|$ at the point $\mathbf{x}$.

Proof. From the embedding theorems between Besov spaces on fractal domains (see, for example, [57]), the RKHS $\mathcal{H}\left(X^{\mathcal{M}}_{s(\cdot)}\right)$ of $X^{\mathcal{M}}_{s(\cdot)}$ is continuously embedded into the space $\mathcal{C}^{\underline{s}-\alpha/2}(\mathcal{M})$ of Hölder continuous functions of order $\underline{s} - \alpha/2$ restricted to $\mathcal{M}$. Therefore, the functions in the RKHS are Hölder continuous of order $\underline{s} - \alpha/2$ on $\mathcal{M}$. Equivalently, the random field solution $\mathcal{X}^{\mathcal{M}}_{s(\cdot)}$ of equation (24) is Hölder continuous of order $\underline{s} - \alpha/2$, in the mean-square sense. Hence, the weak-sense multifractional pseudodifferential representation on fractals derived in Section 3 becomes 'strong-sense'. Note that, from equation (22), the ordinary continuous random field $\mathcal{X}^{\mathcal{M}}_{s(\cdot)}$ is constructed from white noise, in terms of the kernel factorizing its covariance function by self-convolution. This is the kernel of the covariance operator $R_{X^{\mathcal{M}}_{s(\cdot)}}$.

The local exponent of the mean quadratic local variation of the increments of $\mathcal{X}^{\mathcal{M}}_{s(\cdot)}$ coincides with the local Hölder exponent of its covariance function, given by $2[s(\cdot) - \alpha/2]$, with $s(\cdot) - \alpha/2$ denoting, as before, the local Hölder exponent of the functions in the RKHS $\mathcal{H}(X^{\mathcal{M}}_{s(\cdot)})$ of $X^{\mathcal{M}}_{s(\cdot)}$. That is,

$$E\left[\mathcal{X}^{\mathcal{M}}_{s(\mathbf{x}+\mathbf{h})}(\mathbf{x}+\mathbf{h}) - \mathcal{X}^{\mathcal{M}}_{s(\mathbf{x})}(\mathbf{x})\right]^2 \leq C\|\mathbf{h}\|^{2\underline{s}_{\mathbf{h}}(\mathbf{x})-\alpha},$$

for a certain positive constant C.

Remark 1. From Proposition 2, using the results of [1], for any $\epsilon > 0$, the following inequality holds with probability one for the sample paths of a Gaussian random field X with RKHS isomorphic to the trace space $H^{s(\cdot)}(\mathcal{M})$:

$$\sup_{|\mathbf{x}|<\delta} |X(\mathbf{x}_0 + \mathbf{x}) - X(\mathbf{x}_0)| \leq Z\delta^{s(\mathbf{x}_0)-\alpha/2-\epsilon}, \quad \delta \longrightarrow 0, \quad \forall \mathbf{x}_0 \in \mathbb{R}^n, \qquad (25)$$

where Z is an almost surely finite random variable.

Remark 2. The results derived in this section reveal the fact that the heterogeneous Hölder exponent of the random field solution to a multifractional pseudodifferential equation is modified by the fractality order of the domain, in the sense of increasing the heterogeneous local singularity of its sample paths in the Gaussian case. Therefore, chaotic systems are affected by the singularity of the physical law governing them, and by the local fractal dimension (local singularity of the support measure) of the domain of definition.

6 Examples

To illustrate the results established in this paper, in this section, some examples of multifractional Gaussian random fields on fractals are constructed. The corresponding multifractional pseudodifferential equation on a fractal is derived from its covariance factorization.

Example 1.

$$\sum_{\sigma(\cdot)\in\mathcal{C}_F\subset\mathcal{B}^\infty(\mathbb{R}^n)} a^{\mathcal{M}}_{\sigma(\cdot)}(\cdot)\langle D.\rangle^{\sigma(\cdot)}\cdot \mathrm{tr}_{\mathcal{M}}\mathcal{X}_{s(\cdot)} \underset{\mathrm{m.s.}}{=} \varepsilon_{\mu_{\mathcal{M}}}, \tag{26}$$

where $\varepsilon_{\mu_{\mathcal{M}}}$ represents, as before, generalized white noise on $L^2_{\mu_{\mathcal{M}}}(\mathcal{M})$, $a^{\mathcal{M}}_{\sigma(\cdot)}(\cdot)$ denotes a distribution with variable local singularity order $-\sigma$ restricted to the compact set $\mathcal{M}$, $s(\cdot)=\sigma(\cdot)-\frac{n-\alpha}{2}$, and $\mathrm{tr}_{\mathcal{M}}$ denotes the trace operator on the fractal set $\mathcal{M}$. Here, the multifractional operators involved are defined as in equation (34), and $\mathcal{C}_F$ denotes a finite subset of $\mathcal{B}^\infty(\mathbb{R}^n)$. The functional exponent $\sigma(\cdot)$ characterizing the local singularity order of the variogram of the solution on $\mathbb{R}^n$ is modified by he local dimension α of the fractal domain $\mathcal{M}$ where its restriction is defined.

Example 2.

$$(-\Delta)^{\gamma(\cdot)/2}\mathrm{tr}_{\mathcal{M}}\mathcal{X}_{\gamma(\cdot)} \underset{\mathrm{m.s.}}{=} \varepsilon_{\mu_{\mathcal{M}}}, \tag{27}$$

where $(-\Delta)^{\gamma(\cdot)/2}$ is the negative Laplacian of variable order $\gamma(\cdot)/2$, with $\gamma\in\mathcal{B}^\infty(\mathbb{R}^n)$. Here, the continuous solution of this equation is derived pointwise when $\underline{\gamma}>\alpha/2$. In this case, the trace of the covariance operator is finite, and the decay velocity of its eigenvalues depends on the singularity order $\gamma(\cdot)$ and on the fractal local dimension α of the domain $\mathcal{M}$.

Example 3.

$$\langle D.\rangle^{s(\cdot)}(-\Delta)^{\gamma(\cdot)}\mathrm{tr}_{\mathcal{M}}\mathcal{X}_{s(\cdot)+\gamma(\cdot)} \underset{\mathrm{m.s.}}{=} \varepsilon_{\mu_{\mathcal{M}}}, \tag{28}$$

where the multifractional operators of variable order involved are as given in Examples 1 and 2. Note that the multifractional exponent defining the variable fractality order of $\mathcal{X}_{s(\cdot)+\gamma(\cdot)}$ is given in terms of $2(s(\cdot)+\gamma(\cdot))$, and the multifractional exponent defining the slow decay of the covariance function of $\mathcal{X}_{s(\cdot)+\gamma(\cdot)}$ is given in terms of $\gamma(\cdot)$, both of them modified by the local dimension α of the multifractal domain $\mathcal{M}$. This model family is of special interest in the context of heterogeneous anomalous diffusion processes on disordered media.

Example 4.

$$Q_{q(\cdot)}(\mathcal{A})\mathrm{tr}_{\mathcal{M}}\mathcal{X}_{-[q(\cdot)-p(\cdot)]s(\cdot)} = P_{p(\cdot)}(\mathcal{A})\varepsilon_{\mu_{\mathcal{M}}}, \tag{29}$$

where Q/P is an elliptic rational function with variable order $q(\cdot)/2-p(\cdot)/2$ of an elliptic self-adjoint pseudodifferential operator $\mathcal{A}$ of variable order $s(\cdot)$.

This class of models extends the one introduced in [45] to the multifractional context on fractal domains. Indeed, the trace on a fractal domain $\mathcal{M}$ of the pseudodifferential operator $\mathcal{A}$ leads to the compactness of operator

$$\frac{P_{p(\cdot)}(\mathcal{A})}{Q_{q(\cdot)}(\mathcal{A})},$$

and to the compactification of its continuous spectrum, which becomes a pure point spectrum, constituted by the eigenvalues with asymptotic exponent defined in terms of

$$\frac{[q(\cdot) - p(\cdot)]s(\cdot) - \frac{n-\alpha}{2}}{\alpha}.$$

7 Conclusion

In this paper, a class of multifractional random systems on fractals is introduced, in the weak-sense, using the theory of generalized random fields and their associated RKHSs. The characterization of the singular features introduced by the fractal geometry of the domain, altering the exponent of fractal heterogeneity of the random field solution originally defined on $\mathbb{R}^n$, is achieved through the application of trace theorems of Besov spaces on fractal domains, and the pseudoduality condition introduced in [50]. The identification of a new family of RKHSs allows to derive the local regularity properties of the solution. In the Gaussian case, the regularity properties of the sample paths of the solution are also derived.

Since the study is developed in the context of fractional Sobolev spaces of variable order, the smoothness of the functions defining such orders does not allow the introduction of multifractality or of a solution class with a nontrivial singularity spectrum. This is one of the limitations of this framework. In return, the explicit determination of the regularity properties of the solution, which is non-stationary and displays heterogeneous fractality, is obtained. Additionally, in the Gaussian case, it is proved that the fractal geometry of the domain affects the local regularity of the sample paths of the solution, increasing their erraticity.

This work confirms the conjecture that complex systems, like the ones introduced in terms of multifractional pseudodifferential equations, become highly singular when their restrictions to a fractal domain is considered, increasing the local singularity of the support measure (defining the geometry of the domain). This fact is reflected in both the spectral and sample-path properties of the solution, in the Gaussian case.

Acknowledgement. This work has been supported in part by projects MTM2009-13250 and MTM2009-13393 of the SGPI, and P08-FQM-3834 and P09-FQM-5052 of the Andalusian CICE, Spain.

References

1. Adler, R.J.: The Geometry of Random Fields. Wiley, New York (1981)
2. Anh, V.V., Leonenko, N.N., Shieh, N.-R.: Multifractality of products geometric Ornstein-Uhlenbeck type processes. Adv. Appl. Prob. 40, 1129–1156 (2008)
3. Anh, V.V., Leonenko, N.N., Shieh, N.-R.: Multifractal scaling of products of birth-death processes. Bernoulli. 15, 508–531 (2009)
4. Anh, V.V., Leonenko, N.N., Shieh, N.-R.: Multifractal products of stationary diffusion processes. Stoch. Anal. Appl. 27, 475–499 (2009)
5. Ayache, A.: The generalized multifractional Brownian motion can be multifractal. Tech. Rep. LSP-2000-22, Laboratoire de Statistique et Probabilités, UMR C5583 Université Paul Sabatier (2000)
6. Ayache, A., Lévy-Véhel, J.: Generalized multifractional Brownian motion: Definition and preliminary results. In: Dekking, M., Lévy-Véhel, J., Lutton, E., Tricot, C. (eds.) Fractals: Theory and Applications in Engineering, pp. 17–32. Springer, Heidelberg (1999)
7. Bacry, E., Muzy, J.E.: Log-infinitely divisible multifractal processes. Comm. Math. Phys. 236, 449–475 (2003)
8. Barral, J., Mandelbrot, B.B.: Multifractal products of cylindrical pulses. Prob. Th Rel. Fields 124, 409–430 (2002)
9. Benassi, A., Jaffard, S., Roux, D.: Elliptic Gaussian random processes. Rev. Mat. Iberoam 13, 19–90 (1997)
10. Benzi, R., Paladin, G., Parisi, G., Vulpiani, A.: On multifractal nature of fully developed turbulence and chaotic systems. J. Phys. A - Math. Gen. 17, 3521–3531 (1984)
11. Boudafel, M.C., Lu, S., Molz, F.J., Lavallée, D.: Multifractal scaling of the intrinsic permeability. Water Resour. Res. 36, 3211–3222 (2000)
12. Carl, B.: Entropy numbers of diagonal operators with an application to eigenvalue problems. J. Approx. Theory 32, 135–150 (1981)
13. Edmund, D., Triebel, H.: Function Spaces, Entropy Numbers, Differential Operators. Cambridge University Press, Cambridge (1996)
14. Falconer, K.: The Geometry of Fractal Sets. Cambridge University Press, Cambridge (1985)
15. Falconer, K.: Fractal Geometry: Mathematical Foundations and Applications. Wiley, Chichester (1990)
16. Folorunso, O.A., Puente, C.E., Rolston, D.E., Pinzon, J.E.: Statistical and fractal evaluation of the spatial characteristics of soil surface strength. Soil Sci. Soc. Am. J. 58, 284–294 (1994)
17. Frisch, U., Parisi, G.: On the singularity structure of fully developed turbulence. In: Ghil, M., Benzi, R., Parisi, G. (eds.) Turbulence and Predictability in Geophysical Fluid Dynamics and Climate Dynamics, pp. 84–87. North-Holland, New York (1985)
18. Giona, M., Roman, H.E.: Fractional diffusion equation on fractals: one-dimensional case and asymptotic behaviour. J. Phys. A - Math. Gen. 25, 2093 (1992)
19. Grout, H., Tarquis, A.M., Wiesner, M.R.: Multifractal analysis of particle size distributions in soil. Environ. Sci. Technol. 32, 1176–1182 (1998)
20. Harte, D.: Multifractals: Theory and Applications. Chapman & Hall/CRC, Boca Raton (2001)

21. Halsey, T.C., Jensen, M.H., Kadanoff, L.P., Procaccia, I., Shraiman, B.I.: Fractal measures and their singularities: the characterization of strange sets. Phys. Rev. A 33, 1141–1151 (1986)
22. Hentschel, H.G.E., Procaccia, I.: The infinite number of generalized dimensions of fractals and strange attractors. Physica D 8, 435–444 (1983)
23. Jacob, N., Leopold, H.-G.: Pseudo-differential operators with variable order of differentiation generating Feller semigroups. Integr. Equat. Oper. Th. 17, 544–553 (1993)
24. Jaffard, S.: Multifractal formalism for functions part I: results valid for all functions. SIAM J. Math. Anal. 28, 944–970 (1997)
25. Jaffard, S.: Multifractal formalism for functions part II: self-similar functions. SIAM J. Math. Anal. 28, 971–998 (1997)
26. Kikuchi, K., Negoro, A.: Pseudodifferential operators and Sobolev spaces of variable order of differentiation. Rep. Fac. Liberal Arts, Shizuoka University, Sciences 31, 19–27 (1995)
27. Kikuchi, K., Negoro, A.: On Markov process generated by pseudodifferential operator of variable order. Osaka J. Math. 34, 319–335 (1997)
28. Kravchenko, A., Boast, C.W., Bullock, D.G.: Multifractal analysis of soil spatial variability. Agron. J. 91, 1033–1041 (1999)
29. Kropp, J., Block, A., Bloh, W.V., Klenke, T., Schellnhuber, H.J.: Characteristic multifractal element distributions in recent bioactive marine sediments. In: Kruhl, J.H. (ed.) Fractals and Dynamic Systems in Geoscience. Springer, Berlin (1994)
30. Lau, K.S.: Iterated function systems with overlaps and multifractal structure. In: Kono, N., Shieh, N.R. (eds.) Trends in Probability and Related Analysis. World Scientific, River Edge (1999)
31. Lavallée, D., Lovejoy, S., Schertzer, D., Ladoy, P.: Nonlinear variability and landscape topography: analysis and simulation. In: De Cola, L., Lam, N. (eds.) Fractals in Geography. Prentice Hall, Englewood Cliffs (1993)
32. Leopold, H.-G.: Embedding of function spaces of variable order of differentiation. Czech Math. J. 49, 633–644 (1999)
33. Lévy-Véhel, J., Peltier, R.F.: Multifractional Brownian motion: definitions and preliminary results. Tech. Rep. RR-2645 INRIA Rocquencourt (1995)
34. Liu, H.H., Molz, F.J.: Multifractal analyses of hydraulic conductivity distributions. Water Resour. Res. 33, 2483–2488 (1997)
35. Mandelbrot, B.: Intermittent turbulence in self-similar cascades; divergence of high moments and dimension of the carrier. J. Fluid Mech. 62, 331–358 (1974)
36. Mandelbrot, B.: Multifractal measures, especially for the geophysicist. Pure Appl. Geophys. 131, 5–42 (1989)
37. Marsan, D., Bean, J.C.: Multiscaling nature of sonic velocities and lithology in the upper crystalline crust: evidence from the KTB Main Borehole. Geophys. Res. Lett. 26, 275–278 (1999)
38. Mattila, P.: Geometry of Sets and Measures in Euclidean Spaces. Cambridge University Press, Cambridge (1995)
39. Müller, J., Mccauley, J.L.: Implication of fractal geometry for fluid flow properties of sedimentary rocks. Transport Porous Med. 8, 133–147 (1992)
40. Ngai, S.-M.: A dimension result arising from the L_q-spectrum of a measure. P. Am. Math. Soc. 125, 2943–2951 (1997)
41. Pecknold, S., Lovejoy, S., Schertzer, D.: Universal multifractals and anisotropic scale invariance in aeromagnetic fields. Geophys. J. Int. 145, 127–144 (2001)

42. Posadas, A.N.D., Giménez, D., Bittelli, M., Vaz, C.M.P., Flury, M.: Multifractal characterization of soil particle-size distributions. Soil Sci. Soc. Am. J. 65, 1361–1367 (2001)
43. Riedi, R.H.: Multifractal processes. In: Doukhan, P., Oppenheim, G., Taqqu, M.S. (eds.) Theory and Applications of Long-Range Dependence. Birkhäuser, Basel (2002)
44. Riedi, R., Mandelbrot, B.B.: Exceptions to the multifractal formalism for discontinuous measures. Math. P. Camb. Phil. Soc. 123, 133–157 (1998)
45. Ramm, A.G.: Random Fields Estimation Theory. Logman Scientific & Technical - Wiley, New York (1990)
46. Ruiz-Medina, M.D., Angulo, J.M., Anh, V.V.: Stochastic fractional-order differential models on fractals. Theor. Prob. Math. Stat. 67, 130–146 (2002)
47. Ruiz-Medina, M.D., Angulo, J.M., Anh, V.V.: Fractional generalized random fields on bounded domains. Stoch. Anal. Appl. 21, 465–492 (2003)
48. Ruiz-Medina, M.D., Angulo, J.M., Anh, V.V.: Spatial and spatiotemporal Karhunen-Loève-type representations on fractal domains. Stoch. Anal. Appl. 24, 195–219 (2006)
49. Ruiz-Medina, M.D., Anh, V.V., Angulo, J.M.: Stochastic fractional-order differential models with fractal boundary conditions. Stat. Prob. Lett. 54, 47–60 (2001)
50. Ruiz-Medina, M.D., Anh, V.V., Angulo, J.M.: Fractional generalized random fields of variable order. Stoch. Anal. Appl. 22, 775–799 (2004)
51. Ruiz-Medina, M.D., Anh, V.V., Angulo, J.M.: Fractional random fields on domains with fractal boundary. Infin. Dimens Anal. Quantum Prob. Rel. Top 7, 395–417 (2004)
52. Samko, S.G.: Differentiation and integration of variable (fractional) order. Dokl. Akad. Nauk. 342, 458–461 (1995)
53. Schwartz, L.: Théorie des Distributions. Hermann, Paris (1966)
54. Strichartz, R.S.: Differential Equations on Fractals: A Tutorial. Princeton University Press, Princeton (2006)
55. Tennekoon, L., Boufadel, M.C., Lavallée, D., Weaver, J.: Multifractal anisotropic scaling of the hydraulic conductivity. Water Resour. Res. 39, 1193 (2003)
56. Triebel, H.: Interpolation Theory, Function Spaces, Differential Operators. North-Holland, Amsterdam (1978)
57. Triebel, H.: Fractals and Spectra. Birkhäuser, Basel (1997)
58. Weissel, J.K., Pratson, L.F., Malinverno, A.: The length-scaling properties of topography. J. Geophys. Res. 99, 13997–14012 (1994)
59. Young, L.S.: Dimension, entropy and Lyapunov exponents. Ergod. Theor. Dyn. Syst. 2, 109–124 (1982)

Appendix

The main functional tools applied in the development of this paper consist of elements of the theory of fractional Sobolev spaces of variable order, and fractal measures with support given by a compact d-set, $0 < d < n$, as well as the trace theorems of functional spaces on these sets. In this Appendix, we collect all this material to facilitate the reading and understanding of this paper. First, we introduce the main definitions and results from the theory of fractional Sobolev spaces of variable order, and multifractional psedodifferential operators.

Let δ and ρ be real numbers with $0 \leq \delta < \rho \leq 1$, and let σ be a real-valued function in $\mathcal{B}^\infty(\mathbb{R}^n)$, the space of all C^∞-functions on $\mathbb{R}^n$ whose derivatives of all orders are bounded. We say that a function $p(\mathbf{x}, \boldsymbol{\xi}) \in \mathcal{B}^\infty\left(\mathbb{R}^n_{\mathbf{x}} \times \mathbb{R}^n_{\boldsymbol{\xi}}\right)$ belongs to $S^\sigma_{\rho,\delta}$ if and only if for any multi-indices α and β there exists some positive constant $C_{\alpha,\beta}$ such that

$$|D^\alpha_{\boldsymbol{\xi}} D^\beta_{\mathbf{x}} p(\mathbf{x}, \boldsymbol{\xi})| \leq C_{\alpha,\beta} \langle \boldsymbol{\xi} \rangle^{\sigma(\mathbf{x}) - \rho|\alpha| + \delta|\beta|}, \tag{30}$$

where $D^\alpha_{\boldsymbol{\xi}}$ and $D^\beta_{\mathbf{x}}$ respectively denote the derivatives with respect to $\boldsymbol{\xi}$ and $\mathbf{x}$, and $\langle \boldsymbol{\xi} \rangle = (1 + |\boldsymbol{\xi}|^2)^{1/2}$. The following semi-norm is considered for the elements of $S^\sigma_{\rho,\delta}$:

$$|p|^{(\sigma)}_l = \max_{|\alpha+\beta|\leq l} \sup_{(\mathbf{x},\boldsymbol{\xi}) \in \mathbb{R}^n \times \mathbb{R}^n} \left\{ |D^\alpha_{\boldsymbol{\xi}} D^\beta_{\mathbf{x}} p(\mathbf{x}, \boldsymbol{\xi})| \langle \boldsymbol{\xi} \rangle^{-\sigma(\mathbf{x}) + \rho|\alpha| - \delta|\beta|} \right\}.$$

Definition 5. *([26], [27]) For $u \in \mathcal{S}(\mathbb{R}^n)$, the set of rapidly decreasing Schwartz functions, and $p \in S^\sigma_{\rho,\delta}$, let $P : \mathcal{S}(\mathbb{R}^n) \longrightarrow \mathcal{S}(\mathbb{R}^n)$ be defined as*

$$Pu(\mathbf{x}) = (2\pi)^{-n} \int_{\mathbb{R}^n} e^{i\mathbf{x}\boldsymbol{\xi}} p(\mathbf{x}, \boldsymbol{\xi}) \hat{u}(\boldsymbol{\xi}) d\boldsymbol{\xi}, \tag{31}$$

where $\hat{u}(\boldsymbol{\xi}) = \int_{\mathbb{R}^n} e^{-i\mathbf{x}\boldsymbol{\xi}} u(\mathbf{x}) d\mathbf{x}$ is the Fourier transform of u. We refer to $P = p(\mathbf{x}, D_{\mathbf{x}})$ as a pseudodifferential operator of variable order with symbol $p \in S^\sigma_{\rho,\delta}$. The set of all pseudodifferential operators with symbol p of the class $S^\sigma_{\rho,\delta}$ is denoted by $\boldsymbol{S}^\sigma_{\rho,\delta}$.

A pseudodifferential operator $P \in \boldsymbol{S}^\sigma_{\rho,\delta}$ is elliptic if there exist $c > 0$ and $M > 0$ such that

$$|p(\mathbf{x}, \boldsymbol{\xi})| \geq c \langle \boldsymbol{\xi} \rangle^{\sigma(\mathbf{x})}, \quad |\boldsymbol{\xi}| \geq M. \tag{32}$$

Furthermore, $Q \in \boldsymbol{S}^\infty_{\rho,\delta} = \bigcup_{m \in \mathbb{R}} \boldsymbol{S}^m_{\rho,\delta}$ is said to be a left (resp. right) parametrix of P if there exists $R_L \in \boldsymbol{S}^{-\infty}_{\rho,\delta} = \bigcap_{m \in \mathbb{R}} \boldsymbol{S}^m_{\rho,\delta}$ (resp. $R_R \in \boldsymbol{S}^{-\infty}_{\rho,\delta} = \bigcap_{m \in \mathbb{R}} \boldsymbol{S}^m_{\rho,\delta}$) such that

$$QP = I + R_L \quad (\text{resp.} \quad PQ = I + R_R),$$

where I denotes the identity operator. A pseudodifferential operator Q is a parametrix of P if Q is simultaneously a left and right parametrix of P.

Definition 6. *Let σ be a real-valued function in $\mathcal{B}^{\infty}(\mathbb{R}^n)$. The Sobolev space of variable order σ on $\mathbb{R}^n$ is defined as*

$$H^{\sigma(\cdot)}(\mathbb{R}^n) = \left\{ u \in H^{-\infty} = \bigcup_{s\in\mathbb{R}} H^s(\mathbb{R}^n) : \ \langle D.\rangle^{\sigma(\cdot)} u \in L^2(\mathbb{R}^n) \right\}, \tag{33}$$

where

$$\langle D_{\mathbf{x}}\rangle^{\sigma(\mathbf{x})} u = (2\pi)^{-n} \int_{\mathbb{R}^n} \exp(i\mathbf{x}\boldsymbol{\xi})\langle\boldsymbol{\xi}\rangle^{\sigma(\mathbf{x})}\hat{u}(\boldsymbol{\xi})d\boldsymbol{\xi}, \tag{34}$$

with $\langle\boldsymbol{\xi}\rangle = (1+|\boldsymbol{\xi}|^2)^{1/2}$, as before, and

$$H^s(\mathbb{R}^n) = \left\{ u \in \mathcal{S}'(\mathbb{R}^n) : \ \langle D_{\mathbf{x}}\rangle^s u \in L^2(\mathbb{R}^n) \right\}.$$

In the following we write $\underline{\sigma} = \inf_{\mathbf{x}\in\mathbb{R}^n} \sigma(\mathbf{x})$.

Proposition 3. ([27]) *The above-introduced fractional Sobolev spaces of variable order satisfy the following properties:*
(i) If $u \in H^{\sigma(\cdot)}(\mathbb{R}^n)$, then, for $P \in \boldsymbol{\mathcal{S}}^{\sigma}_{\rho,\delta}$, $Pu \in L^2(\mathbb{R}^n)$.
(ii) Let σ_1 and σ_2 be functions in $\mathcal{B}^{\infty}(\mathbb{R}^n)$ with $\sigma_1(\mathbf{x}) \geq \sigma_2(\mathbf{x})$ for each $\mathbf{x} \in \mathbb{R}^n$. Then $H^{\sigma_1(\cdot)}(\mathbb{R}^n) \subset H^{\sigma_2(\cdot)}(\mathbb{R}^n)$. In particular, $H^{\sigma(\cdot)}(\mathbb{R}^n) \subset H^{\underline{\sigma}}(\mathbb{R}^n)$.
(iii) $H^{\sigma(\cdot)}(\mathbb{R}^n)$ is a Hilbert space with the inner product

$$\begin{aligned}\langle u, v\rangle_{H^{\sigma(\cdot)}(\mathbb{R}^n)} &= \int_{\mathbb{R}^n} \left(\langle D_{\mathbf{x}}\rangle^{\sigma(\mathbf{x})} u\right)(\mathbf{x})\overline{\left(\langle D_{\mathbf{x}}\rangle^{\sigma(\mathbf{x})} v\right)(\mathbf{x})}d\mathbf{x} \\ &+ \int_{\mathbb{R}^n} \left(\langle D_{\mathbf{x}}\rangle^{\underline{\sigma}} u\right)(\mathbf{x})\overline{\left(\langle D_{\mathbf{x}}\rangle^{\underline{\sigma}} v\right)(\mathbf{x})}d\mathbf{x}.\end{aligned} \tag{35}$$

Moreover, $\mathcal{S}(\mathbb{R}^n)$ is dense in $H^{\sigma(\cdot)}(\mathbb{R}^n)$.
(iv) Let σ and τ be functions in $\mathcal{B}^{\infty}(\mathbb{R}^n)$. Suppose that $P \in \boldsymbol{\mathcal{S}}^{\sigma}_{\rho,\delta}$. Then, there exist some constant $C > 0$ independent of P and some positive integer l depending only on $\sigma, \tau, \rho, \delta$, and n such that

$$\|Pu\|_{H^{\tau(\cdot)}(\mathbb{R}^n)} \leq C|p|_l^{(\sigma)} \|u\|_{H^{\sigma(\cdot)+\tau(\cdot)}(\mathbb{R}^n)},$$

for $u \in H^{\sigma(\cdot)+\tau(\cdot)}(\mathbb{R}^n)$, which provides the continuity of P from $H^{\sigma(\cdot)+\tau(\cdot)}(\mathbb{R}^n)$ into $H^{\tau(\cdot)}(\mathbb{R}^n)$.

Theorem 1. ([27]) *Let $P \in \boldsymbol{\mathcal{S}}^{\sigma}_{\rho,\delta}$ be elliptic. Then*

$$H^{\sigma(\cdot)}(\mathbb{R}^n) = \left\{ u \in H^{-\infty}(\mathbb{R}^n) : \ Pu \in L^2(\mathbb{R}^n) \right\} \tag{36}$$

as a set. Moreover, the norm $\|u\|_{H^{\sigma(\cdot)}}(\mathbb{R}^n)$ is equivalent to the norm

$$\|u\|_{H^{\sigma(\cdot),P}}(\mathbb{R}^n) := \left(\|Pu\|^2_{L^2(\mathbb{R}^n)} + \|u\|^2_{H^{\underline{\sigma}}(\mathbb{R}^n)}\right)^{1/2}. \tag{37}$$

The following statement on embeddings and lifting properties for fractional Sobolev spaces of variable order on $L^p(\mathbb{R}^n)$ holds (see [23]).

Theorem 2. *Let $1 < p < \infty$ and $j \in \mathbb{N}$, and let $\sigma(\mathbf{x}) = s + \psi(\mathbf{x})$, with $\psi \in \mathcal{S}(\mathbb{R}^n)$, satisfy $0 < m' \leq \sigma(\mathbf{x}) \leq m \leq 2$ for all $\mathbf{x} \in \mathbb{R}^n$. Then the following assertions hold:*
(i) The space

$$H_p^{j,\sigma(\cdot)}(\mathbb{R}^n) = \left\{ f \in \mathcal{S}'(\mathbb{R}^n) : \langle D_{\mathbf{x}} \rangle^{j\sigma(\mathbf{x})} f \in L^2(\mathbb{R}^n) \right\}$$

is a Banach space and $C_0^\infty(\mathbb{R}^n)$ is dense in this space.
(ii) For $m'j > n/p$, the embedding of $H_p^{j,\sigma(\cdot)}(\mathbb{R}^n)$ into $C^\infty(\mathbb{R}^n)$ is continuous.

The main elements and results of the theory of fractal compact $d-$sets and trace theorems of Sobolev spaces on such sets are now introduced.

Definition 7. *([57], pp. 1-5) Let Γ be a set in $\mathbb{R}^n$ ($n \in \mathbb{N}$). Then Γ is called a $d-$set, with $0 \leq d \leq n$, if there exists a Borel measure μ_Γ in $\mathbb{R}^n$ with the following two properties:*

(i) supp $\mu_\Gamma = \Gamma$;

(ii) there are two constants $c_1 > 0$ and $c_2 > 0$ such that, for all $\gamma \in \Gamma$ and all r with $0 < r < 1$, $c_1 r^d \leq \mu_\Gamma (B(\gamma, r) \cap \Gamma) \leq c_2 r^d$, where $B(\gamma, r)$ is the closed ball in $\mathbb{R}^n$ centred at γ and with radius r.

In the development below we consider compact $d-$sets in the sense of the above definition.

We first consider the classical definitions of fractional Sobolev spaces on $\mathbb{R}^n$ and on a domain S of $\mathbb{R}^n$. We denote by $\mathcal{D}(\mathbb{R}^n)$ the space of infinitely differentiable functions with compact support contained in $\mathbb{R}^n$, and by $\mathcal{S}(\mathbb{R}^n)$ the subspace of $C^\infty-$functions with rapid decay at infinity. Their dual spaces are respectively denoted by $\mathcal{D}'(\mathbb{R}^n)$, the space of distributions on $\mathbb{R}^n$, and by $\mathcal{S}'(\mathbb{R}^n)$, the space of tempered distributions. Similarly, $\mathcal{D}(S)$, with $S \subseteq \mathbb{R}^n$, represents the space of infinitely differentiable functions with compact support contained in S, and $\mathcal{D}'(S)$ the space of distributions on S.

Definition 8. *For $s \in \mathbb{R}$, $H^s(\mathbb{R}^n)$ is the space of tempered distributions u such that*

$$(1+ \mid \boldsymbol{\xi} \mid^2)^{s/2} \hat{u}(\boldsymbol{\xi}) \in L^2(\mathbb{R}^n), \quad \boldsymbol{\xi} \in \mathbb{R}^n.$$

In this space the following inner product is considered:

$$(u, v)_s = \int_{\mathbb{R}^n} (1+ \mid \boldsymbol{\xi} \mid^2)^s \hat{u}(\boldsymbol{\xi}) \hat{v}(\boldsymbol{\xi}) d\boldsymbol{\xi},$$

with associated norm $\| u \|_s = \left(\int_{\mathbb{R}^n} (1+ \mid \boldsymbol{\xi} \mid^2)^s \mid \hat{u}(\boldsymbol{\xi}) \mid^2 d\boldsymbol{\xi} \right)^{1/2}$, where $\hat{}$ stands for the Fourier transform.

For $s \in \mathbb{R}$, the Hilbert spaces $H^s(\mathbb{R}^n)$ and $H^{-s}(\mathbb{R}^n)$ are dual.

Fractional Sobolev spaces on domains are introduced as factor spaces of the above fractional Sobolev spaces on $\mathbb{R}^n$ when the domains under consideration satisfy the following extension property:

Definition 9. *A domain $S \subset \mathbb{R}^n$ is said to satisfy an $s-$*extension property *if there exists a bounded extension operator $E : H^s(S) \longrightarrow H^s(\mathbb{R}^n)$, $s \geq 0$, satisfying $Ef = f$ on S.*

Definition 10. *([56], p. 310) Let $S \subset \mathbb{R}^n$ be an arbitrary (bounded or unbounded) domain. Then $H^s(S)$ is the restriction to S of $H^s(\mathbb{R}^n)$. That is,*

$$H^s(S) = \{f \in \mathcal{D}'(S) : \exists F \in H^s(\mathbb{R}^n) \text{ such that } f = F_S\},$$

where F_S denotes the restriction of F to S. With the quotient norm $\|f\|_{H^s(S)} = \inf_{F; F_S = f} \|F\|_{H^s(\mathbb{R}^n)}$, $H^s(S)$ is a Hilbert space.

For S a bounded domain, $\overline{H}^s(S)$ represents the set of functions in $H^s(\mathbb{R}^n)$ with support contained in $\overline{S}$. That is,

$$\overline{H}^s(S) = \left\{u \in H^s(\mathbb{R}^n) : \text{supp } u \subseteq \overline{S}\right\} = \overline{\mathcal{D}(S)}^{\|\cdot\|_{H^s(\mathbb{R}^n)}}.$$

Remark 3. Note that $\overline{H}^s(S) \subseteq \overline{\mathcal{D}(S)}^{\|\cdot\|_{H^s(S)}}$ ([56], p. 318).

The definition of fractional Sobolev spaces on fractal sets via trace operators allows the connection between fractal geometry and functional regularity properties. That is, the weak-sense regularity properties of functions in these spaces depend on the Hausdorff dimension of their fractal domains and on the fractional regularity order of the functions whose traces define them. Since the existence of a Radon measure $\mu_\Gamma \in \mathbb{R}^n$ with support equal to Γ is always guaranteed for a compact $d-$set in the class introduced in Definition 1, the trace on Γ makes sense pointwise for every function in $\mathcal{S}(\mathbb{R}^n)$ (see [57], p. 138). The following definitions of tempered distributions with compact fractal support and fractional Sobolev spaces on compact fractal sets are then introduced.

Definition 11. *The space of tempered distributions $\mathcal{S}'^{\,\Gamma}(\mathbb{R}^n)$ is defined as*

$$\mathcal{S}'^{\,\Gamma}(\mathbb{R}^n) = \{f \in \mathcal{S}'(\mathbb{R}^n) : f(\varphi) = 0 \quad \text{if} \quad \varphi \in \mathcal{S}(\mathbb{R}^n) \quad \text{and} \quad \text{tr}_\Gamma(\varphi) = 0\}.$$

This space is referred to as the space of tempered distributions with compact fractal support Γ, *since it defines the dual space of the factor space $\mathcal{S}(\Gamma)$ constituted by the pointwise traces on Γ of functions in the space $\mathcal{S}(\mathbb{R}^n)$.*

Definition 12. *([57], pp. 192-193) Let Γ be a compact $d-$set in $\mathbb{R}^n$, with $0 < d < n$, and with associated fractal measure μ_Γ (see Definition 1). Then, the Sobolev space $H^s(\Gamma)$ is defined as the trace on Γ of the fractional Sobolev space $H^{s+\frac{n-d}{2}}(\mathbb{R}^n)$. That is,*

$$H^s(\Gamma) = \text{tr}_\Gamma\left(H^{s+\frac{n-d}{2}}(\mathbb{R}^n)\right), \quad s > 0,$$

equipped with the norm

$$\|\varphi\|_{H^s(\Gamma)} = \inf \|\phi\|_{H^{s+\frac{n-d}{2}}(\mathbb{R}^n)},$$

where the infimum is taken over all the functions $\phi \in H^{s+\frac{n-d}{2}}(\mathbb{R}^n)$ *such that* $\mathrm{tr}_\Gamma(\phi) = \varphi$.

The following orthogonal decomposition of $H^{s+\frac{n-d}{2}}(\mathbb{R}^n)$ is then obtained (see [57], p. 193):

$$H^{s+\frac{n-d}{2}}(\mathbb{R}^n) = \left\{\phi \in H^{s+\frac{n-d}{2}}(\mathbb{R}^n) : \mathrm{tr}_\Gamma(\phi) = 0\right\} \oplus H^s(\Gamma).$$

The spaces $H^s(\Gamma)$, $s > 0$, are densely embedded in $L^2(\Gamma)$, which is interpreted as the set of tempered distributions f on $\mathbb{R}^n$ defined by

$$f(\varphi) = \int_\Gamma f(\gamma)\mathrm{tr}_\Gamma(\varphi)(\gamma)\mu_\Gamma(d\gamma), \quad \varphi \in \mathcal{S}(\mathbb{R}^n), \tag{38}$$

where $\mathrm{tr}_\Gamma(\varphi)$ is the pointwise trace of φ on Γ (see [57], pp. 135-141).

Definition 13. *([57], pp. 125, 147) Let* Γ *be a compact* $d-$*set in the sense of Definition 1. The space* $H^{s,\Gamma}(\mathbb{R}^n)$ *is defined as the space of tempered distributions in* $H^s(\mathbb{R}^n)$ *with support contained in* Γ. *That is,*

$$H^{s,\Gamma}(\mathbb{R}^n) = \{f \in H^s(\mathbb{R}^n) : \ f(\varphi) = 0 \text{ if } \varphi \in \mathcal{S}(\mathbb{R}^n) \text{ and } \mathrm{tr}_\Gamma(\varphi) = 0\}, \ s \in \mathbb{R}.$$

The distribution dimension of Γ, denoted by $\dim_D\Gamma$, is then defined as

$$\dim_D\Gamma = \sup\left\{d : \ H^{-\frac{n-d}{2},\ \Gamma}(\mathbb{R}^n) \text{ is nontrivial}\right\},$$

and coincides with the Hausdorff dimension of Γ.

Note that for the trace operator considered in Definition 12, there exists a (non-linear) bounded extension operator from $H^s(\Gamma)$ into $H^{s+\frac{n-d}{2}}(\mathbb{R}^n)$ (see [57], pp. 159-168).

For a bounded linear operator $T : \mathcal{A} \to \mathcal{B}$, with $\mathcal{A}$ and $\mathcal{B}$ being quasi-Banach spaces, its entropy numbers $\{e_n(T)\}_{n\in\mathbb{N}}$ are defined as

$$e_k(T) = \inf\left\{\varepsilon > 0 : \ T(\mathcal{U}_\mathcal{A}) \subset \bigcup_{j=1}^{2^{k-1}} (b_j + \varepsilon\mathcal{U}_\mathcal{B}) \text{ for some } b_1, \ldots, b_{2^{k-1}} \in \mathcal{B}\right\}, \tag{39}$$

where $\mathcal{U}_\mathcal{H}$ stands for the unit ball in $\mathcal{H}$. The following Carl's inequality connects the spectral properties of a compact operator T with the geometry properties of such an operator, described in terms of its entropy numbers (see [12], for the Banach space case, and [13], for the quasi-Banach space case):

$$|\mu_k(T)| \leq (2)^{1/2} e_k(T), \quad k \in \mathbb{N}, \tag{40}$$

where $\{\mu_n(T)\}_{n\in\mathbb{N}}$ is the sequence of all non-zero eigenvalues of T, repeated according to algebraic multiplicity and ordered so that

$$|\mu_1(T)| \geq |\mu_2(T)| \geq \ldots \longrightarrow 0,$$

and $\{e_n(T)\}_{n\in\mathbb{N}}$ represents, as before, the corresponding sequence of entropy numbers.

Approximation numbers constitute another important tool in the characterization of spectral properties of fractal pseudodifferential operators. In general, for a bounded operator T defined between quasi-Banach spaces $\mathcal{A}$ and $\mathcal{B}$, its sequence of approximation numbers $\{a_n(T)\}_{n\in\mathbb{N}}$ is defined as follows:

$$a_k(T) = \inf \{\|T - L\| : \ L \in \mathcal{L}(\mathcal{A}, \mathcal{B}), \quad \text{rank } L < k\}, \quad k \in \mathbb{N},$$

where rank L is the dimension of the range of L, and $\mathcal{L}(\mathcal{A}, \mathcal{B})$ denotes the space of bounded operators from $\mathcal{A}$ to $\mathcal{B}$. In the particular case where T is a compact self-adjoint operator on a Hilbert space H, the following equality holds:

$$|\mu_k(T)| = a_k(T), \quad k \in \mathbb{N}. \tag{41}$$

This equality, considered in the case where $H = L^2(\Gamma)$, is very useful in the characterization of spectral properties of fractal pseudodifferential operators and fractional differential operators defined on compact fractal sets (see Chapters IV and V of [57]). In our framework, Equation (41) is used in the characterization of the spectral properties of the covariance operator of the fractal dual, or, equivalently, of the operator generating the bilinear form which defines the inner product in the RKHS of the trace generalized random field.

The spectral properties of compact embeddings between fractional Besov spaces on fractals are characterized in terms of the corresponding sequences of entropy numbers ([57], pp. 162-170). We consider here two particular cases of such embeddings which are applied in the derivation of the results presented in this paper.

Theorem 3. *Let Γ be a compact $d-$set in $\mathbb{R}^n$, with $0 < d < n$. For $0 \leq s_2 < s_1 < \infty$,*
(i) the embedding $id : H^{s_1}(\Gamma) \longrightarrow H^{s_2}(\Gamma)$ is compact and there exists a positive constant c such that for the related entropy numbers the following inequality holds:

$$e_k \left(id : \ H^{s_1}(\Gamma) \longrightarrow H^{s_2}(\Gamma)\right) \leq ck^{-\frac{s_1 - s_2}{d}}, \quad k \in \mathbb{N};$$

(ii) if $s_1 - s_2 - d/2 > 0$, then, the embedding $id : H^{s_1}(\Gamma) \longrightarrow \mathcal{C}^{s_2}(\Gamma)$ is compact and there exists a positive constant $\widetilde{c}$ such that for the related entropy numbers the following inequality holds:

$$e_k\,(id:\ H^{s_1}(\Gamma) \longrightarrow \mathcal{C}^{s_2}(\Gamma)) \leq \widetilde{c}k^{-\frac{s_1-s_2}{d}}, \quad k \in \mathbb{N},$$

where $\mathcal{C}^{s_2}(\Gamma) = B^{s_2}_{\infty,\infty}(\Gamma)$ *represents the Hölder-Zigmund space of fractional order* s_2 *on* Γ.

From part (ii) of the above theorem, examples of distributions that define measures with compact fractal support Γ (in the sense established in [53]) can be constructed considering the spaces $H^{-\alpha,\Gamma}(\mathbb{R}^n)$, with $\alpha > n/2$. We refer to such distributions as distributions with finite order $\alpha - n/2 = s - d/2$ and compact fractal support Γ.

On the Transient Behavior of the Maximum Level Length in Structured Markov Chains*

Jesús R. Artalejo

Faculty of Mathematics, Complutense University,
Madrid 28040, Spain
jesus_artalejo@mat.ucm.es

Summary. This paper studies the transient behavior of the maximum level length for general block structured continuous-time Markov chains ($CTMC$). The approach is presented for the bidimensional case, however, it still holds for multi-dimensional chains. The results can also be easily modified to cover the discrete-time case. This work complements the busy period analysis by Neuts [12] and the asymptotic approach by Serfozo [14]. Some illustrative examples (SIR epidemic model, retrial queue) including numerical implementations are presented.

Keywords: maximum level length, $M/M/c$ retrial queue, SIR epidemic model, structured Markov chains, transient behavior.

1 Introduction

This paper deals with structured Markov chains (see the recent book by Li [10] and the references therein) which are useful for representing a great variety of stochastic models arising from epidemics and population process, manufacturing, queueing, reliability, etc. An important issue is the study of the maximum level length because this descriptor provides an excellent measure of the system congestion.

Extreme values of stochastic processes can be investigated following different approaches. Serfozo [14] studies the asymptotic behavior of maximum values of birth and death processes over large intervals whereas Neuts [12] concentrates on the distribution of extreme values during a busy period (i.e., the elapsed time between two successive visits to a given initial state). Both methods have been widely used in the literature. Our contribution in this paper is to investigate the transient behavior of the maximal level of the structured chain visited during $[0, t]$. As far as the author knows, this paper is the first attempt to deal with the transient version of the maximum level length

* The author would like to dedicate this paper to the memory of Professor Marisa Menéndez.

of a general structured $CTMC$. To reach our goal, we only need to combine some classical existing techniques (extended Markov chain, Laplace transforms, numerical inversion) but the proposed approach has a straightforward novel use in applications. Our study includes two methods to compute the distribution of the highest level reached before time t and discussion on the related computational issues. In a recent paper, Artalejo *et al.* [4] presented an algorithm for computing the maximum number of infected individuals in transient regime for the SIS (susceptible → infected → susceptible) stochastic epidemic model. As partially related work, we also mention the paper by Artalejo and Chakravarthy [2] where an algorithmic analysis of the maximum level length during a busy period is performed.

The organization of the paper is as follows. In Section 2, our transient analysis is presented for the bidimensional case. In Section 3, the approach is illustrated by applications to the stochastic SIR epidemic model and the $M/M/c$ retrial queue. Finally, a few comments on possible generalizations and concluding remarks are given in Section 4.

2 Transient Behavior Analysis in the Bidimensional Case

We consider a regular $CTMC$ $Z = \{(X(t), Y(t));\ t \geq 0\}$ with state space given by $S = \{(i,j); i \geq 0, 0 \leq j \leq L_i\}$. In the state (i,j), the first coordinate i is called the level of the state. The number of states in each level, $l(i) = L_i + 1$, is assumed to be finite.

The infinitesimal generator of the $CTMC$, $\mathbf{Q} = [q_{(i,j)(i',j')}]$, is of the form

$$\mathbf{Q} = \begin{pmatrix} \mathbf{Q}_{00} & \mathbf{Q}_{01} & \mathbf{Q}_{02} & \cdots \\ \mathbf{Q}_{10} & \mathbf{Q}_{11} & \mathbf{Q}_{12} & \cdots \\ \vdots & \vdots & \vdots & \ddots \end{pmatrix},$$

with block components $\mathbf{Q}_{ii'}$ describing the infinitesimal motion between levels i and i'.

In the above description, the level process $X(t)$ is assumed to have an unbounded state space, but obviously this not excludes the finite case, as the example in Subsection 3.1 shows. On the other hand, the number of phases per level may vary, and the generator $\mathbf{Q}$ has a general block matrix structure, so that any $CTMC$ can be written in this form provided that $l(i)$ is finite.

2.1 The Extended Chain Method

In this subsection, we follow a natural approach by adding a third component to the state description.

Define an extended chain $\widehat{Z} = \{(X(t), Y(t), M(t));\ t \geq 0\}$ where $M(t)$ is the maximum level length visited by the chain Z during $[0,t]$. Now the state space is $\widehat{S} = \{(i,j,k);\ 0 \le i \le k,\ 0 \le j \le L_i\}$. The generator $\widehat{\mathbf{Q}} = [\widehat{q}_{(i,j,k)(i',j',k')}]$ is related to the original generator $\mathbf{Q}$ through the following relationships:

$$\widehat{q}_{(i,j,k)(i',j',k)} = q_{(i,j)(i',j')},\ 0 \le i' \le k,\ 0 \le j' \le L_{i'},$$
$$\widehat{q}_{(i,j,k)(k',j',k')} = q_{(i,j)(k',j')},\ k < k',\ 0 \le j' \le L_{k'},$$
$$\widehat{q}_{(i,j,k)(i',j',k')} = 0,\ \text{otherwise}.$$

The consideration of the extended chain $\widehat{Z}$ is the key for the computational analysis of the transient version of the maximum level length.

For each $t \ge 0$, denote the transient probabilities

$$p_{ijk}(t) = P\{X(t) = i,\ Y(t) = j,\ M(t) = k\},$$

for $(i,j,k) \in \widehat{S}$, and the initial probabilities $p_{ijk}(0) = \delta_{(i,j,k)(i_0,j_0,k_0)}$, where δ_{ab} stands for the Kronecker's function defined by

$$\delta_{ab} = \begin{cases} 1, \text{ if } a = b, \\ 0, \text{ otherwise.} \end{cases}$$

Let $p^*_{ijk}(s)$ be the Laplace transform of the probability $p_{ijk}(t)$; that is, $p^*_{ijk}(s) = \int_0^\infty e^{-st} p_{ijk}(t)dt$, for $\text{Re}(s) \ge 0$.

We notice that the primes notation in the definition of $\widehat{\mathbf{Q}}$ is used to indicate the state after transitions, while later in Theorem 1 the primes are used to indicate the state before transitions. The motivation to adopt this notation is to assign the no-prime notation to the state playing the main role in the corresponding expression under consideration. In this sense, it is natural to describe a generator in terms of its rows, while each forward Kolmogorov equation in Theorem 1 is associated with a final state (i,j,k).

The next result provides the system of equations governing the dynamics of the Laplace transforms.

Theorem 1. *For each $k \ge k_0$, the Laplace transforms $p^*_{ijk}(s)$ satisfy the following system:*

$$sp^*_{ijk}(s) - \delta_{(i,j,k)(i_0,j_0,k_0)} = \sum_{i'=0}^{k} \sum_{j'=0}^{L_{i'}} q_{(i',j')(i,j)} p^*_{i'j'k}(s)$$
$$+\delta_{ik} \sum_{k'=k_0}^{k-1} \sum_{i'=0}^{k'} \sum_{j'=0}^{L_{i'}} q_{(i',j')(i,j)} p^*_{i'j'k'}(s),$$

for $0 \le \max\{i, k_0\} \le k$ and $0 \le j \le L_i$.

Proof. The forward Kolmogorov equations governing the infinitesimal dynamics of process $\widehat{Z}$ are given by

$$\frac{d}{dt}p_{ijk}(t) = \sum_{i'=0}^{k}\sum_{j'=0}^{L_{i'}} q_{(i',j')(i,j)}p_{i'j'k}(t)$$
$$+\delta_{ik}\sum_{k'=k_0}^{k-1}\sum_{i'=0}^{k'}\sum_{j'=0}^{L_{i'}} q_{(i',j')(i,j)}p_{i'j'k'}(t),$$

for $0 \leq \max\{i, k_0\} \leq k$ and $0 \leq j \leq L_i$.

Taking Laplace transforms and using

$$\int_0^\infty e^{-st}\frac{d}{dt}p_{ijk}(t)dt = sp^*_{ijk}(s) - \delta_{(i,j,k)(i_0,j_0,k_0)},$$

yields the Laplace transform equations given in the statement. □

From the scalar formulation given in the theorem, it is clear the fact that, for each fixed $k \geq k_0$, the computation of $p^*_{ijk}(s)$, for $0 \leq \max\{i, k_0\} \leq k$ and $0 \leq j \leq L_i$, involves the unknowns corresponding to the previous indices $k' \in \{0, ..., k-1\}$. An alternative matrix form formulation is given in Corollary 1. The matrix formulation provides a compact expression and helps to construct numerical codes.

First, we introduce some notation. The following vectors comprise the Laplace transforms partitioned according to the levels:

$$\mathbf{p}^*_{ik}(s) = \left(p^*_{i0k}(s), ..., p^*_{iL_ik}(s)\right),\ 0 \leq \max\{i, k_0\} \leq k,$$
$$\mathbf{p}^*_{k}(s) = \left(\mathbf{p}^*_{0k}(s), ..., \mathbf{p}^*_{kk}(s)\right),\ k \geq k_0.$$

Let $\mathbf{a}^0_k$ denote a row vector of dimension $\bar{l}(k) = \sum_{i=0}^{k} l(i)$ with 1 in the (i_0, j_0)th position and 0 elsewhere. The vector $\mathbf{0}_r$ denotes a row vector of zeroes of dimension r. The identity matrix of dimension r is denoted by $\mathbf{I}_r$.

Corollary 1. *For each $k \geq k_0$, the Laplace vector $\mathbf{p}^*_k(s)$ verifies the block structured system*

$$\mathbf{p}^*_k(s)\left(\mathbf{Q}_k - s\mathbf{I}_{\bar{l}(k)}\right) = \mathbf{v}_k,\ k \geq k_0,$$

where $\mathbf{Q}_k$ is the submatrix of $\mathbf{Q}$ corresponding to the states in levels $i = 0, ..., k$, and

$$\mathbf{v}_k = -\mathbf{a}^0_k\delta_{kk_0} - (1-\delta_{kk_0})\mathbf{b}_k,\ k \geq k_0,$$
$$\mathbf{b}_k = \left(\mathbf{0}_{l(0)}, ..., \mathbf{0}_{l(k-1)}, \mathbf{c}_k\right),\ k \geq k_0,$$

where $\mathbf{c}_k$ is a row vector of dimension $l(k)$ and the jth component is equal to $\sum_{k'=k_0}^{k-1}\sum_{i'=0}^{k'}\sum_{j'=0}^{L_{i'}} q_{(i',j')(k,j)}p^*_{i'j'k'}(s)$, for $0 \leq j \leq L_k$.

Once the Laplace transforms have been computed, the transient probabilities $p_{ijk}(t)$ can be obtained by numerical inversion. We propose to carry out the numerical inversion by using Fourier series methods (see Cohen [7]). In particular, the maximum level length distribution $p_{..k}(t) = P\{M(t) = k\}$, for $k \geq k_0$, follows by inverting numerically the sum $\sum_{i=0}^{k}\sum_{j=0}^{L_i} p^*_{ijk}(s)$.

2.2 The Absorbing Macro-state Method

In the introduction, the computation of the maximal level visited in a busy period [12], [2] was mentioned as a related problem. Although the transient analysis (i.e., the subject matter of this paper) is typically a more involved problem, it might seem natural to explore whether or not the approach used when the study is based on a busy period can be successfully extended to the transient framework. This consideration gives the initial motivation for the absorbing macro-state method investigated in this subsection.

Let us assume that the busy period is defined as the elapsed time between two successive visits to a certain state 0^* of the level 0. For a given initial state (i_0, j_0), we note that $P\{M(t) < k\}$, for $k > i_0$, corresponds to the probability of the event that the $CTMC$ Z hits the state 0^* before hitting level k. Hence, the computation of the distribution of the maximal level reached during a busy period reduces to finding the probability of absorption by 0^* in an auxiliary finite state chain with two absorbing states, say, 0^* and k^*, where k^* represents a macro-state comprising all states of level k or above. A minor variant of the above idea can be translated to the transient context.

Theorem 2. *The distribution function of $M(t)$, for $t \geq 0$, is given by*

$$P\{M(t) \leq k\} = \mathbf{a}_k^0 \exp\{\mathbf{Q}_k t\}\mathbf{e}_{\overline{l}(k)}, \ k \geq k_0,$$

where $\mathbf{e}_r$ denotes a column vector of dimension r with all entries equal to 1.

Proof. Instead of considering two absorbing states (i.e., 0^* and k^*), we now deal with the auxiliary chain $\widetilde{Z}_k$ with the absorbing macro-state k^*. Then, it is clear that $P\{M(t) \geq k_0\} = 1$ and

$$P\{M(t) \geq k\} = P\{\widetilde{X}(t) = k^*\}, \ k > k_0, \ t \geq 0,$$

that is; the probability $P\{M(t) \geq k\}$ in the original chain amounts to the probability $P\{\widetilde{X}(t) = k^*\}$ in the auxiliary chain. The infinitesimal generator $\widetilde{\mathbf{Q}}_k$ associated with $\widetilde{Z}_k$ follows easily from $\mathbf{Q}_k$ (see formula (3.2) in [2]).

We recall that the distribution of the time until absorption in a finite Markov chain is said a phase-type (PH) distribution, and it reduces to a matrix exponential [10]. Thus, we have

$$P\{M(t) \geq k\} = 1 - \mathbf{a}_{k-1}^0 \exp\{\mathbf{Q}_{k-1} t\}\mathbf{e}_{\overline{l}(k-1)}, \ k > k_0,$$

which amounts to the desired expression for the distribution function. □

At a first glance the use of the absorbing macro-state method avoids the numerical inversion of the Laplace transforms, which is replaced by the computation of a matrix exponential. However, finding accurate methods to compute the matrix exponential is a non-trivial matter which is still an open problem in numerical analysis. Following [11], we remark that "*the exponential of a matrix can be computed in many ways. In practice, consideration of computational stability and efficiency indicates that some of the methods are preferable to others, but that none are completely satisfactory*". Kulkarni [9] presents four methods for dealing with the transient behavior of a $CTMC$. One of them is based on the Laplace transform method, but it essentially implies to come back to the first method described in Subsection 2.1. In fact, the Laplace transform of $P\{M(t) \leq k\}$ is given by

$$M_k^*(s) = \mathbf{a}_k^0 \left(s\mathbf{I}_{\bar{l}(k)} - \mathbf{Q}_k \right)^{-1} \mathbf{e}_{\bar{l}(k)}, \ k \geq k_0.$$

In principle, a detailed comparison among different methods of numerical computation is not our aim in this paper. Here, we simply mention that both methods rely on the $\mathbf{Q}_k$ matrix, so if a well-posed structure of $\mathbf{Q}_k$ (e.g. existence of zero blocks) might be exploited in the extended chain method, then it also might be exploited in the absorbing macro-state method, and vice versa. Limitations with respect to the dimensionality of the models could affect to the two methods.

However, we can mention a distinguished feature giving support to the use of the extended chain method. The point is that the numerical inversion of the Laplace transforms leads to the probabilities $p_{ijk}(t)$ of the whole tridimensional vector $(X(t), Y(t), M(t))$, while the absorbing macro-state method only gives the marginal distribution of $M(t)$. Due to this fact, our implementation of the numerical examples in the next section is based on the extended chain method.

3 Illustrative Examples

3.1 The Stochastic SIR Epidemic Model

The maximum size of an epidemic is the largest number of infective individuals ever present during its course. Recently, Artalejo *et al.* [4], [3] pointed out that the maximum population size is an interesting descriptor worthy of some extra attention in biological applications. We next deal with the SIR (susceptible $\rightarrow$ infected $\rightarrow$ removed) stochastic epidemic model [1]. For this model, Neuts and Li [13] propose an algorithm for computing the maximum size distribution reached before the absorption. Here, the study is extended by applying the approach developed in Subsection 2.1 for the transient analysis.

Let $(X(t), Y(t))$ be the bidimensional $CTMC$ describing the epidemic. At time t, the population consists of $X(t) = i$ infectives and $Y(t) = j$ susceptible. The initial condition is $(X(0), Y(0)) = (m, n)$. The population state changes either if a susceptible individual is infected, which occurs at rate λ_{ij}, or if an infective is removed at rate μ_i. The state space of the SIR epidemic model is $S = \{(i,j);\ 0 \leq i \leq m+n,\ 0 \leq j \leq \min\{n, m+n-i\}\}$ with $l(i) = \min\{n, m+n-i\} + 1$. We notice that states $(0, j)$, for $0 \leq j \leq n$, are absorbing states. As a result, the set of transient states $S \setminus \{(0,j); 0 \leq j \leq n\}$ is a reducible set. This fact influences the calculation of the quasi-stationary distribution [15], [6].

For $1 \le i \le m$, $\mathbf{Q}_{i,i-1}$ is a square block of dimension $n+1$. If $m+1 \le i \le m+n$, then $\mathbf{Q}_{i,i-1}$ has dimension $l(i) \times l(i)+1$. In both cases, its elements are given by

$$q_{(i,j)(i-1,j')} = \begin{cases} \mu_i, & \text{if } 0 \le j \le \min\{n, m+n-i\},\ j' = j, \\ 0, & \text{otherwise.} \end{cases}$$

The blocks $\mathbf{Q}_{ii}$, for $1 \le i \le m+n$, are square matrices of dimension $l(i)$ with elements

$$q_{(i,j)(i,j')} = \begin{cases} -((1-\delta_{0j})\lambda_{ij} + \mu_i), & \text{if } 0 \le j \le \min\{n, m+n-i\},\ j' = j, \\ 0, & \text{otherwise.} \end{cases}$$

On the other hand, $\mathbf{Q}_{i,i+1}$ for $1 \le i \le m-1$, is a square block of dimension $n+1$, whereas $\mathbf{Q}_{i,i+1}$ for $m \le i \le m+n-1$, has dimension $l(i) \times l(i)-1$. The elements are as follows

$$q_{(i,j)(i+1,j')} = \begin{cases} \lambda_{ij}, & \text{if } 1 \le j \le \min\{n, m+n-i\},\ j' = j-1, \\ 0, & \text{otherwise.} \end{cases}$$

Finally, the blocks $\mathbf{Q}_{0i'}$, for $0 \le i' \le m+n$, $\mathbf{Q}_{ii'}$, for $1 \le i \le m+n-1$ and $i' \notin \{i-1, i, i+1\}$, and $\mathbf{Q}_{m+n,i'}$, for $i' \notin \{m+n-1, m+n\}$, are defined as null blocks of the appropriate dimension. This completes the description of the structured generator.

The transient analysis for the SIR epidemic model is illustrated in Figures 1-3. We assume that $\lambda_{ij} = i^{\alpha} j \beta$, for $0 < \alpha \le 1$, and $\mu_i = i\eta$. Following Neuts and Li [13], we notice that the parameter α can be interpreted as the degree of interaction between infectives and susceptibles.

In Figure 1, we take $\beta = \eta = 1.0$ and the initial condition $(m, n) = (15, 15)$. Then, we plot the probability of having none infectives at time t, $P\{X(t) = 0\}$, as a function of the interaction parameter α. In agreement with the fact that $(0, j)$, for $0 \le j \le n$, are absorbing states, we observe that $P\{X(t) = 0\}$ tends to 1, as $t \to \infty$.

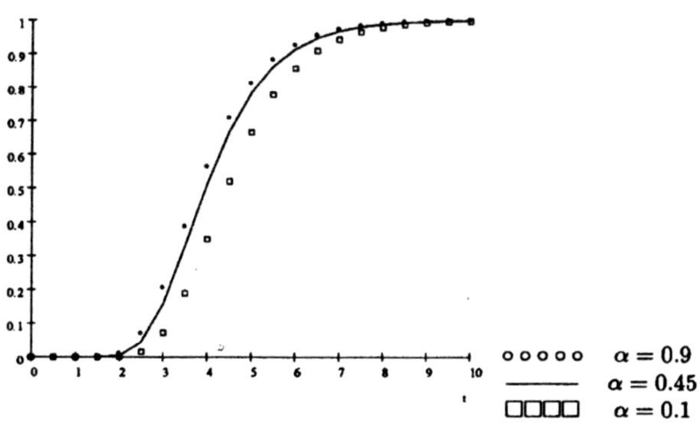

Fig. 1. $P\{X(t) = 0\}$ versus α

The expected number of infectives at time t, $E[X(t)]$, is plotted in Figure 2 against η. The rest of parameters are $\alpha = 0.95$, $\beta = 3.0$ and $(m, n) = (15, 15)$. Curves associated with $\eta = 0.2$ and $\eta = 1.0$ show an initial increment resulting in a peak close to the time origin. In contrast, in the case $\eta = 2.0$, the curve decreases from the initial value $m = 15$ to 0.

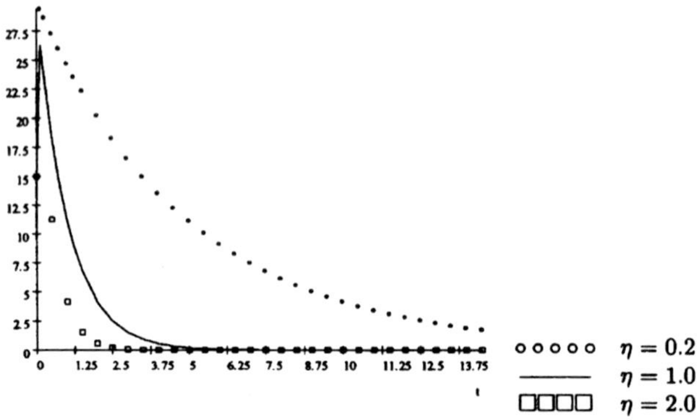

Fig. 2. $E[X(t)]$ versus η

The numerical inversion of the equations given in Theorem 1 allows us to compute the distribution function $P\{M(t) \leq k\}$, for $k_0 \leq k \leq m + n$, at any given time t. In Figure 3, we take $\alpha = \beta = \eta = 1.0$, $k_0 = m$ and $t = 5.0$. Three curves corresponding to different choices of the initial pair (m, n) are plotted. In the three cases, we assume that $m + n = 30$.

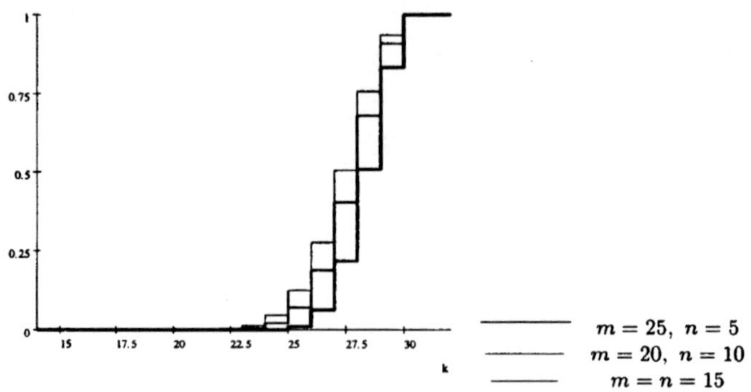

Fig. 3. $P\{M(5) \leq k\}$ versus (m, n)

3.2 The $M/M/c$ Retrial Queue

Retrial queues have been widely used to model telephone systems and other telecommunication and computer networks. We next describe the main retrial queue of $M/M/c$-type (see, for example, Artalejo and Gómez-Corral [5]). Primary customers arrive according to a Poisson process with rate λ. Service is rendered by c identical servers with service times exponentially distributed with rate ν. The system does not have a waiting space. Customers who find at least one server free upon arrival immediately occupy a position and leave the system after service. In contrast, any arriving customer finding all servers busy will enter into a retrial orbit. From there, the retrial customers will compete for service. The inter-retrial times are assumed to be exponentially distributed with rate μ. Moreover, we assume that the process of primary arrivals, services and retrial times are mutually independent.

At any time t, the state of the process is represented by the bidimensional process $(X(t), Y(t))$, where $X(t)$ denotes the number of customers in orbit and $Y(t)$ is the number of busy servers. The state space of the resulting $CTMC$ is $S = \{(i,j);\ i \geq 0,\ 0 \leq j \leq c\}$. Thus, all the matrices $\mathbf{Q}_{ii'}$ are square blocks of dimension $l(i) = c+1$. They are given by

$$q_{(i,j)(i-1,j')} = \begin{cases} i\mu, \text{ if } 0 \leq j \leq c-1,\ j' = j+1, \\ 0, \text{ otherwise,} \end{cases}$$

$$q_{(i,j)(i,j')} = \begin{cases} \lambda, & \text{if } 0 \leq j \leq c-1,\ j' = j+1, \\ j\nu, & \text{if } 1 \leq j \leq c,\ j' = j-1, \\ -(\lambda + j\nu + (1-\delta_{jc})i\mu), & \text{if } 0 \leq j \leq c,\ j' = j, \\ 0, & \text{otherwise,} \end{cases}$$

$$q_{(i,j)(i+1,j')} = \begin{cases} \lambda, \text{ if } j = j' = c, \\ 0, \text{ otherwise.} \end{cases}$$

The rest of matrices $\mathbf{Q}_{ii'}$, for $i \geq 0$ and $i' \notin \{i-1, i, i+1\}$, are blocks of zeroes.

In the next numerical illustration, we take $\nu = \mu = 1.0$, $c = 5$ and the initial state $(i_0, j_0, k_0) = (0,0,0)$. First, we set $k = 0$. Then, the numerical inversion of the vector $p_0^*(s)$ (see Theorem 1) leads to the probability of having no customers in orbit before time t. The probability $P\{M(t) = 0\}$ is plotted in Figure 4 for various choices of the traffic intensity $\rho = \lambda/c\nu$.

At this point, it should be pointed out that the analysis of the maximum level length can be carried out independently of whether or not a stationary distribution exist. In fact, the lowest curve in the figure correspond to the non stable case $\rho = 1.2$ (i.e., ρ is greater than 1.0).

The numerical inversion of $p_k^*(s)$ can be iterated for $k \geq 1$. In this way, we may calculate the probability mass function of the maximal number of customers in orbit during $[0,t]$. In Table 1, we consider the case $\rho = 0.9$, $\nu = 1.0$, $c = 5$ and $(i_0, j_0, k_0) = (0,0,0)$. Each entry in the table contains the 99th

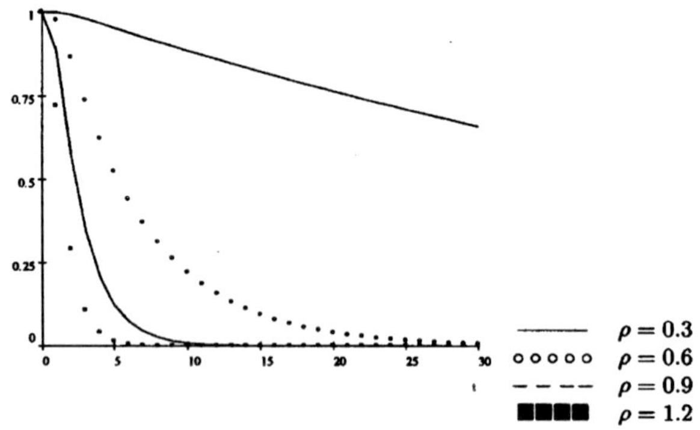

Fig. 4. $P\{M(t) = 0\}$ versus ρ

Table 1. k_{99} versus ρ and μ

	$\mu = 0.1$	$\mu = 1.0$	$\mu = 10.0$	$\mu = 100.0$
$\rho = 0.3$	3	3	3	3
$\rho = 0.6$	13	11	10	10
$\rho = 0.9$	30	26	24	15
$\rho = 1.2$	51	47	45	45

percentile, denoted by k_{99}, of the probability mass function $P\{M(15) = k\}$, for $k \geq 0$. In agreement with the intuitive expectations, the system congestion measured through the percentile k_{99}, increases with ρ but it decreases with increasing values of the retrial rate μ.

4 Concluding Remarks

In this paper, we have investigated the transient distribution of the maximum level length in general block structured bidimensional $CTMC$. Two methods of computation have been presented in Section 2. The transform approach combined with the numerical inversion lead to a reasonable route of evaluation. Unlike the study of the stationary system state, the analysis of the transient maximum level length can be done under very general non restrictive conditions. The SIR epidemic model studied in Subsection 3.1 shows a situation where the underlying Markov chain is finite and reducible whereas the Markov chain of the $M/M/c$ retrial queue in Subsection 3.2 is infinite and irreducible.

The analysis can be extended in several directions. For example, the methodology can be used to study the maximum level length in many other stochastic models including more complicated retrial queues [5] and biological models subject to killing or catastrophes [3], [8]. The transient behavior of the minimum level length visited by the process in $[0, t]$ is the dual problem to the maximum level length considered in this paper. For the sake of easiness, we have focused only on the bidimensional case. However, it is clear that the result in Theorem 1 can be extended to the multidimensional case. The key is just to employ the lexicographical order where the first component (i.e., the level of the process) is reserved for the characteristic which maximum level length is under study. Finally, we notice that the approach can be easily updated to the discrete-time case. Similar arguments to those given in Subsection 2.1 lead to the following recursive equations for the probabilities $p_{ijk}^{(n)} = P\{X_n = i,\ Y_n = j,\ M_n = k\}$, for $n \geq 1$, $0 \leq \max\{i, k_0\} \leq k$ and $0 \leq j \leq L_i$:

$$p_{ijk}^{(1)} = \delta_{kk_0} P_{(i_0,j_0)(i,j)} + (1 - \delta_{kk_0})\delta_{ik} P_{(i_0,j_0)(i,j)},$$

$$p_{ijk}^{(n)} = \sum_{i'=0}^{k} \sum_{j'=0}^{L_{i'}} p_{i'j'k}^{(n-1)} P_{(i',j')(i,j)}$$

$$+\delta_{ik} \sum_{k'=k_0}^{k-1} \sum_{i'=0}^{k'} \sum_{j'=0}^{L_{i'}} p_{i'j'k'}^{(n-1)} P_{(i',j')(i,j)},$$

where (X_n, Y_n, M_n) denotes the discrete-time version of $(X(t), Y(t), M(t))$. In particular, M_n represents the maximum level length visited by the discrete-time Markov chain during the first n time epochs. Obviously, $\mathbf{P} = [P_{(i,j)(i',j')}]$ is the one step transition probability matrix of the original chain $Z = \{(X_n, Y_n);\ n \geq 0\}$. After an appropriate recursive computation, we easily find the desired probabilities $P\{M_n = k\} = \sum_{i=0}^{k} \sum_{j=0}^{L_i} p_{ijk}^{(n)}$, for $k \geq k_0$.

Acknowledgement. This research was supported by the Spanish Government (Department of Science and Innovation) and the European Commission through project MTM2008-01121.

References

1. Allen, L.J.S.: An Introduction to Stochastic Processes with Applications to Biology. Pretince Hall, New Jersey (2003)
2. Artalejo, J.R., Chakravarthy, S.R.: Algorithmic analysis of the maximal level length in general-block two-dimensional Markov processes. Math. Probl. Eng., Article ID 53570, 15 (2007)
3. Artalejo, J.R., Economou, E., López-Herrero, M.J.: Evaluating growth measures in an immigration process subject to binomial and geometric catastrophes. Math. Biosci. Eng. 4, 573–594 (2007)
4. Artalejo, J.R., Economou, A., López-Herrero, M.J.: The maximum number of infected individuals in SIS epidemic models: Computational techniques and quasi-stationary distributions. J. Comput. Appl. Math. 223, 2563–2574 (2010)
5. Artalejo, J.R., Gómez-Corral, A.: Retrial Queueing Systems: A Computational Approach. Springer, Berlin (2008)

6. Artalejo, J.R., López-Herrero, M.J.: Quasi-stationary and ratio of expectations distributions: A comparative study. J. Theor. Biol. 266, 264–274 (2010)
7. Cohen, A.M.: Numerical Methods for Laplace Transform Inversion. Springer, New York (2007)
8. Coolen-Schrijner, P., van Doorn, E.A.: Quasi-stationary distributions for a class of discrete-time Markov chains. Methodol. Comput. Appl. Probab. 8, 449–465 (2006)
9. Kulkarni, V.G.: Modeling and Analysis of Stochastic Systems. Chapman & Hall, London (1995)
10. Li, Q.L.: Constructive Theory in Stochastic Models with Applications: The *RG*-Factorizations. Springer, Tsinghua University Press, Berlin, Beijing (2009)
11. Moler, C., van Loan, C.: Nineteen dubious ways to compute the exponential of a matrix, twenty-five years later. SIAM Rev. 45, 3–49 (2003)
12. Neuts, M.F.: The distribution of the maximum length of a Poisson queue during a busy period. Oper. Res. 12, 281–285 (1964)
13. Neuts, M.F., Li, J.M.: An algorithmic study of S-I-R stochastic epidemic models. In: Heyde, C.C., Prohorov, Y.V., Pyke, R., Rachev, S.T. (eds.) Athens Conference on Applied Probability and Time Series, vol. 1, pp. 295–306. Springer, Heidelberg (1996)
14. Serfozo, R.F.: Extreme values of birth and death processes and queues. Stoch. Process. Appl. 27, 291–306 (1998)
15. van Doorn, E.A., Pollet, P.K.: Survival in a quasi-death process. Linear Algebra Appl. 429, 776–791 (2008)

On Record-Like Observations: Asymptotic Analysis Using Martingale Tools*

Raúl Gouet[1], F. Javier López[2], and Gerardo Sanz[2]

[1] Departamento de Ingeniería Matemática y Centro de Modelamiento Matemático, Universidad de Chile, UMI-CNRS-2807, Casilla 170-3, Correo 3, Santiago, Chile
rgouet@dim.uchile.cl

[2] Dpto. Métodos Estadísticos and BIFI, Facultad de Ciencias, Universidad de Zaragoza. C/ Pedro Cerbuna, 12 50009 Zaragoza, Spain
{javier.lopez,gerardo.sanz}@unizar.es

Summary. We consider a general definition of record-like observations and present a methodology based on martingales to describe the asymptotic behavior of the corresponding counting process. Our definition includes not only the well-known records and near-records, from continuous or discrete distributions. It also contains, as particular cases of interest, weak records and geometric records. We provide concrete examples and discuss possible extensions and alternative approaches. We also propose some problems for future work.

1 Introduction

Our starting point is a remarkable fact, apparently discovered by Rényi [29], about the indicators of record observations. Suppose $X_n, n \geq 1$, is a sequence of independent and identically distributed (iid) random variables (rv), with common continuous distribution function (df) F. Define the record indicators as $I_1 = 1$ and $I_n = 1_{\{X_n > M_{n-1}\}}$, for $n \geq 2$, where $M_n = \max\{X_1, \ldots X_n\}$, $n \geq 1$. Then the I_n are independent random variables, with expectations $E[I_n] = 1/n, n \geq 1$. This simple structure of record indicators is perfect for obtaining asymptotic results, such as the law of large numbers (LLN) or the central limit theorem (CLT), for the counting process of records, defined as $N_n = \sum_{k=1}^{n} I_k, n \geq 1$. This process is of central importance in record theory, because of its relationship with the so-called record times. See [1, 20, 26] for the theory and statistical applications of records.

If the continuity of F is dropped both the independence of indicators and their distribution freeness is lost. The complexity of the dependence structure of record indicators, when F has (infinitely many) discontinuities,

* To Marisa. We thank the editors for giving us the opportunity to participate in this tribute to Marisa Manéndez.

has not been satisfactorily assessed. However, it turns out that martingale tools are successful when dealing with the LLN and the CLT for N_n, for discrete distributions F (purely atomic) concentrated on the nonnegative integers, see [8, 9].

On the other hand, a variety of definitions of record-like objects have been introduced, some of them motivated by potential applications in insurance or in stress-testing in material science ([4, 16, 22]). One significant example in this category are near-records, which were first defined in [4]. Given a parameter $a > 0$, an observation is considered to be a near-record if it is not a record but it is distance less than a of M_{n-1}, the current maximum. More precisely, the indicator of X_n being a near-record is given by $I_n = 1_{\{X_n \in (M_{n-1}-a, M_{n-1})\}}, n \geq 2$. One may hope that, when F is continuous, near-record indicators have a simple structure, such as that of record indicators, but this is not the case. Again, the dependence of the indicators is not well understood but martingale tools succeed in extracting a LLN or a CLT for the corresponding counting process N_n, in a wide range of cases ([14, 15]).

The above examples show that the value of N_n is obtained by sequentially testing, on each observation, a condition that only depends on the current maximum. More specifically, X_n is declared to be a record or a near-record if it belongs to a random interval whose randomness only depends on M_{n-1}. As we are interested in the analysis of (upper) records and (upper) record-like objects, from now on we assume that the df F of the observations is concentrated on $[0, \infty)$ and has infinite right-endpoint. We introduce a general definition for record-like observations as follows.

Definition 1. *Let $(a_n, b_n), n \geq 1$, be a sequence of random intervals of real numbers such that $a_n = a(M_n)$ and $b_n = b(M_n)$, with $a, b : [0, \infty) \to [-\infty, \infty]$ nondecreasing functions such that $a(x) < b(x)$, for all $x \geq 0$, and $\lim_{x\to\infty} a(x) = \lim_{x\to\infty} b(x) = \infty$. The observation X_n is defined to be record-like if $X_n \in (a_{n-1}, b_{n-1}), n \geq 1$, with $(a_0, b_0) = \mathbb{R}$ for convenience.*

This definition includes as particular cases, among others, usual records $a(x) = x$, $b(x) = \infty$; weak records for integer valued distributions $a(x) = x - 1/2$, $b(x) = \infty$; near-records $a(x) = x - a$, $b(x) = x$; δ-records, [10, 16], $a(x) = x + \delta$, $b(x) = \infty$; geometric records, [7, 17], $a(x) = kx$, $b(x) = \infty$, and ties for the maximum in integer-valued distributions, [13], $a(x) = x - 1/2$, $b(x) = x + 1/2$.

Our first objective in this paper is to present, in a unified way, a martingale analysis of the counting process of record-like observations, focusing on the LLN and CLT. We give two general results and then survey in some detail several particular cases. The analysis is based on the application of classical convergence theorems to two martingales, which are constructed by centering N_n using either a predictable or a nonpredictable process. Another objective of the paper is to show extensions and propose problems for future work.

The plan of the paper is the following. In Section 2 we introduce the two fundamental martingales and show how they can be used to obtain LLN and

CLT for the counting process of record-like observations. In Section 3 we apply the results of Section 2 to several record-like statistics such as records from discrete observations, near-records and geometric records. In Section 4 we present some extensions and open problems, including the use of our techniques to analyze the sum of record-like observations, the characterization of distributions using martingales related to record-like observations and a different approach to the CLT, for the number of record-like observations, based on point processes theory.

Parts of this paper (especially Section 3) can be seen as a survey of our results on the matter. There are also new results: in particular, Propositions 1, 2, 3 of Section 2 are new (although they have been proved in some particular cases) and Propositions 4, 5 of Section 4 are also new.

2 Martingales

Consider the natural filtration $\mathbb{F} = \{\mathcal{F}_n, n \geq 0\}$, related to the sequence $X_n, n \geq 1$, that is, $\mathcal{F}_n = \sigma(X_1, \ldots, X_n)$ is the σ-algebra generated by the first n observations, $n \geq 1$, and $\mathcal{F}_0$ is the trivial σ-algebra. Assuming that the X_n are iid, with common df F, and letting $I_n = \mathbf{1}_{\{X_n \in (a_{n-1}, b_{n-1})\}}$ be the indicators of record-like observations, as defined in Definition 1, it is clear that $E[I_n \mid \mathcal{F}_{n-1}] = \int_{(a_{n-1}, b_{n-1})} dF(t) = \phi(M_{n-1})$, for $n \geq 1$, where ϕ is given by $\phi(x) = \overline{F}(a(x)) - \overline{F}^{-}(b(x))$, with $F^{-}(x) = \lim_{t \to x^{-}} F(t)$ and $\overline{F}(x) = 1 - F(x)$.

An immediate consequence is the following.

Proposition 1. *The process*

$$N_n - \sum_{k=1}^{n} \phi(M_{k-1}),\ n \geq 1, \tag{1}$$

is a square-integrable $\mathbb{F}$-martingale.

The martingale in (1) can be used to derive a LLN for N_n. This is due to a well known law of large numbers, [21, 25], which states that, for any filtration $\mathbb{G} = \{\mathcal{G}_n, n \geq 0\}$ and any sequence of 0-1 variables $J_n, n \geq 1$, adapted to $\mathbb{G}$, the series $\sum_{k=1}^{\infty} J_k$ and $\sum_{k=1}^{\infty} E[J_k|\mathcal{G}_{k-1}]$ converge or diverge simultaneously almost surely (a.s.). And, when both diverge, the following holds, as $n \to \infty$:

$$\frac{\sum_{k=1}^{n} J_k}{\sum_{k=1}^{n} E[J_k|\mathcal{G}_{k-1}]} \to 1 \quad \text{a.s.}$$

This result implies for record-like observations, that the asymptotic behavior of N_n and that of $\sum_{k=1}^{n} \phi(M_{k-1})$ are tightly related.

Proposition 2. *Let N_n, ϕ and M_k be as defined above. Then $\lim_{n\to\infty} N_n = \infty$ if and only if $\sum_{k=1}^{\infty} \phi(M_{k-1}) = \infty$, a.s. Furthermore, under divergence,*

$$\frac{N_n}{\sum_{k=1}^{n} \phi(M_{k-1})} \to 1 \quad \text{a.s.} \tag{2}$$

As a consequence of (2), the LLN for N_n can now be reduced to the study of partial sums of $\phi(M_{k-1})$, but the task looks difficult because the $\phi(M_{k-1})$ are highly dependent random variables. However, in many cases of interest, as we will see later, the function ϕ is decreasing and hence $\phi(M_k) = \min\{\phi(X_1), \ldots, \phi(X_k)\}$. So, $\sum_{k=1}^{n} \phi(M_{k-1})$ is in fact a sum of partial minima of nonnegative, iid random variables $\phi(X_i)$. Such process was studied in detail by several authors but was not linked to record-like observations. In particular, the weak and strong LLN, as stated by Deheuvels in [6], are of central interest for us. For a version of Deheuvels' results, adapted to our applications, see Theorem A1 in [8] or Theorem 2 in [9].

The arguments above stress the role of martingale (1) in the derivation of the LLN for N_n. Martingale (1) is also well suited for other limit theorems, such as the CLT, because of its bounded increments and the simple expression of its conditional variances. However, in those cases where the CLT for N_n can be obtained from (1), the result is not satisfactory because it has a random centering process, which cannot be replaced by a deterministic sequence in a simple way. This motivates the search for a new martingale, with nonpredictable compensator, hoping that this feature would lead to a CLT with deterministic centering sequence. We have the following:

Proposition 3. *The process*

$$N_n - \left[\int_{[0,M_n]} \frac{dF(a(t))}{\overline{F}^{-}(t)} - \int_{[0,M_n)} \frac{dF(b(t))}{\overline{F}^{-}(t)} \right], \ n \geq 1, \tag{3}$$

is an $\mathbb{F}$-martingale.

Proof. Let $I_n^1 = \mathbf{1}_{\{X_n > a(M_{n-1})\}}$ and $I_n^2 = \mathbf{1}_{\{X_n \geq b(M_{n-1})\}}$. Then $N_n = \sum_{k=1}^{n} I_k^1 - \sum_{k=1}^{n} I_k^2$. Let us prove that

$$\sum_{k=1}^{n} I_k^1 - \int_{[0,M_n]} \frac{dF(a(t))}{\overline{F}^{-}(t)} \tag{4}$$

is a martingale. It can be easily checked that $\int_{[0,M_n]} \frac{dF(a(t))}{\overline{F}^{-}(t)}$ is integrable. For the martingale property notice that the increment of (4) is given by

$$I_k^1 - \int_{(M_{k-1},M_k]} \frac{dF(a(t))}{\overline{F}^{-}(t)}$$

and that $E[I_k^1 \mid \mathcal{F}_{k-1}] = P[X_k > a(M_{k-1})] = \overline{F}(a(M_{k-1}))$. On the other hand,

$$E\left[\int_{(M_{k-1},M_k]} \frac{dF(a(t))}{\overline{F}^-(t)} \mid \mathcal{F}_{k-1}\right] = \int_{(M_{k-1},\infty)} \left(\int_{(M_{k-1},x]} \frac{dF(a(t))}{\overline{F}^-(t)}\right) dF(x)$$

$$= \int_{(M_{k-1},\infty)} \left(\int_{[t,\infty)} dF(x)\right) \frac{dF(a(t))}{\overline{F}^-(t)} = \int_{(M_{k-1},\infty)} dF(a(t)) = \overline{F}(a(M_{k-1})).$$

The proof of $\sum_{k=1}^n I_k^2 - \int_{[0,M_n)} \frac{dF(b(t))}{\overline{F}^-(t)}$ being a martingale is analogous. □

Particular instances of (3) can be analyzed using one of the classical versions of the martingale CLT, from which one finally obtains a CLT for N_n, with non-random centering sequence. It is also remarkable that, in some examples, the study of the conditional variance process and the Lyapunov condition for the CLT can be reduced to the analysis of sums of minima. This reveals another aspect of the relationship between record-like observations and sums of partial minima.

To keep this paper self-contained, we provide the specific version of the martingale CLT, used in the applications below. See [21].

Theorem 1. *Let $\mathcal{G}_k, k \geq 0$, be a filtration and $\xi_k, k \geq 1$, a square-integrable and adapted sequence of random variables, such that $E[\xi_k \mid \mathcal{G}_{k-1}] = 0$, for $k \geq 1$. If there exists a non-random, positive and increasing sequence $b_n, n \geq 1$, with $b_n \to \infty$, such that*
(i) $b_n^{-2} \sum_{k=1}^n E[\xi_k^2 \mid \mathcal{G}_{k-1}] \to 1$ in probability and
(ii) $b_n^{-2} \sum_{k=1}^n E[\xi_k^2 \mathbf{1}_{\{|\xi_k| > \epsilon b_n\}} \mid \mathcal{G}_{k-1}] \to 0, \forall \epsilon > 0$, in probability .
Then $b_n^{-1} \sum_{k=1}^n \xi_k \to N(0,1)$.

If the ξ_i are cubic-integrable, condition (ii) can be replaced by the Lyapunov-type condition
(iii) $b_n^{-3} \sum_{k=1}^n E[\xi_k^3 \mid \mathcal{G}_{k-1}] \to 0$ in probability.

3 Applications

Here we present the LLN and the CLT for particular instances of record-like observations, based on martingales (1) and (3). This section is mainly a survey of results in [8, 9, 14, 15, 17]. Sketches of proofs are given in some cases.

3.1 Records from Discrete Distributions[1]

As mentioned in the Introduction, the discontinuities of the underlying df F induce dependencies among record indicators I_n and their structure is not easily described anymore. For simplicity we consider discrete df concentrated on $\mathbb{Z}_+$ (the set of nonnegative integers). Extension to discrete df with general sets of atoms (that do not accumulate) is straightforward. The theory

[1] Technical details and proofs of results in this subsection can be found in [8, 9].

of records from discrete distributions has attracted relatively little attention. Two important pioneering papers are [30], where the process of record values for general df is described, and [32], containing a variety of limiting results. Later, researchers from computer science considered records from the geometric distribution, because of their relationship with certain data structures ([3, 28]).

Recall that we assume $F(x) < 1$, for all $x \geq 0$. This is necessary to avoid trivialities because, if F has a terminal atom, the total number of records in the whole sequence is a.s. finite. It is interesting to observe that the normalizing sequence in the LLN for N_n depends on the hazard rates (also known as failure rates) $r_k, k \geq 0$, which are defined as $r_k = P[X = k]/P[X \geq k]$, where X is a generic rv with df F. Another important ingredient is the quantile function defined as $m(t) = \min\{k \in \mathbb{Z}_+; \overline{F}(k) < 1/t\}$.

The following theorem is quite general but has restrictions when F is light-tailed (see case (ii) below). It states essentially that N_n obeys a strong LLN with normalizing sequence $\eta_n = \sum_{k=0}^{m(n)} r_k, \quad n \geq 1$.

Theorem 2. *(i) If* $\lim r_k = 0$, *then* $N_n / \log n \to 1$ *a.s. and if* $\lim r_k = r$, *then*

$$\frac{N_n}{\log n} \to \frac{r}{-\log(1-r)} \quad a.s.$$

(ii) If $0 < \limsup r_k < 1$, *then* $N_n/\eta_n \to 1$ *a.s. and*
(iii) If $\lim r_k = 1$ *and if either the* $r_k, k \geq 0$, *is an increasing sequence or there exists a constant* $C > 0$ *such that* $\overline{F}(k) > e^{-e^{kC}}$, *for sufficiently large* k, *then* $N_n/m(n) \to 1$ *a.s.*

Proof. We use Proposition 2. In this case $\phi(x) = \overline{F}(x)$ is non-increasing. So, $\phi(M_k) = \min\{\phi(X_1), \ldots, \phi(X_k)\}$, with the random variables $Y_i := \phi(X_i) = \overline{F}(X_i)$ taking the values $\overline{F}(k)$ with respective probabilities $p_k := P[X = k], k \geq 0$. The asymptotic behavior of $\sum_{k=1}^{n} \min\{Y_1, \ldots, Y_k\}$ is analyzed using Deheuvels' theorem and the result follows from (2). See [8]. □

We show some examples of strong LLN. The Zeta or discrete Pareto distribution is defined by $p_k = C_d k^{-d}$, for $k \geq 1$, where $d > 1$ is a parameter and C_d a positive constant. In this case $r_k = (d-1)k^{-1} + o(k^{-1}) \to 0$ and so, $N_n/\log n \to 1$. The geometric distribution has $p_k = (1-p)^k p$, for $k \geq 0$, with $p \in (0,1)$. In this case $r_k = p$, for all $k \geq 0$, hence $N_n/\log n \to -p/\log(1-p)$. Finally, consider the Poisson distribution, with parameter $\lambda > 0$, where $p_k = e^{-\lambda}\lambda^k/k!$. It can be shown that $r_k = 1 - \lambda k^{-1} + o(k^{-1})$ and $m(n) \sim \log n/\log\log n$, hence $N_n \log\log n/\log n \to 1$.

The CLT is derived from martingale (3), which takes the form

$$N_n - \sum_{k=0}^{M_n} r_k \tag{5}$$

and is shown to be square integrable.

Theorem 3. *Let* $y_k = \overline{F}(k) = \sum_{i>k} p_i$, $z_k = \sum_{i>k} r_i y_i$ *and* $b_n^2 = \sum_{k=0}^{m(n)} z_k r_k / y_k$, $k, n \in \mathbb{Z}_+$.

(i) If $\sum_{k=0}^{\infty}(1 - r_k) = \infty$ *and* $\limsup r_k - \liminf r_k < 1$ *then*

$$b_n^{-1}(N_n - \eta_n) \to N(0,1).$$

(ii) If $\sum_{k=0}^{\infty}(1 - r_k) < \infty$ *then* $N_n - m(n)$ *is tight.*

Proof. Letting ξ_k denote the increments of (5), we find that $E[\xi_k^2 \mid \mathcal{F}_{k-1}] = \sum_{i>M_{k-1}} p_i(1 - r_i)$, which is a decreasing function of M_{k-1}. Then (i) of Theorem 1 is a LLN for sums of minima . The checking of condition (ii) in Theorem 1 is more involved. See [9]. □

An interesting first conclusion is that N_n is not asymptotically Normal if $\sum_{k=0}^{\infty}(1 - r_k) < \infty$; this contrasts with the continuous case where N_n has always an asymptotically Normal distribution.

The companion CLT for the LLN in the examples above are as follows. For the Zeta distribution, $(N_n - \log n)/\sqrt{\log n} \to N(0,1)$; for the geometric distribution, $(\log n)^{-1/2}(N_n + p \log n / \log(1-p)) \to N(0, -p(1-p)/\log(1-p))$ and, for the Poisson distribution, $(N_n - m(n) + \lambda \log(m(n)))/\sqrt{\log\log n} \to N(0, \lambda)$.

Theorems 2 and 3 show that the limiting behavior of N_n in discrete distributions may differ significantly from that in the continuous case, especially for light-tailed distributions. This is a warning for practitioners expecting to see order of $\log n$ records in a sample of size n, from a continuous distribution, because the instrumental precision is finite and so the observations are necessarily discrete. For instance, if $\overline{F}(x)$ decreases as e^{-x^2}, when $x \to \infty$, the discretized version of the distribution would have $N_n \sim \sqrt{\log n}$ records instead of $\log n$.

3.2 Near Records[2]

Near-records are another relevant example of record-like observation, that fits in the context of Definition 1 and, as the name suggests, they fall short of being records. Given a parameter $a > 0$, X_n is said to be near-record if $X_n \in (M_{n-1} - a, M_{n-1})$. In the notation of Definition 1, we choose $a(x) = x - a$ and $b(x) = x$. Near-records, which have interesting properties and applicability in statistics, were introduced by [4] and have been studied by several authors [14, 15, 23, 27]. In what follows we present the results of the asymptotic analysis of the counting process N_n of near-records, additionally assuming that F is absolutely continuous, with continuous density f and hazard function $\lambda(x) = f(x)/\overline{F}(x)$. Observe that ϕ of Proposition 2 is given by $\phi(x) = \overline{F}(x - a) - \overline{F}(x)$ but this function is not decreasing in general and so, we cannot invoke Deheuvels' theorem. This problem is fixed by assuming

[2] Technical details and proofs of results in this subsection can be found in [14, 15].

also that f is ultimately decreasing because this implies ϕ ultimately decreasing and this is sufficient for the asymptotic analysis. Since most important densities have this property, the assumption is not too restrictive.

Let $\eta_n = \int_a^{m(n)} (\overline{F}(x-a) - \overline{F}(x))\lambda(x)/\overline{F}(x)dx,\ n \geq 1$. For the strong LLN we have the following.

Theorem 4. *If* $\int_0^\infty \lambda^2(x)dx < \infty$ *then* $N_\infty < \infty$ *a.s. If* $\int_0^\infty \lambda^2(x)dx = \infty$ *and any of the following conditions hold:*
(i) $C_1 \leq \lambda(x) \leq C_2$, *for all* $x \geq 0$ *and some positive constants* C_1, C_2*;*
(ii) λ *is decreasing;*
(iii) λ *is differentiable, tends to* ∞ *as* $x \to \infty$ *and has* $|\lambda'(x)| < x^{-r}$, *for some* $r > 1/2$ *and all* x *large enough.*

Then $N_n/\eta_n \to 1$ *a.s.*

Proof. As for records, we apply Proposition 2 (see [14]). □

A first result derived from our analysis is that the number of near-record in the whole sequence may be finite. This result tells that heavy-tailed distributions, with square-integrable hazard function, will almost surely stop showing near-records in any iid sequence of observations. The reason for this behavior is that large values appear so easily in such distributions that near-records have only a dim chance of existing. It is also interesting to mention that, from the structure of the near-record-values process, one can derive the distribution of N_∞, which is geometric when $\overline{F}(x) = 1/x, x > 1$ (see [18]). Condition (i) or (iii) in Theorem 4 can be relaxed to λ or λ' bounded above but then we can only prove weak convergence. Let us see in detail the normalizing sequence η_n, for some distributions. For a heavy-tailed example, let F with $\lambda(x) = 1/\sqrt{x}$, Theorem 4(ii) can be applied and we have $\eta_n \sim 2a \log\log n$. For the exponential distribution with parameter μ we have $\lambda(x) = \mu$ and Theorem 4(i) applies, with $\eta_n = (e^{a\mu} - 1)\log n$. Finally, take the normal distribution $N(0,1)$, which has $\lambda(x) \sim x$, as $x \to \infty$. In this case Theorem 4 does not apply and only a weak LLN for N_n is obtained, with $\eta_n = a^{-1}e^{-a^2/2}e^{am(n)}m(n)$.

A CLT is obtained under hypotheses close to those of the weak LLN.

Theorem 5. *If* $\int_0^\infty \lambda^2(x)dx = \infty$ *and any of the following conditions hold:*
(i) $\lambda(x) \leq C$, *for all* $x \geq 0$ *and some positive constant* C*;*
(ii) λ *is differentiable,* $\lim_{x\to\infty} \lambda(x) = \infty$ *and* $\lim_{x\to\infty} \lambda'(x) = \alpha$, *for some* $\alpha \geq 0$.

Then there exist sequences a_n, b_n *such that* $(N_n - a_n)/b_n \to N(0,1)$.

Proof. The CLT of Theorem 1 is applied to the martingale (3), which simplifies to $N_n - \int_0^{M_n} \frac{f(x-a)-f(x)}{\overline{F}(x)}dx$. See [15]. □

We exhibit the normalizing sequences for the distributions considered after Theorem 4. If $\lambda(x) = 1/\sqrt{x}$, $a_n = 2a\log\log n$ and $b_n^2 = a_n$. For the exponential distribution, $a_n = (e^{a\mu} - 1)\log n$ and $b_n^2 = (2e^{a\mu} - 1)a_n$ and, for the normal, we have $a_n = a^{-1}e^{-a^2/2}e^{am(n)}(m(n) - a - a^{-1})$ and $b_n^2 = a^{-1}e^{-a^2}e^{2am(n)}m(n)$.

3.3 Geometric Records[3]

Suppose one is interested in record breaking observations that beat all previous observations by a factor $k > 1$. These objects are called geometric records and were introduced in [7]. Geometric records are also amenable to asymptotic analysis, taking $a(x) = kx$ and $b(x) = \infty$ in Definition 1 and, again, martingale tools allow to exhibit precise rates for the counting process N_n.

The definition of geometric record is quite stringent and we can expect to observe, in general, very few of them. In fact, using our martingale-based analysis, we can show that a necessary condition for $N_n \to \infty$ is that $\lambda(x) \to 0$, as $x \to \infty$. Results for the LLN are presented below. We previously define $r(x) = \overline{F}(kx)/\overline{F}(x)$, for $x > 0$, and $\eta_n = \int_0^{m(n)} \lambda(x)r(x)dx, n \geq 1$.

Theorem 6. *(i) If* $\inf_{x>0} r(x) > 0$ *then* $\eta_n = O(\log n)$ *and* $N_n/\eta_n \to 1$, *a.s. Moreover, if* $r(x) \to \rho$, *as* $x \to \infty$, *then* $N_n/\log n \to \rho$, *a.s.*
(ii) $\inf_{x>0} r(x) = 0$ *then, if* η_n *is bounded,* N_n *is also bounded, a.s. Otherwise, if* $\eta_n \to \infty$ *and* $r(x)$ *decreases to 0, as* $x \to \infty$, *then* $N_n/\eta_n \to 1$, *a.s.*

Proof. Follows from Proposition 2 since $\phi(x) = \overline{F}(kx)$ is decreasing. See [17].□

Let us consider, for example, $\overline{F}(x) = 1/x, x \geq 1$. Then, $\lambda(x) = 1/x, r(x) = 1/k$ and $\int_0^{\infty} \lambda(x)r(x)dx = \infty$. Hence, Theorem 6(i) can be applied to obtain $N_n/\log n \to k^{-1}$ *(a.s)*. For the exponential distribution we have, according to Theorem 6(ii), that N_n is bounded a.s. The same conclusion is obtained for all distributions with tails thinner than the exponential, such as the normal. Speeds other than $\log n$ are possible in the LLN. For example, taking $\overline{F}(x) = e^{-\log x \log\log x}$, for $x > e$, and $k = e$, we obtain $\eta_n = (\log\log n)^2/2$ but, if $k > e$, N_n is bounded a.s.

The CLT is derived from Theorem 1 applied to the general martingale (3), which can be simplified to $N_n - k\int_0^{M_n} \lambda(kx)r(x)dx$, (see [17]).

Theorem 7. *If* $\lim_{x\to\infty} r(x) = \rho \geq 0$ *and any of the following conditions hold:*
(i) $\rho > 0$;
(ii) $\rho = 0$, λ, r *are decreasing functions and* $\lim \eta_n = \infty$.

Then $(N_n - a_n)/b_n \to N(0,1)$, *with* $a_n = k\int_0^{m(n)} \lambda(kx)r(x)dx$, *and* $b_n^2 = (1 + 2\rho\log\rho)\eta_n, n \geq 1$ *(with* $0\log 0 = 0$*).*

4 Extensions

In this Section we analyze some extensions of our results and techniques and propose some problems for future research.

[3] Technical details and proofs of results in this subsection can be found in [17].

4.1 Sums of Record-Like-Values

In some applications it is of interest to consider sums of values of record-like observations instead of simply counting them. The martingale approach is also useful here to obtain limit theorems. Let φ be a positive payoff function and consider the total payoff process

$$S_n = \sum_{k=1}^{n} \varphi(X_k) \mathbf{1}_{\{X_k \in (a(M_{k-1}), b(M_{k-1}))\}}, \tag{6}$$

corresponding to record-like observations up to X_n. Note that $N_n = S_n$ if $\varphi(x) = 1$. An interesting particular case, first studied in [2], is the cumulative process of record-values, which is obtained by setting $a(x) = x$, $b(x) = +\infty$ and $\varphi(x) = x$. Notice that (6) differs from the process studied in [2] because the sum in (6) is up to the n-th observation while in [2], it is up to the n-th record time. Another interesting example is $a(x) = x - a$, $b(x) = x$, $\varphi(x) = x$, yielding the sum of near-record-values among the first n observations. With this choice of $a(x)$, $b(x)$ and $\varphi(x)$, S_n can be interpreted in actuarial sciences as the sum of claims at a distance less than a of being a record. See [4] for properties and applications of sums of near-record values.

The structure of (6) suggests that a martingale analysis, analogous to that of N_n, can be carried out. The task is to find the right martingales for the LLN or the CLT. We will develop only the CLT under the following assumptions: F has density f and $E[\varphi^2(X_1)] < \infty$. We also restrict the class of possible record-like observations (see Definition 1) by assuming $a(x)$ differentiable, with $a(x) \leq x, x \geq 0$, and $b = \infty$. We have the following

Proposition 4. *Let $g(t) = \varphi(a(t))f(a(t))a'(t), t \geq 0$. The process*

$$Z_n = S_n - \int_0^{M_n} \frac{g(t)}{\overline{F}(t)} dt, \; n \geq 1, \tag{7}$$

is an $\mathbb{F}$-martingale. Moreover, if

$$\lim_{x \to \infty} \overline{F}(x) \left(\int_0^x \frac{g(t)}{\overline{F}(t)} dt \right)^2 = 0 \quad \text{and} \quad \int_0^\infty \frac{g(x)}{(\overline{F}(x))^{1/2}} dx < \infty, \tag{8}$$

the martingale is square integrable and, denoting $\xi_k = Z_k - Z_{k-1}$,

$$E\left[\xi_k^2 \mid \mathcal{F}_{k-1}\right] = \int_{a(M_{k-1})}^\infty \varphi^2(t) f(t) dt + 2 \int_{M_{k-1}}^\infty \int_{a(t)}^t \varphi(x) f(x) dx \frac{g(t)}{\overline{F}(t)} dt, \tag{9}$$

Proof. Notice first that the integrability of $\varphi(X_1)$ implies the integrability of (7). We now check $E[\xi_k \mid \mathcal{F}_{k-1}] = 0, k \geq 1$. On the one hand

$$E\left[\varphi(X_k) \mathbf{1}_{\{X_k > a(M_{k-1})\}} \mid \mathcal{F}_{k-1}\right] = \int_{a(M_{k-1})}^\infty \varphi(t) f(t) dt.$$

Also, using the change of variable $u = a(t)$,

$$E\left[\int_{M_{k-1}}^{M_k} \frac{g(t)}{\overline{F}(t)}dt \;\middle|\; \mathcal{F}_{k-1}\right] = \int_{M_{k-1}}^{\infty} \left(\int_{M_{k-1}}^{x} \frac{g(t)}{\overline{F}(t)}dt\right) f(x)dx$$
$$= \int_{M_{k-1}}^{\infty} g(t)dt = \int_{a(M_{k-1})}^{\infty} \varphi(u)f(u)du.$$

The square integrability of the martingale follows from (8) and the integrability of $\varphi^2(X_1)$. To compute $E\left[\xi_k^2 \mid \mathcal{F}_{k-1}\right]$ note that

$$\xi_k^2 = \varphi^2(X_k)\mathbf{1}_{\{X_k > a(M_{k-1})\}} - 2\varphi(X_k)\mathbf{1}_{\{X_k > a(M_{k-1})\}} \int_{M_{k-1}}^{M_k} \frac{g(t)}{\overline{F}(t)}dt + \left(\int_{M_{k-1}}^{M_k} \frac{g(t)}{\overline{F}(t)}dt\right)^2.$$

Now $E\left[\varphi^2(X_k)\mathbf{1}_{\{X_k > a(M_{k-1})\}} \mid \mathcal{F}_{k-1}\right] = \int_{a(M_{k-1})}^{\infty} \varphi^2(t)f(t)dt$. As $a(t) \leq t$,

$$\varphi(X_k)\mathbf{1}_{\{X_k > a(M_{k-1})\}} \int_{M_{k-1}}^{M_k} \frac{g(t)}{\overline{F}(t)}dt = \varphi(X_k) \int_{M_{k-1}}^{M_k} \frac{g(t)}{\overline{F}(t)}dt,$$

$$E\left[\varphi(X_k) \int_{M_{k-1}}^{M_k} \frac{g(t)}{\overline{F}(t)}dt \;\middle|\; \mathcal{F}_{k-1}\right] = \int_{M_{k-1}}^{\infty} \varphi(x) \left(\int_{M_{k-1}}^{x} \frac{g(t)}{\overline{F}(t)}dt\right) f(x)dx$$
$$= \int_{M_{k-1}}^{\infty} \left(\int_{t}^{\infty} \varphi(x)f(x)dx\right) \frac{g(t)}{\overline{F}(t)}dt.$$

On the other hand, using integration by parts formula, we obtain

$$E\left[\left(\int_{M_{k-1}}^{M_k} \frac{g(t)}{\overline{F}(t)}dt\right)^2 \;\middle|\; \mathcal{F}_{k-1}\right] = \int_{M_{k-1}}^{\infty} \left(\int_{M_{k-1}}^{x} \frac{g(t)}{\overline{F}(t)}dt\right)^2 f(x)dx$$
$$= 2\int_{M_{k-1}}^{\infty} \left(\int_{M_{k-1}}^{x} \frac{g(t)}{\overline{F}(t)}dt\right) g(x)dx = 2\int_{M_{k-1}}^{\infty} \left(\int_{a(t)}^{\infty} \varphi(y)f(y)dy\right) \frac{g(t)}{\overline{F}(t)}dt,$$

and the result is proved. □

From the proposition above, we see that $E\left[\xi_k^2 \mid \mathcal{F}_{k-1}\right]$ is a decreasing function of M_{k-1}, and therefore, $\sum_{k=1}^{n} E\left[\xi_k^2 \mid \mathcal{F}_{k-1}\right]$ is a sum of minima of iid random variables Y_k, defined by

$$Y_k = \int_{a(X_k)}^{\infty} \varphi^2(t)f(t)dt + 2\int_{X_k}^{\infty} \left(\int_{a(t)}^{t} \varphi(x)f(x)dx\right) \frac{g(t)}{\overline{F}(t)}dt.$$

Note also that, when $a(t) = t$, expression (9) reduces to $\int_{M_{k-1}}^{\infty} \varphi^2(t)f(t)dt$.

Thus, results about sums of minima can be used to find the normalizing sequence of the CLT for S_n. Following the approach in [10], a Lyapunov condition must be verified; it is expectable that the conditional third moment of ξ_k can be written in terms of a sum of minima. Last, a change in the centering sequence from $\int_0^{M_n} g(t)/\overline{F}(t)dt$ to a deterministic sequence is required.

There remains the task of completing the program outlined above to obtain the asymptotic normality for S_n. It would also be interesting to find martingales similar to (7), for other choices of a and b and prove a CLT. In particular, for the sum of near-record-values, where $a(x) = x - a, b(x) = x, \varphi(x) = x$.

4.2 Characteristic Martingales

The rather complicated expression of martingale (3) makes it hard to believe that it sprang as result of intuitive considerations. In fact, (3) is a refined version of an initial martingale, discovered when looking for an alternative to (1). For the counting process of records N_n, we found that $N_n - cN_n$, where c is a positive constant, is an $\mathbb{F}$-martingale if F is either the exponential or the geometric distribution. In [11] we characterize all distributions F such that $N_n - cM_n$ is a martingale, showing that, in the continuous and the discrete case, the exponential and geometric distributions are, up to a simple transformation, the only distributions with that property.

We may consider an extension of the above result to other record-like statistics. For instance, let N_n be the number of observations greater than the previous maximum minus a constant $a > 0$; that is $a(t) = t - a, b(t) = +\infty$ (these observations are called δ-records in [10, 16], with $\delta = -a$, and include the case of weak records for integer-valued random variables, taking $a = 1/2$).

In the following proposition we define a process, related to $N_n - cM_n$, which turns out to be a martingale for the exponential and geometric distributions.

Proposition 5. *Let N_n be the number of record-like observations with $a(t) = t - a$, $b(t) = +\infty$. Then*

$$N_n - c(M_n \vee a),\ n \geq 1, \tag{10}$$

is a martingale if and only if

$$\overline{F}(t-a) = c \int_t^{\infty} \overline{F}(u)du, \qquad \text{for all } t \geq a \tag{11}$$

in the support of F and $c = 1/\int_a^{\infty} \overline{F}(u)du$. In particular, (10) is a martingale when F is the exponential distribution or the geometric distribution (in the latter case a is supposed to be a positive integer number).

Proof. The proof follows the lines of Lemma 2.1 in [11]. The increment of (10) is given by $\mathbf{1}_{\{X_n > M_{n-1} - a\}} - c((M_n \vee a) - (M_{n-1} \vee a))$. Clearly $E\left[\mathbf{1}_{\{X_n > M_{n-1} - a\}} \mid \mathcal{F}_{n-1}\right] = \overline{F}(M_{n-1} - a)$. On the other hand, the value of

$E[(M_n \vee a) - (M_{n-1} \vee a) \mid \mathcal{F}_{n-1}]$ depends on whether $M_{n-1} < a$ or $M_{n-1} \geq a$. In the first case

$$\begin{aligned} E[(M_n \vee a) - (M_{n-1} \vee a) \mid \mathcal{F}_{n-1}] &= E[(X_n \vee a) - a \mid \mathcal{F}_{n-1}] \\ &= \int_0^\infty P[X_n > t + a]dt = \int_a^\infty \overline{F}(u)du. \end{aligned}$$

In the second case, we have

$$E[(M_n \vee a) - (M_{n-1} \vee a) \mid \mathcal{F}_{n-1}] = \int_0^\infty P[X_n - M_{n-1} > t]dt = \int_{M_{n-1}}^\infty \overline{F}(u)du.$$

Since $\overline{F}(t-a) = 1$ for $t < a$, the first part of the proposition is proved.

For the exponential distribution, it is immediate that $\overline{F}(x) = e^{-\lambda x}$ satisfies equation (11). In the case of the geometric distribution (starting at 1), with survival function $\overline{F}(x) = (1-p)^{\lfloor x \rfloor}$, we have $\int_a^\infty \overline{F}(x)dx = (1-p)^a/p$, so $c = p/(1-p)^a$; then, letting $t = a + i$, with $i = 0, 1, 2, \ldots$,

$$c\int_{a+i}^\infty \overline{F}(x)dx = c\sum_{k=a+i}^\infty (1-p)^k = c(1-p)^{a+i}/p = (1-p)^i = \overline{F}(i). \quad \square$$

The natural continuation is to find all distributions such that (10) is a martingale, with a and b defined in Proposition 5. It would also be interesting to study the characteristic martingale problem for other choices of a and b.

4.3 Limits without Martingales

We have shown how useful martingales are in obtaining asymptotic results for the counting process of record-like observations. One may wonder whether it is possible to analyze $N_n = \sum_{k=1}^n \mathbf{1}_{\{X_k \in (a(M_{k-1}), b(M_{k-1}))\}}$ directly, as a sum of Bernoulli random variables, applying results on laws of large numbers and central limit theorem for dependent 0-1 variables, thus circumventing the use of martingales and sums of minima. This approach seems to be difficult, since the dependence of the indicators is not well understood, except for the particular case of records from continuous distributions. However, in some cases, another approach can be used. It consists of studying the sequence of record-like values, finding the asymptotic behavior of this sequence and obtaining the results for N_n via a random change of time. We now show how to apply this procedure to prove a CLT for the number of records in a sequence of iid random variables with general distribution F.

Let ξ be the point process of record values on $[0, \infty)$; that is, $\xi(x)$ is the cardinal of the set $\{t \in [0, x] : t \text{ is a record value of the sequence } X_n, n \geq 1\}$. Assuming that the atoms of F do not accumulate, ξ is a point process on $[0, \infty)$, with independent increments, which admits the representation $\xi = \xi_c + \xi_d$, where ξ_c and ξ_d are the independent continuous and discrete components of ξ. The continuous component is non-homogeneous Poisson and the discrete component is a Bernoulli process. It can be proved, under some conditions on the tail of F, that $\xi(x_n)$ satisfies a CLT, where $x_n, n \geq 1$, is any sequence of real numbers increasing to ∞. Then we use the identity $N_n = \xi(M_n)$ to translate this result into a CLT for N_n. We call the replacement of x_n by M_n a random change of time and, of course, it is not clear if the new process will obey the CLT. A result of [24] can be applied to justify the replacement and obtain the CLT for $\xi(M_n)$. In [19], we characterize the distributions F such that N_n is asymptotically normal. This result improves our previous CLT for discrete distributions, in [9], and solves a conjecture posed in [3].

It would be interesting to use this approach in the study of other record-like statistics. However, in doing this, several difficulties arise. The first is that we need a good description of the point process ξ of the record-like values under consideration. This is known in some examples other than records. For instance, the case of weak records, which satisfy the condition $X_n \geq M_{n-1}$, was described in [31]. In this situation, ξ is an independent increment process, with $\xi(x) = \sum_{j\geq 1} Z_j \mathbf{1}_{\{a_j \leq x\}}$, where the Z_j are independent with geometric distribution of parameter $1 - r_j$. Another situation where ξ is well described is the case of near-records, as defined in Subsection 3.2. In this case it can be shown that ξ is a cluster Poisson process where the center process is non-homogeneous Poisson and each center x casts a geometric number of points in $(x - a, x)$, whose locations have density $f(x)/(\overline{F}(x - a) - \overline{F}(x))$ (see [18]). Finally, geometric records, considered in Subsection 3.3, can be translated, via a logarithmic transformation of the observations, into δ-records, which are record-like observations satisfying $X_n > M_{n-1} + a$. In this situation ξ is a type II counter process, with a nonhomogeneous Poisson input process (See [5, 17]). Once the corresponding process ξ has been successfully described, a CLT for $\xi(x_n)$ must be found. This is not very difficult if ξ has independent increments (as in the case of weak records) but it is likely to be harder if the increments of ξ are dependent, as is the case of near-records or geometric records. Another problem with this approach is that the identity relating N_n, ξ and M_n for records ($N_n = \xi(M_n)$) does not hold for other record-like statistics. For instance $N_n \neq \xi(M_n)$ for weak records, since $\xi(M_n)$ may count weak records which occur after time n. It is expectable, however, that the difference $\xi(M_n) - N_n$ is negligible when compared with N_n. The formula does not hold for near records either, although again the differences may be small. Last, the result of [24], used in [19], requires ξ to be written as a sum of independent random variables. This representation is available for records and weak records, but not for other record-like statistics, such as near-records or geometric records. A more general CLT with random change of time is needed in those cases.

Overall, we believe that this procedure, without martingales, is adequate for records or weak records, but may be difficult to apply for other cases of record-like statistics.

Acknowledgement. This research was supported by project MTM2007-63769 and MTM2010-15972 of MICINN (Spain), PFB-03-CMM project and FONDECYT Grant 1090216 (Chile). The authors are members of the research group Modelos Estocásticos (DGA).

References

1. Arnold, B.C., Balakrishnan, N., Nagaraja, H.N.: Records. Wiley, New York (1998)
2. Arnold, B.C., Villaseñor, J.A.: The asymptotic distribution of sums of records. Extremes 1, 351–363 (1998)
3. Bai, Z., Hwang, H., Liang, W.: Normal approximation of the number of records in geometrically distributed random variables. Random Struct. Alg. 13, 319–334 (1998)
4. Balakrishnan, N., Pakes, A.G., Stepanov, A.: On the number and sum of near-record observations. Adv. Appl. Probab. 37, 765–780 (2005)
5. Cox, D.R., Isham, V.: Point processes. Chapman & Hall, London (1980)
6. Deheuvels, P.: Valeurs extrémales d'échantillons croissants d'une variable aléatoire réelle. Ann. Inst. H Poincaré X, 89–114 (1974)
7. Eliazar, I.: On geometric record times. Phys. A 348, 181–198 (2005)
8. Gouet, R., López, F.J., San Miguel, M.: A martingale approach to strong convergence of the number of records. Adv. Appl. Probab. 33, 864–873 (2001)
9. Gouet, R., López, F.J., Sanz, G.: Central limit theorems for the number of records in discrete models. Adv. Appl. Probab. 37, 781–800 (2005)
10. Gouet, R., López, F.J., Sanz, G.: Asymptotic normality for the counting process of weak records and δ-records in discrete models. Bernoulli. 13, 754–781 (2007)
11. Gouet, R., López, F.J., Sanz, G.: A characteristic martingale related to the counting process of records. J. Theor. Probab. 20, 443–455 (2007)
12. Gouet, R., López, F.J., Sanz, G.: Laws of large numbers for the number of weak records. Stat. Probabil. Lett. 78, 2010–2017 (2008)
13. Gouet, R., López, F.J., Sanz, G.: Limit laws for the cumulative number of ties for the maximum in a random sequence. J. Stat. Plan. Infer. 139, 2988–3000 (2009)
14. Gouet, R., López, F.J., Sanz, G.: Limit theorems for the counting process of near-records. To appear in J. Stat. Plan. Infer (2010)
15. Gouet, R., López, F.J., Sanz, G.: Central limit theorem for the number of near-records. To appear in Comm. Statist.-Theor. Meth (2010)
16. Gouet, R., López, F.J., Sanz, G.: On δ-record observations: asymptotic rates for the counting process and elements of maximum likelihood estimation (2010) (submitted)
17. Gouet, R., López, F.J., Sanz, G.: On the number of geometric records (2010) (preprint)
18. Gouet, R., López, F.J., Sanz, G.: On the structure of near-record values (2010) (preprint)
19. Gouet, R., López, F.J., Sanz, G.: Asymptotic normality for the number of records from general distributions. To appear in Advances in Applied Probability (2011)

20. Gulati, S., Padgett, W.J.: Parametric and Nonparametric Inference from Record-Breaking Data. Lect. Notes in Statist., vol. 172. Springer, New York (2003)
21. Hall, P., Heyde, C.C.: Martingale Limit Theory and its Applications. Academic Press, New York (1980)
22. Hashorva, E.: On the number of near-maximum insurance claim under dependence. Insur. Math. Econ. 32, 37–49 (2003)
23. Hashorva, E., Hüsler, J.: Estimation of tails and related quantities using the number of near-extremes. Comm. Statist.-Theor. Meth. 34, 337–349 (2005)
24. Kubacki, K.S., Szynal, D.: On the rate of convergence in a random central limit theorem. Probab. Math. Statist. 9, 95–103 (1988)
25. Neveu, J.: Martingales à Temps. Discret. Masson, Paris (1972)
26. Nevzorov, V.B.: Records: Mathematical theory. Translations of Mathematical Monographs, vol. 194. American Mathematical Society, Providence (2001)
27. Pakes, A.G.: Limit theorems for numbers of near-records. Extremes 10, 207–224 (2007)
28. Prodinger, H.: Combinatorics of geometrically distributed random variables: left-to-right maxima. Discrete Math. 153, 253–270 (1996)
29. Rényi, A.: Théorie des éléments saillants d'une suite d'observations. Ann. Fac. Sci. Univ. Clermont-Ferrand 8, 7–13 (1962)
30. Shorrock, R.W.: On record values and record times. J. Appl. Probab. 9, 316–326 (1972)
31. Stepanov, A.V.: Limit theorems for weak records. Theor. Prob. Appl. 37, 570–574 (1992)
32. Vervaat, W.: Limit theorems for records from discrete distributions. Stoch. Proc. Appl. 1, 317–334 (1973)

p-Symmetric Measures: Definition, Properties and Perspectives*

Pedro Miranda[1] and Susana Martínez[2]

[1] Complutense University of Madrid, Spain
pmiranda@mat.ucm.es
[2] I.E.S.O. Arturo Plaza, Spain
susanamartinezsuarez@hotmail.com

Summary. In this paper we give a review of the main properties of p-symmetric measures. This subfamily of fuzzy measures has an appealing representation and can be defined with a reduced number of coefficients. Moreover, the corresponding Choquet integral is related to internal-external coverings and its expression has an intuitive meaning. Next, the polytope of p-symmetric measures is an example of order polytope; from this property, it is possible to derive many results applying the structure of the polytope of p-symmetric measures. We finish with the conclusions and highlighting several situations in which p-symmetric measures could be an appealing choice.

Keywords: fuzzy measures, p-symmetry, order polytopes, Choquet integral.

1 Motivation and Basic Concepts

Consider a referential set of n elements $X = \{x_1, \ldots, x_n\}$. Subsets of X are denoted A, B and so on, and also $A_1, A_2, \ldots$ The set of subsets of X is denoted by $\mathcal{P}(X)$.

A **fuzzy measure** [26] (or **capacity** [4] or **non-additive measure** [7]) over X, is a set function $\mu : \mathcal{P}(X) \to [0, 1]$ satisfying

- $\mu(\emptyset) = 0, \mu(X) = 1,$
- $\mu(A) \leq \mu(B)$ for all $A, B \in \mathcal{P}(X)$ such that $A \subseteq B$.

As it can be seen from the above definition, fuzzy measures are a generalization of probability distributions on X, where we have removed additivity and we have imposed monotonicity instead. Note on the other hand that $n-1$ values suffice to define a probability measure, while $2^n - 2$ coefficients are

* This paper has been done to pay homage to Prof. Marisa Menéndez. As soon as we met her, she offered us her support, her valuable advices and her friendship. These are the kind of things that make the difference between a friend and a colleague. Dear friend, we will never forget all you have done for us; we miss you very much.

needed for fuzzy measures. As it will become apparent below, this exponential complexity is the *Achilles heel* of the Theory of Fuzzy Measures.

Next step in the Theory of Fuzzy Measures is to define an extension of the concept of expected value that can be applied to any fuzzy measure. This is done through the so-called Choquet integral [4].

Consider a non-negative function $f : X \to \mathbb{R}^+$. The **Choquet integral** for finite referentials is defined by

$$\mathcal{C}_\mu(f) := \sum_{i=1}^{n} (f(x_{(i)}) - f(x_{(i-1)}))\mu(B_i),$$

where parenthesis mean a permutation such that $0 = f(x_{(0)}) \leq f(x_{(1)}) \leq \ldots \leq f(x_{(n)})$ and $B_i = \{x_{(i)}, \ldots, x_{(n)}\}$. Another equivalent expression is

$$\mathcal{C}_\mu(f) := \sum_{i=1}^{n} f(x_{(i)})(\mu(B_i) - \mu(B_{i+1})) \tag{1}$$

with $B_{n+1} = \emptyset$.

For general functions, not necessarily non-negative, there are two possible extensions of Choquet integral [7]: the *asymmetric Choquet integral*, usually known as Choquet integral, and the *symmetric Choquet integral*, also known as *Šipoš integral.*

In general, Choquet integral is complex to compute, this complexity inherited from the complexity of fuzzy measures. However, there are two special cases in which Choquet integral can be computed very quickly. One of them is the case of additive measures (i.e. probabilities); in this case, Choquet integral reduces to a weighed mean (indeed a expected value). The other one appears when μ is *symmetric*, i.e. the values of μ do not depend on the subset but on its cardinality; from a mathematical point of view, a fuzzy measure μ is symmetric if $\mu(A) = \mu(B)$ if $|A| = |B|$. When μ is symmetric, it can be seen [18] that Choquet integral coincides with an OWA operator [27] and Eq. (1) reduces to

$$\mathcal{C}_\mu(f) = \sum_{i=1}^{n} (\mu_{n-i} - \mu_{n-i-1}) f(x_{(i)}), \tag{2}$$

where μ_i denotes the value of the fuzzy measure for any subset whose cardinality is i.

Fuzzy measures, together with Choquet integral, have been applied in many different situations in which probabilities are too restrictive. For example, fuzzy measures have become a powerful tool in Decision Theory (see e.g. [12], [23] and [2]), where the Choquet Expected Utility model generalizes the Expected Utility one; this model offers a simple theoretical foundation for explaining phenomena that cannot be accounted for in the framework of Expected Utility Theory, as the well-known Ellsberg's and Allais' paradoxes (see [2] for a survey about this topic). Other fields related to non-additive measures are Combinatorics, Pseudo-Boolean functions, Welfare Theory and many others (see [11] for a review of theoretical and practical applications of fuzzy measures). Moreover, they are included in the field of Aggregation Operators, that constitutes a major research topic nowadays [10].

As we have seen, despite of the many advantages of fuzzy measures, their practical use has to face with the hurdle of an increment in the complexity. In order to reduce this complexity, several subfamilies have been defined. The basic idea is to include some additional constraints in the definition, so that the complexity reduces while the modeling capability of the subfamily is kept as rich as possible. Among all these subfamilies, surely the most famous is the subfamily of k-additive measures, introduced by Grabisch in [9] and that is defined below. Previously, we introduce the Möbius transform of a fuzzy measure [22].

Let μ be a fuzzy measure on X. The **Möbius transform (or inverse)**[1] of μ is a set function on X defined by

$$m(A) := \sum_{B\subseteq A} (-1)^{|A\setminus B|}\mu(B), \ \forall A \subseteq X.$$

The Möbius transform given, the original fuzzy measure can be recovered via the *Zeta transform* [3]:

$$\mu(A) = \sum_{B\subseteq A} m(B).$$

Therefore, the Möbius transform is an alternative representation of fuzzy measures. The value $m(A)$ represents the strength of the subset A in any coalition in which it appears. It is also known as *dividends* in Game Theory [13]. Choquet integral in terms of m is given by (see [3])

$$\mathcal{C}_\mu(f) = \sum_{T\subset X} m(T)\left[\bigwedge_{x_i\in T} f(x_i)\right], \quad f \in [0,1]^n. \tag{3}$$

For additive measures, it can be seen that the Möbius transform vanishes for subsets that are not singletons. This inspires the concept of k-additive measures.

A fuzzy measure μ is said to be **k-additive** if its Möbius transform vanishes for any $A \subseteq X$ such that $|A| > k$ and there exists at least one subset A with exactly k elements such that $m(A) \neq 0$.

In this sense, a probability measure is just a 1-additive measure. Thus, k-additive measures generalize probability measures and they fill the gap between probability measures and general fuzzy measures. As the Möbius transform is an alternative representation of a fuzzy measure, for a k-additive measure the number of coefficients is reduced to

---

[1] The Möbius transform also applies if μ is just a set function satisfying $\mu(\emptyset) = 0$. This is the case in Game Theory.

$$\sum_{i=1}^{k} \binom{n}{i} - 1,$$

as one coefficient is completely determined by

$$1 = \mu(X) = \sum_{B \subseteq X} m(B).$$

Similarly, the complexity of a Choquet integral with respect to k-additive measure reduces a great deal, as it can be seen from Eq. (3).

The goal of p-symmetric measures is to provide a concept, similar to k-additive measures, bridging the gap between symmetric fuzzy measures and general fuzzy measures. The rest of the paper is organized as follows: In next section we introduce p-symmetric measures; Section 3 shows some properties of p-symmetric measures and their corresponding Choquet integral. Section 4 studies the polytope of p-symmetric measures. Section 5 outlines some situations where p-symmetry could be an interesting tool.

2 *p*-Symmetric Measures

Let us consider an OWA operator (Eq. 2). If we look at the definition, we can see that only the order in the scores is important, i.e. we are interested in the scores, but we do not care about which criterium each score has been obtained. Mathematically, this means that the fuzzy measure defining the OWA operator only depends on the cardinality of the subsets, and not in the elements of the subset themselves.

Thus, all criteria have the same importance or, in other words, we have a "subset of indifference" (X itself). Then, it makes sense to define 2-symmetric measures as those measures for which we have two subsets of indifference, 3-symmetric measures as those with three subsets of indifference, and so on. Let us now translate this idea.

Definition 1. *[17] Given two elements x_i, x_j of the universal set X, we say that x_i and x_j are* **indifferent elements** *if and only if*

$$\forall A \subseteq X \backslash \{x_i, x_j\},\ \mu(A \cup x_i) = \mu(A \cup x_j).$$

This definition translates the idea that we do not care about which element, x_i or x_j is in the coalition; that is, we are indifferent between x_i and x_j. This concept can be generalized for subsets of more than two elements, as shown in the following definition:

Definition 2. *[17] Given a subset A of X, we say that A is a* **subset of indifference** *if and only if*

$$\forall B_1, B_2 \subseteq A,\ |B_1| = |B_2|,\ \forall C \subseteq X \backslash A,\ \mu(B_1 \cup C) = \mu(B_2 \cup C).$$

Definition 3. *[17] Given a fuzzy measure μ, we say that μ is a* p**-symmetric measure** *if and only if the coarsest partition of the universal set in subsets of indifference is $\{A_1, \ldots, A_p\}, A_i \neq \emptyset, \forall i \in \{1, \ldots, p\}$.*

The existence and unicity of this partition has been proved in [16]. We will denote by $\mathcal{FM}(A_1, \ldots, A_p)$ the set of fuzzy measures for which A_i, $i = 1, \ldots, p$, is a subset of indifference (but not necessarily being p-symmetric! Indeed, any symmetric measure belongs to $\mathcal{FM}(A_1, \ldots, A_p)$).

3 Basic Properties

In this section we study some basic properties of p-symmetric measures. Detailed proofs can be found in [17].

Let us start with the representation of p-symmetric measures. As all elements in a subset of indifference have the same behavior, when dealing with a fuzzy measure in $\mathcal{FM}(A_1, \ldots, A_p)$, we only need to know the number of elements of each A_i that belong to a given subset C of the universal set X. Therefore, the following result holds:

Lemma 1. *If $\{A_1, \ldots, A_p\}$ is a partition of X, then in order to define a fuzzy measure in $\mathcal{FM}(A_1, \ldots, A_p)$, any $C \subseteq X$ can be identified with a p-dimensional vector $(c_1, \ldots, c_p)$ with $c_i := |C \cap A_i|$.*

In other words, we can write $\mu(C) \equiv \mu(|A_1 \cap C|, \ldots, |A_p \cap C|)$, and the same applies for the Möbius transform. As a consequence:

Proposition 1. *Let μ be a p-symmetric measure with respect to the partition $\{A_1, \ldots, A_p\}$. Then, the number of values that are needed in order to determine μ is*

$$\left[(|A_1| + 1) \times \cdots \times (|A_p| + 1)\right] - 2.$$

For example, if we are dealing with the 2-symmetric measure with subsets of indifference A_1, A_2, we obtain that the number of coefficients needed is $(|A_1| + 1) \times (|A_2| + 1)$. These coefficients can be represented in a $(|A_1| + 1) \times (|A_2| + 1)$ matrix:

$$\begin{pmatrix} \mu(0,0) & \mu(0,1) & \ldots & \mu(0,|A_2|-1) & \mu(0,|A_2|) \\ \mu(1,0) & \mu(1,1) & \ldots & \mu(1,|A_2|-1) & \mu(1,|A_2|) \\ \vdots & \vdots & \ddots & \vdots & \vdots \\ \mu(|A_1|-1,0) & \mu(|A_1|-1,1) & \ldots & \mu(|A_1|-1,|A_2|-1) & \mu(|A_1|-1,|A_2|) \\ \mu(|A_1|,0) & \mu(|A_1|,1) & \ldots & \mu(|A_1|-1,|A_2|-1) & \mu(|A_1|,|A_2|) \end{pmatrix}$$

Let us now turn to Choquet integral. As an application of Eq. (3) and Proposition 1, the following holds:

Proposition 2. *Let μ be a p-symmetric measure with respect to the partition $\{A_1, \ldots, A_p\}$. Given a function $f : X \to \mathbb{R}^+$, the Choquet integral is given by*

$$\sum_{i=1}^{n} f(x_{(i)}) \sum_{c_k \leq b_k^{i-1}, \forall k} m(c_1, \ldots, c_j + 1, \ldots, c_p) \prod_{k=1}^{p} \binom{b_k^{i-1}}{c_k}$$

with $x_{(i)} \in A_j$, and where $(b_1^{i-1}, \ldots, b_p^{i-1}) \equiv B_{(i-1)} = \{x_{(1)}, \ldots, x_{(i-1)}\}$.

As a subset $C \equiv (|C \cap A_1|, \ldots, |C \cap A_p|)$, we can find all possible families of coefficients for Choquet integral finding all possible paths from $(0, \ldots, 0)$ to $(|A_1|, \ldots, |A_p|)$. For example, if we are dealing with the 2-symmetric case, this result can be depicted in Figure 1.

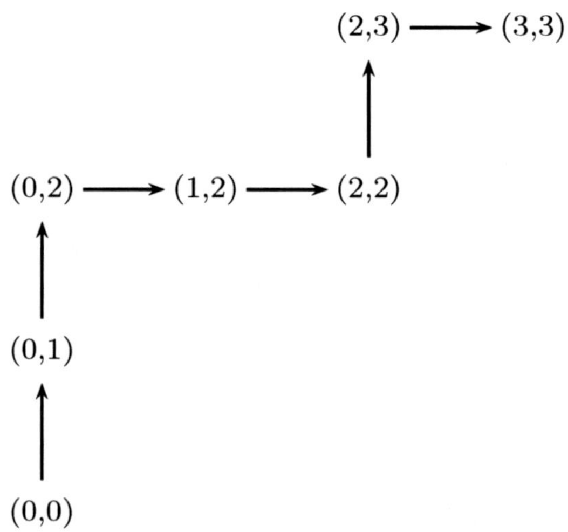

Fig. 1. Possible path from (0,0) to (3,3) when $|A_1| = 3$ and $|A_2| = 3$

Moreover, the Choquet integral can be decomposed in a sum of p Choquet integrals plus a correction term.

Proposition 3. *Let μ be a p-symmetric measure with respect to the partition $\{A_1, \ldots, A_p\}$, and suppose $\mu(A_i) > 0, \forall i$. Then, the Choquet integral is given by*

$$\sum_{i=1}^{p} \mu(A_i) \mathcal{C}_{\mu_i}(f) + \sum_{B \not\subseteq A_j, \forall j} m(B) \bigwedge_{x_i \in B} f(x_i),$$

where μ_i is defined by its Möbius transform

$$m_i(C) = \begin{cases} \frac{m(C)}{\mu(A_i)} & \text{if} C \subseteq A_i \\ 0 & \text{otherwise} \end{cases}$$

The last summand in this proposition represents the part of the Choquet integral that cannot be assigned to any subset in the partition. Note that the integrals in the first part are just OWA's, as μ_i is a symmetric measure restricted to A_i, so that they can be computed very quickly. This result is related to the internal-external coverings of Murofushi *et al.* [19], [20].

4 p-Symmetric Measures as Order Polytopes

It can be easily seen that $\mathcal{FM}(A_1, \ldots, A_p)$ is a convex polytope for a fixed partition $\{A_1, \ldots, A_p\}$. In this section we show that $\mathcal{FM}(A_1, \ldots, A_p)$ belongs to a special class of polytopes, the so-called order polytopes, and derive some properties from this result.

Let us recall the basic notions about order polytopes. Consider a finite poset $(P, \preceq)$ (or P for short) of p elements. We will denote the subsets of P by capital letters $A, B, \ldots$ and also $A_1, A_2, \ldots$; elements of P are denoted a, b, and so on. If A is a subset of P, it inherits a structure of poset from the restriction of $\preceq$ to A. In this case, we say that A is a **subposet** of P. If two elements a, b of P satisfy $a \preceq b$ or $b \preceq a$, we say that they are *comparable.* A subposet $(A, \preceq)$ is a **chain** if for any $a, b \in A$, either $a \preceq b$ or $b \preceq a$. A subposet $(A, \preceq)$ is an **antichain** if for any $a, b \in A$, neither $a \preceq b$ nor $b \preceq a$.

Given a poset $(P, \preceq)$, we define the **dual** poset $(\overline{P}, \preceq')$ as another poset with the same underlying set and satisfying

$$a \preceq b \text{ in } P \Leftrightarrow b \preceq' a \text{ in } \overline{P}.$$

If $(P, \preceq)$ is isomorphic to $(\overline{P}, \preceq')$, we say that P is **autodual**.

A subset I of P is an **ideal** or **downset** if for any $a \in I$ and any $b \in P$ such that $b \preceq a$, it follows that $b \in I$. Notice that with this definition the empty set is an ideal. The dual notion of an ideal is a **filter** or **upset**, i.e., a set that contains all upper bounds of its elements. Remark that for a given antichain in P, we can build an ideal whose minimal elements are the elements in the antichain. Reciprocally, we can identify any ideal with the antichain of its minimal elements. Thus, the number of ideals and the number of antichains is the same.

Let us now turn to order polytopes. Given a poset $(P, \preceq)$, it is possible to associate to P, in a natural way, a polytope $O(P)$ in $\mathbb{R}^p$, called the **order polytope** of P (cf. [25]). The polytope $O(P)$ is formed by the p-uples f of real numbers indexed by the elements of P satisfying

- $0 \leq f(a) \leq 1$ for every a in P,
- $f(a) \leq f(b)$ whenever $a \preceq b$ in P.

Thus, the polytope $O(P)$ consists in (the p-uples of images of) the order-preserving functions from P to $[0, 1]$. It is a well-known fact [25] that $O(P)$ is a 0/1-polytope, i.e. its extreme points are all in $\{0, 1\}^p$. In fact, it is easy to see that the extreme points of $O(P)$ are exactly (the characteristic functions of) the filters of P.

The set $\mathcal{FM}(A_1, \ldots, A_p)$ can be seen as the order polytope of the poset $(P(A_1, \ldots, A_p), \preceq)$, where

$$P(A_1, \ldots, A_p)$$
$$:= \{(i_1, \ldots, i_p) : i_j \in \{0, \ldots, |A_j|\}, i, j \in \mathbb{Z}\} \backslash \{(0, \ldots, 0), (|A_1|, \ldots, |A_p|)\},$$

and $\preceq$ is given by $(c_1, \ldots, c_p) \preceq (b_1, \ldots, b_p) \Leftrightarrow c_i \leq b_i, i = 1, \ldots, p$.

Next Figure shows the poset for $\mathcal{FM}(A_1, A_2)$ when $|A_1| = 2, |A_2| = 1$.

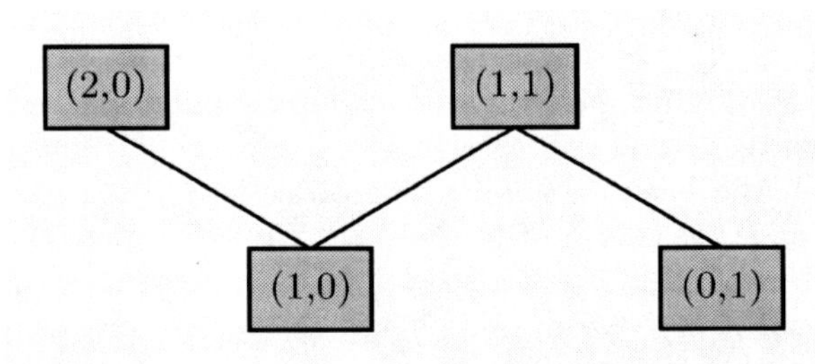

Fig. 2. Hasse diagram of the poset for $\mathcal{FM}(A_1, A_2)$ when $|A_1| = 2, |A_2| = 1$

The importance of this result will become apparent below, where we show some properties of $\mathcal{FM}(A_1, \ldots, A_p)$ as an order polytope; all of them are given in terms of the subjacent poset, thus simplifying a great deal the corresponding proofs.

As a first consequence, the vertices of $\mathcal{FM}(A_1, \ldots, A_p)$ are included in the set of $\{0, 1\}$-valued measures, thus recovering the results obtained in [15]. We can also give the values of the number of vertices of $\mathcal{FM}(A_1, \ldots, A_p)$ for small values of p (the general case is a long standing open problem in Combinatorics [1]). We only have to note that these numbers are, in fact, two less (we have to exclude the antichains $\{(|A_1|, \ldots, |A_p|)\}$ and $\{(0, \ldots, 0)\}$) than the numbers of antichains in the product of chains of sizes $|A_1| + 1, \ldots, |A_p| + 1$. Then, from the results in [1, 24] we can deduce the following result.

Theorem 1

- *The number of vertices of 1-symmetric measures on a referential set of n elements is n.*
- *The number of vertices in $\mathcal{FM}(A, B)$ is given by*

$$\binom{a+b}{a} - 2$$

where $a = |A| + 1$ and $b = |B| + 1$.

- *The number of vertices in* $\mathcal{FM}(A, B, C)$ *is*

$$\left(\Pi_{j=0}^{a-1} \frac{\binom{c+b+j}{b}}{\binom{b+j}{b}} \right) - 2$$

where $a = |A| + 1$, $b = |B| + 1$ *and* $c = |C| + 1$.
- *The number of vertices of* $\mathcal{FM}(A, B, C, D)$ *with* $|B| = |C| = |D| = 1$ *is*

$$48\binom{a+8}{8} - 96\binom{a+7}{7} + 63\binom{a+6}{6} - 15\binom{a+5}{5} + \binom{a+4}{4} - 2$$

where $a = |A| + 1$.

It must be noticed that the number of vertices of p-symmetric measures is very reduced if we compare it with the number of vertices of general fuzzy measures, as it can be seen in next table

Table 1. General measures vs. p-symmetric measures. Number of vertices.

$\|X\|$	General measures	symmetry	2-symmetry	3-symmetry
3	18	3	8	18
4	166	4	18	48
5	7,579	5	33	173
6	7,828,352	6	68	978

For this table, we are considering for the 2-symmetric and 3-symmetric cases, the choice of subsets of indifference leading to a maximal number of vertices.

As the set of fuzzy measures and the set $\mathcal{FM}(A_1, \ldots, A_p)$ are polytopes, they can be given in terms of their vertices. The reduction in the number of the vertices shown in Table 1 is one of the main advantages of p-symmetric measures and simplifies ulterior studies.

Let us now treat the problem of adjacency. As proved in [14, 21], the problem of determining non-adjacency of vertices of a polytope is, in some cases, NP-complete (see [8] for a definition of NP-complete problems and related notions). However, in [5], the following has been proved for order polytopes.

Lemma 2. *A necessary condition for* F_1 *and* F_2 *to be adjacent vertices in* $O(P)$ *is that either* $F_1 \subset F_2$ *or* $F_2 \subset F_1$.

Theorem 2. *If* F_1 *and* F_2 *are filters of* P *and* $F_1 \subset F_2$, *then* F_1 *and* F_2 *are adjacent vertices in* $O(P)$ *if and only if* $F_2 \setminus F_1$ *is a connected subposet of* P.

This result is important not only because it characterizes adjacency, but because it also allows to derive the following corollary.

Corollary 1. *Checking whether two filters F_1 and F_2 are adjacent can be done in quadratic time in the number of elements of P.*

In particular, the adjacency structure of $\mathcal{FM}(A_1, \ldots, A_p)$ can be checked in quadratic time. Figure 3 (which has been drawn with the help of the Pigale computer program[2]) depicts the adjacency structure for $\mathcal{FM}(A_1, A_2, A_3)$ when $|A_1| = |A_2| = |A_3| = 1$, i.e. the adjacency structure of general fuzzy measures when $|X| = 3$. In this figure, filters are noted by their corresponding antichains.

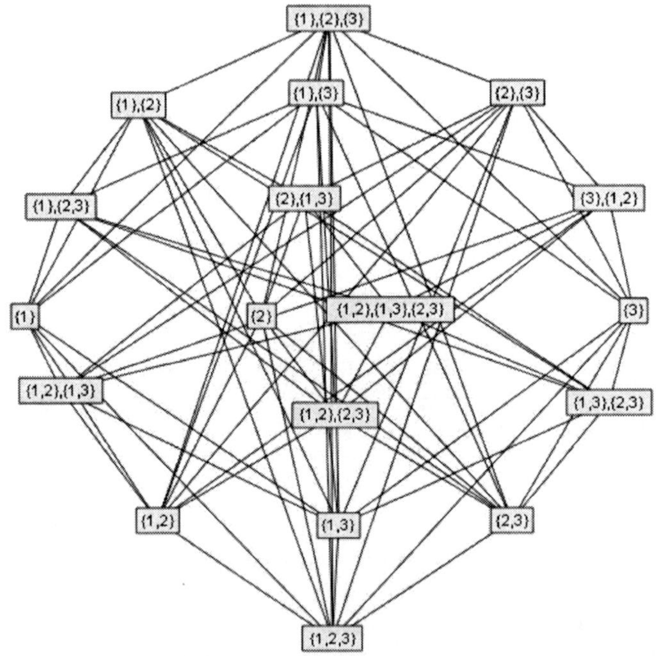

Fig. 3. Adjacency structure of $\mathcal{FM}(A_1, A_2, A_3)$ when $|A_1| = |A_2| = |A_3| = 1$

More results related to the adjacency structure of $\mathcal{FM}(A_1, \ldots, A_p)$ can be found in [5].

Let us now deal with the problem of obtaining the group of isometries on $\mathcal{FM}(A_1, \ldots, A_p)$.

Suppose $f : P \to P$ is a bijection such that there exist two disjoint filters F, F' (we allow one of them to be empty) such that $P = F \cup F'$ and

1. $f(F)$ and $f(F')$ are filters in P.
2. If $i, j \in F$, then $i \preceq j$ if and only if $f(i) \preceq f(j)$.
3. If $i, j \in F'$, then $i \preceq j$ if and only if $f(j) \preceq f(i)$.

---

[2] PIGALE: Public Implementation of a Graph Algorithm Library and Editor, H. de Fraysseix and P. Ossona de Mendez. http://pigale.sourceforge.net/

That is, f is isotone on F and antitone on F'. For a given f in these conditions, we define $h_{f,F,F'} : O(P) \to O(P)$ by

$$h_{f,F,F'}(a_1, \ldots, a_p) := (b_1, \ldots, b_p),$$

where

$$b_{f(i)} := \begin{cases} a_i & \text{if } i \in F \\ 1 - a_i & \text{if } i \in F' \end{cases}.$$

It has been proved in [6] that $h_{f,F,F'}$ is an isometry on $O(P)$, called **isometry induced by f and the filters** F, F'. Now, the following can be shown:

Theorem 3. *[6] Let $h : O(P) \to O(P)$ be an isometry; then, there exists a bijection $f : P \to P$ and two filters F, F' determining a partition on P and satisfying conditions 1-3 such that $h = h_{f,F,F'}$.*

Let us now consider the isometries satisfying $F = P$; we will denote this set of isometries by H_0. From Theorem 3, H_0 is given by the mappings f such that $h_{f,P,\emptyset}$ is an isometry. Note also that H_0 is never empty, as the identity map determines an isometry in H_0. Moreover, it is easy to check that H_0 is a subgroup of the group of isometries. Indeed, the group H_0 is isomorphic to the group of order automorphisms of P (isomorphisms preserving the order structure of P). For the particular case of connected posets, the following result can be stated.

Proposition 4. *[6] If P is connected, then:*

1. *H_0 is a normal subgroup of the set of isometries on $O(P)$.*
2. *If P is autodual, then H_0 is a subgroup of index 2.*
3. *If P is not autodual, then H_0 covers the whole set of isometries on $O(P)$.*

For the case of $\mathcal{FM}(A_1, \ldots, A_p)$, the corresponding poset is connected and autodual. Then, we can apply the previous results to completely characterize the set of isometries on $\mathcal{FM}(A_1, \ldots, A_p)$.

Lemma 3. *[6] If a poset P is a product of p chains of sizes $a_1, \ldots, a_p$ except the top and bottom elements, then the group of automorphisms of P is generated by the functions $f_{j,k}$ given by:*

$$f_{j,k}(c_1, \ldots, c_j, \ldots, c_k, \ldots, c_p) = (c_1, \ldots, c_k, \ldots, c_j, \ldots, c_p),$$

where j, k are such that $a_j = a_k$. We call this mapping the **transposition** *between the chains j and k.*

Define $g : P(A_1, \ldots, A_p) \to P(A_1, \ldots, A_p)$ by

$$g((c_1, \ldots, c_p)) = (|A_1| - c_1, \ldots, |A_p| - c_p).$$

Then, g is an order-reversing isomorphism between $P(A_1, \ldots, A_p)$ and its dual poset. We will call such mapping g the **dual application**.

Theorem 4. *[6] The group of isometries on $\mathcal{FM}(A_1, \ldots, A_p)$ is generated by the isometries induced by transpositions between subsets of indifference of the same cardinality, and by the isometry induced by the dual application.*

5 Conclusions and Perspectives

In this paper we have introduced the definition and some properties of p-symmetric measures. The main point of this subfamily is that it provides a gradation between symmetric measures and general fuzzy measures; therefore, we can choose the degree of symmetry in terms of the concrete problem and the desired complexity. Moreover, they share almost the same properties as general fuzzy measures; this is a property that others subfamilies do not fulfill, as for example k-additive measures. The reason relies in the fact that general fuzzy measures and p-symmetric measures can be both treated as order polytopes; this allows the study in terms of the corresponding poset, thus reducing the complexity and giving a deeper insight in the structure of the subfamily.

There are many circumstances under which p-symmetry could be an interesting tool. As we have seen, p-symmetric measures allow a reduction in the complexity of computing Choquet integral. When we have many criteria, we could think of using a p-symmetric measure *near* the real measure, so that we reduce the complexity while obtaining similar Choquet values.

Another example comes from Multicriteria Decision Making. In many real problems of Multicriteria Decision Making there are many criteria; in this case, it is usual to decompose the problem in several sub-problems by introducing a hierarchy of criteria. In this case, we could think in decompose the problem in several sub-problems based on symmetry or 2-symmetry. Thus, the problem of identifying the fuzzy measure for each sub-problem is easy to solve and the overall complexity reduces.

Acknowledgement. This paper has been supported in part by grant numbers MTM2007-61193, MTM2009-10072 and BSCH-UCM910707.

References

1. Berman, J., Koehler, P.: Cardinalities of finite distributive lattices. Mitteilungen aus dem Mathematischen Seminar Giessen 121, 103–124 (1976)
2. Chateauneuf, A., Cohen, M.: Choquet Expected Utility Model: A new approach to individual behaviour under uncertainty and to Social Welfare. In: Grabisch, M., Murofushi, T., Sugeno, M. (eds.) Fuzzy Measures and Integrals, pp. 289–314. Physica-Verlag, Heidelberg (2000)
3. Chateauneuf, A., Jaffray, J.-Y.: Some characterizations of lower probabilities and other monotone capacities through the use of Möbius inversion. Math. Soc. Sci. 17, 263–283 (1989)
4. Choquet, G.: Theory of capacities. Ann. de l'Institut. Fourier 5, 131–295 (1953)
5. Combarro, E.F., Miranda, P.: Adjacency on the order polytope with applications to the theory of fuzzy measures. Fuzzy Sets and Systems 180, 384–398 (2010)
6. Combarro, E.F., Miranda, P.: Characterizing isometries on the order polytope with an application to the theory of fuzzy measures. Inform. Sci. 180, 384–398 (2010)

7. Denneberg, D.: Non-additive measures and integral. Kluwer Acad. Pub., Dordrecht (1994)
8. Garey, M., Johnson, D.: Computers and Intractability: A Guide to the Theory of NP-Completeness. Mc Graw Hill, New York (1979)
9. Grabisch, M.: k-order additive discrete fuzzy measures and their representation. Fuzzy Sets and Systems 92, 167–189 (1997)
10. Grabisch, M., Marichal, J.-L., Mesiar, R., Pap, E.: Aggregation Functions. Cambridge Univ. Press, Cambridge (2009)
11. Grabisch, M., Murofushi, T., Sugeno, M.: Fuzzy Measures and Integrals-Theory and Applications. Studies in Fuzziness and Soft Computing, vol. 40. Physica–Verlag, Heidelberg (2000)
12. Grabisch, M., Roubens, M.: Application of the Choquet integral in Multicriteria Decision Making. In: Grabisch, M., Murofushi, T., Sugeno, M. (eds.) Fuzzy Measures and Integrals, pp. 348–375. Physica-Verlag, Heidelberg (2000)
13. Harsanyi, J.C.: A simplified bargaining model for the n-person cooperative game. Int. Econom. Rev. 4, 194–220 (1963)
14. Matsui, T.: NP-completeness of non-adjacency relations on some 0-1-polytopes. Lect. Notes Oper. Res. 1, 249–258 (1995)
15. Miranda, P., Combarro, E.F., Gil, P.: Extreme points of some families of non-additive measures. European J. Oper. Res. 33(10), 3046–3066 (2006)
16. Miranda, P., Grabisch, M.: p-symmetric bi-capacities. Kybernetika 40(4), 421–440 (2004)
17. Miranda, P., Grabisch, M., Gil, P.: p-symmetric fuzzy measures. Int J Uncert. Fuzz. Know.-Based Syst. 10(suppl.), 105–123 (2002)
18. Murofushi, T., Sugeno, M.: Some quantities represented by the Choquet integral. Fuzzy Sets and Systems 56, 229–235 (1993)
19. Murofushi, T., Sugeno, M., Fujimoto, K.: Canonical separated hierarchical decomposition of the Choquet integral over a finite set. Int. J. Uncert. Fuzz. Know.-Based Syst. 6(3), 257–271 (1997)
20. Murofushi, T., Sugeno, M., Fujimoto, K.: Separated hierarchical decomposition of Choquet integral. Int. J. Uncert. Fuzz. Know.-Based Syst. 5, 563–585 (1997)
21. Papadimitriou, C.: The adjacency relation on the travelling salesman polytope is NP-complete. Math. Program 14(3), 312–324 (1978)
22. Rota, G.C.: On the foundations of combinatorial theory I. Theory of Möbius functions. Zeitschrift für Wahrscheinlichkeitstheorie und Verwandte Gebiete (2), 340–368 (1964)
23. Schmeidler, D.: Integral representation without additivity. Proc. Amer. Math. Soc. 97(2), 255–261 (1986)
24. Stanley, R.: Ordered Structures and Partitions. Mem. Amer. Math. Soc. 119 (1972)
25. Stanley, R.: Two poset polytopes. Discrete Comput. Geom. 1(1), 9–23 (1986)
26. Sugeno, M.: Theory of Fuzzy Integrals and its Applications. PhD Thesis, Tokyo Institute of Technology (1974)
27. Yager, R.R.: On ordered weighted averaging aggregation operators in multicriteria decision making. IEEE Trans. Syst. Man Cybern. 18, 183–190 (1988)

Part IX

Robust and Soft Methods in Statistics

Robustification of the MLE without Loss of Efficiency

Biman Chakraborty[1], Sahadeb Sarkar[2], and Ayanendranath Basu[3]

[1] School of Mathematics, University of Birmingham, Edgbaston, Birmingham B15 2TT, United Kingdom
B.Chakraborty@bham.ac.uk

[2] Operations Management Group, Indian Institute of Management, Joka, Kolkata 700 104, India
sahadeb@iimcal.ac.in

[3] Bayesian and Interdisciplinary Research Unit, Indian Statistical Institute, 203 B.T. Road, Kolkata 700 108, India
ayanbasu@isical.ac.in

Summary. A robust procedure, which produces the maximum likelihood estimator when the data are in conformity with the parametric model, and generates the outlier deleted maximum likelihood estimator under the presence of extreme outliers, has obvious intuitive appeal to the practising scientist. None of the currently available robust estimators achieves this automatically. Here we propose a density-based divergence belonging to the family of disparities ([7]) where the corresponding weighted likelihood estimator ([10], [11]) exhibits this desirable behavior for proper choices of tuning parameters. Some properties of the corresponding estimation procedure are discussed and illustrated through examples.

Keywords: Hellinger distance, outlier deleted maximum likelihood estimator, residual adjustment function, weighted likelihood estimation.

1 Introduction

The maximum likelihood estimator is the cornerstone of classical parametric inference. However, it is notoriously non-robust to deviations, even very small ones, from the parametric conditions in many common models. In much of the robustness literature, the main focus has been to modify the methods based on maximum likelihood in order to make them more stable under model misspecifications. While the robustness issue requires careful attention, the universal appeal of the classical parametric methods such as the maximum likelihood estimation is something we would hate to lose. In real data situations with extreme outliers, many data analysts still naively apply standard techniques routinely after simply deleting the outlying observations – an approach which is theoretically unsatisfactory for obvious reasons. However, a robust method which automatically (rather than through subjective deletion) generates the outlier deleted maximum likelihood estimator under extreme outliers and exactly matches the ordinary maximum likelihood estimator when the data are generally concordant with the model will have immediate intuitive appeal to the practitioner. None of the currently available robust methods manage that in practise.

Using a density-based minimum divergence approach, [5] showed that the conflicting issues of efficiency and robustness can be reconciled using the *minimum Hellinger distance estimator.* [15], [13] and [7], among others, pursued this line of research. The latter work extended the scope of application to a general class of density based divergences called *disparities.* The corresponding minimum disparity estimators are all first order efficient while several of them have remarkable robustness properties. [12] provides a general description of many of these methods.

It may be noted that the class of disparities is a reformulation of the class of ϕ-divergences (see, eg., [6] and [2]). However [5] appears to be the first who seriously dealt with the robustness issue based on such divergences and Lindsay's ([7]) work stands out in that he identified the geometrical properties needed by the divergence to inherit strong robustness properties.

In this paper we have followed the set up and approach of [7] and considered minimization procedures based on the *robustified likelihood disparity* and the corresponding weighted likelihood version; this family is a modification of the *likelihood disparity* ([7]). The proposed estimators remove the non-robustness of the maximum likelihood estimator in a very natural way. Under proper choice of tuning parameters the procedure has the remarkable property that the corresponding weighted likelihood estimators ([10], [11]) are *exactly* equal to the maximum likelihood estimator when the data generally follow the model, while being exactly equal to the outlier deleted maximum likelihood estimator when the data contain some extreme outliers. This is the novelty of this method – it is not just another minimum disparity estimation technique generating robust and efficient estimators simultaneously.

Throughout the paper, we will (a) denote the true distribution by G and the model family by $\mathcal{F}_\Theta = \{F_\theta, \theta \in \Theta \subseteq \mathbb{R}^p\}$, (b) assume that $\mathcal{F}_\Theta$ and G are both contained in $\mathcal{G}$, the class of all distributions having probability density functions with respect to (w.r.t.) a dominating measure, and (c) write lower case letters to represent the probability density functions and upper case letters for the corresponding distribution functions (e.g., write g, f_θ for the densities of G and F_θ).

2 Disparities and Weighted Likelihood

For ease of presentation we first consider the discrete case. Let $X_1, X_2, \ldots, X_n$ represent a random sample from a discrete distribution F_θ having a countable support. Let $d_n(x)$ denote the proportion of X_j's in the sample having the value x. [7] defined the *Pearson residual* $\delta(x)$ at a value x by $\delta(x) = (d_n(x) - f_\theta(x))/f_\theta(x)$. An x-value is called an *outlier* if $\delta(x)$ is a relatively large positive number. Let $C(\cdot)$ be a real–valued, thrice differentiable convex function on $[-1,\infty)$ with $C(0)=0$. Then the *disparity* ρ_C between d_n and f_θ is defined as

$$\rho_C(d_n, f_\theta) = \sum_x C\big(\delta(x)\big) f_\theta(x). \tag{1}$$

The minimizer of (1) with respect to θ, provided one exists, is called the *minimum disparity estimator* of θ corresponding to the disparity ρ_C.

Letting ∇ denote the gradient with respect to θ, the estimating equation, under differentiability of the model, takes the form

$$-\nabla\rho_C = \sum_x A(\delta(x))\nabla f_\theta(x) = 0, \tag{2}$$

where

$$A(\delta) \equiv (\delta+1)C'(\delta) - C(\delta), \tag{3}$$

where $C'(\delta)$ represents the derivative of $C(\delta)$ with respect to its argument. The corresponding derivative of $A(\delta)$ will be denoted by $A'(\delta)$. Without changing the estimating properties of the disparity ρ_C, $C(\delta)$ can be redefined so that its $A(\delta)$ function satisfies $A(0)=0$ and $A'(0)=1$. This standardized function $A(\delta)$ is called the *residual adjustment function* of the disparity. The maximum likelihood estimator of θ minimizes the *likelihood disparity* $LD(d_n, f_\theta)$ with $C_{LD}(\delta) = (\delta+1)\log(\delta+1)-\delta$ and the residual adjustment function has the form $A_{LD}(\delta)=\delta$; here log represents the natural logarithm.

Let $u_\theta(x) = \nabla \log f_\theta(x)$ denote the maximum likelihood score function. A little algebra shows that equation (2) can be written as

$$\sum_{x:d_n(x)\neq 0} \frac{A(\delta(x))+1}{\delta(x)+1} d_n(x)u_\theta(x) + \sum_{x:d_n(x)=0} [A(-1)+1]\nabla f_\theta(x) = 0.$$

For divergences with $A(-1)=-1$, the above equation becomes

$$\sum_{d_n(x)\neq 0} w(x)d_n(x)u_\theta(x) = \frac{1}{n}\sum_{i=1}^n w(X_i)u_\theta(X_i) = 0. \tag{4}$$

where

$$w(x) = w(\delta(x)) \equiv (A(\delta(x))+1)/(\delta(x)+1). \tag{5}$$

The last equation in (4) can be viewed as a weighted likelihood score equation; it equates a weighted sum of the likelihood scores to zero. The solution to this equation will be referred to as the weighted likelihood estimator. If $w(x)=1$ identically in x, equation (4) reduces to the maximum likelihood score equation. For fixed weights w, a closed form solution of equation (4) is usually available. Hence, (4) can be solved via a fixed point iteration algorithm similar to iteratively reweighted least squares. One can directly construct weighted likelihood estimating equations as in (4) starting from the residual adjustment function of any appropriate disparity; however, the method will correspond exactly to the minimization of a disparity only if $A(-1) = -1$ for the corresponding function. Notice that in case of the likelihood disparity $A(-1) = -1$, and the weights are all identically equal to 1.

Large outliers are manifested through large positive values of δ. However, the negative side of the δ axis is not a robustness concern, and one convenient modification is to put $w(x) = w(\delta(x)) = 1$ for $\delta(x) < 0$. This does not affect the robustness or efficiency properties of the corresponding estimator, but removes the intuitively confusing possibilities of having negative weights or weights greater than 1. This may be viewed as applying the weighted likelihood methodology to the combined disparity obtained by combining the likelihood disparity on the inlier side to the disparity in question on the outlier side (see [9]). This branch of weighted likelihood estimation was developed by Markatou et al. ([10], [11]), and more details about these methods can be found in their papers. [1] have also developed efficient testing procedures based on the weighted likelihood idea.

In the above weighted likelihood scheme, our objectives therefore are that (i) the observations generally following the model should get weights *exactly* equal to 1 (as in the case of the likelihood disparity), and (ii) the outlying data points (with relatively large positive values of δ) should be smoothly downweighted, with the degree of downweighting increasing as the observations become more and more aberrant; wildly discrepant observations should get weights practically equal to zero. With these objectives in mind we present the following proposal. We modify the residual adjustment function of the likelihood disparity with numbers c and c^* satisfying $-1 \leq c^* < 0 < c < \infty$, so that

$$A_{c^*,c}(\delta) = \begin{cases} c^* & \text{for } -1 \leq \delta \leq c^* \\ \delta & \text{for } c^* < \delta < c \\ c & \text{for } \delta \geq c. \end{cases} \tag{6}$$

By solving the differential equation (3) for the above residual adjustment function one gets the defining equation of the disparity to be

$$C_{c^*,c}(\delta) = \begin{cases} (\delta+1)\log(c^*+1) - c^* & \text{for } -1 \leq \delta \leq c^* \\ (\delta+1)\log(\delta+1) - \delta & \text{for } c^* < \delta < c \\ (\delta+1)\log(c+1) - c & \text{for } \delta \geq c. \end{cases} \tag{7}$$

Clearly, $C_{c^*,c}(\delta)$ is convex on (c^*, c) and linear on $[-1, c^*]$ and $[c, \infty)$ with the slopes at $\delta = c^*$ and $\delta = c$ well defined. Hence $C_{c^*,c}(\delta)$ is convex on the entire interval $[-1, \infty)$. We will refer to the disparity generated by $C_{c^*,c}$ as the *robustified likelihood disparity* (with tuning parameters c^* and c). The minimizer of the robustified likelihood disparity $RLD_{c^*,c}(d_n, f_\theta)$ over $\theta \in \Theta$ is the *robustified maximum likelihood estimator*. Similarly the *robustified weighted likelihood estimators* is the solution of the estimating equation (4) where the weight function uses the form of the residual adjustment function as given in (6). However in the latter case we restrict c^* to equal -1, so as to stick to the convention that $w(x) = 1$ for $\delta(x) < 0$.

Now consider the case of a continuous distribution G, modeled by a continuous family of distributions $\mathcal{F}_\theta$. Let $X_1, \ldots, X_n$ be a random sample from G. In extending the minimum disparity ideas to the estimation of θ we are faced with the additional complication that the data are discrete but the model is continuous. In this case we let

$$\hat{g}_n(x) = \frac{1}{n}\sum_{i=1}^{n} k\left(x, X_i, h_n\right) = \int k(x, y, h_n) dG_n(y) \tag{8}$$

denote a nonparametric density estimator where $k(\cdot, y, h_n)$ is a smooth kernel function with bandwidth h_n and G_n is the empirical distribution function. Then, we can estimate θ by minimizing the disparity

$$\rho_C(\hat{g}_n, f_\theta) = \int C(\delta(x)) f_\theta(x) dx, \quad \delta(x) = (\hat{g}_n(x) - f_\theta(x))/f_\theta(x), \tag{9}$$

between $\hat{g}_n$ and f_θ. Under differentiability of the model the estimating equation now has the form

$$-\nabla \rho_C = \int_x A(\delta(x)) \nabla f_\theta(x) dx = 0.$$

The solution of this equation, when one uses the form of the residual adjustment function as in equation (6), represents the robustified maximum likelihood estimator in this case.

When one tries to determine a weighted likelihood estimating equation as in the discrete case, one ends up with the equation

$$\int w(x) u_\theta(x) d\hat{G}_n(x) = 0, \tag{10}$$

where $w(x) = (A(\delta(x)) + 1)/(\delta(x) + 1)$ and $\hat{G}_n(x)$ is the cumulative distribution function corresponding to $\hat{g}_n$. While this may also be solved by a fixed point iteration method, this will still require numerical solutions of integrals at every stage (multiple integrals if the data are multivariate). However, in analogy with the discrete case, if one considers the unsmoothed version of the above equation by removing the smoothing from $\hat{G}_n$, one gets the following form

$$\int w(x) u_\theta(x) dG_n(x) = 0, \quad \text{or} \quad \frac{1}{n}\sum_{i=1}^{n} w(X_i) u_\theta(X_i) = 0, \tag{11}$$

which is a sum over the data points, and not an integral over the entire sample space. The weights $w(x)$ have interpretations similar to those given for the discrete models. The solutions of these estimating equations will represent the robustified weighted likelihood estimators for this case. However, one additional modification we consider in this case is the smoothing of the model in the construction of the weights. Thus we compute

$$\hat{g}_n(x) = \int k(x, y, h_n) dG_n(y), \quad \text{and } f_\theta^*(x) = \int k(x, y, h_n) dF_\theta(y). \tag{12}$$

Then one computes $\delta(x)$ as $\delta(x) = (\hat{g}_n(x) - f_\theta^*(x))/\hat{g}_n(x)$, and the weights are computed as before. This approach of smoothing the model in the context of minimum disparity estimation has been discussed in [3].

Henceforth we will refer to the estimator obtained by minimizing the RLD as the *robustified maximum likelihood estimator* ($RMLE$), and that obtained by solving the weighted likelihood equations as *robustified weighted likelihood estimator* ($RWLE$), with subscripts c^* and c if necessary. Unless otherwise mentioned, c^* will be assumed to be equal to -1 for the $RWLE$.

3 Asymptotic Efficiency

Asymptotic results of [3] cover general disparities in the continuous case when the model densities are smoothed. But in this case the asymptotic efficiencies are dependent on the availability and the use of model specific kernels. In the present paper we primarily focus on the minimum disparity approach which minimizes (9) as in [5]. For the weighted likelihood case, however, we consider the smoothed model version as would follow from the Basu and Lindsay approach, since in this case model specific kernel choices are not critical for asymptotic efficiency. In addition it allows one to use a fixed bandwidth, which does not need to go to zero as the function of the sample size, and the experimenter is spared the difficulty of having to negotiate the complicated bandwidth selection problem.

Unlike the Basu and Lindsay approach, however, there is no general theory in the continuous case for the minimum disparity estimators which minimize (9) as in the Beran approach, and here we specifically deal with the case of the RLD. Let us define the RLD estimation functional $T_{c^*,c} : \mathcal{G} \to \Theta$ satisfying

$$RLD_{c^*,c}(g, f_{T_{c^*,c}(G)}) = \min_{\theta \in \Theta} RLD_{c^*,c}(g, f_\theta), \tag{13}$$

provided such a minimum exists. For an appropriate kernel density estimator $\hat{g}_n$, the $RMLE_{c^*,c}$ of θ is $T_{c^*,c}(\hat{G}_n)$. We will write $\hat{\theta}_{c^*,c}$ for $T_{c^*,c}(\hat{G}_n)$. The existence, consistency, and asymptotic normality of $\hat{\theta}_{c^*,c}$ follow with slight modifications of the proofs of [4]. Here we simply state the main result:

Theorem 1. *Let $u_\theta(x) = \nabla \log f_\theta(x)$ be the p-dimensional likelihood score function. The Fisher information matrix is defined as $I(\theta) \equiv \int u_\theta u_\theta^T f_\theta$. Consider any fixed c^*, c satisfying $-1 < c^* < 0 < c < \infty$. Suppose the true distribution $G = F_{\theta_0} \in \mathcal{F}_\Theta$. Then, under the vector-parameter generalizations of* *the conditions given in [4], the asymptotic distribution of* $n^{1/2}(\hat{\theta}_{c^*,c} - \theta_0)$ *is* $N(0, I^{-1}(\theta_0))$ *where* $I(\theta)$ *is the information matrix for* f_θ.

Actually, the proof of Theorem 1 establishes more than just the asymptotic optimality of the estimators concerned. It also shows that $\hat{\theta}_{c^*,c}$ and the MLE $\hat{\theta}_{ML}$ are asymptotically equivalent, both being equal to

$$\theta_0 + I(\theta_0)^{-1} n^{-1} \sum_{i=1}^{n} u_{\theta_0}(X_i) + o_p(n^{-1/2}).$$

Thus when the model is correct, the two estimators may be expected to behave very similarly for large values of n.

4 Robustness

It is easy to verify that the robustified maximum likelihood estimator and the robustified weighted likelihood estimator both have influence functions identical to the influence function of the maximum likelihood estimator at the model. While this implies that the the influence function is not useful in characterizing the robustness of these estimators, they do indicate their asymptotic efficiency. Several authors including [5] and [7] have discussed the inadequacy of the influence function in describing the robustness of such estimators. When dealing with the minimum Hellinger distance estimator, Beran described its robustness through the boundedness of the "α influence function", while Lindsay considered the second order term in the expansion of the bias (influence function is linked to the first order term). The works of these authors (as well as those of several others) have demonstrated that the minimum disparity estimators have strong robustness properties which are not captured by the influence function analysis.

The robustness of the robustified maximum likelihood estimator and robustified weighted likelihood estimator can be partly understood by the markedly dampened response it provides to outlying observations. Large outliers are characterised by large positive values of the Pearson residual $\delta(x)$; as $\delta(x)$ becomes larger than c, the estimating equation of the robustified maximum likelihood estimator will begin to downweight such observations. Similarly in the weighted likelihood scenario, the weights in (5) are going to be strictly smaller than 1 when $\delta(x) > c$, and will tend to zero as the value of δ increases, so that for an extreme outlier the estimating equation (4) is expected to behave as if the observation is simply deleted from the data set.

The breakdown point of a statistical functional is roughly the smallest percentage of contaminated values in the data that may result in an arbitrarily extreme value of the estimate. Here we establish the breakdown point of the minimum robustified likelihood disparity functional $T_{c^*,c}(\cdot)$ under the following setup. In order to have a clear focus, we consider θ to be a scalar for the rest of the discussion in this section. For $\epsilon \in (0,1)$, consider the contamination model, $H_{\epsilon,m} = (1-\epsilon)G + \epsilon K_m, \quad m \geq 1$, where G is the true distribution, $\{K_m\}$ is a sequence of contaminating distributions, and $h_{\epsilon,m}$, g and k_m are the corresponding densities w.r.t. the dominating measure Q. For a given sequence $\{K_m\}$, we will say that breakdown in $T_{c^*,c}$ occurs for ϵ level of contamination if

$$\lim_{m\to\infty} |T_{c^*,c}(H_{\epsilon,m})| = \infty. \tag{14}$$

We are interested in the smallest ϵ for which there exists a sequence $\{K_m\}$ such that (14) holds. From a practical point of view, we will restrict ϵ to be in (0, 1/2] in the following discussion.

For our analysis we make the following assumptions, which put a structure on the model and on the contamination sequence and enable us to determine the disparities under extreme forms of contaminations.

Assumptions: *The true density g, the model density $\{f_\theta\}$, and the contamination density $\{k_m\}$ satisfy the following*:

A1. $\int \min\{g, k_m\} \to 0$ *as* $m \to \infty$, *that is the contamination distribution becomes asymptotically singular to the true distribution.*

A2. $\int \min\{f_\theta, k_m\} \to 0$ *as* $m \to \infty$ *uniformly for* $|\theta| \leq M$, *for any fixed* $M > 0$. *That is, the contamination distribution becomes asymptotically singular to specified models.*

A3. $\int \min\{g, f_{\theta_m}\} \to 0$ *as* $m \to \infty$, *if* $|\theta_m| \to \infty$ *as* $m \to \infty$. *That is, for large values of the parameter* θ, *model distributions become asymptotically singular to the true distribution.*

Contamination sequences satisfying assumptions **A1** and **A2** will be called outlier sequences. Intuitively, these outlier sequences represent the worst possible type of contamination, and hence it seems natural to study the breakdown properties of the functional under such sequences. Assumption **A3** formalizes the expected behavior of the model.

Theorem 2. *Consider any fixed* c^*, c *satisfying* $-1 \leq c^* < 0 < c < \infty$. *Given a contamination level* ϵ, *let* θ_ϵ^* *be the minimizer of* $RLD((1-\epsilon)g, f_\theta)$, *which will be assumed to exist. Let* $b(\epsilon) = C_{c^*,c}(\epsilon - 1) + (1-\epsilon)\log(c+1)$, *and* $\gamma(\epsilon) = RLD((1-\epsilon)g, f_{\theta_\epsilon^*}) + \epsilon \log(c+1)$ *and* $\epsilon^* = \inf\{\epsilon : b(\epsilon) \leq \gamma(\epsilon)\}$. *Then, breakdown does not occur as long as* $\epsilon < \epsilon^*$. *In particular, when the true distribution* G *is in the model, breakdown does not occur if* $\epsilon < 1/2$.

Proof. Let $T_{c^*,c}(H_{\epsilon,m}) = \theta_{\epsilon,m}$. Given $0 < \epsilon < 1/2$, suppose breakdown occurs, i.e., there exists $\{K_m\}$ such that $|\theta_{\epsilon,m}| \to \infty$ as $m \to \infty$. Then, $RLD(h_{\epsilon,m}, f_{\theta_{\epsilon,m}}) - RLD(\epsilon k_m, f_{\theta_{\epsilon,m}})$ is equal to:

$$\begin{aligned}
&\int_{h_{\epsilon,n}\le(c^*+1)f_{\theta_{\epsilon,m}}}[h_{\epsilon,m}\log(c^*+1)-c^*f_{\theta_{\epsilon,m}}] \\
&-\int_{\epsilon k_m\le(c^*+1)f_{\theta_{\epsilon,m}}}[\epsilon k_m\log(c^*+1)-c^*f_{\theta_{\epsilon,m}}] \\
&+\int_{(c^*+1)f_{\theta_{\epsilon,m}}<h_{\epsilon,m}<(c+1)f_{\theta_{\epsilon,m}}}[h_{\epsilon,m}\log\tfrac{h_{\epsilon,m}}{f_{\theta_{\epsilon,m}}}-(h_{\epsilon,m}-f_{\theta_{\epsilon,m}})] \\
&-\int_{(c^*+1)f_{\theta_{\epsilon,m}}<\epsilon k_m<(c+1)f_{\theta_{\epsilon,m}}}[\epsilon k_m\log\tfrac{\epsilon k_m}{f_{\theta_{\epsilon,m}}}-(\epsilon k_m-f_{\theta_{\epsilon,m}})] \\
&+\int_{h_{\epsilon,m}\ge(c+1)f_{\theta_{\epsilon,m}}}[h_{\epsilon,m}\log(c+1)-cf_{\theta_{\epsilon,m}}] \\
&-\int_{\epsilon k_m\ge(c+1)f_{\theta_{\epsilon,m}}}[\epsilon k_m\log(c+1)-cf_{\theta_{\epsilon,m}}].
\end{aligned} \tag{15}$$

Define the set $B_m = \{x : g(x) > f_{\theta_{\epsilon,m}}(x)\}$. Then the probabilities of $B_m \cap \{x : h_{\epsilon,m} \le (c^*+1)f_{\theta_{\epsilon,m}}\}$ w.r.t. G, K_m and $F_{\theta_{\epsilon,m}}$ converge to zero, and those of $B_m \cap \{x : \epsilon k_m \le (c^*+1)f_{\theta_{\epsilon,m}}\}$ w.r.t. K_m and $F_{\theta_{\epsilon,m}}$ also converge to zero as $m \to \infty$. By assumption **A3**, the probability of B_m^c converges to zero w.r.t. G as $m \to \infty$. Note that $I_{B_m^c}(x)(h_{\epsilon,m}(x) - \epsilon k_m(x))$ converges to zero almost surely as $m \to \infty$. Thus, the difference between first two integrals in (15) converges to zero. Similarly, by considering the sets $B_m \cap \{x : (c^*+1)f_{\theta_{\epsilon,m}} < h_{\epsilon,m} < (c+1)f_{\theta_{\epsilon,m}}\}$ and $B_m \cap \{x : (c^*+1)f_{\theta_{\epsilon,m}} < \epsilon k_m < (c+1)f_{\theta_{\epsilon,m}}\}$, we see the difference between the third and fourth integrals of (15) converges to zero as $m \to \infty$. Now, write the last two integrals in (15) as

$$\int_{\{h_{\epsilon,m}\ge(c+1)f_{\theta_{\epsilon,m}};\epsilon k_m<(c+1)f_{\theta_{\epsilon,m}}\}}\{h_{\epsilon,m}\log(c+1)-cf_{\theta_{\epsilon,m}}\}$$
$$+\int_{\epsilon k_m\ge(c+1)f_{\theta_{\epsilon,m}}}(1-\epsilon)g\log(c+1).$$

Again, the probability of the set $B_m^c \cap \{x : h_{\epsilon,m} \ge (c+1)f_{\theta_{\epsilon,m}}, \epsilon k_m < (c+1)f_{\theta_{\epsilon,m}}\}$ w.r.t. G, K_m and $F_{\theta_{\epsilon,m}}$ converges to zero as $m \to \infty$ from our assumptions, and the probability of the set $B_m \cap \{x : h_{\epsilon,m} \ge (c+1)f_{\theta_{\epsilon,m}}, \epsilon k_m < (c+1)f_{\theta_{\epsilon,m}}\}$ w.r.t. K_m and $F_{\theta_{\epsilon,m}}$ converges to zero. Then the difference between last two integrals in (15) converges to $(1-\epsilon)\log(c+1)$. Thus, under the existence of an outlier sequence which causes breakdown,

$$\liminf_{m\to\infty} RLD(h_{\epsilon,m}, f_{\theta_{\epsilon,m}}) = \liminf_{m\to\infty} RLD(\epsilon k_m, f_{\theta_{\epsilon,m}}) + (1-\epsilon)\log(c+1). \tag{16}$$

Notice that from Jensen's inequality, the right hand side of (16) is bounded below by

$$b(\epsilon) = C_{c^*,c}(\epsilon - 1) + (1-\epsilon)\log(c+1). \tag{17}$$

We will get a contradiction to our assumption of the existence of this outlier sequence $\{K_m\}$ for which breakdown occurs at contamination level ϵ if we can show that there exists a bounded value θ of the parameter such that

$$\limsup_{m\to\infty} RLD(h_{\epsilon,m}, f_\theta) < b(\epsilon) \tag{18}$$

since in this case the sequence $\{\theta_{\epsilon,m}\}$ cannot minimize $RLD(h_{\epsilon,m}, f_\theta)$ for every m. Given a level ϵ of contamination, and an outlier sequence $\{K_m\}$, let θ be any bounded value of the parameter. Using assumption **A2**, and similar arguments as before, we obtain

$$\begin{aligned}\limsup_{m\to\infty} RLD(h_{\epsilon,m}, f_\theta) &= \limsup_{m\to\infty} RLD((1-\epsilon)g, f_\theta) + \epsilon\log(c+1)\\ &= RLD((1-\epsilon)g, f_\theta) + \epsilon\log(c+1).\end{aligned} \tag{19}$$

Since $RLD((1-\epsilon)g, f_\theta)$ is minimized by $\theta = \theta^*_\epsilon$ among all fixed and bounded values of θ,

$$\begin{aligned}\limsup_{m\to\infty} RLD(h_{\epsilon,m}, f_\theta) &= RLD((1-\epsilon)g, f_\theta) + \epsilon\log(c+1) \qquad (20)\\ &\geq RLD((1-\epsilon)g, f_{\theta^*_\epsilon}) + \epsilon\log(c+1) = \gamma(\epsilon).\end{aligned}$$

Thus an outlier sequence $\{K_m\}$ cannot cause breakdown for values of ϵ satisfying $\gamma(\epsilon) < b(\epsilon)$. Hence breakdown does not occur for $\{K_m\}$ as long as $\epsilon < \epsilon^*$, where $\epsilon^* = \inf\{\epsilon : b(\epsilon) \leq \gamma(\epsilon)\}$.

Now consider the case $G = F_{\theta_0}$. Then, $RLD((1-\epsilon)g, f_{\theta_0}) = RLD((1-\epsilon)f_{\theta_0}, f_{\theta_0}) = C_{c^*,c}(-\epsilon)$, which is also the lower bound (over θ) of $RLD((1-\epsilon)g, f_\theta)$ by Jensen's inequality. Hence when $G = F_{\theta_0}$, $\theta^*_\epsilon = \theta_0$ and $\limsup_{m\to\infty}$ $RLD(h_{\epsilon,m}, f_{\theta^*_\epsilon}) = C_{c^*,c}(-\epsilon) + \epsilon\log(c+1) = a(\epsilon)$, say, and there is no breakdown for ϵ level contamination if $a(\epsilon) < b(\epsilon)$. Note that $a(\epsilon)$ and $b(\epsilon)$ are strictly increasing and decreasing, respectively, in ϵ, with $a(1/2) = b(1/2)$. Thus, when G is in the model there is no breakdown for $\epsilon < 1/2$. □

Remark 1. The proof does more than simply establishing the breakdown point. When breakdown does not occur under a given outlier sequence, $RLD(h_{\epsilon,m}, f_\theta)$ is minimized in the limit by θ^*_ϵ among all bounded values of θ, and hence θ^*_ϵ is the minimum disparity estimator of θ in the limit. At the model, in particular, $\theta^*_\epsilon = \theta_0$ so that a contamination proportion $\epsilon < 1/2$ does not have any limiting impact at all.

5 Examples and Simulations

5.1 Examples

Example 1. (*Drosophila Data*) We consider a part of an experimental data analyzed by [13]. The data are in the form of frequencies of daughter flies (drosophila) carrying a recessive lethal mutation on the X-chromosome, where the male parents have been exposed to a certain dose of a chemical. Approximately 100 daughter flies were sampled for each male. This particular experiment resulted in $(x_i, f_i) = (0, 23), (1, 7), (2, 3), (91, 1)$, where x_i is the number of daughters carrying the recessive lethal mutation and f_i is the number of male parents having x_i such daughters. Notice that the last pair is highly discordant with the rest of the observations. We have fitted a *Poisson*(θ) model to these data and estimate θ using robustified maximum likelihood estimation and robustified weighted likelihood estimation for various values of c and c^*. While all the robust estimators perform very well in discounting the large outlier, the robustified weighted likelihood estimators (which are the same as the corresponding robustified maximum likelihood estimators with $c^* = -1$) are *exactly* equal to the outlier deleted maximum likelihood estimator for all c between 1 and 50. Note that the $MHDE$ (0.364), which also effectively ignores the large outlier, does not *exactly* equal the outlier deleted MLE.

Table 1. The robustified maximum likelihood estimates ($RMLE$s) and robustified weighted likelihood estimates ($RWLE$s) under the *Poisson* model for the Drosophila Data, with the maximum likelihood estimate (MLE) and the outlier deleted maximum likelihood estimate ($OD - MLE$), obtained after the subjective deletion of the large outlier 91, presented for comparison

	$RMLE$				$RWLE$
c^*	-0.5	-0.8	-0.9	-1.0	$c^* = -1.0$
$c = 0.2$	0.3292	0.3217	0.3194	0.3172	0.3172
$c = 1.0$	0.4054	0.3983	0.3961	0.3939	0.3939
$c = 50.0$	0.4054	0.3983	0.3961	0.3939	0.3939
MLE	-	-	-	3.0588	-
$OD - MLE$	-	-	-	0.3939	-

Example 2. (*Short's data*) We consider Short's data ([14], Dataset 2) for the determination of the parallax of the sun, the angle subtended by the earth's radius, as if viewed and measured from the surface of the sun. For our estimation, we have used the kernel density defined in (8), with the normal kernel, and bandwidth $h_n = 0.75186\hat{\sigma}n^{-1/5}$, where $\hat{\sigma} = \text{median}(|X_i - \text{median}(X_i)|)/0.674$. Table 2 gives the values of our robust estimates of μ and σ^2 for various values of c^* and c under the normal model, as well as the maximum likelihood estimate for all the observations and that after deleting the outlying value at 5.76. All the robust estimators exhibit very strong outlier resistance. The robustified weighted likelihood estimates coincide with the outlier deleted maximum likelihood estimate for $c = 5$ and 10; however by the time $c = 50$ the tuning parameter is too liberal and treats the outlier as a regular observation, and the estimate settles on the maximum likelihood estimate for all observations.

Figure 1 shows the fit of the kernel density estimate and normal densities using the maximum likelihood estimate and the robustified weighted likelihood estimate for $c = 10$, exhibiting the robust fit of the latter curve.

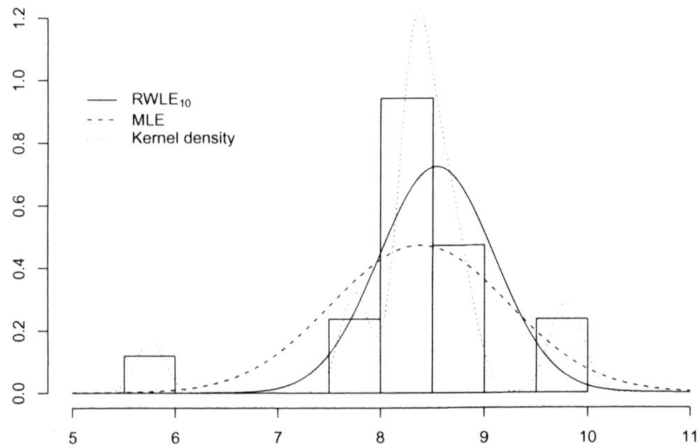

Fig. 1. Density estimates for Short's determination of the parallax of the sun

Table 2. Fits of a $N(\mu, \sigma^2)$ model to Short's data using the robustified maximum likelihood estimates ($RMLE$s) and the robustified weighted likelihood estimates with the normal kernel. The maximum likelihood estimate for the full data (MLE) and the same without the outlier ($OD - MLE$) are also presented for comparison.

	$RMLE$				$RWLE$
c^*	-0.5	-0.8	-0.9	-1.0	$c^* = -1.0$
$c = 0.2$	$(8.437, 0.068)$	$(8.437, 0.068)$	$(8.437, 0.068)$	$(8.437, 0.068)$	$(8.433, 0.055)$
$c = 1.0$	$(8.386, 0.111)$	$(8.365, 0.116)$	$(8.363, 0.116)$	$(8.363, 0.116)$	$(8.363, 0.093)$
$c = 2.0$	$(8.459, 0.053)$	$(8.385, 0.111)$	$(8.366, 0.117)$	$(8.366, 0.117)$	$(8.363, 0.093)$
$c = 5.0$	$(8.570, 0.357)$	$(8.552, 0.339)$	$(8.551, 0.340)$	$(8.600, 0.289)$	$(8.541, 0.305)$
$c = 10.0$	$(8.570, 0.357)$	$(8.552, 0.337)$	$(8.553, 0.334)$	$(8.542, 0.331)$	$(8.541, 0.305)$
$c = 50.0$	$(8.570, 0.357)$	$(8.552, 0.337)$	$(8.553, 0.334)$	$(8.542, 0.331)$	$(8.378, 0.715)$
MLE	-	-	-	$(8.378, 0.715)$	-
$OD - MLE$	-	-	-	$(8.541, 0.305)$	-

5.2 Simulation Results

In Table 3 we present a small simulation example to give more insight into the behavior of the proposed weighted likelihood estimators. The simulation is done with sample sizes of 20, 50 and 100, and uses 1000 replications for each experiment. We have used the $Poisson$ model, and calculated the robustified weighted likelihood estimates of the mean parameter θ. Data are generated both from the $Poisson(5)$ distribution and the $0.9Poisson(5) + 0.1Poisson(12)$ mixture. The empirical means of each of these estimators are presented in Table 3, as are the empirical mean square errors. When the data come from the $Poisson$ mixture, the empirical mean square error is computed against 5, the mean of the larger component. The results illustrate the role of the tuning parameter c in controlling the tradeoff between robustness and efficiency and also show that our robust methods corresponding to, say $c = 1.0$, are competitive with or better than the minimum Hellinger distance estimator ($MHDE$) in terms of performance in this case. Similar results are obtained in simulation with the $N(\mu, \sigma^2)$ model, which are not presented here for brevity.

Table 3. Empirical means of the robustified weighted likelihood estimates ($RWLE$s), the maximum likelihood estimates (MLEs) and the minimum Hellinger distance estimates ($MHDE$s) for sample sizes 20, 50 and 100 under the *Poisson* model with empirical mean square errors of the estimates given in parentheses

	Poisson(5)					
Sample	$RWLE$					
Size	$c = 0.2$	$c = 1.0$	$c = 10.0$	$c = 50.0$	MLE	$MHDE$
$n = 20$	4.7766	4.8913	4.9962	5.0052	5.0080	4.8544
	(0.4266)	(0.3137)	(0.2514)	(0.2472)	(0.2465)	(0.3012)
$n = 50$	4.8714	4.9464	4.9994	5.0044	5.0059	4.9092
	(0.1464)	(0.1099)	(0.0958)	(0.0954)	(0.0955)	(0.1147)
$n = 100$	4.9096	4.9660	4.9951	4.9977	4.9986	4.9338
	(0.0703)	(0.0518)	(0.0481)	(0.0482)	(0.0482)	(0.0553)
	0.9 *Poisson*(5)+ 0.1 *Poisson*(12)					
$n = 20$	4.9856	5.1100	5.3620	5.5129	5.7288	5.1507
	(0.5504)	(0.4732)	(0.5825)	(0.7648)	(1.0325)	(0.4784)
$n = 50$	5.0750	5.1847	5.4084	5.5431	5.7157	5.2563
	(0.1877)	(0.1860)	(0.3365)	(0.4933)	(0.7192)	(0.2331)
$n = 100$	5.1314	5.2225	5.4413	5.5654	5.7101	5.3338
	(0.1087)	(0.1280)	(0.2878)	(0.4228)	(0.6075)	(0.1991)

6 Concluding Remarks

In this paper we have considered an inference technique based on a naturally-robust version of the maximum likelihood procedure. Recently, there has been quite a bit of activity in the area of density-based minimum divergence methods. Admittedly, there can be many choices of the function $C(\cdot)$ defined in (1) [or alternatively $A(\cdot)$ defined in (3)] that can generate reasonably robust yet efficient methods. Generally, it is difficult to single out some of these functions to be particularly more desirable than others although estimators such as the $MHDE$ have received special attention in the literature. So the natural question in connection with the proposed methods is: 'What makes them stand out?' The proposed functions $C_{c^*,c}$ and $A_{c^*,c}$ are special as they provide a natural modification of the likelihood disparity; and unless the value of c is very small, the MLE and $RWLE$ are exactly equal under the model with high probability. On the other hand when the data contain extreme outliers, the $RWLE$s are practically equivalent to the outlier deleted MLE

for moderate values of c, although larger choices of c will eventually force the estimators to coincide with the MLE for the full data. This is what we have, in fact, observed in many examples, including several that we have not presented here. The $RMLE$ also shows a similar behavior although it generally does not equal the MLE exactly except in discrete models. As the second derivative of the $A_{c^*,c}(\delta)$ function equals zero at $\delta = 0$, the estimators are automatically second order efficient in the sense described by [7] and [3]. The proposed estimators are naturally inlier robust as well (see, eg. [8], unlike the $MHDE$, although we have not done a detailed inlier analysis in this paper. The simulation results and real-life examples clearly show the desirable properties of the new inference methods.

References

1. Agostinelli, C., Markatou, M.: Test of hypotheses based on the Weighted Likelihood Methodology. Statistica Sinica 11, 499–514 (2001)
2. Ali, S.M., Silvey, S.D.: A general class of coefficients of divergence of one distribution from another. J. Royal Statist. Soc. Ser. B 28, 131–142 (1966)
3. Basu, A., Lindsay, B.G.: Minimum disparity estimation in the continuous case: efficiency, distributions, robustness. Ann. Inst. Statist. Math. 46, 683–705 (1994)
4. Basu, A., Sarkar, S., Vidyashankar, A.N.: Minimum negative exponential disparity estimation in parametric models. J. Statist. Plan. Infer. 58, 349–370 (1997)
5. Beran, R.J.: Minimum Hellinger distance estimates for parametric models. Ann. Statist. 5, 445–463 (1977)
6. Csiszár, I.: Eine informationstheoretische Ungleichung und ihre Anwendung auf den Beweis der Ergodizitat von Markoffschen Ketten. Publ. Math. Inst. Hungar. Acad. Sci. 3, 85–107 (1963)
7. Lindsay, B.G.: Efficiency versus robustness: the case for minimum Hellinger distance and related methods. Ann. Statist. 22, 1081–1114 (1994)
8. Mandal, A.: Minimum Disparity Inference: Strategies for Improvement in Efficiency. PhD Thesis, Indian Statistical Institute, Kolkata, India (2010)
9. Mandal, A., Bhandari, S.K., Basu, A.: Minimum disparity estimation based on combined disparities: Asymptotic results. J. Statist. Plan. Infer. 141, 701–710 (2011)
10. Markatou, M., Basu, A., Lindsay, B.G.: Weighted likelihood estimating equations: the discrete case with applications to logistic regression. J. Statist. Plan. Infer. 57, 215–232 (1997)
11. Markatou, M., Basu, A., Lindsay, B.G.: Weighted likelihood equations with bootstrap root search. J. Amer. Statist. Assoc. 93, 740–750 (1998)
12. Pardo, L.: Statistical Inference Based on Divergence Measures. Chapman & Hall/CRC, New York (2006)
13. Simpson, D.G.: Minimum Hellinger distance estimation for the analysis of count data. J. Amer. Statist. Assoc. 82, 802–807 (1987)
14. Stigler, S.M.: Do robust estimators work with real data? Ann. Statist. 5, 1055–1098 (1977)
15. Tamura, R.N., Boos, D.D.: Minimum Hellinger distance estimation for multivariate location and covariance. J. Amer. Statist. Assoc. 81, 223–229 (1986)

Hotelling's T^2-Test with Multivariate Normal Mixture Populations: Approximations and Robustness

Alfonso García-Pérez

Departamento de Estadística, Investigación Operativa y Cálculo Numérico,
Universidad Nacional de Educación a Distancia (UNED),
Madrid, Spain
agar-per@ccia.uned.es

Summary. Many classical multivariate tests, such as the one-sample Hotelling's T^2-test, are based in considering a multivariate normal distribution as underlying model because, in this situation, we can use the usual $F_{(a,b)}$-distribution to compute p-values and critical values. In this contributed paper we obtain good analytic approximations for these elements in a close to normal situation which allow us to analyze the robustness of this test against small departures from normality.

1 Introduction

Many parametric tests were obtained assuming a normal distribution as underlying model. This assumption can be avoided, in some cases, when we deal with univariate problems. Nevertheless, in a multivariate context, this dependence on the multivariate normal distribution is even stronger.

In García-Pérez [5] a method for obtaining good analytic approximations to the elements of a test was proposed, method that has been successfully applied, with univariate observations, to χ^2-tests in García-Pérez [6], to t-tests in García-Pérez [7] and to F-tests in García-Pérez [8].

In this paper we extend this methodology to multivariate observations, obtaining good analytic approximations for the p-value and the critical value of the Hotelling's T^2-test, when the underlying model is close but different from the multivariate normal distribution. In particular, we consider a mixture of two multivariate normal populations.

The exact density (but not the tail probability) of Hotelling's T^2-statistic, under a mixture of two multivariate normal, was been previously studied by Srivastava and Awan [17] and Gupta and Kabe [9] but the elements involved there are very hard to interpret and the density very unpleasant to apply (really they have to use numerical integration to analyze the robustness of Hotelling's T^2-test). On the contrary, the (very accurate) analytic approximations for the tail probability and the critical value that we obtain here allow a direct interpretation of the elements involved, are easier to apply, and its robustness nicer to interpret. Moreover, the approximations obtained can be applied to other multivariate models not necessarily a mixture of normals.

The idea of the method we expose, consists in considering the elements for which we obtain the approximations (the p-value and the critical value) as functionals of the model distribution. Then, to use the first two terms of their von Mises expansion that depends on (an obvious generalization of) the Tail Area Influence Function. This is finally approximated in two different ways: with a saddlepoint approximation and with an expansion provided by Fujikoshi [4].

The von Mises approximations are showed in Section 2 and the approximations for the TAIF are obtained in Section 3.

The approximations so established are applied for multivariate normal mixtures in Section 4 and some examples are included there.

The range of validity of the approximations obtained in the paper is settled in Section 5 with the Breakdown Condition.

Finally, the robustness of the one-sample Hotelling's T^2-test is analyzed in Section 6, concluding that, for small deviations of the multivariate normal model, this test has Robustness of Validity with symmetric or nearly symmetric underlying models but that this is lost with asymmetric distributions. Moreover, in this last situation, the lack of robustness gets worse as the dimension m of the observable random vector increases (maybe another *dimensionality curse*), because for the t-test ($m = 1$) we have Robustness of Validity also in this situation (García-Pérez [7]).

These conclusions are in accordance with the obtained, only performing simulations, by Everitt [2], Nachtsheim and Johnson [13] or Rencher ([15], pp. 97).

1.1 Preliminaries

Let $\mathcal{X}$ be the space where takes values the observable random vector $\mathbf{X}$, contained in (usually equal to) $\mathbb{R}^m$ and T a functional defined on a convex set $\mathcal{F}$ of distribution functions on $\mathcal{X}$, taking values in $\mathbb{R}^p$ (i.e., m is the dimension of the observable random vector and p the dimension of the parameter space; here, it will be $p = 1$ although we shall consider, mainly, two different functionals, the p-value and the critical value). All over the paper $\delta_{\mathbf{x}}$ will denote the probability measure which puts mass 1 at the point $\mathbf{x} \in \mathcal{X}$.

If there exists the Influence Function of the functional T at $G \in \mathcal{F}$, $IF(\mathbf{x}; T, G)$, the first-order von Mises expansion of T evaluated at the distribution $F \in \mathcal{F}$ is (see Withers [19], García-Pérez [5])

$$T(F) = T(G) + \int_{\mathcal{X}} IF(\mathbf{x}; T, G)\, dF(\mathbf{x}) + Rem \tag{1}$$

where

$$Rem = \frac{1}{2}\int_{\mathcal{X}}\int_{\mathcal{X}} T^{(2)}(\mathbf{x},\mathbf{y};T,G_F)\,d[F(\mathbf{x})-G(\mathbf{x})]\,d[F(\mathbf{y})-G(\mathbf{y})] \tag{2}$$

$$= \frac{1}{2}\int_{\mathcal{X}}\int_{\mathcal{X}} \frac{\partial}{\partial\epsilon} IF(\mathbf{x};T,G_F^{\epsilon,\mathbf{y}})\Big|_{\epsilon=0} d[F(\mathbf{x})-G(\mathbf{x})]\,d[F(\mathbf{y})-G(\mathbf{y})]$$

$$+\frac{1}{2}\int_{\mathcal{X}} IF(\mathbf{y};T,G_F)\,d[F(\mathbf{y})-G(\mathbf{y})]$$

and where $G_F = (1-t)F+tG$ for some $t\in[0,1]$, and $G_F^{\epsilon,\mathbf{y}} = (1-\epsilon)G_F+\epsilon\delta_{\mathbf{y}}$.

If F is close to G, the remainder term Rem will be close to zero. If we restrict our attention to ϵ-contamination models, $F=(1-\epsilon)G+\epsilon H$ (for instance, multivariate normal mixtures) we can bound the remainder term (2) by $O(\epsilon^2)$. Others authors, such as Hampel et al. ([10], chap. 3) or Ronchetti [16] consider ϵ-contamination models $F=(1-\epsilon/\sqrt{n})G+\epsilon/\sqrt{n}\,H$; in these cases, the remainder term can be bounded by $O(\epsilon^2/n)$ and the error can be then, controlled with the sample size n ($> m+1$ to be $\mathbf{S}$ nonsingular).

2 Multivariate Tests with an $F_{(a,b)}$-Distribution

Let us consider now multivariate tests in which the test statistic T_n follows an $F_{(a,b)}$-distribution under the null hypothesis, when the observations follow a multivariate normal distribution, $\mathbf{X}_i \sim \Phi_{\boldsymbol{\mu},\boldsymbol{\Sigma}} \equiv N(\boldsymbol{\mu},\boldsymbol{\Sigma})$. ("$X_i \sim F$" stands for "$X_i$ is distributed as F".)

Our aim is to obtain an analytic approximation for the p-value and the critical value of this kind of tests when the model distribution is not the multivariate normal but another one close to this.

If T is the functional p-value of a test based on the test statistic T_n, i.e., $p_n^F = P_{\mathbf{X}_i\sim F}\{T_n > t\}$, or the functional critical value k_n^F, and G is the multivariate normal distribution, the VOM approximations given by equation (1) are,

$$p_n^F \simeq p_n^{\Phi} + \int_{\mathcal{X}} \overset{\bullet}{p_n^{\Phi}}(\mathbf{x})\,dF(\mathbf{x}) \tag{3}$$

$$k_n^F \simeq k_n^{\Phi} + \int_{\mathcal{X}} \overset{\bullet}{k_n^{\Phi}}(\mathbf{x})\,dF(\mathbf{x}) \tag{4}$$

that will be more accurate as distribution F is closer to $\Phi \equiv \Phi_{\boldsymbol{\mu},\boldsymbol{\Sigma}}$.

Influence functions $\overset{\bullet}{p_n^{\Phi}}$ and $\overset{\bullet}{k_n^{\Phi}}$ are related with (an obvious multivariate generalization of) the Tail Area Influence Function defined by Field and Ronchetti [3] in the univariate context. Here, if $\mathbf{x}\in\mathcal{X}\subset\mathbb{R}^m$, we define

$$\text{TAIF}(\mathbf{x};t;T_n,H) = \frac{\partial}{\partial \epsilon} P_{H^{\epsilon,\mathbf{x}}}\{T_n > t\}\Big|_{\epsilon=0}$$

where $H^{\epsilon,\mathbf{x}} = (1-\epsilon)H + \epsilon\,\delta_{\mathbf{x}}$. In fact,

$$\overset{\bullet}{p}{}_n^{\Phi}(\mathbf{x}) = \text{TAIF}(\mathbf{x};t;T_n,\Phi) \qquad \text{and} \qquad \overset{\bullet}{k}{}_n^{\Phi}(\mathbf{x}) = \frac{\text{TAIF}(\mathbf{x};k_n^{\Phi};T_n,\Phi)}{f_{(a,b)}(k_n^{\Phi})}$$

where $f_{(a,b)}$ is the density function of an $F_{(a,b)}$-distribution with (a,b) degrees of freedom, i.e., of T_n when $\mathbf{X}_i \sim \Phi_{\boldsymbol{\mu},\boldsymbol{\Sigma}}$.

3 Approximations for the TAIF

To obtain VOM approximations (3) and (4) we need to compute the TAIF at the normal model; however, in most of the cases, this must be approximated before to be included there. In the paper we shall use two approximations of it. The first one is based on a saddlepoint approximation and the second one, on a specific approximation for the Hotelling's T^2-statistic. This statistic is defined, as usual, by

$$T^2 = n(\overline{\mathbf{x}} - \boldsymbol{\mu})'\mathbf{S}^{-1}(\overline{\mathbf{x}} - \boldsymbol{\mu})$$

if the $\mathbf{X}_i$'s have mean vector $\boldsymbol{\mu}$ and covariance matrix $\boldsymbol{\Sigma}$, and where $\overline{\mathbf{x}} = n^{-1}\sum_{i=1}^n \mathbf{X}_i$ and $\mathbf{S} = (n-1)^{-1}\sum_{i=1}^n (\mathbf{X}_i - \overline{\mathbf{x}})(\mathbf{X}_i - \overline{\mathbf{x}})'$, for which we know that $T_n = (n-m)T^2/[m(n-1)] \sim F_{(m,n-m)}$ if $\mathbf{X}_i \sim \Phi_{\boldsymbol{\mu},\boldsymbol{\Sigma}}$.

3.1 SAD Approximation for the TAIF

The first approximation that we shall obtain is based on the saddlepoint approximation for the tail probability given by the Lugannani and Rice formula (see Lugannani and Rice [12] or, better, Daniels [1]). T_n can be expressed as a ratio of sums of squares of independent variables (Rao [14], pp. 541) as

$$T_n = \frac{(n-m)T^2}{m(n-1)} = \frac{(n-m)}{m(n-1)}(n-1)\frac{n(\overline{\mathbf{x}} - \boldsymbol{\mu})'\boldsymbol{\Sigma}^{-1}(\overline{\mathbf{x}} - \boldsymbol{\mu})}{\frac{(\overline{\mathbf{x}} - \boldsymbol{\mu})'\boldsymbol{\Sigma}^{-1}(\overline{\mathbf{x}} - \boldsymbol{\mu})}{(\overline{\mathbf{x}} - \boldsymbol{\mu})'[(n-1)\mathbf{S}]^{-1}(\overline{\mathbf{x}} - \boldsymbol{\mu})}}$$

that, under a $\Phi \equiv \Phi_{\boldsymbol{\mu},\boldsymbol{\Sigma}}$ model follows an $F_{(m,n-m)}$-distribution. Under this distribution we have that

$$P_{\Phi}\{T_n > t\} = P_{\Phi}\{Y_1 - t\,Y_2 > 0\}$$

where $Y_1 \sim \gamma(\frac{m}{2},\frac{m}{2})$ distribution and $Y_2 \sim \gamma(\frac{n-m}{2},\frac{n-m}{2})$. Hence, using Lugannani and Rice formula for a sample of size one of the random variable $W = Y_1 - t\,Y_2$ with cumulant generating function

$$K(v) = m \log M_{\gamma}(v/m) + (n-m)\log M_{\gamma}(-tv/(n-m)) \tag{5}$$

where

$$M_\gamma(v) = \int_{-\infty}^{\infty} e^{v\,z^2}\, d\Phi_s(z) = \int_{-\infty}^{\infty} \dots \int_{-\infty}^{\infty} e^{v\,(u_c-\mu_c)^2/\sigma_c^2}\, d\Phi_{\boldsymbol{\mu},\boldsymbol{\Sigma}}(u_1,\dots,u_m) \tag{6}$$

being μ_c and σ_c the marginal mean and standard deviation of the cth component of the multivariate normal distribution, we have that

$$p_n^{\Phi} = P_{\Phi}\{T_n > t\} = P\{W > 0\} = 1 - \Phi_s(w) + \phi_s(w)\left\{\frac{1}{r} - \frac{1}{w} + O(1)\right\} \tag{7}$$

where Φ_s and ϕ_s are the cumulative distribution and density functions of the univariate standard normal distribution, $w = sign(z_0)\sqrt{-2K(z_0)}$, $r = z_0\sqrt{K''(z_0)}$ and z_0 the saddlepoint solution of $K'(z_0) = 0$.

In this way we express the functional *tail probability under a normal model*, $P_{\Phi}\{T_n > t\} = p_n^{\Phi}$ explicitly depending on this normal model $\Phi_{\boldsymbol{\mu},\boldsymbol{\Sigma}}$. Now, changing this by the contaminated model $\Phi^{\epsilon,\mathbf{x}} = (1-\epsilon)\Phi_{\boldsymbol{\mu},\boldsymbol{\Sigma}} + \epsilon\,\delta_{\mathbf{x}}$ and differentiating the resultant functional with respect to ϵ at $\epsilon = 0$ we obtain (see García-Pérez [8] for details),

$$\widehat{\mathrm{TAIF}}_1\left(\mathbf{x}; t; T_n, \Phi_{\boldsymbol{\mu},\boldsymbol{\Sigma}}\right) \simeq C_1 \left\{\dot{K}\, C_2 - \frac{\dot{z}_0}{z_0} - \frac{\dot{K}''}{2K''}\right\}$$

where C_1 and C_2 are the constants

$$C_1 = \frac{e^K}{\sqrt{2\pi}z_0\sqrt{K''}} \quad , \quad C_2 = \left[1 - \frac{z_0\sqrt{K''}}{(-2K)^{3/2}}\right].$$

In (6) we select the cth component that makes the pivotal multivariate normal closest to the distribution F (see examples below).

Now, making the same computations than in García-Pérez [8] we obtain the following approximations because in a Hotelling's T^2-test is $T_n = (n-m)T^2/[m\;(n-1)] \sim F_{(m,n-m)}$ if $\mathbf{X}_i \sim \Phi_{\boldsymbol{\mu},\boldsymbol{\Sigma}}$, and the first degree of freedom does not depend on the sample size,

$$\begin{aligned} p_n^F \simeq p_n^{\Phi} + A_1 \int_{\mathcal{X}} \Bigg\{ \left(A_2\,m - \frac{3\,t+1}{4\,(t-1)}\right) t^{-1/2}\, e^{\frac{t-1}{2t}\frac{(x_c-\mu_c)^2}{\sigma_c^2}} \\ + \frac{3t-1}{2(t-1)}\, t^{-3/2}\, \frac{(x_c-\mu_c)^2}{\sigma_c^2}\, e^{\frac{t-1}{2t}\frac{(x_c-\mu_c)^2}{\sigma_c^2}} - \frac{1}{4}\, t^{-5/2}\, \frac{(x_c-\mu_c)^4}{\sigma_c^4}\, e^{\frac{t-1}{2t}\frac{(x_c-\mu_c)^2}{\sigma_c^2}} \\ - \frac{t}{t-1}\, \frac{(x_c-\mu_c)^2}{\sigma_c^2} + \frac{t}{t-1} - A_2\,m \Bigg\}\, dF(x_1,\dots,x_m) \end{aligned} \tag{8}$$

where

$$A_1 = e^{-m(t-1)/2} \; \frac{t^{m/2}}{\sqrt{\pi}\,\sqrt{m}\,(t-1)} \qquad , \qquad A_2 = 1 - \frac{t-1}{m\,\sqrt{2}\,(t-1-\log t)^{3/2}}.$$

For the critical value we obtain

$$k_n^F \simeq t + \frac{A_1}{f_{(m,n-m)}(t)} \int_{\mathcal{X}} \left\{ \left(A_2\, m - \frac{3\,t+1}{4\,(t-1)} \right) t^{-1/2}\, e^{\frac{t-1}{2t}\frac{(x_c-\mu_c)^2}{\sigma_c^2}} \right.$$
$$+ \frac{3t-1}{2(t-1)}\, t^{-3/2}\, \frac{(x_c-\mu_c)^2}{\sigma_c^2}\, e^{\frac{t-1}{2t}\frac{(x_c-\mu_c)^2}{\sigma_c^2}} - \frac{1}{4}\, t^{-5/2}\, \frac{(x_c-\mu_c)^4}{\sigma_c^4}\, e^{\frac{t-1}{2t}\frac{(x_c-\mu_c)^2}{\sigma_c^2}}$$
$$\left. - \frac{t}{t-1}\, \frac{(x_c-\mu_c)^2}{\sigma_c^2} + \frac{t}{t-1} - A_2\, m \right\} dF(x_1, ..., x_m) \qquad (9)$$

where, in this approximation (9), $t = F_{(m,n-m);\alpha}$ is the $(1-\alpha)$-quantile of an $F_{(m,n-m)}$-distribution and $f_{(m,n-m)}$ the density function of this distribution.

Let us observe that if $F \sim \Phi_{\boldsymbol{\mu},\boldsymbol{\Sigma}}$ it is $\int_{\mathcal{X}} \widehat{\text{TAIF}}_1 \left(\mathbf{x}; t; T_n, \Phi_{\boldsymbol{\mu},\boldsymbol{\Sigma}} \right) d\Phi_{\boldsymbol{\mu},\boldsymbol{\Sigma}}(x_1, ..., x_n) = 0$, and so, $p_n^F = p_n^\Phi$ and $k_n^F = F_{(m,n-m);\alpha}$. Hence, (8) and (9) are *linear* generalizations of the usual values used under a normal model.

3.2 Fujikoshi Approximation for the TAIF

Instead of using a general saddlepoint approximation to the tail probability as the starting point for obtaining the approximation of the TAIF, in this section we propose to use an existing good analytic approximation for the tail probability of the Hotelling's T^2-statistic at the normal model, as the starting point in the obtaining of the approximation to the TAIF. Namely, the approximation given in Fujikoshi ([4], pp. 189-190),

$$p_n^\Phi = P_\Phi\{T_n > t\} = P_\Phi\{T^2 > y\} = 1 - H_m(y) - \frac{1}{n}[\beta_0\, H_m(y) + \beta_1\, H_{m+2}(y)$$
$$+ \beta_2\, H_{m+4}(y) + \beta_3\, H_{m+6}(y)] + o(n^{-1}) \qquad (10)$$

where $y = m(n-1)t/(n-m)$ and H_p the cumulative distribution function of a χ_p^2-distribution.

Because we are considering this tail probability at the normal distribution and this is an elliptical distribution $E_m(\boldsymbol{\mu}, \boldsymbol{\Sigma}, \psi)$ where the *characteristic generator* is $\psi(u) = \exp(-u/2)$, the coefficients β_i that appear in (10) are (see Iwashita [11])

$$\beta_0 = -\tfrac{1}{4}m[m+(m+2)k] \quad , \quad \beta_1 = -\tfrac{1}{2}m[1-(m+2)k]$$

$$\beta_2 = \tfrac{1}{4}m(m+2)(1-k) \quad , \quad \beta_3 = 0$$

where k is the *kurtosis parameter* defined by $k = \psi''(0)/(\psi'(0))^2 - 1$.

Let us observe that the dependence on (i.e., the influence of) the underlying distribution (normal in our case) lies only in the β_i coefficients, not on the H_p, which are common to all F models. Hence,

$$P_\Phi\{T_n > t\} \simeq 1 - H_m(y) - \frac{1}{n}\left[\beta_0\, H_m(y) + \beta_1\, H_{m+2}(y) + \beta_2\, H_{m+4}(y)\right].$$

Now, changing $\Phi_{\boldsymbol{\mu},\boldsymbol{\Sigma}}$ by the contaminated model $\Phi^{\epsilon,\mathbf{x}} = (1-\epsilon)\Phi_{\boldsymbol{\mu},\boldsymbol{\Sigma}} + \epsilon\,\delta_{\mathbf{x}}$ and differentiating with respect to ϵ at $\epsilon = 0$, we obtain (the approximation of) the TAIF at the normal model,

$$\widehat{\mathrm{TAIF}}_2(\mathbf{x};t;T_n,\Phi_{\boldsymbol{\mu},\boldsymbol{\Sigma}}) \simeq -\frac{1}{n}\left[\dot{\beta_0}\; H_m(y) + \dot{\beta_1}\; H_{m+2}(y) + \dot{\beta_2}\; H_{m+4}(y)\right]$$

$$= \frac{m\,(m+2)}{4\,n}\;\dot{k}\;\left[H_m(y) - 2\,H_{m+2}(y) + H_{m+4}(y)\right]$$

because $\dot{\beta_0} = \dot{\beta_2} = -m(m+2)\,\dot{k}\,/4$ and $\dot{\beta_1} = m(m+2)\,\dot{k}\,/2$.

Contaminating also functional k we obtain $k_\epsilon = \psi''(0)_\epsilon/(\psi'(0)_\epsilon)^2 - 1$ and

$$\dot{k} = \frac{\dot{\psi''}(0)\;[\psi'(0)]^2 - 2\,\psi'(0)\;\dot{\psi'}(0)\;\psi''(0)}{[\psi'(0)]^4} = 4\left[\dot{\psi''}(0) + \dot{\psi'}(0)\right]$$

because $\psi'(0) = -1/2$ and $\psi''(0) = 1/4$.

Finally, after some computations, we obtain

$$\dot{\psi'}(0) = \frac{1}{2}\left[1 - \left(\frac{x_c-\mu_c}{\sigma_c}\right)^2\right] \qquad\qquad \dot{\psi''}(0) = \frac{1}{12}\left[\left(\frac{x_c-\mu_c}{\sigma_c}\right)^4 - 3\right]$$

where μ_c and σ_c are the marginal mean and standard deviation of the cth component of the multivariate normal distribution, that make the pivotal multivariate normal closest to the distribution F (see examples below). Hence,

$$\widehat{\mathrm{TAIF}}_2\left(\mathbf{x};t;T_n,\Phi_{\boldsymbol{\mu},\boldsymbol{\Sigma}}\right) \simeq \frac{m\,(m+2)}{4\,n}\left[1 - 2\left(\frac{x_c-\mu_c}{\sigma_c}\right)^2 + \frac{1}{3}\left(\frac{x_c-\mu_c}{\sigma_c}\right)^4\right]$$

$$\cdot\left[H_m(y) - 2\,H_{m+2}(y) + H_{m+4}(y)\right]$$

and so,

$$p_n^F \simeq p_n^{\Phi} + \frac{m\,(m+2)}{4\,n}\;[H_m(y) - 2\,H_{m+2}(y) + H_{m+4}(y)]$$

$$\cdot \left[1 + \int_{\mathcal{X}} \left\{ \frac{1}{3}\frac{(x_c-\mu_c)^4}{\sigma_c^4} - 2\frac{(x_c-\mu_c)^2}{\sigma_c^2} \right\} dF(x_1,...,x_m) \right]. \quad (11)$$

And for the critical value,

$$k_n^F \simeq t + \frac{m\,(m+2)}{4\,n\,f_{(m,n-m)}(t)}\;[H_m(y) - 2\,H_{m+2}(y) + H_{m+4}(y)]$$

$$\cdot \left[1 + \int_{\mathcal{X}} \left\{ \frac{1}{3}\frac{(x_c-\mu_c)^4}{\sigma_c^4} - 2\frac{(x_c-\mu_c)^2}{\sigma_c^2} \right\} dF(x_1,...,x_m) \right] \quad (12)$$

where, in approximation (12), $t = F_{(m,n-m);\alpha}$ is the $(1-\alpha)$-quantile of an $F_{(m,n-m)}$-distribution and $f_{(m,n-m)}$ the density function of this distribution.

Let us observe again that if $F \sim \Phi_{\boldsymbol{\mu},\boldsymbol{\Sigma}}$ it is $\int_{\mathcal{X}} \widehat{\mathrm{TAIF}}_2\left(\mathbf{x}; t; T_n, \Phi_{\boldsymbol{\mu},\boldsymbol{\Sigma}}\right) d\Phi_{\boldsymbol{\mu},\boldsymbol{\Sigma}}(x_1,...,x_n) = 0$, and so, $p_n^F = p_n^{\Phi}$ and $k_n^F = F_{(m,n-m);\alpha}$ obtaining again *linear* extensions of the usual normal values.

4 Hotelling's T^2-Test with a Multivariate Normal Mixture Population

From here, we shall consider only p-values, as model F a Location/Scale Contaminated Normal (LSCN) distribution

$$F = (1-\lambda)\,\Phi_{\boldsymbol{\mu_1},\boldsymbol{\Sigma_1}} + \lambda\,\Phi_{\boldsymbol{\mu_2},\boldsymbol{\Sigma_2}}$$

and, as pivotal distribution, the multivariate normal $\Phi_{\boldsymbol{\mu_1},\boldsymbol{\Sigma_1}}$. Because the integral of the TAIF with respect the pivotal distribution is zero, using a left-superscript on the marginal mean and variance to distinguish between the two normals of the mixture, we have that approximations (8) and (11) are now: The **SAD approximation**,

$$p_n^F \simeq p_n^{\Phi} + \lambda\,A_1 \int_{\mathcal{X}} \left\{ \left(A_2\,m - \frac{3\,t+1}{4\,(t-1)} \right) t^{-1/2}\, e^{\frac{t-1}{2t}\frac{(x_c-{}^1\mu_c)^2}{{}^1\sigma_c^2}} \right.$$

$$+ \frac{3t-1}{2(t-1)}\,t^{-3/2}\,\frac{(x_c-{}^1\mu_c)^2}{{}^1\sigma_c^2}\,e^{\frac{t-1}{2t}\frac{(x_c-{}^1\mu_c)^2}{{}^1\sigma_c^2}}$$

$$- \frac{1}{4}\,t^{-5/2}\,\frac{(x_c-{}^1\mu_c)^4}{{}^1\sigma_c^4}\,e^{\frac{t-1}{2t}\frac{(x_c-{}^1\mu_c)^2}{{}^1\sigma_c^2}}$$

$$\left. - \frac{t}{t-1}\,\frac{(x_c-{}^1\mu_c)^2}{{}^1\sigma_c^2} + \frac{t}{t-1} - A_2\,m \right\} d\Phi_{\boldsymbol{\mu_2},\boldsymbol{\Sigma_2}}(x_1,...,x_m) \quad (13)$$

and the **Fujikoshi approximation**

$$p_n^F \simeq p_n^\Phi + \frac{m\,(m+2)}{4\,n}\,[H_m(y) - 2\,H_{m+2}(y) + H_{m+4}(y)]$$

$$\lambda\left[1 + \int_{\mathcal{X}} \left\{\frac{1}{3}\frac{(x_c - {}^1\mu_c)^4}{{}^1\sigma_c^4} - 2\frac{(x_c - {}^1\mu_c)^2}{{}^1\sigma_c^2}\right\} d\Phi_{\boldsymbol{\mu_2},\boldsymbol{\Sigma_2}}(x_1, ..., x_m)\right] \quad (14)$$

Approximations (13) and (14) can easily be computed because

$$\int_{\mathcal{X}} e^{\frac{t-1}{2t}\frac{(x_c - {}^1\mu_c)^2}{{}^1\sigma_c^2}}\, d\Phi_{\boldsymbol{\mu_2},\boldsymbol{\Sigma_2}}(x_1, ..., x_m)$$

$$= \frac{A_3\, {}^1\sigma_c\, \sqrt{t}}{\sqrt{t\ {}^1\sigma_c^2 - {}^2\sigma_c^2\ (t-1)}}$$

$$\int_{\mathcal{X}} \frac{(x_c - {}^1\mu_c)^2}{{}^1\sigma_c^2}\, e^{\frac{t-1}{2t}\frac{(x_c - {}^1\mu_c)^2}{{}^1\sigma_c^2}}\, d\Phi_{\boldsymbol{\mu_2},\boldsymbol{\Sigma_2}}(x_1, ..., x_m)$$

$$= \frac{A_3\ \sqrt{t}}{{}^1\sigma_c \sqrt{t\ {}^1\sigma_c^2 - {}^2\sigma_c^2\ (t-1)}}\,[\sigma^2 + (\mu - {}^1\mu_c)^2]$$

$$\int_{\mathcal{X}} \frac{(x_c - {}^1\mu_c)^4}{{}^1\sigma_c^4}\, e^{\frac{t-1}{2t}\frac{(x_c - {}^1\mu_c)^2}{{}^1\sigma_c^2}}\, d\Phi_{\boldsymbol{\mu_2},\boldsymbol{\Sigma_2}}(x_1, ..., x_m)$$

$$= \frac{A_3\ \sigma}{{}^1\sigma_c^4\ {}^2\sigma_c}\,[\mu_4 + 4(\mu - {}^1\mu_c)\,\mu_3 + 6(\mu - {}^1\mu_c)^2\,\sigma^2 + (\mu - {}^1\mu_c)^4]$$

$$\int_{\mathcal{X}} \frac{(x_c - {}^1\mu_c)^2}{{}^1\sigma_c^2}\, d\Phi_{\boldsymbol{\mu_2},\boldsymbol{\Sigma_2}}(x_1, ..., x_m) = \frac{1}{{}^1\sigma_c^2}\,[{}^2\sigma_c^2 + ({}^2\mu_c - {}^1\mu_c)^2]$$

$$\int_{\mathcal{X}} \frac{(x_c - {}^1\mu_c)^4}{{}^1\sigma_c^4}\, d\Phi_{\boldsymbol{\mu_2},\boldsymbol{\Sigma_2}}(x_1, ..., x_m) =$$

$$\frac{1}{{}^1\sigma_c^4}\,[{}^2\mu_{4c} + 4({}^2\mu_c - {}^1\mu_c)\,{}^2\mu_{3c} + 6({}^2\mu_c - {}^1\mu_c)^2\ \ {}^2\sigma_c^2 + ({}^2\mu_c - {}^1\mu_c)^4]$$

where

$$A_3 = \exp\left\{\frac{(t-1)({}^1\mu_c - {}^2\mu_c)^2}{2(t\ {}^1\sigma_c^2 - {}^2\sigma_c^2\ (t-1)}\right\}\,,\ \mu = \frac{{}^2\mu_c\ {}^1\sigma_c^2\ t - {}^1\mu_c\ {}^2\sigma_c^2\ (t-1)}{t\ {}^1\sigma_c^2 - {}^2\sigma_c^2\ (t-1)}\,,$$

$$\sigma = \frac{\sqrt{t}\ {}^1\sigma_c\ {}^2\sigma_c}{\sqrt{t\ {}^1\sigma_c^2 - {}^2\sigma_c^2\ (t-1)}}\,,\ \mu_3 = \int_{-\infty}^{\infty} (x-\mu)^3\, d\Phi_{\mu,\sigma}(x)\,,$$

$$\mu_4 = \int_{-\infty}^{\infty} (x-\mu)^4\, d\Phi_{\mu,\sigma}(x)\,,\ {}^2\mu_{3c} = \int_{-\infty}^{\infty} (x - {}^2\mu_c)^3\, d\Phi_{{}^2\mu_c,{}^2\sigma_c}(x)\,,$$

$${}^2\mu_{4c} = \int_{-\infty}^{\infty} (x - {}^2\mu_c)^4\, d\Phi_{{}^2\mu_c,{}^2\sigma_c}(x).$$

4.1 Scale Contaminated Normal (SCN) population

If the underlying model is $F = (1-\lambda)\,\Phi_{\boldsymbol{\mu},\boldsymbol{\Sigma}} + \lambda\,\Phi_{\boldsymbol{\mu},b\boldsymbol{\Sigma}}$ then, it is ${}^2\mu_c = {}^1\mu_c$ and ${}^2\sigma_c = \sqrt{b}\;{}^1\sigma_c$ in expressions (13) and (14). The approximations appear in the next example.

Example 1. If the pivotal distribution is the standard multivariate normal distribution, $N(\mathbf{0},\mathbf{I})$, it is ${}^1\mu_c = {}^2\mu_c = 0$ and ${}^2\sigma_c = \sqrt{b}\;{}^1\sigma_c = \sqrt{b}$ and the approximations will finally be: the **SAD approximation**, if $t>1$ and $b<t/(t-1)$,

$$p_n^F \simeq p_n^{\Phi} + \lambda\,A_1 \left\{ \left(A_2\,m - \frac{3\,t+1}{4\,(t-1)} \right) [t-b(t-1)]^{-1/2} + \frac{3t-1}{2(t-1)}\,b \right.$$
$$\left. \cdot [t-b(t-1)]^{-3/2} - \frac{3}{4}\,b^2\,[t-b(t-1)]^{-5/2} + \frac{t}{t-1}\,(1-b) - A_2\,m \right\}$$

and the **Fujikoshi approximation**,

$$p_n^F \simeq p_n^{\Phi} + \frac{m\,(m+2)}{4\,n}\,[H_m(y) - 2\,H_{m+2}(y) + H_{m+4}(y)]\;\lambda\,(1-b)^2.$$

Let us observe that if $b=1$ or $\lambda=0$, i.e., if we have a standard multivariate normal distribution instead of a scale contaminated normal distribution, the second term in the previous approximations is zero and we have the usual p-value.

In this example, the selection of the cth component that makes the pivotal distribution closest to the underlying model F, is irrelevant because all the components have the same mean and the same variance but, if this would not be the case, we should choose the cth component such that $|1-\sigma_c^2| = \min_{i=1,\ldots,m}\{|1-\sigma_i^2|\}$.

To give some numerical values of the previous approximations, we show in Table 1 the "*exact*" values (really obtained with the package R and a simulation of 80.000 samples), of the p-value of a test based on a Hotelling's T^2-statistic where $m=4$ and $n=49$, considering here, the SCN, $0.95\,N(\mathbf{0},\mathbf{I}) + 0.05\,N(\mathbf{0},1.2\cdot\mathbf{I})$, instead of the usual multivariate standard normal distribution.

Table 1. *Exact*, VOM+SAD and VOM+Fujikoshi approximate p-values under a SCN model

t	"exact"	VOM+SAD approx.	VOM+Fujikoshi approx.
2	0.1108	0.1102	0.1107
2.5	0.0557	0.0560	0.0557
3	0.0276	0.0282	0.0281
3.5	0.0141	0.0141	0.0143
4	0.0073	0.0069	0.0073
4.5	0.0036	0.0032	0.0038

From this table we see that both approximations are quite good. In the previous situation, i.e., under a $0.95\,N(\mathbf{0},\mathbf{I}) + 0.05\,N(\mathbf{0}, 1.2\cdot\mathbf{I})$ model, the test statistic T_n does not follow now an $F_{(4,45)}$-distribution. Really, using the previous formulas, the (approximate) actual level of the classical 0.05 $F_{(4,45)}$-test is 0.0503 with a SAD approximation and 0.049987 with the Fujikoshi approximation, under this model, showing signs of stability in the level with SCN models.

4.2 Location Contaminated Normal (LCN) Population

In this situation, the model distribution F for which we want to approximate the p-value is $F = (1-\lambda)\,\Phi_{\boldsymbol{\mu}_1,\boldsymbol{\Sigma}} + \lambda\,\Phi_{\boldsymbol{\mu}_2,\boldsymbol{\Sigma}}$. Hence, we have to make ${}^2\sigma_c = {}^1\sigma_c$ in expressions (13) and (14). The approximations obtained appear in the next example.

Example 2. If the pivotal distribution is the standard multivariate normal distribution, $N(\mathbf{0},\mathbf{I})$, it is ${}^1\mu_c = 0$, ${}^2\mu_c = \theta$ and ${}^2\sigma_c = {}^1\sigma_c = 1$. The approximations will be then: The **SAD approximation**, if $t > 1$,

$$p_n^F \simeq p_n^\Phi + \lambda\,A_1 \left\{ \left(e^{(t-1)\theta^2/2} - 1 \right) \left(A_2\,m + \frac{t\theta^2}{t-1} \right) - \frac{\theta^4\,t^2}{4}\,e^{(t-1)\theta^2/2} \right\}$$

and the **Fujikoshi approximation**,

$$p_n^F \simeq p_n^\Phi + \frac{m\,(m+2)}{4\,n}\,\left[H_m(y) - 2\,H_{m+2}(y) + H_{m+4}(y)\right]\,\lambda\,\frac{\theta^4}{3}.$$

Let us observe that if $\theta = 0$ or $\lambda = 0$, i.e., if we have a standard multivariate normal distribution instead of a location contaminated normal distribution, the second term in all the previous approximations is zero and we have the usual p-value. In this example, we should choose the cth component if $|\theta_c| = \min_{i=1,\dots,m}\{|\theta_i|\}$.

In Table 2 we show the "exact" values (obtained with the package R and a simulation of 80.000 samples), of the p-value of a test based on a Hotelling's T^2-statistic where $m = 4$ and $n = 49$, considering here, the LCN, $0.95\,N(\mathbf{0},\mathbf{I}) + 0.05\,N(\boldsymbol{\theta},\mathbf{I})$, where $\boldsymbol{\theta}' = (\theta, ..., \theta) = (1, ..., 1)$ instead of the usual multivariate standard normal distribution.

From this table we see again that both approximations are quite good. If we used it to obtain, in the same setting, the (approximate) actual level of the 0.05 $F_{(4,45)}$-test based on $T_n = (49-4)T^2/(4\cdot 48)$ under a $0.95\,N(\mathbf{0},\mathbf{I}) + 0.05\,N(\mathbf{1},\mathbf{I})$ model, for which T_n does not follow now an $F_{(4,45)}$-distribution, we should obtain that the (approximate) actual level is 0.0511, using the SAD approximation, and 0.049987, using the Fujikoshi approximation, showing again robustness of validity also with location contaminated multivariate normal models.

Table 2. *Exact*, VOM+SAD and VOM+Fujikoshi approximate p-values under a LCN model

t	"exact"	VOM+SAD approx.	VOM+Fujikoshi approx.
2	0.1186	0.1072	0.1106
2.5	0.0599	0.0568	0.0556
3	0.0310	0.0287	0.0280
3.5	0.0157	0.0141	0.0142
4	0.0080	0.0068	0.0073
4.5	0.0037	0.0032	0.0038

5 Breakdown Condition

As far as we move away from the pivotal distribution (the multivariate normal in the paper), the approximations are getting worse until they "break down". Indeed, this is the concept of breakdown point: the limit up to the VOM approximations (and so, the VOM+SAD and VOM+Fujikoshi approximations) are valid because they are based on an Influence Function. Nevertheless, this concept must be suited to the context, as Staudte and Sheather ([18], pp. 41) say. We consider the *Breakdown Condition*

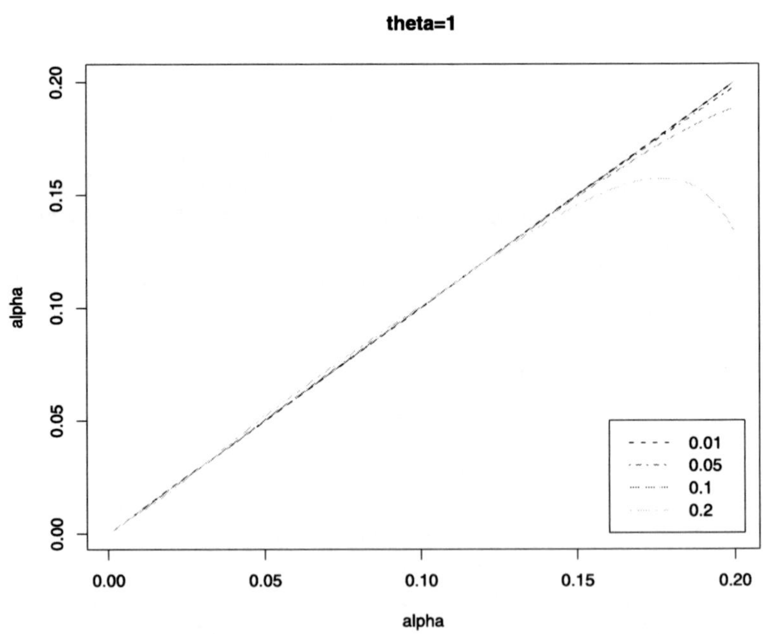

Fig. 1. α versus $P_F\{T_n > k_n^F\}$ for $\lambda = 0.01\,,\, 0.05\,,\, 0.1\,,\, 0.2$

$$P_F\{T_n > k_n^F\} \approx \alpha \tag{15}$$

where the tail probability and the critical value in this equation are computed with the previous approximations.

Let us observe that this k_n^F is the actual (approximate) critical value, not the nominal one obtained with the nominal level. Hence, the Breakdown Condition is valid not only for robust tests but for any kind of test.

Moreover, we remark that the proposed approximations are local in nature because they are based on an Influence Function. Then, their use will be limited to a reduced neighborhood around the pivotal distribution. Hence, the Breakdown Condition depends on the class of distributions considered; i.e., the limit distribution for which the condition breaks down, depends on the problem at issue.

Let us consider, for instance, VOM+SAD approximations and the LCN model considered in Example 4.2, $(1-\lambda)\,N(\mathbf{0},\mathbf{I}) + \lambda\,N(\boldsymbol{\theta},\mathbf{I})$.

If we use graphics to analyze where condition (15) breaks down, we can observe in Figure 1 different representations of this condition: α versus $P_F\{T_n > k_n^F\}$, for $\lambda = 0.01$, $\lambda = 0.05$, $\lambda = 0.1$, $\lambda = 0.2$, and $\boldsymbol{\theta}' = (\theta, ..., \theta) = (1, ..., 1)$. We can deduce from this picture that the Breakdown Condition is satisfied very well. Nevertheless, if we now move the location parameter to $\theta = 1.4$, we should observe that condition (15) should break down when we pass from $\lambda = 0.01$ to $\lambda = 0.05$. Then, for a range of contamination $0 \le \lambda \le 0.2$, the location parameter θ can vary between $-1 \le \theta \le 1$ (negative values of θ have the same effect because this parameter appears to the square in the approximations of Example 4.2).

Hence, for LCN models, the VOM+SAD approximations can be used for models F in the class $\{F = (1-\lambda)\,N(\mathbf{0},\mathbf{I}) + \lambda\,N(\boldsymbol{\theta},\mathbf{I}) \;\; 0 \le \lambda \le 0.2\,,\, -1 \le \theta \le 1\}$.

Of course we can increase (decrease) the value of λ and decrease (increase) θ if we want to restructure the class of distributions in which the VOM+SAD approximations work well.

In the same way, for SCN models, the VOM+SAD approximations can be used for models F in the class $\{F = (1-\lambda)\,N(\mathbf{0},\mathbf{I}) + \lambda\,N(\mathbf{0}, b\cdot\mathbf{I}) \;\; 0 \le \lambda \le 0.2\,,\, 0 < b \le 1.3\}$.

Similarly, for SCN models, the VOM+Fujikoshi approximations can be used for models F in the class $\{F = (1-\lambda)\,N(\mathbf{0},\mathbf{I}) + \lambda\,N(\mathbf{0}, b\cdot\mathbf{I}) \;\; 0 \le \lambda < 0.4\,,\, 0 < b \le 4\}$.

And finally, for LCN models, the VOM+Fujikoshi approximations can be used for models F in the class $\{F = (1-\lambda)\,N(\mathbf{0},\mathbf{I}) + \lambda\,N(\boldsymbol{\theta},\mathbf{I}) \;\; 0 \le \lambda < 0.4\,,\, -2.5 \le \theta \le 2.5\}$. Hence, the Fujikoshi approximations are wider than the SAD ones.

6 Robustness of Hotelling's T^2-Test

In an α-test based on the test statistic $T_n(X_1, ..., X_n)$ (that rejects H_0 for large values), α is the nominal level of the test and $P_{X_i \sim F}\{T_n > c_\alpha\}$ the actual level of the test under the model F, if c_α is the critical value of the test under a multivariate normal model Φ.

We can say that the test has a complete robustness of validity under a distribution F, with respect to the normal, if $\forall\,\alpha,\; P_{X_i \sim F}\{T_n > c_\alpha\} = \alpha$.

We can display this robustness of validity, plotting, for different $\alpha' s$, the nominal level versus the actual level of the test; i.e., plotting the pairs of points $(\alpha, P\{T_n > c_\alpha\})$. We call the resultant diagram, "Robustness of Validity Plot".

Obviously, because for all α, it is $P_{X_i \sim \Phi}\{T_n > c_\alpha\} = \alpha$, a completely robust test will be represented, in this figure, on the diagonal line. As far as we move away from this line, the test will be less robust and points over and under this line will indicate over and under actual levels of the test.

With the previous VOM+Fujikoshi approximations of the p-values now it is possible to compute and represent these lines very easily, obtaining so a quick idea of the robustness of validity of a test, especially if we want to compare several overprinted tests.

As we saw in the previous Section, the VOM+Fujikoshi approximations are valid just in a small neighborhood of the target distribution and so, we can study the robustness in this neighborhood.

If we consider several LCN $(1-\lambda)\,N(\mathbf{0},\mathbf{I}) + \lambda\,N(\boldsymbol{\theta},\mathbf{I})$ models, moving (θ,λ), the "Robustness of Validity Plot" of Hotelling's test appears in Figure 2. From this we obtain the same conclusion that Everitt [2], Nachtsheim

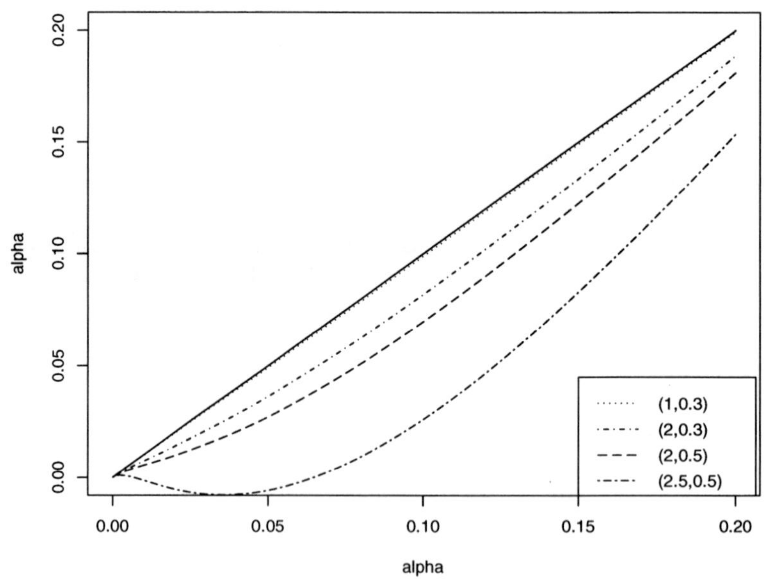

Fig. 2. Nominal levels versus Actual levels with a LCN model for different values of (θ, λ)

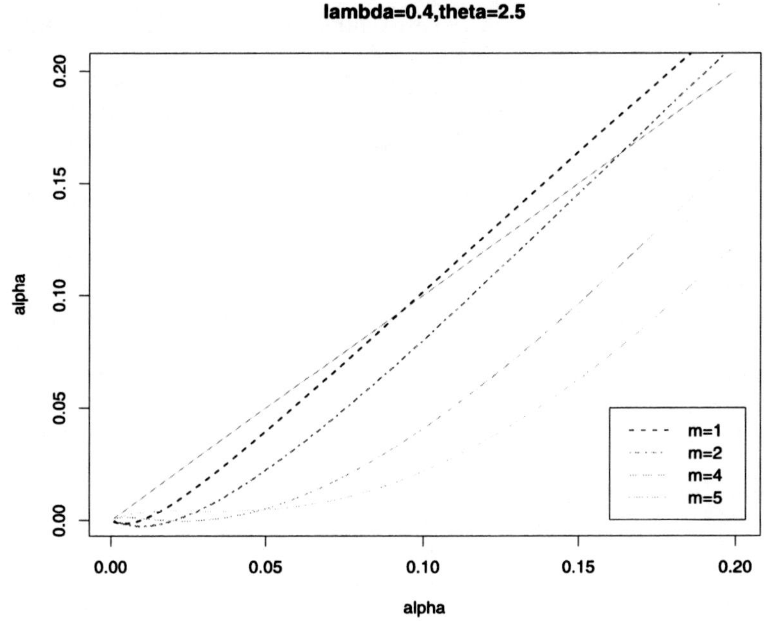

Fig. 3. Nominal levels versus Actual levels with the LCN model ($\theta = 2.5, \lambda = 0.4$) for different m's

and Johnson [13] or Rencher ([15], pp. 96-97) obtained with simulations: the one-sample Hotelling's T^2-test is somewhat robust to departures from the assumptions of multivariate normality if the underlying model is symmetric or nearly symmetric, but that this is lost if the underlying model is asymmetric.

Studying now what happens as we increase the dimension m of the observable random vector $\mathbf{X}$, first with a symmetric or nearly symmetric underlying model (a LCN with $\theta = 2$ and $\lambda = 0.2$), we observe that the actual levels are acceptably close to the nominal ones.

Nevertheless, if we consider an asymmetric underlying model (a LCN with $\theta = 2.5$ and $\lambda = 0.4$), the lack of Robustness of Validity increases as m increases as is showed in Figure 3. This is in accordance with Rencher ([15], pp. 97) but here, we obtain this conclusion using analytic expressions and not with simulations.

Acknowledgement. This work was partially supported by Grant MTM2009-10072.

References

1. Daniels, H.E.: Saddlepoint approximations for estimating equations. Biometrika 70, 89–96 (1983)
2. Everitt, B.S.: A Monte Carlo investigation of the robustness of Hotelling's one and two-sample T^2 statistic. J. Amer. Statist. Assoc. 74, 48–51 (1979)

3. Field, C.A., Ronchetti, E.: A tail area influence function and its application to testing. Sequent. Anal. 4, 19–41 (1985)
4. Fujikoshi, Y.: An asymptotic expansion for the distribution of Hotelling's T^2-statistic under nonnormality. J. Multivar. Anal. 61, 187–193 (1997)
5. García-Pérez, A.: Von Mises approximation of the critical value of a test. TEST 12, 385–411 (2003)
6. García-Pérez, A.: Chi-square tests under models close to the normal distribution. Metrika 63, 343–354 (2006)
7. García-Pérez, A.: t-tests with models close to the normal distribution. In: Balakrishnan, N., Castillo, E., Sarabia, J.M. (eds.) Advances in Distributions, Order Statistics, and Inference, pp. 363–379. Birkhäuser, Boston (2006)
8. García-Pérez, A.: Approximations for F-tests which are ratios of sums of squares of independent variables with a model close to the normal. TEST 17, 350–369 (2008)
9. Gupta, A.K., Kabe, D.G.: Distributions of Hotelling's T^2 and multiple and partial correlation coefficients for the mixture of two multivariate Gaussian populations. Statistics 32, 331–339 (1999)
10. Hampel, E.R., Ronchetti, E.M., Rousseeuw, P.J., Stahel, W.A.: Robust Statistics: The Approach Based on Influence Functions. Wiley, New York (1986)
11. Iwashita, T.: Asymptotic null and nonnull distribution of Hotelling's T^2-statistic under the elliptical distribution. J. Statist. Plan Infer. 61, 85–104 (1997)
12. Lugannani, R., Rice, S.: Saddle point approximation for the distribution of the sum of independent random variables. Adv. Appl. Probab. 12, 475–490 (1980)
13. Nachtsheim, C.J., Johnson, M.E.: A new family of multivariate distributions with applications to Monte Carlo studies. J. Amer. Statist. Assoc. 83, 984–989 (1988)
14. Rao, C.R.: Linear Statistical Inference and its Applications, 2nd edn. Wiley, New York (1973)
15. Rencher, A.C.: Multivariate Statistical Inference and Applications. Wiley, New York (1998)
16. Ronchetti, E.: Robust inference by influence functions. J. Statist. Plan. Infer. 57, 59–72 (1997)
17. Srivastava, M.S., Awan, H.M.: On the robustness of Hotelling's T^2-test and distribution of linear and quadratic forms in sampling from a mixture of two multivariate normal populations. Comm. Statist.-Theor. Meth. 11, 81–107 (1982)
18. Staudte, R.G., Sheather, S.J.: Robust Estimation and Testing. Wiley, New York (1990)
19. Withers, C.S.: Expansions for the distribution and quantiles of a regular functional of the empirical distribution with applications to nonparametric confidence intervals. Ann. Statist. 11, 577–587 (1983)

Interval and Fuzzy-Valued Approaches to the Statistical Management of Imprecise Data*

Norberto Corral, María Ángeles Gil, and Pedro Gil

Departamento de Estadística e I.O. y D.M., Universidad de Oviedo, Spain
norbert@uniovi.es, magil@uniovi.es, pedro@uniovi.es

Summary. In real-life situations experimental data can arise which do not derive from exact measurements or observations, but they correspond to ranges, judgements, perceptions or ratings often involving imprecision and subjectivity. These data are usually formalized with (and treated as) grouped or categorical/qualitative data for which the statistical analysis techniques to be applied are rather limited.

However, many of these data could be alternatively and suitably identified with either interval- or fuzzy number-valued data. This approach offers in fact mathematical languages/scales allowing us to express many imprecise data related either to ranges/fluctuations or to judgements/perceptions/ratings, and to capture the underlying imprecision, subjectivity and variability. Besides capturing the information surrounding the imprecision, subjectivity and variability (which is frequently ignored in dealing with grouped or categorical data), the use of the rich interval and fuzzy scales enables to state distances between data with a meaning similar to that for numerical ones. Moreover, it will possible to develop statistical methods based on these distances and exploiting the added information.

This paper aims to review the key ideas in this approach as well as some of the existing techniques for the statistical analysis.

Keywords: fuzzy data, interval data, metrics between imprecise data, random fuzzy set, random interval.

1 Introduction

Most of the traditional statistical data analysis assume precise experimental data are available, though in reality many random experiments cannot provide us with exact data.

* This paper has been written as a tribute to our beloved friend Marisa Menéndez. She touched us with her friendship and warm hospitality, and we have had the opportunity to share with her many wonderful personal meetings and fruitful scientific discussions. She has been always ready to help us in many respects, and we will always feel indebted to her. Thank you, Marisa, for your affection and care; you will always occupy a very special place in our hearts.

Sometimes, imprecision is due to the available information not being enough to quantify exactly the values a real-valued random variable takes on. In this way, existing numerical data can be known either only to lie within bounded intervals or only up to some categories. For instance: the information supplied by most of the National Statistical Offices concerning annual income usually appears grouped in intervals or income ranges; in reporting somebody on the price of a given item one can say it is RATHER CHEAP. In other words, we refer to a type of imprecision/uncertainty affecting the knowledge about variable values, although values are actually numerical.

In other cases, the imprecision is due to the fact that the considered variables are associated with judgements, perceptions or ratings which are intrinsically imprecisely-valued and involve subjectivity. In these cases, existing data are non-numerical and this is a consequence of the nature of the variable instead of a consequence of the knowledge on it. For instance: the random variable salary range for a specific job at different companies is itself interval-valued; rating the quality of life or the degree of agreement with a given policy, etc. lead to random attributes which could be labeled as ordinal categorical ones, and often appropriately formalized by means of fuzzy number-valued variables.

When available information correspond to imprecise data, they are often treated as either grouped or categorical ones, and the statistical analysis techniques to be applied are rather limited. Many of these procedures are based on the frequencies of different variable values and maybe their rank in case of ordinal ones, but the values themselves are usually ignored. The approach to be recalled in this paper suggests the use of either random intervals or random fuzzy numbers as possible ways to model many of the random mechanisms supplying these imprecise data. The approach is supposed to be ultimately addressed to draw statistical conclusions on the distributions of these imprecisely-valued random elements, irrespectively of the situation data come from involving or not an underlying real-valued random variable.

If imprecise data can be properly described in terms of either intervals or fuzzy numbers, a valuable and wide statistical methodology can be developed on the basis of some versatile and intuitive metrics between these data. In this way, many statistics can be defined by considering such distances, these statistics exploiting relevant information which could not be exploited in case of using categorical data, so that the accuracy of the derived conclusions will be improved.

On the other hand, the suggested approach can be carried out in two different ways, namely, by prefixing a list of intervals or categories (which are usually assumed to be mutually exclusive and exhaustive) so that each imprecise data is one-to-one converted into a value in such a list, or by enabling a full freedom in describing data. Whenever it is possible to apply the last way, it allows us to capture both the subjectivity involved in the judgements/perceptions/ratings and the variability in describing them much more accurately than by using prefixed lists. Thus, individuals to which one could assign the label HIGH QUALITY, if a short list of labels concerning their quality is pre-established, would be probably associated with different fuzzy numbers (depending on both the individual the rating refers to and the person rating the quality) if the person rating the quality is free to describe imprecise data.

In this paper we first present the preliminary concepts and results concerning the modeling of imprecise data by means of intervals and fuzzy numbers, the arithmetic and distances between them, the formalization of the random mechanisms supplying these data and some associated summary measures of the distribution of these random elements. Later we will briefly present some existing methods to estimate and test hypotheses about the distribution of these random elements.

2 Preliminaries

The statistical analysis of imprecise data requires an adequate probabilistic setting so that statistical developments, and especially inferential ones, can be well-formalized. Furthermore, it would be convenient this setting enables that most of the concepts and ideas in usual statistical techniques with numerical data can be preserved.

The space of *interval values* for data to be considered in the approach is the class

$$\mathcal{K}_c(\mathbb{R}) = \big\{[a,b] \,:\, a \leq b\big\}.$$

The basic arithmetical operations to develop statistics with interval data are the sum and the product by a real number, which are usually defined as the image of the involved intervals through the corresponding operation in $\mathbb{R}$. Thus, given intervals $K, K' \in \mathcal{K}_c(\mathbb{R})$ and a scalar $\gamma \in \mathbb{R}$, the *sum of* K *and* K' is defined as the interval in $\mathcal{K}_c(\mathbb{R})$

$$K + K' = \text{Minkowski sum of } K \text{ and } K' = \big\{x + y \,:\, x \in K, y \in K'\big\},$$

and the *product of* K *by* γ is defined as the interval in $\mathcal{K}_c(\mathbb{R})$ such that

$$\gamma \cdot K = \big\{\gamma \cdot x \,:\, x \in K\big\}.$$

$(\mathcal{K}_c(\mathbb{R}), +, \cdot)$ has not a linear (but a semilinear-conical) structure, since $K + (-1) \cdot K \neq [0,0] =$ neutral element of the Minkoswski sum in $\mathcal{K}_c(\mathbb{R})$, but in case K reduces to a singleton.

The space of *fuzzy numbers* for data to be considered in the approach is the class

$$\mathcal{F}_c(\mathbb{R}) = \big\{\widetilde{U} : \mathbb{R} \to [0,1] \,:\, \widetilde{U}_\alpha \in \mathcal{F}_c(\mathbb{R}) \text{ for all } \alpha \in (0,1]\big\}$$

where $\widetilde{U}_\alpha = \{x \in \mathbb{R} : \widetilde{U}(x) \geq \alpha\}$. A fuzzy number $\widetilde{U} \in \mathcal{F}_c(\mathbb{R})$ models an ill-defined property on (or subset of) $\mathbb{R}$, so that for each real number x the value $\widetilde{U}(x)$ can be interpreted as the 'degree of compatibility' of x with the property 'defining' $\widetilde{U}$ (or 'degree of truth' of the assertion "x is $\widetilde{U}$", or 'degree of membership' of x to $\widetilde{U}$). Of course, $\mathcal{K}_c(\mathbb{R}) \subset \mathcal{F}_c(\mathbb{R})$ since indicator functions of intervals are examples of fuzzy numbers.

The basic arithmetical operations to develop statistics with fuzzy data are usually assumed to be based on Zadeh's (also called the maximum-minimum) extension principle (Zadeh [28]) or, equivalently and based on the results by Nguyen [22], as the level-wise extension of the usual interval arithmetic. That is, given two fuzzy numbers $\widetilde{U}, \widetilde{V} \in \mathcal{F}_c(\mathbb{R})$ and a scalar $\gamma \in \mathbb{R}$, the *sum of* $\widetilde{U}$ *and* $\widetilde{V}$ is defined as the fuzzy number $\widetilde{U} + \widetilde{V} \in \mathcal{F}_c(\mathbb{R})$ such that for each $\alpha \in (0,1]$:

$$(\widetilde{U} + \widetilde{V})_\alpha = \widetilde{U}_\alpha + \widetilde{V}_\alpha,$$

and the *product of* $\widetilde{U}$ *by* γ as the fuzzy number $\gamma \cdot \widetilde{U} \in \mathcal{F}_c(\mathbb{R})$ such that for each $\alpha \in (0,1]$:

$$(\gamma \cdot \widetilde{U})_\alpha = \gamma \cdot \widetilde{U}_\alpha = \{\gamma \cdot y \,:\, y \in \widetilde{U}_\alpha\}.$$

$(\mathcal{F}_c(\mathbb{R}), +, \cdot)$, has not a linear (but a semilinear-conical) structure.

'Location' and 'imprecision/shape' are two key features in characterizing both interval and fuzzy values. In fact, interval and fuzzy values can be represented by means of their so-called t-vector and support function, respectively, which will serve us to reformulate the spaces $\mathcal{K}_c(\mathbb{R})$ and $\mathcal{F}_c(\mathbb{R})$ along with the corresponding arithmetics within certain Hilbert spaces.

On one hand, any interval $K \in \mathcal{K}_c(\mathbb{R})$ can be characterized by the so-called *t-vector* (or *mid/spread characterization*) of K, $t_K = (\text{mid}\, K, \text{spr}, K)$, where $\text{mid}\, K$ = mid-point/center of K and $\text{spr}\, K$ = spread/radius of K. This characterization enables to embed $\mathcal{K}_c(\mathbb{R})$ into $\mathbb{R}^2$ through the *t-vector function*

$$t : \mathcal{K}_c(\mathbb{R}) \to \mathbb{R}^2 \text{ s.t. } t(K) = t_K$$

(see Blanco [4]).

The t-vector function preserves the semilinearity of $\mathcal{K}_c(\mathbb{R})$ since

$$t(K + K') = t(K) + t(K'), \quad t(\lambda \cdot K)) = \lambda \cdot t(K),$$

for all $K, K' \in \mathcal{K}_c(\mathbb{R})$ and $\lambda \geq 0$.

This function allows us to induce a family of L^2 metrics on $\mathcal{K}_c(\mathbb{R})$ from a family of L^2 distances on $\mathbb{R}^2$ like the one given for $\mathbf{x}, \mathbf{y} \in \mathbb{R}^2$ (with $\mathbf{x} = (x_1, x_2)$ and $\mathbf{y} = (y_1, y_2)$) by

$$\mathfrak{d}_\theta(\mathbf{x}, \mathbf{y}) = \|\mathbf{x} - \mathbf{y}\|_\theta = \sqrt{\langle \mathbf{x} - \mathbf{y}, \mathbf{x} - \mathbf{y} \rangle_\theta} = \sqrt{(x_1 - y_1)^2 + \theta \cdot (x_2 - y_2)^2},$$

with $\theta > 0$.

This induction can be carried out so that the t-vector enables to embed isometrically the space $\mathcal{K}_c(\mathbb{R})$ onto the cone $\mathbb{R} \times [0, +\infty)$ of $\mathbb{R}^2$. In this respect, and based on the ideas in Gil *et al.* [6] and Blanco [4], and more generally Trutschnig *et al.* [27] and González-Rodríguez *et al.* [8], the following family of metrics can be introduced:

Theorem 1. *Given* $\theta \in (0, +\infty)$, *the mapping* $d_\theta : \mathcal{K}_c(\mathbb{R}) \times \mathcal{K}_c(\mathbb{R}) \to [0, +\infty)$ *such that for any* $K, K' \in \mathcal{K}_c(\mathbb{R})$

$$d_\theta(K, K') = \mathfrak{d}_\theta(t_K, t_{K'}) = \sqrt{(\text{mid}\, K - \text{mid}\, K')^2 + \theta \cdot (\text{spr}\, K - \text{spr}\, K')^2}$$

satisfies that

- d_θ *is an* L^2*-type metric on* $\mathcal{K}_c(\mathbb{R})$ *(enabling to weight through the choice of* θ *the influence of the distance between the location -in terms of the centers- of interval data in contrast to the distance between the imprecision -in terms of the spreads-).*
- $(\mathcal{K}_c(\mathbb{R}), d_\theta)$ *is a separable metric space.*
- *The t-vector function* $t : \mathcal{K}_c(\mathbb{R}) \to \mathbb{R}^2$ *states an isometrical embedding of* $\mathcal{K}_c(\mathbb{R})$ *(with the interval arithmetic and* d_θ*) onto a closed convex cone of* $\mathbb{R}^2$, $\mathbb{R} \times [0, +\infty)$ *(with the vectorial arithmetic and the distance* $\mathfrak{d}_\theta$*).*

On the other hand, any fuzzy number $\widetilde{U} \in \mathcal{F}_c(\mathbb{R})$ can be characterized by the so-called (Minkowski) *support function* of $\widetilde{U}$ (see Puri and Ralescu [24]) extends level-wise the notion of the support function of a set and is given by the mapping $s_{\widetilde{U}} : \{-1, 1\} \times (0, 1] \to \mathbb{R}$ defined so that

$$s_{\widetilde{U}}(-1, \alpha) = -\inf \widetilde{U}_\alpha, \quad s_{\widetilde{U}}(1, \alpha) = \sup \widetilde{U}_\alpha$$

for all $\alpha \in (0, 1]$. This characterization enables to embed $\mathcal{F}_c^2(\mathbb{R}) = \{\widetilde{U} \in \mathcal{F}_c(\mathbb{R}) : s_{\widetilde{U}} \in \mathcal{H}_1\}$, with $\mathcal{H}_1$ = space of the L^2-type real-valued functions on $\{-1, 1\} \times (0, 1]$ w.r.t. the corresponding uniform probability measures, into $\mathcal{H}_1$ through the *support function*

$$s : \mathcal{F}_c^2(\mathbb{R}) \to \mathcal{H}_1 \ \text{ s.t. } \ s(\widetilde{U}) = s_{\widetilde{U}}.$$

The support function preserves the semilinearity of $\mathcal{F}_c^2(\mathbb{R})$ since

$$s(\widetilde{U} + \widetilde{V}) = s(\widetilde{U}) + s(\widetilde{V}), \quad s(\gamma \cdot \widetilde{U})) = \gamma \cdot s(\widetilde{U}),$$

for all $\widetilde{U}, \widetilde{V} \in \mathcal{F}_c^2(\mathbb{R})$ and $\gamma \geq 0$.

This function allows us to induce a family of L^2 metrics on $\mathcal{F}_c^2(\mathbb{R})$ from a family of L^2 distances on $\mathcal{H}_1$ like the one given for $f, g \in \mathcal{H}_1$ by

$$\mathfrak{D}_\theta^\varphi(f, g) = \|f - g\|_\theta^\varphi = \sqrt{\langle f - g, f - g\rangle_\theta^\varphi}$$

$$= \sqrt{\int_{(0,1]} \left([\mathfrak{mid}\, f(1, \alpha) - \mathfrak{mid}\, g(1, \alpha)]^2 + \theta\, [\mathfrak{spr}\, f(1, \alpha) - \mathfrak{spr}\, g(1, \alpha)]^2 \right) d\varphi(\alpha)},$$

where

$$\mathfrak{mid}\, f(-1, \alpha) = -\mathfrak{mid}\, f(1, \alpha) = \frac{f(-1, \alpha) - f(1, \alpha)}{2},$$

$$\mathfrak{spr}\, f(-1, \alpha) = \mathfrak{spr}\, f(1, \alpha) = \frac{f(-1, \alpha) + f(1, \alpha)}{2},$$

with $\theta > 0$ and φ being a weighting measure which is formalized by means of an absolutely continuous probability measure on the measurable space $\big((0,1], \mathcal{B}_{(0,1]}\big)$ with the mass function being positive in $(0,1)$.

This induction can be carried out so that the support function enables to embed isometrically the space $\mathcal{F}^2_c(\mathbb{R})$ onto a cone in $\mathcal{H}_1$. In this respect, and based on the ideas in Montenegro [18] and more generally Trutschnig *et al.* [27] and González-Rodríguez *et al.* [8], the following family of metrics can be introduced:

Theorem 2. *Given $\theta \in (0, +\infty)$ and an absolutely continuous probability measure φ on $\big((0,1], \mathcal{B}_{(0,1]}\big)$ with the mass function being positive in $(0,1)$, the mapping $D^\varphi_\theta : \mathcal{F}^2_c(\mathbb{R}) \times \mathcal{F}^2_c(\mathbb{R}) \to [0, +\infty)$ such that for any $\widetilde{U}, \widetilde{V} \in \mathcal{F}^2_c(\mathbb{R})$*

$$D^\varphi_\theta(\widetilde{U}, \widetilde{V}) = \mathfrak{D}^\varphi_\theta(s_{\widetilde{U}}, s_{\widetilde{V}}) = \sqrt{\int_{(0,1]} \left[\mathfrak{d}_\theta(\widetilde{U}_\alpha, \widetilde{V}_\alpha)\right]^2 \, d\varphi(\alpha)}$$

$$= \sqrt{\int_{(0,1]} \left(\left[\operatorname{mid} \widetilde{U}_\alpha - \operatorname{mid} \widetilde{V}_\alpha\right]^2 + \theta \left[\operatorname{spr} \widetilde{U}_\alpha - \operatorname{spr} \widetilde{V}_\alpha\right]^2\right) d\varphi(\alpha)}$$

satisfies that

- *D^φ_θ is an L^2-type metric on $\mathcal{F}^2_c(\mathbb{R})$ (enabling to weight through the choice of θ and φ the influence of the distance between the location in contrast to the distance between the shape, as well as the relevance of different levels α, respectively).*
- *$\big(\mathcal{F}^2_c(\mathbb{R}), D^\varphi_\theta\big)$ is a separable metric space.*
- *The support function $s : \mathcal{F}^2_c(\mathbb{R}) \to \mathcal{H}_1$ states an isometrical embedding of $\mathcal{F}^2_c(\mathbb{R})$ (with the fuzzy arithmetic and D^φ_θ) onto a closed convex cone of $\mathcal{H}_1$ (with the functional arithmetic and the distance $\mathfrak{D}^\varphi_\theta$).*

The embeddings in Theorems 1 and 2 can be used to induce the notions of $\mathcal{K}_c(\mathbb{R})$- and $\mathcal{F}^2_c(\mathbb{R})$-valued random elements and associated relevant parameters of its distribution from the concepts of random elements and associated relevant parameters.

In dealing with interval and fuzzy data for statistical purposes (especially for inferential ones), there is a need for modeling the random mechanisms producing these data within a probabilistic setting. In this way, by considering the isometrical embedding in Theorem 1, the notion of a *random interval* can be immediately induced from that of random element as follows:

Definition 1. *Given a probability space $(\Omega, \mathcal{A}, P)$, a mapping $X : \Omega \to \mathcal{K}_c(\mathbb{R})$ is said to be a* ***random interval*** *(RI for short) if $t_X : \Omega \to \mathbb{R}$ is a two-dimensional random vector, that is, a Borel measurable mapping w.r.t. $\mathcal{A}$ and the Borel σ-field generated by the topology induced by $\mathfrak{d}_\theta$ on $\mathbb{R}^2$.*

Equivalently, RIs can be formalized in any of the following ways:

- X *is a compact convex random set, that is, it is a Borel measurable mapping w.r.t.* $\mathcal{A}$ *and the Borel* σ*-field generated by the topology induced by the Hausdorff metric on* $\mathcal{K}_c(\mathbb{R})$.
- X *is a Borel measurable mapping w.r.t.* $\mathcal{A}$ *and the Borel* σ*-field generated by the topology induced by the metric* d_θ *on* $\mathcal{K}_c(\mathbb{R})$.
- X *is a random interval, that is, the real-valued functions* $\inf X : \Omega \to \mathbb{R}, \sup X : \Omega \to \mathbb{R}$ *are real-valued random variables.*
- X *is a random interval, that is, the real-valued functions* $\operatorname{mid} X : \Omega \to \mathbb{R}, \operatorname{spr} X : \Omega \to [0, +\infty)$ *are real-valued random variables.*

Analogously, by considering the isometrical embedding in Theorem 2, the notion of a *random fuzzy number* (or random fuzzy set) can be immediately induced from that of random element as follows:

Definition 2. *Given a probability space* $(\Omega, \mathcal{A}, P)$*, a mapping* $\mathcal{X} : \Omega \to \mathcal{F}_c^2(\mathbb{R})$ *is said to be a* ***random fuzzy number*** *(or, more generally, a* ***random fuzzy set****) (RFN for short) if* $s_X : \Omega \to \mathcal{H}_1$ *is an* $\mathcal{H}_1$*-valued random element, that is, a Borel measurable mapping w.r.t.* $\mathcal{A}$ *and the Borel* σ*-field generated by the topology induced by the metric* $\mathfrak{D}_\theta^\varphi$ *on* $\mathcal{H}_1$.

Equivalently, RFNs can be formalized in any of the following ways:

- $\mathcal{X}$ *is a fuzzy random variable as intended by Puri and Ralescu [25], that is, for all* $\alpha \in (0, 1]$ *the interval-valued mapping* $\mathcal{X}_\alpha : \Omega \to \mathcal{K}_c(\mathbb{R})$ *is a random interval.*
- $\mathcal{X}$ *is a Borel measurable mapping w.r.t.* $\mathcal{A}$ *and the Borel* σ*-field generated by the topology induced by the metric* D_θ^φ *on* $\mathcal{F}_c^2(\mathbb{R})$.

The Borel measurability of RIs and RFNs guarantee that one can properly refer in this setting to concepts like the *distribution induced by either an RI or an RFN*, the *stochastic independence of either RIs or RFNs*, and so on, which are crucial for inferential developments.

In analyzing interval and fuzzy data from an RI and an RFN, respectively, two relevant *summary measures/parameters* are to be considered, namely, the mean value and the Fréchet variance, both induced from the expectation and Fréchet's variance of a random element. Thus,

Definition 3. *Let* $(\Omega, \mathcal{A}, P)$ *be a probability space.*

If $X : \Omega \to \mathcal{K}_c(\mathbb{R})$ *is an associated RI such that* $t_X \in L^1(\Omega, \mathcal{A}, P)$*, the* ***mean value of RI*** X *(in Aumann's sense [3]) is the interval* $E[X] \in \mathcal{K}_c(\mathbb{R})$ *such that* $t_{E[X]} = E(t_X)$*, i.e.,* $E[X]$ = *Aumann integral of* X *w.r.t.* P*, which in this one-dimensional case is such that* $\operatorname{mid} E[X] = E(\operatorname{mid} X)$ *and* $\operatorname{spr} E[X] = E(\operatorname{spr} X)$*, that is,* $E[X] = [E(\inf X), E(\sup X)]$.

If $\mathcal{X} : \Omega \to \mathcal{F}_c^2(\mathbb{R})$ *is an associated RFN such that* $s_\mathcal{X} \in L^1(\Omega, \mathcal{A}, P)$*, the* ***mean value of RFN*** $\mathcal{X}$ *(in Puri and Ralescu's sense [25]) is the fuzzy number* $\widetilde{E}(\mathcal{X}) \in \mathcal{F}_c^2(\mathbb{R})$ *such that* $s_{\widetilde{E}(\mathcal{X})} = E(s_\mathcal{X})$*, i.e., for all* $\alpha \in (0, 1]$ *we have that* $\left(\widetilde{E}(\mathcal{X})\right)_\alpha = E[\mathcal{X}_\alpha]$.

The mean values of RIs and RFNs in Aumann's and Puri-Ralescu's senses, respectively, preserve the main properties of the mean value of a random variable (see, for instance, Blanco [4] and González-Rodríguez *et al.* [8]). Thus,

Proposition 1. *The mean values of RIs and RFNs as stated in Definition 3 satisfy the following properties:*

- *They are coherent with the usual interval and fuzzy arithmetics. If X is an RI associated with $(\Omega, \mathcal{A}, P)$ such that $t_X \in L^1(\Omega, \mathcal{A}, P)$, and X is 'discrete' (i.e., $X(\Omega) = \{K_1, K_2, \dots\}$, and $\Omega_i = X^{-1}(K_i)$ with $\{\Omega_i\}_i$ being a countable partition of Ω), then*

$$E[X] = P(\Omega_1) \cdot K_1 + P(\Omega_2) \cdot K_2 + \dots.$$

If $\mathcal{X}$ is an RFN associated with $(\Omega, \mathcal{A}, P)$ such that $s_{\mathcal{X}} \in L^1(\Omega, \mathcal{A}, P)$, and $\mathcal{X}$ is 'discrete' (i.e., $\mathcal{X}(\Omega) = \{\widetilde{x}_1, \widetilde{x}_2, \dots\}$, and $\Omega_i = \mathcal{X}^{-1}(\widetilde{x}_i)$ with $\{\Omega_i\}_i$ being a countable partition of Ω), then

$$\widetilde{E}(\mathcal{X}) = P(\Omega_1) \cdot \widetilde{x}_1 + P(\Omega_2) \cdot \widetilde{x}_2 + \dots$$

- *They are the Fréchet expectations in the metric spaces $(\mathcal{K}_c(\mathbb{R}), d_\theta)$ and $\left(\mathcal{F}_c^2(\mathbb{R}), D_\theta^\varphi\right)$. If X is an RI such that $t_X \in L^1(\Omega, \mathcal{A}, P)$ and $d_\theta(X, E[X]) \in L^2(\Omega, \mathcal{A}, P)$, then*

$$E[X] = \arg\min_K E\left(\left(d_\theta(X, K)\right)^2\right),$$

the minimum being considered on the class of intervals $K \in \mathcal{K}_c(\mathbb{R})$ for which $E\left(\left(d_\theta(X, K)\right)^2\right)$ exists. If $\mathcal{X}$ is an RFN such that $s_{\mathcal{X}} \in L^1(\Omega, \mathcal{A}, P)$ and $D_\theta^\varphi(\mathcal{X}, E(\mathcal{X})) \in L^2(\Omega, \mathcal{A}, P)$, then

$$\widetilde{E}(\mathcal{X}) = \arg\min_{\widetilde{U}} E\left(\left(D_\theta^\varphi(\mathcal{X}, \widetilde{U})\right)^2\right),$$

the minimum being considered on the class of fuzzy numbers $\widetilde{U} \in \mathcal{F}_c^2(\mathbb{R})$ for which $E\left(\left(D_\theta^\varphi(\mathcal{X}, \widetilde{U})\right)^2\right)$ exists.

- *They are equivariant by linear operations and transformations. If $\gamma, \nu \in \mathbb{R}$, $K \in \mathcal{K}_c(\mathbb{R})$, and X and Y are two RIs such that $t_X, t_Y \in L^1(\Omega, \mathcal{A}, P)$, then*

$$E[\gamma \cdot X + \nu \cdot Y + K] = \gamma \cdot E[X] + \nu \cdot E[Y] + K.$$

If $\gamma, \nu \in \mathbb{R}$, $\widetilde{U} \in \mathcal{F}_c^2(\mathbb{R})$, and $\mathcal{X}$ and $\mathcal{Y}$ are two RFNs such that $s_{\mathcal{X}}, s_{\mathcal{Y}} \in L^1(\Omega, \mathcal{A}, P)$, then

$$\widetilde{E}[\gamma \cdot \mathcal{X} + \nu \cdot \mathcal{Y} + \widetilde{U}] = \gamma \cdot \widetilde{E}(\mathcal{X}) + \nu \cdot \widetilde{E}(\mathcal{Y}) + \widetilde{U}.$$

The mean values of RIs and RFNs in Aumann's and Puri-Ralescu's senses, respectively, are also supported by *Strong Laws of Large Numbers* (cf., Artstein and Vitale [2], Puri and Ralescu [23] or Molchanov [17] for RIs, and

Colubi *et al.* [5], Molchanov [16], Terán [26], etc. for RFNs). The mean value is the almost sure limit (w.r.t. different metrics) of the 'sample fuzzy mean'. The result for the metrics d_θ and D_θ^φ could be alternatively derived on the basis of the above mentioned embeddings from that for Hilbert space-valued random elements (see, for instance, Ledoux and Talagrand [12]).

Based on the second conclusion in Proposition 1, the variance of RIs and RFNs could be appropriately formalized by considering Fréchet's approach. This approach looks at the variance as a measure of the 'error' in approximating/estimating the values of the random element through the corresponding mean value, this error being quantified in terms of a squared metric. Thus (see, for instance, Lubiano *et al.* [14]),

Definition 4. *Let $(\Omega, \mathcal{A}, P)$ be a probability space.*

*If $X : \Omega \to \mathcal{K}_c(\mathbb{R})$ is an associated RI such that $\|t_X\|_\theta \in L^2(\Omega, \mathcal{A}, P)$, the θ-**Fréchet variance** of X is the real number $\sigma_X^2(\theta)$ (or simply σ_X^2) such that $\sigma_X^2 = \mathrm{Var}(t_X) = E\left([\|t_X - E(t_X)\|_\theta]^2\right)$ in $(\mathcal{K}_c(\mathbb{R}), d_\theta)$, i.e.,*

$$\sigma_X^2 = E\left(\left(d_\theta(X, E[X])\right)^2\right).$$

*If $\mathcal{X} : \Omega \to \mathcal{F}_c^2(\mathbb{R})$ is an associated RFN such that $\|s_\mathcal{X}\|_\theta^\varphi \in L^2(\Omega, \mathcal{A}, P)$, the (θ,φ)-**Fréchet variance** of $\mathcal{X}$ is the real number $\sigma_\mathcal{X}^2(\theta,\varphi)$ (or simply $\sigma_\mathcal{X}^2$) such that $\sigma_\mathcal{X}^2 = \mathrm{Var}(s_\mathcal{X}) = E\left([\|s_\mathcal{X} - E(s_\mathcal{X})\|_\theta^\varphi]^2\right)$ in $\left(\mathcal{F}_c^2(\mathbb{R}), D_\theta^\varphi\right)$, i.e.,*

$$\sigma_\mathcal{X}^2 = E\left(\left(D_\theta^\varphi(\mathcal{X}, \widetilde{E}(\mathcal{X}))\right)^2\right) = \mathrm{Var}(\mathrm{mid}\,\mathcal{X}) + \theta\,\mathrm{Var}(\mathrm{spr}\,\mathcal{X})$$

the Var *being intended w.r.t. $P \times \varphi$.*

The (θ,φ)-Fréchet variance of RIs and RFNs preserves de valuable properties for this concept, such as

Proposition 2. *The Fréchet variance of RIs and RFNs as stated in Definition 4 satisfy the following properties:*

- *$\sigma_X^2 \geq 0$ with $\sigma_X^2 = 0$ if, and only if, there exists $K \in \mathcal{K}_c(\mathbb{R})$ s.t. $X = \widetilde{K}$ a.s.$[P]$.*
  $\sigma_\mathcal{X}^2 \geq 0$ with $\sigma_\mathcal{X}^2 = 0$ if, and only if, there exists $\widetilde{U} \in \mathcal{F}_c^2(\mathbb{R})$ s.t. $\mathcal{X} = \widetilde{U}$ a.s.$[P]$.
- *If $\gamma \in \mathbb{R}$, $K \in \mathcal{K}_c(\mathbb{R})$ and X, Y are two independent RIs associated with probability space $(\Omega, \mathcal{A}, P)$ s.t. $\|t_X\|_\theta, \|t_Y\|_\theta \in L^2(\Omega, \mathcal{A}, P)$, then*

$$\sigma_{\gamma\cdot X+K}^2 = \gamma^2 \cdot \sigma_X^2, \quad \sigma_{X+Y}^2 = \sigma_X^2 + \sigma_Y^2.$$

If $\gamma \in \mathbb{R}$, $\widetilde{U} \in \mathcal{F}_c^2(\mathbb{R}^p)$ and $\mathcal{X}, \mathcal{Y}$ are two independent RFNs associated with $(\Omega, \mathcal{A}, P)$ and such that $\|s_\mathcal{X}\|_\theta^\varphi, \|s_\mathcal{Y}\|_\theta^\varphi \in L^2(\Omega, \mathcal{A}, P)$, then

$$\sigma_{\gamma\cdot\mathcal{X}+\widetilde{U}}^2 = \gamma^2 \cdot \sigma_\mathcal{X}^2, \quad \sigma_{\mathcal{X}+\mathcal{Y}}^2 = \sigma_\mathcal{X}^2 + \sigma_\mathcal{Y}^2.$$

3 Statistical Analysis of Imprecise Data

In this section we will briefly review some ideas in a recent statistical methodology which has been developed to analyze interval and fuzzy data from an inferential perspective. This methodology is based on the metrics recalled in Section 2.

It should be emphasized both, the lack of realistic and operational 'parametric' families of probability distribution models for RIs and RFNs and the lack of Central Limit Theorems for these special random elements which could directly applicable for inferential purposes. However, a crucial role is played in this framework by the existence of Central Limit Theorems for Hilbert space-valued random elements (see, for instance, Laha and Rohatgi [11]), and bootstrapped versions (see Giné and Zinn [7]). Alternatively, and because of the embeddings of the spaces of values these random elements take on, some methods from Multivariate or Functional Data Analysis could be particularized to deal with RIs and RFNs, respectively; anyway, (see, González-Rodríguez *et al.* [8]) care should be taken to guarantee that in applying Hilbert space-valued results we are always moving within the corresponding cones.

The aim of the statistical methods we are reviewing is to draw conclusions about the distribution of either RIs or RFNs over populations, on the basis of the information supplied by samples of observations from the random elements. In this respect, some inferential procedures with RIs and RFNSs have been developed, like

- Inferential statistics about the population mean values of RIs and RFNs;
- Inferential statistics about the population Fréchet variances or other summary measures of the distribution of RIs and RFNs;
- Other statistical developments involving RIs or RFNs, like the interval or fuzzy arithmetics-based linear regression problem, and so on.

Since RIs are special instances of RFNs we will constrain along the following ideas and results to the fuzzy number-valued case. Many of the developed studies using this methodology is based on the metrics recalled in Section 2 concern the mean values of the involved RIs or RFNs. Among these studies we can outline those referred to 'point'/'interval' (imprecise) estimation of the mean value and testing 'two-sided' hypotheses about mean values.

Regarding testing about the means of RFNs, two-sided tests have been established. More concretely, by assuming as available sample information:

1. *One-sample case:* a realization from a simple random sample $(\mathcal{X}_1, \dots, \mathcal{X}_n)$ from the RFN $\mathcal{X}$;
2. *Two-sample case:* realizations from two simple random samples $(\mathcal{X}_1, \dots, \mathcal{X}_{n_1})$ and $(\mathcal{Y}_1, \dots, \mathcal{Y}_{n_2})$ from $\mathcal{X}$ and $\mathcal{Y}$, respectively (independent/linked samples);

k. *k-sample case:* realizations from k simple random samples $(\mathcal{X}_{11} \ldots, \mathcal{X}_{1n_1})$, $\ldots$, $(\mathcal{X}_{k1}, \ldots, \mathcal{X}_{kn_k})$ from $\mathcal{X}_1 \ldots \mathcal{X}_k$, respectively (independent/linked samples).

the following corresponding 'two-sided' null hypotheses (equalities of fuzzy numbers) have been tested:

1. $H_0 : \widetilde{E}(\mathcal{X}) = \widetilde{U} \in \mathcal{F}_c^2(\mathbb{R})$
2. $H_0 : \widetilde{E}(\mathcal{X}) = \widetilde{E}(\mathcal{Y})$
k. $H_0 : \widetilde{E}(\mathcal{X}_1) = \ldots = \widetilde{E}(\mathcal{X}_k)$

these hypotheses being equivalent, respectively, to the following ones (expressed as equalities of real numbers):

1. $H_0 : D_\theta^\varphi(\widetilde{E}(\mathcal{X}), \widetilde{U}) = 0$
2. $H_0 : D_\theta^\varphi(\widetilde{E}(\mathcal{X}), \widetilde{E}(\mathcal{Y})) = 0$
k. $H_0 : \sum_{i=1}^{k} \left[D_\theta^\varphi \left(\widetilde{E}(\mathcal{X}_i), \widetilde{E}\Big(\frac{1}{k} \cdot [\mathcal{X}_1 + \ldots + \mathcal{X}_k]\Big)\right)\right]^2 = 0.$

The aim of these tests would be that of concluding (at a given significance level) whether or not H_0 should be rejected on the basis of the available sample of observations. It should be noted that to preserve the usual statistical notation, and because there will not be confusion, α will denote in the sequel the nominal significance level.

In connection with the two-sided tests about the mean of an RFN (***one-sample case***), studies have been developed including:

- An exact test for 'normal' RFNs (in accordance with Puri and Ralescu's view of normality for an RFN, i.e, $\mathcal{X} = \widetilde{V} + \mathcal{N}(0,1)$ for a certain $\widetilde{V} \in \mathcal{F}_c^2(\mathbb{R})$) (cf. Montenegro *et al.* [20]);
- Asymptotic tests for RFNs;
- Bootstrap test for general RFNs.

Although the exact test leads to and easy-to-apply method, the assumption of $\mathcal{X}$ being normal in Puri and Ralescu's sense [24] is quite restrictive and unrealistic. Moreover, the asymptotic general method based on the Central Limit Theorem for Banach space-valued random elements by Araujo and Giné [1] (cf. Körner [10]) is easy-to-apply when $\mathcal{X}$ takes on a finite number of different values (cf. Montenegro *et al.* [20]), but the asymptotic distribution of the statistic can involve unknown parameters or elements out of the cone, and large sample sizes would be required.

Empirical preliminary studies based on simulations have indicated that estimating some terms in the asymptotic distribution often entails a substantial loss of precision w.r.t. the nominal significance level. Motivated by this assertion, we have considered the use of D_θ^φ and the generalized bootstrapped Central Limit Theorem by Giné and Zinn [7] which allows us to consider bootstrap techniques in this context, and to conclude that

Theorem 3. *Let $\mathcal{X} : \Omega \to \mathcal{F}_c(\mathbb{R}^p)$ be an RFN associated with $(\Omega, \mathcal{A}, P)$ such that*

- $\|s_{\mathcal{X}}\|_\theta^\varphi \in L^2(\Omega, \mathcal{A}, P)$,
- $(\mathcal{X}_1, \ldots, \mathcal{X}_n)$ *is a simple random sample from* $\mathcal{X}$,
- $(\mathcal{X}_1^*, \ldots, \mathcal{X}_n^*)$ *is a bootstrap simple random sample sample from* $(\mathcal{X}_1, \ldots, \mathcal{X}_n)$.

To test $H_0 : \widetilde{E}(\mathcal{X}) = \widetilde{U} \in \mathcal{F}_c^2(\mathbb{R}^p)$ at the nominal significance level $\alpha \in [0,1]$, H_0 should be rejected whenever

$$T_n = \frac{\left[D_\theta^\varphi(\overline{\mathcal{X}}_n, \widetilde{U})\right]^2}{\widehat{S}_n^2} > z_\alpha,$$

where $z_\alpha = 100(1-\alpha)$ fractile of the bootstrap distribution of

$$T_n^* = \left[D_\theta^\varphi(\overline{\mathcal{X}^*}_n, \overline{\mathcal{X}}_n)\right]^2 \Big/ \widehat{S}_n^{*\,2}$$

with

$$\overline{\mathcal{X}_n^*} = \sum_{i=1}^n \mathcal{X}_i^*/n, \qquad \widehat{S}_n^{*\,2} = \sum_{i=1}^n \left[D_\theta^\varphi(\mathcal{X}_i^*, \overline{\mathcal{X}^*}_n)\right]^2/(n-1),$$

(for which the distribution can be approximated by Monte Carlo method).

The probability of rejecting the null hypothesis under alternative assumptions converges to 1 as $n \to \infty$ (i.e., both the asymptotic and the bootstrap tests are consistent).

Comparative simulation studies have shown that for small/medium samples, the bootstrap method performs and behaves usually much better than the asymptotic one, whereas for large sample sizes (over 300), the improvement is not that remarkable, but the bootstrap approach still provides the best approximation to the nominal significance level.

For the ***two-sample case***, some studies have been also developed for both independent (cf. Montenegro et al. [19]) and dependent samples (cf. González-Rodríguez *et al.* [9]) leading to conclusions similar to those for the one-sample case.

Two-sided tests about the means of k RFNs (i.e., the ***k-sample case*** or ANOVA for independent or dependent samples) have been also established. In this way we have obtained (see González-Rodríguez *et al.* [8]) that for the independent case

Theorem 4. *Let $\mathcal{X}_1, \ldots, \mathcal{X}_k$ be independent RFNs for which the existence of the associated fuzzy means, Fréchet variances and the covariance functions of their support functions is assumed. Consider k independent realizations, each of them of size n_i from a simple random sample $(\mathcal{X}_{i1}, \ldots, \mathcal{X}_{in_i})$ from $\mathcal{X}_i$ $(i = 1, \ldots, k)$, and let $(\mathcal{X}_{i1}^*, \ldots, \mathcal{X}_{in_i}^*)$ denote the bootstrap sample randomly chosen from $\{\mathcal{X}_{i1}, \ldots, \mathcal{X}_{in_i}\}$ $(i = 1, \ldots, k)$.*

To test $H_0 : \widetilde{E}(\mathcal{X}_1) = \ldots = \widetilde{E}(\mathcal{X}_k)$ at the nominal significance level $\alpha \in [0,1]$, H_0 should be rejected whenever

$$T_n^{\kappa} = \sum_{i=1}^{k} n_i \left[D_\theta^\varphi \left(\overline{\mathcal{X}_{i\cdot}}, \overline{\mathcal{X}_{\cdot\cdot}} \right) \right]^2 > z_\alpha^*,$$

where $z_\alpha^* = 100(1-\alpha)$ fractile of the distribution of

$$T_n^{\kappa *} = \sum_{i=1}^{k} n_i \left[D_\theta^\varphi \left(\overline{\mathcal{X}_{i\cdot}^*} + \overline{\mathcal{X}_{\cdot\cdot}}, \overline{\mathcal{X}_{i\cdot}} + \overline{\mathcal{X}_{\cdot\cdot}^*} \right) \right]^2$$

(for which the distribution can be approximated by Monte Carlo method), where

$$\overline{\mathcal{X}_{i\cdot}} = \frac{1}{n_i} \left[\mathcal{X}_{i1} + \ldots + \mathcal{X}_{in_i} \right], \qquad \overline{\mathcal{X}_{i\cdot}^*} = \frac{1}{n_i} \left[\mathcal{X}_{i1}^* + \ldots + \mathcal{X}_{in_i}^* \right],$$

$$\overline{\mathcal{X}_{\cdot\cdot}} = \frac{1}{n_1 + \ldots + n_k} \left[\mathcal{X}_{11} + \ldots + \mathcal{X}_{1n_1} + \ldots + \mathcal{X}_{k1} + \ldots + \mathcal{X}_{kn_k} \right]$$

$$\overline{\mathcal{X}_{\cdot\cdot}^*} = \frac{1}{n_1 + \ldots + n_k} \left[\mathcal{X}_{11}^* + \ldots + \mathcal{X}_{1n_1}^* + \ldots + \mathcal{X}_{k1}^* + \ldots + \mathcal{X}_{kn_k}^* \right].$$

The probability of rejecting H_0 under alternative assumptions converges to 1 as $n = n_1 + \ldots + n_k \to \infty$ (i.e., the test is consistent).

Example 1. To illustrate the procedure above described, consider the problem of rating the quality of trees in a reforestation carried out in Huerna Valley (Asturias-León, Spain). The study has been developed by researchers in the Institute of Natural Resources and Land Management of the University of Oviedo (INDUROT) and full dataset is full property of the Institute, so only a few data have been shown.

Instead of considering usual Lickert's 1-5 or 1-7 codings for rating the quality of trees, field researchers have been informed about the possibility of using fuzzy numbers (for instance, fuzzy trapezoidal ones with general support $[0, 100]$ where 0 = lowest quality, 100 = highest quality) and the meaning for their assessment. The available sample information for $n_1 = 133$ birches (*Betula celtiberica*), $n_2 = 109$ sessile oaks (*Quercus petraea*), and $n_3 = 37$ rowans (*Sorbus aucuparia*) is gathered in the following table (only one datum per species has been shown due to confidentiality). Let $\mathcal{X}_1$, $\mathcal{X}_2$ and $\mathcal{X}_3$ denote the quality for the birches, sessile oaks and rowans, respectively.

Betula celtiberica	*Quercus petraea*	*Sorbus aucuparia*
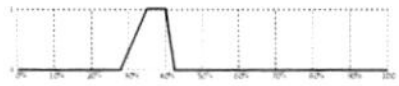	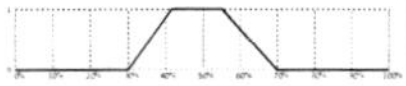	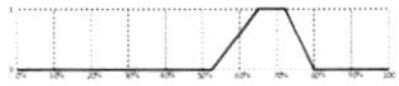
⋮	⋮	⋮

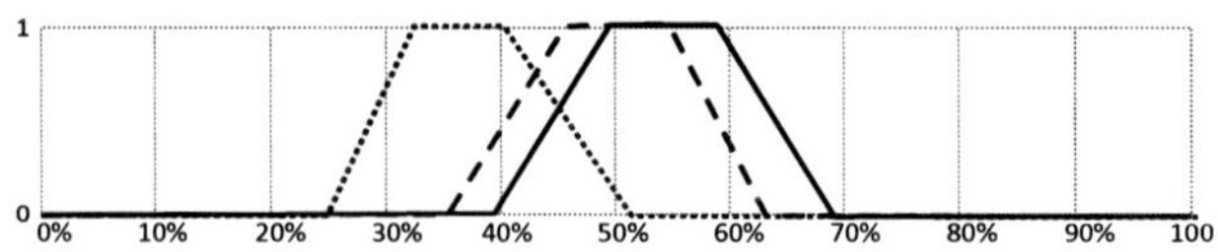

Fig. 1. Mean values of quality ratings; solid line for the birches, doted line for the sessile oaks, and dashed line for the rowans

To test the null hypothesis $H_0 : \widetilde{E}(\mathcal{X}_1) = \widetilde{E}(\mathcal{X}_2) = \widetilde{E}(\mathcal{X}_3)$ on the basis of the available fuzzy data, we should consider different sample means (see Figure 1) and apply the above-described bootstrap ANOVA test.

The application of the above-described bootstrap approach (with $\theta = 1/3$, $\varphi =$ Lebesgue measure on $(0, 1]$, and $10,000$ replications) leads to a bootstrap p-value equal to .000, which means that at almost any significance level the mean quality is significantly different for the three species.

For the dependent case a variation in the statistics has been carried out leading to a slightly different test (see Montenegro *et al.* [21]).

4 Concluding Remarks

RIs and RFNs are well-formalized notions within the probabilistic setting and such that using them to model imprecise data enables to preserve all the key concepts and ideas in Statistical Reasoning. Based on some tools from Interval and Fuzzy Set Computations an integral methodology to develop statistical inferences on the population mean values an other 'parameters' is being carried out. This methodology shows convenient properties (like strong consistency and others). It has been mostly generalized to higher dimension, that is, to random compact convex sets and random fuzzy sets (e.g., González-Rodríguez *et al.* [8]).

An R-package called SAFD (Statistical Analysis of Fuzzy Data) have been recently designed by Lubiano and Trutschnig (see, for instance, [15]) to perform computations with RFNs.

A lot of theoretical developments on the statistical analysis of interval and fuzzy data remain to be performed. Regarding the empirical studies, only preliminary though rather promising studies have been carried out (e.g., to consider a sensitivity analysis w.r.t. the choice of the metric for the first type of studies).

A wide list of related literature on the topic in this paper can be found at the website `http://bellman.ciencias.uniovi.es/SMIRE/Publications.html`.

Acknowledgement. This research has been partially supported by/benefited from the Spanish Ministry of Science and Innovation Grants MTM2009-09440-C02-01 and the COST Action IC0702. Their financial support is gratefully acknowledged.

References

1. Araujo, A., Giné, E.: The Central Limit Theorem for Real and Banach Valued Random Variables. Wiley, New York (1980)
2. Artstein, Z., Vitale, R.A.: A strong law of large numbers for random compact sets. Ann. Probab. 3, 879–882 (1975)
3. Aumann, R.J.: Integrals of set-valued functions. J. Math. Anal. Appl. 12, 1–12 (1965)
4. Blanco, A.: Análisis estadístico de un nuevo modelo de regresión lineal flexible para intervalos aleatorios. PhD Thesis, University of Oviedo, Spain (2009), `http://bellman.ciencias.uniovi.es/SMIRE/Archivos/Teseo-ABlanco.pdf`
5. Colubi, A., López-Díaz, M., Domínguez-Menchero, J.S., Gil, M.A.: A generalized strong law of large numbers. Prob. Theor. Rel. Fields 114, 401–417 (1999)
6. Gil, M.A., Lubiano, M.A., Montenegro, M., López-García, M.T.: Least squares fitting of an affine function and strength of association for interval data. Metrika 56, 97–111 (2002)
7. Giné, E., Zinn, J.: Bootstrapping general empirical measures. Ann. Probab. 18, 851–869 (1990)
8. González-Rodríguez, G., Colubi, A., Gil, M.A.: Fuzzy data treated as functional data. A one-way ANOVA test approach. Comp. Statist. Data Anal (2010), doi:10.1016/j.csda.2010.06.013 (in press)
9. González-Rodríguez, G., Colubi, A., Gil, M.A., D'Urso, P.: An asymptotic two dependent samples test of equality of means of fuzzy random variables. In: Proc 17th Conf IASC-ERS (COMPSTAT 2006), Roma, pp. 689–695 (2006)
10. Körner, R.: An asymptotic α-test for the expectation of random fuzzy variables. J. Stat. Plann. Infer. 83, 331–346 (2000)
11. Laha, R.G., Rohatgi, V.K.: Probability Theory. Wiley, New York (1979)
12. Ledoux, M., Talagrand, M.: Probability in Banach Spaces: Isoperimetry and Processes. Springer, Berlin (1991)
13. Lubiano, M.A.: Medidas de Variación de Elementos Aleatorios Imprecisos. PhD Thesis, University of Oviedo, Spain (1999), `http://www.tesisenred.net/TESIS_UOV/AVAILABLE/TDR-0209110-122449//UOV0067TMALG.pdf`
14. Lubiano, M.A., Gil, M.A., López-Díaz, M., López-García, M.T.: The $\overrightarrow{\lambda}$-mean squared dispersion associated with a fuzzy random variable. Fuzzy Sets and Systems 111, 307–317 (2000)
15. Lubiano, M.A., Trutschnig, W.: ANOVA for Fuzzy Random Variables Using the R-package SAFD. In: Borgelt, C., González-Rodríguez, G., Trutschnig, G., Lubiano, M.A., Gil, M.A., Grzegorzewski, P., Hryniewicz, O. (eds.) Combining Soft Computing and Statistical Methods in Data Analysis. AICS, vol. 77, pp. 449–456. Springer, Heidelberg (2010)
16. Molchanov, I.: On strong laws of large numbers for random upper semicontinuous functions. J. Math. Anal. Appl. 235, 349–355 (1999)
17. Molchanov, I.: Theory of Random Sets. Springer, Berlin (2005)
18. Montenegro, M.: Estadística con datos imprecisos basada en una métrica generalizada. PhD Thesis, University of Oviedo, Spain (2003), `http://www.tesisenred.net/TESIS_UOV/AVAILABLE/TDR-0209110-120109//UOV0066TMMH.pdf`
19. Montenegro, M., Casals, M.R., Lubiano, M.A., Gil, M.A.: Two-sample hypothesis tests of means of a fuzzy random variable. Inform. Sci. 133, 89–100 (2001)

20. Montenegro, M., Colubi, A., Casals, M.R., Gil, M.A.: Asymptotic and Bootstrap techniques for testing the expected value of a fuzzy random variable. Metrika 59, 31–49 (2004)
21. Montenegro, M., López-García, T., Lubiano, M.A., González-Rodríguez, G.: A dependent multi-sample test for fuzzy means. In: Abst. 2nd Workshop ERCIM Working Group on Computing & Statistics (ERCIM 2009), Limassol, Cyprus, vol. 102 (2009)
22. Nguyen, H.T.: A note on the extension principle for fuzzy sets. J. Math. Anal. Appl. 64, 369–380 (1978)
23. Puri, M.L., Ralescu, D.A.: Strong law of large numbers for Banach space-valued random sets. Ann. Probab. 11, 222–224 (1983)
24. Puri, M.L., Ralescu, D.A.: The concept of normality for fuzzy random variables. Ann. Probab. 11, 1373–1379 (1985)
25. Puri, M.L., Ralescu, D.A.: Fuzzy random variables. J. Math. Anal. Appl. 114, 409–422 (1986)
26. Terán, P.: A strong law of large numbers for random upper semicontinuous functions under exchangeability conditions. Statist. Prob. Lett. 65, 251–258 (2003)
27. Trutschnig, W., González-Rodríguez, G., Colubi, A., Gil, M.A.: A new family of metrics for compact, convex (fuzzy) sets based on a generalized concept of mid and spread. Inform. Sci. 179, 3964–3972 (2009)
28. Zadeh, L.A.: The concept of a linguistic variable and its application to approximate reasoning, Part 1. Inform. Sci. 8, 199–249 (1975); Part 2. Inform. Sci. 8, 301–353; Part 3. Inform. Sci. 9, 43–80

Part X

Modelling in Biological and Medical Problems

Autocorrelation Measures and Independence Tests in Spike Trains

Aldana González-Montoro[1], Ricardo Cao[1], Nelson Espinosa[2], Jorge Mariño[2], and Javier Cudeiro[2]

[1] Department of Mathematics, Universidade da Coruña, Facultad de Informática, Campus Elviña, 15071 A Coruña, Spain
agonzalezmo@udc.es, rcao@udc.es

[2] Neuroscience and Motor Control Group (NEUROcom), Department of Medicine, Universidade da Coruña, Campus de Oza, 15006 A Coruña, Spain
nespinosa@udc.es, xurxo@udc.es, jcud@udc.es

Summary. In the nervous system, neurons convey information by means of electric pulses called action potentials or spikes. The information is encoded in sequences of these pulses, called spike trains. In neurophysiological experiments, spike trains are recorded and analyzed statistically. Time intervals between action potentials is the key feature of spike trains.

One of the statistical techniques available for studying spike trains is the autocorrelation. In the neuroscientific literature the term autocorrelation is used to denote frequency histograms of time intervals between every pair of the spikes generated by a single neuron for a period of time. The autocorrelation function is very useful for characterizing spike trains. The shape of the autocorrelation indicates the nature of the dependence among time intervals between consecutive spikes for a specified time window. In this work we propose two statistics to test the hypothesis of independence between time intervals. The bootstrap method is used to calibrate the null distribution of these tests. The tests are applied to real spike trains recorded from the primary visual cortex of anesthetized cats, both during spontaneous activity and after electric stimulation-induced activity.

1 Introduction

Neurons are, together with glial cells, the basic structural and functional units of the nervous system. One of the most important characteristics of neuronal cells is their ability to propagate big quantities of information, at fairly fast speeds throughout neural networks. The information is conveyed along the nervous system in the form of electrical signals called action potentials (AP) or spikes traveling through cable-like cellular extensions called *axons*.

Neurons are highly specific cells whose structure and physiology make possible the generation and transmission of AP. Roughly speaking, when the input signals that continuously arrive to the initial segment of the axon reach a certain threshold, an abrupt change in the cells electrical membrane potential takes place, giving rise to an AP. Those electrical pulses travel along the membrane of the axon, and their trains are used as a binary code for the transfer of information between cells. Since AP are sharp potential changes, they are relatively easy to record. Researchers do so by placing electrodes close to, or inside, the neurons. In mammals APs have an approximate amplitude of 100 mV and a duration of 1 ms. The shape of the spikes remains practically unchanged while they travel through the axon, so the information they carry must be coded not in each AP but in a sequence of them.

A sequence of APs is called a *spike train*. Spike trains are very important because a great part of the information transmitted by neurons is coded in them. Spike trains are the object of study of the present work.

The global brain activity, i.e., the level of arousal and attentiveness, is modulated by the so called *activating ascending pathways*, constituted by neuronal nuclei situated in the brainstem (*bs*) and the basal forebrain (*bf*). This modulation is responsible of the main changes of activity that take place during the sleep-wake cycle. During the slow sleep state, neural activity is highly synchronized, which is reflected in the electroencephalogram (EEG) by waves of more amplitude and less frequency than in the wake state. The anesthetized state is very similar to and mimics the slow sleep state that occurs under physiological conditions, characterized by low frequency oscillatory activity. The arousal state, i.e., the transition from sleep to awake, can be induced by stimulating either the *bs* or the *bf*. In this work, simultaneous neuronal recordings made in the primary visual cortex of anesthetized cats were used to study the effects induced by the electrical stimulation of both pathways. Discussion on spike trains and modeling of neural systems can be found in [2] and a complete description of the state of the art on statistical analysis tools for multiple spike train data can be found in [1].

The following sections of the paper are organized as follows. Section 2 describes the experimental data. Two correlation measures for spike trains are introduced in Section 3. Section 4 deals with the statistics proposed to test independence between interspike intervals and include the results obtained using these methods. Finally, Section 5 contains the main conclusions.

2 Experimental Data

The experiments which yielded the data were performed in anesthetized cats. An eight-point multielectrode was introduced in the primary visual cortex in order to make simultaneous extracellular recordings of several neurons. Concurrently, an EEG was made and two other electrodes were introduced for electrical stimulation at *bs* and *bf*. The stimuli were electric pulses of 2 s of duration (trains of 0.05 s micro-pulses at a frequency of 50 Hz delivered for 2 s) which were applied differentially in the areas under study according to the following protocol. First, a group of neurons was identified and isolated using the multielectrode, and their spontaneous activity was recorded for 2 minutes. Then, electrical stimulation was delivered either to *bs* or *bf* (the sequence of stimulation of those areas was randomized) for two seconds and, after another period of time (8 min), enough for the neurons to return to their spontaneous activity, the other region (*bs* or *bf*) was stimulated following the same procedure. Finally, the recording continued for another amount of time that allowed the neurons to return to spontaneous activity again.

Each of the eight electrodes of the multielectrode device may lay close enough to no, one or more than one neurons. Hence, under favorable experimental conditions, it is possible to record more than eight neurons simultaneously. In this work we deal with the simultaneous recording of seven neurons. We used three different recordings (called trials) for each stimulus and each neuron.

In our context, one trial is the recording of one neuron during the spontaneous activity followed by the application of the stimulus (either *bs* or *bf*) and a final time period for recovery. Each trial had a duration of around 600 seconds and stimuli were applied approximately after 120 seconds of the beginning of the recording, but not all of them occurred at the same instant. Original data are presented as time instants that indicate spike events (with a millisecond precision). The neurons are denoted N1, N3a, N3b, N4a, N4b, N5 and N7.

Since the aim of the neurophysiological work was to characterize the differential effects of the stimuli on neuronal activity, the recording of each trial was separated in three periods: before stimulation (called *pre*), immediately after the stimulus (called *post*) and the final period of recovery (called *final*). It is worth mentioning that the spontaneous neuronal activity recovers gradually and not suddenly, but these three intervals will be considered as a first approach. In the current experimental preparation, the spontaneous activity (measured as firing rate) of the studied neurons is fairly low, as can be seen in Figure 1.

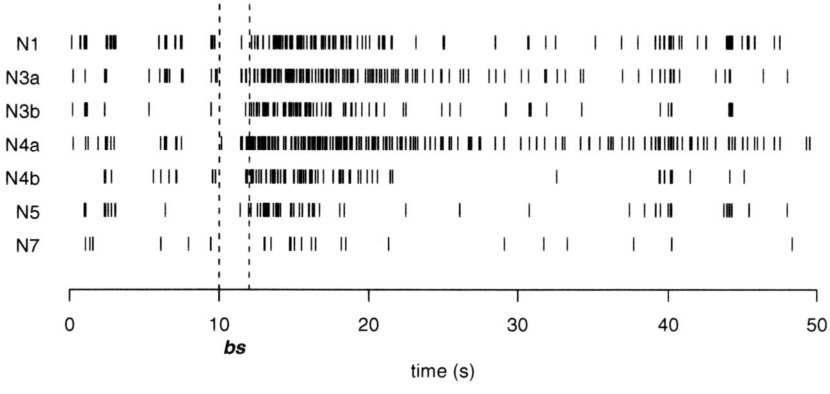

Fig. 1. Raster plots of 50 s recordings of one trial for each neuron. The short vertical lines represent the moments when the spikes occurred. The dotted lines represent the moments when the *bs* stimulus begins and ends.

3 Correlation Measures in Spike Trains

From a statistical viewpoint spike trains can be described as the sequence of AP generated by a point process, $\{T_i\}$, where T_i is the instant when the i-th AP occurred. In particular, if we have a T seconds duration neural activity recording and $(n+1)$ AP have occurred, we have a spike train $\{T_i\}_{i=1}^{n+1}$ with $0 \le T_1 \le T_2 \le \ldots \le T_{n+1} \le T$. We can also define the inter spike intervals (ISI), $\{S_i\}_{i=1}^{n}$, as the elapsed times between consecutive spikes, where $S_i = T_{i+1} - T_i$, $i = 1, 2, \ldots, n$.

One useful feature to characterize neuronal activity is the correlation in spike trains. In the next subsections, two different methods to measure autocorrelation in spike trains are introduced.

3.1 Autocorrelation

Given the ISI of an observed spike train, $S_1, \ldots, S_n$, we can estimate the serial autocovariance function as

$$\hat{\gamma}(h) = \frac{1}{n} \sum_{i=1}^{n-h} (S_{i+h} - \bar{S})(S_i - \bar{S}), \qquad 0 \le h < n$$

where $\bar{S} = \frac{1}{n} \sum_{i=1}^{n} S_i$ is the sample mean. Then, the serial autocorrelation function is estimated by

$$\hat{\rho}(h) = \frac{\hat{\gamma}(h)}{\hat{\gamma}(0)}, \qquad 0 \le h < n\,.$$

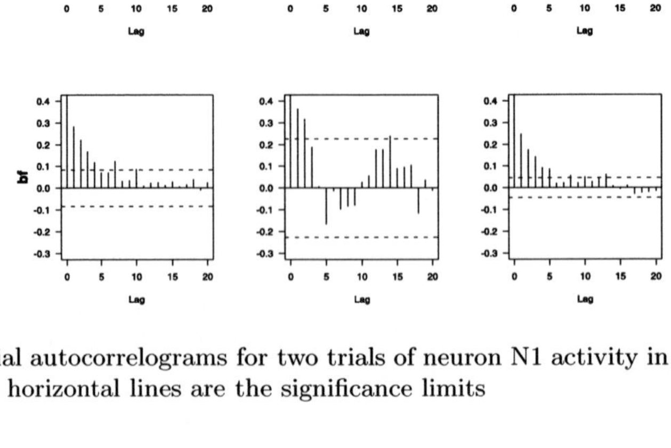

Fig. 2. Serial autocorrelograms for two trials of neuron N1 activity in each period. The dashed horizontal lines are the significance limits

In Figure 2 the autocorrelograms for two trials of neuron N1 can be observed. In this example, the structure of the autocorrelograms of the two trials are dissimilar. In the first one, there is serial autocorrelation of order up to nine for the *pre* period and there is no autocorrelation in the *post* period while in the second trial there is only correlation up to lower orders in *pre* and *final* and there exists a moderate autocorrelation in the *post* stage.

In general, autocorrelograms vary form trial to trial and there is no fixed structure for the autocorrelation of each neuron. Even though, it can be seen that the serial autocorrelation is low (most of the times not significant) in the *post* part of the recordings and that there exists autocorrelation of high orders in *pre* and *final*. A particular case is the one of neuron N5, which has a very low correlation of order one during spontaneous activity and absolutely no correlation during the effect of the stimulus. On the other hand, in the autocorrelograms for neuron N4a a very well defined alternation pattern between negative and positive coefficients can be observed. The autocorrelograms for neurons N4a and N5 can be seen in Figure 3.

For almost every neuron, the serial autocorrelation drops in the *post* period. This autocorrelation decrease is presumably due to the disruption of the spontaneous slow oscillatory activity induced by the electrical stimulation.

In [4] different pattern types of firing rates are discussed, such as, among others, cyclic rates or periods of local growth (or decrease). This patterns lead to different autocorrelogram shapes. Some of these patterns are shown by simulation studies and, for others, references are presented. One particular

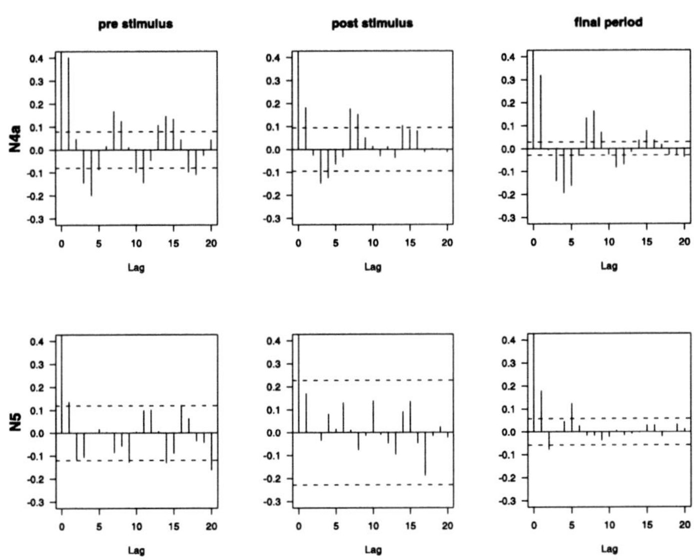

Fig. 3. Serial autocorrelograms for one trial of the *bs* stimulus of neurons N4a and N5 activity in each period. The dashed horizontal lines are the significance limits.

example of what is discussed in [4] is the pattern exhibited by neuron N4a: the alternation between negative and positive autocorrelation coefficients may proceed from an alternation between small and large ISIs.

3.2 Higher Order Interspike Autocorrelation

One may take into account, not only the elapsed times between consecutive firings but also the elapsed times between any two spikes. In fact this is the correlation measure mostly used in neuroscience.

The idea consists in breaking the total time of recording, T, in $Q = \left[\frac{T}{q}\right] + 1$ intervals of length q. Let A_i be the i-th interval, $A_i = [(i-1)q, iq)$. Here it is convenient to note that, if only one spike is intended to fall in each interval, q must be sufficiently small, 1 ms for example. This is caused, for instance, by the refractory period that a neuron needs to recover before firing again. Let us define the new series $\{V_i\}_{i=1}^{Q}$:

$$V_i = \sum_{j=1}^{n+1} I(T_j \in A_i) \tag{1}$$

where $I(A)$ is the indicator function of the set A. We can estimate its autocovariance function by

$$\hat{\gamma}_V(h) = \frac{1}{Q} \sum_{i=1}^{Q-h} (V_{i+h} - \bar{V})(V_i - \bar{V}), \quad \bar{V} = \frac{1}{Q} \sum_{i=1}^{Q} V_i \, .$$

Now, if $h << Q$ the following approximations, $Q - h \approx Q$ and $\sum_{i=1}^{Q-h} V_{i+h} \approx \sum_{i=1}^{Q} V_i \approx \sum_{i=1}^{Q-h} V_i$ can be used to obtain, after some algebra,

$$\hat{\gamma}_V(h) \approx \frac{1}{Q} \sum_{i=1}^{Q-h} V_{i+h} V_i - \bar{V}^2 \, . \tag{2}$$

Since $\bar{V}$ does not depend on h, we can concentrate in $\hat{\gamma}_V^*(h) = \hat{\gamma}_V(h) + \bar{V}^2 = \frac{1}{Q} \sum_{i=1}^{Q-h} V_{i+h} V_i$ without modifying the shape of the function $\hat{\gamma}_V(h)$.

Now,

$$V_{i+h} V_i = \sum_{j=1}^{n+1} \sum_{l=1}^{n+1} I(T_j \in A_i, T_l \in A_{i+h})$$

and then, $\hat{\gamma}_V^*(h)$, as a function of h, is the histogram of these frequencies, though divided by Q.

Observe that in the case of a small enough q (q=1 ms for example), $V_{i+h} V_i = 0$, except when $V_{i+h} = V_i = 1$, in which case the product is 1. This is why, for each h:

$$\hat{\gamma}^*_V(h) = \frac{1}{Q}\sum_{i=1}^{Q-h} I((V_{i+h}, V_i) : (V_{i+h}, V_i) = (1,1)) ,$$

which is easy to think in terms of time: $V_{i+h} = V_i = 1$ means that there are two spikes that are separated by a distance of, at least $h-1$ and at most $h+1$. From this follows that $Q\hat{\gamma}^*_V(h)$ counts the number of spike pairs (although some might be missing) that are separated by a distance between $h-1$ and $h+1$. This is the main idea to define the autocorrelation as it follows.

Higher order interspike autocorrelation, as it is used in neuroscience, is very similar to $\hat{\gamma}^*_V(h)$ but it is built in an alternative way. Actually, it is defined as the histogram of relative frequencies (or absolute sometimes) of the elapsed time between any two spikes of a train that do not surpass a certain $w_{\max}$ chosen by the researcher. This $w_{\max}$ is usually much smaller than T, which allows us to compare with the serial covariance function of $\{V_i\}_{i=1}^{Q}$, since the approximations in (2) are valid.

Given a spike train $\{T_i\}_{i=1}^{n+1}$, let the set of distances between any two spikes be $\{D_m\}_{m=1}^{M} = \{T_i - T_j / i, j \in \{1, \ldots, n+1\}, i \neq j\}$, such that $-w_{\max} \leq D_m \leq w_{\max}$. Moreover, we need to choose b, where $2b$ is the width of the histograms intervals. In this context, we define the higher order interspike autocorrelation (HOISA) of a spike train at the distance d, by:

$$\hat{g}(d) = \frac{1}{M}\sum_{m=1}^{M} I(d - b \leq D_m \leq d + b) .$$

Here b plays a similar role to q in (2) and, in fact, this histogram is very similar to that obtained from the serial autocovariance of the series $\{V_i\}_{i=1}^{Q}$. Some differences might arise from discretization and normalization. Interestingly, for $\hat{\gamma}^*_V(h)$ the discretization is done before calculating the distances, while in the definition of $\hat{g}(d)$ the discretization is carried out when constructing the histogram. On the other hand, to obtain $\hat{\gamma}^*_V(h)$ the absolute frequencies are divided by Q while for $\hat{g}(d)$ the denominator is M. Then the results should be almost proportional. In Figure 4 we can observe the degree of similarity of these histograms for three lengths of q and b, $q = b = 0.01, 0.1$ and $1\,\text{s}$ for the activity in the *pre* period of one trial of neuron N1. The functions $\hat{\gamma}^*_V(h)$ and $\hat{g}(d)$ have been multiplied by Q and M respectively, so that the similarities are better shown up. The histograms are practically the same though there are some differences for the three sizes. These differences grow with the size of q and b and when $q = b = 1$ these are noticeable.

Note that, in fact, the HOISA is just an estimate of the probability density of time between any two spikes. To get a smoother estimate, a nonparametric kernel estimator is proposed:

$$\tilde{g}(d) = \frac{1}{Mh}\sum_{m=1}^{M} K\left(\frac{d - D_m}{h}\right) = \frac{1}{M}\sum_{m=1}^{M} K_h(d - D_m) ,$$

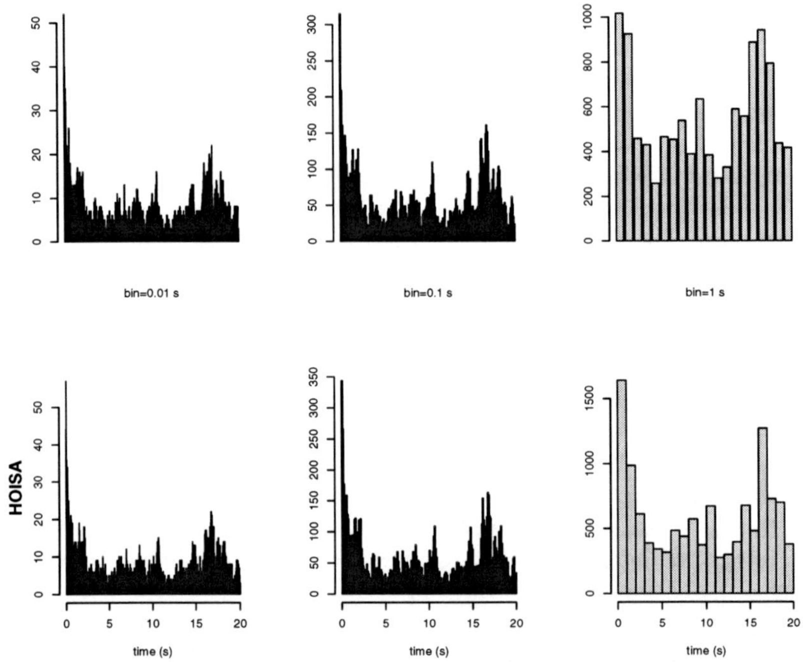

Fig. 4. Comparison between $Q\gamma_V^*(h)$ (top panel) and $M\hat{g}(d)$ (bottom panel) for three bin sizes, $q = b = 0.01, 0.1, 1$

where K is a kernel function and $h > 0$ a smoothing parameter (see [3], [5], [7] and [8] for details). We have used the gaussian kernel function and the Sheather and Jones "plug in" method for bandwidth selection (see [6]).

Figure 5 shows the autocorrelation for the *pre* and *post* periods for two trials, one for each stimulus, of neuron N1 activity. We have used a $w_{\max}$ of 10 s. In the *pre* period Figure 5 shows that, given that there is a spike in time t, there is a high probability density for another spike to occur within the next few seconds (less than 2 s). Also, differences between *pre* and *post* can be observed and also between the different trials in the *post* period.

Next, Figure 6 shows the estimates for a trial of each stimulus and for the three stages of the record of four different neurons. As we had observed before for neuron N1, in *pre* there is a high probability of two spikes occurring very close in time and there also exist some other probability peaks.

Many of the plots in Figure 6 exhibit secondary peaks. It is interesting to analyze what these secondary peaks mean. As indicated above, under sleep states or, as in this case, anesthesia, most cortical neurons display an oscillatory activity. Some rhythms have been characterized neurophysiologically in cats, as the slow rhythm (< 1 Hz), the delta rhythm (1–5 Hz) and the spindle oscillation (7–14 Hz). These rhythms are designated as *slow sleep*

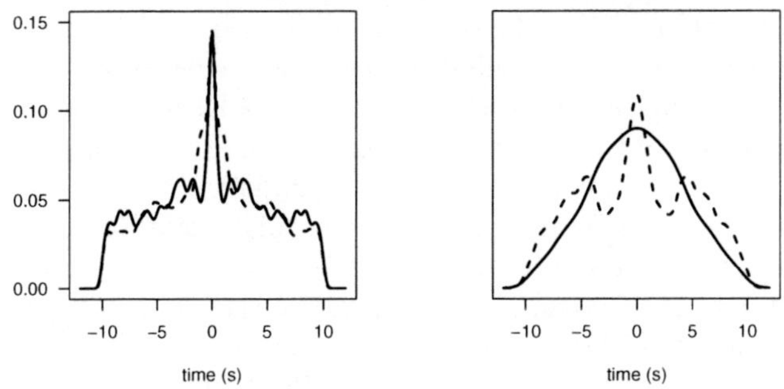

Fig. 5. Kernel estimates of the higher order interspike autocorrelation for one trial of stimulus *bs* (solid line) and one trial of stimulus *bf* (dashed line) of neuron N1 in the *pre* (left panel) and *post* (right panel) periods

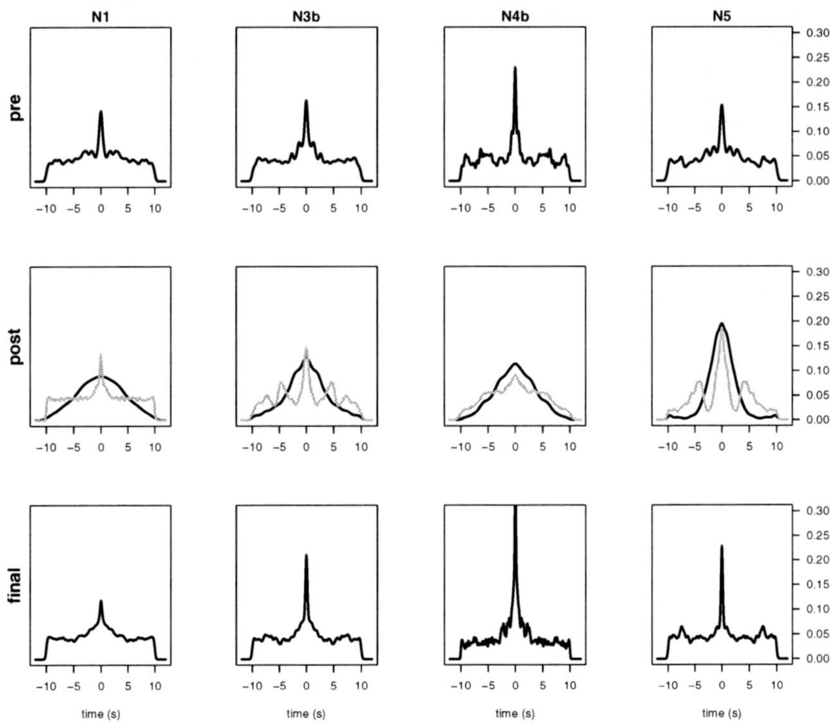

Fig. 6. Kernel estimates of the higher order interspike autocorrelation for one trial of each neuron in each period. Black lines correspond to trials of the *bs* stimulus and grey lines to the *bf* stimulus

oscillations. On the other hand, neuronal spike responses are grouped into what are called *bursts*. These features are sequences of action potentials fulfilling certain characteristics, including: a) consecutive spikes within a burst are not separated one from another in more than certain distance, and b) between one burst and another there is, at least, a certain distance. The distances in this definition may change for different areas of the cortex. If there is an oscillation, for example a delta oscillation of 2 Hz, what happens is that after a neuron generates a burst, it is quite likely that the next burst will occur after about 500 ms. In anesthetized cats it is common to record oscillations of about 0.1 Hz (belonging to the so termed *slow rhythm*). Thus, a slow oscillatory activity of 0.1 Hz could be the cause of the peaks at 10 s. If larger values of $w_{\max}$ were chosen, peaks at around 20 and 30 s could be observed in the HOISA.

Regarding the HOISA functions in the *post* stage. The differences between the autocorrelation functions are mainly found in their dispersion. For this period of the recorded trials, most of the histograms are unimodal, but there are some trials in which conspicuous secondary peaks can be observed; these are supposed to reflect stimulation-induced oscillations. In the *post* period it makes sense to compare the estimates obtained for each of the two stimuli. In several neurons, autocorrelations for the stimulus *bs* present more dispersion than those for the stimulus *bf*.

The estimates of the autocorrelation function for the *final* stage of the study are very similar to the corresponding ones of the *pre* condition. The main peak of the probability density remains at zero. There are also other peaks as in *pre*. In the *post* period, distances between spikes were mainly small but when the effect of the stimulus is over, the distances return to the behavior they had before the stimulus was applied.

These correlation measures, counting the distances along the entire train, do not take into account the possible lack of stationarity on trains and therefore assume that the correlation is stationary. Often, this stationarity is not easy to justify. When neurons are under the effect of a stimulus, they can adapt to it over time, or they can lose stationarity because the stimulus vary in time and the neuronal response varies with it. To check the lack of stationarity and see how the autocorrelation changes over time, the HOISA functions can be computed using sliding windows. Thus, autocorrelation is calculated at each instant using information of a neighborhood of each point. The problem with this approach is that it takes a lot of data for the estimates to be accurate. From the neurobiological perspective it is interesting to calculate these functions in the *post* period, to reveal the autocorrelation dynamics after the stimulus is applied. However, with the amount of data we had for this analysis no accurate estimates could be computed. Instead, in Figure 7 we see an example of how this functions behave in the *pre* stage, where they could, or not, be stationary.

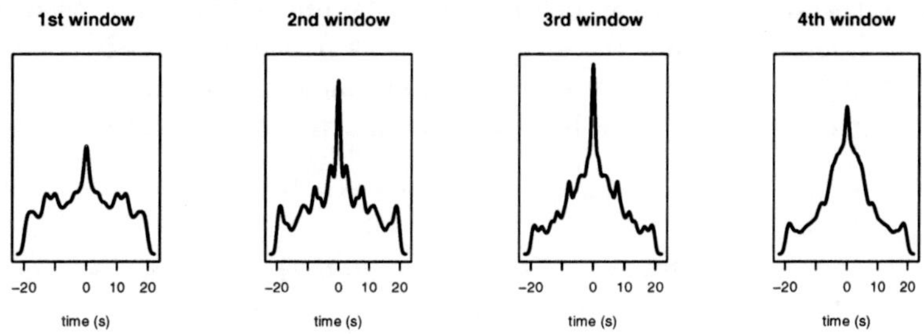

Fig. 7. HOISA functions in the *pre* part of the first trial of neuron N1. A sliding window of 48 s has been used centered at times: 24 s, 48 s, 72 s and 96 s.

4 Testing Independence for Interspike Intervals

In this section we will study the existence of dependence among the elapsed times between consecutive spikes. So, we will test the null hypothesis H_0 : the ISIs are independent, versus the alternative H_1 : they are not independent.

Two different tests will be proposed. If the ISIs are dependent, this situation will influence the shape of the HOISA. These functions will be used to build the first test. The estimated autocorrelation function for the original train will be compared with another one obtained from independent spike trains. On the other hand, the Kolmogorov-Smirnov goodness-of-fit test will be used to compare the distribution of the elapsed times between consecutive spikes in the original train with the distribution of the times of independent trains.

To obtain the sample of independent ISIs, a random shuffle is performed in the original ISIs. A new sample $\{S_i^*\}_{i=1}^n$ is obtained from $\{S_i\}_{i=1}^n$, destroying all the possible serial dependence but preserving any other possible features. With this new sample a new spike train is built: $T_1^* = 0$ y $T_i^* = \sum_{j=1}^n S_j^*$, whose times between consecutive spikes are independent. The differences between the HOISA function of this independent train and the one of the original train will show how far from independence the train under study is. In [4] this methodologies for independence tests are discussed.

The first hypothesis test is carried out as follows. The HOISA function of a registered spike train will be compared with the one obtained from a shuffled train. More specifically, N shuffled trains are used, their HOISA functions are computed and averaged to avoid falling in a case that is not representative. This *average HOISA function* is denoted $\bar{g}(t)$. The test statistic is defined as the L_1 distance:

$$T_{HOISA} = \int |\tilde{g}(x) - \bar{g}(x)| \mathrm{d}x \ .$$

H_0 will be rejected for large values of T_{HOISA}.

For the second test, the empirical distributions of the D_m for both populations, the original, $\tilde{F}$, and the shuffled one, which we call $\bar{F}$, are needed. To estimate the second one, as before, several shuffled trains are used, say N. From each shuffled train, the set of distances between spikes is constructed, and $\bar{F}$ is defined as the empirical distribution of the sample of all these N sets put together. Then, the Kolmogorov-Smirnov test statistic is used:

$$T_{KS} = \sup_x |\tilde{F}(x) - \bar{F}(x)| \ .$$

To calibrate the distributions of the test statistics a bootstrap method is proposed. The steps for the first test are the following:

1. Sample from $\{s_i = t_{i+1} - t_i\}_{i=1}^n$ to obtain a resample $\{s_i^*\}_{i=1}^n$ of distances between consecutive spikes and build a bootstrap train: $t_1^* = 0$, and $t_i^* = \sum_{j=1}^{i-1} s_j^*$, for $i = 2, \ldots, n+1$.
2. Calculate $\tilde{g}^*(t)$ for this bootstrap train.
3. Resample N times from s^* to obtain: $s^{**(i)}$, $i = 1, \ldots, N$ as before. Build $t^{**(i)}$ as in Step 1 and calculate $\tilde{g}^{**(i)}$ for each train $t^{**(i)}$. Then define $\bar{g}^* = \frac{1}{N}\sum_{i=1}^N \tilde{g}^{**(i)}$.
4. Obtain $T_{HOISA}^* = \int |\tilde{g}^* - \bar{g}^*|$.
5. Repeat Steps 1-4 B times to get $T_{HOISA,1}^*, \ldots, T_{HOISA,B}^*$ and use them to estimate the desired quantiles of the T_{HOISA} distribution or the p-value for T_{HOISA}^{obs}.

For the Kolmogorov-Smirnov test the procedure is very similar:

1. Build the independent spike train t^* as before.
2. Calculate the distances between consecutive spikes for the bootstrap train: $\{d_m^*\}_{m=1}^M$.
3. Resample N times from t^*, to build N trains $\{t^{**(j)}, j = 1, \ldots, N\}$ and for each train build the set of distances: $\{d_m^{**(j)}\}_{m=1}^{M_j}$.
4. Calculate T_{KS}^* as the Kolmogorov-Smirnov statistic for the samples $\{d_m^*\}_{m=1}^M$ y $(d_1^{**(1)}, \ldots, d_{M_1}^{**(1)}, d_1^{**(2)}, \ldots, d_{M_2}^{**(2)} \ldots, d_1^{**(N)}, \ldots, d_{M_N}^{**(N)})$.
5. Repeat Steps 1-4 B times to obtain $T_{KS,1}^*, \ldots, T_{KS,B}^*$ and use them to estimate the desired quantiles of the T_{KS} distribution or the p-value for T_{KS}^{obs}.

In general, differences can be observed between the HOISA function of the original train and the one obtained with the resamples. Roughly speaking, the density of the resampled data is more uniformly distributed and then the main peak is lower than in the case of the real data. It is also very common the absence of secondary peaks in the HOISA functions of the resampled trains.

In Table 1 the results of the tests for four neurons, N1, N3a, N3b and N4b and three different recordings (one in the *pre* part and two in the *post* part, one for each stimulus) can be observed. The p-values obtained with each test were calculated using the bootstrap method described above. A total number of 500 bootstrap resamples and $N = 80$ shuffles were used for each bootstrap train in the *pre* part and $N = 100$ in the *post* part. Also, a Ljung-Box (T_{LB}) test was implemented to compare the results.

Table 1. p-values for the independence tests T_{HOISA}, T_{KS} and T_{LB}, constructed using the distances between two spikes

neuron	N1			N3a			N3b			N4b		
period	*pre*	*post*		*pre*	*post*		*pre*	*post*		*pre*	*post*	
stimulus		*bs*	*bf*		*bs*	*bf*		*bs*	*bf*		*bs*	*bf*
T_{HOISA}	0.000	0.080	0.204	0.006	0.004	0.126	0.000	0.006	0.308	0.000	0.254	0.574
T_{KS}	0.002	0.036	0.668	0.266	0.002	0.456	0.002	0.004	0.634	0.001	0.148	0.514
T_{LB}	0.000	0.205	0.012	0.000	0.000	0.018	0.000	0.000	0.980	0.002	0.320	0.911

These results show that, in the *pre* period, the distances between consecutive spikes are not independent (but one case: N3a KS test). This fact is not true in most of the cases for the *post* period. The null hypothesis is rejected in the *post* part for the *bs* stimulus in neurons N3a and N3b (and for *bs* in N1 using the KS test and in *bf* in N1 and N3a using the Ljung-Box test) but it is not rejected for the *bf* stimulus, showing a difference between stimuli.

Figure 8 shows the HOISA functions of three original trains, both *pre* and *post* (*bs* stimulus) periods, and the average HOISA function for the independent case, averaged over 100 shuffles of the original train. In Figure 8 it is easy to recognize the cases where independence is rejected (neurons N1, N3b and the *pre* period of neuron N4b) and the one in which it is not (*post* of neuron N4b).

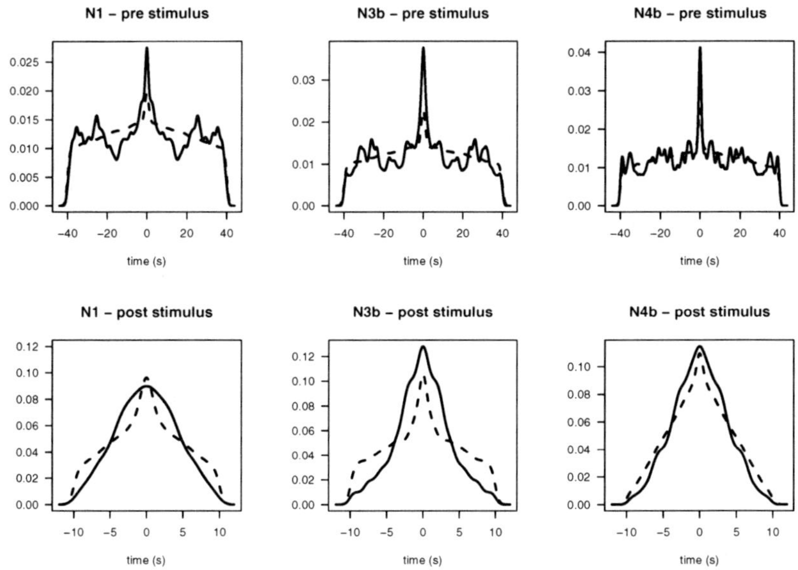

Fig. 8. Comparison of the HOISA function for the original trains (solid line) and the average HOISA function for independent trains (dashed line). First trial of neurons N1, N3b and N4b (*bs* stimulus).

5 Conclusions

Two correlation measures for spike trains have been introduced. First of all, the serial autocovariance has been discussed. On the other hand, the higher order interspike autocorrelation (HOISA) was presented. This autocorrelation measure is commonly used in neuroscience. The spontaneous activity of neurons is characterized by the existence of dependence among spikes. In this work, a test for independence based on the HOISA function was proposed. As this function is constructed in the basis of a histogram, another statistic, based on the empiric distribution function, was discussed. The distribution of these statistics under the null hypothesis was calibrated with a bootstrap procedure. Finally, a Ljung-Box statistic was also used for comparison. This last statistic has the inconvenience of being based on the serial autocorrelation which, varies very much from one trial to another. In general, it can be observed that dependence exists in the *pre* part, reflecting the highly synchronized neuronal oscillatory activity. This dependence is present for some neurons after the *bs* stimulus while it does not appear after the *bf* stimulus for most of the neurons. In some cases, the T_{KS} and T_{LB} statistics present values that are not consistent with the ones obtained with the other tests. This does not happen with the T_{HOISA} statistic, what makes it more robust. Our results indicate that the HOISA-based test for independence is a useful method for the characterization and analysis of the dynamics of the neuronal oscillatory activity.

References

1. Brown, N.E., Kass, R.E., Mitra, P.: Multiple neural spike train data analysis: state-of-the-art and future challenges. Nature Neurosci. 7(5), 456–461 (2004)
2. Dayan, P., Abbott, L.F.: Theoretical Neuroscience: Computational and Mathematical Modeling of Neural Systems. The MIT Press, Cambridge (2001)
3. Parzen, E.: On estimation of a probability density function and mode. Ann. Math. Statist. 33, 1065–1076 (1962)
4. Perkel, D.H., Gerstein, G.L., Moore, G.P.: Neuronal spike trains and stochastic processes. I. The single spike train. Biophys. J. 7(4), 391–418 (1967)
5. Rosenblatt, M.: Remarks on some nonparametric estimates of a density function. Ann. Math. Statist. 27, 832–837 (1956)
6. Sheather, S.J., Jones, M.C.: A reliable data-based bandwidth selection method for kernel density-estimation. J. Royal Statist. Soc. Ser. B 53(3), 683–690 (1991)
7. Silverman, B.W.: Density Estimation. Chapman & Hall, London (1986)
8. Wand, M.P., Jones, M.C.: Kernel Smoothing. Chapman & Hall, London (1995)

Multiple Comparison of Change Trends in Cancer Mortality/Incidence Rates Taking with Overlapping Regions and Time-Periods*

Nirian Martín[1] and Yi Li[2]

[1] Dept. Statistics, Carlos III University of Madrid, Spain
nirian.martin@uc3m.es
[2] Dept. Biostatistics – Harvard School of Public Health and
Dep. Biostatistics & Computational Biology – Dana Farber Cancer Institute, USA
yili@jimmy.harvard.edu

Summary. When analyzing trends in cancer rates, it is common to rely on the so-called Annual Percent Change (APC). For dealing with such a measure of trend, directly age-adjusted rates are usually considered. Classical methods such as pooled t-tests are often applied for comparing APCs of two groups of individuals in a simple way under independence assumption. In practice, it is quite common to find groups of interest for which the independence assumption fails because their regions or periods of time overlap. No one of the papers that deal this problem consider the case where more than two APCs are compared. In this work we propose a Wald-type test-statistic which is not difficult to compute once we provide the estimators of two or more APCs to be compared. These estimators are the minimum power divergence estimators that cover as special case those obtained by maximum likelihood.

1 Introduction

According to the World Health Statistics 2009, published by WHO, the different types of cancer jointly cardiovascular diseases constituted the main causes of death during year 2004, specially in upper-middle and high income countries. It is well known that as income level increases, the percentage of non-communicable deaths (NCDs) increases, in other words, as people live longer, the risk of NCD grows. In such a study it is shown that in high-income countries during year 2004, 77% of all the deaths were NCDs. Cardiovascular diseases and cancers, as main part of NCDs, provided values of

* This work is related to the stay of Nirian Martín in Harvard University (September 2008 - August 2009), supported by the Real Colegio Complutense. She met Marisa just the first day after arriving in Spain, September 1. In the last conversation Marisa told Nirian that both will meet soon again. Due to her optimistic character it was impossible to imagine they both could not meet again. This paper has been written in memory to Marisa.

age-standardized mortality rates equal to 408 per 100,000 and 164 per 100,000 respectively. In order to implement prevention and control plans against these diseases, it is very important for health authorities to analyze whether the mortality (incidence) rates related to such diseases have increased or decreased in a set of successive years and in what degree, that is the trends in mortality (incidence) are taken into account. The National Surveillance, Epidemiology and End Results (SEER [2]) Program of the National Cancer Institute (NCI) constitutes nowadays one of the most prestigious epidemiological surveillance program in the world that allows the scientific community to do research on cancer data based on the US population, actually the SEER provides a database and statistical software, `SEER*Stat`. The new techniques that are needed to cover different problems related to the age-adjusted cancer rates are being continuously implemented by the SEER and particularly the Annual Percent Change (APC) for studying trends of cancer rates is considered.

Statistically, the trend in cancer rates is an average rate of change per year in a given period of time framework when constant change along the time has been assumed. Let r_i be the expected mean of the cancer rate associated with the i-th time point in a sequence of ordered I time points $\{t_i\}_{i=1}^{I}$, for example each time point could be a year ($t_i = i$, $i = 1, ..., I$). The cancer rates can be referred to either cancer incidence rates or cancer mortality rates which demonstrate the risk of developing cancer or dying from cancer respectively. The annual percent change (APC) is a measure that allows us to compare trends in age-adjusted rates that are taken year by year. Regarding the age-adjusted rates, if we apply directly standardized rates, the expected mean of the cancer rate associated with the rate of the k-th region ($k = 1, ..., K$) at the time-point t_{ki} ($i = 1, ..., I_k$), or the i-th year ($t_{ki} = i$), is given by $r_{ki} = \sum_{j=1}^{J} \omega_j r_{kji}$, where J is the number of age-groups, $\{\omega_j\}_{j=1}^{J}$ is the age-distribution of the Standard Population ($\sum_{j=1}^{J} \omega_j = 1$, $\omega_j > 1$, $j = 1, ..., J$) and $\{r_{kji}\}_{j=1}^{J}$ is the set of rates associated with the k-th region at the time-point i in each of the age-groups ($j = 1, ..., J$). The SEER Program applies directly standardized rates on the US population of year 2000 with $J = 19$ age-groups $[0, 1)$, $[1, 5)$, $[5, 10)$, $[10, 15)$, ..., $[80, 85)$, $[85, *)$. The APC associated with the rates $\{r_{kji}\}_{j=1}^{J}$ of the k-th region is defined as the percentage

$$\mathrm{APC}_k = 100(\theta_k - 1), \tag{1}$$

where $\theta_k = r_{k2}/r_{k1} = ... = r_{kI_k}/r_{k,I_k-1}$ (constant change assumption along the whole period of time in region k).

Let n_{kji} the population at risk in the k-th region, j-th age-group, at the time-point t_{ki} and d_{kji} the number of deaths (or incidences) in the k-th region, j-th age-group, at the time-point t_{ki}. The r.v.s that generate d_{kji}, D_{kji}, are considered to be mutually independent with Poisson distribution $D_{kji} \sim \mathcal{P}(m_{kji})$, where $m_{kji} \equiv \mathrm{E}[D_{kji}] = n_{kji} r_{kji}$.

The Age-stratified Poisson Regression model, introduced by Li *et al.* [6], is useful for modelizing $\{r_{kji}\}_{j=1}^{J}$ under constant change assumption. For m_{kji} (or for r_{kji}) it is considered to hold

$$\log \frac{m_{kji}}{n_{kji}} = \beta_{0kj} + \beta_{1k} t_{ki} \quad \text{or} \quad \log r_{kji} = \beta_{0kj} + \beta_{1k} t_{ki}. \tag{2}$$

The parameter of the model is $\boldsymbol{\beta}_k \equiv (\beta_{0k1}, ..., \beta_{0kJ}, \beta_{1k})^T$ and its parameter space $\Theta_k \equiv \mathbb{R}^{J+1}$. Note that

$$\theta_k = \exp(\beta_{1k}), \tag{3}$$

with β_{1k} being the parameter we would like to estimate in the practice. The same model in matrix notation is given by

$$\log\left(\mathrm{Diag}^{-1}(\boldsymbol{n}_k)\, \boldsymbol{m}_k(\boldsymbol{\beta}_k)\right) = \boldsymbol{X}_k \boldsymbol{\beta}_k \quad \text{or} \quad \boldsymbol{m}_k(\boldsymbol{\beta}_k) = \mathrm{Diag}(\boldsymbol{n}_k) \exp(\boldsymbol{X}_k \boldsymbol{\beta}_k), \tag{4}$$

where $\boldsymbol{m}_k(\boldsymbol{\beta}_k) \equiv (m_{k11}(\boldsymbol{\beta}_k), ..., m_{kJI_k}(\boldsymbol{\beta}_k))^T$, $\mathrm{Diag}(\boldsymbol{n}_k)$ is a diagonal matrix of individuals at risk $\boldsymbol{n}_k \equiv (n_{k11}, ..., n_{kJI_k})^T$ and

$$\boldsymbol{X}_k = \begin{pmatrix} \mathbf{1}_{I_k} & & \boldsymbol{t}_k \\ & \ddots & \vdots \\ & \mathbf{1}_{I_k} & \boldsymbol{t}_k \end{pmatrix}_{JI_k \times (J+1)} = (\mathbf{I}_J \otimes \mathbf{1}_{I_k}, \mathbf{1}_J \otimes \boldsymbol{t}_k),$$

with $\boldsymbol{t}_k \equiv (t_{k1}, ..., t_{kI_k})^T$, is a full rank $M_k \times (J+1)$ design matrix, with $M_k \equiv JI_k$.

If we focus on model (4), for making statistical inference with sample $\boldsymbol{D}_k \equiv (D_{k11}, ..., D_{kJI_k})^T$ there is no problem. However, if we take jointly $k = 1, ..., K$ regions, $\boldsymbol{D} = (\boldsymbol{D}_1, ..., \boldsymbol{D}_K)^T$ is not a simple random sample, because it is possible to find individual at risk at the same time in different regions. In order to overcome this difficulty in Martín and Li [3] it were made some assumptions with regard to the overlapping region associated exclusively to a couple of regions k and $h \neq k$, which is applied in the case where both regions have the same APC: if $D_{kji}^{(h)} \sim \mathcal{P}(m_{kji}^{(h)})$ is the number of deaths (incidences) in the overlapping part "in space" between regions k and h, then for $n_{kji}^{(h)} > 0$ (population at risk) it holds

$$m_{kji}^{(h)} = \frac{n_{kji}^{(h)}}{n_{kji}} m_{kji}. \tag{5}$$

In addition, in the complementary part with respect to the overlapping part "in space" between regions k and h, it holds $D_{kji}^{\prime(h)} \sim \mathcal{P}(m_{kji}^{\prime(h)})$, which is independent with respect to $D_{kji}^{(h)}$, and $m_{kji}^{\prime(h)} = (n_{kji}^{\prime(h)}/n_{kji}) m_{kji}$.

In this paper the most important characteristics for the minimum power divergence estimators of the Age-stratified Poisson Regression model, when overlapping group of individuals are taking into account, are revised in Section 2. As main part of the paper Wald-type test-statistics for testing the equality of APCs of K regions with possible overlapping group of individuals are presented in Section 3. A numerical result is shown in Section 4.

2 Minimum Power Divergence Estimators for the Age-Stratified Poisson Regression Model with a Possible Overlapping Group of Individuals

Based on the likelihood function of a Poisson sample associated exclusively to the k-th region, $\boldsymbol{D}_k$, and using a single index for vectors rather than triple index, the kernel of the log-likelihood function is given by $\ell_{\boldsymbol{\beta}_k}(\boldsymbol{D}_k) = \sum_{s=1}^{M_k} D_s \log m_s(\boldsymbol{\beta}_k) - \sum_{s=1}^{M_k} m_s(\boldsymbol{\beta}_k)$, and thus the MLE of $\boldsymbol{\beta}_k$ is $\widehat{\boldsymbol{\beta}}_k = \arg\max_{\boldsymbol{\beta}_k \in \Theta_k} \ell_{\boldsymbol{\beta}_k}(\boldsymbol{D}_k)$. Focussed on a multinomial contingency table it is intuitively understandable that a good estimator of the probabilities of the cells should be such that the discrepancy with respect to the empirical distribution or relative frequencies is small enough. The oldest discrepancy or distance measure we know is the Kullback divergence measure, actually the estimator which is built from the Kullback divergence measure is the MLE. By considering the unknown parameters of a Poisson contingency table, the expected values, rather than probabilities and the observed frequencies rather than relative frequencies, we are going to show how is it possible to carry out statistical inference for Poisson models through power divergence measures. According to the Kullback divergence measure, the discrepancy or distance between the Poisson sample $\boldsymbol{D}_k$ and its vector of means $\boldsymbol{m}_k(\boldsymbol{\beta}_k)$ is given by

$$\mathrm{d}_{\mathrm{Kull}}(\boldsymbol{D}_k, \boldsymbol{m}_k(\boldsymbol{\beta}_k)) = \sum_{s=1}^{M_k} \left(D_s \log \frac{D_s}{m_s(\boldsymbol{\beta}_k)} - D_s + m_s(\boldsymbol{\beta}_k) \right). \quad (6)$$

Observe that $\mathrm{d}_{\mathrm{Kull}}(\boldsymbol{D}_k, \boldsymbol{m}_k(\boldsymbol{\beta}_k)) = -\ell_{\boldsymbol{\beta}_k}(\boldsymbol{D}_k) + C(k)$, where $C(k)$ does not depend on parameter $\boldsymbol{\beta}_k$. Based on such a relationship we can define the MLE of $\boldsymbol{\beta}_k$ as the minimum Kullback divergence estimator $\widehat{\boldsymbol{\beta}}_k = \arg\min_{\boldsymbol{\beta}_k \in \Theta_k} \mathrm{d}_{\mathrm{Kull}}(\boldsymbol{D}_k, \boldsymbol{m}_k(\boldsymbol{\beta}_k))$, and the MLE of $\boldsymbol{m}_k(\boldsymbol{\beta}_k)$ functionally as $\boldsymbol{m}_k(\widehat{\boldsymbol{\beta}}_k)$, due to the invariance property of the MLEs. The power divergence measures are a family of measures defined as

$$\mathrm{d}_{\lambda}(\boldsymbol{D}_k, \boldsymbol{m}_k(\boldsymbol{\beta}_k)) = \frac{1}{\lambda(1+\lambda)} \sum_{s=1}^{M_k} \left(\frac{D_s^{\lambda+1}}{m_s^{\lambda}(\boldsymbol{\beta}_k)} - D_s(1+\lambda) + \lambda m_s(\boldsymbol{\beta}_k) \right), \quad (7)$$

such that from each possible value for subscript $\lambda \in \mathbb{R} - \{0, -1\}$ a different way to quantify the discrepancy between $\boldsymbol{D}_k$ and $\boldsymbol{m}_k(\boldsymbol{\beta}_k)$ arises. In case of $\lambda \in \{0, -1\}$, it is defined $\mathrm{d}_{\lambda}(\boldsymbol{D}_k, \boldsymbol{m}_k(\boldsymbol{\beta}_k)) = \lim_{\ell \to \lambda} \mathrm{d}_{\ell}(\boldsymbol{D}_k, \boldsymbol{m}_k(\boldsymbol{\beta}_k))$, and in this manner the Kullback divergence appears as special case of power divergence measures when $\lambda = 0$, $\mathrm{d}_0(\boldsymbol{D}_k, \boldsymbol{m}_k(\boldsymbol{\beta}_k)) = \mathrm{d}_{\mathrm{Kull}}(\boldsymbol{D}_k, \boldsymbol{m}_k(\boldsymbol{\beta}_k))$ and on the other hand case $\lambda = -1$ is obtained by changing the order of the arguments for the Kullback divergence measure, $\mathrm{d}_{-1}(\boldsymbol{D}_k, \boldsymbol{m}_k(\boldsymbol{\beta}_k)) = \mathrm{d}_{\mathrm{Kull}}(\boldsymbol{m}_k(\boldsymbol{\beta}_k), \boldsymbol{D}_k)$. The estimator of $\boldsymbol{\beta}_k$ obtained on the basis of (7) is the so-called minimum power divergence estimator (MPDE) and it is defined for each value of $\lambda \in \mathbb{R}$ as

$$\widehat{\boldsymbol{\beta}}_{k,\lambda} = \arg \min_{\boldsymbol{\beta}_k \in \Theta_k} \mathrm{d}_\lambda(\boldsymbol{D}_k, \boldsymbol{m}_k(\boldsymbol{\beta}_k)), \tag{8}$$

and the MPDE of $\boldsymbol{m}_k(\boldsymbol{\beta}_k)$ functionally as $\boldsymbol{m}_k(\widehat{\boldsymbol{\beta}}_{k,\lambda})$ due to the invariance property of the MPDEs. For more details about such a estimator see Pardo and Martín [5].

In order to obtain the MPDE of (1), $\widehat{\mathrm{APC}}_{k,\lambda} = 100(\exp(\widehat{\beta}_{1k,\lambda}) - 1)$, we need to compute the estimator of the parameter of interest by following the next result. In Martín and Li [3] it was established the next result.

Proposition 1. *The MPDE of β_{1k}, $\widehat{\beta}_{1k,\lambda}$, is the solution of the nonlinear equation*

$$f(\widehat{\beta}_{1k,\lambda}) = \sum_{i=1}^{I_k} t_{ki} \Upsilon_{ki} = 0,$$

with

$$\Upsilon_{ki} = \sum_{j=1}^{J} m_{kji}(\widehat{\boldsymbol{\beta}}_\lambda) \left(\varphi_{kji} - 1\right),$$

$$m_{kji}(\widehat{\boldsymbol{\beta}}_\lambda) = n_{kji} \exp(\widehat{\beta}_{0kj,\lambda}) \exp(\widehat{\beta}_{1ki,\lambda} t_{ki}) \quad \textit{and} \quad \varphi_{kji} = \left(\frac{D_{kji}}{m_{kji}(\widehat{\boldsymbol{\beta}}_\lambda)} \right)^{\lambda+1},$$

$$\exp(\widehat{\beta}_{0kj,\lambda}) = \left(\sum_{s=1}^{I_k} p_{kjs} \psi_{kjs}^{\lambda+1} \right)^{\frac{1}{\lambda+1}}, \quad j = 1, ..., J,$$

$$p_{kjs} = \frac{n_{kjs} \exp(\widehat{\beta}_{1k,\lambda} t_{ks})}{\sum_{h=1}^{I_k} n_{kjh} \exp(\widehat{\beta}_{1k,\lambda} t_{kh})} \quad \textit{and} \quad \psi_{kjs} = \frac{D_{kjs}}{n_{kjs} \exp(\widehat{\beta}_{1k,\lambda} t_{ks})}.$$

For the asymptotic distributions some assumptions have to be made:
a) $m^*_{kji}(\boldsymbol{\beta}^0_k) = m_{kji}(\boldsymbol{\beta}^0_k)/N_k$ remains constant as $N_k \equiv \sum_{s=1}^{M_k} m_s(\boldsymbol{\beta}_k)$ increases, that is $m_{kji}(\boldsymbol{\beta}^0_k)$ increases at the same rate as N_k.
b) $N^*_k \equiv \frac{N_k}{N}$ $(k = 1, ..., K)$ is constant as $N \equiv N_1 + ... + N_K$ increases, that is N increases at the same rate as N_k.

For details about the asymptotic distribution of parameters see Martín and Li [3].

3 Wald-Type Test-Statistics for the Age-Stratified Poisson Regression Model with a Possible Overlapping Group of Individuals

The Z-test provided in Li *et al.* [6],

$$Z_{\lambda}(\mathrm{r};\mathrm{s}) = \frac{\widehat{\beta}_{1r,\lambda} - \widehat{\beta}_{1s,\lambda}}{\sqrt{\widehat{\mathrm{Var}}(\widehat{\beta}_{1r,\lambda} - \widehat{\beta}_{1s,\lambda})}} \tag{9}$$

with $\lambda = 0$, for testing $\mathcal{H}_0(\mathrm{r};\mathrm{s}) : \mathrm{APC}_r = \mathrm{APC}_s$ (or $\mathcal{H}_0(\mathrm{r};\mathrm{s}) : \beta_{1r} - \beta_{1s} = 0$) based on the age-adjusted Poisson Regression model is useful only if we want to compare the APCs of two specific regions r and s. This test-statistic is not valid only for MLEs ($\lambda = 0$) but also for MPDEs, actually all the MPDEs of $\boldsymbol{\beta}$ have the same asymptotic distribution. The asymptotic distribution of $Z(\mathrm{r};\mathrm{s})$ is standard normal under $\mathcal{H}_0(\mathrm{r};\mathrm{s})$. Based directly on (9), in Martín and Li [3] $K-1$ test hypotheses separately were performed, $\mathcal{H}_0(1;\mathrm{k}) : \mathrm{APC}_1 = \mathrm{APC}_k$, $k = 2, ..., K$. For the moment there is no paper concerned in making comparisons between more than two regions at the same time, $\mathcal{H}_0 : \mathrm{APC}_1 = \mathrm{APC}_2 = ... = \mathrm{APC}_K$. In order to do that we reformulate the hypotheses

$$\mathcal{H}_0(1;2,...,\mathrm{K}) : \boldsymbol{\theta} = (\theta_1, ..., \theta_{K-1})^T = \mathbf{0}, \tag{10}$$

against $\mathcal{H}_1(1;2,...,\mathrm{K}) : \exists b \in \{1, ..., K-1\} : \theta_b \neq 0$, where $\theta_b \equiv \beta_{11} - \beta_{1b+1}$, $b = 1, ..., K-1$. The following test-statistic is applied

$$W_{\lambda}(1;2,...,\mathrm{K}) = \widehat{\boldsymbol{\theta}}_{\lambda}^T \widehat{\mathrm{Var}}^{-1}(\widehat{\boldsymbol{\theta}}_{\lambda})\widehat{\boldsymbol{\theta}}_{\lambda}, \tag{11}$$

$\widehat{\boldsymbol{\theta}}_{\lambda} = (\widehat{\theta}_{1,\lambda}, ..., \widehat{\theta}_{K-1,\lambda})^T$, where $\widehat{\theta}_{b,\lambda} \equiv \widehat{\beta}_{11,\lambda} - \widehat{\beta}_{1b+1,\lambda}$, $b = 1, ..., K-1$. In particular, when there are no individuals at risk shared by regions to be compared, its expression is given by

$$W_{\lambda}(1;2,...,\mathrm{K}) = \sum_{b=1}^{K-1} \frac{\widehat{\theta}^2_{b,\lambda}}{\widehat{\sigma}^2_{1,b+1,\lambda}} - \frac{1}{\sum_{k=1}^{K} \frac{1}{\widehat{\sigma}^2_{1k,\lambda}}} \sum_{b=1}^{K-1} \sum_{b'=1}^{K-1} \frac{\widehat{\theta}^2_{b,\lambda}\widehat{\theta}^2_{b',\lambda}}{\widehat{\sigma}^2_{1,b+1,\lambda}\widehat{\sigma}^2_{1,b'+1,\lambda}}. \tag{12}$$

Note that in practice when $K = 2$ test-statistic (12) is equivalent to (9) with $\widehat{\mathrm{Var}}(\widehat{\beta}_{1r,\lambda} - \widehat{\beta}_{1s,\lambda}) = \widehat{\sigma}^2_{1r,\lambda} + \widehat{\sigma}^2_{1s,\lambda}$ and $r = 1$, $s = 2$, because $W_{\lambda}(1;2) = Z^2_{\lambda}(1;2)$.

Our aim in this paper is to find the asymptotic distribution of $\widehat{\boldsymbol{\theta}}_{\lambda}$ first and the distribution of (11) later. When comparing regions b and b', we must consider a new reference point for time index i, denoted by $\bar{I}_{bb'}$, such that $t_{b\bar{I}_{bb'}}$ represents the time point within $\{t_{bi}\}_{i=1}^{I_b}$ where the time series associated with region b' is about to start, i.e. we have $\{t_{b'i}\}_{i=1}^{I_{b'}}$ such that $t_{b'1} = t_{b\bar{I}_{bb'}} + 1$. Taking into account Martín and Li [3] it is easy to prove the next result.

Theorem 1. *Under the hypothesis that* $\theta_b = 0$ *or* $\beta_{11} = \beta_{1b+1}$, $b = 1, ..., K-1$, *the asymptotic distribution of* $\widehat{\boldsymbol{\theta}}_\lambda$ *is central Normal with*

$$\mathrm{Var}(\widehat{\theta}_{b,\lambda}) = \sigma_{11}^2 + \sigma_{1b+1}^2 - 2\sigma_{11}^2\sigma_{1b+1}^2\xi_{1,b+1},$$

$$\mathrm{Cov}(\widehat{\theta}_{b,\lambda}, \widehat{\theta}_{b',\lambda}) = \sigma_{11}^2 + \sigma_{1b+1}^2\sigma_{1b'+1}^2\xi_{b+1,b'+1} - \sigma_{11}^2\sigma_{1b+1}^2\xi_{1b+1} - \sigma_{11}^2\sigma_{1b'+1}^2\xi_{1b'+1},$$

where σ_{1k}^2 *is equal to*

$$\sigma_{1k}^2 = \left(\sum_{j=1}^{J}\sum_{i=1}^{I_k} m_{kji}(\boldsymbol{\beta}^0)(t_{ki} - \widetilde{t}_{kj}(\boldsymbol{\beta}_k^0))^2 \right)^{-1}, \tag{13}$$

with $\widetilde{t}_{kj}(\boldsymbol{\beta}_k^0) = \left(\sum_{i=1}^{I_k} m_{kji}(\boldsymbol{\beta}_k^0)t_{ki}\right) / \sum_{i=1}^{I_k} m_{kji}(\boldsymbol{\beta}_k^0)$ *and*

$$\xi_{bb'} = \sum_{j=1}^{J}\sum_{i=1}^{I_b - \bar{I}_{bb'}} \frac{n_{b'ji}^{(b)}}{n_{b'ji}} m_{b'ji}(\boldsymbol{\beta}^0)(t_{b'i} - \widetilde{t}_{bj}(\boldsymbol{\beta}^0))(t_{b'i} - \widetilde{t}_{b'j}(\boldsymbol{\beta}^0)). \tag{14}$$

Proof. It follows by the bilinear property of covariance and taking into account the asymptotic distribution of $\widehat{\boldsymbol{\beta}}_\lambda$ that was established in Martín and Li [3]: The asymptotic distribution of $\widehat{\boldsymbol{\beta}}_\lambda$ is central Normal with $\mathrm{Var}(\widehat{\beta}_{1k,\lambda}) = \sigma_{1k}^2$ and $\mathrm{Cov}(\widehat{\beta}_{1b,\lambda}, \widehat{\beta}_{1b',\lambda}) = \sigma_{1b}^2\sigma_{1b'}^2\xi_{bb'}$. □

We need to obtain the MPDEs of σ_{1k}^2 and $\xi_{bb'}$, $\widehat{\sigma}_{1k,\lambda}^2$ and $\widehat{\xi}_{bb',\lambda}$ respectively. A way to proceed is based on replacing $\boldsymbol{\beta}^0$ by the most efficient MPDE. For example, for $\widehat{\sigma}_{1k,\lambda}^2$ we shall use

$$\widehat{\boldsymbol{\beta}}_\lambda^0 \equiv \begin{cases} \widehat{\boldsymbol{\beta}}_{1,\lambda}^0, \text{ if } N_1 \geq N_k \\ \widehat{\boldsymbol{\beta}}_{k,\lambda}^0, \text{ if } N_1 < N_k \end{cases}.$$

Theorem 2. *The asymptotic distribution of* $W_\lambda(1; 2, ..., \mathrm{K})$ *is* χ_{K-1}^2 *under (10).*

Proof. Let $\boldsymbol{L}^T = (\boldsymbol{1}_{K-1}, -\boldsymbol{I}_{K-1})$ with $\boldsymbol{1}_{K-1}$ being a $(K-1)$-dimensional vector of 1's and $\boldsymbol{I}_{K-1}$ is the $(K-1)$-dimensional identity matrix. Taking into account that $\widehat{\boldsymbol{\beta}}_\lambda = \boldsymbol{L}^T\widehat{\boldsymbol{\theta}}_\lambda$ and based on (14) and Corollary 2.1 in Dik and de Gunst [1] it is concluded that $W_\lambda(1; 2, ..., \mathrm{K})$ is χ_g^2 under (10), where $g = \mathrm{rank}(\boldsymbol{L}) = K - 1$. □

To determine whether N is large, it can be used the rule of thumb of the average of events per cell $\eta \equiv N/(KJI)$. Data sets with $\eta < 5$ are considered to be small, but the length of the whole period of time I is also an important factor to take into account. Because for $\lambda = 0$, $\widehat{N}_{k,0} = \sum_{j=1}^{J}\sum_{i=1}^{I_k} D_{kji}$, it is also held $\widehat{N}_0 = \sum_{k=1}^{K}\widehat{N}_{k,0} = \sum_{k=1}^{K}\sum_{j=1}^{J}\sum_{i=1}^{I_k} D_{kji}$. Hence in practice $\widehat{\eta} \equiv \sum_{k=1}^{K}\sum_{j=1}^{J}\sum_{i=1}^{I_k} D_{kji}/(KJI)$ is useful to estimate whether N is large.

4 Real Data Example

We have considered thyroid cancer mortality in three regions, Western (W) US population (compounded by Arizona, New Mexico and Texas), South Western (SW) US population (compounded by Arizona, California and Nevada) and West Coast (WC) US population (companied by California, Oregon and Washington). Note that Arizona is shared by W and SW, and California by SW and WC. Different periods of time, 1969-1983, 1977-1991 and 1990-1999 are taken for W, SW and WC respectively. The third one differs from the rest in the sense that it considers a shorter period of time for its study. In Martín and Li [3] it was analyzed the performance of the minimum chi-square estimator (MCSE), that is the minimum power divergence estimator obtained with $\lambda = 1$, and it was concluded that it behaves more accurately than the MLEs ($\lambda = 0$) for small data sets. We have chosen thyroid cancer because it is a rare cancer, and both estimators have been computed in order to analyze the data. The rates are expressed per $100,000$ individuals at risk. Taking into account $\exp(\beta_{0k}) = \sum_{j=1}^{J}\omega_j \exp(\beta_{0kj})$, in Figure 1 the fitted models are plotted and from them it seems at first sight that there is a decreasing trend for Thyroid cancer in WC and SW, and null or decreasing trend in W.

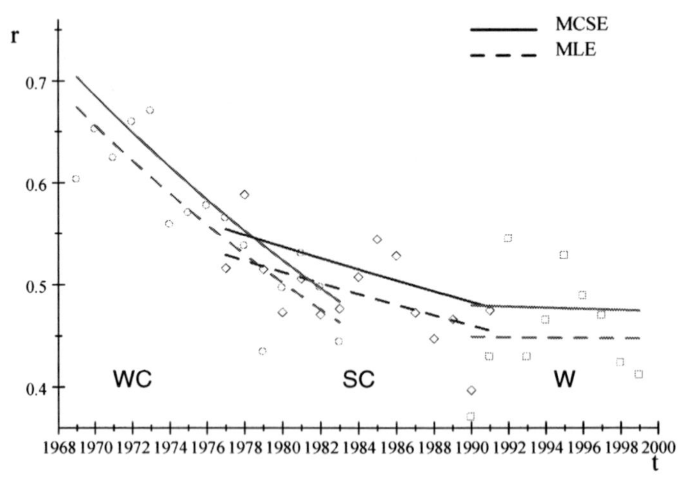

Fig. 1. Fitted models to data

The specific values for estimates and test-statistics $W_\lambda(1;2,3)$, for $\lambda = 0, 1$, are summarized in Table 1.

Table 1. Thyroid cancer mortality trends comparison among WC, SW and W during 1969-1983, 1977-1991 and 1990-1999 respectively: Maximum Likelihood Estimators and Minimum Chi-Square Estimators

Region	k	λ	$\widehat{\beta}_{0k,\lambda}$	$\widehat{\beta}_{1k,\lambda}$	$\widehat{APC}_{k,\lambda}$
WC	1	0	−0.3680	−0.0267	−2.639
	1	1	−0.3241	−0.0268	−2.646
SW	2	0	−0.5404	−0.0107	−1.064
	2	1	−0.4943	−0.0106	−1.053
W	3	0	−0.7939	0.0003	0.031
	3	1	−0.7084	−0.0012	−0.117
WC vs. SW: $Z_{12,0} = -1.85$; $Z_{12,1} = -1.92$					
SW vs. W: $Z_{23,0} = -0.85$; $Z_{23,1} = -0.75$					
$W_0(1;2,3) = 6.02$; $W_1(1;2,3) = 6.01$					

Apart from the Wald-type test-statistic, we have included test-statistics $Z_{12,\lambda}$, $Z_{23,\lambda}$ ($\lambda = 0, 1$) for couples of regions. For their computation are essential the values of three matrices:

$$(\widehat{\xi}_{bb',0})_{b,b' \in \{1,2,3\}} = \begin{pmatrix} * & -8684.09 & 0 \\ -8684.09 & * & -827.56 \\ 0 & -827.56 & * \end{pmatrix},$$

$$\left(\mathrm{Cov}(\widehat{\beta}_{1b,0}, \widehat{\beta}_{1b',0})\right)_{b,b' \in \{1,2,3\}}$$
$$= \begin{pmatrix} 2.92354 \times 10^{-5} & -0.77292 \times 10^{-5} & 0 \\ -0.77292 \times 10^{-5} & 3.04440 \times 10^{-5} & -0.32915 \times 10^{-5} \\ 0 & -0.32915 \times 10^{-5} & 13.06460 \times 10^{-5} \end{pmatrix},$$

$$\mathrm{Var}(\widehat{\boldsymbol{\theta}}_0) = \begin{pmatrix} 7.51380 \times 10^{-5} & 3.68340 \times 10^{-5} \\ 3.68340 \times 10^{-5} & 15.98815 \times 10^{-5} \end{pmatrix}.$$

$$(\widehat{\xi}_{bb',1})_{b,b' \in \{1,2,3\}} = \begin{pmatrix} * & -9125.47 & 0 \\ -9125.47 & * & -866.18 \\ 0 & -866.18 & * \end{pmatrix},$$

$$\left(\mathrm{Cov}(\widehat{\beta}_{1b,1}, \widehat{\beta}_{1b',1})\right)_{b,b' \in \{1,2,3\}}$$
$$= \begin{pmatrix} 2.78598 \times 10^{-5} & -0.73429 \times 10^{-5} & 0 \\ -0.73429 \times 10^{-5} & 2.88823 \times 10^{-5} & -0.31074 \times 10^{-5} \\ 0 & -0.31074 \times 10^{-5} & 12.42113 \times 10^{-5} \end{pmatrix},$$

$$\mathrm{Var}(\widehat{\boldsymbol{\theta}}_1) = \begin{pmatrix} 7.14279 \times 10^{-5} & 3.50927 \times 10^{-5} \\ 3.50927 \times 10^{-5} & 15.20711 \times 10^{-5} \end{pmatrix}.$$

Because p-value($Z_{12,0}$) = 0.064, p-value($Z_{12,1}$) = 0.054, p-value($Z_{23,0}$) = 0.39, p-value($Z_{23,1}$) = 0.45, p-value($W_0(1;2,3)$) = 0.049, p-value($W_1(1;2,3)$) = 0.049, the hypothesis of equal APCs is rejected with 0.05 significance level for the three regions when using the Wald-type test, while cannot be rejected using Z -test-statistics for couples of regions.

Acknowledgement. This work was carried out during the stay of the first author as Visiting Scientist at Harvard University and Dana Farber Cancer Institute, supported by the Real Colegio Complutense and Grant MTM2009-10072. Li's work was partially supported by R01CA95747 and 1P01CA134294.

References

1. Dik, J.J., de Gunst, M.C.M.: The distribution of general quadratic forms in normal variables. Statistica Neerl. 39, 14–26 (1985)
2. Horner, M.J., Ries, L.A.G., Krapcho, M., Neyman, N., Aminou, R., Howlader, N., Altekruse, S.F., Feuer, E.J., Huang, L., Mariotto, A., Miller, B.A., Lewis, D.R., Eisner, M.P., Stinchcomb, D.G., Edwards, B.K. (eds.): SEER Cancer Statistics Review. National Cancer Institute, Bethesda (1975-2006)
3. Martín, N., Li, Y.: A new class of minimum power divergence estimators with applications to cancer surveillance. J. Multivar. Anal (2011) (in press)
4. Pardo, L.: Statistical Inference Based on Divergence Measures. Chapman & Hall/CRC, New York (2006)
5. Pardo, L., Martin, N.: Homogeneity/Heterogeneity Hypotheses for Standardized Mortality Ratios Based on Minimum Power-divergence Estimators. Biomet. J. 51, 819–836 (2009)
6. Li, Y., Tiwari, R.C., Zou, Z.: An age-stratified model for comparing trends in cancer rates across overlapping regions. Biomet. J. 50, 608–619 (2008)

On the Existence of Solutions of a Mathematical Model of Morphogens*

J. Ignacio Tello

Departamento de Matemática Aplicada, E.U. Informática,
Universidad Politécnica de Madrid, Spain
jtello@eui.upm.es

Summary. We consider a simple mathematical model of distribution of morphogens (signaling molecules responsible for the differentiation of cells and the creation of tissue patterns) similar to the model proposed by Lander, Nie and Wan in 2002. The model consists of a system of two equations: a PDE of parabolic type modeling the distribution of free morphogens with a dynamic boundary condition and an ODE describing the evolution of bound receptors. Three biological processes are taken into account: diffusion, degradation and reversible binding. We study the existence and uniqueness of solutions.

1 Introduction

Morphogenesis is the biological process whereby the cells of the embryo differentiate into specialized cells to create tissues. Morphogenesis is one of the most important process of developmental biology and responsible of the creation of shapes and organs. The process consists of the production and distribution of signaling molecules called Morphogens. Morphogens are synthesized at signaling localized sites and spread into the body. We consider the transport of morphogens from the localized points of synthesis to the cells surface by diffusion. Recently several reports suggest the important roll of transcytosis in the transport trough the cells. In transcytosis, vesicles are employed to intake the morphogen on one side of the cell, draw them across the cell, and eject them on the other side. Once the morphogens arrive to the surface of the cell, they bind the receptors situated in the surface and a process of internalization of the complex morphogen-receptors occurs. The response of the cell consists of the activation of particular genes which then determine the pattern of cell differentiation.

The transport of morphogens and other crucial issues, as how morphogens coordinate growth inside the cell, are not well understood and different theories have been proposed in the last decades to describe them.

---

* This paper is dedicated to the memory of Professor María Luisa Menéndez with deep gratitude and affection.

Morphogenesis has been studied from the early 20th century. The first mathematical model was proposed by Turing in the 1950s (see [19]) where a reaction-diffusion system of differential equations modelized the process.

Lander, Nie and Wan [9] studied numerically several mathematical models and focused on the *Drosophila* wing disc. They obtain (by using recent experimental data) that diffusive mechanisms of morphogen transport may produce gradients of morphogens and show that those mechanisms are much more plausible than the non-diffusive ones. They propose several mathematical models, one of them, the diffusion-reversible binding model with degradation is the model which has been analyzed in the following sections. One of the main novelties of that model arises in the peculiar dynamic boundary condition at $x = 0$ (see formula (4)).

Lander, Nie, Vargas and Wan [8] and Lander, Nie and Wan [10] proposed several models of differential equations. The models consider a PDE of parabolic type to describe the evolution of morphogens and a set of ODE′s to model the receptor and the bound-receptor. They study the steady states and the linear stability of them under the action of a source in a region of the domain.

Merkin and Sleeman [14] have studied the system proposed by Lander, Nie and Wan [9] with degradation and without it. They provide an analysis of the models under the assumption of constant concentration of morphogens at the boundary $x = 0$ and gradient of morphogens equals to zero at infinity. The authors prove that the case where the bound morphogen complex is not degraded, the free morphogen profile is essentially linear and spreads as a square root law.

Recently Merkin, Needham and Sleeman [13] have considered a mathematical model with diffusion and have included a chemosensitivity term to describe morphogen concentration. They have presented results on the existence and uniqueness of classical solutions and self-similarity. Their numerical simulations have showed periodic pulse solutions.

Lou, Nie and Wan [12] consider a model with two species of morphogens. The system consists of three PDE′s of parabolic type and one ODE. They study the steady states and numerical simulation for the evolution problem.

In Tello [18] a similar system is studied in the whole domain for a particular boundary conditions given by a system of ODE′*s* at $x = 0$. The existence and uniqueness of solutions is proved by using a Banach fixed point argument. In [18] the asymptotic behavior of the boundary conditions is studied and used to prove the behavior of the solution (u, v) which satisfies

$$\lim_{t\to\infty} u(x,t) = \phi(x), \qquad \lim_{t\to\infty} v(x,t) = \xi(x)$$

in $L^p(\Omega)$ for $1 \le p < \infty$ where (ϕ, ξ) is a steady state of the problem.

The mathematical model proposed by Lander, Nie and Wan [10] is considered in Muñoz and Tello [15] under structural simplifications which reduce a system of 5 equations to a model of 3. The resulting mathematical model does not consider the effects of the processes in the interior of the cell and it is similar to the model described by Lander, Nie, Vargas and Wan [8]. In [15] the steady states of the problem are considered before the existence and uniqueness of solutions. The numerical simulations of the problem show a rich behavior depending of the range of the parameters.

Another system of equations modeling morphogenesis has been proposed in Bollenbach, Kruse, Pantazis, González-Gaitán and F. Julicher [1] where a chemotaxis term is introduced in the system. The problem is reduced to a system of two equations, one of parabolic type coupled to an ODE. A rigorous mathematical analysis is applied to study a system in Stinner, Tello and Winkler [16] for a range of parameters. In [16] global existence of solutions and uniform bounds in $L^\infty(\Omega)$ are given by using an iterative argument. A regularization method is considered to study the problem, where diffusion is included for the non-diffusive equation. Uniform bounds are obtained and the approximated solution converges to the solution of the limit problem when the diffusion coefficient goes to 0.

In this work we consider the case of diffusive transport of morphogens. We first describe the mathematical model proposed by Lander, Nie and Wan [9] where a dynamic boundary condition governs the production of morphogens at $x = 0$. In the third section we study the steady states and the boundary conditions. Existence and uniquenes of solutions are studied in the last two sections by using a Schauder fixed point argument.

2 The Mathematical Model

Different models of distribution of morphogens have been introduced by several authors in the last decade. A simple mathematical model is presented in Lander, Nie and Wan [9] in an unbounded domain $(0, \infty)$. We study the mathematical model in a one dimensional bounded domain. Lander et al [9] consider the evolution of Decapentaplegic (Dpp) one of the morphogens present in Drosophila larvae wing disc. We denote by L the morphogen Dpp (the lingand), by R the receptor per unit of extracellular space, by LR the complex ligand-receptor and their respective concentrations by $[L]$, $[R]$ and $[LR]$. The following formula express the processes of formation of the complex

$$L + R \underset{\mathrm{k}_{off}}{\overset{\mathrm{k}_{on}}{\rightleftharpoons}} LR,$$

where k_{on} and k_{off} are the binding and dissociation rate constants. We assume that the number of receptors (free and bound) is constant on time, i.e.

$$[R] = R_{tot} - [LR], \tag{1}$$

where R_{tot} is the total receptor concentration per unit of extracellular space. Assumption [1] reduces the number of equations and simplifies the problem.

We consider linear diffusion of $[L]$ with diffusion constant d. Then, $[L]$ satisfies the equation:

$$\frac{\partial}{\partial t}[L] - d\frac{\partial^2}{\partial x^2}[L] = -k_{on}\left(R_{tot} - [LR]\right)[L] + k_{off}[LR].$$

We consider degradation of the complex $[LR]$. Let k_{deg} be the degradation rate constant, then $[LR]$ satisfies the equation:

$$\frac{\partial}{\partial t}[LR] = k_{on}R_{tot}[L] - k_{on}[L][LR] - k_{off}[LR] - k_{deg}[LR].$$

Let u and v be the normalized concentrations of morphogen and complex morphogen-receptor, then after normalization of the equations we arrive to the system

$$\frac{\partial u}{\partial t} - \frac{\partial^2}{\partial x^2}u = -u(1-v) + v, \tag{2}$$

$$\frac{\partial v}{\partial t} = \lambda\left[u(1-v) - v\right] - \mu v. \tag{3}$$

The system (2), (3) is proposed by Lander, Nie and Wan [9] in the domain $x \in (0,\infty)$, $t \in (0,\infty)$ with the boundary conditions

$$\frac{\partial u}{\partial t} = \nu - u(1-v) + v, \quad \text{at} \quad x = 0, \quad t > 0, \tag{4}$$

$$\lim_{x\to\infty} u(x,t) = 0, \qquad t > 0, \tag{5}$$

and initial data:

$$u(x,0) = v(x,0) = 0, \qquad x \geq 0. \tag{6}$$

The existence of solutions to the system (2)-(6) has been study in Tello [18] by using Banach fixed point Theorem. In the following sections we study the system in the domain $x \in (0,1)$, $t \in (0,\infty)$. The problem we study is the following

$$\frac{\partial u}{\partial t} - \frac{\partial^2}{\partial x^2}u = -u(1-v) + v, \quad x \in (0,1), \quad t > 0 \tag{7}$$

$$\frac{\partial v}{\partial t} = \lambda\left[u(1-v) - v\right] - \mu v, \quad t > 0, \tag{8}$$

$$\frac{\partial u}{\partial t} = g(u,v), \quad \text{at} \quad x = 0, \quad u(1,t) = 0, \quad t > 0, \tag{9}$$

$$u(x,0) = v(x,0) = 0, \quad x \in (0,1). \tag{10}$$

The parameters λ and μ are positives and g satisfies

$$g \in C^1(\mathbb{R}^2_+), \tag{11}$$

there exists $k_0 > 0$ such that

$$g(k_0, v) < -\beta < 0, \qquad \text{for any} \quad v \in [0,1]. \tag{12}$$

$$g(0, v) > 0 \qquad \text{for any } v \in [0,1]. \tag{13}$$

Through the following sections we use the notation:

$$I := (0,1); \qquad I_T := I \times (0,T).$$

3 Boundary Conditions and Steady States

We consider the system of ordinary differential equations

$$\frac{\partial \overline{u}}{\partial t} = g(\overline{u}, \overline{v}), \quad t > 0, \tag{14}$$

$$\frac{\partial \overline{v}}{\partial t} = \lambda \left[\overline{u}(1-\overline{v}) - \overline{v}\right] - \mu\overline{v}, \quad t > 0, \tag{15}$$

with the initial data

$$\overline{u})0) = \overline{v}(0) = 0. \tag{16}$$

Lemma 1. *$\overline{u}$ and $\overline{v}$ satisfy*

$$\overline{u} \geq 0; \quad 0 \leq \overline{v} \leq 1.$$

Proof. We denote by $f : \mathbb{R}^2_+ \to \mathbb{R}$ the function defined by

$$f(\overline{u}, \overline{v}) := \lambda \left[\overline{u}(1-\overline{v}) - \overline{v}\right] - \mu\overline{v}.$$

Then the problem is given by

$$\frac{\partial \overline{u}}{\partial t} = g(\overline{u}, \overline{v}), \quad t > 0,$$

$$\frac{\partial \overline{v}}{\partial t} = f(\overline{u}, \overline{v}), \quad t > 0,$$

$$\overline{u}(0) = \overline{v}(0) = 0.$$

Since $f \in C^1(\mathbb{R}^2)$ and $g \in C^1(\mathbb{R}^2)$, we have the existence of a unique solution in $(0,T)$ for $T \leq \infty$ as far as the solution remains bounded and non-negative. The positivity of $\overline{u}$ and $\overline{v}$ is a consequence of

$$\left.\frac{d\overline{u}}{dt}\right|_{\overline{u}=0} > 0 \text{ for } \overline{v} \in [0,1]$$

and

$$\left.\frac{d\overline{v}}{dt}\right|_{v=0} \geq 0 \text{ for } u \geq 0.$$

Since $f(\overline{u}, 1) < 0$ and $\overline{v}(0) \leq 1$ we have

$$\overline{v} \leq 1, \tag{17}$$

which ends the proof. □

Lemma 2. *$\overline{u}$ and $\overline{v}$ are uniformly bounded.*

Proof. We consider $\overline{u}$ and $\overline{v}$ the solution to (14) – (16). By previous lemma we have that $\overline{v} \in [0,1]$ and therefore $\overline{v}$ is bounded. Since $\overline{u} \geq 0$ and $\overline{u}_t < 0$ for $\overline{u} = k_0$ and any $\overline{v} \in [0,1]$ we conclude the proof. □

Proposition 1. *There exists a unique solution $(\overline{u}, \overline{v}) \in [C^1(0,\infty)]^2$ to (14)-(16).*

Proof. Since g and f are $C^1(\mathbb{R}^2_+)$, there exists $T > 0$ and a unique solution $(\overline{u}, \overline{v}) \in [C^1(0,T)]^2$ to (14)-(16) as far as the solution remains bounded and non-negative. Lemma 1 and Lemma 2 imply $T = \infty$. Regularity of solutions is a consequence of the regularity of f and g. □

3.1 Steady States

In this subsection we consider the steady states of the problem given by the solutions of the system:

$$-\frac{\partial^2}{\partial x^2}u = -u(1-v) + v, \quad x \in (0,1), \tag{18}$$

$$\lambda\left[u(1-v) - v\right] - \mu v = 0, \quad x = 0 \tag{19}$$

with the boundary conditions:

$$0 = g(u,v), \text{ at } x = 0, \qquad u(1) = 0. \tag{20}$$

Thanks to (19) we have

$$v = \frac{\lambda u}{\mu + \lambda + \lambda u}$$

and therefore

$$-\frac{\partial^2}{\partial x^2}u + \frac{\mu u}{\lambda u + \lambda + \mu} = 0, \qquad \text{for} \quad x \in (0,1). \tag{21}$$

Lemma 3. *Under assumptions (11)-(13) there exists at least a solution $\alpha \in (0, k_0)$ such that*

$$g\left(\alpha, \frac{\lambda\alpha}{\mu + \lambda + \lambda\alpha}\right) = 0.$$

Proof. Since $\frac{\lambda k_0}{\mu+\lambda+\lambda k_0} \in (0,1)$ we have

$$g\left(k_0, \frac{\lambda k_0}{\mu+\lambda+\lambda k_0}\right) < 0 \quad \text{and} \quad g(0,0) > 0.$$

Continuity of g and Bolzano′s Theorem end the proof. □

Previous lemma ensures that there exists at least an stationary state at the boundary, that we denote by α. Then, the boundary conditions of the problem are given by

$$u(0) = \alpha, \qquad u(1) = 0. \tag{22}$$

The problem (21), (22) has been studied in several works, for readers convenience we include the following result concerning the steady states.

Lemma 4. *For every λ and μ satisfying (11) and for any $\alpha \in (0, k_0)$ there exists a unique solution $\phi_\alpha \in C^1(I)$ to (21) and (22). Moreover the solution is a positive and monotone decreasing function.*

The proof of the Lemma 4 can be found in Lander, Nie, Vargas and Wang [8] where a monotonicity argument is used. In [15] a different method to prove the existence of solutions is considered.

Lander, Nie, Vargas and Wang [8] have considered the problem with a source term and nonlinear mixed boundary conditions. The problem with nonlinear boundary condition is more complicated and a monotone method is used to prove the existence of solutions. Uniqueness is also treated in [8]. In [8], the existence and uniqueness of solutions is obtained by using a monotonicity method of Amman and Sattinger based on upper and lower solutions. The monotonicity of solutions is also studied in [12] and [8].

It is also possible to obtain a parametric series expansion in powers of a new parameter defined by ν/λ .

4 Existence of Solutions

We consider the problem (7)-(8) with the boundary conditions

$$u(0,t) = \overline{u}(t), \qquad u(1,t) = 0, \tag{23}$$

$$v(0,t) = \overline{v}(t), \qquad v(1,t) = 0, \tag{24}$$

and the initial data

$$u(x,0) = v(x,0) = 0. \tag{25}$$

We first introduce some preliminary results concerning a priori estimates of the solution.

Lemma 5. $0 \le u$ *and* $0 \le v \le 1$.

Proof. We consider $(v-1)$ which satisfies

$$(v-1)_t + (\lambda + \mu + \lambda u)(v-1) = -(\lambda+\mu), \quad t > 0.$$

By integration, we obtain

$$\begin{array}{l} v(x,t) = 1 - \int_0^t (\lambda+\mu) exp\left\{\int_t^\tau (\lambda+\mu+\lambda u(x,s))ds\right\} d\tau - \\ exp\left\{ -\int_0^t (\lambda+\mu+\lambda u(x,s))d\tau \right\} \end{array} \tag{26}$$

which implies

$$v \le 1. \tag{27}$$

We introduce the functions ψ_ϵ, ψ, Ψ_ϵ and $\Psi : \mathbb{R} \to \mathbb{R}$ defined by

$$\psi_\epsilon(s) := \begin{cases} -1, \ s \le -\epsilon \\ \frac{1}{\epsilon}s, \ -\epsilon < s < 0 \\ 0, \quad s \ge 0, \end{cases} \qquad \psi(s) := \begin{cases} -1, \ s < 0 \\ 0, \quad s \ge 0, \end{cases} \tag{28}$$

$$\Psi_\epsilon(s) := \begin{cases} -s - \frac{\epsilon}{2}, \ s \le -\epsilon \\ \frac{1}{2\epsilon}s^2, \qquad -\epsilon < s < 0 \\ 0, \qquad\quad s \ge 0, \end{cases} \qquad \Psi(s) := \begin{cases} -s, \ s \le 0 \\ 0, \quad s \ge 0. \end{cases} \tag{29}$$

Notice that $\Psi'_\epsilon = \psi_\epsilon$ and

$$\lim_{\epsilon\to 0} \psi_\epsilon = \psi; \qquad \lim_{\epsilon \to 0} \Psi_\epsilon = \Psi,$$

$s\psi(s) = \Psi(s)$ for $s \in \mathbb{R}$. We multiply (7) by $\psi_\epsilon(u)$, integrate over I to obtain:

$$\frac{d}{dt}\int_I \Psi_\epsilon(u)dx + \int_I \psi'_\epsilon |u_x|^2 dx = \int_I \left(-(u(1-v)\psi_\epsilon(u) + v\psi_\epsilon(u)\right) dx, \tag{30}$$

for $t > 0$. Since $\psi'_\epsilon \ge 0$, we have

$$\frac{d}{dt}\int_I \Psi_\epsilon(u)dx \le \int_I \left(-u(1-v)\psi_\epsilon(u) + v\psi_\epsilon(u)\right) dx. \tag{31}$$

Notice that

$$\lim_{\epsilon\to 0} -u(1-v)\psi_\epsilon(u) = -u(1-v)\psi(u) = -(1-v)\Psi(u) \qquad a.e.$$

$$\lim_{\epsilon \to 0} v\psi_\epsilon(u) \le v\psi(u) \le \Psi(v)$$

and take limits as $\epsilon \to 0$ in (31) to obtain

$$\frac{d}{dt}\int_I \Psi(u)dx \le -\int_I (1-v)\Psi(u)dx + \int_I \Psi(v)dx, \quad t > 0. \tag{32}$$

In the same way we multiply (8) by $\psi_\epsilon(v)$ and integrate over I

$$\frac{d}{dt}\int_I \Psi_\epsilon(v)dx = \int_I \left(\lambda(u(1-v)\psi_\epsilon(v) - \mu v\psi_\epsilon(v)\right) dx. \tag{33}$$

We take limits as $\epsilon \to 0$ and it results:

$$\frac{d}{dt}\int_I \Psi(v)dx = \lambda\int_I u(1-v)\psi(v)dx - \mu\int_I \Psi(v)dx. \tag{34}$$

We apply the inequalities

$$(1-v)\psi(v) \le 0,$$

$$u(1-v)\psi(v) \le -\Psi(u)(1-v)\psi(v) \le (1-v)\Psi(u),$$

to (34) and it becomes

$$\frac{d}{dt}\int_I \Psi(v)dx \le \lambda\int_I (1-v)\Psi(u)dx - \mu\int_I \Psi(v)dx. \tag{35}$$

We multiply equation (32) by λ and add to equation (35) to get

$$\frac{d}{dt}\left(\lambda\int_I \Psi(u)dx + \int_I \Psi(v)dx\right) \le (\lambda-\mu)\int_I \Psi(v)dx, \quad t>0. \tag{36}$$

We apply Growall′s lemma to (36) to end the proof. □

Lemma 6. *$u \le \phi$ and $v \le \xi$.*

Proof. Let (ϕ, ξ) the steady state of the problem. Then, $u-\phi$ and $v-\xi$ satisfies:

$$(u-\phi)_t-(u-\phi)_{xx} = -(u-\phi)(1-v)+(v-\xi)(1+\phi), \quad x\in(0,1), \quad t>0 \tag{37}$$

and

$$(v-\xi)_t = \lambda(u-\phi)(1-v) - (v-\xi)(\lambda+\mu+\lambda\phi), \qquad t>0. \tag{38}$$

We introduce the functions θ_ϵ, θ, Θ_ϵ and $\Theta : \mathbb{R} \to \mathbb{R}$ defined by

$$\theta_\epsilon(s) := \begin{cases} 0, & s\le 0, \\ \frac{1}{\epsilon}s, & 0<s<\epsilon, \\ 1, & s\ge\epsilon, \end{cases} \qquad \theta(s) := \begin{cases} 0, \; s<0, \\ 1, \; s\ge 0, \end{cases} \tag{39}$$

$$\Theta_\epsilon(s) := \begin{cases} 0, & s\le 0, \\ \frac{1}{2\epsilon}s^2, & 0<s<\epsilon, \\ s-\frac{1}{2}\epsilon & s\ge\epsilon, \end{cases} \qquad \Theta(s) := \begin{cases} 0, \; s\le 0, \\ s, \; s\ge 0. \end{cases} \tag{40}$$

Notice that $\Theta_\epsilon' = \theta_\epsilon$, $s\theta(s) = \Theta(s)$ for $s \in \mathbb{R}$ and

$$\lim_{\epsilon \to 0} \theta_\epsilon = \theta, \qquad \lim_{\epsilon \to 0} \Theta_\epsilon = \Theta.$$

We multiply equation (37) by $\theta_\epsilon(u-\phi)$, integrate over I and take limits when $\epsilon \longrightarrow 0$ to get:

$$\begin{aligned} \frac{d}{dt}\int_I \Theta(u-\phi)dx \leq \int_I -\Theta(u-\phi)(1-v)dx + \\ \textstyle\int_I (v-\xi)(1+\phi)\theta(u-\phi)dx, \text{ for } t>0. \end{aligned} \tag{41}$$

Notice that

$$\begin{aligned} (v-\xi)(1+\phi)\theta(u-\phi) \leq (v-\xi)(1+\phi)\theta(u-\phi)\theta(v-\xi) = \\ (1+\phi)\theta(u-\phi)\Theta(v-\xi) \leq (1+\phi)\Theta(v-\xi), \end{aligned}$$

we obtain, from (41)

$$\frac{d}{dt}\int_I \Theta(u-\phi)dx \leq -\int_I \Theta(u-\phi)(1-v)dx + \int_I \Theta(v-\xi)(1+\phi)dx. \tag{42}$$

In the same way we multiply (37) by $\lambda^{-1}\theta_\epsilon(v-\xi)$. We integrate over I and take limits when $\epsilon \to \infty$ to get

$$\begin{aligned} \lambda^{-1}\frac{d}{dt}\int_I \Theta(v-\xi)dx = \\ \int_I (u-\phi)(1-v)\theta(v-\xi)dx - \int_I \Theta(v-\xi)\left(1+\frac{\mu}{\lambda}+\phi\right)dx \end{aligned} \tag{43}$$

Since

$$(u-\phi)(1-v)\theta(v-\xi) \leq (u-\phi)(1-v)\theta(v-\xi)\theta(u-\phi) \leq \Theta(u-\phi)(1-v)$$

(43) becomes

$$\begin{aligned} \lambda^{-1}\frac{d}{dt}\int_I \Theta(v-\xi)dx \leq \\ \int_I \Theta(u-\phi)(1-v)dx - \int_I \Theta(v-\xi)(1+\frac{\mu}{\lambda}+\phi)dx. \end{aligned} \tag{44}$$

By (42) and (44) it results

$$\frac{d}{dt}\left(\int_I \Theta(u-\phi)dx + \lambda^{-1}\int_I \Theta(v-\theta)dx\right) \leq -\frac{\mu}{\lambda}\int_I \Theta(v-\xi)dx \leq 0. \tag{45}$$

Then, Gronwall's lemma implies

$$\int_I \Theta(u-\phi)dx = \int_I \Theta(v-\xi)dx = 0,$$

which ends the proof. □

Theorem 1. *There exists at least one solution to the problem (7)-(11).*

Proof. We consider the new unknown $z := u - \overline{u}\sigma(x)$, for $\sigma(x) \in C^\infty(I)$ monotone decreasing function satisfying $\sigma(x) = 1$ in $(0, \frac{1}{4})$ and $\sigma(x) = 0$ for $\frac{1}{2} < x < 1$, then z satisfies

$$\frac{\partial z}{\partial t} - \frac{\partial^2 z}{\partial x^2} + (1-v)z = -\overline{u}_t\sigma - \overline{u}\sigma_{xx} - \overline{u}\sigma(1-v) + v, \qquad x \in I,\ t > 0. \quad (46)$$

where v is the solution of the equation

$$\frac{\partial v}{\partial t} = \lambda\left[(z + \overline{u}\sigma)(1-v) + v\right] - \mu v, \qquad t > 0,$$

with the boundary conditions

$$z(0,t) = 0, \quad (47)$$

and initial data

$$z(x,0) = v(x,0) = 0. \quad (48)$$

By (26) v is given by

$$v(x,t) = 1 - \textstyle\int_0^t (\lambda+\mu) exp\left\{\int_t^\tau (\lambda + \mu + \lambda z(s) + \lambda\overline{u}(s)\sigma(x))ds\right\} d\tau - \\ exp\left\{-\textstyle\int_0^t (\lambda + \mu + \lambda z(x,s) + \lambda\overline{u}(s)\sigma(x))d\tau\right\}. \quad (49)$$

We denote by $f(t,v)$ the right side part of (46), then the problem (46)-(48) became:

$$\begin{cases} \dfrac{\partial z}{\partial t} - \dfrac{\partial^2}{\partial x^2} z + (1-v)z = f(t,v),\ t > 0, & x \in I \\ z(x,0) = 0, & x \in I, \\ z(0,t) = 0, & t > 0. \end{cases} \quad (50)$$

We use a fixed point argument to prove the existence of solutions of (50). Let A be the subset of continuous functions defined by

$$A := \{z \in L^2(0,T : H_0^1(I));\ 0 \le z + \overline{u}\sigma(x) \le \phi,\ z(0,t) = 0\}$$

and

$$J(\hat{z}) := z$$

where z is the solution to (50) for v defined by

$$1 - \textstyle\int_0^t (\lambda+\mu) exp\left\{\int_t^\tau (\lambda + \mu + \lambda\hat{z}(x,s) + \lambda\overline{u}(s)\sigma(x))ds\right\} d\tau - \\ exp\left\{-\textstyle\int_0^t (\lambda + \mu + \lambda\hat{z}(x,s) + \lambda\overline{u}(s)\sigma(x))d\tau\right\}. \quad (51)$$

Notice that, since $\hat{z} \in L^2(0,T : H_0^1(I)) \cap L^\infty([0,T] \times I)$ and $\sigma \in C^\infty(\Omega)$, we have that

$$v, v_t \in L^\infty([0,T] \times I). \tag{52}$$

In order to apply Schauder fixed point theorem, we consider the following:

i. There exists a unique solution z to (50) and z, z_t, z_x and $z_{xx} \in L^p(0,T : L^p(I))$ for $1 \leq p < \infty$.

ii. $0 \leq z + \overline{u}\sigma \leq \phi$:
We consider the function $z + \overline{u}\sigma - \phi$ which satisfies the equation

$$(z+\overline{u}\sigma-\phi)_t-(z+\overline{u}\sigma-\phi)_{xx}+(z+\overline{u}\sigma-\phi)(1-v) = (v-\xi)(1+\phi), \tag{53}$$

and

$$(v-\xi)_t = \lambda(z+\overline{u}\sigma-\phi)(1-v)-(v-\xi)(\lambda+\mu+\lambda\phi). \tag{54}$$

Consider the functions (39)-(40) and proceed as in Lemmas 5 and 6 to obtain

$$0 \leq z + \overline{u}\sigma \leq \phi. \tag{55}$$

iii. $J : A \to L^2(0,T : H_0^1(I))$ is well defined:
We multiply equation (50) by z and integrate by parts to obtain

$$\begin{gathered} \tfrac{1}{2}\int_I z^2 dx\Big|_{t=T} + \int_0^T \int_I |z_x|^2 dx = \\ \int_0^T \int_I z(f(x,t) + (1-v)\overline{u}\sigma - \overline{u}_t\sigma z) dx dt. \end{gathered} \tag{56}$$

Since v and f are bounded and $0 \leq z + \overline{u}\sigma \leq \phi \leq \alpha$ we have that

$$\frac{1}{2}\int_I z^2 dx\Big|_{t=T} + \int_0^T \int_I |z_x|^2 dx \leq k_0 T \tag{57}$$

which implies $z \in L^2(0,T : H_0^1(I))$. Uniqueness of solution has been discuss in (i).

iv. J is a continuous function:
it is a consequence of the continuity of v and f as functions of z.

v. $J(A) \subset A$:
by (*ii*) and (*iii*) we deduce (*v*).

vi. $z \in L^2(0,T : H_0^1(I))$:
see (*iii*) formula (57).

vii. $z \in L^2(0,T : H^2(I) \cap H_0^1(I))$ and $z_t \in L^2(0,T : L^2(I))$.
Since $f(x,t) + (1-v)\overline{u}\sigma - \overline{u}_t\sigma z \in L^\infty((0,T) \times I)$ we have the result. See for instance Brezis [2].

viii$J(A)$ is a precompact subset of $L^2(0,T:H_0^1(I))$:

We define the space

$$W=\{z;\ z\in L^2(0,T:H^2(I)\cap H_0^1(I)),\ z_t=\frac{dz}{dt}\in L^2(0,T:L^2(I))\},$$

with the norm

$$\|v\|_W=\|v\|_{L^2(0,T:H^2(I))}+\|v'\|_{L^2(0,T:L^2(I))}.$$

Since

-. $H^2(I)\cap H_0^1(I)\hookrightarrow H_0^1(I)$ is a compact embedding.

-. $H_0^1(I)\subset L^2(I)$.

We obtain that $W\hookrightarrow L^2(0,T:H_0^1(I))$ is a compact embedding (see Lions [11], Theorem 5.1). Hence we have that $J(A)$ is a precompact subset of A.

Thanks to (*iv*), (*v*) and (*viii*) we apply Schauder's fixed point theorem to J to obtain the existence of at least a solution to (7)-(11). □

4.1 Uniqueness of Solutions

Proposition 2. *There exists at most one solution to (7)-(11).*

Proof. By the contrary we assume there exists two different solutions (u_1,v_1) and (u_2,v_2) to the system (7)-(11). We consider $U=u_1-u_2$ and $V=v_1-v_2$, then (U,V) satisfies

$$\begin{aligned}&\tfrac{\partial U}{\partial t}-\tfrac{\partial^2}{\partial x^2}U=-U(1-v_1)+V(1-u_2), && x\in I,\quad t>0,\\ &\tfrac{\partial V}{\partial t}=\lambda\left[U(1-v_1)-V(1-u_2)\right]-\mu V, && x\in I,\quad t>0.\end{aligned}\tag{58}$$

$$U(0,t)=U(1,t)=0.\tag{59}$$

and initial data:

$$U(x,0)=V(x,0)=0,\qquad x\in I.\tag{60}$$

Multiply the system (58) by (U,V) and integrate over I to obtain:

$$\frac{d}{dt}\frac{1}{2}\int_I U^2dx+\int_I\left|\frac{\partial}{\partial x}U\right|^2dx=-\int_I U^2(1-v_1)dx+\int_I UV(1-u_2)dx,\tag{61}$$

and

$$\frac{d}{dt}\frac{1}{2}\int_I V^2dx=\lambda\int_I VU(1-v_1)dx-\int_I V^2(\lambda+\mu-\lambda u_2)dx.\tag{62}$$

Since $|v_1|\le\beta$ and $|u_2|\le\alpha$, we have that:

$$-\int_I U^2(1-v_1)dx+\int_I UV(1-u_2)dx\le c_1\int_I U^2dx+c_2\int_I V^2dx;\tag{63}$$

and

$$\lambda \int_I VU(1-v_1)dx - \int_I V^2(\lambda+\mu-\lambda u_2)dx \leq c_3 \int_I U^2 dx + c_4 \int_I V^2 dx, \quad (64)$$

for

$$c_1 := \beta - 1 + \tfrac{|1-\alpha|}{2}; \ c_2 := \tfrac{|1-\alpha|}{2};$$

$$c_3 := \tfrac{\lambda}{2}|1-\beta|; \qquad c_4 := \lambda\left(\alpha + \tfrac{|1-\beta|}{2} - 1\right) - \mu.$$

Hence, by (61) – (64), we obtain

$$\frac{d}{dt}\frac{1}{2}\int_I \left(U^2+V^2\right)dx + \int_I \left|\frac{d}{dx}U\right|^2 dx \leq c_5 \int_I \left(U^2+V^2\right)dx; \qquad (65)$$

for $c_5 := \max\{c_1+c_3, c_2+c_4\}$. We apply Growall's lemma to (65) and the proof ends. □

Remark 1. The asymptotic behavior of the solution for

$$g(u,v) = \nu - u(1-v) + v$$

and $\nu > 0$ is similar to case studied in [18] where the solution goes to the steady state as t goes to ∞. In that case, $u(t,0) = \overline{u}(t)$ is a monotone increasing function and

$$\lim_{t\to\infty} \overline{u}(t) = \alpha > 0.$$

The solution u goes to ϕ_α as t goes to ∞, where ϕ_α is the solution to the stationary problem

$$-\frac{\partial^2}{\partial x^2}u + \frac{\mu u}{\lambda u + \lambda + \mu} = 0, \qquad \text{in } x \in (0,1),$$

with the boundary condition

$$u(0) = \alpha, \qquad u(1) = 0.$$

Acknowledgement. The author is supported by project MTM2009-13655 of DIGISPI.

References

1. Bollenbach, T., Kruse, K., Pantazis, P., González-Gaitán, M., Julicher, F.: Morphogen transport a a u in epithelia. Physical Rev. E 75, 11901 (2007)
2. Brezis, H.: Analyse Fonctionnelle: Théorie et applications. Masson, Paris (1983)
3. Entchev, E.V., Schwabedissen, A., González-Gaitán, M.: Gradient formation of the TGF-beta homolog Dpp. Cell 103, 981–991 (2000)

4. Entchev, E.V., González-Gaitán, M.: Morphogen gradient formation and vesicular trafficking. Traffic 3, 98–109 (2002)
5. Friedman, A.: Partial Differential Equations of Parabolic Type. Prentice-Hall, Englewood Cliffs (1964)
6. Kerszberg, M., Worpert, L.: Mechanism for positional signalling by morphogen transport: A theoretical study. J. Theor. Biol. 191, 103–114 (1998)
7. Krzyzanowski, P., Laurençot, P., Wrzosek, D.: Well-posedness and convergence to the steady state for a model of morphogen transport. SIAM J. Math. Anal. 40, 1725–1749 (2009)
8. Lander, A.D., Nie, Q., Vargas, B., Wan, F.Y.N.: Aggregation of a distributed source in morphogen gradient formation. Stud. Appl. Math. 114, 343–374 (2005)
9. Lander, A.D., Nie, Q., Wan, F.Y.M.: Do morphogen gradients arise by diffusion? Dev. Cell 2, 785–796 (2002)
10. Lander, A.D., Nie, Q., Wan, F.Y.M.: Internalization and end flux in morphogen gradient formation. J. Comp. Appl. Math. 190, 232–251 (2006)
11. Lions, J.L.: Quelques Méthodes de Résolution de Problemes aux Limites Non Linéaires. Dunod-Gauthier Villars, Paris (1967)
12. Lou, Y., Nie, Q., Wan, F.Y.N.: Effects of sog DPP-Receptor binding. SIAM J. Appl. Math. 65, 343–374 (2005)
13. Merking, J.H., Needham, D.J., Sleeman, B.D.: A mathematical model for the spread of morphogens with density dependent chemosensitivity. Nonlinearity 18, 2745–2773 (2005)
14. Merking, J.H., Sleeman, B.D.: on the spread of morphogens. J. Math. Biol. 51, 1–17 (2005)
15. Muñoz, A.I., Tello, J.I.: (2010) (submitted)
16. Stinner, C., Tello, J.I., Winkler, M.: (2010) (in progress)
17. Teleman, A., Cohen, S.: Dpp gradient formation in the Drosophila wing imaginal disc. Cell 103, 971–980 (2000)
18. Tello, J.I.: Mathematical analysis of a model of morphogenesis. Discr. Contin. Dynam. Syst. - Ser. A 25, 343–361 (2009)
19. Turing, A.M.: The chemical basis for morphogenesis. Phil Trans. Royal Soc. London, Ser. B Biol. Sci. 237, 37–72 (1952)
20. Wolpert, L.: Positional information and the spatial pattern of cellular differentiation. J. Theor. Biol. 25, 1–47 (1969)

Author Index

Lightning Source UK Ltd.
Milton Keynes UK

176220UK00005B/1/P

9 783642 208522